6th International Conference on Advanced Materials and Ecological Environment (AMEE 2024)

Journal of Physics: Conference Series Volume 2941

Wuhan, China
20-21 November 2024

ISBN: 979-8-3313-1428-6
ISSN: 1742-6588

Printed from e-media with permission by:

Curran Associates, Inc.
57 Morehouse Lane
Red Hook, NY 12571

CURRAN ASSOCIATES INC. proceedings.com

Some format issues inherent in the e-media version may also appear in this print version.

Institute of Physics
Dirac House, Temple Back
Bristol BS1 6BE UK

Phone: 44 1 17 929 7481
Fax: 44 1 17 920 0979

techtracking@iop.org

Additional copies of this publication are available from:

Curran Associates, Inc.
57 Morehouse Lane
Red Hook, NY 12571 USA
Phone: 845-758-0400
Fax: 845-758-2633
Email: curran@proceedings.com
Web: www.proceedings.com

TABLE OF CONTENTS

Journal of Physics: Conference Series **2941** (2025) 011001 doi:10.1088/1742-6596/2941/1/011001

Preface

The 2024 6th International Conference on Advanced Materials and Ecological Environment (AMEE 2024), organized by the Hubei Zhongke Institute of Geology and Environment Technology, was successfully conducted virtually from November 20th, 2024, due to travel restrictions.

The AMEE 2024 proceedings aim to compile the latest, most comprehensive, and globally cutting-edge knowledge on Advanced Materials and Ecological Environment. Each paper accepted has undergone rigorous peer-review by 1-2 expert referees and has been selected for its originality, significance, and clarity, aligning with the conference's objectives. The conference program is exceptionally rich and profound, featuring presentations of selected papers with high impact, as well as additional contributions that are timely and groundbreaking. Director and Taishan Distinguished Professor Qing Wang from Taishan University, along with other professors, delivered enlightening speeches at the conference. We sincerely hope that the conference will not only provide participants with a broad overview of the latest research outcomes in related fields but also serve as a significant platform for academic networking and exchange.

The Technical Program Committee has been diligently working to meet the review deadline. Over the course of the conference, we have received more than 200 submissions. The final conference program features about 100 papers, which are organized into three distinct parts: keynote speeches, oral presentations, and electronic poster presentations.

We wish to convey our heartfelt gratitude to the editors and all members of the Technical Program Committee for their diligent efforts, invaluable time, and unwavering commitment in preparing for the conference. Our profound appreciation extends to the volunteers and staff for their tireless dedication and the generosity they have bestowed upon the conference. Last but certainly not least, we are grateful to each and every author, speaker, and participant for their significant contributions to the success of AMEE 2024.

Changbo Cheng

Hubei Zhongke Institute of Geology and Environment Technology
Organizer
Hubei Zhongke Institute of Geology and Environment Technology

Technical Program Committee
Chair
Prof. Qing Wang, Taishan University, China
Program Chair
Prof. Leonardo Pagnotta, University of Calabria, Italy
Prof. Ioannis Karapanagiotis, Aristotle University of Thessaloniki, Greece
Prof. Xianqiang Pei, Lanzhou Institute of Chemical Physics, Chinese Academy of Sciences, China

Committee Member
Prof. Michele Penza, ENEA - Italian National Agency for New Technologies, Energy, and Economic Sustainable Development, Italy
Asst. Prof. Hasan Dilbas, Yuzuncu Yil University, Turkey
Prof. Satya Narayan, Kumamoto University, Japan
Assoc. Prof. Mohd Hasmizam Razali, Universiti Malaysia Terengganu, Malaysi
Dr. Carlos E. Flores Chaparro, Instituto Potosino de Investigacion Cientificay Tecnologica, Mexico
Dr. Enrico Zacchei, Senior Engineer and Principal Investigator at ITECONS Institute, Portugal
Dr. Grzegorz Woroniak, Bialystok University of Technology, Poland
Assoc. Prof. Dhanesh G. Mohan, University of Sunderland, United Kingdom
Dr. Lin Feng Ng, Universiti Teknologi Malaysia, Malaysia
Assoc. Prof. Mohsen Padervand, University of Maragheh, Iran
Asst. Prof. Qian Wang, Joining and Welding Research Institute, Osaka University, Japan
Prof. Leonardo Pagnotta, University of Calabria, Italy
Dr. Raúl Martín Lozano, University of Castilla La Mancha, Spain
Assoc. Prof. Abu Zahrim Yaser, Universiti Malaysia Sabah, Malaysia
Dr. Milad Khatib, University of Balamand, Lebanon
Asst. Prof. Biao Hu, Shenzhen University, China
Asst. Prof. Ammar Hamed Abdelsalam Elsheikh, Tanta University, Egypt
Assoc. Prof. Khor Shing Fhan, Universiti Malaysia Perlis, Malaysia
Prof. Tarek Aly Elsayed, Helwan University, Egypt
Asst. Prof. Supachok Tanpichai, King Mongkut's University of Technology Thonburi, Thailand
Prof. Babu P. Sekhar, Siddhartha Institute of Engineering & Technology, India
Dr. Rohit Gupta, Dr. Bhimrao Ambedkar University Agra, India

Dr. Amuthakkannan Pandian, PSR Engineering College, India

Journal of Physics: Conference Series **2941** (2025) 011001 doi:10.1088/1742-6596/2941/1/011001

Peer Review Statement

All papers published in this volume have been reviewed through processes administered by the Editors. Reviews were conducted by expert referees to the professional and scientific standards expected of a proceedings journal published by IOP Publishing.

- **Type of peer review:** Double Anonymous
- **Conference submission management system:** Morressier
- **Number of submissions received:** 245
- **Number of submissions sent for review:** 231
- **Number of submissions accepted:** 104
- **Acceptance Rate (Submissions Accepted / Submissions Received × 100):** 42.4
- **Average number of reviews per paper:** 2
- **Total number of reviewers involved:** 33
- **Contact person for queries:**
  Name: Amanda Chen
  Email: ameeconfer@163.com
  Affiliation: Hubei Zhongke Institute of Geology and Environment Technology - 2024 6th International Conference on Advanced Materials and Ecological Environment

Utilization of River Snail Shells as Formaldehyde Absorbents

Xiaofeng Tan[1,2], Chao Tian[2,*], Wenyi Li[2] and Xinrui Yu[1]

[1] School of Material Science and Chemical Engineering, Hubei University of Technology, Wuhan 430068, Hubei, China.
[2] The Research Center of Circular Economy Development in Hubei, Hubei University of Technology, Wuhan 430068, Hubei, China.
Email: 2939699553@qq.com; *charles_tian@126.com; 3195975350@qq.com; 3318193136@qq.com

Abstract. This study addresses aquaculture waste management challenges by proposing a pollution control and resource utilization strategy through the calcination of discarded river snail shells (RSC) at different temperatures for formaldehyde absorption. The findings indicate that calcining RSC at 900°C transforms the material into calcium oxide with a porous structure, resulting in excellent adsorption capacity. Formaldehyde adsorption tests show that RSC calcined at both 900°C and 1000°C perform well in adsorption, with the 900°C calcined RSC demonstrating optimal effectiveness considering both economic and energy-saving factors. This approach presents an innovative solution for enhancing indoor air quality by incorporating shell powder into surface coatings.
Keywords: Resource utilization; river snail shell; formaldehyde adsorption.

1. Introduction

Numerous waste shells are discarded after the consumption of seafood and other food products. The seafood industry alone generates a substantial amount of this waste, with around 10 million tonnes produced annually in China, and the majority of these shells are disposed of in landfills [1]. This massive accumulation of shells occupies substantial land resources and causes severe environmental pollution. Fresh piles of waste shells left in open areas can become breeding grounds for microorganisms, attracting flies that may carry diseases as the leftover residue inside the shells decomposes, leading to public health concerns [2]. Over time, these accumulations release harmful gases, such as ammonia and hydrogen sulfide, and generate leachate rich in organic matter and heavy metals, which can seep into the groundwater and contaminate local water sources if untreated, posing a serious threat to environmental safety [3]. Therefore, shell waste should be disposed of and utilized more effectively to support the sustainable development of the aquaculture industry. Shells primarily consist of about 95% calcium carbonate, approximately 5% organic matter, and trace amounts of inorganic salts, including elements like potassium, sodium, magnesium, iron, zinc, and selenium [4]. Since shells have a porous and fibrous double helix structure after calcination, they can be used for adsorbing harmful gases [5]. The use of water-based coatings with shell powder as a filler enhances the formaldehyde purification rate, which increases with the amount of shell powder added, reaching up to 68.3% [6].

The river snail (Viviparidae) is a sizable, operculated freshwater snail belonging to the order Architaenioglossa and is typically found in various waterways [7]. The river snail is a traditional aquatic product in China, valued for its nutritional and health benefits, and is increasingly popular in people's lives [8]. River snail shells (RSC) are primarily composed of calcium and have been found to be reusable for environmentally benign purposes [9]. RSC has been used as a component for catalysts, concrete, and purifiers [9-11]. However, there is a lack of publications on the use of RSC for formaldehyde absorption. This study systematically analyzed the compositional characteristics and adsorption properties of waste RSC. The formaldehyde adsorption performance test revealed its strong capability of effectively removing formaldehyde pollution from indoor air.

2. Experimental

2.1. Raw Materials and Preparation

The RSC was collected as waste from a garden and farm in Wuhan, Hubei Province, China. They were thoroughly washed with water several times and air-dried in an oven at 100°C for 24 hours. The dried shells were then crushed into a fine powder and filtered through a 60-mesh stainless steel sieve.

2.2. Calcination and Characterization

To enhance their adsorption capabilities, the RSC was calcined to remove organic impurities and volatile substances. At temperatures exceeding 900°C, the calcium carbonate in the shells fully decomposes into calcium oxide, which increases the specific surface area of the material [12]. An SXL-1200C muffle furnace, capable of reaching a maximum calcination temperature of 1150°C, was used for this process. The initial calcination temperature was set at 255°C, with a heating rate of 10 °C /min, and maintained at the target temperature for 2 hours before being allowed to cool naturally to room temperature. Calcination temperatures of 400°C, 500°C, 600°C, 700°C, 800°C, 900°C, 1000°C, and 1100°C were tested. The calcined RSC was stored in a sealed container to prevent reactions with air. Thermal stability and compositional analyses were performed using thermogravimetric analysis (TGA) and X-ray diffraction (XRD) patterns.

2.3. *Formaldehyde Absorption Tests*

To identify the optimal calcination temperature for RSC in terms of formaldehyde absorption performance, a standardized experimental procedure was conducted.

2.3.1. Experimental Preparation. Calcined RSC powder was evenly distributed in Petri dishes. Two sealed glass chambers were prepared, each with dimensions of 1250 mm (length) × 800 mm (width) × 1000 mm (height) and a total volume of 1 m^3 . Glass thickness ranged between 8 mm and 10 mm. One chamber, containing the calcined RSC samples, served as the experimental sample chamber, while the other chamber, containing uncalcined RSC samples, acted as a control.

A 20 ml analytical-grade formaldehyde solution was placed in a beaker inside each chamber, which were then sealed. The chambers were kept at a controlled temperature of (20 ± 2)°C and relative humidity of (50 ± 10)% throughout the experiment.

2.3.2. Experimental Procedure. Both chambers' RSC samples were exposed to formaldehyde for 48 hours. After this period, colorimetric phenol reagents were added to the Petri dishes with RSC powder to develop color changes indicating formaldehyde absorption levels. Color intensity was recorded as a qualitative measure of RSC's absorption performance.

Following this, the RSC samples were retrieved for Fourier Transform Infrared Spectroscopy (FTIR) analysis, which examined molecular vibrations and structural changes associated with formaldehyde absorption by focusing on absorption bands corresponding to functional groups' fundamental vibrational frequencies.

2.3.3. Data Collection and Analysis. Colorimetric observations and FTIR spectra were documented and analyzed for shifts in absorption bands indicating structural changes in RSC samples. The performance of RSC at various calcination temperatures was compared based on colorimetric and FTIR results, allowing the optimal calcination temperature for formaldehyde absorption to be determined in a consistent and controlled manner.

2.4. *Moisture Absorption Tests*

To evaluate the moisture absorption and desorption capabilities of uncalcined and calcined RSC at 900°C, both RSC samples were placed in a humidity chamber maintained at 80% relative humidity (RH). The mass of each sample was recorded periodically to monitor moisture absorption over time. This process continued until the samples reached equilibrium, defined as a constant mass with no further moisture gain. The equilibrated, moist samples were then transferred to a drying oven set to a low-humidity environment to initiate desorption. The mass of each sample was again recorded at regular intervals until equilibrium was reached, indicating a stable, dry state.

3. Results and Discussions

3.1. *Optimal Calcination Temperature*

TGA was conducted on the RSC to determine the weight loss ratio during heating. Weight loss rate = (Initial mass - Current mass) / Initial mass * 100%. As illustrated in figure 1, the heating process of the RSC could be divided into two distinct phases. In the first phase, the physical desorption of water and dehydration of the shell powder occurred, along with the decomposition of organic matter present within the powder. This was followed by a significant mass loss, primarily due to the decomposition of calcium carbonate into calcium oxide during high-temperature calcination. At a calcination temperature of 900°C, the weight loss ratio of the RSC stabilized, indicating the complete decomposition of the calcium carbonate.

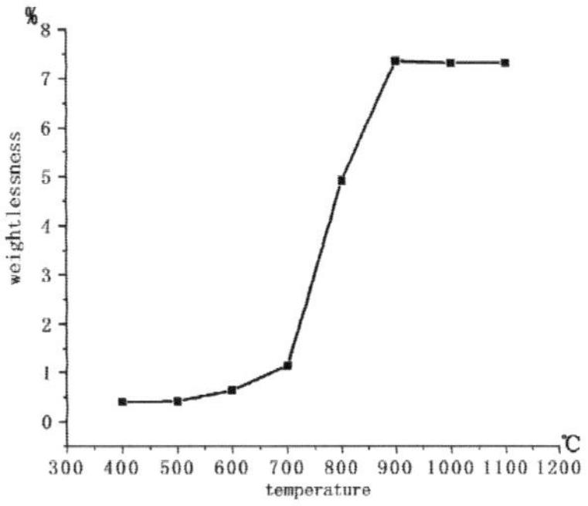

Figure 1. TGA of RSC powder.

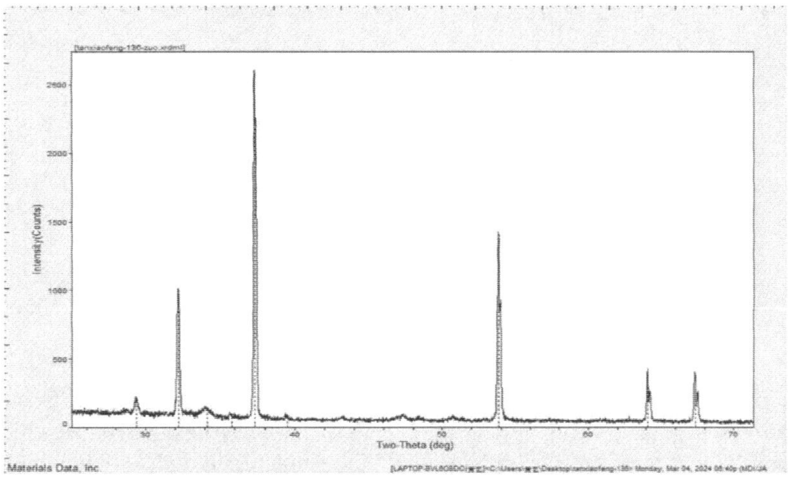

Figure 2. XRD of RSC calcined at 900°C.

XRD analysis was performed on the RSC powder calcined at 900°C. As depicted in figure 2, the XRD pattern confirms the transformation of the RSC powder into calcium oxide. The average grain size of the calcined powder was approximately 900 nm, and the structure appeared relatively loose, contributing to its adsorption capacity. Comparison with the characteristic peaks of calcium oxide revealed that the RSC powder exhibited high crystalline quality, indicating its high purity. This high crystalline quality and purity enhance the RSC powder's strong adsorption capacity.

3.2. Absorbents Performance

The results of the MBTH spectrophotometric method are presented in figure 3. The color change of the colorimetric phenol reagents was more pronounced when the RSC was calcined at 900°C and 1000°C, indicating superior absorption performance at these temperatures.

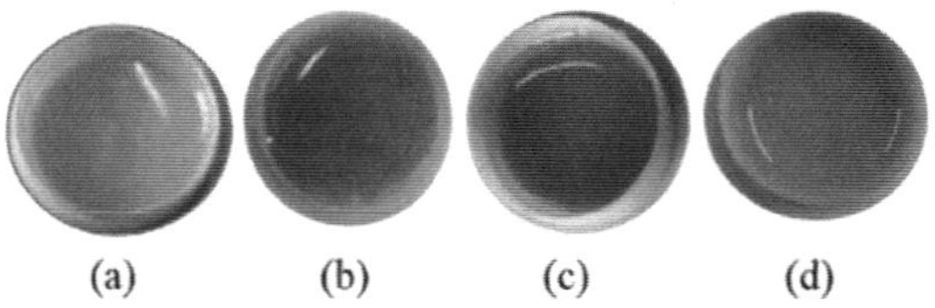

Figure 3. MBTH spectrophotometric analysis of uncalcined RSC (a), calcined RSC at 900°C (b), 1000°C (c), and 1100°C (d).

Figure 4 illustrates the FTIR analysis results of RSC calcined at 600°C, 800°C, 900°C, 1000°C, and 1100°C, respectively. A prominent characteristic peak was observed in the mid-wavenumber region (around 1600 cm^{-1}), which was indicative of the presence of a double bond (C=O). The RSC samples calcined at 600°C, 800°C, 900°C, and 1000°C all demonstrated the ability to adsorb formaldehyde. The peak intensity suggested that RSC calcined at 900°C and 1000°C exhibited the best adsorption performance.

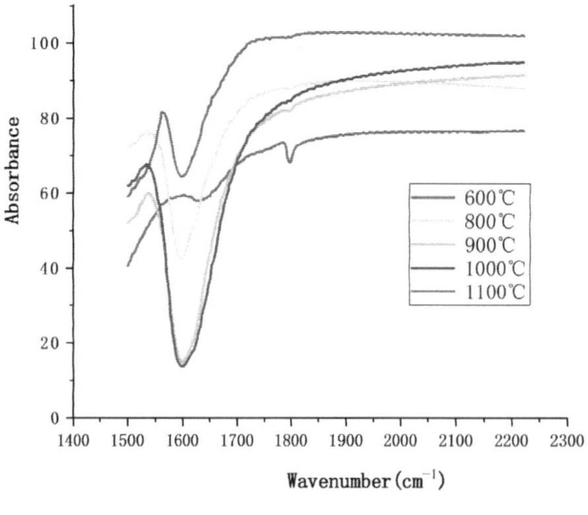

Figure 4. FTIR analysis of calcinated RSC.

Figure 5(a) demonstrates that both the uncalcined and 900°C calcined RSC samples reached moisture saturation within approximately 5 hours. The moisture absorption of the calcined RSC was notably superior to that of the uncalcined RSC, indicating that calcination enhances the moisture-holding capacity. Figure 5(b) shows the desorption performance of both samples, which reached equilibrium within about 20 hours. The calcined RSC performed slightly better in desorption, suggesting that calcination at 900°C not only enhances absorption but also supports more effective moisture release in a low-humidity environment.

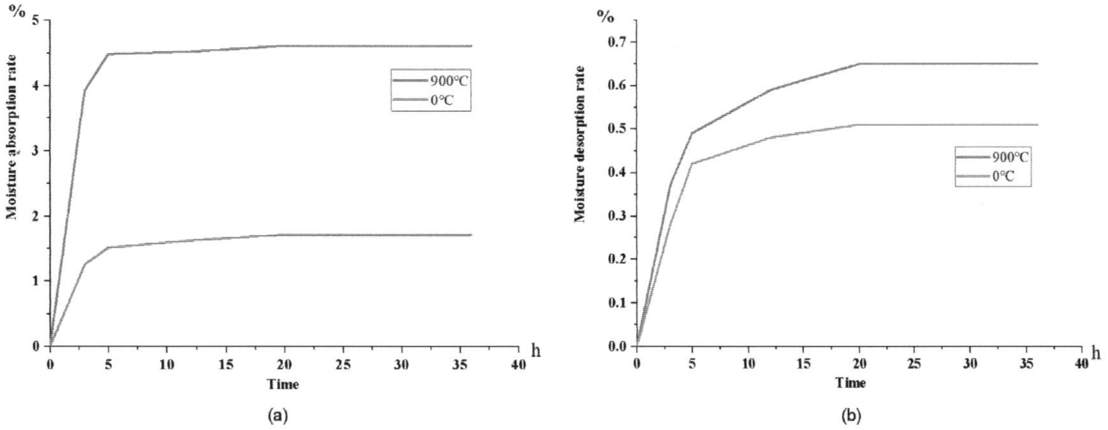

Figure 5. Moisture absorption rate (a) and moisture desorption rate(b).

3.3. Feasibility
The use of RSC as a raw material offers significant cost advantages, given its abundance as aquaculture waste and near cost-free availability, which greatly reduces procurement expenses. The production process for calcined RSC powder is relatively straightforward, requiring minimal investment in equipment and maintenance. Furthermore, the collection and utilization of carbon dioxide produced during calcination can further lower production costs and even generate additional revenue. With increasing environmental awareness, the market demand for eco-friendly building materials is on the rise. Preliminary estimates suggest that recycling 1 million tons of discarded RSC annually to produce activated powder for environmentally friendly coatings could generate a market value of approximately 500 million yuan. Considering raw material costs, production expenses, and market potential, the economic benefits of this plan are substantial.

4. Conclusion
In this paper, RSC powder was prepared and subjected to calcination at different temperatures to determine the optimal calcination temperature. TGA identified 900°C as the most suitable calcination temperature. XRD analysis revealed that the RSC calcined at 900°C transformed into calcium oxide with a porous structure, demonstrating good adsorption capacity. Formaldehyde adsorption experiments indicated that RSC calcined at 900°C and 1000 °C exhibited the best adsorption performance. FTIR analysis further confirmed the effectiveness of the shell powder in adsorbing formaldehyde at these temperatures. Considering both economic and energy-saving factors, RSC calcined at 900°C demonstrated the optimal performance for formaldehyde adsorption, along with superior moisture absorption and desorption capabilities. These properties make it well-suited for incorporation into coatings designed to enhance indoor air quality, offering an effective and sustainable solution for pollutant reduction. This plan is a feasible approach for recycling aquaculture solid waste and reusing it as a resource.

Acknowledgments
This paper is supported by the College Students' Innovation and Entrepreneurship Training Program (S202310500085).

Journal of Physics: Conference Series **2941** (2025) 012001 doi:10.1088/1742-6596/2941/1/012001

5. Reference

[1] Mo K H, Alengaram U J, Jumaat M Z, Lee S C, Goh W I and Yuen C W 2018 Recycling of seashell waste in concrete: A review *Construction and Building Materials* **162** 751-764.

[2] Sawai J 2011 Antimicrobial characteristics of heated scallop shell powder and its application *Biocontrol Science* **16** 95-102.

[3] Hart A 2020 Mini-review of waste shell-derived materials' applications *Waste Management & Research* **38(5)** 514-527.

[4] Li T, Xin R, Wang D, Yuan L, Wu D and Wu X 2023 Research progress on the applications of seashell adsorption behaviors in cement-based materials *Buildings* **13(5)** 1289.

[5] Zhan J, Lu J and Wang D 2022 Review of shell waste reutilization to promote sustainable shellfish aquaculture *Reviews in Aquaculture* **14(1)** 477-488.

[6] Wang Z 2020 Research on *Water-Based Seashell Powder Coating* (Changchun: Jilin University of Architecture) Master's Thesis.

[7] Kaewdang S and Nirunsin R 2019 Synthesis of calcium oxide from river snail shell as a catalyst in production of biodiesel *Applied Environmental Research* **41(1)** 31-37.

[8] Liu X H, Bai H X, Zhu D J and Cao G 2011 Green catalyzing transesterification of soybean oil with methanol for biodiesel based on the reuse of waste river-snail shell *Advanced Materials Research* **148** 794-798.

[9] Roschat W, Siritanon T, Kaewpuang T, Yoosuk B and Promarak V 2016 Economical and green biodiesel production process using river snail shells-derived heterogeneous catalyst and co-solvent method *Bioresource Technology* **209** 343-350.

[10] Maftukhah S, Agustine D, Heri S M and Arrizal H 2022 Application of golden snail shell powder as a river water purifier *The 1st International Conference on Social, Science, and Technology* （Tangerang) vol 1 .(Tangerang :Indonesia) P24.

[11] Stel'makh S A, Shcherban' E M, Beskopylny A N, Mailyan L R, Meskhi B, Beskopylny N, et al. 2022 Nanomodified concrete with enhanced characteristics based on river snail shell powder *Applied Sciences* **12(15)** 7839.

[12] Kaewdaeng S, Sintuya P and Nirunsin R 2017 Biodiesel production using calcium oxide from river snail shell ash as catalyst *Energy Procedia* **138** 937-942.

Preparation and Thermal Aging Properties of Nano-Modified Alkylbenzene Insulating Oil

Jianping Liao[1], Huo Yang[2], Zhifeng Liu[1], Fan Gao[1], Yongye Xu[1] and Zhiping Zhu[2,*]

[1] Electric Power Research Institute of EHV Transmission Company, China Southern Power Grid Co., Ltd., Guangzhou 510405, Guangdong, China.
[2] Hunan Provincial Key Laboratory of Materials Protection for Electric Power and Transportation, School of Chemistry and Chemical Engineering, Changsha University of Science and Technology, Changsha 410114, Hunan, China.
Email: 81425128@qq.com; 1094365410@qq.com; 1412149786@qq.com; 416327881@qq.com; 461431507@qq.com; *zhuzp@csust.edu.cn

Abstract. The implementation of submarine cables is more and more widespread with the rapid economic development of the island economy. Insulating oil is one of the core materials of submarine cables, but domestic alkylbenzene insulating oil is limited to 220 kV, the alkylbenzene insulating oil used for the ultra-high voltage submarine cables above 500 kV is imported. In order to replace imported oil, nano-modified alkylbenzene insulating oil was prepared by modifying SiO_2 nanoparticles with 3-aminopropyltriethoxysilane (APTES) in this study. A comparative analysis was conducted to evaluate the thermal aging properties of this oil and imported T3788 oil. The results show that the nano-modified insulating oil has a lower acid number, dielectric loss factor after aging, and the decrease of interfacial tension is the smallest. This indicates that the heat aging resistance of nano-modified insulating oil is stronger than that of imported T3788 oil, which offers a new reference for the selection of insulating oil for submarine cables in the future.
Keywords. Submarine cable; insulating oil; thermal aging; nano-modification.

1. Introduction

With the development of the world's energy and the growing demand for electricity, the demand for trans-sea power transmission has increased sharply. A submarine cable is a cable laid on the sea floor, the world's first submarine cable appeared in the mid-19th century, between Britain and France. In recent years, with the increasing energy demand in coastal areas, submarine cables have evolved into an integral part of the transmission of electrical power between terrestrial and insular locations [1-3].

There are three main types of submarine cable: oil-filled cable, impregnated paper-insulated cable, and cross-linked polyethylene cable. Although impregnated paper-insulated cables have more applications in ultra-high voltage DC networking projects, and cross-linked polyethylene cables are gradually developing into ultra-high voltage fields, oil-filled cables have advantages in terms of safety and service life. At present, 500 kV AC ultra-high voltage submarine cables in the world are all oil-filled cables [4,5].

Oil-filled cables are a class of power cables that use supplemental impregnants to prevent air pockets from forming within the insulating material in order to increase the cable's working field strength [6]. Oil-filled cables boast significant technical superiority in terms of safety and durability, making them the preferred choice for ultra-high voltage submarine cables. Research into this area is of paramount importance [7].

In the context of oil-filled cables, insulating oil is one of the indispensable materials. Its physical properties, chemical properties, electrical properties, etc. are instrumental in ensuring the stable work of oil-filled cables [8]. In the early days, the insulating oil was mainly low viscosity mineral oil. Until the 1980s, alkylbenzene insulating oil became the main insulating oil used in oil-filled cables because of its excellent low-temperature properties, gas precipitation, and kinematic viscosity, replacing the traditional mineral insulating oil [9].

Currently, domestically produced alkylbenzene insulating oils are mainly limited to oil-filled cable applications at voltage levels of 220 kV and below. Therefore, it is particularly important to seek a solution that can both enhance the performance of domestically produced alkylbenzene oil and reduce the cost. Nanotechnology, as a cutting-edge technology, has a wide range of application prospects in the field of material modification [10,11].

Cong et al. [12] investigated the properties of Fe_3O_4, TiO_2, and Al_2O_3 nanoparticles in vegetable insulating oils and found the particles exhibiting a concentration of 0.01‰ demonstrated the most favourable effect in terms of modification and the nano-Fe_3O_4 was superior to the other two. Wu et al. [13] investigated the impact of modifying SiO_2 particles with KH550 to enhance the performance associated with insulating oil and found that modified SiO_2 particles have a highly effective adsorbing capacity for particles in the oil, which can improve its dielectric properties and have the effect of delaying aging. Qin et al. [14] modified natural ester insulating oils using silane coupling agent grafted nanohexagonal boron nitride (APTES-hBN) thereby enhancing the resistance of the insulating oil to thermal aging. This was attributed to the ability of APTES-hBN to engage in hydrogen bonding with acids, inhibiting the hydrolysis process of the oil.

In this study, the domestic dodecylbenzene insulating oil was modified by nano-SiO_2 modified with silane coupling agent 3-aminopropyltriethoxysilane (APTES). Secondly, thermal aging tests were carried out on the nano-modified alkylbenzene insulating oil as well as the imported T3788 oil to compare the changes in acid value, dielectric loss factor, and interfacial tension during thermal aging of different oils.

2. Experimental

2.1. Specimen Preparation

In the course of this study, two distinct types of oils are employed. One is domestic dodecylbenzene insulating oil and the other is imported T3788 oil. The domestic dodecylbenzene insulating oil is used as the base oil for adding nanoparticles, and the T3788 oil is used as the comparison oil in the heat aging test. The base oils were vacuum dried at 90°C for 24h before the start of the test to avoid moisture in the oils from interfering with the test results.

The nanoparticles utilized were procured as SiO_2 powders, exhibiting a dimension of approximately 15 nanometres, a specific area of greater capacity of over 150 m^2/g, and a pureness of over 99.9%. The silane coupling agent: 3-aminopropyltriethoxysilane (APTES) was chosen to modify the SiO_2 nanoparticles.

2.2. Preparation of Nano-Modified Insulating Oil

To ensure that the dimensions of SiO_2 nanoparticles corresponded to the specified requirements, it was subjected to milling prior to utilization and subsequently dried at 90°C, with the objective of eliminating any adsorbed water. 1 g of the treated nano-SiO_2 was added to a solution of 20 mL of ethanol. The SiO_2 nanoparticles were treated using an ultrasound technique for a period of 30 minutes, with the objective of ensuring uniform dispersion of the nanoparticles in the solution. Subsequently, 0.6 mL of APTES was added to the solution, while stirring. The reflux reaction was then conducted in a stable temperature environment, maintained at 80°C for a period of 30 minutes. After the reaction solution was cooled down, the mixture was centrifuged and the lower precipitate was separated from the solution.

The precipitate was subjected to vacuum dehydration in an oven maintained at 90°C for a period of 24 hours, with the objective of achieving the modification of SiO_2 nanoparticles with APTES. A nano-modified insulating oil with a SiO_2 nanoparticle content reaching 0.12 g/L was prepared and treated with ultrasonication for 30 minutes. This process ensured the uniform dispersion of APTES-modified nanoparticles in the oil. The prepared nano-modified insulating oil was named as A-SiO_2 oil and the insulating oil formulated with pure SiO_2 nanoparticles was named as O-SiO_2 oil. The specific preparation process is shown in figure 1.

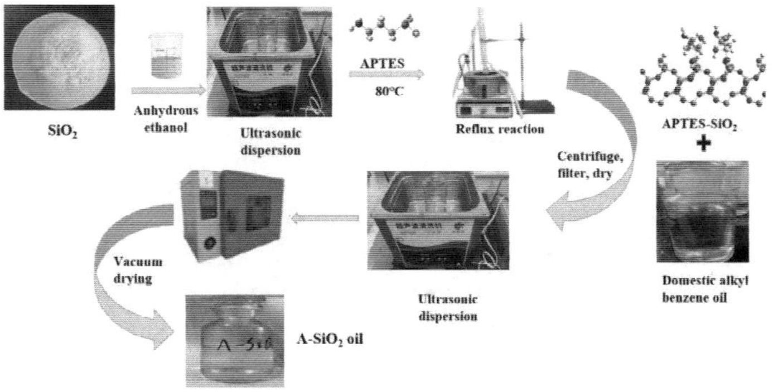

Figure 1. The procedure for nano-modifying insulating oil.

2.3. Thermal Aging Test

To explore the difference between A-SiO_2 oil and imported oil in thermal aging resistance, thermal aging tests were conducted on A-SiO_2 oil and imported T3788 oil. Before the aging test, the insulating oil was subjected to a vacuum oven drying process at 90°C for a duration of 24 hours. Following this, thermal aging experiments were conducted in a drying oven at 120°C for 20 days with the A-SiO_2 oil and the imported T3788 oil.

During this period, the sampling interval was 5 days and the samples taken were tested for a range of parameters such as dielectric loss factor, interfacial tension, and acid number. The dielectric loss factor is a measure of the dielectric properties of insulating oil in response to an electric field and thus provides insight into the loss characteristics of insulating oil. The acid number can enable the acid content contained in the oil, which is an integral chemical parameter for assessing new oil and gauging the extent of deterioration in existing oil. The interfacial tension is conceptualized as the tension at the interface between insulating oil and other substances. This concept can elucidate the magnitude of interaction between insulating oil and other substances.

3. Results and Discussion

3.1. Characterization of Modified Nanoparticles

To illustrate the efficacy of APTES grafting on nano-SiO_2, infrared spectroscopy (FT-IR) and TEM tests were performed on pure SiO_2 nanoparticles and APTES-modified SiO_2 nanoparticles, respectively. As shown in figure 2, in the pure nano-SiO_2 sample, a broad peak exists in the wavelength range of 3000 cm^{-1} - 3500 cm^{-1}, reflecting the characteristics of the stretching vibration associated with the hydroxyl groups located on the surface of the nano-SiO_2 particles. However, after the modification treatment with APTES, the peaks in this region were significantly weakened, which is strong evidence that part of the surface hydroxyl groups were effectively replaced by APTES.

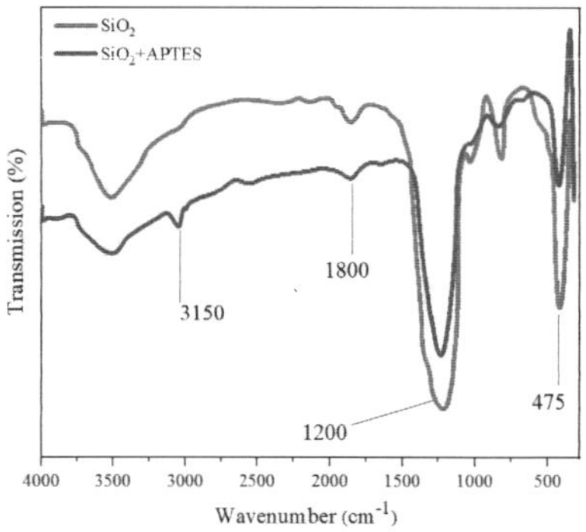

Figure 2. FT-IR plots of SiO_2 and APTES-modified SiO_2.

Notably, the APTES-modified nano-SiO_2 showed characteristic peaks about organic groups at 3150 cm^{-1}. This indicates that the modification treatment of the nano-SiO_2 surface by APTES not only effectively reduced the number of polar hydroxyl moieties on the exterior, but also successfully introduced organic groups into the nanoparticles. According to the principle of similar solubility, these newly introduced organic groups will help to achieve a more stable and homogeneous dispersion of nano-SiO_2 in organic insulating oils, and thus effectively reduce the agglomeration of nanoparticles.

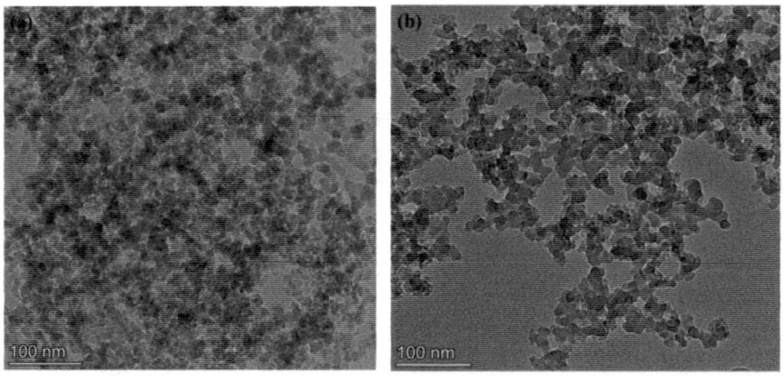

Figure 3. Transmission electron micrographs of (a) SiO_2 nanoparticles without any treatment and (b) SiO_2 nanoparticles modified by APTES.

The results of the transmission electron microscopy (TEM) analysis are presented in figure 3. As illustrated in figure 3(a), it is evident that the SiO_2 nanoparticles have an obvious agglomeration phenomenon and their edges are blurred. In figure 3(b), the agglomeration phenomenon of the SiO_2 nanoparticles after APTES modification has been significantly improved. The edges of the SiO_2 nanoparticles become clear and transparent, showing good dispersion. The results demonstrate that the modification of SiO_2 nanoparticles on the surface by APTES effectively replaces the hydroxyl groups on the surface of SiO_2 nanoparticles with organic groups, thereby facilitating the uniform dispersion of SiO_2 nanoparticles in the oil and preventing the formation of agglomerates.

3.2. Thermal Aging Test

The acid numbers of the examined insulating oil samples exhibited a gradual increase with the number of days dedicated to the aging process, as shown in figure 4(a). During the inaugural phase of the aging process, the growth of acid value was relatively gentle; however, as the aging process progresses, the growth rate of acid value accelerates significantly. The accumulation of acids can exert a deleterious effect on the electrical characteristics of insulating oils. In the early stage of aging, the difference between A-SiO_2 oil O-SiO_2 oil, and T3788 oil is not significant; however, in the late stage of aging, The acid number of T3788 oil was markedly higher in comparison to that of A-SiO_2 oil. This indicates the A-SiO_2 oil has excellent thermal stability and helps to retard the acidic hydrolysis process of the insulating ester.

As illustrated in figure 4(b), the dielectric loss factor of A-SiO_2 oil was consistently kept at the lowest level throughout the test. However, after aging for a certain period of time, there was a notable increase in the dielectric loss factor of the insulating oil, and this change coincides with the growth trend of the acid value, which further reveals the impact of acids upon the performance of the oils during the aging process. This phenomenon may be attributed to the fact that the modified nano-SiO_2 on the surface of APTES is able to effectively adsorb water and acid in the oil, thus attenuating the promotional effect of water and acid on the aging process, which in turn effectively slows down the aging rate of the insulating oil.

The interfacial tension can reflect the quantity of hydrophilic polar molecules present among the oil's constituents. As shown in figure 4(c), A-SiO_2 oil has the smallest decrease in interfacial tension, which is better than that of O-SiO_2 oil. It indicates that A-SiO_2 oil has better anti-aging performance, which is due to the fact that the APTES-modified SiO_2 nanoparticles can better dissolve in dodecylbenzene oil, which produces fewer polar molecules. In conclusion, the anti-aging performance of A-SiO_2 oil is superior to that of imported T3788 oil.

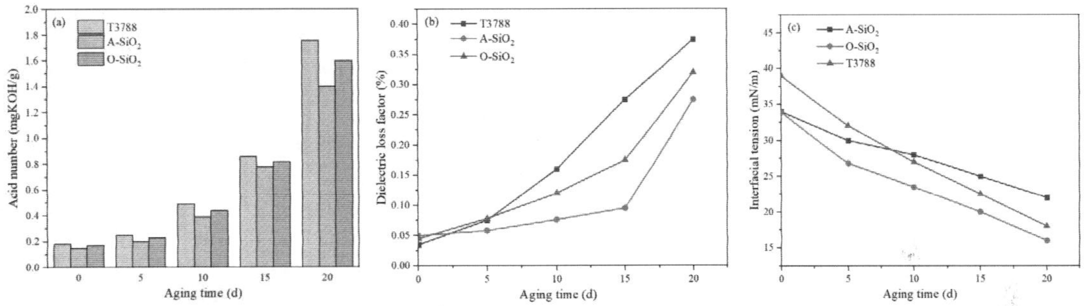

Figure 4. Changes in (a) acid value, (b) dielectric loss factor, and (c) interfacial tension of various types of oils during the aging process.

4. Conclusion

In this study, SiO_2 nanoparticles modified by APTES were successfully prepared, and these nanoparticles had better solubility in the oil and could be dispersed more uniformly, which significantly reduced the agglomeration of the nanoparticles.

The comparisons of acid value, dielectric loss factor, and interfacial tension changes during the aging process of different oils strongly indicated that the nano-modified alkylbenzene insulating oil exhibited better thermal stability and aging resistance than imported T3788 oil. Compared with the pure SiO_2 nanoparticles, SiO_2 nanoparticles modified by APTES had a better modification effect, which effectively improved the performance of alkylbenzene insulating oil in all aspects and inhibited the aging process of alkylbenzene insulating oil. The successful formulation of nano-modified alkylbenzene insulating oil brings a new kind of choice for the use of insulating oil in the future, which is of great consequence to the safe and stable operational performance of current and future power systems.

Acknowledgments

This work was supported by the Technology Research Program supported by the Electric Power Research Institute of EHV Transmission Company (CGYKJXM20220080).

5. References

[1] Tang W, Brown K, Mitchell D, Blanche J and Flynn D 2023 Subsea power cable health

management using machine learning analysis of low-frequency wide-band sonar data *Energies* **16(17)** 6172.

[2] Taormina B, Bald J, Want A, Thouzeau G, Lejart M, Desroy N and Carlier A 2018 A review of potential impacts of submarine power cables on the marine environment: Knowledge gaps, recommendations and future directions *Renewable and Sustainable Energy Reviews* **96** 380-391.

[3] Wang T J, Moran B and Zukerman M 2024 Capacity-aware undersea cable system design *Journal of Lightwave Technology* **42(8)** 2648-56.

[4] Zhang T, Shi G R, Zhang N, Du A Q and Wu Q 2024 Analysis and aging evaluation method of frequency domain dielectric spectroscopy for high-voltage XLPE submarine cables *Electrical Engineering* **106** 3403-13.

[5] Gao C F, He D X, Zhou Y F, Wang W and Wang P 2019 A study on the space charge characteristics of AC sliced XLPE cables *IEEE Access* **7** 20531-37.

[6] Sugimoto S and Nakade M 2018 Copper compound deposition for decreasing electrical insulation in aged oil-filled cables *IEEE Transactions on Dielectrics and Electrical Insulation* **25(5)** 1660-67.

[7] Gustavsen B, Høidalen H K and Ohnstad T M 2014 Field measurement and simulation of electromagnetic transients on 132kV oil-filled submarine cables *Electric Power Systems Research Volume* **115** 43-49.

[8] Wang Y, Zhang X, Li L, Du J and Gao J 2019 Design of partial discharge test environment for oil-filled submarine cable terminals and ultrasonic monitoring *Energies* **12(24)** 4774.

[9] Hosier I L, Koilraj J E A and Vaughan A S 2016 Effect of aging on the physical, chemical and dielectric properties of dodecylbenzene *IEEE Transactions on Dielectrics and Electrical Insulation* **23(6)** 3389-96.

[10] Chen J, Sun P, Sima W, Shao Q, Ye L and Li C 2019 A promising nano-insulating-oil for industrial application: electrical properties and modification mechanism *Nanomaterials* **9(5)** 788.

[11] Liao J P, Chu J W, Gao F, Chen Y F, Zhang C, Yu S M, Li H Q and Zhong L S 2021 Research progress on the performance of alkylbenzene insulating oil for submarine cables *Insulation materials* **54(10)** 20-25.

[12] Cong H X, Shao H M, Du Y L, Hu X F, Zhao W J and Li Q M 2022 Influence of nanoparticles on long-term thermal stability of vegetable insulating oil *IEEE Transactions on Dielectrics and Electrical Insulation* **29(5)** 1642-50.

[13] Wu Y R, Chen R, Zhang J Y and Tang C 2024 The influence of doping with different surface modified SiO_2 nanoparticles on the dielectric properties of natural ester insulating oil *Renewable Energy* **228** 120671.

[14] Qin J S, Peng X, Qiu Q P and Tang C 2022 A new type of nano APTES-hBN modified palm oil as natural ester insulating oil with upgraded thermal aging characteristics *Renewable Energy* **200** 743-750.

Moisture Absorbing Properties of Alkylbenzenes with Experiment and Molecular Dynamics Simulation Calculation

Zhifeng Liu[1], Shangming Zhou[2], Jianping Liao[1], Fan Gao[1], Yufei Chen[1] and Zhiping Zhu[2,*]

[1] Electric Power Research Institute of EHV Transmission Company, China Southern Power Grid Co., Ltd., Guangzhou 510405, Guangdong, China.
[2] School of Chemistry and Chemical Engineering, Changsha University of Science and Technology, Changsha 410114, Hunan, China.
Email: 1412149786@qq.com; 469157890@qq.com; 81425128@qq.com; 416327881@qq.com; 181997300@qq.com; *zhuzp@csust.edu.cn

Abstract. The presence of moisture accelerates the aging of insulating oil thus greatly reducing the insulating performance. In this study, in order to investigate their absorb moisture characteristics, comparison experiments and molecular dynamics simulation calculations at 1wt% oil-water mixing were carried out for the alkylbenzene insulating oil, mineral oil and vegetable oil. The experiment results show that vegetable oil has the strongest moisture absorption ability, alkylbenzene insulating oil is the second, and mineral oil is the weakest at temperature 15-45°C and relative humidity 90%, as the temperature rises from 15°C to 30°C, the growth of the moisture absorption rate is significant, but as temperature rises from 30°C to 45°C the increase tendency is smaller; the results of molecular dynamics simulation the show that with the increase of temperature, the molecular motion is intensified, the interaction energy decreases, and the free volume and diffusion coefficient increase. These results help to understand the moisture absorption mechanism of alkylbenzene insulating oil and provide a theoretical basis for its storage and use.
Keywords. ReaxFF-MD; insulating oil; model construction; moisture absorption property.

1. Introduction

The stable operation of power transmission and transformation equipment usually depends on the protective effect of transformer insulating oil on transmission and transformation system, with the renewal of the power system and the unceasing innovation of transformer technology, the performance needs of the insulating oil are also more and more stringent, not only need to have a low viscosity to ensure smooth operation inside the cable, but also must have excellent electrical properties to guarantee the safety and stability of the cable.

The insulating substance used inside oil-filled cables is insulating oil, usually for the convenience of daily management and strategic reserve, the management organization usually stores a large amount of insulating oil, but in the process of long-term placement, aging or deterioration may occur due to environmental factors, etc., and its physicochemical and electrical properties will be reduced, in addition to the temperature, the humidity is considered to be the most important factor leading to the degradation of the transformer. And during the operation of transformers, humidity, and temperature will have a synergistic effect, accelerating the aging of the oil-paper insulation system [1]. Therefore, an in-depth study of the effect of environmental factors on the aging performance of alkylbenzene insulating oil is essential.

Under the current situation of rapid development of computer technology, it has begun to be broadly used in the field of chemistry and materials. Molecular simulation technology can be used to explore the simulation of processes of chemical reactions [2], chemical bonding [3] properties, and other content, these calculations can be corroborated with the results of macroscopic experiments, but also predict the change rule of macroscopic experiments [4], so its application in the realm of insulating oil technology research is also more and more. The use of molecular simulation to study the diffusion process of an insulating oil system [5] can greatly reduce the manpower and material resources and can analyze the molecular motion behavior from a microscopic point of view, which can provide theoretical underpinning support for the study of insulating oils in the new type of high-voltage power transmission devices.

Therefore, in this study, the moisture absorption experiments of mineral oil, alkylbenzene insulating oil, and vegetable oil were carried out, and the changes of property parameters in the oil-water mixture system during the moisture absorption process were simulated by molecular dynamics [6] as a means of explaining the moisture absorption variability of the three kinds of oil-water mixtures from a microscopic point of viewpoint and providing certain theoretical guidance for selecting a high-standard insulating oil for transmission and substation equipment that meets the development rate of the power system.

2. Experimental

2.1. Insulation Oil Moisture Absorption Experiment

The experimental program for the study of moisture absorption characteristics of three oils, naphthenic mineral oil (25#), alkylbenzene insulating oil (DDB), and vegetable oil (FR3), is as follows:

(1) Vacuum drying treatment is carried out under the condition of a vacuum drying oven with air pressure -0.1GPa and temperature 70°C for 24 hours to remove moisture, and after drying, the initial moisture content is measured after 6 hours of natural cooling in the vacuum drying oven.

(2) The oil after initial determination was placed in a constant temperature and humidity chamber with a relative humidity of 90% to accelerate the moisture absorption rate of the insulating oil. Three experimental temperature points were set, namely 15°C, 30°C, and 45°C, and eight sampling time points were 0, 2, 6, 12, 18, 24, 30, and 36 h, totaling 24 sampling time points. At each sampling time point, the moisture content was measured three times in parallel to take the average value.

2.2. Molecular Dynamics Experiment

In this paper, the SPCE model was chosen, which is suitable for water vapor simulation, and the values of the diffusion coefficient and other values calculated by this model are well correlated with the experimental values. After the geometry optimization of each molecule, the AC module of MS is used to construct a mixing model with a water molecule mass fraction of 1%, and the obtained oil mixing model is firstly geometrically optimized, and then annealed and pressurized decompression calculations are carried out respectively after convergence to obtain the final oil-water mixing model with a stable density of 0.84 g/cm^3.

The model building of the other two insulating oils is the same process as that of mineral oil-water, in which the final density of the DDB-water mixing model is 0.85 g/cm^3 and that of the FR3-water mixing model is 0.897 g/cm^3, which is close to the density of the actual oil samples, and the energies of the models eventually tend to be stabilized, which proves the success of the modeling, and can be used for the subsequent molecular dynamics calculations.

The above hybrid model was used to simulate the molecular dynamics of the NVT system with a total duration of 300ps. The COMPASS II force field was used throughout the simulation, and the simulated temperatures were 15°C, 30°C, and 45°C, respectively, with a step size of 1fs, and the output of one frame every 1ps, and the resulting trajectory files were analyzed for the molecular dynamics properties: interaction energy, free volume, and diffusion coefficient.

3. Result Analysis

3.1. Analysis of Moisture Absorption Experiment Results

The moisture absorption experiments at different temperatures are shown in figure 1, and from figure 1(c), it can be found that the initial water content of vegetable oil can only be at 100 μg/mL in the same water removal method, and in addition, the water content of vegetable oil shows a linear increase at 30°C as well as 45°C, and does not reach the maximum for a long time [7,8]. The other moisture absorption data were fitted according to the moisture absorption equation, and the fitting calculation formula is shown in equation (1):

$$y = y_0 + A_1 e^{-x/t_1} \tag{1}$$

where A_1 is the moisture absorption rate constant of the oil sample at that temperature, t_1 is the moisture absorption time constant at that temperature, and y_0 is the saturated water content of the oil at that temperature, and the results of the fitting are shown in table 1.

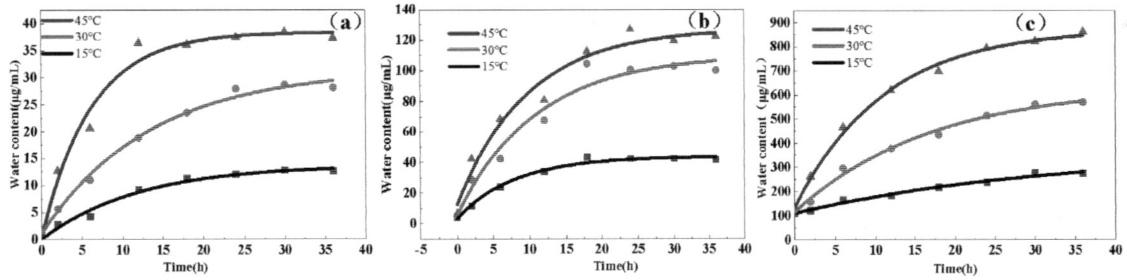

Figure 1. Moisture absorption diagram of insulating oil at different temperatures (a) naphthenic mineral oil (b) DDB insulating oil (c) FR3 insulating oil.

Table 1. Fitting results of moisture absorption curves of three insulating oils.

	25#			DDB			FR3		
	15°C	30°C	45°C	15°C	30°C	45°C	15°C	30°C	45°C
A_1	13.71	31.34	37	40.41	104.3	114.85	251.52	522.96	742.08
t_1	11.87	13.45	6.15	8	9.7	10.96	32	17.02	11.09
R^2	0.98	0.99	0.98	0.98	0.96	0.97	0.98	0.98	0.99

From figure 1 and table 1, it can be seen that the moisture absorption process of mineral oil, alkylbenzene, and vegetable oils is close to the same law, and the water content of the water content showed a swift surge in the early stage and a slow increase in the later stage of the experimental time, and the maximum moisture absorption of the mineral oil has been at a low level, while the saturated water content of alkylbenzene was 60 μg/mL at the temperature of 15°C to 30°C, and only 20 μg/mL increase in saturated water content at the temperature from 30°C to The saturated water content of vegetable insulating oils increased linearly from 15°C to 30°C, and the saturated water content of alkylbenzene increased by about 290 μg/mL at a temperature of 30°C to 45°C, while the saturated water content of vegetable insulating oils increased linearly by about 290 μg/mL at a temperature of 15°C, and the level of moisture present in vegetable-based insulating oils increased by about 290 μg/mL at a temperature of 30°C to 45°C, and the level of moisture present in vegetable-based insulating oils increased by about 290 μg/mL at a temperature of 45°C, and the level of moisture present in vegetable-based insulating oils increased linearly at a temperature of 15°C.

3.2. Interaction Energy

The interaction energy can indicate how close the fragments are to each other, the smaller the interaction energy represents the greater the adsorption between two fragments and the more stable their structure. The formula is shown in equation (2):

$$E_{Inter} = E_{total} - (E_A + E_B) \quad (2)$$

where E_{Inter} denotes the interaction energy between the media, and E_A and E_B denote the potential energy of segment A and segment B, respectively. In addition, since there will be differences in size between each model, the interaction energy needs to be processed using the normalization method to calculate the interaction energy per unit area, which is shown in equation (3):

$$E_{int\ Area} = \frac{E_{int}}{E_{surface}} \quad (3)$$

where $E_{int\ Area}$ denotes the interaction energy per unit area, E_{int} denotes the interaction energy between fragments, and $E_{surface}$ denotes the contact area between two fragments when they are combined.

Table 2. The interaction energy between oil and water in different models.

	25#			DDB			FR3		
	15°C	30°C	45°C	15°C	30°C	45°C	15°C	30°C	45°C
E_{Inter} (kcal/mol)	-8.033	-7.758	-7.67	-18.34	-14.99	-13.68	-15.89	-13.14	-12.94
$E_{intArea}$ (kcal·mol^{-1}·Å^{-2})	-0.0119	-0.0115	-0.0115	-0.0274	-0.0224	-0.0205	-0.0223	-0.0202	-0.0199

From the interaction energy table 2, It is observable that there exists interaction energy between oil and water in these three oil-water mixtures decreases with increasing temperature, which is due to the fact that the intermolecular motion is intensified by increasing temperature, resulting in weakening of the force between the water molecules and the oil molecules. At the same temperature, the comparison of the interaction energy is $E_{DDB} > E_{25\#} > E_{FR3}$. From the data of $E_{int\ Area}$, it can be seen that the mineral oil is less sensitive to the change, while the DDB and FR3 reflect a stronger sensitivity to the change of temperature. Vegetable oils are mainly due to the presence of hydrogen bonding, a stronger specific bond, in the vegetable oil system. The alkylbenzene system, on the other hand, is due to the presence of a large number of benzene ring π-π bonds and [9] alkylbenzene insulating oils form π-H bonds with water molecules that enter by moisture absorption. The $E_{int\ Area}$ of alkylbenzene is greater than that of vegetable oils because the number of alkylbenzene molecules in the model is greater than that of vegetable oils, and the magnitude of the interaction energy is positively related to the number of molecules.

3.3. Free Volume

The larger the free volume, the more favorable the diffusion and displacement of particles in the system. According to the free volume theory of Fox and Flory, the total volume (V_t) of a system is divided into the volume occupied by molecules (Occupied volume (V_o)) and the free volume of unoccupied "holes" (free volume (V_F)). The ratio of free volume to total volume is the free volume fraction (FFV) [10]. The formula is shown in equation (4):

$$FFV = \frac{V_F}{V_F + V_O} \quad (4)$$

Different molecular sizes have different free volumes in the same model, and in this paper, we calculate the free volume of water molecules in the system by creating Connolly surfaces as a way to calculate the V_F of the interfacial region with a hard-sphere probe, and the water molecules have a van der Waals force radius of 1.4 Å.

Table 3. Free volume fractions in different models.

	25#			DDB			FR3		
	15°C	30°C	45°C	15°C	30°C	45°C	15°C	30°C	45°C
V_F (Å^3)	1268.6	1462.6	1486.9	1718.9	1759.3	1915.9	1429.5	1446.2	1475.1
V_O (Å^3)	16330	16270	16246	15568	15529	15372	15178	15073	15133
FFV (%)	0.0721	0.0824	0.0839	0.0995	0.1018	0.1108	0.0861	0.0875	0.0888

Based on the graphical representation of table 3 on free volume and the calculation of the free volume fraction. It can be seen that the free volume fraction increases with an increase in temperature. The free volume of alkylbenzene is close to that of mineral oil, which is 1.16 times more than that of vegetable oil, which is due to the fact that alkylbenzene is close to the viscosity of mineral oil, whereas the viscosity of vegetable oil is higher than that of alkylbenzene. Although alkylbenzene and mineral oil provide more space for the diffusion of water molecules, the smaller interaction force between mineral oil and water molecules will lead to the intensification of the migratory movement of water molecules in mineral oil between mineral oils, which makes it more difficult for other water molecules to enter into the interior of mineral oils, and thus leads to a weaker moisture-absorption capacity of mineral oils.

3.4. Diffusion Coefficient

The diffusion behavior of water molecules can be analyzed by the mean-square displacement (MSD) line in the oil-water mixture system, which is analyzed as a whole in a multimolecular system to reflect the law of molecular motion. In this study, the mean square displacement during molecular simulation is calculated and analyzed by using Einstein's formula, which is a conversion of the microscopic MSD into a macroscopic diffusion coefficient [11,12], where the integral part is equivalent to the ratio of the MSD curve to the time differential, i.e., the slope k in the MSD curve, and thus the diffusion coefficient can be replaced by equation (5):

$$D = k/6 \tag{5}$$

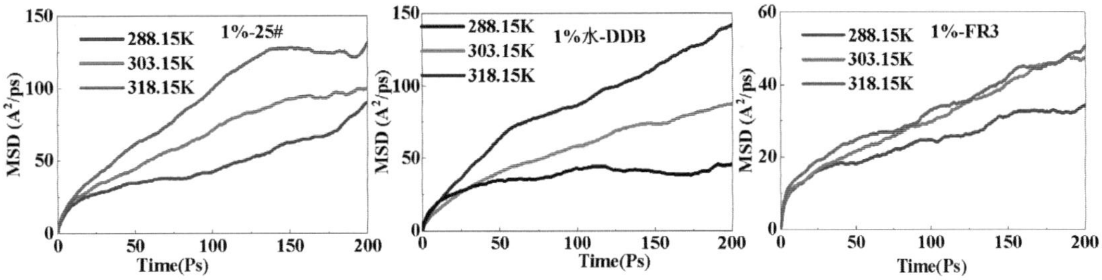

Figure 2. MSD curves of different water-oil systems

Table 4. Diffusion coefficients of different water-oil systems at different temperatures.

	25#			DDB			FR3		
Temperature	15°C	30°C	45°C	15°C	30°C	45°C	15°C	30°C	45°C
k	0.216	0.516	0.793	0.112	0.377	0.533	0.108	0.179	0.187
D (Å^2/ps)	0.036	0.086	0.132	0.0187	0.0628	0.0888	0.018	0.0298	0.0312

During the calculation of the MSD curves, 50ps before and after were discarded, and the MSD curves for the middle 200ps were investigated to avoid the error of simulating the unadapted and fatigued phenomena in the early stage. From figure 2, it is obvious that the diffusion coefficients of all three water-oil systems gradually increase with increasing temperature, and the MSDs of mineral oils are 1.37-1.96 and 2-4.23 times higher than those of alkylbenzene as well as vegetable oils, respectively, which is attributed to the fact that the higher the temperature is, the higher the kinetic energy of the molecules is, and the molecular Brownian motion is intensified. From the change of diffusion coefficient in table 4, it can be seen that the diffusion coefficient of alkylbenzene increases by three times when the temperature changes (from 15°C to 30°C) at low temperatures, however, from 30°C to 45°C, the increase is weak, which is the same as that of the interaction energy and the change rule of the saturated water content in the moisture absorption experiments before, which further illustrates the reliability of simulation calculations.

4. Conclusion

In this study, moisture absorption experiments were conducted on alkylbenzene, mineral oil, and vegetable oil, and a 1% water-oil model was developed using molecular dynamics, and the following conclusions were obtained.

(1) The saturated water contents of mineral oil, alkylbenzene, and vegetable oils were 13.08-38.28 μg/mL, 36-124 μg/mL, and 273.9-845 μg/mL at 90% relative humidity from 15°C to 45°C. The moisture uptake rates of mineral oil and alkylbenzene all increased substantially when the temperature was raised from 12°C to 30°C, whereas the rate change was smaller when the temperature was raised from 30°C to 45°C. The rate change was small.

(2) The results of molecular dynamics simulation show that with the increase of temperature, the intermolecular motion is intensified, which leads to the decrease of interaction energy and the increase of free volume and diffusion coefficient, and matches with the pattern of moisture absorption experimental data, which verifies the reasonableness of the simulation analysis.

Acknowledgment

This work was supported by the Technology Research Program supported by the Electric Power Research Institute of EHV Transmission Company (CGYKJXM20220080).

5. References

[1] Hao J, Chen G, Liao R J, Yang L J and Tang C 2012 Influence of moisture on space charge dynamics in multilayer oil-paper insulation *IEEE Transactions on Dielectrics and Electrical Insulation* **19(4)** 1456-1464.

[2] Xin L Y, Liu C, Liu Y, Huo E G, Li Q B, Wang X R and Cheng Q L 2020 Thermal decomposition mechanism of some hydrocarbons by ReaxFF-based molecular dynamics and density functional theory stud *Fuel* **275** 117885.

[3] Yan H W, Nie B S, Peng C, Liu P J, Wang X T, Yin F F, Gong J, Wei Y Y and Lin S 2021 Molecular model construction of low-quality coal and molecular simulation of chemical bond energy combined with materials studio *Energy & Fuels* **35(21)** 17602-17616.

[4] Chen Y Z, Wang Z T, Li B, Yu K, Wang H, Wang J, Huo Y P and Wang J F 2023 ReaxFF molecular dynamics study on the microscopic mechanism for kerogen pyrolysis *Langmuir* **39(50)** 18581-18593.

[5] Zuo H C, Huang Z Y, Wang Q, Li J and Chen X Y 2023 Molecular dynamics investigation on the diffusion of POSS in insulating oil under external electric field *IEEE Transactions on Dielectrics and Electrical Insulation* **30(5)** 1991-2000.

[6] Zhou J, Huang Z Y, Wang F P, Wang Q, Zhang K and Zuo H C 2023 Molecular-level evaluation of thermal field confinement effects on moisture distribution at the interface of natural ester oil-paper insulation: An investigation of electric field correlation *Journal of Molecular Liquid* **387** 122661.

[7] Wang L H, Yin F, Shen Y, Tang C 2020 Reactive molecular dynamics research on influences of water on aging characteristics of PMIA insulation paper *Journal of Applied Physics* **127(10)** 105107

[8] Arsad S R, Ker P J, Jamaludin M Z, Choong P Y, Lee H J, Thiviyanathan V A and Yang Ghazali Y Z **2023** Water content in transformer insulation system: a review on the detection and quantification methods *Energies* **16** 1920.

[9] Gierszal K P, Davis J G, Hands M D, Wilcox D S, Slipchenko L V and Ben-Amotz D 2011 π - hydrogen bonding in liquid water The Journal of Physical Chemistry Letter 2(22) 2930-2933.

[10] Jiang T Y, Yang J, Yang X, Yang L, Bi M Q and Zhou X 2022 Diffusion mechanisms of dissolved gases in transformer oil influenced with moisture based on molecular dynamics simulation *ACS Omega* **7(44)** 39812-39821.

[11] Qiu Q P, Zhang J W, Yang L, Du D Y and Tang C 2021 Differences analysis of water molecular diffusion behaviors in vegetable oil and mineral oil under temperature field *Journal of Molecular Liquids* **323** 115030.

[12] Qiu Q P, Tian W X, Yang L, Ding Z Y, Zhang J W and Tang C 2021 Simulation effect of SiO_2 nanoparticles on the water molecules diffusion inside insulating oil at different temperatures *Colloids and Surfaces A: Physicochemical and Engineering Aspects* **610** 125738.

Improved Microstructures and Performance of Public Filler Aggregate and Recycled Mortar with Additional Tannic Acid

Chen Cheng[1], Kingsley Chiang[1], Xinxin Wang[1], Wenjie Zhang[1], Yazhi Zhu[2,*] and Hui Luo[3]

[1] China State Construction Engineering (Hong Kong) Limited, Hongkong 999077, China.
[2] Department of Structural Engineering, Tongji University, Shanghai 200092, China.
[3] School of Civil Engineering, Jiangsu Ocean University, Lianyungang 222005, Jiangsu, China.
Email: chen_cheng@cohl.com; kingsleychiang@cohl.com; zjwangxinxin@cohl.com; wenjie.zhang@cohl.com; *yzzhu@tongji.edu.cn; water@njfu.edu.cn

Abstract. The enhancement method of modifying public filler (PF) with tannic acid (TA) for the preparation of modified fully recycled aggregate mortar (FRAM) was introduced in this research, aiming to improve its strength and durability and further enhance its application in construction. The improvement effect of TA on the microstructure and physical properties of public fine aggregate (PFA) at different concentrations and immersion times was analyzed. Besides, the physical properties and durability of FRAM were investigated and the strengthening mechanism of TA on FRAM was analyzed by micro-technical methods. The results revealed that the loose cement mortar on the surface of PFA could be removed after TA modification. TA could react with calcium hydroxide and calcite to generate nanoparticles in situ which promote cement hydration to improve the mechanical strength and durability of FRAM. Compared with the unmodified PFA, the compressive strength was increased by 24.62% at 28 d, and the capillary water absorption and electrical flux were reduced by 54.12% and 16.56%, respectively. Microscopic analysis further indicated that TA could improve the interfacial bonding effect of PFA, and enhance the cohesion of the hardened cementitious material, as well as that TA adsorbed on the surface of PFA could be dispersed to the freshly mixed cement mortar to improve the internal structure. The above results indicated that TA can significantly improve the performance of PFA, which has a positive effect on improving the mechanical strength and durability of FRAM and is conducive to saving natural aggregates (NA) and recycling construction resources.
Keywords. Public filler (PF); fully recycled aggregate mortar (FRAM); tannic acid (TA); microscopic analysis; strengthening mechanism.

1. Introduction

A considerable amount of construction and demolition waste (CDW) is generated as urbanization progresses, posing challenges to the environment and economy [1-4]. It has been reported that over 500 million tons of CDW are produced annually worldwide [5]. In contrast, the extensive consumption of natural resources, particularly using natural aggregates (NA) in construction, imposes considerable environmental burdens. [6]. Therefore, reducing natural resource consumption and promoting CDW recycling and reuse have become global challenges [7-10]. More than 90% of CDW is composed of inert materials, commonly referred to as public fillers (PF). Utilizing PF to produce recycled mortar effectively offers a practical solution to the challenges related to construction materials, energy consumption, and environmental impact. This approach is aligned with the increasing demand for environmentally sustainable construction materials, offering promising potential [11-14]. However, recycled aggregate mortar (RAM) often falls short of engineering standards due to the presence of various internal defects [15]. It is generally believed that the presence of aged mortar adhered to the surface of recycled aggregates is a source of defects. Many researchers have reported that the presence of aged mortar on PF surfaces significantly affects the bond between recycled concrete aggregates (RCA) and fresh mortar [16,17]. Therefore, pre-treating PF is of great importance to enhance its performance.

Tannic acid (TA) is a polyphenolic compound derived from plants, characterized by its cost-effectiveness, broad accessibility, and favorable biocompatibility, making it a promising material for various applications. [18]. Oh et al. [19] found that tannic acid can capture calcium and induce localized mineralization. However, tannic acid forms precipitates with alkaline substances in cement, hindering the reaction between water and cement and thereby impeding hydration. Consequently, the direct addition of tannic acid to concrete may result in decreased compressive strength. Singh et al. [20] incorporated surface-modified bamboo fibers into glass fiber-reinforced plastic (GFRP), increasing the modulus of elasticity by 55% and providing a low-cost, efficient green surface modification agent for modified porous natural fibers. Wang et al. [21] used a nanoindentation instrument to test the elastic modulus of the interfacial transition zone (ITZ) between TA-modified recycled fine aggregate (RFA) and fresh mortar. Their study demonstrated that TA-modified RFA produced a denser microstructure at the interface with fresh mortar, promoting cement hydration and exhibiting a higher elastic modulus. Therefore, TA serves as an effective and multifunctional filler modifier, capable of comprehensively enhancing the performance of concrete, including mechanical properties, durability, and workability.

Therefore, public fine aggregate (PFA) was modified with various concentrations of tannic acid (TA) solution through the "pre-wetting + air-drying" method. The fully recycled aggregate mortar (FRAM) was produced with TA-modified PFA. Various techniques were applied to examine the microstructure, physical properties, and durability of samples, aiming to analyze the enhancement effect of TA on PFA and the reinforcement mechanism of FRAM. This approach offers both theoretical and technical perspectives on utilizing TA to improve the performance of recycled mortar.

2. Materials and Methods

2.1. Materials

The raw materials were obtained from a demolished residential area in Guangzhou and were mainly composed of brick aggregates. These aggregates underwent the process of crushing and screening, producing PFA with particle sizes ranging from 0.15 mm to 4.75 mm. Figure 1 shows the microstructure and mineral composition of PFA, the PF presented a molten state after high-temperature calcination. Besides, the PF had larger internal pores compared to NA and most of them were interconnected, resulting in strong water absorption capabilities. In this experiment, ordinary Portland cement (P·O 42.5) and TA reagent with a 95% purity were selected. The water was municipal tap water and the water reducer was a high-efficiency polycarboxylate superplasticizer with a water-reducing rate of 22.5%.

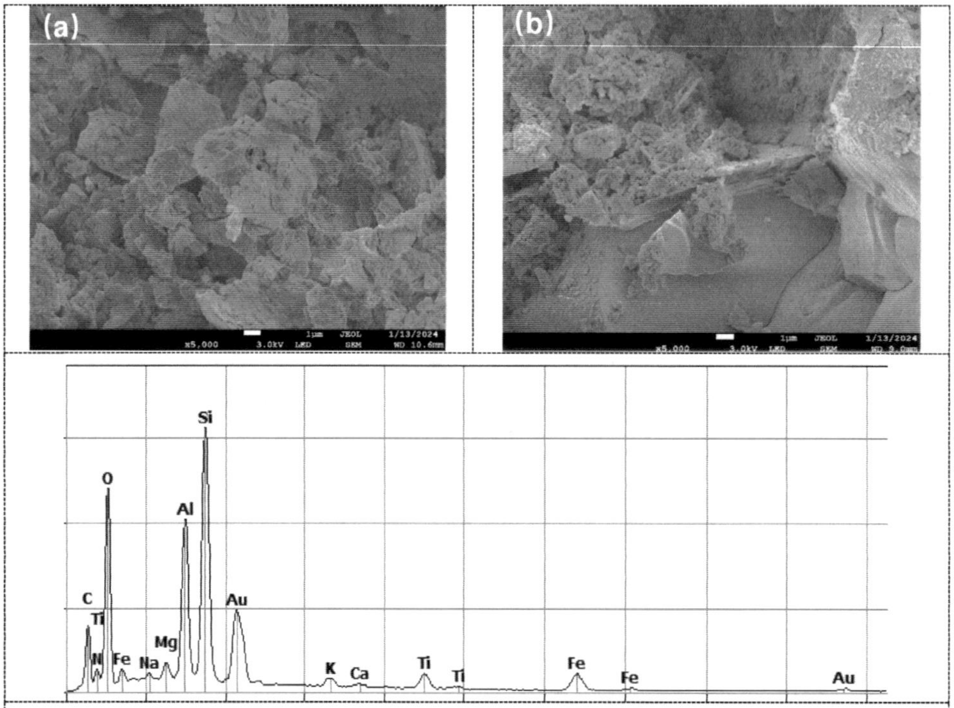

Figure 1. Microstructure and mineral composition of PFA.

2.2. Preparation of TA-modified PFA and FRAM

2.2.1. PFA Modification. The "pre-wetting + air-drying" method was used to enhance PFA with TA, following a three-step process: (1) pre-treatment of PFA, (2) TA adsorption, and (3) air-drying. First, the aggregates were placed in a 105°C oven and dried until they reached a constant weight, then cooled at 23 ± 2°C for 24 h to remove excess moisture. Then PFA was soaked in TA solutions of varying concentrations (0.2%, 0.4%, 0.6%, 0.8%, and 1.0%) for 24 h, with a solid-liquid ratio of 1:10. The mixture was then washed, re-soaked for 2 h, and then dried at 70°C for 8 h.

2.2.2. FRAM Preparation. In this experiment, the mix proportion for all specimens consisted of 472 kg/m³ of cement, 1451 kg/m³ of fine aggregate, 261 kg/m³ of water, and 1.62 kg/m³ of water reducer. Natural sand (NS) replaced PFA and was labeled NA. The unmodified PFA (0% TA) was labeled as the Control group, and the PFA modified with 0.2%, 0.4%, 0.6%, 0.8%, and 1.0% concentrations of TA solution was labeled as TA-PFA-0.2, TA-PFA-0.4, TA-PFA-0.6, TA-PFA-0.8, and TA-PFA-1, respectively. The measured cement and PFA were first mixed for 1 minute in a forced planetary mixer. Then, water along with the water reducer was added and the mixture was stirred for 3 minutes. Once the materials were thoroughly combined, the mixture was poured into 70.7 × 70.7 × 70.7 mm molds. The molds were vibrated to eliminate air bubbles, and the surface was smoothed. After 24 hours of curing, the specimens were demolded and placed in a curing box set at 23 ± 2°C with 95% humidity for 28 d before conducting strength tests.

2.3. Testing and Analytic Methods

2.3.1. Microscopic Characterization. The nitrogen adsorption-desorption test (Micromeritics ASAP 2020) was conducted to analyze the pore volume and pore size distribution of the samples. Each sample weighed approximately 5 g and degassed at 60°C for 12 h before testing. Scanning electron microscopy (SEM, JSM-7200F) was employed to investigate the surface morphology of the samples before and after

Journal of Physics: Conference Series **2941** (2025) 012004 doi:10.1088/1742-6596/2941/1/012004

modification, while energy-dispersive X-ray spectroscopy (EDS) was used to analyze their elemental composition. The samples were immersed in isopropanol for 48 h to remove any residual water and then dried in a vacuum dryer at 50°C for 48 h before testing.

2.3.2. Mineral Composition. The phase composition of both unmodified and TA-modified PFA was examined by X-ray diffraction (XRD, Ultima IV). The measurements were conducted in a 0°-100° range, with 10° intervals, and the data were analyzed using the International Crystal Structure Database (ICSD). Additionally, a Fourier transform infrared spectrometer (FTIR, Nicolet iS50) was applied to validate the attenuated total reflectance (ATR) FTIR spectra of the TA-modified PFA.

2.3.3. Physical Properties. According to GB/T 14685-2011, the physical properties of PFA were tested, including water absorption, crushing value, and bulk density. A chloride ion rapid migration test (RCM) was also conducted to examine the effect of RA on chloride ion diffusion and concrete's resistance to chloride-induced corrosion. Following GB/T 50082-2009, the chloride ion migration depth was measured for concrete mixtures with different proportions of recycled aggregates.

3. Results and Discussions

3.1. Physical Properties of TA-modified PFA

3.1.1. Water Absorption Capacity. Figure 2 demonstrates the changes in the water absorption capacity of TA-modified PFA. After modifying PFA with TA, the water absorption rates of TA-modified PFA were lower than those of unmodified PFA (Control), which exhibited the highest water absorption rate (11.3%). Moreover, with the concentration of TA solution increased, the water absorption rates of TA-PFA-0.2, TA-PFA-0.4, TA-PFA-0.6, TA-PFA-0.8, and TA-PFA-1 decreased to 10.1%, 9.6%, 9.1%, 8.3%, and 8.1%, respectively. These findings proved the positive effect of TA in reducing water absorption, which attributed to the dissolution of some loose hydration products on the surface of PFA by TA solution, thereby reducing the content of cement mortar and consequently lowering the water absorption rate of PFA.

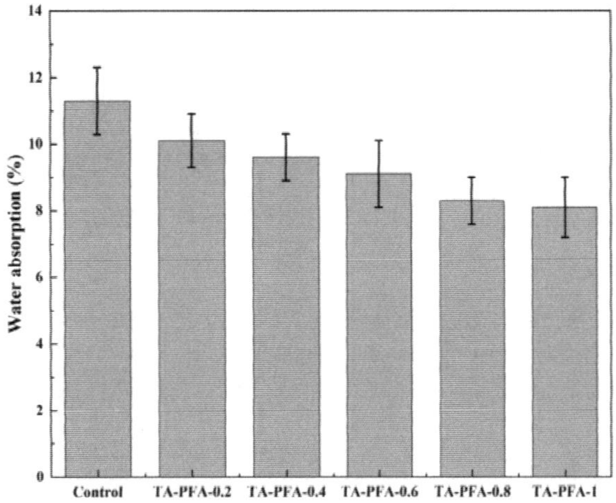

Figure 2. Water absorption of the unmodified and TA-modified PFA.

In addition, the improvement effect on water absorption weakened after the solution concentration reached a certain level. This situation might be because TA doping will react with CH and calcite in PFA to produce calcium tannate, which dissolves some loose hydration products in the reaction process and reduces the porosity of PFA. However, the content of CH (calcium hydroxide) and calcite in PFA was low, and as the solution concentration and soaking time increased, these substances were gradually consumed. This may affect the physical and chemical properties of PFA and its performance in applications. Therefore, the reduction in water absorption rate in the later stage was primarily due to TA filling the pores of PFA, which altered the pore structure and water penetration pathways of the TA-modified PFA.

3.1.2. Apparent Density. There is a large amount of irregular old cement mortar adhered to the surface of the PF. As the apparent density increases, both the proportion of attached mortar and the material's porosity decrease, indicating a denser, more compact structure that typically correlates with enhanced durability and overall superior quality. It can be seen in figure 3 that the apparent density of PFA showed an increasing trend with higher concentrations of TA and longer immersion times. PFA had the best treatment effect when immersed in 1% TA solution with an immersion time of 72 h, and the apparent density increased from the initial 1729 kg/m^3 to 2158 kg/m^3. Under the same concentration, the improvement of apparent density was slower with the increase of solution concentration and immersion time, and the increase of apparent density from 24 h to 48 h was much larger than that from 48 h to 72 h; under the same immersion time, the concentration of the solution from 0.8~1% had very little effect on the apparent density compared with that from 0.2~0.4%, 0.4~0.6%, and 0.6~0.8%. This might be because the PFA surface had more holes and gaps with the increase of immersion time in TA and the pore filling tends to saturation, continuing to increase the immersion time for the improvement of apparent density had no obvious effect. Nanomaterials had the effect of agglomeration and the continued increase of the concentration of the TA solution might lead to the TA blockage of the pores of the regenerated aggregates, which was not conducive to the filling of the pore interior.

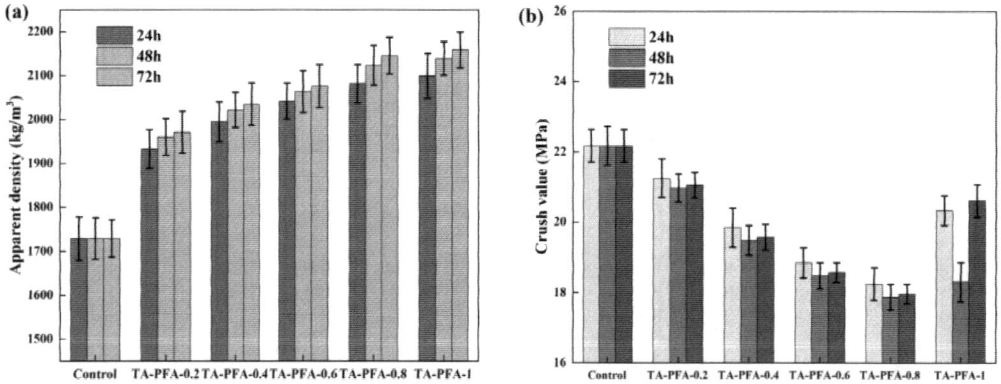

Figure 3. Physical properties of unmodified and TA-modified PFA: (a) Apparent density and (b) Crushing value.

3.1.3. Crushing Value. The changes in the crushing value of PFA modified with different TA solutions are shown in figure 3(b). The crushing value of TA-modified PFA was enhanced and gradually increased with higher solution concentrations and longer immersion times. The best improvement was achieved at the concentration of 0.8% and the immersion time of 48 h, and the crushing value was reduced from 22.1% to 18.2%. Besides, it can be found that the immersion time from 24h to 48h showed the best improvement effect on PFA under the same TA concentration. When the soaking time was raised to 72h, the crushing value produced the phenomenon of rising, in different modified solution concentrations that exist in this situation, and the concentration of more than 0.8%, the crushing value of PFA rose more. The observed phenomenon might be because of the filling effect of TA, which effectively enhanced the crushing value of PFA. However, as TA particles gradually filled the gaps between the aggregates, the improvement in the crushing value became less significant once the TA reached a saturation point in the aggregate structure.

3.1.4. Pore Structure. The comparison between the pore structure of unmodified PFA (Control) and TA-modified PFA is shown in figure 4. When the TA solution concentration was 0.6% and 1.0%, the cumulative pore volume of PFA was reduced by 43% and 66%, respectively. The reduction in porosity was primarily attributed to the reaction of TA with cement hydration products, which produced nanoparticles that partially filled the nanopores. It can be seen in figure 4(b) that the majority of the poles in PFA were smaller than 40 nm, which indicates they were mostly capillary or gel pores. Treatment of PFA with a 0.2% TA solution resulted in a slight reduction of these pores. In addition, it can be seen that the reduction of pores was higher as the concentration of TA solution increased.

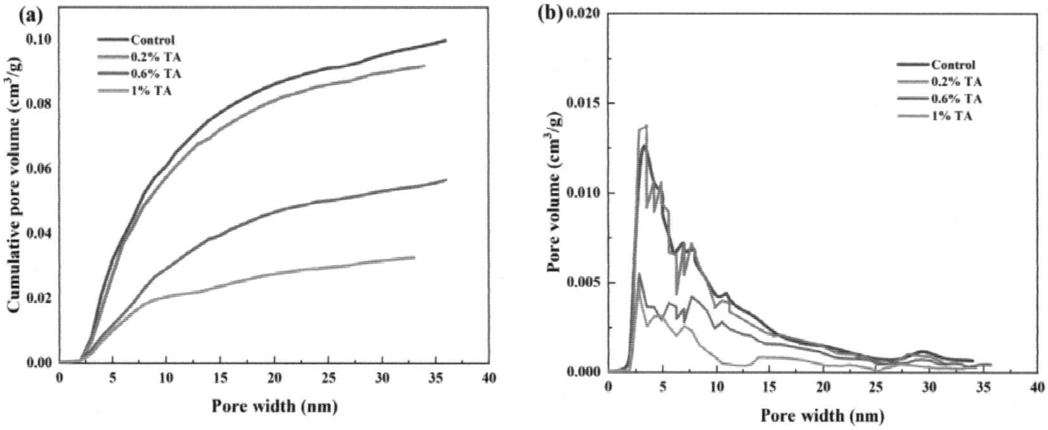

Figure 4. Pore structure of unmodified and TA-modified PFA: (a) Cumulative pore volume and (b) Pore volume.

3.2. Microstructure Characterization of TA-modified PFA

3.2.1. SEM. Figure 5 shows the microstructure of the TA-modified PFA observed under SEM. With increasing PEI concentration, the coverage of TA on the PFA surface increases, forming uneven calcium carbonate patches. The new precipitation of $CaCO_3$ is unevenly arranged, and the defects and gaps in the interface decrease, thus improving the bonding performance between the fine aggregates and the matrix. When the TA concentration increased to 0.2%, the texture of PFA became more undulated, with carbonates stacked onto each other, forming complex intersected rough segments. Moreover, the internal structure of TA-PFA-1 exhibits a smooth and continuous surface, with uniform distribution and minimal geometric shape variation, forming a smoother and significantly enhanced plate compared to TA-PFA-0.6. Therefore, lower concentrations of TA may lead to aggregation or clustering between fine aggregates, while higher concentrations of TA can more effectively disperse the fine aggregates and prevent their aggregation.

Journal of Physics: Conference Series **2941** (2025) 012004 doi:10.1088/1742-6596/2941/1/012004

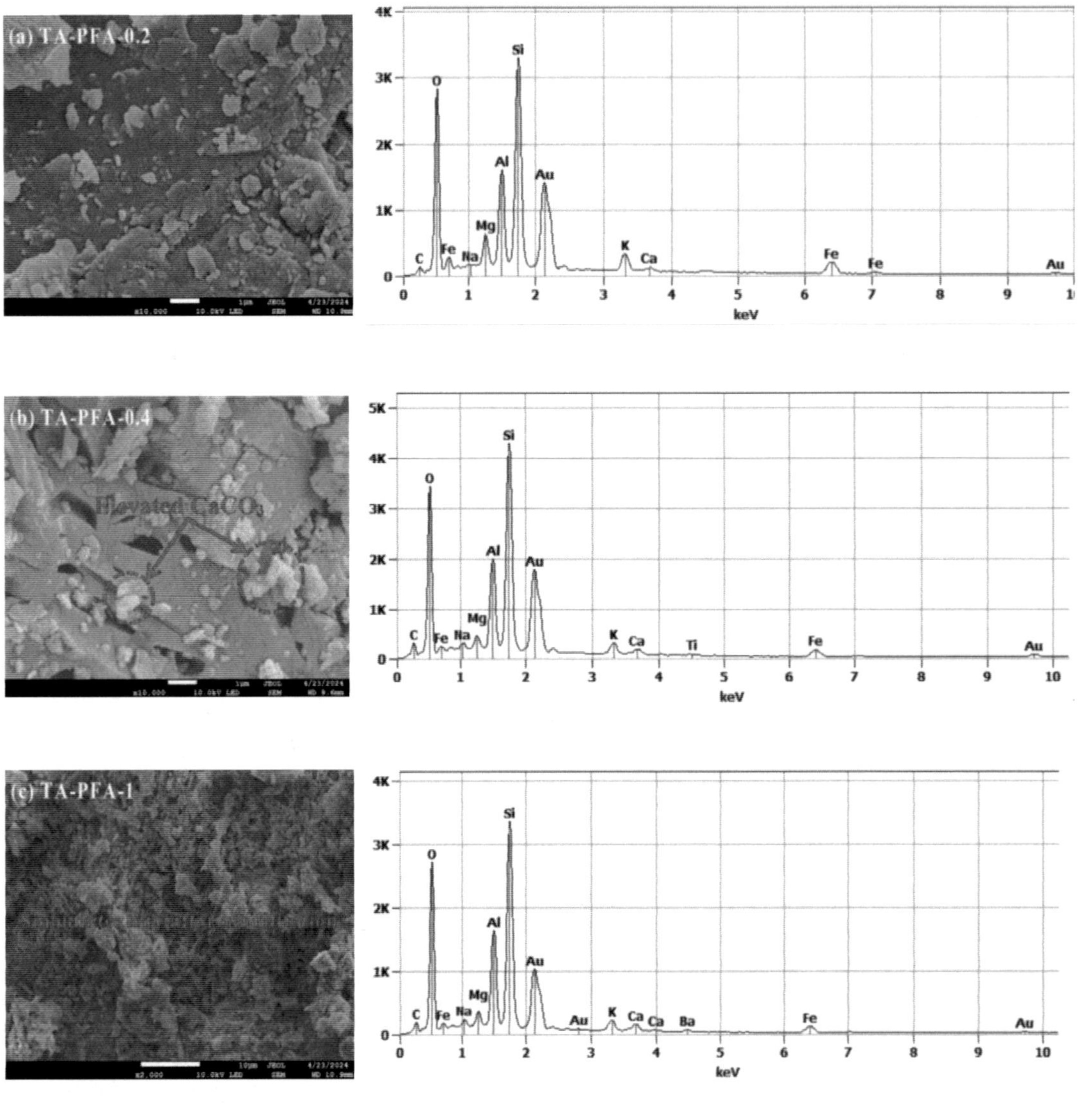

Figure 5. SEM-EDS images of TA-modified PFA.

3.2.2. FITR. Figure 6 presents the FTIR spectra of unmodified PFA (control) and TA-modified PFA. The modified PFA was named after the concentration of TA solution used for treatment. The peaks observed at 694 cm^{-1}, 779 cm^{-1}, and 1049 cm^{-1} can be attributed to specific vibrational modes of the carbonate (CO_3) group in calcite. Specifically, the peak at 694 cm^{-1} is associated with the out-of-plane bending of CO_3, the one at 779 cm^{-1} corresponds to the in-plane bending vibration of the CO_3^{2-} ion, and the peak at 1049 cm^{-1} represents the asymmetric stretching vibration of CO_3 in calcite. The peak at 3440 cm^{-1} was associated with the asymmetric O-H bond stretching vibrations. Besides, as the concentration of TA increased, the characteristic peaks at 694 cm^{-1} and 3440 cm^{-1} gradually diminished, suggesting that the reaction between PFA and TA led to the consumption of calcite and CH. This observation was consistent with the findings from the XRD analysis.

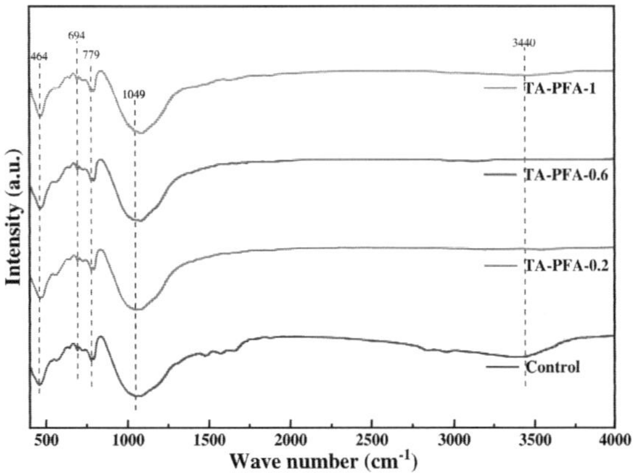

Figure 6. FTIR Spectra of unmodified and TA-modified PFA.

3.2.3. XRD. The XRD spectra of unmodified and TA-modified PFA are shown in figure 7. The main crystal compositions of unmodified PFA were quartz, calcite, gypsum, and dolomite. The dolomite was derived from NA in construction waste. The presence of calcite in unmodified PFA was caused by the carbonation of CH on the PFA surface in the air. The peaks of CH and some peaks of calcite disappeared after treatment with 0.6% TA solution, indicating that they were consumed in the chemical reaction with TA. Consequently, some of the cement paste was removed, exposing more natural sand to the PFA surface and the peak value of treated quartz increased.

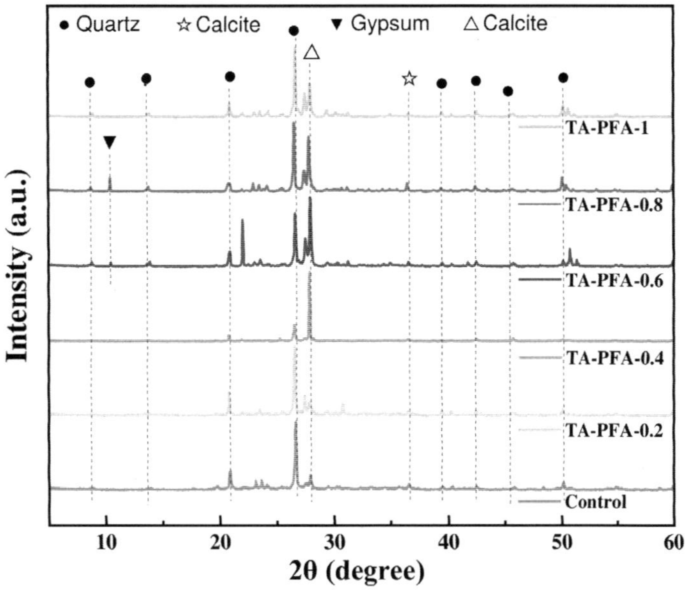

Figure 7. XRD pattern of unmodified and TA-modified PFA.

3.3. Influence of TA-modified PFA on the Properties of FRAM

3.3.1. Compressive Strength. Figure 8 reveals the effect of TA-modified PFA on the compressive strength of FRAM. The compressive strength of the NA group was all higher than that of the Control group because the surface of PFA was wrapped by a layer of hardened cement mortar, which led to higher porosity, larger crushing value, poor solidity, irregular surface shape, and many internal defects of PFA. The compressive strength of PFA was lower than that of NA and many tiny cracks during crushing were produced to reduce its mechanical properties. Moreover, the ITZ between the old and new mortar was weak and the bonding performance was poor, resulting in insufficient absorption of more energy when subjected to external loads, leading to lower compressive strength. The FRAM prepared with TA-modified PFA had higher compressive strength. The improvement of early compressive strength by TA-modified PFA was small because of the inhibitory effect of TA. With the increase in age, this retarding effect gradually weakened at the later stage, and the 28d compressive strength was improved compared to the control. Modification of PFA with 0.8% concentration of TA solution resulted in the greatest improvement in compressive strength. Therefore, the compressive strength of FRAM increased by 24.62% at 28d.

In addition, modifying PFA with higher concentrations of TA solution improved the strength even more and this trend continued until the TA concentration reached 1%. Under the condition of higher TA solution concentration (1%), too much TA remained in PFA and the inhibitory effect of TA considerably slowed the hydration rate of FRAM. Consequently, the compressive strength of FRAM with 1% TA-modified PFA was considerably lower compared to the other groups. Therefore, it is suggested that the TA concentration for PFA modification should be limited to a maximum of 0.8% to avoid an excessive inhibition of cement hydration.

Moreover, the compressive strengths of FRAM were also compared with those of the NA group. As expected, the compressive strength of FRAM samples prepared with modified PFA was lower than that of the NA group at all ages. After modification with TA solution, the compressive strength of FRAM in the test group was significantly increased at 3 d, 7 d, and 28 d, and was only slightly lower than that of the NA group. At 90 d, the compressive strength of FRAM prepared with TA-PFA-0.6 and TA-PFA-0.8 exceeded that of the NA group, indicating the improvement effect of TA on the performance of PFA and FRAM. This was because the higher concentration of TA solution can remove the loose cement particles on the surface of PFA and improve the densification of its ITZ, thus improving the compressive strength of FRAM.

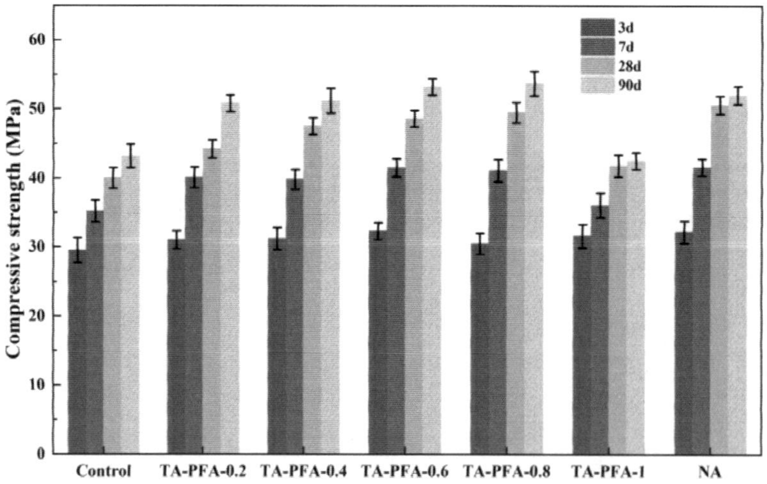

Figure 8. Compressive strength of FRAM prepared with various TA-modified PFA.

3.3.2. Capillary Water Absorption. Figure 9 reveals the changes in cumulative pore volume with water absorption time of FRAM prepared with various TA-modified PFA. Overall, the capillary water absorption mass increased significantly with the increase of water absorption time in all groups. In comparison to the NA group, the capillary water absorption mass of the Control group increased significantly within 8 d. The Control group had the same capillary water absorption mass as the NA group. This was due to the generation of more cracks in the PFA during the crushing process, along with a fragile and porous interface between the old and new cement mortar. These pores and microcracks created an interconnected network, which promoted the rapid transportation of water. Moreover, the capillary water absorption mass of the TA-modified PFA groups was dramatically decreased, and this result closely correlated with the pore characterization tests. The total porosity of FRAM prepared with various TA-modified PFA was reduced, as the nanoparticles formed by the reaction of TA with calcium hydroxide and calcite covered the pore surfaces. This coverage reduced the pore diameter, thereby limiting the water transport channels and decreasing the water transport rate. At the water absorption time of 831s$^{1/2}$, the capillary water absorption mass of each group of FRAM was 16.01g (NA), 60.88g (Control), 47.14g (0.2% TA), 27.93g (0.6% TA), and 31.12g (1% TA), respectively. The capillary water absorption mass of FRAM was reduced by 22.57% after the modification of TA solution, 54.12%, and 48.89%, indicating that PFA had a better waterproofing effect after modification with TA solution. In addition, the water absorption process of NA exhibited a tendency to gradually level off after an initial phase of rapid increase. In the early stage, the interior of the specimen was relatively dry, and water primarily entered through capillary suction. As the internal water content increased, capillary suction progressively diminished, and the water absorption rate gradually stabilized due to the combined influence of gravity and the diminishing capillary forces. However, the Control group had a higher porosity and increased water absorption, which showed a linear growth trend during the test cycle, and the capillary suction process of the FRAM modified with TA solution was in between. Besides, the TA molecule contained phenolic hydroxyl groups which are hydrophilic, leading to an increase in capillary suction.

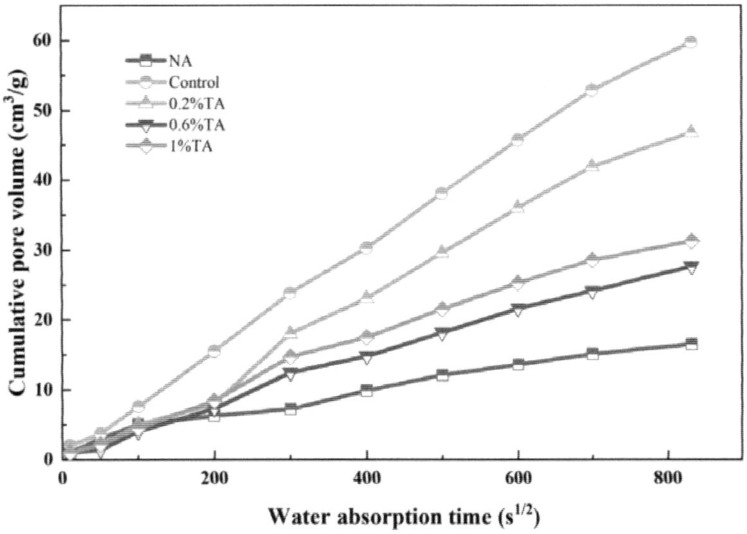

Figure 9. Capillary water absorption mass of FRAM prepared with various TA-modified PFA.

3.3.3. Electric Flux. Durability is the first requirement to be met by the structure, most of the erosion causing deterioration damage to concrete occurs through water immersion, so low permeability is the first line of defense for concrete. Concrete durability is affected by the homogeneity and stability of the mix, the denseness after hardening, the interface structure, and the quality of raw materials, while permeability is mainly determined by two factors, the porosity and pore connectivity within the mortar and the microcracks within the concrete, especially at the interfacial joints. Duan et al. [20] observed that the electric flux increased linearly with the replacement rate of recycled aggregate. This was attributed to the higher porosity caused by the old mortar adhered to the surface of the recycled fine aggregate, along with the less dense structure of the ITZ, which formed a transmission pathway.

Figure 10 reveals the electrical flux on FRAM prepared with various TA-modified PFA. It was evident that the FRAM prepared with TA-modified PFA significantly strengthened the protection against chloride ion infiltration. The electric fluxes for the 0.2% TA, 0.4% TA, and 1% TA groups decreased by 13.23%, 14.81%, and 16.56%, respectively, compared to the control group. This improvement might stem from the formation of nanoparticles through the reaction between TA, CH, and calcite. These substances coated the pore surfaces, resulting in a decrease in both pore size and the channels for chloride ion transport. In addition, the incorporation of TA-modified PFA improved the bonding between the new mortar and the interfacial transition zone (ITZ), leading to a denser ITZ and further strengthening the protection against chloride ion infiltration.

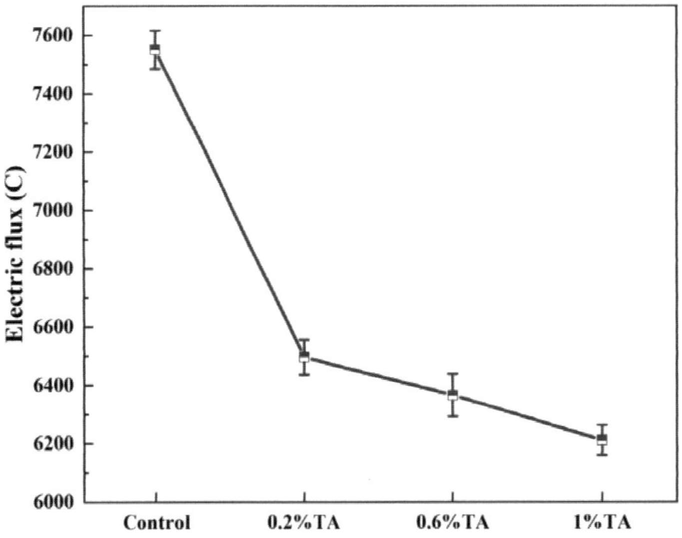

Figure 10. The electric flux of FRAM prepared with various TA-modified PFA.

4. Conclusions

This research enhances public fine aggregate (PFA) with tannic acid (TA) to produce modified fully recycled aggregate mortar (FRAM), aiming to improve its strength, durability, and broader application in construction. The effects of TA on the macroscopic physical properties of PFA at varying concentrations and immersion times were examined and its impact on the physical properties and durability of FRAM was explored. The detailed conclusions are as follows:

The water absorption rates of TA-modified PFA decreased with the increasing concentration of TA solution, which proved the positive effect of TA in reducing the content of cement mortar and consequently lowering the water absorption rate of PFA.

The microstructure tests indicated that the internal structure of TA-PFA-1 exhibits a smooth and continuous surface. The main crystal compositions of unmodified PFA were quartz, calcite, gypsum, and dolomite.

The compressive strength of FRAM prepared with different TA-modified PFA was increased by 24.62% at 28 d, and it is suggested that the TA concentration for PFA modification should be limited to a maximum of 0.8% to avoid an excessive inhibition of cement hydration.

5. References

[1] Ahmad W, Ahmad A, Ostrowski K A, Aslam F and Joyklad P 2021 A scientometric review of waste material utilization in concrete for sustainable construction *Case Studies in Construction Materials* **15** e00683.

[2] Liu J, Zhang S and Wagner F 2018 Exploring the driving forces of energy consumption and environmental pollution in China's cement industry at the provincial level *Journal of Cleaner Production* **184** 274-285.

[3] Ghisellini P, Ji X, Liu G and Ulgiati S 2018 Evaluating the transition towards cleaner production in the construction and demolition sector of China: A review *Journal of Cleaner Production*

195 418-434.
[4] Chen, J L, Zhou W J and Li W 2015 Precondition and problems in reutilization of building wastes *Architecture Technology* **12** 1114-1116.
[5] Younis K H and Pilakoutas K 2013 Strength prediction model and methods for improving recycled aggregate concrete *Construction and Building Materials* **49** 688-701.
[6] Gao C, Huang L, Yan L, Jin R and Chen H 2020 Mechanical properties of recycled aggregate concrete modified by nano-particles *Construction and Building Materials* **241** 118030.
[7] Liu J, Nie J and Yuan H 2020 Interactive decisions of the waste producer and the recycler in construction waste recycling *Journal of Cleaner Production* **256** 120403.
[8] Bao Z, Lee W M and Lu W 2020 Implementing on-site construction waste recycling in Hong Kong: Barriers and facilitators *Science of the Total Environment* **747** 141091.
[9] He L and Yuan H 2020 Investigation of construction waste recycling decisions by considering consumers' quality perceptions *Journal of Cleaner Production* **259** 120928.
[10] Ibrahim M I M 2016 Estimating the sustainability returns of recycling construction waste from building projects *Sustainable Cities and Society* **23** 78-93.
[11] Lu W and Tam V W 2013 Construction waste management policies and their effectiveness in Hong Kong: A longitudinal review *Renewable and Sustainable Energy Reviews* **23** 214-223.
[12] Choudhary J, Kumar B and Gupta A 2021 Potential utilization of construction wastes in asphalt pavements as fillers using ranking framework *Construction and Building Materials* **277** 122262.
[13] Lam P T, Ann T W, Wu Z and Poon C S 2019 Methodology for upstream estimation of construction waste for new building projects *Journal of Cleaner Production* **230** 1003-1012.
[14] Sormunen P and Kärki T 2019 Recycled construction and demolition waste as a possible source of materials for composite manufacturing *Journal of Building Engineering* **24** 100742.
[15] Adessina A, Fraj A B, Barthélémy J F, Chateau C and Garnier D 2019 Experimental and micromechanical investigation on the mechanical and durability properties of recycled aggregates concrete *Cement and Concrete Research* **126** 105900.
[16] Shaban W M, Yang J, Su H, Mo K H, Li L and Xie J 2019 Quality improvement techniques for recycled concrete aggregate: A review *Journal of Advanced Concrete Technology* **17(4)** 151-167.
[17] Meng T, Zhang J, Wei H and Shen J 2020 Effect of nano-strengthening on the properties and microstructure of recycled concrete *Nanotechnology Reviews* **9(1)** 79-92.
[18] Hagerman A E, Riedl K M, Jones G A, Sovik K N, Ritchard N T, Hartzfeld P W and Riechel T L 1998 High molecular weight plant polyphenolics (tannins) as biological antioxidants *Journal of Agricultural and Food Chemistry* **46(5)** 1887-1892.
[19] Oh D X, Prajatelistia E, Ju S W, Jeong Kim H, Baek S J, Joon Cha H, Soo Hwang D, et al. 2015 A rapid, efficient and facile solution for dental hypersensitivity: The tannin–iron complex *Scientific Reports* **5(1)** 10884.
[20] Singh A S, Halder S, Wang J, Imam M A and Chen P 2019 Tannic acid intermediated surface functionalization of bamboo micron fibers to enhance mechanical performance of hybrid GFRP *Composites Part B: Engineering* **177** 107322.
[21] Wang L, Wang J, Qian X, Fang Y, Chen P and Tuinukuafe A 2020 Tea stain-inspired treatment for fine recycled concrete aggregates *Construction and Building Materials* **262** 120027.

Thermodynamic Study of CoCrFeNiX (X=Cu, Ti) High-Entropy Alloy

Zhennan Liu[1], Cong Liu[1], Ying Sun[1], Songyuan Zhang[1], Yong Cheng[1,2] and Chunling Yao[1,*]

[1] Faculty of Metallurgy and Mining, Kunming Metallurgy College, Kunming, Yunnan, China
[2] School of Earth Sciences, Yunnan University, Kunming, Yunnan, China
Email: 547430739@qq.com; 56956541@qq.com; 5109787081@qq.com; 87718493@qq.com; cheng_yong1988@163.com; *yaochunling2015@sina.com

Abstract. Thermodynamic properties play a crucial role in the design of high-entropy alloys, and using reliable thermodynamic models for theoretical calculations to obtain key thermodynamic data of alloys has become an important and effective approach. This study predicts the mixing Gibbs free energy, mixing enthalpy, mixing entropy, excess Gibbs free energy, and excess entropy of the solid solution of the transition metal high-entropy alloy CoCrFeNi-X (X=Cu, Ti) system using binary infinite dilution activity coefficients, Miedema theory, and modified M-MIVM, while considering the difference in atomic size. From a thermodynamic perspective, the possibility, phase stability, and structural stability composition of these high-entropy alloys are evaluated. The research results show that the CoCrFeNi-X (X=Cu, Ti) system meets the thermodynamic criteria for high-entropy alloys and can form a five-component high-entropy alloy. For the CoCrFeNiCu alloy system, with appropriately balanced component content, the system's mixing free energy is more negative, and the alloy structure is more stable. This rule guides the alloy composition design, resulting in a better scheme for the $Co_{0.20}Cr_{0.20}Fe_{0.23}Ni_{0.22}Cu_{0.15}$ alloy. For the CoCrFeNiTi alloy system, with appropriately high Ni and low Ti conditions, the system's mixing free energy is more negative, and the alloy structure is more stable. This rule guides the alloy composition design, resulting in a better scheme for the $Co_{0.20}Cr_{0.17}Fe_{0.21}Ni_{0.28}Ti_{0.14}$ alloy. At high temperatures, the mixing free energy of these alloy systems significantly decreases, enhancing the stability of the alloys. This study provides reliable thermodynamic data and theoretical guidance for the design of CoCrFeNi-X (X=Cu, Ti) high-entropy alloys.
Keywords. Thermodynamic; high-entropy alloys; Miedema theory; mixing free energy.

1. Introduction

High-entropy alloys (HEAs) are a type of alloy composed of five or more elements in equimolar or near-equimolar proportions. They have high entropy values in thermodynamics and form simple FCC or BCC solid solution phases rather than complex intermetallic compound phases during solidification. This unique design concept endows high-entropy alloys with excellent mechanical and functional properties, such as high strength, high toughness, high plasticity, high corrosion resistance, and excellent electromagnetic properties. They have broad application prospects in industrial manufacturing and are a new type of metal material with great development potential [1-4]. High-entropy alloys are mainly divided into three categories: one is based on transition elements such as Co, Cr, Fe, Ni, etc., by adjusting the content of each component and adding alloying elements such as Al,

Cu, Mn, Ti, etc. to obtain high-performance high-entropy alloys; one is based on refractory elements such as Mo, Nb, Zr, Ti, Hf, etc., by adjusting the alloy composition to improve their properties; one is based on lightweight elements such as Al, Mg, Ti, etc., such as AlMgLiZnCu, AlMgZnCuSi, AlZrTiNbMo, etc. [3, 4]. In recent years, a large amount of research has been conducted on CoCrFeNi-based transition element high-entropy alloys, which are now close to engineering applications. As typical transition element high-entropy alloys, these alloys usually have a single solid solution structure of FCC or BCC phase, showing good comprehensive mechanical properties [4-8]. However, single-phase high-entropy alloys have not completely solved the trade-off problem between high strength and high toughness of traditional single-phase metal materials. For example, the fracture elongation rate of CoCrFeNiMn-based high-entropy alloys at room temperature can reach 50%, while the yield strength is only about 410MPa [9]. In engineering applications, it is common to add other elements such as Al, Cu, Mn, Ti, etc. to CoCrFeNi high-entropy alloys to change the phase structure and regulate their mechanical and physical properties [10, 11]. Therefore, seeking a new design concept for high-entropy alloys to achieve an effective combination of high strength and high toughness is a difficult problem that needs to be solved in the field of high-entropy alloys [12].

The performance of alloys depends on their chemical composition. Finding the optimal chemical composition to improve their performance remains a key issue. However, the design space of high-entropy alloy compositions is vast. Predicting alloy properties solely through traditional "trial and error" methods inevitably leads to the occasional and inaccurate results of research, increasing research costs [3, 10]. Therefore, from a theoretical perspective, using less experimental data and thermodynamic models with a solid physical foundation to calculate the thermodynamic properties of alloys is expected to quickly and specifically screen out components with stable structures and excellent performance, greatly improving the research efficiency of high-entropy alloys. Researchers have used thermodynamic calculation software to simulate the phase diagrams of CoCrFeNiTa and AlCoCrFeNi high-entropy alloys [13, 14], providing theoretical references for designing specific alloy compositions. Researchers [15] derived expressions for the configurational partition function, excess Gibbs energy, and thermodynamic activity coefficients of solid solutions based on the free volume theory, lattice model, and Scatchard-Hildebrand theory from the perspective of statistical thermodynamics, using modified molecular interaction model (M-MIVM) [16, 17]. This model has been applied and validated in multiple binary, ternary, quaternary, and penta-component solid solution alloy systems (mainly composed of Co, Cr, Fe, Ni, Mn), achieving good results [15, 18, 19]. The modified M-MIVM can simultaneously express the contribution of molecular configuration's microscopic state number (entropy) and molecular interaction (enthalpy) to excess Gibbs free energy, has a solid physical foundation, satisfies thermodynamic consistency, and is suitable for thermodynamic research on simple solid solution phases of high-entropy alloys. Based on previous work, this study uses binary infinite dilution activity coefficients, Miedema theory, and the modified M-MIVM to predict the mixing Gibbs free energy, mixing enthalpy, mixing entropy, excess Gibbs free energy, and excess entropy of CoCrFeNi-X (X=Cu, Ti) high-entropy alloy solid solutions, considering atomic size differences, and evaluating the possibility, phase stability, and structural stability composition of this high-entropy alloy formation from a thermodynamic perspective, providing reliable thermodynamic data and theoretical guidance for high-entropy alloy design.

2. Calculation Method

2.1. Model Introduction

For the modified M-MIVM, the expression for the molar excess Gibbs free energy of solid solution alloy is : [15]

$$\frac{G_{\mathrm{m}}^{\mathrm{E}}}{RT}=\sum_{i=1}^{C}x_i\ln\left(\frac{V_{si}}{\sum_{j=1}^{C}x_jV_{sj}B_{ji}}\right)+\sum_{i=1}^{C}x_i\left[\frac{\sum_{j=1}^{C}x_jB_{ji}(\lambda_j-\lambda_i)}{\sum_{l=1}^{C}x_lB_{li}}\right]+\sum_{i=1}^{C-1}\sum_{j>i}^{C}x_ix_j\left[\frac{A_{ij}}{\sum_{l=1}^{C}x_l\frac{V_{sl}}{V_{si}}B_{li}}+\frac{A_{ji}}{\sum_{l=1}^{C}x_l\frac{V_{sl}}{V_{sj}}B_{lj}}\right] \tag{1}$$

In equation (1), *G*E m is the molar excess Gibbs free energy of the system; *R* is the ideal gas constant; *T* is the absolute temperature; x_i is the mole fraction of component *i*; *C* is the number of components; *i*, *j*, *l* are natural numbers from 1 to *C*, representing each component in the system; A_{ij} and A_{ji} are the energy parameters of the *i*-*j* binary system at temperature *T*, when $i=j$, $A_{ij}=A_{ji}=0$; in solid solution alloy systems, the energy parameters are related to temperature, and their relationship can be expressed as:

$$TA_{ij}=T'A'_{ij},\quad TA_{ji}=T'A'_{ji} \tag{2}$$

In equation (2), *A' ij* and *A' ji* represent the energy parameters of the *i-j* binary system at temperature T'. In equation (1), V_{si} and V_{sj} are the molar void volumes of components *i* and *j* at temperature *T*, respectively, and their calculation formulas are:

$$V_{si}=V_{mi}-V_{mi0},\quad V_{sj}=V_{mj}-V_{mj0} \tag{3}$$

In equation (3), V_{mi} and V_{mi0} represent the molar volume and molar hard-sphere volume of pure solid *i*, respectively; V_{mj} and V_{mj0} represent the molar volume and molar hard-sphere volume of pure solid *j*, respectively. In equation (1), λ_i and λ_j represent the void numbers of pure solids *i* and *j*, respectively, and their calculation formula is:

$$\lambda_i=\frac{V_{mi}}{V_{mi}-V_{mi0}},\quad \lambda_j=\frac{V_{mj}}{V_{mj}-V_{mj0}} \tag{4}$$

In equation (1), B_{ij} and B_{ji} are the volume parameters of the *i-j* binary system at temperature *T*. When $i=j$, $B_{ij}=B_{ji}=1$. This parameter is related to temperature, and its relationship can be expressed as:

$$T\ln B_{ij}=T'\ln B'_{ij},\quad T\ln B_{ji}=T'\ln B'_{ji} \tag{5}$$

In equation (5), *B' ij* and *B' ji* represent the volume parameters of the *i-j* binary system at temperature T'.

Based on the standard thermodynamic formulas and the relationship between molar and partial molar quantities, the expression for the activity coefficient of any component *i* in a solid solution alloy is derived as [15]:

$$\begin{aligned}\ln\gamma_i=&1+\ln\left(\frac{V_{si}}{\sum_{j=1}^{C}x_jV_{sj}B_{ji}}\right)-\sum_{k=1}^{C}\left(\frac{x_kV_{si}B_{ik}}{\sum_{j=1}^{C}x_jV_{sj}B_{jk}}\right)\\&+\frac{\sum_{j=1}^{C}x_jB_{ji}(\lambda_j-\lambda_i)}{\sum_{k=1}^{C}x_kB_{ki}}+\sum_{j=1}^{C}\frac{x_jB_{ij}}{\sum_{k=1}^{n}x_kB_{kj}}\left[\left(\lambda_i-\lambda_j\right)-\frac{\sum_{t=1}^{C}x_tB_{tj}\left(\lambda_t-\lambda_j\right)}{\sum_{k=1}^{C}x_kB_{kj}}\right]+\sum_{j\neq i}^{C}x_j\left(\frac{A_{ij}}{\sum_{l=1}^{C}x_l\frac{V_{sl}}{V_{si}}B_{li}}+\frac{A_{ji}}{\sum_{t=1}^{C}x_t\frac{V_{st}}{V_{sj}}B_{tj}}\right)\\&-\sum_{j=1}^{C-1}\sum_{k>j}^{C}x_jx_k\left(\frac{A_{jk}}{\sum_{l=1}^{C}x_l\frac{V_{sl}}{V_{sj}}B_{lj}}+\frac{A_{kj}}{\sum_{t=1}^{C}x_t\frac{V_{st}}{V_{sk}}B_{tk}}\right)-\sum_{j=1}^{C-1}\sum_{k>j}^{C}x_jx_k\left[\frac{A_{jk}\left(\frac{V_{si}}{V_{sj}}B_{ij}-\sum_{t=1}^{C}x_t\frac{V_{st}}{V_{sj}}B_{tj}\right)}{\left(\sum_{l=1}^{C}x_l\frac{V_{sl}}{V_{sj}}B_{lj}\right)^2}+\frac{A_{kj}\left(\frac{V_{si}}{V_{sk}}B_{ik}-\sum_{f=1}^{C}x_f\frac{V_{sf}}{V_{sk}}B_{fk}\right)}{\left(\sum_{g=1}^{C}x_g\frac{V_{sg}}{V_{sk}}B_{gk}\right)^2}\right]\end{aligned} \tag{6}$$

In equation (6), γ_i represents the activity coefficient of component *i*, and the standard state is the solid phase of pure *i*. For the binary system *i-j*, when x_i or x_j approaches 0, the expressions for the infinite dilution activity coefficients $\gamma\infty\ i$ and $\gamma\infty\ j$ are derived from the binary expansion of equation (6) as follows:

$$\ln\gamma_i^{\infty} = 1+\ln\left(\frac{V_{si}}{V_{sj}B_{ji}}\right)-\frac{V_{si}}{V_{sj}}B_{ij}+\left(1-B_{ij}\right)\left(\lambda_j-\lambda_i\right)+\frac{V_{si}A_{ij}}{V_{sj}B_{ji}}+A_{ji} \tag{7}$$

$$\ln\gamma_j^{\infty} = 1+\ln\left(\frac{V_{sj}}{V_{si}B_{ij}}\right)-\frac{V_{sj}}{V_{si}}B_{ji}+\left(1-B_{ji}\right)\left(\lambda_i-\lambda_j\right)+\frac{V_{sj}A_{ji}}{V_{si}B_{ij}}+A_{ij} \tag{8}$$

From equations (1) and (6), it can be seen that the modified M-MIVM is a four-parameter model. For a multi-*component* system, if the infinite dilution activity coefficient or activity values of its sub-binary systems are known, the parameters A_{ij}, A_{ji}, B_{ij}, and B_{ji} can be determined, and the thermodynamic properties of the multi-component system can be predicted.

2.2. Basic Data Acquisition

A significant advantage of the modified M-MIVM is that it can describe and predict the thermodynamic properties of multicomponent solid solutions using only binary infinite dilution activity coefficients or binary activity data. Therefore, in order to determine the model parameters A_{ij}, A_{ji}, B_{ij}, and B_{ji}, it is necessary to organize data such as atomic radius, molar volume, and molar hard-sphere volume of each pure component in the CoCrFeNi-X (X=Cu, Ti) system; at the same time, collect and obtain infinite dilution activity coefficients corresponding to the 14 sub-binary systems Co-Cr, Co-Fe, Co-Ni, Cr-Fe, Cr-Ni, Fe-Ni, Co-X (X=Cu, Ti), Cr-X (X=Cu, Ti), Fe-X (X=Cu, Ti), Ni-X (X=Cu, Ti) in the CoCrFeNi-X (X=Cu, Ti) system.

In this study, reliable activity data for binary systems were primarily collected from established thermodynamic data manuals for binary solid solution alloys, such as those in Hultgren [20] and SGTE [21]. However, due to the scarcity of activity data for known binary solid alloy systems, parameters for some multi-component systems cannot be calculated from experimental activity values of the corresponding binary systems. This necessitates the development of new methods for obtaining parameters, thereby enabling the modified M-MIVM to be more widely applied to the prediction of thermodynamic properties of multi-component solid alloys. Since the Miedema model can predict the thermodynamic properties of binary alloys from the physicochemical properties of pure substances [22, 23], Miedema theory can be employed to obtain thermodynamic data for binary systems lacking such data. In recent years, Miedema theory has been continuously refined, making it an effective method for obtaining thermodynamic data for binary systems.

This study utilizes the Miedema infinite dilution activity coefficient method to obtain the model parameters for two components. The Miedema model extends the Wigner-Seitz theory from pure metals to *i-j* binary alloys, resulting in the Miedema formula for the infinite dilution mixing enthalpy of component *i* in *i-j* binary alloys, as follows:

$$\Delta H_{i,\mathrm{Mie}}^{\mathrm{M}\infty} = H_i^{\mathrm{E}\infty} = \frac{2pV_i^{2/3}[1+\mu_i(\phi_i-\phi_j)][9.4(n_{wsi}^{1/3}-n_{wsj}^{1/3})^2-(\phi_i-\phi_j)^2-(r/p)]}{n_{wsi}^{-1/3}-n_{wsj}^{-1/3}} \tag{9}$$

Where: p is an empirical constant that divides all elements into two categories: one is transition metals or elements of the sub-group except for group IIB, and the other is elements other than those in the first category. When both components are elements in the first category, p=14.1; when one component is an element in the first category and the other is an element in the second category, p=12.3; when both components are elements in the second category, p=10.6. V_i is the molar volume of pure metal i; μ_i is an empirical constant, where μ_i=0.14 when the valence of component i is 1, μ_i=0.10 when the valence is 2, μ_i=0.07 when the valence is 3, and μ_i=0.04 when the valence is greater than 3 or when i is an element in the first category; ϕ_i and ϕ_j are the electronegativity of components i and j, respectively; n_{wsi} and n_{jsw} are the electron densities of components i and j (the number of electrons per $0.148\times10^{-30}m^3$), respectively; $r/p=(r/p)_i\times(r/p)_j$ is an empirical constant. At temperature T, the Miedema infinite dilution activity coefficient of component i is expressed as:

$$\ln \gamma_{i,\mathrm{Mie}}^{\mathrm{M}\infty} = \frac{\Delta H_{i,\mathrm{Mie}}^{\mathrm{M}\infty}}{RT} \tag{10}$$

Similarly, the Miedema infinite dilution activity coefficient of component j can be obtained. In approximate calculations, we can use the Miedema infinite dilution activity coefficient instead of the experimental infinite dilution activity coefficient to obtain the model parameters. The Miedema parameters of each pure component in the CoCrFeNi-X (X=Cu, Ti) system are shown in table 1.

Table 1. Miedema parameters of pure components [23].

Component i	$V2/3\ i$	μ_i	ϕ_i	$n1/3\ wsi$	$(r/p)_i$
Co	3.5	0.04	5.10	1.75	1.0
Cr	3.7	0.04	4.65	1.73	1.0
Fe	3.7	0.04	4.93	1.77	1.0
Ni	3.5	0.04	5.20	1.75	1.0
Cu	3.7	0.04	4.55	1.47	0.3
Ti	4.8	0.04	3.65	1.47	1.0

2.3. Model Parameter Acquisition

The M-MIVM parameters of the pure components in the CoCrFeNi-X (X=Cu, Ti) system are shown in table 2. The energy parameters A_{ij} and A_{ji}, and volume parameters B_{ij} and B_{ji} of the 14 sub-binary systems contained in this system can be determined by fitting the binary infinite dilution activity coefficients. When reliable binary infinite dilution activity coefficients $\gamma\infty\ i$ and $\gamma\infty\ j$ are known, the simultaneous equations (7) and (8) form an equation system, and the optimal solution is found by nonlinear function fitting. The results are shown in table 3. By substituting the relevant parameters in tables 2 and 3 into the binary expansion of equation (6), the predicted values of the activity coefficients or activities of the two components in the binary solid solution alloy can be obtained. Comparing them with the experimental data, some results are shown in figures 1-8. It can be seen from the results that the modified M-MIVM can well describe the sub-binary solid solution alloy system, its predicted values are in good agreement with the experimental values, and the obtained model parameters are accurate and reliable.

Table 2. M-MIVM parameters of pure components.

Component i	$r_i\,(10^{-8}$cm)	$V_{mi0}\,(cm^3\cdot mol^{-1})$	$V_{mi}(cm^3\cdot mol^{-1})$	Reference sources
Co	1.252	4.950	6.55	[24]
Cr	1.282	5.315	7.12	[24]
Fe	1.274	5.216	7.12	[24]
Ni	1.246	4.880	6.55	[24]
Cu	1.278	5.265	7.10	[25]
Ti	1.460	7.852	10.63	[22]

Note: r_i is the atomic hard sphere radius of component i.

Table 3. M-MIVM parameters and infinite dilution activity coefficients of binary solid solution alloys.

i-j	*T*(K)	A_{ij}	A_{ji}	B_{ij}	B_{ji}	$\gamma\infty\ i$	$\gamma\infty\ j$	Reference sources
Co-Cr	1600	0.8330	-1.0459	0.7870	0.7900	1.316	0.847	[21]
Co-Fe	1650	0.0521	0.0603	0.9596	0.8686	1.293	1.426	[26]
Co-Ni	1200	0.0930	0.0869	1.1759	0.8945	1.172	1.145	[21]
Cr-Fe	1600	0.2175	-0.4050	0.9278	0.4880	2.197	1.609	[20]
Cr-Ni	1600	-1.5370	0.6781	0.4090	0.6098	0.395	3.518	[21]
Fe-Ni	1273	-1.0670	1.2387	1.6376	0.3214	0.104	0.649	[27, 28]
Co-Cu	1200	0.2164	0.3743	0.7146	1.8241	1.054	1.013	Miedema
Cr-Cu	1300	0.2253	0.3726	0.5981	2.2291	1.036	1.031	Miedema
Fe-Cu	1300	0.3237	0.6636	1.3702	1.5293	1.026	1.034	Miedema
Ni-Cu	973	0.2355	0.3718	0.5446	0.8386	3.285	5.898	[20]
Co-Ti	1200	0.0115	1.0249	0.9734	1.9778	1.255	0.758	Miedema
Cr-Ti	1670	0.0312	1.0481	0.9112	0.9329	3.021	6.671	[21]
Fe-Ti	1300	-0.0148	0.9792	1.3585	1.6081	1.171	0.831	Miedema
Ni-Ti	1300	0.0855	1.2543	0.7559	2.8525	1.268	0.750	Miedema

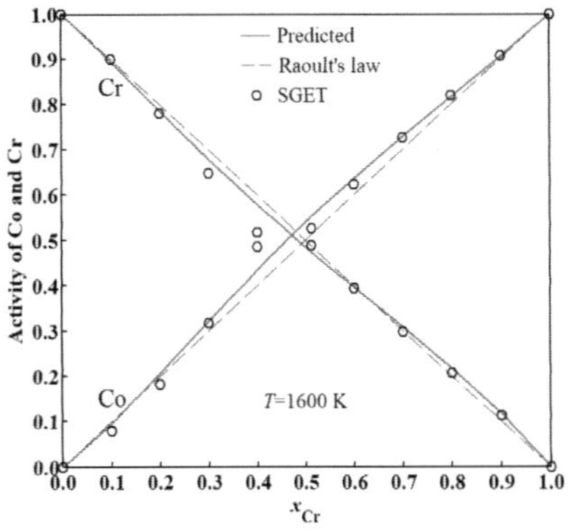

Figure 1. Comparison of predicted values and SGET data [21] of Co and Cr activities in the Co-Cr solid solution at 1600 K.

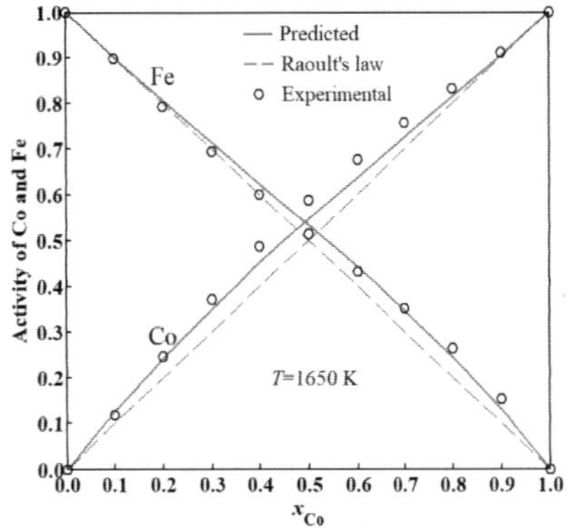

Figure 2. Comparison of predicted values and experimental data [26, 29] of Co and Fe activities in the Co-Fe solid solution at 1650 K.

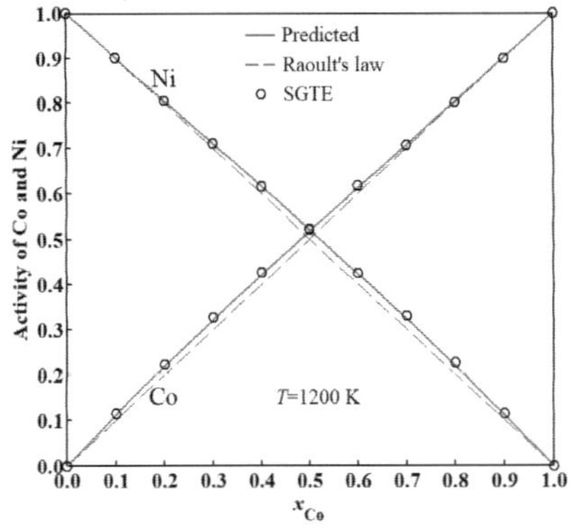

Figure 3. Comparison of predicted values and SGET data [21] of Co and Ni activities in the Co-Ni solid solution at 1200 K.

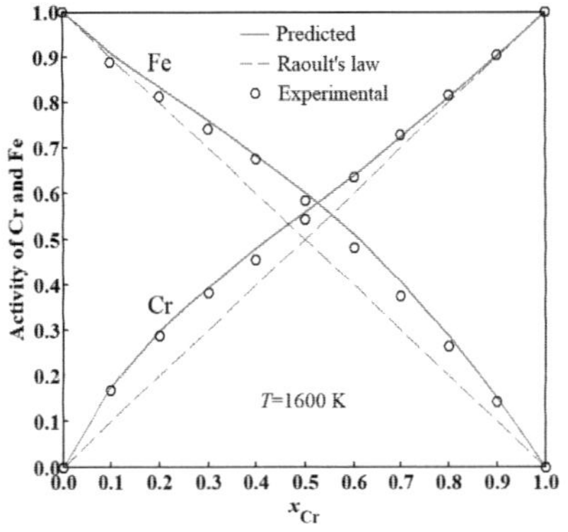

Figure 4. Comparison of predicted values and experimental data [20] of Cr and Fe activities in the Cr-Fe solid solution at 1600 K.

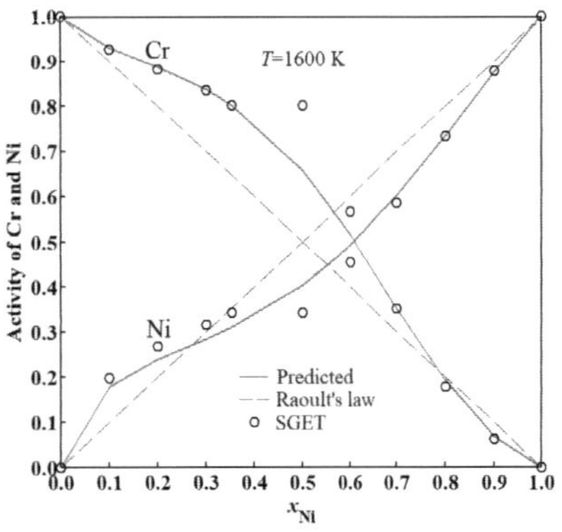

Figure 5. Comparison of predicted values and SGET data [21] of Cr and Ni activities in the Cr-Ni solid solution at 1600 K.

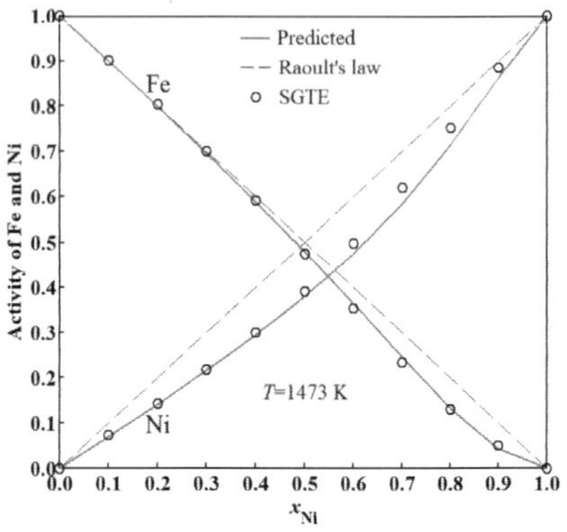

Figure 6. Comparison of predicted values and SGET data [21] of Fe and Ni activities in the Fe-Ni solid solution at 1473 K.

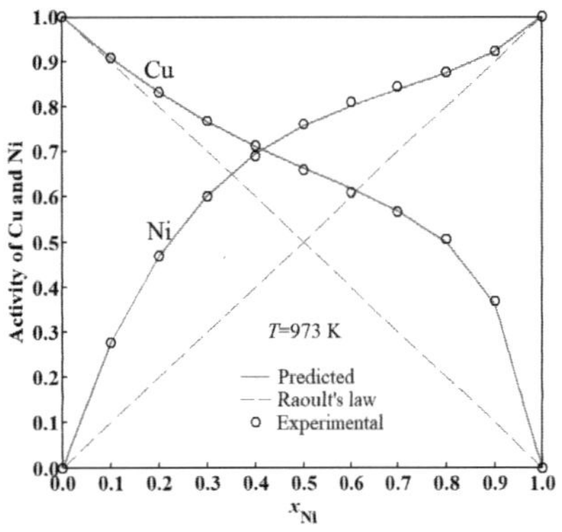

Figure 7. Comparison of predicted values and experimental data [20] of Cu and Ni activities in the Cu-Ni solid solution at 973 K.

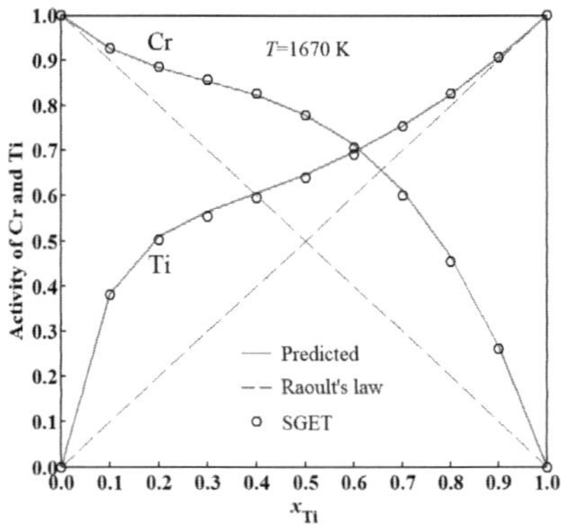

Figure 8. Comparison of predicted values and SGET data [21] of Cr and Ti activities in the Cr-Ti solid solution at 1670 K.

2.4. Calculation of Thermodynamic Properties

For a non-ideal solution system, its thermodynamic properties usually include: mixing properties, excess properties, and partial molar properties and activities of each component. Mixing properties often include mixing Gibbs free energy, mixing enthalpy, and mixing entropy. Excess properties often include excess Gibbs free energy, excess enthalpy, and excess entropy. Theoretically, these properties are interrelated, and as long as one property function is known, all other property functions should be strictly derived.

In this study, the modified M-MIVM was used to predict the thermodynamic properties of the solid solution of CoCrFeNi-X (X=Cu, Ti) high-entropy alloys. By substituting the relevant parameters into the five-component expansion of equations (1) and (6), the excess molar Gibbs free energy *G*E m and the activity a_i (activity coefficient γ_i) of each component in the system can be obtained. The molar excess enthalpy *H*E m, molar mixing enthalpy Δ*H*M m, and molar excess entropy *S*E m of the system can be calculated by the Gibbs-Helmholtz equation, as shown in equations (11) and (12); the molar mixing Gibbs free energy Δ*G*M m of the system is calculated by equation (13); and the molar mixing entropy Δ*S*M m of the system can be calculated by the formula *G=H-TS*.

$$\Delta H_{\mathrm{m}}^{\mathrm{M}} = H_{\mathrm{m}}^{\mathrm{E}} = -RT^2 \left(\frac{\partial (G_{\mathrm{m}}^{\mathrm{E}} / RT)}{\partial T} \right)_{x,P} \tag{11}$$

$$S_{\mathrm{m}}^{\mathrm{E}} = -\left(\frac{\partial G_{\mathrm{m}}^{\mathrm{E}}}{\partial T}\right)_{x,P} \tag{12}$$

$$\Delta G_{\mathrm{m}}^{\mathrm{M}} = RT\sum_{i=1}^{C} x_i \ln \gamma_i x_i \tag{13}$$

Using the above method, the basic thermodynamic properties of CoCrFeNi-X (X=Cu, Ti) high entropy alloy solid solutions under different conditions can be obtained. By analyzing the values of these properties and their variations with conditions, the possibility of forming the high entropy alloy, phase stability, and structural stability composition can be evaluated, providing thermodynamic data and theoretical guidance for high entropy alloy design.

3. Results and Discussion

The mole fraction of each main component of high-entropy alloys ranges from 0.05 to 0.35 [6], and the heat treatment temperature ranges from 600 to 1100°C [30, 31]. This study predicts the mixing Gibbs free energy, mixing enthalpy, mixing entropy, excess Gibbs free energy, and excess entropy of the solid solution of CoCrFeNi-X (X=Cu, Ti) alloy within the above conditions, obtaining a total of 2940 thermodynamic data.

The key to designing high-entropy alloys lies in forming a simple solid solution phase by dissolving components in a certain proportion, while avoiding the formation of intermetallic compounds or other non-metallic inclusions as much as possible. Researchers [32-35] have provided empirical criteria for the formation of solid solution phases in high-entropy alloys, which are: the number of alloy components is not less than 5, the mixing entropy ΔSM m$\geq 1.61R$; the variance of atomic sizes of each component $\delta \leq 6.6\%$; the mixing enthalpy of the high-entropy alloy system satisfies -15 $\mathrm{KJ \cdot mol^{-1}} \leq \Delta H$M m$\leq 5\ \mathrm{KJ \cdot mol^{-1}}$. The mixing entropy in the criteria is determined under the assumption of Boltzmann's theory, considering the solid solution alloy as an ideal solution, and the calculation formula is as follows:

$$\Delta S_{\mathrm{m}}^{\mathrm{M}} = -R\sum_{i=1}^{C} x_i \ln x_i \tag{14}$$

When the molar ratio of each component is equal, the mixing entropy of the alloy system reaches its maximum value, and the calculation formula is as follows:

$$\Delta S_{\mathrm{m}}^{\mathrm{M}} = R\ln C \tag{15}$$

For a five-component alloy system with equal molar ratio, the calculated value of the mixing entropy is $1.61R$. However, the real solution is different from the ideal solution. The mixing entropy is affected not only by the composition but also by temperature, component type, and other factors. The calculation formula [35] for the atomic size mean square deviation δ is as follows:

$$\delta = \sqrt{\sum_{i=1}^{C} x_i\left(1-\frac{r_i}{\bar{r}}\right)^2} \tag{16}$$

$$\bar{r} = \sum_{i=1}^{C} x_i r_i \tag{17}$$

where, $\bar{r}$ is the weighted value of atomic radii of all components.

At the same time, the stability of the phase is closely related to the Gibbs free energy, and the lower the free energy of the system, the more stable the structure. When the alloy components are determined, the composition and temperature of the system have a greater impact on the free energy. Considering the mixing free energy of the solid solution alloy is of great significance for designing

Journal of Physics: Conference Series **2941** (2025) 012005 doi:10.1088/1742-6596/2941/1/012005

high entropy alloys.

3.1. CoCrFeNiCu System

The model-predicted thermodynamic properties of CoCrFeNiCu high-entropy alloy solid solutions at 900°C are shown in table 4 for readers to replicate and verify. The model-predicted values of the mixing entropy of equimolar CoCrFeNiCu high-entropy alloy solid solutions at different temperatures are shown in figure 9, with an average value of 1.50*R*, consistent with the high-entropy effect in thermodynamics. As can be seen from figure 9, the mixing entropy of the system gradually increases with increasing temperature, but the variation is small. The model-predicted values of the mixing enthalpy of CoCrFeNiCu high-entropy alloy solid solutions are shown in figure 10. It can be seen that the mixing enthalpy of the system falls within the range plotted in the figure at different temperatures and component concentrations, between -2.32 KJ·mol^{-1} and 0.78 KJ·mol^{-1}, meeting the mixing enthalpy criterion for the formation of high-entropy alloys. Using equations (16) and (17) and the parameters in table 2, the mean square deviation δ of the atomic sizes of the components in the CoCrFeNiCu alloy system within the concentration range of x_i=0.05~0.35 can be calculated, which is between 1.0% and 1.2%, meeting the δ parameter criterion for the formation of high-entropy alloys. In summary, the results indicate that CoCrFeNiCu alloys meet the relevant criteria and are prone to forming quinary high-entropy alloys, but the alloy formation and phase stability need to be specifically analyzed and judged based on the system's mixing free energy and actual preparation conditions.

Table 4. Model predictions of thermodynamic properties of CoCrFeNiCu high-entropy alloy solid solutions with different compositions at 1173K.

Component *i*	x_i	Δ*G*M m (KJ·mol^{-1})	Δ*H*M m (KJ·mol^{-1})	Δ*S*M m (J·mol^{-1}·K^{-1})	*G*E m (KJ·mol^{-1})	*S*E m (J·mol^{-1}·K^{-1})
	0.05	-14.8407	-1.6542	11.2417	-0.0611	-1.3581
	0.10	-15.1114	-1.0436	11.9930	0.2265	-1.0828
	0.15	-15.1805	-0.6499	12.3875	0.4336	-0.9237
i=Cu	0.20	-15.1127	-0.3834	12.5570	0.5831	-0.8239
	0.25	-14.9344	-0.1954	12.5652	0.6894	-0.7543
	0.30	-14.6590	-0.0586	12.4471	0.7620	-0.6996
	0.35	-14.2944	0.0430	12.2228	0.8074	-0.6517

Note: When the concentration of component *i* is determined, the other components are all equimolar.

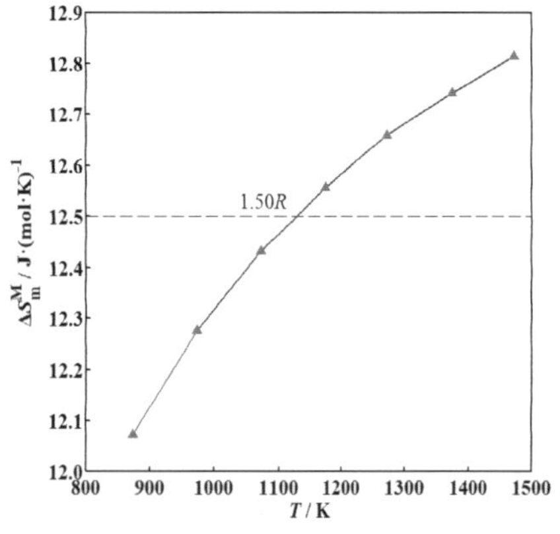

Figure 9. Variation of mixing entropy of equimolar CoCrFeNiCu high-entropy alloy solid solution with temperature.

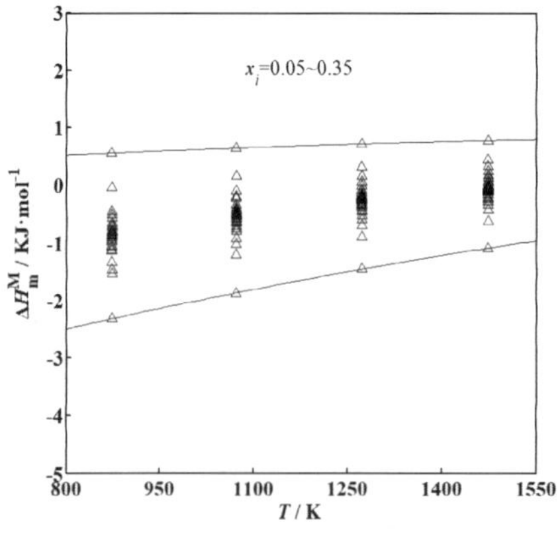

Figure 10. Mixing enthalpy region of the solid solution of CoCrFeNiCu high entropy alloy.

Using the grid search method, the alloy composition that yields the minimum Δ*G*M m under specific conditions has been identified as $Co_{0.20}Cr_{0.20}Fe_{0.23}Ni_{0.22}Cu_{0.15}$. This alloy achieves a Δ*G*M m of -16.44 KJ·mol^{-1} at 1000°C. Conversely, the alloy composition resulting in the maximum Δ*G*M m under certain conditions is $Co_{0.05}Cr_{0.20}Fe_{0.05}Ni_{0.35}Cu_{0.35}$, which attains a Δ*G*M m of -12.54 KJ·mol^{-1} at 1000°C. It can be seen that in the CoCrFeNiCu high entropy alloy, the appropriately balanced component content results in a more negative mixing Gibbs free energy Δ*G*M m of the system, and the alloy structure is more stable. Figure 11 shows the relationship between the mixing Gibbs free energy of the solid solution of CoCrFeNiCu high entropy alloy and temperature. As shown in figure 11, with the increase of temperature, the mixing Gibbs free energy of the system decreases, and the stability of the alloy increases; conversely, with the decrease of temperature, the mixing Gibbs free energy of the system increases, and the stability of the alloy decreases. This conclusion is consistent with experimental studies [30, 31].

3.2. CoCrFeNiTi System

The model-predicted thermodynamic properties of CoCrFeNiTi high-entropy alloy solid solutions with different compositions at 900°C are shown in table 5 for readers to replicate and verify. The model-predicted values of the mixing entropy of equimolar CoCrFeNiTi high-entropy alloy solid solutions at different temperatures are shown in figure 12, with an average value of 1.43*R*, consistent with the high-entropy effect in thermodynamics. As can be seen from figure 12, the mixing entropy of the system gradually increases with increasing temperature, but the magnitude of the change is small. The model-predicted values of the mixing enthalpy of CoCrFeNiTi high-entropy alloy solid solutions are shown in figure 13. It can be seen that the mixing enthalpy of the system falls within the range plotted in the figure at different temperatures and component concentrations, between -4.13KJ·mol^{-1} and 0.70KJ·mol^{-1}, meeting the mixing enthalpy criterion for the formation of high-entropy alloys. Using equations (16) and (17) and the parameters in table 2, the mean square deviation δ of the atomic sizes of the components in the CoCrFeNiTi alloy system within the concentration range of x_i=0.05~0.35 can be calculated, ranging from 3.6% to 7.1%, with some concentration ranges meeting the δ parameter criterion for the formation of high-entropy alloys. In summary, the results indicate that CoCrFeNiTi alloys can meet relevant criteria and form quinary high-entropy alloys, but the formation and phase stability of the alloys need to be analyzed and judged based on the mixing free energy of the system and actual preparation conditions.

Table 5. Model predictions of thermodynamic properties of CoCrFeNiTi high-entropy alloy solid solutions with different compositions at 1173K.

Component *i*	x_i	Δ*G*M m (KJ·mol^{-1})	Δ*H*M m (KJ·mol^{-1})	Δ*S*M m (J·mol^{-1}·K^{-1})	*G*E m (KJ·mol^{-1})	*S*E m (J·mol^{-1}·K^{-1})
	0.05	-15.1452	-2.6218	10.6764	-0.3656	-1.9235
	0.10	-15.5069	-2.2836	11.2731	-0.1690	-1.8027
	0.15	-15.5481	-1.8232	11.7007	0.0660	-1.6106
i=Ti	0.20	-15.3883	-1.3293	11.9855	0.3074	-1.3953
	0.25	-15.0871	-0.8479	12.1391	0.5366	-1.1804
	0.30	-14.6786	-0.4038	12.1695	0.7424	-0.9771
	0.35	-14.1838	-0.0100	12.0834	0.9181	-0.7912

Note: When the concentration of component *i* is determined, the other components are all equimolar.

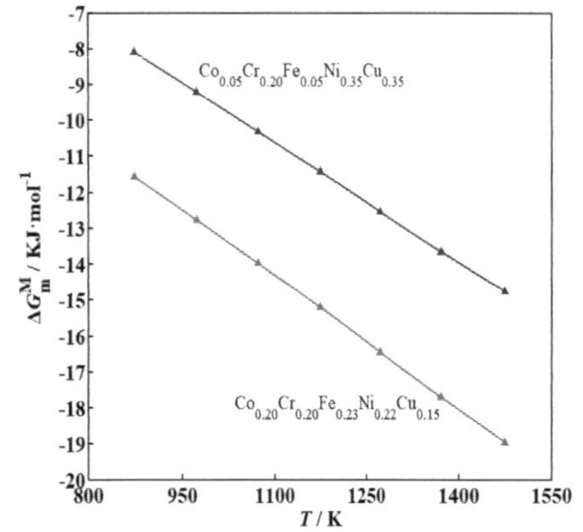

Figure 11. Variation of mixing Gibbs free energy of CoCrFeNiCu high-entropy alloy solid solution with Temperature.

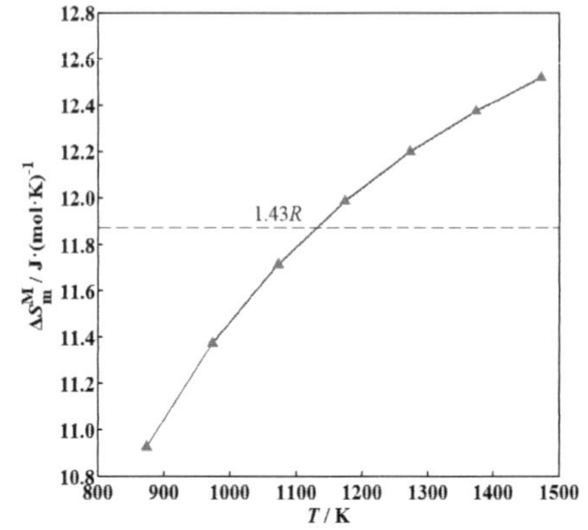

Figure 12. Variation of mixing entropy of equimolar CoCrFeNiTi high-entropy alloy solid. solution with temperature.

Using the grid search method, the alloy composition that yields the minimum ΔGM m under specific conditions has been identified as $Co_{0.20}Cr_{0.17}Fe_{0.21}Ni_{0.28}Ti_{0.14}$. This alloy achieves a ΔGM m of -16.91 KJ·mol^{-1} at 1000°C. Conversely, the alloy composition resulting in the maximum ΔGM m under certain conditions is $Co_{0.05}Cr_{0.35}Fe_{0.21}Ni_{0.05}Cu_{0.34}$, which attains a ΔGM m of -11.73 KJ·mol^{-1} at 1000°C. It can be seen that in the CoCrFeNiTi high-entropy alloy system, appropriate conditions of high Ni and low Ti result in a more negative mixing free energy ΔGM m, leading to a more stable alloy structure. Figure 14 shows the relationship between the mixing Gibbs free energy of the solid solution in the CoCrFeNiTi high-entropy alloy and temperature. As the temperature increases, the mixing Gibbs free energy of the system decreases, enhancing the stability of the alloy; conversely, as the temperature decreases, the mixing Gibbs free energy of the system increases, reducing the stability of the alloy. This conclusion is consistent with experimental studies [30, 31]. In summary, appropriately increasing the processing temperature and reducing the cooling rate can significantly reduce the mixing free energy of the system, reduce the tendency of segregation and ordering, and facilitate the formation of a stable single-phase solid solution with a disordered arrangement.

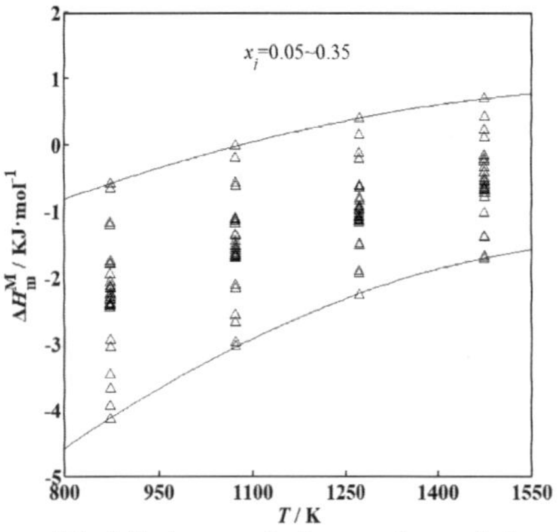

Figure 13. Mixing enthalpy region of the solid solution of CoCrFeNiTi high entropy alloy.

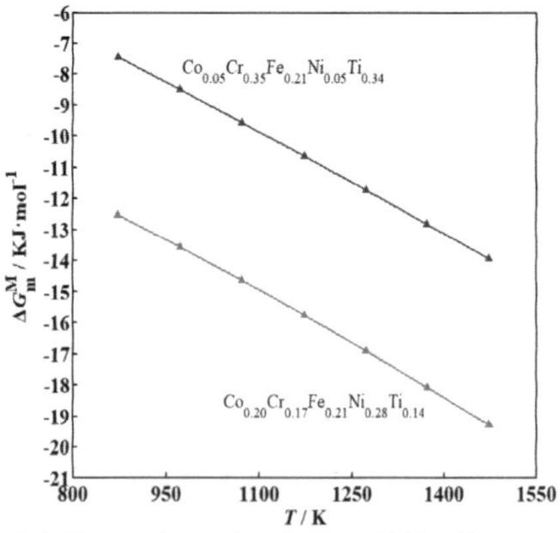

Figure 14. Variation of mixing Gibbs free energy of CoCrFeNiTi high-entropy alloy solid solution with Temperature.

4. Conclusion

(1) Using existing binary thermodynamic data and the modified M-MIVM, the thermodynamic properties of the solid solution of CoCrFeNi-X (X=Cu, Ti) high-entropy alloys, including mixing

Gibbs free energy, mixing enthalpy, mixing entropy, excess Gibbs free energy, and excess entropy, were predicted under the conditions of x_i=0.05~0.35 and T=600~1200°C, resulting in a total of 2940 thermodynamic data points. Based on the good results of previous studies, the model predictions are relatively reliable. The significant advantage of the modified M-MIVM is that it can predict the thermodynamic properties of multicomponent solid solutions using only binary infinite dilution activity coefficients or binary activity data, making it suitable for the thermodynamic study of simple solid solution phases in high-entropy alloys.

(2) By calculating the mean square deviation δ of the atomic sizes of the components in the CoCrFeNi-X (X=Cu, Ti) system, combined with the mixing enthalpy, and based on the criteria of $\delta \leq 6.6\%$ and $-15 KJ\cdot mol^{-1} \leq \Delta H M\ m \leq 5 KJ\cdot mol^{-1}$, the possibility of forming a five-component high-entropy alloy was analyzed from a thermodynamic perspective. It was concluded that within the range of conditions studied, the CoCrFeNi-X (X=Cu, Ti) alloy system satisfies the relevant criteria and can form a five-component high-entropy alloy.

(3) By analyzing the relationship between mixing Gibbs free energy and composition of cocrfeni-x (x=Cu, Ti) system high entropy alloy solid solution, some better alloy composition design programmes can be obtained. For the CoCrFeNiCu alloy system, an appropriately balanced component content results in a more negative mixing free energy and a more stable alloy structure. This law guides alloy composition design, leading to an optimal scheme of $Co_{0.20}Cr_{0.20}Fe_{0.23}Ni_{0.22}Cu_{0.15}$ alloy. For the CoCrFeNiTi alloy system, appropriate conditions of high Ni and low Ti content result in a more negative mixing free energy and a more stable alloy structure. This law guides alloy composition design, leading to an optimal scheme of $Co_{0.20}Cr_{0.17}Fe_{0.21}Ni_{0.28}Ti_{0.14}$ alloy.

(4) By analyzing the relationship between the mixing Gibbs free energy and temperature of the CoCrFeNi-X (X=Cu, Ti) high entropy alloy solid solution, it is concluded that appropriately increasing the processing temperature can significantly reduce the mixing free energy of the system, improve the stability of the alloy, and facilitate the formation of a disordered single-phase solid solution. This conclusion is consistent with experimental results.

Acknowledgments
This work was financially supported by the Scientific Research Fund of the Yunnan Education Department (Grant No. 2022J1303 and Grant No. 2019J0891).

5. References

[1] Yeh J W, Chen S K, Lin S J, et al. 2004 Nanostructured High-entropy alloys with multiple principal elements: novel alloy design concepts and outcomes *Advanced Engineering Materials* **6(5)** 299-303

[2] Zhang Y 2022 Discovery and development of high entropy alloys *Journal of Sichuan Normal University: Natural Science* **45(6)** 711-722, 708

[3] Sng X F and Zhang Y 2022 Progress of high entropy alloys *Powder Metallurgy Technology* **40(5)** 451-457

[4] Zheng H T 2020 *Microstructure Evolution and Mechanical Properties of Directionally Solidified CoCrFeNi System High Entropy Alloy* (Harbin: Harbin Industrial University) Doctor's Thesis

[5] Niu L C, Li J, Zhao S J, et al. 2022 Research progress of deformation mechanism of FeMnCoCrNi high entropy alloy system *The Chinese Journal of Nonferrous Metals* **32(8)** 2316-2326

[6] Liu C X, Zhang L C and Jian Z Y 2019 Research progress of FeNiCrCo-X high-entropy alloy. *Applied Chemical Industry* **48(6)** 1483-1486, 1490

[7] Lu Y, Li C Y and Tian L 2022 Research progress in properties of high-entropy alloy *Rare Metals* **46(10)** 1352-1364

[8] Hang S Y 2019 *Directional Solidification Microstructure and Mechanical Properties of CoCrFeNi Series High Entropy Alloy* (Xi'an: Xi'an Technological University) Master's

Thesis
[9] Gludovatz B, Hohenwarter A, Catoor D, et al. 2014 A fracture-resistant high-entropy alloy for cryogenic applications *Science* **345(6201)** 1153
[10] Meng S, Guo D, Zhao D F, et al. 2023 Research progress of CoCrFeNi high entropy alloy *Journal of Netshape Forming Engineering* **15 (08)** 156-168
[11] Li C W and Zhang Y 2022 Effects of Copper Addition on Microstructure and Properties of CoCrFeNi High Entropy Alloy *Journal of Netshape Forming Engineering* **14(12)** 1-9
[12] An M R, Li S L, Su M J, et al. 2022 Molecular dynamics simulation of size dependent plastic deformation mechanism of CoCrFeNiMn crystalline/amorphous dual-phase high-entropy alloys *Acta Physica Sinica* **71(24)** 152-163
[13] Sha M H 2021 Study on Preparation, Structure and Properties of CoCrFeNi-M High Entropy Alloy *Anshan: University of Science and Technology Liaoning* 30-33
[14] Mukarram M, Mujahid M and Yaqoob K 2021 Design and development of CoCrFeNiTa eutectic high entropy alloys *Journal of Materials Research and Technology* **10** 1243-1249
[15] Liu Z N, Yao C L, Liu C, et al. 2022 A thermodynamic model of solid solutions and its application in solid alloys. Journal of Iron and Steel Research International **29(2)** 263-280
[16] Dai H and Tao D P 2018 A statistical thermodynamic model with strong adaptability for liquid mixtures *Fluid Phase Equilibria* **473** 154-165
[17] Dai H 2019 *Improvement of MIVM and Its Application in Liquid Alloys and Organic Solutions* (Kunming: Kunming University of Science and Technology) Doctor's Thesis
[18] Yao C L, LIU Z N, LIU C, et al. 2022 Application of Modified M-MIVM in ni-based solid alloys *Nonferrous Metals Engineering* **12(5)** 58-66+74
[19] Liu Z N, Sun Y, Yao C Y, et al. 2023 Thermodynamic study of CoCrFeNiMn high-entropy alloy *Mining and Metallurgy* **32(03)** 61-69
[20] Hultgren R, Desai P D, Hawkins D T, et al. 1973 Selected values of the thermodynamic properties of binary alloys
[21] Predel B 2002 *Landolt-Börnstein Group IV Physical Chemistry Volume 19: Thermodynamic Properties of Inorganic Materials* (New York: Springer-Verlag Berlin Heidelberg)
[22] Gao Y F 2005 *Application of MIVM in Iron-Based Solid Alloy System* (Kunming: Kunming University of Science and Technology) Master's Thesis
[23] Yang H W 2012 *Study on Thermodynamic Model of Mixing Enthalpy of Liquid Alloy* (Kunming University of Science and Technology) Doctor's Thesis
[24] Tao D P 2004 A thermodynamic model for solid solutions and its application to the C-Fe-Co, C-Fe-Ni and Mn-Cr-Pt solid dilutions *Materials Science and Engineering A* **366(2)** 239-247
[25] Tao D P 2003 Prediction of the thermodynamic properties of binary continuous solid solutions by infinite dilute activity coefficients *Thermochimica Acta* **408(1)** 67-74
[26] Tomiska J 1986 Determination of thermodynamic mixing effects of solid Fe-Co alloys by means of Knudsen cell mass spectrometry and calculation of the phase diagram *Z. Met.* **77** 97-102
[27] Mazandarany F N and Pehlke R D 1973 Thermodynamic properties of solid alloys of chromium with nickel and iron Metallurgical Transactions **4(9)** 2067-2075
[28] Conard B R, Mcanency T B and Sridhar R 1979 Thermodynamics of iron-nickel alloys by mass spectrometry *Metallurgical & Materials Transactions B* **9(3)** 463-468
[29] Wang J J, Lu X G, Zhu N Q, et al. 2017 Thermodynamic and diffusion kinetic studies of the Fe-Co system *CALPHAD* **58** 82-100
[30] Jia X Q, Xu Z L, Zhou S X, et al. 2023 Effect of annealing temperature on pitting corrosion resistance of CoCrFeMnNi high entropy alloy fabricated by laser additive *Surface Technology* **52(02)** 272-281
[31] Zhang T C, Li J K and Wang X B 2018 Phase stability of CoCrFeMnNi high entropy alloys. Heat *Treatment of Metals* **43(9)** 10-15
[32] Zhang Y, Zhou Y, Lin J, et al. 2008 Solid-Solution Phase Formation Rules for Multi-

component Alloys *Advanced Engineering Materials* **10(6)** 534-538

[33] Yang X and Zhang Y 2012 Prediction of high-entropy stabilized solid-solution in multi-component alloys *Materials Chemistry and Physics* **132(2/3)** 233-238

[34] Zhang Y, Zuo T T, Tang Z, Gao M C, Dahmen K A, Liaw P K and Lu Z P. 2014 Microstructures and properties of high-entropy alloys *Progress in Materials Science* **61** 1-93

[35] Zhang L, Zhen H M, Tao X M, et al. 2019 Thermodynamics study of Al-based high entropy quinary alloys *The Chinese Journal of Nonferrous Metals* **29(11)** 2601-2608

Case Analysis and Influencing Factors Research of Reactor Internals Baffle-Former Bolts on Irradiation Assisted Stress Corrosion Cracking

Wei Qin[1], Chaoliang Xu[2,3,4,*], Xiangbing Liu[2,4], Jian Yin[2,4] and Xiao Jin[2,4]

[1] South China Nuclear and Radiation Safety Supervision Station of the Ministry of Ecology and Environment, Shenzhen, Guangdong, China
[2] Suzhou Nuclear Power Research Institute, Suzhou, Jiangsu, China
[3] School of Mechanical Engineering, Shanghai Jiao Tong University, Shanghai, China
[4] National Engineering Research Center for Nuclear Power Plant Safety & Reliability, Suzhou, Jiangsu, China
Email: qin.wei@scro.mee.gov.cn; *xuchaoliang@cgnpc.com.cn; liuxbing@cgnpc.com.cn; yin.jian@cgnpc.com.cn; jinxiao@cgnpc.com.cn

Abstract: In order to obtain the distribution characteristic of baffle-former bolts (BFBs) cracking due to irradiation assisted stress corrosion cracking (IASCC), the case analysis of 4 nuclear power plant (NPP) is carried out. Then a discussion about the irradiation damage, stress, temperature and water environment on IASCC is given. The analysis shows that a large number of BFBs are cracked. The BFBs in the middle domain are exposed to higher irradiation and cause more BFBs cracking in PB-2 NPP. Higher stress levels in the bottom and top domains with lower irradiation damage result in the BFBs fracture in G NPP. Higher irradiation temperatures will increase the BFBs swelling rate and alter the local water environment around BFBs, leading to more BFBs cracking in the T-1 NPP than in the D-1.
Keywords. Reactor; Baffle-Former bolts; stress corrosion cracking; irradiation damage.

1. Introduction

Reactor internals are key equipment in a nuclear power plant (NPP) and their integrity is related to the reactor operation safety and efficiency. Under pressurized water reactor conditions, internals service in severe environments such as high temperature, high pressure, high neutron irradiation (up to 100 dpa for 40 years of operation) and primary loop water. When the irradiation dose exceeds a certain neutron dose, irradiation-induced material damage will lead to irradiation-accelerated stress corrosion cracking (IASCC) phenomenon and cause internal failure. Internal IASCC failure has been the most important issue for NPP safety operations. In internal components, the most prone to IASCC is the baffle-former bolts (BFBs). The BFBs are an important structural component of the internals and are used to attach the baffle plates to the former plates in the internals. Many BFBs IASCC problems have occurred in NPP [1]. For example, Beach Point-2, Indian point-2, Salem, Ginna, Tihange-1, Doel-1 and so on. According to the NPP operating experience, the BFBs IASCC failure often occurs after an operating time of more than ten equivalent full-power years. With the operation time increasing, the IASCC risk increases gradually. From the point of safe and reliable operation of the NPP, it is necessary to pay close attention to the possibility of IASCC failures. Therefore, the IASCC case analysis and influencing factors studies can provide reference data for the BFBs evaluation and BFB replacement.

Journal of Physics: Conference Series **2941** (2025) 012006 doi:10.1088/1742-6596/2941/1/012006

This study summarizes the experience with BFBs cracking in PB-2 and G (abbr.) NPP in U.S. and T-1 and D-1 (abbr.) NPP in Belgium. Based on this data survey, the influencing factors of irradiation, stress, temperature and water environment are discussed.

2. IASCC Case

BFBs cracking due to IASCC had been detected in PB-2, G, T-1, and D-1. These 4 NPPs were all commercialized in the 1970s. The schematic diagram of the internals is shown in figure 1. The number of formers is 5 layers to 8 layers with different designs. The BFBs are made of austenitic stainless steel. The supplier of the internal is Westinghouse and the power of the reactors ranges from 580 MW to 962 MW.

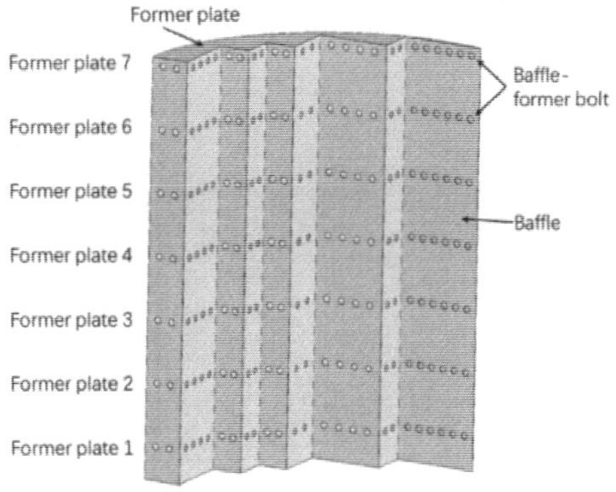

Figure 1. Schematic diagram of the internals and BFBs designed by Westinghouse.

So far, the BFBs inspections have been conducted at T-1 5 times, with a total of 85 IASCC bolts found. In D-1, the BFBs inspections have been conducted 3 times, with a total of 17 IASCC bolts found. In these 2 NPPs, the cumulative number of IASCC bolts has been gradually increasing with operating time as shown in figure 2. PB-2 and G only conducted 1 BFBs inspection and obtained bolt distribution dates in different former plates as shown in figure 2. It is indicated that the IASCC bolts appeared in all former plates, but the distribution characteristics are different. IASCC bolts in PB-2 are mainly concentrated in the middle region of former plates, but mainly concentrated in the top and bottom former plate in G. This indicates that the main influencing factors are different between PB-2 and G.

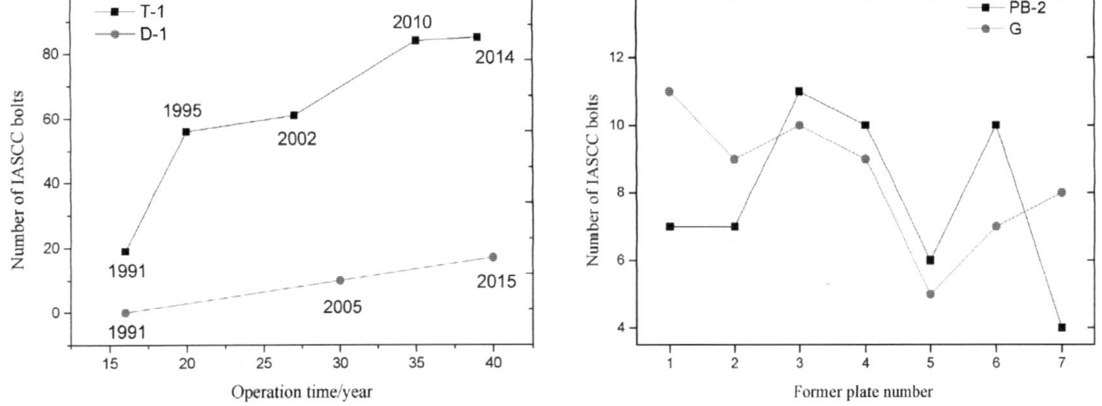

Figure 2. Accumulated IASCC bolts at different operation times in T-1 and D-1(left), IASCC bolts number in different former in PB-2 and G (right).

3. Influencing Factors Analysis

3.1. Irradiation Damage

Due to the BFBs distribution regions difference, BFBs undergo different levels of neutron irradiation damage. Figure 3 shows the neutron flux distribution in different former plate regions in a typical NPP. It can be observed that the BFBs in the middle region are exposed to much more irradiation damage. On the contrary, the BFBs in the top and bottom regions are exposed to lower irradiation damage. The maximum damage difference in BFBs can be up to more than 10 times. Therefore, when analyzing the BFBs IASCC problem from the perspective of irradiation damage, only the BFBs in the middle region of the former plate are considered. The effect of irradiation damage on the BFBs in the top and bottom regions of the former plate is limited.

It can be observed from figure 3 that the number of cracked BFBs in PB-2 shows a significant correlation with irradiation damage. Most of the IASCC bolts are mainly concentrated in the middle region of the former plate (except for the No. 5 former plate). This means that high irradiation damage leads to more IASCC bolts. Therefore, the IASCC in the middle region in PB-2 is mainly caused by irradiation damage. However, a small amount of IASCC bolts can be detected in the top and bottom regions of the former plate. This indicates that there are other influencing factors involved. This will be discussed later.

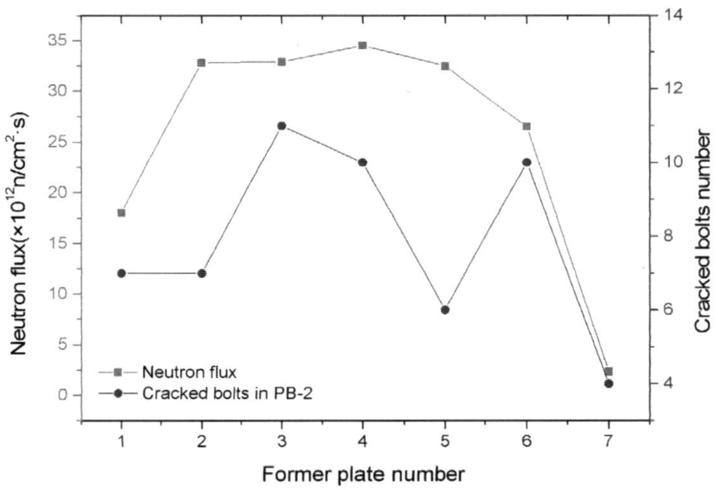

Figure 3. Neutron flux distribution in a typical NPP and IASCC bolts distribution with the former plate number in PB-2.

In order to further analyze the relationship between irradiation damage and IASCC, the distribution of neutron irradiation damage and BFBs cracking number in T-1 is given in figure 4 [2]. It can be found that irradiation damage exhibits different distributions in different BFB columns. Four irradiation damage peaks appear in column 2, column 8, column 13 and column 17. The number of IASCC BFBs is basically consistent with the distribution of irradiation damage. On the whole, high irradiation damage leads to a significant IASCC phenomenon. This indicates that the IASCC is directly related to the irradiation damage. Irradiation damage is a key factor in the occurrence of IASCC.

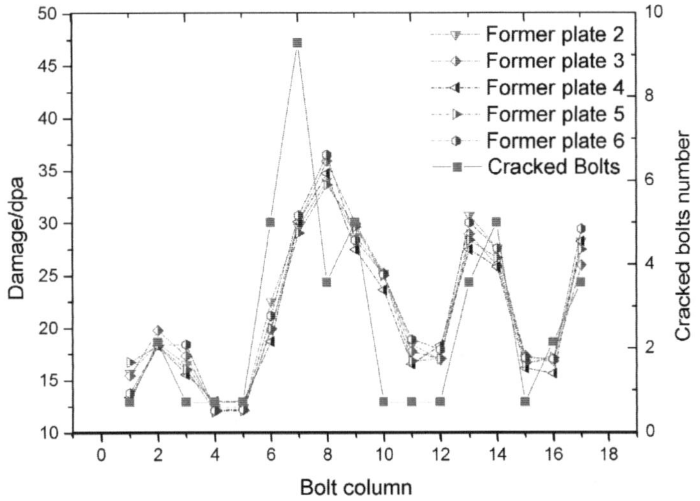

Figure 4. Neutron irradiation damage and IASCC bolts number distribution in T-1 [2].

3.2. Stress

Due to the impacts of internal installation, vibration by fluid impact, irradiation swelling and creep, the stress in BFBs is different. Figure 5 shows the stress distribution in BFBs at different former plates in a typical NPP[3] and cracked BFBs number in G. It can be observed that the actual stress in BFBs after reactor operation shows significant differences. After reactor operation, the stress in the top and bottom BFBs (that is to say, the BFBs in the top and bottom former plate) increases and the stress in BFBs in the middle domain decreases. The actual stress in BFBs exhibits bathtub curve characteristics. Meanwhile, the number of IASCC bolts in former plate 1 and former plate 7 is larger than in the middle former plate in G. According to the discussion in 3.1, the irradiation damage in top and bottom BFBs is weak. Irradiation damage is not the major reason for BFBs cracking. So, under low irradiation damage level conditions, the stress may play the main role in BFB cracking.

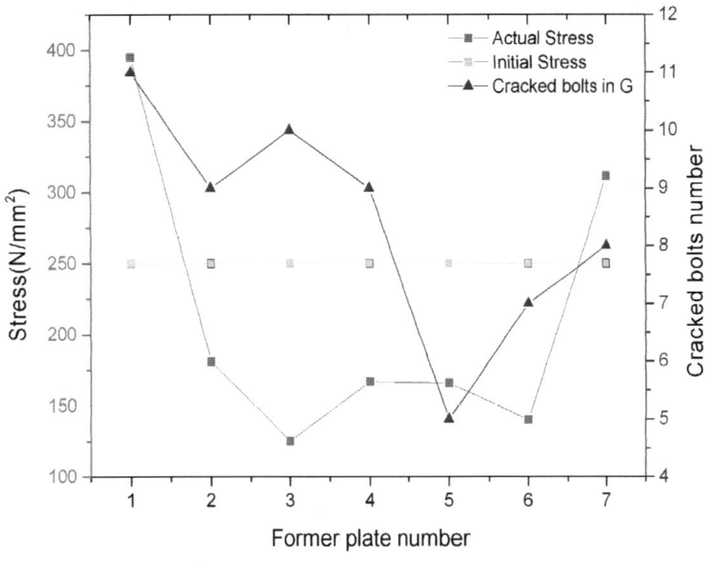

Figure 5. Initial stress and actual stress in the internal former of a typical nuclear power plant [3] and cracked BFBs number in G.

In fact, previous studies have shown that the BFBs in the bottom and top regions are subject to high stresses. For example, a finite element analysis study by Areva and the Electric Power Research Institute (EPRI) has shown that the stresses in top and bottom BFBs continue to increase with increasing operating time and are larger than other BFBs in service life [4]. According to studies by EPRI and Materials Aging Institute (MAI), high stress will increase the IASCC sensitivity and cause the BFBs to crack under a low irradiation damage state [5,6]. In addition, high stress will lead to the formation of the ferrite phase in austenitic stainless steel, and the formation of the ferrite phase will directly increase the IASCC sensitivity [7]. Therefore, because of the continuous increase of stress levels with the increased internal operating time, the stress in the top and bottom BFBs is maintained at a high level, which increases the IASCC sensitivity of the BFBs.

3.3. Temperature

From figure 2, it can be observed that the number of cracked BFBs in T-1 is larger than that in D-1. The highest irradiation damage in T-1 BFBs is about 35 dpa, and the highest damage in D-1 is around 47 dpa. Generally, the high irradiation-damaged BFBs are generally distributed in the middle region of former plates as shown in figure 3, and the influence of stress is relatively weak as shown in figure 5. Therefore, in addition to irradiation damage and stress, there are also other factors affecting IASCC.

Under the fission reactor operation environmental, γ-rays produced by nuclear reaction will cause materials temperature to increase. Analysis [8] shows that the maximum temperature in BFBs is about 340°C in D-1, and the maximum temperature is about 370°C in T-1. The BFBs temperature in D-1 is lower than that in T-1 (figure 6). This temperature difference is due to more coolant holes near the BFB in D-1, which significantly reduces the temperature in BFBs. However, the BFB is not cooled sufficiently in T-1, resulting in higher temperatures during the operation process. It has been shown [9] that the stainless steel irradiated at a temperature higher than 250°C will form gas bubbles and lead to material swelling. The degree of irradiation swelling increases with increasing temperature. Microstructural analysis shows that the size of bubbles formed in the T-1 BFB is 7~8 nm, causing a swelling rate of about 0.24%, while only 2.5 nm bubbles are formed in the D-1 and the swelling rate is basically negligible [8]. The higher swelling rate in T-1 will increase the stress and significantly promote the BFB cracking sensitivity. So higher service temperature in BFBs will lead to a larger number of BFB cracks in T-1 than that in D-1.

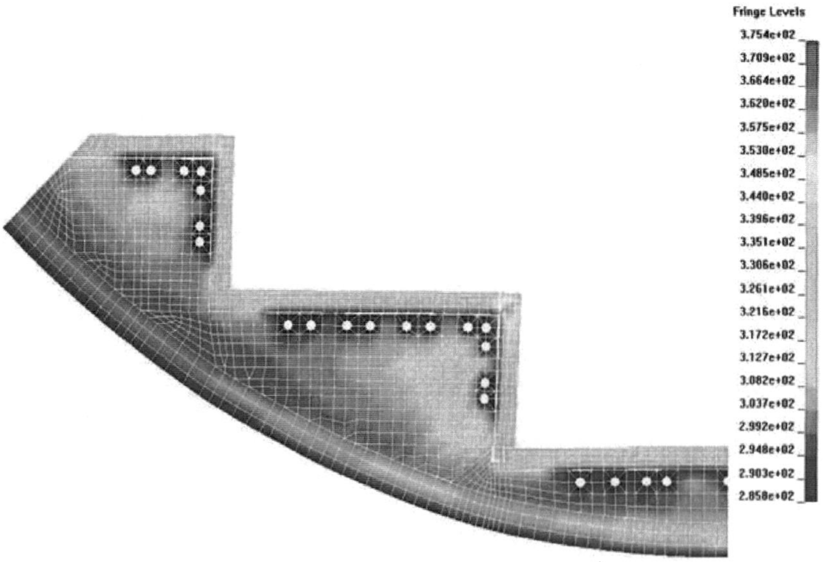

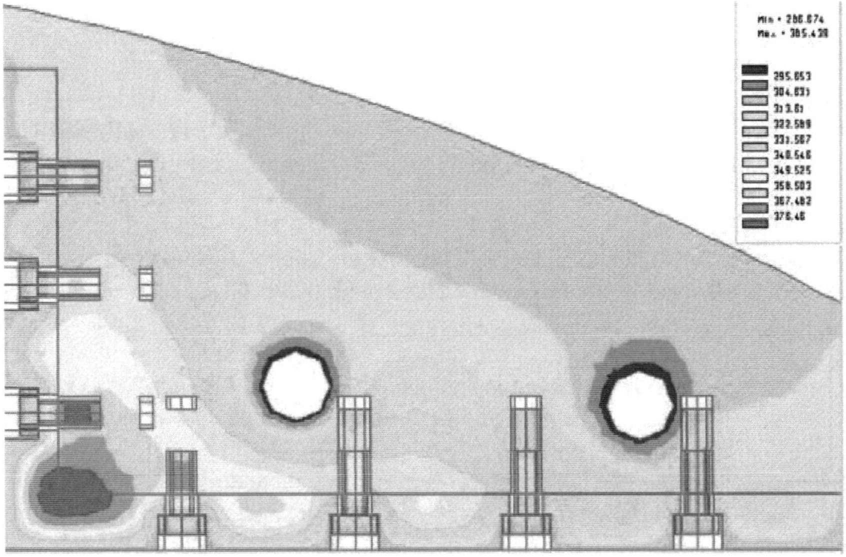

Figure 6. Temperature distribution in D-1 (left) and T-1 (right) former plate [8].

3.4. Water Environment

In the reactor primary circuit water environment, neutron irradiation will produce many irradiation products, such as H^+, OH, H_2O_2, and so on [10,11]. Previous studies have indicated that IASCC sensitivity is related to the concentration of oxidizing substances in the water [12]. Analysis [13] by the MAI further indicates that the γ-heating may result in insufficient cooling of the internals component due to the components' design. This may lead to localized surface water boiling in the BFBs, and then change the irradiation products or oxidizing substances concentration around BFBs. In D-1, a higher number of flowing water holes around BFBs cause a lower possibility of localized boiling. This may be a possible reason why the number of cracked BFBs in D-1 is smaller than that in T-1. However, the current understanding of the localized boiling phenomenon is still limited and further experimental verification is still needed.

4. Conclusions

In order to obtain the distribution characteristic of baffle-former bolts (BFBs) cracking due to irradiation-assisted stress corrosion cracking (IASCC), the case analysis of 4 nuclear power plants (NPP) is carried out. Then a discussion about the irradiation damage, stress, temperature, and water environment on IASCC is given. The analysis shows that a large number of BFBs are cracked. The BFBs in the middle domain are exposed to higher irradiation and cause more BFBs cracking in PB-2 NPP. Higher stress levels in the bottom and top domain with lower irradiation damage result in the BFBs fracture in G NPP. Higher irradiation temperatures will increase the BFBs swelling rate and alter the local water environment around BFBs, leading to more BFBs cracking in the T-1 NPP than in the D-1.

Acknowledgments

This research is supported by the Guangdong Major Project of Basic and Applied Basic Research (grant No. 2019B030302011), the National Key Research and Development Program of China (grant No. 2021YFA1600903, No.2023YFF0716204), the Guangdong Basic and Applied Basic Research Foundation (grant No. 2023B1515120082), the National Natural Science Foundation of China (grant No. U23B2072, 12275200, 12075274, 52101066). Wei Qin (first author) and Chaoliang Xu (co-first author) have the same contribution to this research.

5. References

[1] Makhncnko O V, Kandala S M. Influence of residual process stresses on brittle fracture resistance of wwer-1000 reactor baffle in case of an emergency[J]. The Paton Welding Journal, 2023(1): 55-59.

[2] Robert Gérard, Frédéric Somville, Situation of the Baffle-Former Bolts in Belgian Units, Proceedings of the 17th International Conference on Nuclear Engineering ICONE17, Brussels, 2009.

[3] Yu. C O, Kharchenko V V, Kravchuk L V. Computational Assessment of the Shape Change of the WWER-1000 Reactor Core Baffle Considering Effects of Radiation Swelling, Radiation Creep, and Subcritical Metal Damage Via Ductile Fracture Models[J]. Strength of Materials, 2023(2):55.

[4] Tamás R. Liszkai, Matthew Snyder, Anne Demma. Materials aging degradation of reactor vessel internals part ii: structural evaluation – global finite element models. In: Proceedings of the ASME 2009 Pressure Vessels and Piping Division Conference, Czech Republic, PVP2009-77786, 2009.

[5] Gad-El-Hak I, Mureithi N, Painter K K. Mechanism and risk assessment of baffle-jetting-induced vibrations in fuel rods[J]. Nuclear engineering and design, 2024, 416(Jan.):112747.1-112747.13.

[6] Faïza SEFTA. Proposed revision of the IASCC sensitivity criterion. Materials Ageing Institute, H-B60-2015-00404-EN. 2015.

[7] Chaoliang Xu, Xiangbing Liu, Wangjie Qian, et al. Journal of Nuclear Materials, 2017, 495: 421-424.

[8] F. Somville, R. Gérard, R-W. Bosch, et al. Ageing Management of Baffle Former Bolts in Belgian Nuclear Power Plants. In: 2016 International LWR Material Reliability Conference and Exhibition, Chicago, 2016.

[9] Malcolm Griffiths. Materials, 2021, 14: 2622.

[10] Scott P. Journal of Nuclear Materials, 1994, 211(2): 101-122.

[11] Ishigure K, Nukii T, Ono S. Journal of Nuclear Materials, 2006, 350(1): 56–65.

[12] Yang Wu. Materials for Mechanical Engineering, 1994(2): 16-19.

[13] Thierry C, Salem M, Frédéric D, et al. Superficial oxide examinations on a cracked baffle former bolt, Materials aging institute, 2011, H-B60-2011-02404-EN. 2011.

Method of Detecting Beta Radiation Using 3D-Printed Porous Plastic Scintillator

Haiying Zhang, Yanjun Huang*, Guiyin Guo and Xuefeng Sun

Suzhou Nuclear Power Research Institute, Co., Ltd., China General Nuclear Power Group, Suzhou 215004, Jiangsu, China.
Email: 1052927255@qq.com; *hyj1231@163.com; 273614071@163.com; luckjelly@163.com

Abstract. The novel porous plastic scintillator (PPS) material has potential application prospects in the field of radioactivity detection. In this paper, the PPS columns were printed with Fused Deposition Modeling (FDM) technology, which was applied to β detection in a liquid scintillation counter. The efficiency can reach 3.3% for ^{3}H, 43.5% for ^{14}C, and 67.6% for ^{90}Sr, respectively. The memory effect could be reduced to less than 0.5% with a round of rinsing. It proposes a new solution for β detection, which demonstrates significant value in the field of radionuclide detection in the future.
Keywords. 3D-printing; porous plastic scintillator; radionuclide detection; memory effect.

1. Introduction

Since the early 1950s, liquid scintillation counting (LSC) technology has been developed and improved over the years to become an analytical technique that appeared to be unique for β detection [1]. It was widely used in the field of β detection such as ^{3}H, ^{14}C, and ^{90}Sr in the environment and effluents from nuclear facilities. However, this method uses organic solvents as the matrix for the scintillator, which has disadvantages such as poor stability, toxicity, and non-reusability. Therefore, a scintillator that is safe, environmentally friendly, efficient, and reusable, is practically demanded [2].

Solid plastic scintillators (PS) have been widely used in the fields of radiation detection, security inspection, medical and biological research, industrial testing, and scientific research due to their high detection efficiency, good mechanical properties, radiation resistance, and ease of processing [3]. In recent years, the application of such materials for radionuclide detection received a certain degree of attention [4]. The PS are solid solutions containing organic scintillating substance, similar in composition to liquid scintillators (LS), typically consisting of a matrix plastic material such as polystyrene, Polymethyl Methacrylate (PMMA), and a wavelength-shifting agent, which can efficiently convert the energy of β and other rays into light signals, serving as an alternative to scintillator solutions [3]. How to achieve the substitution of PS for LS is one of the key issues in the field of radionuclide detection.

The 3D-printing (additive manufacturing) technology, with its unique advantages, has a broad application prospect in the personalized customization for porous materials [5]. Applying the technology to the preparation of porous plastic scintillator (PPS) and application of these materials for β detection is clearly forward-looking and feasible. Based on the aforementioned objectives, this paper describes a manufacturing method for PPS based on 3D-printing technology and has conducted detection studies on different β radionuclides. It demonstrated potential technical advantages and value for the promotion and application of the method and the materials.

2. Materials and Methods

2.1. 3D-printing Materials

The 3D printing material used in this study is the ND-SF018S type scintillating fiber (Beijing Nuclear Tai Technology Development Co., Ltd., China (figure 1(a))). The fiber uses polystyrene as the scintillator matrix, with 2,5-diphenyloxazole (PPO) added at a mass concentration of 1wt% and 1,4-bis (5-phenyl-2-oxazolyl) benzene (POPOP) at a mass concentration of 0.02wt%. The fiber has a diameter of 1.75 mm without PMMA cladding. It has an attenuation time of 3ns, a light yield of >5000 photons/MeV, and an emission wavelength of 425 nm.

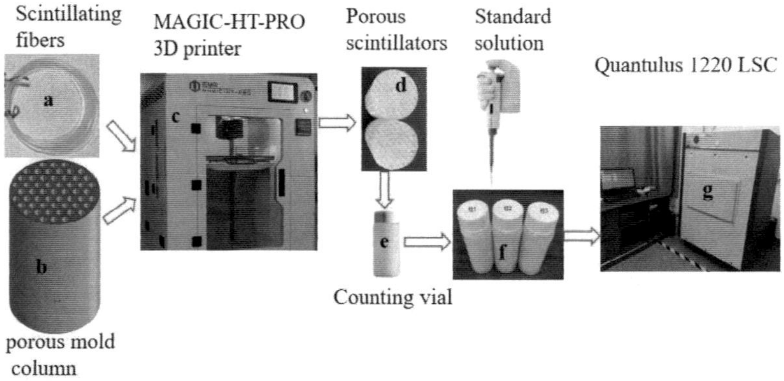

Figure 1. Experimental procedure and method.

2.2. 3D-printing Method

This study employs Fused Deposition Modeling (FDM) technology for the printing of the PPS. In 2020, Berns et al. used FDM technology to melt prefabricated polystyrene scintillating optical fibers and printed PS blocks with good transparency, demonstrating the feasibility of printing polystyrene scintillators for radioactivity detection [6]. FDM is one of the most common additive manufacturing processes, which works by feeding thermoplastic filament material into a nozzle, heating it to a molten state inside the nozzle, and then extruding it from the nozzle tip. Under computer control, the material is deposited layer by layer to eventually construct a three-dimensional object. Compared to other 3D-printing technologies, FDM does not require the development of new compositions. It can directly utilize the optimal polystyrene PS to process plastic scintillating optical fibers. By using FDM technology to melt the scintillating optical fibers in the nozzle, a three-dimensional model is constructed layer by layer according to the designed three-dimensional model, printing out the materials with complex shapes and internal structures.

When designing the internal structure of the PPS, it is necessary to consider the quenching effect caused by the non-homogeneity between the scintillator and the measurement solution. This ensures that radiation particles can reach the surface of the scintillator as much as possible, increasing the contact area between the scintillator and the particles, reducing the distance between the radiation particles and the scintillator, and allowing photons to propagate effectively within the scintillation measurement system with minimal loss. Taking into account the molding precision of FDM technology and the requirements for the internal shape and size of the scintillator, and considering that the scintillator can be directly used for radiation detection performance testing on an LSC spectrometer, the polystyrene scintillator in this study was first designed as a straight-through porous honeycomb column with an inner hole side length of about 1 mm (figure 1(b)). The size of the scintillator is consistent with the inner diameter of a conventional 20 mL polystyrene counting vial (figure 1(e)), with a diameter of 24 mm and a height of 40 mm. The internal holes are hollow tubes with an equilateral hexagonal cross-section of 1 mm side length, with a total of 55 holes. The estimated spatial volume of this structure is 5.7 mL.

In this study, a MAGIC-HT-PRO high-temperature 3D printer (figure 1(c)) was used. The printer nozzle diameter is 0.4 mm, and the nozzle temperature is between 240°C and 260°C. The temperature of the constant temperature chamber is between 60°C and 100°C. The printing speed is between 30 mm/s and 60 mm/s, and the layer thickness is between 0.1 mm and 0.2 mm. After setting the printing parameters, the designed three-dimensional model is imported into the 3D printer, and then the layer-by-layer stacking printing begins.

2.3. *The Experiments for β Detection*

By using the 3D-printed PPS, it is matched with a 20 mL polyethylene counting vial for the conventional LSC spectrometer (figure 1(e)). The vial can hold up to 5.7 mL of solution at most. Before specific application, the neck of the counting vial is cut open to place the PPS, and a small amount of transparent tape is used to seal the body of the vial. The cap is then tightened, and the vial is dark adapted for one day to obtain the background for β detection. Then, 50μL of standard solution (SS) is evenly dropped on the top of the PPS column, and ultrasonic oscillation is performed for 15 minutes to disperse the standard solution as evenly as possible, forming a sample for scintillation counting (figure 1(f)). It is then placed on the Quantulus 1220 ultra-low background LSC spectrometer (PerkinElmer, USA) for β counting (figure 1(g)). After the measurement, 4 mL of pure water is added, and ultrasonic oscillation is performed for 15 minutes. It is then placed on the LSC spectrometer again to determine the detection efficiency of the 3D-printed PPS for beta radionuclides.

When measuring the sample on the LSC spectrometer, different modes are used depending on the properties of the measured nuclides: low-energy mode for 3H, and high-energy mode for 14C and 90Sr.

After the above measurements, the PPS is taken out, washed repeatedly with pure water, ultrasonic oscillated for 15 minutes, and soaked overnight in 100 mL of 0.1M citric acid solution [7]. Then ultrasonic oscillated for another 15 minutes, and finally washed with pure water again. After drying, it is placed back into the 20 mL vial, dark-adapted, and measured again on the LSC spectrometer to calculate the adsorption residue of the PPS. In this study, all standard solutions used are provided by the National Institute of Metrology of China (NIMC).

3. The Results and Discussion

3.1. *The β Detection Efficiencies*

To conduct a comparative analysis of 3D-printed PPS and conventional LSC, this study selected three pure β-radionuclides, ^{3}H, ^{14}C, and ^{90}Sr. Their maximum β energies are 18.5 keV, 156 keV, and 564 keV, respectively, representing low, medium, and high-energy β particles.

The 3D-printed PPS was placed in a counting vial to prepare a background sample, which was then measured on an LSC spectrometer. Background samples were spiked with 50 μL of standard solution, and the β spectrum was measured on an LSC spectrometer to assess the detection efficiency of the 3D-printed PPS and compare it with conventional LSC in terms of detection efficiency and sample spectrum. The comparison of the 3D-printed PPS and the conventional LSC counting background is shown in table 1, the detection efficiency comparison for ^{3}H, ^{14}C, and ^{90}Sr is shown in table 2, and the comparison of the measured sample spectra for ^{3}H, ^{14}C, and ^{90}Sr is shown in figure 2.

Table 1. Comparison of the background for 3D-printed PPS and conventional LS.

The test objects and the test mode	H-3 mode (low-energy)	C-14 mode (High-energy)
3D-printed PPS (This study)	0.51cpm	0.62cpm
8 mL ^{3}H-free water+12 mL Ultima Gold Ullt	0.55cpm	/
9 mL Carbo-sorb E™+9 mL Permafluor E™	/	2.49cpm

When using the PPS column for β detection, in addition to the three common types of quenching in the LSC method (i.e., chemical quenching, colour quenching, and ionization quenching), two new quenching phenomena are considered: distance quenching and optical quenching. The main difference between PPS and LS is the distance between the decay particle generation and the scintillation solvent [4]. For the LSC method, the sample is dissolved or dispersed in the liquid scintillator, with less self-absorption of β particles in the sample, resulting in high detection efficiency. While for PPS the β particles must pass through the medium between the sample liquid in the pores to reach the scintillator surface. The range of β in the medium is related to their energy; the lower the energy, the shorter the range in the medium. Some low-energy β particles complete decay when passing through the sample solution, failing to produce a light signal, which is called distance quenching, also known as particle quenching [8,9]. The fluorescent photons produced by the PPS also need to pass through the solution medium, and even switch from the plastic medium to the solution medium multiple times. During this process, photons may undergo optical processes such as refraction, scattering, and absorption, before finally reaching the photomultiplier tube, reducing the number of photons ultimately detected, which is optical quenching [9].

Table 2. Comparison of detection efficiencies for 3D-printed PPS and conventional LS.

The test objects and the test mode	^3H	^{14}C	^{90}Sr
3D-printed PPS (This study) +0.2 mL SS[a]	3.3%	43.5%	67.6%
0.2 mL SS* +12 mL Ultima Gold Ullt	58.8%	/	/
8 mL SS* +12 mL Ultima Gold Ullt	18.8%	/	/
50uL SS* +9 mL Carbo-Sorb E^{TM}+9 mL Permafluor E^{TM}	/	60.5%	/
50 uL SS* +12 mL Ultima Gold Ullt	/	/	95%
50uL SS* +4 mL water+12 mL Ultima Gold Ullt	/	/	94%

Note: [a]SS, standard solution from the metrology institution.

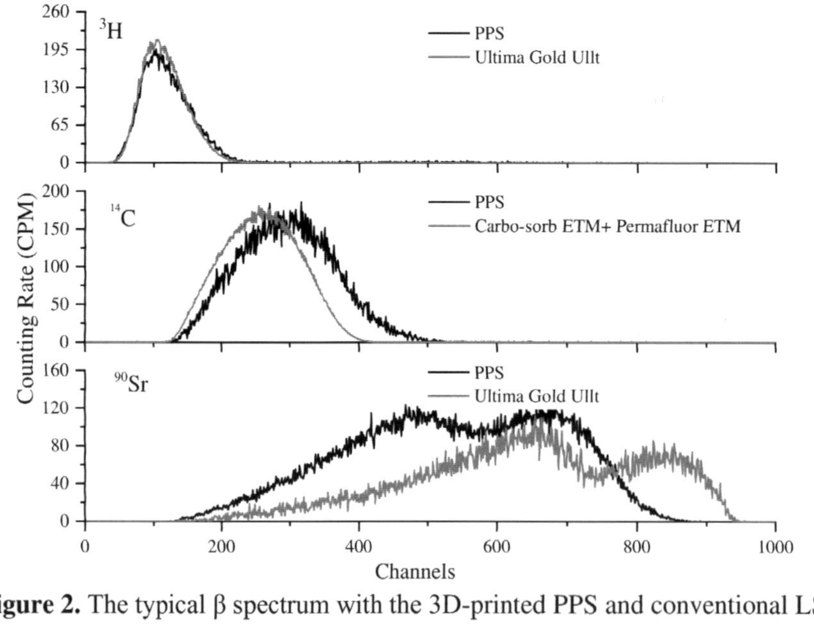

Figure 2. The typical β spectrum with the 3D-printed PPS and conventional LS.

From table 1, it can be seen that the total background of tritium in the low-energy mode is 0.51cpm, which is not much different from the conventional scintillation liquid measurement system of 20 mL; in the high-energy mode, the total background of carbon-14 is 0.62cpm, which is 1/4 of the conventional LS measurement system. From table 2, it can be seen that the best detection efficiencies of the PPS for ^3H, ^{14}C, and ^{90}Sr are 3.3%, 43.5%, and 67.6%, respectively, all significantly lower than the conventional LSC method, and the greater the decrease in detection efficiency, the lower the energy of the β particles.

Figure 2 shows that the PPS measures the β and the conventional LSC sample spectrum are consistent, the peak position difference is not large, and the peak position is not related to the detection efficiency of the β particles.

Distance quenching is the quenching before the production of fluorescence, which mainly occurs in heterogeneous measurement [10] and is the most important reason for energy loss of PPS for β detection. When measuring with a PPS, the radionuclide must be within the radiation particle range distance from the scintillator surface. The lower the energy of the β particles, the more severe the impact of distance quenching, and the lower the detection efficiency obtained. The β emitted by ^{3}H have low energy, and most of the β particles energy loss before reaching the scintillator [11], which also indicates that the detection efficiency of PPS for low-energy β particles needs to be improved, and reducing the size of the scintillator pores may be an effective method.

3.2. The Impact of the Detection Solution

Spiked samples are supplemented with pure water to alter the volume of the measurement solution, and then the β spectrum is measured again on a low-background LS spectrometer to determine the impact of the system's solution volume on detection efficiency and sample external standard spectrum index (SQP(E)). The variation of detection efficiency for β by the 3D-printed PPS with the system's solution volume is shown in figure 3, and the variation of SQP(E) with the system's solution volume is shown in figure 4.

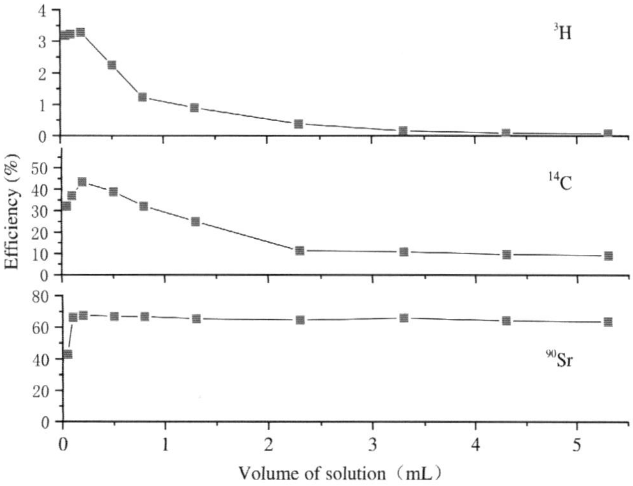

Figure 3. The variations of detection efficiencies for ^{3}H, ^{14}C, and ^{90}Sr with the volume of the solutions in the 3D-printed PPS.

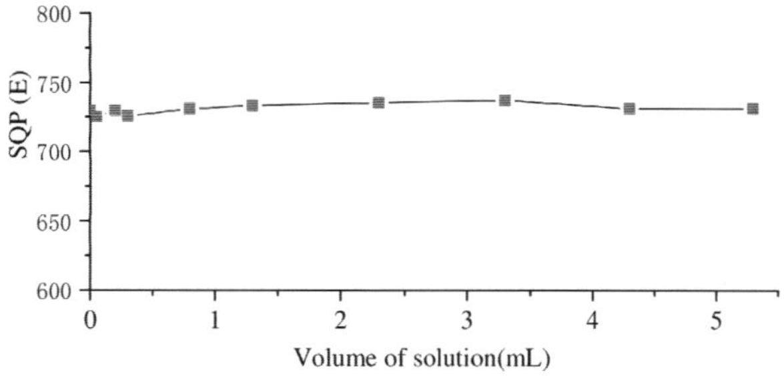

Figure 4. The variation of the SQP(E) with the volume solution for the 3D-printed PPS.

As can be seen from figure 3, the detection efficiencies of the 3D-printed PPS for ^{3}H, ^{14}C, and ^{90}Sr

slightly increase when the solution volume is between 50 μL and 0.2 mL. This is because when the standard substance is dropped at the top of the scintillator, the standard solution often gathers at the top of the scintillator column in the form of small droplets. When pure water is slowly added and the counting vial is agitated, the standard solution disperses within the PPS column along the holes from the top of the scintillator. When the water volume added is about 0.2 mL, the detection efficiency for radionuclides by the PPS reaches its peak. However, as the measurement solution continues to increase, the standard substance precipitates at the bottom of the scintillator column with the water solution, and the influence of solution distance quenching on low and medium energy β particles increases. The detection efficiency for ^{3}H and ^{14}C drops sharply. When the measurement solution reaches 2 mL, the detection efficiency for 3H by the 3D-printed PPS is already below 0.5%, and as the measurement solution approaches the volume of the scintillator column, the detection efficiency for 3H approaches zero, while the detection efficiency for ^{14}C stabilizes at about 9% after the measurement solution exceeds 2.3 mL. In contrast, high-energy β particles are less affected by distance quenching, and once the standard substance is uniformly dispersed within the PPS, the detection efficiency for ^{90}Sr is hardly affected by the volume of the measurement system's solution.

The sample external standard spectrum index SQP(E) is a parameter used in liquid scintillation counting to quantify the degree of sample quenching. SQP(E) is the number of channels (endpoint) that contains 99.5% of the total count of the external standard pulse height spectrum when the sample is irradiated by an external γ source (^{152}Eu inside the Quantulus 1220 ultra-low background liquid scintillation spectrometer) [1]. The distance quenching produced during the measurement with the PPS is different from the quenching in conventional LS counting. As shown in figures 3 and 4, as the volume of the measurement system's solution increases, the detection efficiencies of low and medium energy β change significantly, but SQP(E) remains almost unchanged. Therefore, the classical quenching indicator parameter SQP(E) cannot be used to characterize the quenching degree or changes in the detection efficiency of the measurement system with the 3D-printed PPS.

3.3. The Memory Effect for the β Detection

Plastic scintillators, when applied to β detection, not only do not produce mixed waste but are also reusable. It is essential to verify the adsorption residue of the PPS for β radionuclides, which is defined as the memory effect. As can be seen from table 3, after a single round of the rinsing process (the procedure described in section 2.3), the retested adsorption residues of the 3D-printed PPS for ^{3}H, ^{14}C, and ^{90}Sr are all below 0.5%. The adsorption residue of radionuclides in routine environmental samples has a minimal impact, and if necessary, the number of cleanings can be increased to reduce the adsorption residue. Therefore, the memory effect could be determined as less than 0.5 %.

Table 3. The memory effects of ^{3}H, ^{14}C, and ^{90}Sr on 3D-printed PPS after rinsing.

Radionuclides	The net counts before rinsing	The net counts after rinsing	The memory effect
^{3}H	198.9cpm	0.7cpm	0.05%
^{14}C	1453.1cpm	5.7cpm	0.39%
^{90}Sr	734.5cpm	3.6cpm	0.49%

In summary, the 3D-printed PPS in this project has maximum detection efficiencies of 3.5%, 43.5%, and 67.6% for ^{3}H, ^{14}C, and ^{90}Sr, respectively, and after washing, the adsorption residue of radionuclides is below 0.5%. It is suitable for the measurement of medium to high-energy β radionuclides, especially for high-energy β radionuclides such as ^{90}Sr, whose detection efficiency is not affected by the volume of the measurement solution. To improve the detection efficiency for the measurement of medium and low-energy β radionuclides, it is necessary to control the volume of the measurement solution or to enhance the printing precision, reduce the diameter of the PPS pores, and increase the specific surface area to meet the detection requirements.

4. Conclusions

In this paper, the 3D-printed PPS using FDM technology was obtained. It experimentally verified the effectiveness of β detection. Under optimal measurement conditions, the detection efficiency of the PPS for ^{3}H, ^{14}C, and ^{90}Sr are 3.3%, 43.5%, and 67.6%, respectively, proving for the first time the feasibility of using 3D-printed PPS technology for β detection.

The experiment also demonstrated that the β detection efficiency with the PPS is significantly lower than that with the LS, and is severely affected by distance quenching. Further research is needed on the composition improvement of the PPS, optimization of the 3D printing process, and enhancement of printing precision and the specific surface area of the PPS column.

With the further development and improvement of 3D printing technology, as well as the continuous advancement of research in the field of radionuclide detection, it is believed that more similar studies will proceed further, bringing more innovation and breakthroughs to the fields of radioactivity detection.

Acknowledgments

The authors wish to acknowledge the assistance of Dr. Bo Qian from Suzhou Institute of Nano-tech and Nano-bionics (SINANO), Chinese Academy of Sciences (CAS) for the 3D-printing of the materials.

5. References

[1] L' Annunziata M F, Tarancón A, Bagán H, et al. 2020 Liquid scintillation analysis: principles and practice *Handbook of Radioactivity Analysis* **6** pp 575-801.

[2] Horrocks D 1974 *Applications of Liquid Scintillation Counting*. (America: Academic Press New York and London) pp 2-67.

[3] Hamel M 2021 *Plastic Scintillators* (France: Springer) pp 4-26.

[4] Tarancón A, Bagán H, and García J F 2021 Plastic scintillators in environmental analysis *Plastic Scintillators: Chemistry and Applications* **13** pp 461-508.

[5] Gibson I, Rosen D W, et al. 2021 *Additive Manufacturing Technologies* (Switzerland: Springer). pp. 160-186

[6] Berns S, Boyarintsev A, Hugon S, Kose U, et al. 2020 A novel polystyrene-based scintillator production process involving additive manufacturing *J. Instrum.* **15(10)** P10019

[7] Tarancon A., García J F, and Rauret G 2003 Reusability of plastic scintillators used in beta emitter activity determination *Applied Radiation and Isotopes* **59(5-6)** 373-376.

[8] Santiago L M, Bagán H, Tarancón A, et al. 2013 Systematic study of particle quenching in organic scintillators *Nucl. Instrum. Methods Phys. Res., Sect. A* **698** 26-36.

[9] Santiago L M, Bagán H, Tarancón A, et al. 2013 Synthesis of plastic scintillation microspheres: evaluation of scintillators *Nucl. Instrum. Methods Phys. Res., Sect. A* **698** 106-116

[10] Tarancón A, Bagán H, and García J F 2017 Plastic scintillators and related analytical procedures for radionuclide analysis *J. Radioanal. Nucl. Chem.* **314** 555-572.

[11] Tarancón A, García J F and Rauret G 2004 Determination of beta emitters (^{90}Sr, ^{14}C and ^{3}H) in routine measurements using plastic scintillation beads *Nucl. Instrum. Methods Phys. Res., Sect. A* **516(2-3)** 602-609.

Analysis of the Cryogenic Structure of the Concentric HTS Cable Joint

Liguang Hu[1,*], Jian Cheng[1], Tengbiao Chen[1], Cheng Chen[1], Qi Cheng[1], Chunlong Chen[1] and Bangzhu Wang[2]

[1] Shenzhen Power Supply Bureau, China Southern Power Grid, Shenzhen, Guangdong, China
[2] Centre for Applied Superconductivity, Beijing Jiaotong University, Beijing, China
Email: hlgfox@qq.com; chengjian186@126.com; 13714153288@139.com; 283755920@qq.com; 185338819@qq.com; 846636943@qq.com; ibondrew@gmail.com

Abstract. Due to merits of low loss, large capacity and compactness, concentric high-temperature superconducting (HTS) cable has become an ideal solution for the large-capacity power transmission in a narrow path. As the three phases in the concentric HTS cable are wounded one by one in the same former, the non-superconducting joints are concentrated in a short range. The same cooling method as the normal section will cause the local temperature to rise and subject to hot spot risks. This paper conducts a study on the cryogenic cooling structure design of the joint of the concentric HTS cable. The heat generation mechanism is analysed. Based on the heat transfer enhancement method and computational fluid dynamic (CFD) simulation, the optimal cryogenic structure is proposed. The results are of significance to the safe and stable operation of the joint in the engineering application of the concentric HTS power transmission.
Keywords. Cryogenic structure; cable joint; high-temperature superconducting; heat transfer.

1. Introduction

Concentric type high-temperature superconducting (HTS) cables, characterized by their low losses, high power capacity, and compact design, have emerged as a promising solution for high-capacity power transmission in constrained spaces, such as urban areas in megacities. In recent years, countries such as the United States, Germany, China, and Japan have all undertaken research and engineering demonstration projects for concentric HTS cable, underscoring the mainstream status of this configuration in HTS cable research [1-5].

On September 28th, 2021, Shenzhen Power Supply Co., a subsidiary of China Southern Power Grid, successfully energized a 10 kV/2.5 kA concentric HTS transmission cable, supplying power to the Ping'an Finance Center, a landmark building in the Greater Bay Area [3]. The project did not include a joint as the total length was 400 m. However, it highlights the importance of joint technology for HTS cable system as the joints are crucial for emergency repairs and restoration in the event of localized damage, such as cryogenic fault, vacuum breakdown, or mechanical failure. Furthermore, the development of long-distance HTS cable lines necessitates the use of joints to connect individual cable segments, typically around 500 meters in length due to today's techniques.

The concentric configuration of HTS cables, where the three phases are wound sequentially on the same former, results in a localized concentration of non-superconducting joints. This spatial proximity of the non-superconducting (i.e. resistive) joints poses a significant risk of localized temperature rise and potential hot spots. The concentration of heat dissipation at these joints can lead to a reduction in the critical current of the HTS tapes, potentially impacting the overall performance and reliability of the cable system. Therefore, the design of an effective cooling structure for cable joints, capable of efficiently removing heat from the joint region, is paramount to the reliable operation of superconducting cable system, as shown in Figure 1.

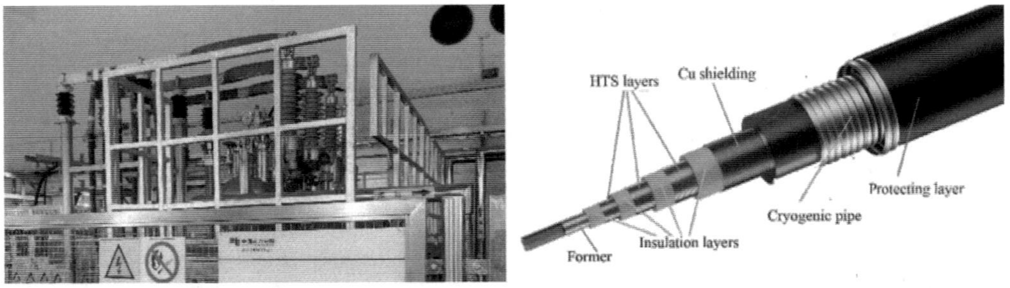

Figure 1. Shenzhen HTS cable project and its concentric HTS cable.

In this paper, we first modeled the joint structure and analyzed the thermal load. Then, we characterized the cryogenic flow of the concentric joint. The optimal cryogenic cooling structure was discussed.

2. Joint Model and Thermal Analysis

A schematic diagram of the joint structure for a concentric HTS cable is shown in figure 2. The joint comprises two parts: the superconducting cable core joint and the cryogenic pipe joint. The cable core joint focuses on connecting the cable's former, splicing the HTS tapes of each phase, and restoring the integrity of the high-voltage insulation. The cryogenic pipe joint, using flanges to the cable's cryostat on both sides, on the other hand, primarily facilitates the connection of external fluid channels and ensures the continuity of the cryostat.

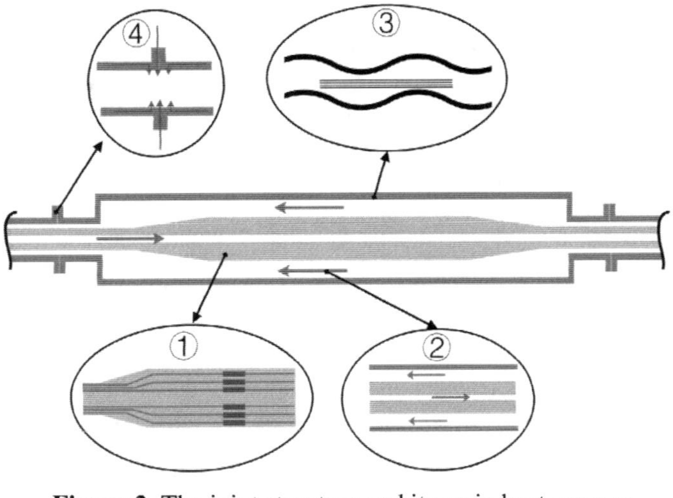

Figure 2. The joint structure and its main heat sources.

The primary heat sources within the joint are also indicated in figure 2, namely: Joule heating from the joint resistance, fluid friction dissipation, heat transfer from the cryostat of the joint, and heat transfer at the connection part between the cryogenic pipe and the joint cryostat. Next, we will delve into a detailed analysis of the characteristics of the four heat loads and provide estimations for each.

2.1. Joule Heating from the Joint Resistance

To date, there is no established method for achieving superconducting joints for YBCO tapes that meets the stringent requirements of practical applications. Engineering solutions currently rely on welding techniques to establish connections, which inherently introduce a resistive element. The heat power *P* generated by the resistive joints is

$$P = I^2R \tag{1}$$

Where *I* is the operating current and *R* is the total resistance.

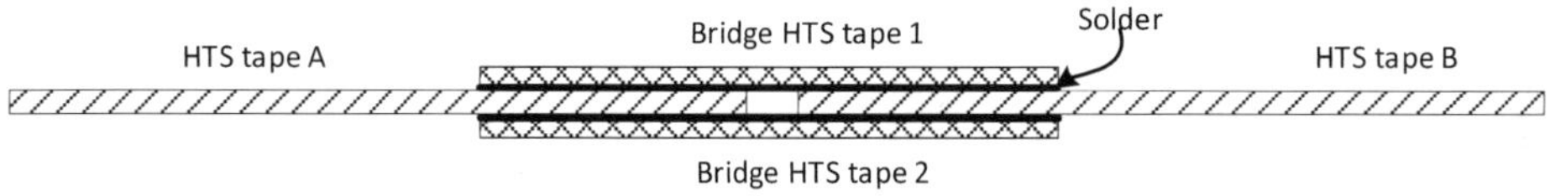

Figure 3. The HTS tape joint structure.

We proposed a bridge type joint method, in which two bridge HTS tapes were soldered upper and down the tapes in the two cable core terminals, as shown in figure 3. The resistance of each joint can be effectively controlled to below 30 nΩ utilizing in-field implementation techniques. With 60 HTS tapes connected in parallel per phase and a rated current of 2.5 kA, the joint heat power at each phase is about 3 W.

2.2. Heat Transfer from the Cryostat

The joint cryostat utilizes vacuum multi-layer insulation (MLI or superinsulation) technology, which is the most effective kind of insulation in cryogenics, minimizing heat transfer through conduction, convection, and radiation to a significantly low level [6]. The basic functioning of MLI technique is shown in Figure 4 below.

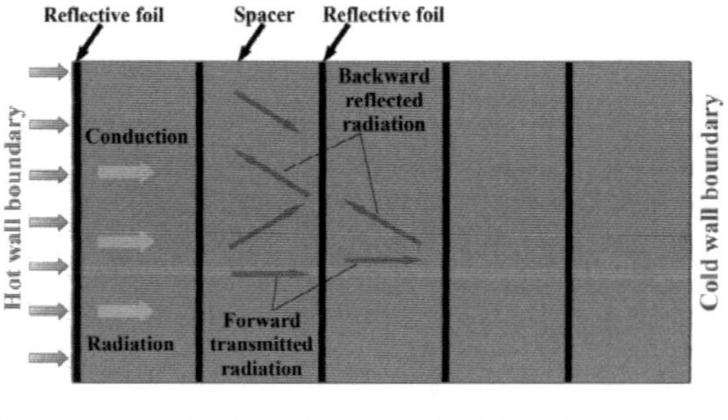

Figure 4. Basic functioning of MLI technique [7].

The cryostat is about 1.5 m long. With the implementation of 40-60 layers of insulation film and a vacuum level of 10^{-3} Pa, the typical total load of the joint cryostat can be reduced to ~2 W.

2.3. Fluid Dissipation and the Heat Transfer at the Connection Part

Fluid viscosity inherently leads to dissipation. The power loss mainly depends on the friction factor *f* and there are several equations proposed to compute it. Moody [8] proposed an explicit formula to compute the frictional factor for $4000 \leq Re \leq 10^8$and $0 \leq \frac{\epsilon}{D} \leq 0.01$

$$f = 0.0055\left[1 + \left(2 \times 10^4\left(\frac{\epsilon}{D}\right) + \left(\frac{10^6}{Re}\right)\right)^{\frac{1}{3}}\right] \tag{2}$$

where ϵ is the roughness of wall and *D* is the hydraulic diameter. However, at the space scale of the joint, these dissipative effects are negligible [9].

The flanged connection between the cryogenic pipe of the superconducting cable and the cryostat of the joint, lacking MLI, exhibits significant thermal conduction. The heat conduction power q can be calculated by Fourier's law:

$$q = -2\pi\lambda r\,\frac{dT}{dr} \tag{3}$$

where λ is thermal conductivity, T is temperature, and r is the radius.

Both the flange and the connected pipes are constructed from stainless steel 304. The ambient temperature is assumed to be room temperature, set at 300 K. The internal temperature is taken as the operating temperature of liquid nitrogen, which is 70 K for calculation purposes. The heat load at this connection part is estimated to be no greater than 0.2 W.

3. Flow Analysis of the Cryogenic Structures

Concentric HTS cable typically employs a counter-flow cooling scheme, characterized by two flow channels with flow in opposite directions. One channel runs along the cable former, while the other is between the cable core and the cryogenic pipe. We employed computational fluid dynamics (CFD) techniques and the widely used CFD software ANSYS FLUENT to conduct a basic cryogenic structure analysis of the two flow channels.

3.1. Flow in the Former Joint

In field welding of corrugated former is not easy. A straight pipe is often used to match the length and radius. Considering that the welding of the former typically occurs in less-than-ideal engineering environments, direct welding could introduce impurities to potentially contaminate the cryogenic fluid. Furthermore, direct welding necessitates precise matching of the straight pipe's length. A bridging method shown in figure 5 is proposed. First, weld pipe 1' to pipe 1 and pipe 2' to pipe 2. Then, clean the inside of both pipes. Finally, due to the gap length, weld pipe 3 to both pipes 1' and 2'. The pipe 3 could be used to match the length and ensure the inside of the pipes are clean.

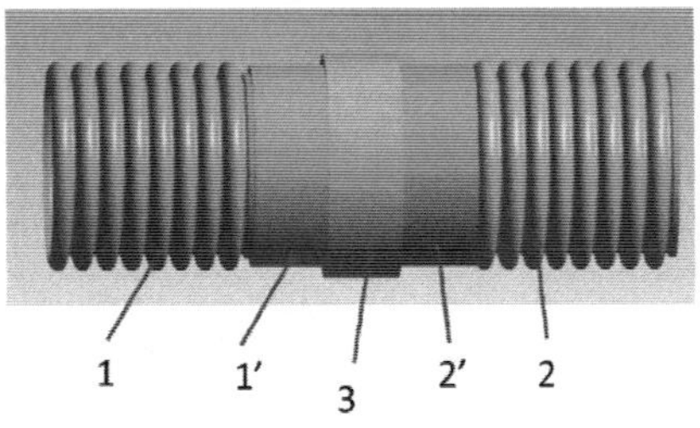

Figure 5. The proposed bridge type former joint.

Figure 6 below illustrates the streamline and velocity of the fluid near the former joint. The introduction of the straight pipe does alter the fluid's motion characteristics. However, considering the short length of the joint section, measuring less than 100 mm, this alteration is essentially negligible.

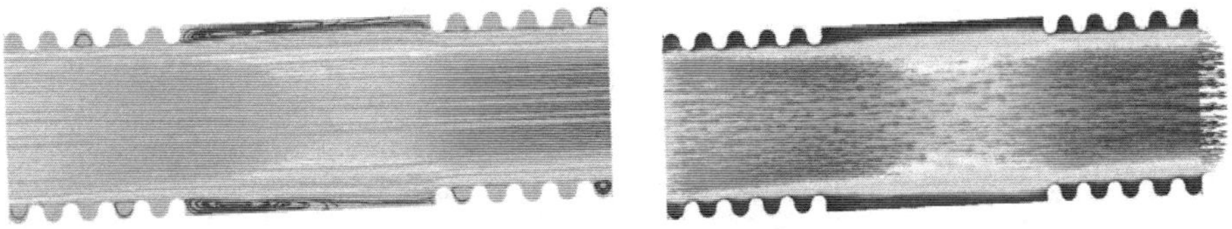

Figure 6. The streamline and velocity of the fluid near the former joint.

3.2. Flow between the Cable Core and the Cryostat

The channel between the cable core and the cryogenic container is not perfectly straight, leading to a slightly more complex flow. Upon entering the joint space, the fluid experiences a change in both flow direction and flow cross-section. Figure 7(a) shows the flow channel between the cable core and the cryostat. The meshing result is shown in figure 7(b).

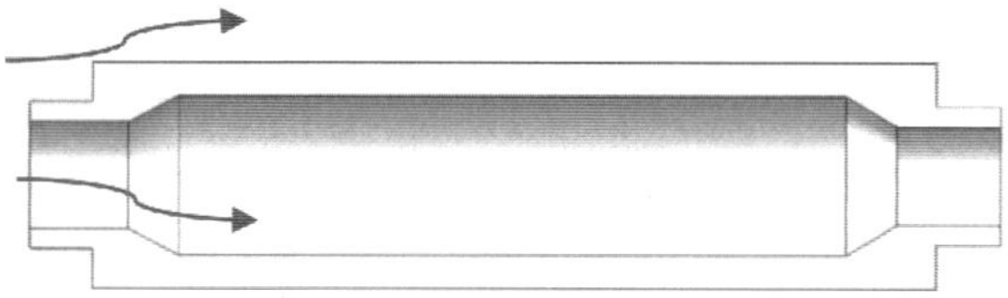

(a)

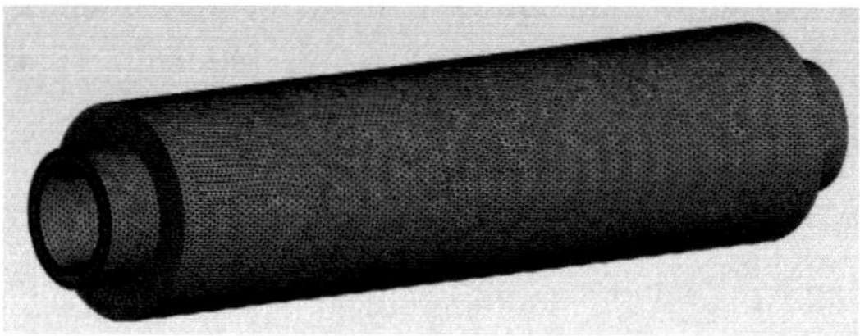

(b)

Figure 7. (a) Flow channel between the cable core and the cryostat and (b) flow region meshing

Figure 8 shows the flow state between the cable core and the joint cryostat under the superconducting cable's rated flow rate of 0.35 kg/s. Figure 8(a) and figure 8(b) are the baseline flow characteristics of the velocity streamline and the turbulent kinetic energy. We can see from the figures that the fluid flow pattern undergoes a transformation upon entering the joint due to the altered geometry of the flow channel. Specifically, a stagnation zone develops at the protruding section of the cable core and the corner part of the joint cryostat in the inlet region. This stagnation zone undoubtedly hinders the heat dissipation capacity of the cable core. To mitigate this negative effect, we optimized the cable core joint's outer shape, creating a more gradual change in diameter to suppress the stagnation zone, as illustrated in Figure 8(c) and figure 8(d).

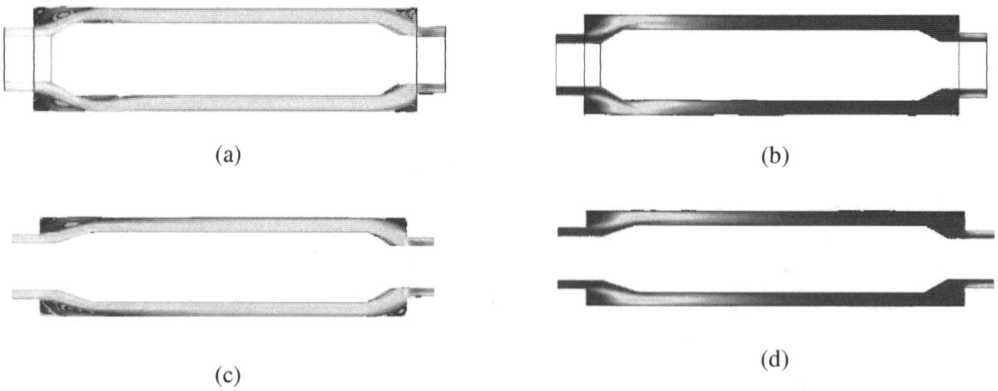

(a) (b)

(c) (d)

Figure 8. Flow between the cable core and the cryostat.

4. Conclusion

The HTS cable joint is a crucial component within the overall system. Due to its unique structure, the thermal load of a concentric HTS cable joint is complex, necessitating specialized design and optimization of its cryogenic structure.

We analyzed the characteristics of the thermal load in the joint of a concentric superconducting cable, providing estimations for each load. Computational fluid dynamics method was employed to analyze the flow of the cryogenic structure of the joint. Results indicate that the former joint has a negligible impact on the flow while the flow channel between the superconducting cable core joint and the cryostat joint requires optimized geometric design to achieve enhanced cooling performance.

Looking ahead, quantitative system analysis and experimental validation are essential for further advancements.

Acknowledgments
This work was financially supported by the China Southern Power Grid Major Science and Technology Program under Grant 090000KK52222197.

5. References

[1] Demko J A, Sauers I, James D R, Gouge M J, Lindsay D, Roden M, Tolbert J, Willen D, Traeholt C and Nielsen C T 2007 Triaxial HTS cable for the AEP Bixby project *IEEE Transactions on Applied Superconductivity* **17** 2047-50

[2] Stemmle M, Merschel F, Noe M and Hobl A 2013 Ampacity project—Worldwide first superconducting cable and fault current limiter installation in a German city center *22nd International Conference and Exhibition on Electricity Distribution (CIRED 2013)* pp 1-4

[3] Tang S, Li M, Xie H, Wang Z, Yu P, Li Z, Wang Z, Chen T and Wang B 2023 Development and demonstration of concentric-type HTS power cable for distribution grid in Shenzhen urban *Front. Energy Effic.* **1** 1160372

[4] Shiohara K, Adachi K, Nakanishi T, Sato M, Mido N, Aoki Y, Hasegawa T, Ogawa M and Ota T 2023 Demonstration test of triaxial superconducting cable system installed in commercial chemical plant grid *IEEE Transactions on Applied Superconductivity* **33** 5400905

[5] Wang B, Wu X, Xie H, Lv Z, Wang Z, Song M, Huang Y, Xia Y, Wang Z, Chen T, Tian B and Dai S 2021 Design, manufacture, and test of a 30 m 10 kV/2.5 kA concentric HTS cable prototype for urban grid *IEEE Access* **9** 120066-77

[6] Kanda M, Matsumoto K and Yamaguchi S 2021 Heat transfer through multi-layer insulation (MLI) *Physica C: Superconductivity and Its Applications* **583** 1353799

[7] Singh D, Pandey A, Singh M K, Singh L and Singh V 2020 Heat radiation reduction in the cryostat with multilayer insulation technique *J. Inst.* **15** 7032

[8] Moody L F 1944 Friction factors for pipe flow *Transactions of the ASME* **66** 671-84

[9] Zajaczkowski B, Giesbers A J M, Holtrust M, Haenen E and den Heijer R 2011 Feasibility of inline cooling in long distance HTS power line *Cryogenics* **51** 180-6

Preparation and Performance Research of Bio-based High Toughness Polylactic Acid Composites

Xinfang Li, Sufen Zhao* and Ziyang Wang

School of Packaging Design, Zhongshan Torch Polytechnic, Zhongshan 528436, Guangdong, China.
Email: xtulxf@163.com; *610230508@qq.com; 1472630528@qq.con

Abstract. The work aims to prepare the polylactic acid composites with high toughness by adding polybutylene succinate (PBS) with good toughening performance and acetyl tributyl citrate (ATBC) as an interface phase solvent and diisopropylbenzene peroxide (DCP) as a compatibilizer and polyamide compounds (TMC) as a nucleating agent. The bio-based polylactic acid composites are prepared by melt blending. The toughening agent and the effects of phase solvents and compatibilizers on the mechanical performances and transparency of PLA blends were studied. The research focused on investigating the effects of the toughening agent, compatibilizer, and interfacial compatibilizer on the mechanical performances and transparency of the PLA blends, with the aim of developing a high-toughness PLA preparation technology. The results indicated that when 0.4-0.6%. TMC was added to the PLA material, the crystallinity reached 32.7%, and the light transmittance exceeded 92%. Evidently, TMC can promote PLA nucleation and accelerate resin crystallization. The optimal mass ratio of each component in the PLA/PBS/ATBC/TMC/DCP blend was 80:20:10:0.4:0.3, resulting in a breaking elongation of 416.5%, which is 71 times higher than that of pure PLA, impact strength of 4.22 kJ/m², an increase of 165% compared to pure PLA, a tensile strength of 44.3 MPa, and a light transmittance of 82%. Clearly, the compatibility of the PLA/PBS blend is optimal, with significant toughening and transparency enhancement, broadening its application in the field of transparent products.
Keywords. Polylactic acid; polybutylene succinate; diisopropylbenzene peroxide; toughness.

1. Introduction

In today's world, plastic waste poses a severe threat to global environmental governance. Against the backdrop of "carbon neutrality" and "plastic restriction orders," the demand for biodegradable packaging materials in our country is set to experience rapid growth in the future [1-3]. Polylactic acid (PLA) boasts high strength and modulus, with tensile strength reaching up to 60 MPa. However, its glass transition temperature is not high (60-65°C), and its elongation at break under normal temperatures is low (usually less than 10%), making it brittle and poor in impact resistance, which limits its application scope [4,5]. Polybutylene succinate (PBS) is a type of biologically degradable material with promising applications, offering good flexibility. Using PBS to toughen PLA results in a toughened PLA material that possesses fully biodegradable properties [6,7]. Nevertheless, due to the differences in optical characteristics between PLA and PBS, the transparency of PLA material decreases after toughening, restricting its use in transparent product areas. This study focuses on examining the relationship between interfacial compatibility and toughening and light-enhancing properties among components of PBS/PLA bio-based composites. By optimizing the composition, key technologies for the preparation of highly tough PLA materials are developed.

2. Experimental

2.1. Materials and Equipment

Polylactic acid (PLA), with a melt flow rate of 10.0 g/10 min, sourced from NatureWorks LLC in the United States; Polybutylene succinate (PBS), with a melt flow rate of 25 g/10 min, supplied by Anhui Xuelang Biotechnology Co., Ltd.; Multiamide nucleating agent (TMC-328, analytical reagent grade), provided by Hangzhou Ximao New Material Technology Company; Diisopropylbenzene peroxide (DCP) and tributyl acetylcitrate (ATBC), both analytical reagent grade, procured from Sinopharm Group Chemical Reagent Co., Ltd.

Internal Mixer (Model X(S)N-3), manufactured by Nanjing Kaitai Machinery Co., Ltd.; Plate Vulcanizer (Model GDSFJM), made by Sanfeng Precision Measuring Instruments (Guangdong) Co., Ltd.; Drying Oven (Model DHG-9030), produced by Shanghai Supu Instrument Co., Ltd.; Differential Scanning Calorimeter (DSC, Model TAQ2000), supplied by American company TA Instruments Inc.; Intelligent Electronic Tensile Tester (Model XLW), provided by Jinan Landmark Electromechanical Technology Co., Ltd.; Impact Tester (Model XCJ-4), manufactured by Chengde Testing Machine Co., Ltd. in Hebei Province; Transmittance/Haze Meter (Model WGT-S), made by Shanghai Precision Scientific Instrument Co., Ltd.

2.2. Preparation of PLA Composites

Dry PLA granules and PBS granules in an oven at a temperature of 80°C for 24 hours. After drying, mix the PLA, and PBS with additives such as ATBC, TMC, DCP, etc., at a certain ratio (mass content) in a Banbury mixer. The mixing temperature is set at 190°C, the rotor speed at 60 rpm, and the mixing time lasts for 10 minutes. While still hot, cut the mixed samples into small particles for later use. Dry these prepared samples in an oven at 80°C for 8 hours. According to different test requirements, the mixed samples are placed in a flat vulcanizing machine to press into sheets of corresponding thickness. The temperature is set at 190°C, preheating for 3 minutes, hot pressing pressure at 20 MPa, hot pressing time for 3 minutes, cold pressing pressure at 20 MPa, and cold pressing time for 10 minutes. Finally, place the prepared sample strips in an oven at a temperature of 30°C to eliminate stress concentration for 12 hours.

2.3. Test Method

The DSC tests are conducted according to the methods specified in GB/T 19466.2-2004 and GB/T 19466.3-2004; the elongation at break and tensile strength are tested according to the method described in GB/T 1040.1-2006; the impact strength is tested according to the method outlined in GB/T 1043.1-2008; and the light transmittance is tested according to the method given in GB/T 2410-2008.

3. Results and Discussion

3.1. Study on the Melting and Crystallization Behavior of PLA/TMC Composites

The effects of multiamide nucleating agent (TMC-328) and its dosage on the melt crystallization properties of PLA were investigated when added to PLA. In the PLA/TMC blends, the mass fraction of TMC is 0.2 wt%, 0.4 wt%, 0.6 wt%, 0.8 wt%, and 1.0 wt%.

The glass transition temperature (Tg), melting temperature (Tm), cold crystallization temperature (Tcc), melting enthalpy (ΔHm), and crystallinity (Xc) of PLA and its blends are listed in table 1. As shown in table 1, with the addition of a nucleating agent, the Tcc of PLA shifts to lower temperatures, and the crystallinity (Xc) increases. This effect is particularly pronounced when the mass fraction of the nucleating agent TMC is between 0.4-0.6 wt%. This occurs because the addition of TMC provides numerous nuclei for PLA crystallization, reducing the activation energy for crystallization, which in turn lowers Tcc and increases Xc. However, the Tg and Tm of PLA do not undergo significant changes due to the addition of the nucleating agent. When the mass fraction of the nucleating agent TMC reaches 0.6 wt%, the crystallinity reaches 32.7%, an increase of 16.5 percentage points over pure PLA, indicating the best nucleation ability for crystallization.

The formula for calculating the degree of crystallinity is adopted.

$$Xc = \frac{\Delta H}{(1-\varphi)\Delta H0} \times 100\% \quad (1)$$

In this formula, ΔH represents the measured enthalpy of fusion for the blend system, $\triangle H_0$ denotes the melting enthalpy of PLA at 100% crystallization, and φ signifies the mass fraction of the nucleating agent in the blend system. $\triangle H_0$ is approximately 93 J/g.

Table 1. DSC characteristics of PLA and PLA/TMC blend.

PLA/TMC	T_g (°C)	T_m (°C)	T_{cc} (°C)	$\triangle H_m$ (J/g)	X_c (%)
100/0	60.3	151.7	120.6	15.2	16.2
99.8/0.2	59.9	152.9	102.6	24.6	26.3
99.6/0.4	59.1	153.6	96.8	28.9	31.0
99.4/0.6	59.6	153.8	96.2	30.4	32.7
99.2/0.8	59.8	154.0	97.6	26.8	28.9
99.0/1.0	59.2	153.7	100.3	25.0	27.0

3.2. Study on the Transparency Performance of PLA/TMC Composites

In the current biodegradable plastics, PLA (Polylactic Acid) stands out as the sole transparent material. Its exceptional transparency endows PLA with a wide range of potential applications. Presently, PLA has been utilized in the production of various transparent packaging containers, films, and daily necessities. Figure 1 displays the effect of the mass fraction of nucleating agent TMC on the light transmittance of PLA. PLA boasts a high light transmittance of 89%. Upon adding 0.4-0.6 wt% TMC, the light transmittance of PLA materials increases, reaching over 92%. However, when 0.8-1.0 wt% TMC is added, the light transmittance of PLA materials decreases, indicating that the addition of 0.4-0.6% TMC in PLA can promote PLA nucleation, refine spherulite size, thereby accelerating resin crystallization and enhancing the transparency of the products.

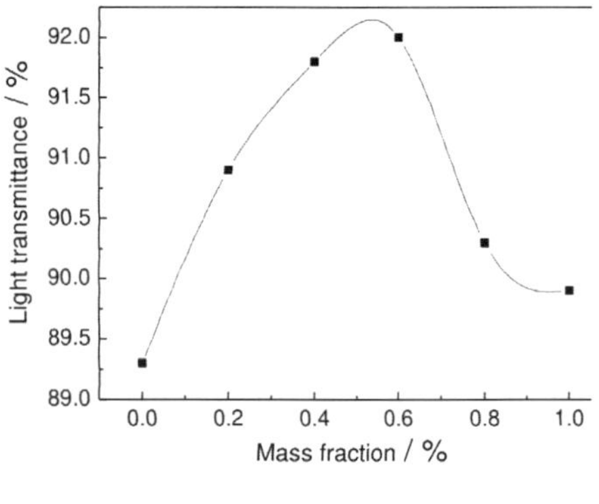

Figure 1. The effect of TMC content on the light transmittance of PLA.

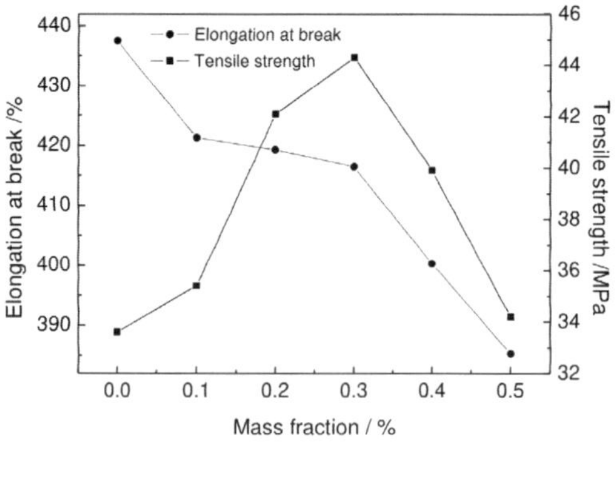

Figure 2. The effect of DCP content on the elongation at break and tensile strength of PLA composites.

3.3. Study on the Mechanical Performances of PLA/PBS/ ATBC/TMC/DCP Blend

During the melt blending process of PLA/PBS/ATBC/TMC, DCP serves as a crucial reactive compatibilizer, effectively improving the compatibility between PLA-PBS interfaces [8-10]. Hence, it is essential to investigate the impact of DCP content on the mechanical properties and transparency of the PLA/PBS/ATBC/TMC blends. The composition is formulated with 80.0% PLA content, 20.0% PBS

Journal of Physics: Conference Series **2941** (2025) 012009 doi:10.1088/1742-6596/2941/1/012009

content, a nucleating agent TMC at 0.4%, a compatibilizer ATBC at 10.0%, and DCP content at 0.1%, 0.2%, 0.3%, 0.4%, and 0.5%, respectively, based on relevant literature selection [11,12].

The influence of DCP content on the elongation at break, tensile strength, and impact strength of PLA/PBS/ATBC/TMC blends is shown in figures 2 and 3. As observed in figure 2, with the increase of DCP content, the elongation at break and tensile strength of the blends exhibits different trends: the elongation at break gradually decreases, while the tensile strength initially increases and then decreases. When the DCP content reaches 0.3%, the tensile strength of the blend reaches its maximum value at 44.3 MPa, which represents a 32% increase compared to the PLA/PBS/ATBC/TMC blend (33.6 MPa) and a 35% decrease compared to pure PLA. Meanwhile, the elongation at break is 419.3%, an increase of 72 times over pure PLA, maintaining high flexibility. This is attributed to the addition of DCP, which leads to increased entanglement of molecular chains in the blend. Upon external force application, deformation becomes more difficult, resulting in a decrease in the elongation at break. On the other hand, the addition of DCP, improves the interfacial interaction between PLA and PBS, enhancing their compatibility and thus increasing the impact strength. However, as the DCP content continues to rise, the compatibility at the PLA-PBS interface decreases, leading to a drop in tensile strength.

As shown in figure 3, with the increase of DCP content, the impact strength of the blend first increases and then decreases. When the DCP content reaches 0.2%, the impact strength of the blend achieves its maximum value of 4.36 kJ/m², which represents a 12% increase compared to the PLA/PBS/ATBC/TMC blend (3.89 kJ/m²) and a 175% increase compared to pure PLA. This is due to the addition of DCP enhances the interfacial adhesion between the PLA and PBS phases, improving interfacial compatibility. Under external force, stress can effectively transfer from the PLA phase to the PBS phase, allowing the PBS phase to dissipate more energy, thereby increasing the impact strength. However, when the DCP content exceeds 0.2%, the interfacial compatibility between PLA and PBS decreases, leading to a drop in the impact strength of the blend.

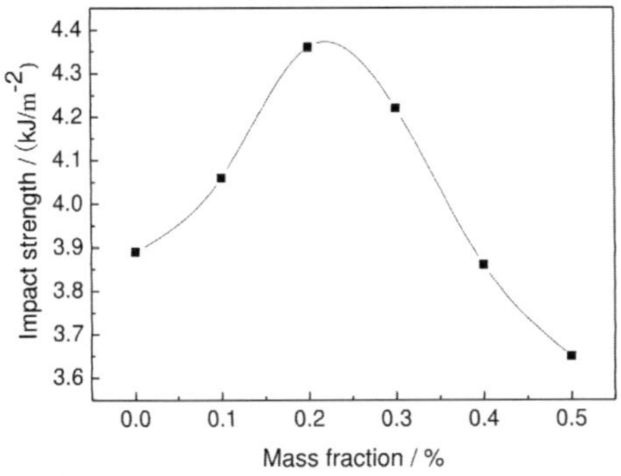

Figure 3. The effect of DCP content on the impact strength of PLA composites.

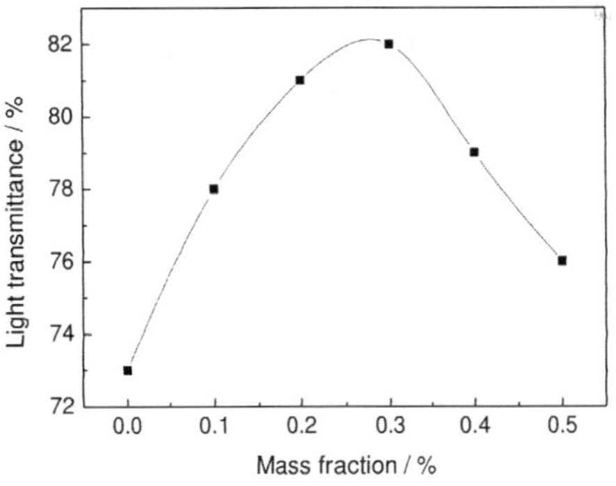

Figure 4. The effect of DCP content on the light transmittance of PLA composites.

3.4. Study on the Transparency Properties of PLA/PBS/ATBC/TMC/DCP Composites

The effect of DCP content on the light transmittance of PLA/PBS/ATBC/TMC/DCP blends is shown in figure 4. The light transmittance of PLA is 89%. Upon the addition of PBS, the light transmittance of the blend significantly decreases. However, with the addition of ATBC, TMC, and DCP, the light transmittance of the PLA/PBS/ATBC/TMC/DCP blend increases. When the DCP content reaches 0.3%, the light transmittance of the blend achieves its maximum value at 82%, which is a 12% increase compared to the PLA/PBS/ATBC/TMC blend (73%). This improvement is attributed to the significant enhancement of interfacial interaction between PLA and PBS by the compatibilizer ATBC and the nucleating agent DCP, which improves the compatibility of the PLA/PBS blend. The nucleating agent TMC promotes the nucleation of PLA and refines the spherulite size, thereby accelerating resin crystallization, and enhancing the transparency of the PLA/PBS blend.

4. Conclusions

When 0.4-0.6% TMC was added to PLA materials, the crystallinity reached 32.7%, and the light transmittance exceeded 92%. It is evident that TMC can promote PLA nucleation and accelerate resin crystallization. Analyzing the effect of DCP content on the mechanical and optical properties of PLA/PBS/ATBC/TMC blends, the optimal composition ratio of each component in the PLA/PBS/ATBC/TMC blend was found to be 80:20:10:0.4:0.3. At this ratio, the elongation at break of the blend was 416.5%, which is 71 times higher than that of pure PLA. The impact strength was 4.22 kJ/m^2, a 165% increase compared to pure PLA, and the tensile strength was 44.3 MPa. The light transmittance was 82%. Clearly, this composition provides the best toughening and transparency enhancement for PLA.

Acknowledgments

The authors like to acknowledge the financial support from the Special Project on Major Fields of Colleges and Universities in Guangdong Provincet (Grant No. 2021ZDZX4068, 2023ZDZX3096), the Innovation Team Project of Colleges and Universities in Guangdong Province (Grant No. 2023KCXDT084), the Science and Technology Program for social benefit of Zhongshan (Grant No. 2022B2001) for the research work.

5. References

[1] Kang H L, Miao X L and Li J H 2021 Synthesis and characterization of biobased thermoplastic polyester elastomers containing poly (butylene 2,5-furandicarboxylate) *RSC Advances* **11(25)** 14932.

[2] Pal A K and Katiyar V 2016 Nano-amphiphilic chitosan dispersed poly (lactic acid) bionanocomposite films with improved thermal, mechanical and gas barrier properties *Biomacromolecules* **17(8)** 2603.

[3] Lambert S and Wagner M 2017 Environmental performance of biobased and biodegrdable plastics: the road ahead *Chemical Society Reviews* **46(22)** 6855.

[4] Sangroniz A, Sangroniz L and Gonzalez 2019 Improving the barrier properties of a biodegradable polyester for packaging applications *European Polymer Journal* **115** 76.

[5] Chen X and Yan N 2020 A brief overview of renewable plastics *Materials Today Sustainability* **7(8)** 100031.

[6] Chen J H and Yang M C 2015 Preparation and characterization of nanocomposite of maleated poly (butylene adipate-co-terephthalate) with organoclay *Materials Science & Engineering C, Materials for Biological Applications* **46** 301.

[7] Wang R, Sun X J and Chen L L 2021 Morphological and mechanical properties of biodegradable poly (glycolic acid)/poly (butylene adipate-co-terephthalate) blends with in situ compatibilization *RSC Advances* **11(3)** 1241.

[8] Semba T, Kitagawak K and Ishiaku U S 2007 Effect of compounding procedure on mechanical properties and dispersed phase morphology of poly (lactic acid)/polycaprolactone blends containing peroxide *Journal of Applied Polymer Science* **103(2)** 1066.

[9] Zhang N and Lu X 2016 Morphology and properties of super-toughened bio-based poly (lactic acid)/poly (ethylene-co-vinyl acetate) blends by peroxide-induced dynamic vulcanization and interfacial compatibilization *Polymer Testing* **56** 354.

[10] Cai X L 2020 *Study on the Structure and Properties of PLA/PBS Blend Modified by Small Molecule Reactive Compatibilization* (Wuhan: Hubei University of Technology Press) p 49.

[11] Chen N, Wang W J, Shan P J, Chen J Z and Liu H Z 2020 Influence of DCP on toughening efficacy of PLA modified via in-situ polymerization of biobased monomers *Packaging Engineering* **41(23)** 51.

[12] Li X F, Huang X P, Hong P and Zhao S F 2024 Study on toughness and transparency of biodegradable polylactic acid composites *New Chemical Materials* **52(7)** 157.

Study on the Preparation Process Optimization and Adsorption Characteristics of Metallic Oxide Composite Modified Quartz Sand

Guohao Li*
Zhaoqing Open University, Zhaoqing, Guangdong, China.
Email: *lgh10112024@126.com

Abstract. The conventional water treatment process is difficult to remove micropollutants. Meanwhile, filtration is the last barrier to this process. One of the important factors of the influence of effluent quality is adsorption and interception on the surface of the filter. The metallic oxide composite modified quartz sand (MCMS) is prepared with raw quartz sands (RQS) as the surface modification material, $FeCl_3$, $AlCl_3$, and ethanol solution of $TiCl_4$ as the modifiers. Test results are as follows: (1) The preparation method is obtained. (2) The optimum preparation conditions are obtained by the method of SEM and EDS detection. (3) The surface characteristic of MCMS had been observed as a little ball shape. The composition of the surface metallic oxide is analyzed which consists of Fe_2O_3, TiO_2, Al_2O_3, SiO_2, and FeO(OH) or AlO(OH). (4) The specific surface area of MCMS is 39.10 times as RQS and 3.38 times as IOCS. The combination of metallic oxide which produces a large number of hydroxyl isα-Fe_2O_3, AlO(OH), and Anatase TiO_2 by the detection of phase identification from XRD. Organic matter removal of MCMS is 1.5 times as IOCS. (5) The average of HA removal of MCMS can reach to 91.56%. The adsorption characteristics of MCMS are both physical adsorption and chemical adsorption, meanwhile being based on chemical adsorption.
Keywords. Modifier of metallic oxide; composite modified quartz sand; process conditions optimization; adsorption characteristic; static adsorption effect.

1. Introduction

With the rapid development of the social economy, the water quality of most drinking water sources has been polluted to varying degrees. The degree of micro pollution of dissolved organic matter has deepened, and the difficulty of degrading organic matter in domestic and industrial wastewater has increased. Water source pollution is becoming increasingly serious [1]. Due to the limitations of traditional filter media such as limited specific surface area and negative charges on the surface under neutral pH conditions, modified filter media has attracted the attention of researchers as an economical and effective filter material [2]. However, in order to be truly applied in practical production, there are still problems such as weak adhesion strength and insufficient adsorption capacity of filter media, which are prone to secondary pollution [3]. Therefore, improving the process conditions for preparing modified quartz sand is one of the main ways to enhance the filtration effect.

Research has shown that the removal efficiency of pollutants by composite modified quartz sand is higher than that of ordinary modified quartz sand [4]. Integrating the pollutant removal preferences of ordinary modified sand, composite modified quartz sand prepared with two or more modifiers possesses special properties and functions that ordinary modified quartz sand does not possess [5]. At present, research on the application of composite modification technology in water purification filter media at home and abroad mainly focuses on using metal iron salts, aluminum salts, manganese salts, etc. as modifiers, and the main implementation methods are the repeated alkaline deposition method under low temperature conditions, single calcination method under high temperature conditions, and surface adhesion method of binders; The purpose of preparing modified filter media is to modify the surface potential of filter material particles and the internal structure of metal media.

Based on the rate of hydroxyl generation of α- Fe_2O_3, γ-Al_2O_3, and TiO_2 in water is relatively fast, which greatly increases the adsorption capacity of quartz sand surface [6]. In this paper, $FeCl_3$, $AlCl_3$, and $TiCl_4$ solutions are used as modifiers, and the preparation process of MCMS is shown in figure 1. The total surface metal loading, modifier adhesion strength, and humic acid removal rate are evaluated as indicators [7-10]. Through orthogonal experiments and continuous experiments, the effects of modifier dosage, stirring strength, drying temperature, and time are studied to obtain the optimal preparation process conditions, in order to improve the shortcomings of modifier adhesion strength, obtain new processes, technologies, and parameter values for preparing composite modified quartz sand, and obtain its adsorption characteristics [11-15].

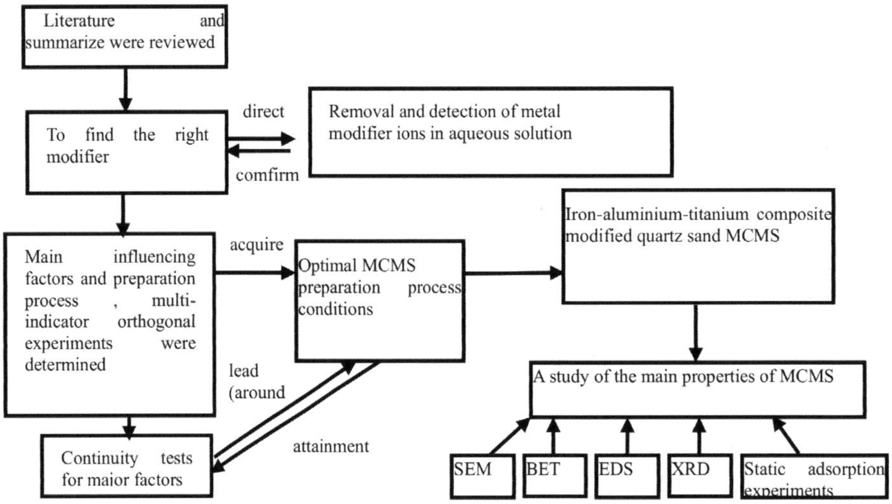

Figure 1. Technology route of the overall research.

2. Major Materials and Methods

2.1. Major Materials

(1) Row quartz sand: acid-soaked in HCl solution of pH=1 for more than 24h, washed to neutral by distilled water, dry the water completely and cooled to room temperature. A diameter of 0.8-1.4 mm was sieved and placed in a sealed polyethylene bottle for use. (2) Modifiers: 2 mol/L $FeCl_3$ solution (analytically pure); 2 mol/L $AlCl_3$ solution (analytically pure); 2 mol/L $TiCl_4$ ethanol solution (analytically pure). (3) Agents used in the humic acid configuration of raw water.

2.2. Major Equipment

T6 UV spectrophotometer (2016, Beijing); 500GWL high temperature calcination furnace (2014, Guangzhou); S3400N scanning electron microscope (2018, Hitachi); SX2-5-12 resistance furnace (2016, Beijing); DHG-9145 drying oven (2016, Shanghai); PHS-3B precision pH meter (2016, Shanghai); ASAP2020 specific surface area tester (2018, USA); X' Pert PRO type X-ray diffractometer (2014, The Netherlands); Perkin Elmer Optima 8000 inductively coupled plasma emission spectrometer (2021, USA).

2.3. Preparation Process of MCMS

The MCMS preparation process is shown in figure 2, where the $FeCl_3$ solution was mixed with unmodified quartz sand at a certain casting ratio *C1* (the ratio of the volume of $FeCl_3$ solution to the mass of RQS, mL·g^{-1}) and placed in a drying oven at 120°C. The processed mixture was then quickly placed in a muffle furnace at a certain temperature (350-390°C) for a high temperature heat treatment for a certain time (2.5-3.5h), removing the sand. After being cooled to room temperature and pounded loose, the $AlCl_3$ solution and the ethanol solution of $TiCl_4$ were fully mixed with the sand according to a certain casting ratio of *C2*, *C3* and sand, respectively, and the mixture was placed in a drying oven at 120°C for drying, and then the processed mixture was rapidly placed in a muffle furnace at a certain temperature *T* (200-360°C) for high temperature calcination for a certain period of time *t* (0-3.0h), MCMS was rinsed clean, dried for moisture, obtained the MCMS and sealed in polyethylene bottles for use.

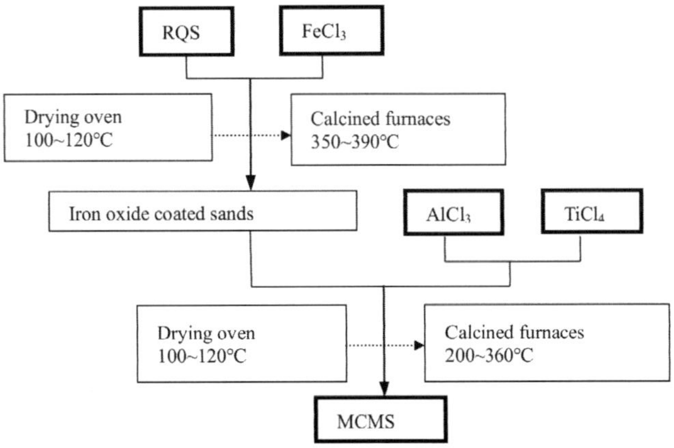

Figure 2. The preparation flow chart of MCMS.

3. Results and Discussion

3.1. Orthogonal Experimental Analysis

In this paper, five factors such as $FeCl_3$ solution, $AlCl_3$ solution and $TiCl_4$ ethanol solution dosing ratios *C1* (mL/g), *C2* (mL/g), *C3* (mL/g) and calcination temperature *T*, calcination time *t* were selected. The total surface metal loading *M* (μg/g), adhesion rate of the surface modifiers β (%), and the humic acid removal rate η (%) were used as the evaluation indexes to determine the process conditions for the preparation of MCMS. The factor level table L_{16} (4^5) of the orthogonal test for MCMS was developed, as shown in table 1.

Table 1. Factors and levels of orthogonal experiment on MCMS.

Factor	Unit (of measure)	Values for each level			
		1	2	3	4
$FeCl_3$ solution dosing ratio *C1*	mL·g^{-1}	0.12	0.14	0.16	0.18
$AlCl_3$ solution dosing ratio *C2*	mL·g^{-1}	0.10	0.12	0.14	0.16
$TiCl_4$ solution dosing ratio *C3*	mL·g^{-1}	0.04	0.06	0.08	0.10
Calcination temperature *T*	°C	200	250	300	350
Calcination time *t*	h	2.0	2.5	3.0	3.5

The orthogonal test procedure was carried out according to the MCMS preparation process and the test results are shown in table 2.

Table 2. Result and analysis of orthogonal test on MCMS.

No.	Factor					Evaluation indicators		
	C_1 /(mL·g^{-1})	C_2 /(mL·g^{-1})	C_3 /(mL·g^{-1})	T /(°C)	t /(h)	M /(mg·g^{-1})	β /(%)	η /(%)
1	0.12	0.10	0.04	200	2.0	870.937	95.96%	93.57%
2	0.12	0.12	0.06	250	2.5	908.873	81.39%	95.54%
3	0.12	0.14	0.08	300	3.0	923.781	94.50%	92.41%
4	0.12	0.16	0.10	350	3.5	1101.011	90.50%	90.98%
5	0.14	0.10	0.06	300	3.5	1006.951	91.59%	93.68%
6	0.14	0.12	0.04	350	3.0	899.251	94.35%	95.54%
7	0.14	0.14	0.10	200	2.5	923.628	91.57%	92.41%
8	0.14	0.16	0.08	250	2.0	968.204	94.96%	96.91%
9	0.16	0.10	0.08	350	2.5	905.165	93.89%	92.41%
10	0.16	0.12	0.10	300	2.0	920.189	90.24%	95.54%
11	0.16	0.14	0.04	250	3.5	973.654	95.88%	95.32%
12	0.16	0.16	0.06	200	3.0	981.261	93.75%	97.11%
13	0.18	0.10	0.10	250	3.0	921.001	90.75%	93.97%
14	0.18	0.12	0.08	200	3.5	923.511	93.52%	85.46%
15	0.18	0.14	0.06	350	2.0	982.457	95.43%	95.24%
16	0.18	0.16	0.04	300	2.5	872.718	94.33%	93.97%

From the data in table 2, the level effect values are calculated, and extreme variance R-values for each column are in table 3.

Table 3. Analysis of the result of orthogonal test on MCMS.

Factor	Total surface metal loading *M* (μg/g)					Modifier adhesion rate *β* (%)					Humic acid removal rate *η* (%)				
	C1	C_2	C_3	T	t	C_1	C_2	C_3	T	t	C_1	C_2	C_3	T	t
$\overline{K_1}$	951	926	904	924	935	90.5	93.0	95.1	93.7	94.1	93.1	93.4	94.6	92.1	95.3
$\overline{K_2}$	950	913	970	942	903	93.1	89.9	90.5	90.7	90.3	94.6	93.0	95.4	95.4	93.6
$\overline{K_3}$	945	951	930	931	931	93.4	94.3	94.2	92.7	93.3	95.1	93.8	91.8	93.9	94.8
$\overline{K_4}$	925	981	966	972	1001	93.5	93.4	90.8	93.5	92.9	92.2	94.7	93.2	93.5	91.4
R	26.2	67.8	65.7	47	98.7	2.92	4.47	4.59	2.96	3.85	2.94	1.72	3.60	3.30	3.96

As can be seen from table 3, for the total surface metal loading *M*, the R-value of factor t is 98.7 is the largest. Therefore, the evaluation indexes of the MCMS preparation process are ranked by the influence of each factor as follows: Total MCMS surface metal loading *M* (μg·g^{-1}): *t*>*C2*>*C3*>*T*>*C1*; MCMS surface modifier adhesion rate *β* (%): *C3*> *C2*>*t*> *T*> *C1*; MCMS removal rate of humic acid *η* (%): *t*>*C3*>*T*>*C1*>*C2*. Through orthogonal experimental analysis and calculation, the optimal combination of level values was determined to be *C1* = 0.16 mL·g^{-1} RQS, *C2* = 0.14 mL·g^{-1} RQS, *C3* = 0.04 mL·g^{-1} RQS, *T* = 250°C, and *t* = 2.0h.

3.2. One-factor Continuity Test

This chapter aims to use single-factor continuous experiments to study the impact of *M*(mg/mL), *β* (%), and *η* (%) of different preparation process conditions.

3.2.1. Effect of $FeCl_3$ Solution Dosing Ratio C1. The unchanged preparation process conditions are set as $C2$ = 0.14 mL·g^{-1}, $C3$ = 0.04 mL·g^{-1}, T = 250°C, t = 2h, different levels of $C1$. Results are shown in figure 3.

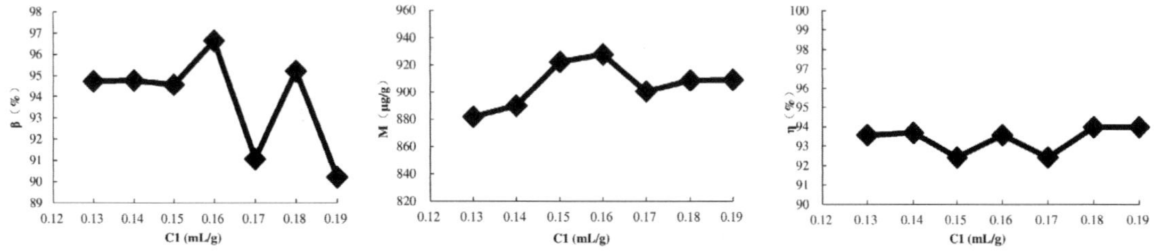

Figure 3. Effect of $C1$ on MCMS performance.

Results are shown in figure 3: M increased faster when $C1$ was small, and then became slower, M reached a maximum value of 928 μg/g when $C1$ was increased to 0.16 mL·g^{-1} RQS, and continued to increase $C1$, M decreased and then tended to stabilise near 909 μg/g. β was stable near 95% when $C1$ was small, when $C1$=0.16, β appeared maximum value of 96.65%. Continued to increase $C1$, β began to fluctuate between 90.2% and 95.2%. Whereas η was relatively stable with the increase of $C1$ between 92.4% to 93.6%.

The surface of RQS will reach a state of loading saturation with the increase in the amount of modifier $FeCl_3$ solution added. When $C1$ is greater than 0.16 mL·g^{-1}, the total metal loading tends to be saturated, the overload metal will be easy to be detached under the high-intensity vibration.

3.2.2. Effect of $AlCl_3$ Solution Dosing Ratio C2. The unchanged preparation process conditions are set as $C1$ = 0.16 mL·g^{-1}, $C3$ =0.04 mL·g^{-1}, T = 250°C, t = 2h, different levels of $C2$. Results are shown in figure 4.

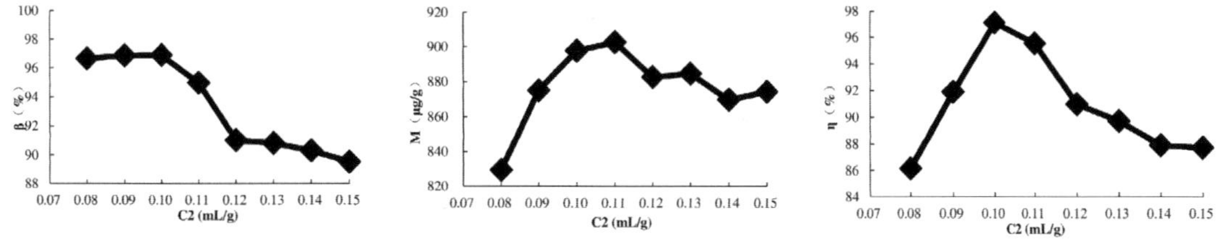

Figure 4. Effect of $C2$ on MCMS performance.

Results are shown in figure 4: M increased faster when $C2$ was small, and then became slower, M reached a maximum value of 903 μg/g when $C2$ was increased to 0.11 mL·g^{-1} RQS and continued to increase $C2$, the overall trend of M is decreasing. When $C2$>0.10 mL·g^{-1}, β and η began to decline continuously, meanwhile, the peak value β=96.8% and η=97.11% appeared in $C2$=0.10 mL·g^{-1}.

The alumina metal is attached to the modified sand with an iron oxide layer encapsulated on the surface. When $C2$ is higher than the load saturation state value, the excess $AlCl_3$ particles with smaller molecular particles will enter into the gap between iron oxide and RQS and undergo a chemical reaction at high temperatures, destroying the originally relatively stable iron oxide crystal layer and causing it to flake off, decreasing the adhesion rate of the modifier. Therefore, when $C2$ reaches 0.10 mL-g^{-1}, there is no need to increase $C2$.

3.2.3. Effect of $TiCl_4$ ethanol solution dosing ratio C3. The unchanged preparation process conditions are set as $C1$ = 0.16mL·g^{-1}, $C2$=0.10mL·g^{-1}, T = 250°C, t = 2h, different levels of $C3$. Results are shown in figure 5.

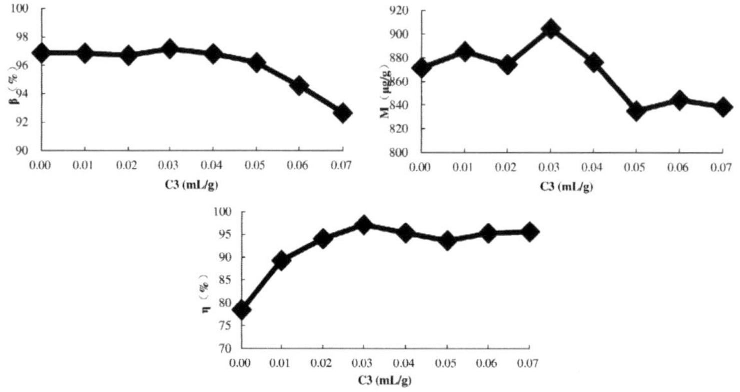

Figure 5. Effect of *C3* on MCMS performance.

Results are shown in figure 5: When $C3$=0.03 mL·g^{-1}, M appears to be the maximum value of 905 μg/g. When the value of $C3$ continues to increase, M begins to decrease and the speed is fast. When $C3$ <0.03 mL·g^{-1}, β stabilised near 96.8%, when $C3$ = 0.03 mL·g^{-1}, β and η reached a maximum value of 97.16% and 97.11%, when $C3$ > 0.03 mL·g^{-1}, β value began to decline and the rate of decline was increasing, η balanced around 95%. Similar to the principle of $AlCl_3$ particles destroying the structure of iron oxide crystal layers, when $C3$ is higher than the saturation value of the load, titanium metal oxides are more likely to destroy the iron oxide crystal layer and reduce the adhesion rate of the modifier. During the process of increasing $C3$ from 0 to 0.03 mL·g^{-1}, η increased by about 20%, indicating that $C3$ is a key factor in organic matter removal.

3.2.4. Effect of Calcination Temperature T. The surface of modified sand was examined by EDS energy spectrum analyser, the lower the percentage of exposed Si elements on the surface of MCMS to the total number of atoms on the surface, the higher the degree of modifier coverage and the better the modification effect. The unchanged preparation process conditions are set as $C1$=0.16 mL·g^{-1}, $C2$=0.10 mL·g^{-1}, $C3$=0.03 mL·g^{-1}, t=2h, different levels of T. Results are shown in table 4.

Table 4. Effect of surface-exposed on MCMS performance.

Sample	T/°C	Percentage of surface-exposed Si elements to total surface atoms/%	M /mg·g^{-1}	β /%	Surface appearance of modified sand after calcination
1	200	11.69	881.758	94.75	Sand covered with a viscous oily substance with a bismuth-yellow surface and an aromatic odour
2	240	0.42	902.617	97.23	Sand with a small amount of clay on the surface, with an earthy yellow surface and an aromatic odour
3	280	2.34	927.727	96.64	Clumps on the outer surface of the sand layer with a tawny surface and powder at the bottom
4	320	4.01	838.624	92.64	Clumps of sand, need to be mashed hard, brownish red surface, and more powder at the bottom

Sample	T/°C	Percentage of surface-exposed Si elements to total surface atoms/%	M /mg·g^{-1}	β /%	Surface appearance of modified sand after calcination
5	360	4.33	857.500	88.72	Clumps of sand, need to be mashed hard, brownish red surface, and more powder at the bottom
0	25	32.29	\	\	Unchanged

When T > 240°C, β decreases continuously with increasing temperature, and the percentage of exposed Si elements on the surface of MCMS to the total number of surface atoms gradually increases, indicating that the modifier surface coating is decreasing. When the T = 240°C, the metal oxides on the surface of MCMS accounted for 99.58% of the total number of atoms on the surface of the modified sand, and the modifier attachment was good, β reached the maximum value of 97.23%.

3.2.5. Effect of Calcination Time t. The unchanged preparation process conditions are set as $C1$=0.16 mL·g^{-1}, $C2$=0.10 mL·g^{-1}, $C3$=0.03 mL·g^{-1}, T=240°C, different levels of t. Results are shown in figure 6.

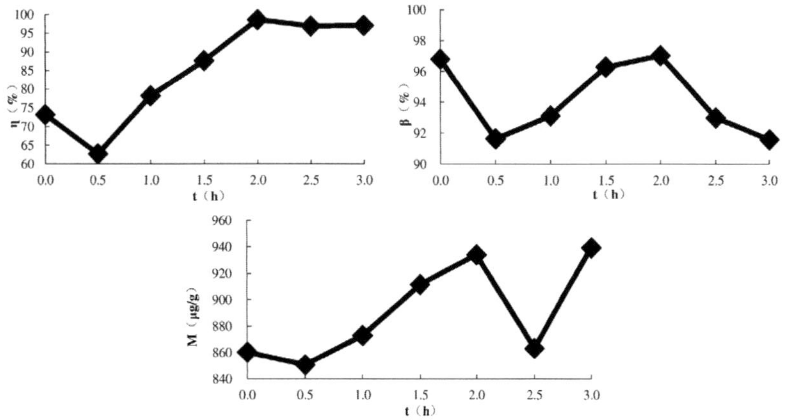

Figure 6. Effect of t on MCMS performance.

When t between 0.5 to 2 hours, all the indexes increase gradually; when t=2h, M reaches a peak of 934 μg/g, β and η reaches the maximum value of 97.04% and 98.67%; as t continues to increase, β began to decrease rapidly, while η tended to be in a stable state near 97%.

When t < 0.5h, the mixture of $AlCl_3$ and $TiCl_4$ changed the structure of the iron oxide layer at high temperature; with the t increasing, the mixture gradually solidified on the iron oxide layer as a carrier; M reached a saturated state at t = 2h, and the continuation of high-temperature heating would damage the structure of the metal oxide layer on the surface of the modified sand and reduce the adhesion rate of the modifier of MCMS.

In summary, the optimal preparation process conditions for MCMS are: $C1$=0.16 mL·g^{-1} RQS, $C2$=0.10 mL·g^{-1} RQS, $C3$=0.03 mL·g^{-1} RQS, T=240°C and t=2.0h.

3.3. MCMS Surface Morphology and Properties Study.

Sand surface morphology was detected by using SEM and EDS observation. Results are shown in figure 7 and figure 8.

Journal of Physics: Conference Series **2941** (2025) 012010 doi:10.1088/1742-6596/2941/1/012010

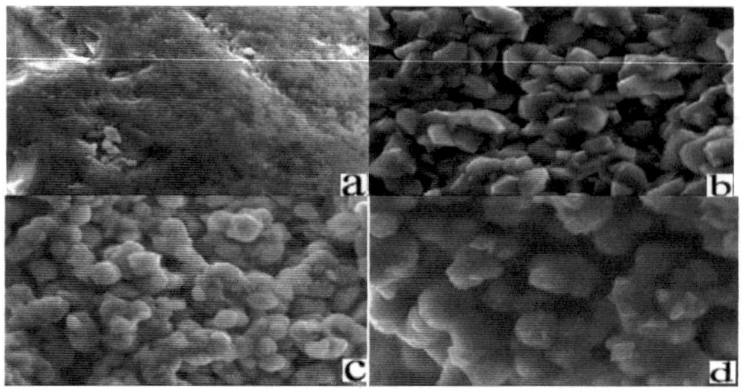

Figure 7. Surface morphological structures of SEM photo.
Note: (a) RQS,5000x; (b) Iron oxide layer of MCMS, 10000x; (c) MCMS, 7500x; (d) MCMS, 10000x.

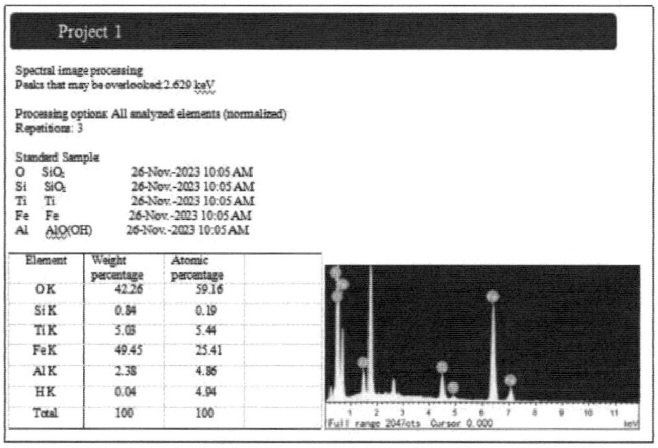
Project 1

Spectral image processing
Peaks that may be overlooked:2.629 keV

Processing options: All analyzed elements (normalized)
Repetitions: 3

Standard Sample
O SiO2 26-Nov.-2023 10:05 AM
Si SiO2 26-Nov.-2023 10:05 AM
Ti Ti 26-Nov.-2023 10:05 AM
Fe Fe 26-Nov.-2023 10:05 AM
Al AlO(OH) 26-Nov.-2023 10:05 AM

Element	Weight percentage	Atomic percentage
OK	42.26	59.16
SiK	0.84	0.19
TiK	5.03	5.44
FeK	49.45	25.41
AlK	2.38	4.86
HK	0.04	4.94
Total	100	100

Figure 8. EDS semi-quantitative analysis of the optimum preparation MCMS.

As shown in figure 7(a), the surface of RQS is flat, with a small specific surface area and adsorption capacity. Figure 7(b) is a chisel-like iron oxide crystal layer magnified 10,000 times the image, EDS results show that the iron oxide compounds for α-Fe_2O_3. The observation of the surface has a very clear and dense pore, the pore size is below 15 μm, and the crystal distribution is uniform, which provides a good carrier surface condition for the mixed modifier of $AlCl_3$ and $TiCl_4$ for re-modification. Figures 7(c) and 7(d) show that the mixed modifier of $AlCl_3$ and $TiCl_4$ was successfully modified on the surface of the iron oxide crystal layer, and the surface crystals were in the shape of tumor balls, which is the final morphology of the successful modification of the spherical small crystals covering the chiseled iron-shaped crystal surface uniformly so that MCMS has a larger specific surface area.

The results of EDS spectroscopy are shown in figure 8, the various substances on the surface are possibly Fe_2O_3, TiO_2, Al_2O_3, SiO_2, and FeO(OH) or AlO(OH). The surface pores of MCMS are massive and complex in structure, with pore sizes below 5 μm, and most of the pore inner diameters are in the nanoscale, which can adsorb a large number of organic matters with different particle sizes and improve the adsorption capacity of the filter media. Therefore, the method of preparing MCMS is reasonable and the effect of RQS surface modification is remarkable.

4. Conclusion

(1) The optimal preparation process conditions for MCMS were obtained: the dosing ratios of $FeCl_3$ solution, $AlCl_3$ solution, and $TiCl_4$ ethanol solution to RQS, *C1*, *C2*, and *C3*, were 0.16 mL·g^{-1} RQS, 0.10 mL·g^{-1} RQS, and 0.03 mL·g^{-1} RQS, calcination temperature T = 240 °C, calcination time t = 2.0 h.

(2) MCMS prepared under the optimal preparation process conditions had a total surface metal loading of 934 μg/g, a modifier attachment rate of 97.04%, and a humic acid removal rate of up to

98.67%.

(3) The SEM electron microscope scanning results of MCMS showed that the surface crystals of MCMS were in the shape of tumefaction spheres, and the metal crystals uniformly covered the surface of the carrier filter media, which was stable and had a large specific surface area. (4) The results of the EDS energy spectroscopy analysis and X-ray diffractometer showed that the composition of the surface oxides of MCMS might be Fe_2O_3, TiO_2, Al_2O_3, SiO_2, and FeO(OH) or AlO(OH), the surface oxides combinations were α-Fe_2O_3, AlO(OH), and Anatase TiO_2. (5) The adsorption process of MCMS was fitted using two adsorption curves, Langmuir and Freundlich, and the correlation coefficient values were all greater than 0.9, but the correlation coefficient value of the former was relatively larger, indicating that the adsorption process of MCMS on micro-pollutants co-existed with both physical and chemical adsorption, but chemical adsorption was dominant.

5. References

[1] Ebrahimzadeh A, Jajarmi A and Baleanu D 2024 Enhancing water pollution management through a comprehensive fractional modeling framework and optimal control techniques *Journal of Nonlinear Mathematical Physics* **31(1)** 48-48.

[2] He L, Yang Z Q, Yuan X H, et al. 2023 Analysis of cleaning, regeneration, and filtration performance of quartz sand filter material in a certain water plant *Water Purification Technology* **42(12)** 72-79.

[3] Ye T J, Liu B, Li D M, et al. 2013 Research on the treatment effect of new filter material iron oxide modified sand on non-typical water source *Proceedings of the 8th International Symposium on Urban Water Development in China* 1-10.

[4] Huang T F 2015 *Research on the Filtration Effect of Two New Modified Sands on HA and Optimization of Backwashing Conditions* (Guangdong: Guangdong University of Technology) Master's Thesis.

[5] Luo B 2017 *Preparation and Adsorption Performance of Graphene Oxide Ferric Chloride Composite Modified Zeolite* (Guangdong: Guangdong University of Technology) Master's Thesis.

[6] Sheng L 2005 *Study on the Manganese Removal Efficiency of Composite Modified Filter Media* (Shanghai: Tongji University) Doctor's Thesis.

[7] Guangdong University of Technology. A novel nano metal oxide iron, Aluminum, Titanium Composite Modified Quartz Sand Filter Material and Its Preparation Method: CN201410243159.6 2014-10-01.

[8] Guangdong University of Technology A Preparation Method for Graphene Based Iron Chloride Composite Modified Filter Material and Its Prepared Modified Filter Material: CN201811108081.1 2019-02-12.

[9] de Freitas Rosa Remiro Paula, de Sousa Cristina Paiva, et al. 2021 In situ evaluation of filter media modified by biocidal nanomaterials to control bioaerosols in internal environments *Water, Air, & Soil Pollution* **232(5)** 176-176.

[10] Jaber L, et al. 2024 Recent trends in surface impregnation techniques on activated carbon for efficient pollutant removal from wastewater *Desalination and Water Treatment* 100562-100562.

[11] Xie F X, Cherng S J, Lu S M, et al. 2014 Functions of self-assembled ultrafine tio_2 nanocrystals for high efficient dye-sensitized solar cells *ACS Applied Materials & Interfaces* **6(8)** 5367-5373.

[12] Vaishya R C and Gupta S K 2011 Optimization of oxide coating process on quartz sand for arsenic (III) removal from groundwater *Journal of Water Supply: Research and Technology—AQUA* **60(2)** 109-120.

[13] Yang T Z and Zhang W M 2019 Effects of ionic strength, inorganic ions and humic acids on U (VI) removal by HAP modified quartz sand *2019 6th International Conference on Machinery, Mechanics, Materials and Computer Engineering (MMMCE 2019)* 830-836.

[14] Hassani E M S, et al. 2024 Elaboration of an innovative plant biomaterial for its valorization in the treatment of wastewater *Bioresources and Bioprocessing* **11(1)** 58-58.
[15] Li G H 2024 Experimental study on the treatment of domestic waste leachate by intensive coagulation and sedimentation method. *Contemporary Chemical Industry* **53(03)** 530-534.

Progress in Separation Technology for Retired Photovoltaic Modules and Suggestions for Recycling

Yanfeng Li[1], Zilin Liu[2], Dehai Lin[2], Enfu Zhang[1], Zhan Liu[3], Yang Dong[2], Xiaoming Sheng[1], Zihan Li[3], Li Li[3] and Faquan He[3,*]

[1] Liaoning Longyuan Renewables Development Co. Ltd, Shenyang, Liaoning, China.
[2] National Institute of Clean-and-Low-Carbon Energy, Beijing, China.
[3] China Energy Longyuan Environmental Protection Co. Ltd, Beijing, China.
Email: 12102931@ceic.com; zilin.liu.a@chnenergy.com.cn; dehai.lin@chnenergy.com.cn; yang.dong@chnenergy.com.cn; 12109055@ceic.com; 20049954@ceic.com; 12071005@ceic.com; 20099003@ceic.com; 12053407@ceic.com; *17240111@ceic.com

Abstract. With the "double carbon" target and energy structure upgrading, the PV industry in China is increasing rapidly, and retired PV modules are becoming one of the key limiting factors for the green development of new energy. The recycling of PV modules is becoming more and more important, and module separation is the focus of research at home and abroad. Based on the investigation of domestic and foreign literature, the principle, research progress, and development potential of separation technology by mechanical method, pyrolysis method, and chemical method are analyzed, which provides reference and suggestion for the development of PV module recycling from the direction of green cycle.
Keywords. Photovoltaic module recycling; separation technologies; mechanical methods; pyrolysis methods; chemical methods; green recycling; dual carbon targets; energy structural upgrade.

1. Introduction

Against the backdrop of the dual-carbon goals, photovoltaic (PV) power generation is becoming increasingly significant in the global energy mix. From January to August 2023, China's installed PV power capacity reached 505.41 million kilowatts, an increase of 44.4% year-on-year, making it a driving force in the transformation of the energy structure. The large-scale decommissioning of PV panels will become a major constraint on the healthy development of new energy. By the end of 2020, China's cumulative decommissioned PV modules reached approximately 160,000 tons, and it is expected to reach 1.5 million tons between 2025 and 2030, with a potential waste volume of 20 million tons by 2050. If not properly handled, this could result in massive resource wastage and introduce new environmental issues, posing a potential crisis that could hinder the development of PV technology. Currently, the recycling mechanisms for PV modules both domestically and internationally are not well-established. Moreover, due to the significant regional differences in climate and sunlight between China's eastern and western regions, the distribution of PV installations is highly localized. Researching recycling technologies and industries suitable for China's specific conditions is of great importance for the sustainable development of photovoltaics.

2. Research Progress

At present, crystalline silicon cells dominate the market with a 90% share. Their basic structure includes solar cells, an organic encapsulation layer, glass, a back sheet, a frame, and a junction box. PV modules are bonded by the organic encapsulation layer (figure 1).

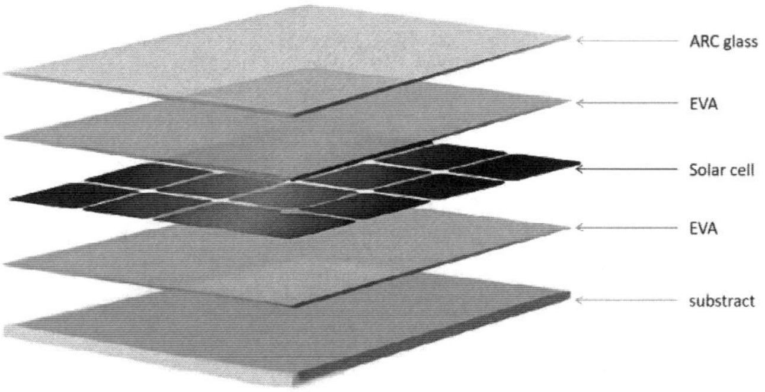

Figure 1. Structure of crystalline silicon solar cells.

2.1. Industry Demand and Current Status

China's PV recycling is in the early exploratory stage. It is estimated that the total revenue from recycling a single PV module is approximately 56.53 RMB, while the cost is about 75 RMB. Low-cost, green recycling has gradually become a significant technological demand in the PV sector (figure 2).

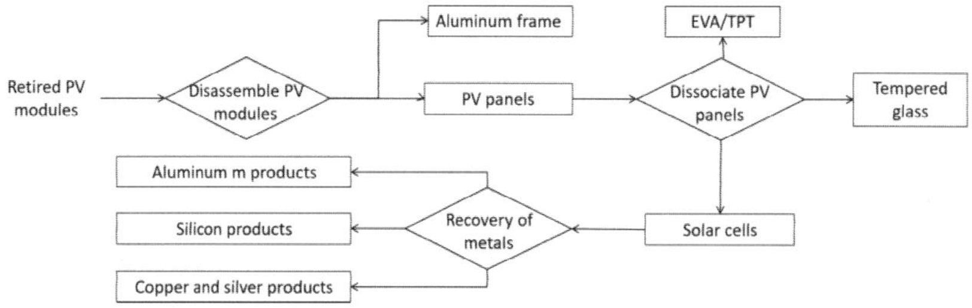

Figure 2. Process flow of crystalline silicon cell recycling.

The number of PV recycling-related papers published domestically and internationally has been steadily increasing from 2006 to 2024, with a sharp rise after 2019, indicating the growing importance of PV recycling. Since 2023, most published papers have focused on PV modules and recycling, suggesting that the separation and recycling of modules are still the main research priorities in China. Module separation is the most critical first step in PV recycling and thus remains a hot research area. This paper comprehensively analyzes the current separation technologies for PV modules and provides an outlook on the prospects of PV recycling (figures 3 and 4).

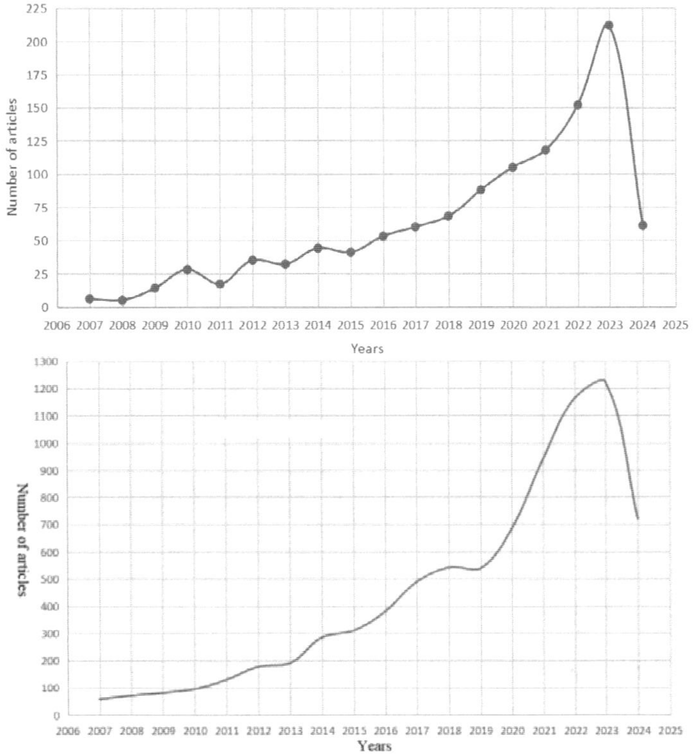

Figure 3. Publication status of crystalline silicon cell recycling papers domestically and internationally.

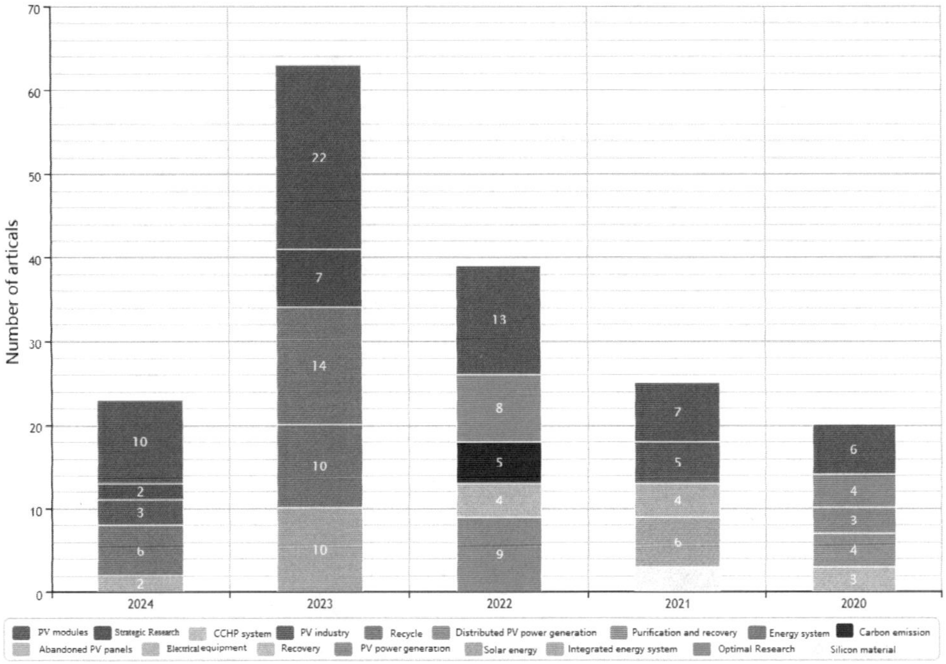

Figure 4. Research areas of crystalline silicon cell recycling in China.

Journal of Physics: Conference Series **2941** (2025) 012011 doi:10.1088/1742-6596/2941/1/012011

2.2 Technological Progress

2.2.1 Physical Methods. The physical method involves using mechanical equipment to crush and recycle PV modules by applying impact, shearing, and compressive forces [2]. This method does not produce wastewater or exhaust gas, is simple to operate, and has relatively high efficiency, but it may damage the cells and result in incomplete separation.

(1) Direct Crushing Method.

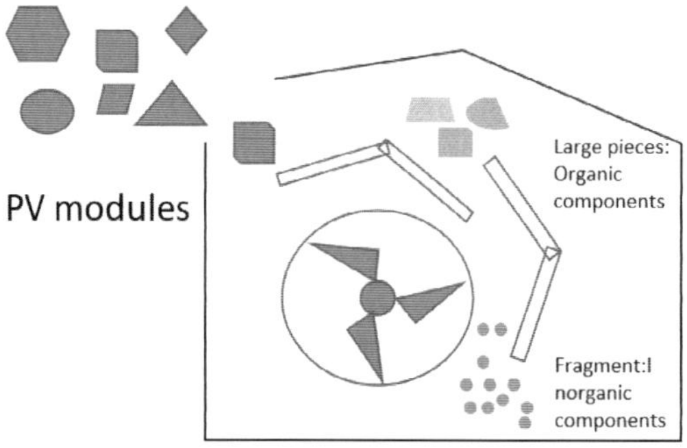

Figure 5. Schematic diagram of the direct crushing method for PV modules.

The direct crushing method leverages the deformation differences between inorganic glass, silicon wafers, and organic EVA (ethylene-vinyl acetate) back sheet materials to achieve component separation (figure 5). Granata and Pagnanelli [3,4] used rotor crushers and hammer mills to crush PV modules multiple times, where the glass components were relatively easy to enrich, facilitating direct recycling. TOKORO et al. [5] utilized the hardness difference between glass and EVA materials, employing selective eccentric grinding to separate EVA from glass. Under conditions of 2500 r/min and 5 minutes, the EVA content in the recovered glass was less than 1%.

(2) High-Voltage Pulse Method. The high-voltage pulse method involves subjecting materials to extremely rapid impact, compression, and fragmentation through a high-voltage electric field, simultaneously performing electro-fracturing (ED) and electro-hydraulic fracturing (EHD). Solid materials are selectively broken along material boundaries by high-voltage breakdown, while shock waves generated in the solution physically destroy the material, achieving selective separation (figure 6).

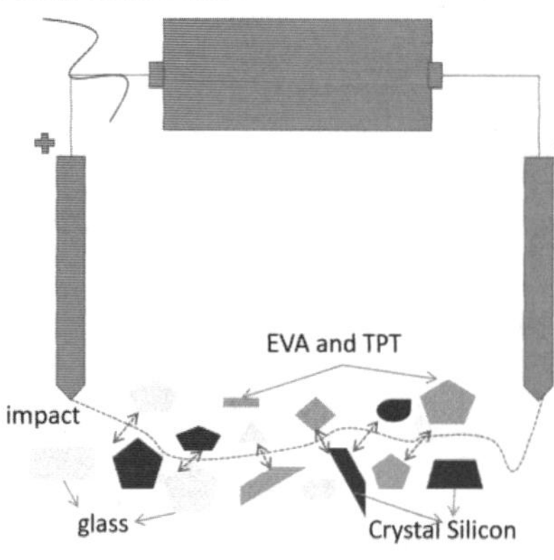

Figure 6. Schematic diagram of the high-voltage pulse crushing method for PV modules.

Duan [6] used high-voltage pulses to separate printed circuit boards (PCBs), achieving narrow-fraction metal enrichment. When the pulse number reached 400, the crushing performance peaked, with 97.92% of the copper concentrated in 2mm-sized products. AKIMOTO et al. [7] used 110 kV pulse voltage to complete the initial crushing of modules with 20 pulses, followed by 250 pulses at 90 kV to separate glass, cells, metals, and EVA. However, silver, silicon powder, and other small particles were significantly mixed. To address this, Azeumo et al. [8] used sodium tungstate heavy medium separation, achieving 100% recovery of metals with about 67% purity and 76% recovery of glass with nearly 100% purity.

Song et al. [9] improved the high-voltage pulse technology using water as the medium, achieving an average particle size of 4.1mm at 160 kV, 300 pulses, and an energy consumption of 192.99 J/g. 95% of the copper and 96% of the silver were concentrated in particles smaller than 1mm, while 85% of the aluminum, 85% of the lead, and 87% of the tin were concentrated in particles smaller than 0.5mm.

Zhao et al. [10] investigated the selective separation of different components, with the order being silver > silicon > glass. High-value elements had a purity greater than 98% in particles between 0.5-4mm. The particle distribution of PV modules is shown in figure 7.

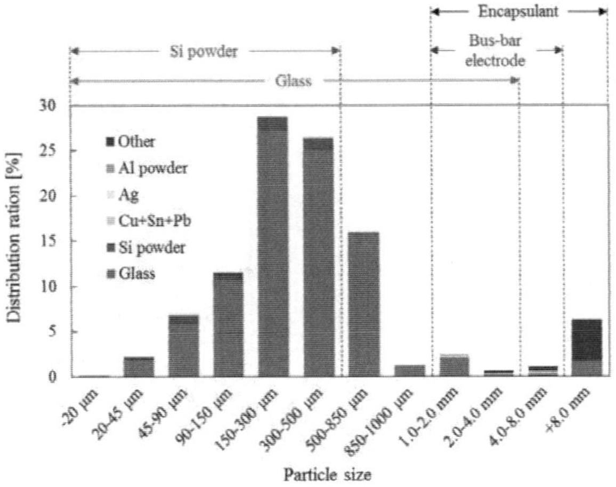

Figure 7. Schematic diagram of the high-voltage pulse crushing method for PV modules [11].

(3) Cryogenic Crushing Method. The principle of the cryogenic crushing method is to reduce the temperature, which lowers the activity of organic polymers, reduces intermolecular forces, makes composite materials brittle and hard, and decreases elasticity and viscosity, making them easier to break off. The brittleness of the back sheet and glass increases, making them easier to crush and remove (figure 8).

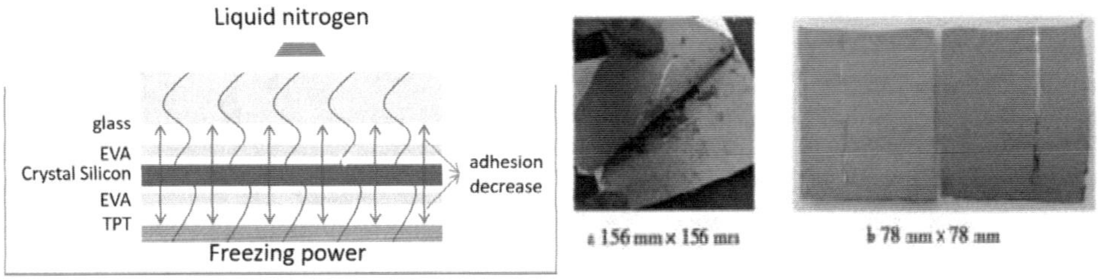

Figure 8. Schematic diagram and results of the cryogenic crushing method for PV modules [12].

Zhang et al. [13] used liquid nitrogen to cryogenically modify PV modules. Compared with untreated samples, the EVA adhesion significantly decreased, brittleness increased, and it became easier to break off. After 8 seconds of crushing, the silicon enrichment rate was 72%, and the glass removal rate was 70%. Wei et al. [12] studied primary crushing-cryogenic treatment-secondary crushing, finding that cryogenic-secondary crushing resulted in higher PV glass recovery (85%) compared to conventional crushing, with a glass removal rate of 99.6%. Silver grid and busbar metals were also recoverable.

Yingli Energy (China) Co., Ltd. employed a deep cryogenic grinding method, which made EVA easier to remove, and the back sheet and glass easier to crush. Screening resulted in enriched products of glass, silicon, back sheet, and EVA [14]. Zhang et al. [15] explored the effects of high-voltage electrostatics, including voltage, rotation speed, electrode gap, and corona electrode on sorting. Under optimal process conditions, the silicon recovery rate reached 99.72%, and the PET recovery rate reached 99.8%.

2.2.2 Pyrolysis Method. The pyrolysis method involves the high-temperature melting and decomposition of organic matter, converting it into gas or liquid, thereby achieving the separation of components with minimal impact on inorganic materials such as metals [16]. The key to component separation is the removal of EVA [17], and pyrolysis can effectively remove EVA [18,19]. The pyrolysis process is the simplest to operate, but it emits fluorine-containing exhaust gases.

(1) Direct Pyrolysis.

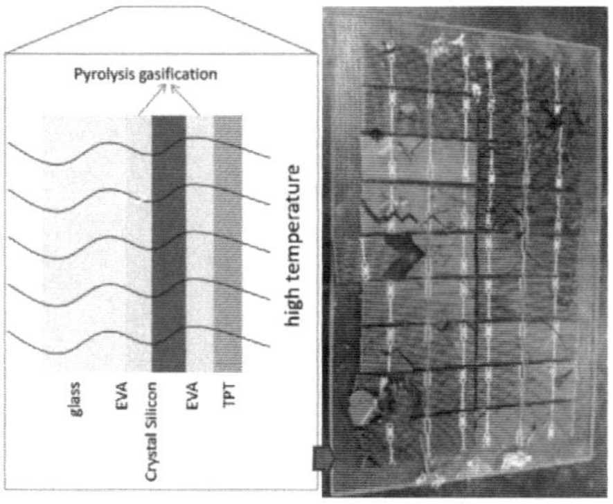

Figure 9. Schematic diagram of the pyrolysis method for PV modules.

Tao, Wang, Fiandra, et al. [20-22] adopted a two-stage pyrolysis technique, removing over 99% of the polymers in PV modules. Tammaro et al. [23] heated PV modules to 600°C, completely degrading the EVA (figure 9). Li et al. [24] found that EVA pyrolysis occurs in two stages: in the first stage, the ester bonds in the EVA polymer break, releasing acetic acid; in the second stage, the residual polymer undergoes further bond breaking and re-polymerization, generating aromatic volatiles and aliphatic hydrocarbons. When the pyrolysis temperature is 500°C, the mass loss rate of EVA exceeds 99% (figure 10).

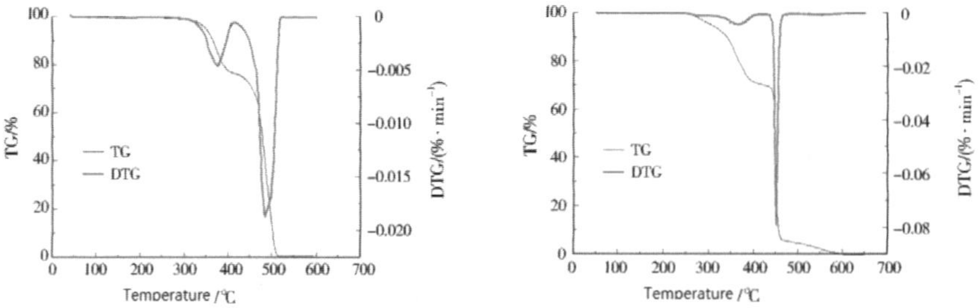

Figure 10. Thermogravimetric analysis of the pyrolysis method for PV modules [24].

(2) Pyrolysis in an Inert Atmosphere. Zhang et al. [25] conducted pyrolysis of PV modules in a nitrogen atmosphere. Under conditions of a flow rate of 0.5 L/min, a temperature of 500°C, and a holding time of 30 minutes, the EVA was almost entirely decomposed. Zhang et al. [26] also found that under conditions of 773K, 30 minutes, and an N2 flow rate of 0.5 L/min, the organic matter degradation rate was close to 100%. Farrell et al. [27] compared recovered and virgin polymers, finding that recovered EVA had a relatively high calorific value of 39.51 MJ.Kg-1, which remained unchanged at 39.87 MJ.Kg-1, confirming the energy recovery potential of EVA. Feng et al. [28] under pyrolysis conditions of 500°C, a holding time of 30 minutes, an airflow time of 90 seconds, and an airflow speed of 85m³/h, achieved a solar cell recovery rate and concentration of 91.09% and 84.4%, respectively, for particle sizes larger than 4mm. Under conditions of an airflow speed of 75m³/h and an airflow time of 90 seconds, the recovery rate and concentration for particles in the 2-4mm size range were 82.29% and 80.52%, respectively.

(3) Hot Knife Method. DONI et al. [29] used a heated plate to heat PV modules. Under conditions of 400 W power and 15 minutes, most of the glass could be recovered. Based on this, a heated blade can be designed to cut the EVA adhesive area, achieving separation of the glass and the cells.

Japan's NPC Corporation designed a hot knife machine [30], the "J-BOX Separation Device," where a 300°C hot knife cuts the adhesive area between the glass and the cells, enabling rapid separation of PV glass and modules. This method is only applicable to intact PV modules, and the glass remains coated with EVA, requiring further processing (figure 11).

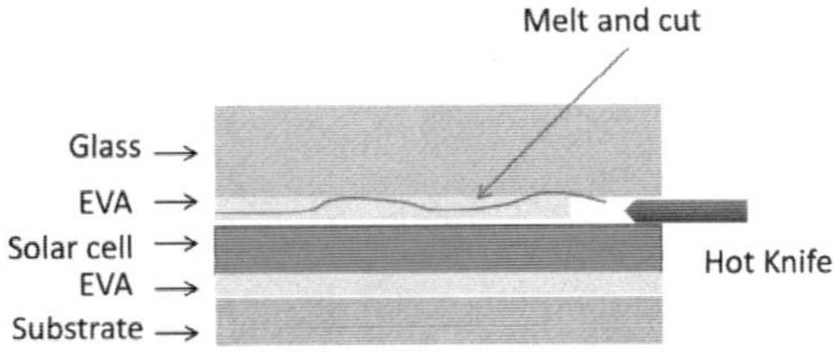

Figure 11. Schematic diagram of the hot knife method for PV module separation.

(4) Other Pyrolysis Techniques. DONI et al. [29] used radiofrequency heating. When the PV module material molecules are exposed to sufficient electromagnetic waves, the molecules are excited into intense motion, causing internal heating of the material. ERCOLE et al. [31] used an infrared heater to heat PV modules to a temperature of 70~150°C, softening the EVA and separating the glass and cells. This separation method is relatively simple, but the EVA removal rate is not high.

The pyrolysis method is highly efficient but generates toxic and harmful gases such as hydrogen fluoride [32,33], requiring consideration of exhaust gas treatment. However, pyrolysis remains an important method for recycling waste PV modules [34].

2.2.3 Chemical Method. The chemical method separates components by dissolving the EVA adhesive, achieving complete separation. Due to the cross-linking and coupling of EVA in PV modules, the adhesion is greatly enhanced, and the chemical separation rate is generally low.

(1) Organic Swelling Method. The molecular structure of EVA consists of a non-polar crystalline ethylene chain and a polar amorphous vinyl acetate chain. Compared to inorganic reagents, organic solvents have better-dissolving effects on EVA (figure 12).

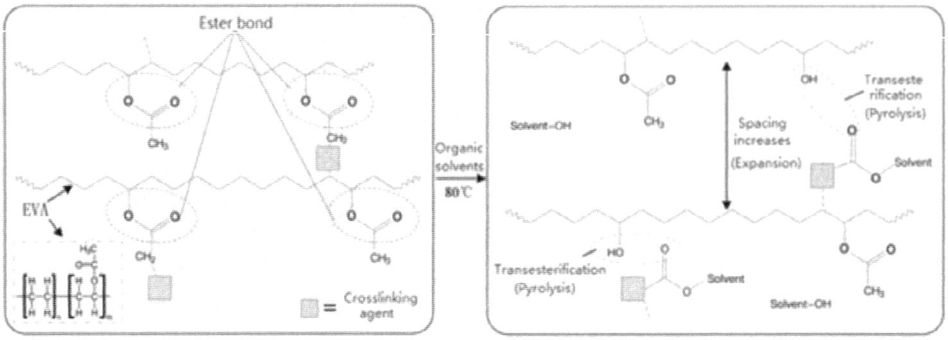

Figure 12. Schematic diagram of the organic solvent method for PV module swelling.

Kang et al. [35] studied the dissolution of EVA with isopropanol, toluene, etc. Using a 90°C toluene solution, separation could be completed in two days. The mechanism of EVA dissolution and swelling was studied in-depth, as shown in figure 13. Approximately 35% of EVA is cross-linked, resulting in the coexistence of cross-linked and non-cross-linked structures. Organic reagents primarily dissolve the non-cross-linked structure, while the cross-linked part mainly undergoes a swelling reaction, leading to an increase in resin volume and stress, causing the solar cell to break. Doi et al. [36] studied the dissolving effects of six organic solvents, including acetone, toluene, and trichloroethylene, with trichloroethylene having the best effect, dissolving both cross-linked and non-cross-linked structures with a dissolution time of 7-10 days. Kang et al. [37] indicated that during the dissolution of EVA, the swollen part still adhered to the glass and silicon wafer surfaces, leading to incomplete EVA removal, requiring further processing. Jiao et al. [38] used o-dichlorobenzene as a solvent. In conditions of a leaching temperature of 160°C, a leaching time of 240 minutes, a solid-liquid ratio of 4:100, and a stirring speed of 800 r/min, the dissolution and separation effect of PV modules was the best.

(2) Enhanced Organic Swelling Method. Pang et al. [39] used microwaves to enhance the EVA swelling reaction. Under conditions of 4 mol/L trichloroethylene concentration, a reaction temperature of 70°C, and a solid-liquid ratio of 50g/L, effective separation of PV modules could be achieved in 2 hours. Azeumo et al. [40] studied an optimized recovery process using ultrasound, under conditions of 60°C, a reaction time of less than 60 minutes, ultrasound intensity of 200W, and 50kHz. Among six different organic solvents, toluene dissolved EVA at the fastest rate. KIM et al. [41] completely dissolved EVA within 30 minutes under conditions of 3 mol/L o-dichlorobenzene, a dissolution temperature of 70°C, and an ultrasound power of 900W. However, EVA might swell and cause cell breakage. PANG et al. [42] indicated that within the temperature range of 30-75°C, the difference in thermal expansion coefficients between glass and EVA could promote interlayer gap formation, accelerating solvent penetration. Under conditions of 4 mol/L trichloroethylene, a solid-liquid ratio of 50g/L, and a reaction temperature of 70°C, EVA could be completely dissolved within 2 hours.

(3) Inorganic Solvent Method. Bruton et al. [43] were the first to use nitric acid to dissolve the EVA adhesive film. The oxidation and decomposition efficiency of nitric acid on EVA was low, with a slow interlayer penetration rate. It took 7-14 days at 60°C, and the reaction produced NOX. Residual EVA remained on the cell surface, resulting in poor removal effectiveness. Institutions such as Trina Solar studied hydrothermal reaction separation of PV modules [44], where chemical reactions occurred in high-temperature water under high pressure of 2-50 MPa and temperatures of 150-600°C. Xu Yang et al. [45] indicated that under conditions of 270°C, 1 mol/L NaHCO3, and a reactor water filling rate of 50%, reacting for 3 hours, partial EVA underwent hydrolysis, effectively reducing the bonding strength of the upper EVA layer.

(4) Other Chemical Methods. Axel Briand et al. [46] used supercritical CO2 foaming to induce delamination of PV modules. The effects of CO2 temperature, pressure, and depressurization rate on EVA deformation were studied. When the temperature exceeded the melting point and the initial pressure was not less than 150 bar, increasing the depressurization rate could enhance EVA layer deformation, achieving module separation and targeted recovery. Li et al. [47] used the green reagent DMPU (C6H12N2O) to separate PV modules, maintaining their original dimensions. Additionally, Li et al. [48] used DBE (C21H36O12) to inhibit EVA swelling, ensuring that the module separation process did not result in excessive cracking. Min et al. [49] used EGDA (C6H8O4) to separate PV modules. Compared to water, EGDA more easily penetrated the glass-EVA gap, increasing the chance of contact with EVA, thus improving separation efficiency (figure 13).

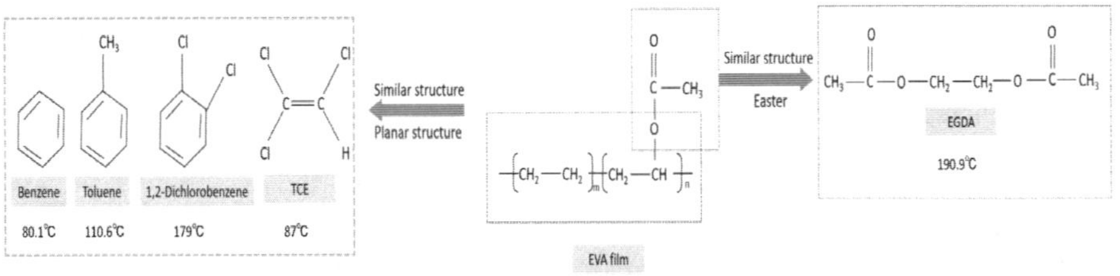

Figure 13. Schematic diagram of the EGDA swelling mechanism for PV modules.

Li et al. [50] used DMPU to separate PV modules, studying the effects of different modules, reaction temperatures, solid-liquid ratios, and ultrasound power on the separation ratio, glass recovery rate, and back sheet peeling time (figure 14).

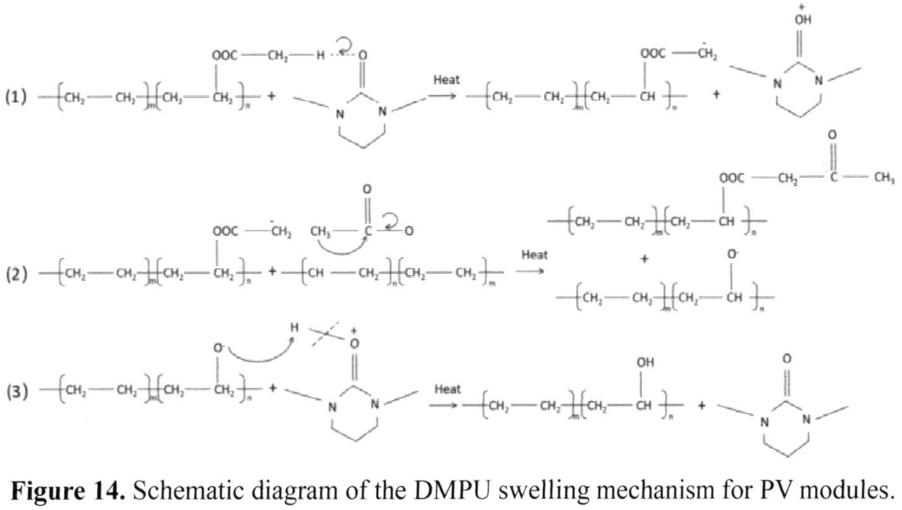

Figure 14. Schematic diagram of the DMPU swelling mechanism for PV modules.

3. Conclusion

With the increasing proportion of PV power generation in the new power system, the potential for PV modules recovery is huge, and the blue ocean market has emerged. The demand for recycling technology and industry is also increasing. Based on the analysis of current technological progress and industrial development needs, the following prospects are proposed for recycling technology, equipment, and industrialization. (1) The future is promising for high-efficiency, complete separation technologies. The current industry faces major bottlenecks, including limited application scenarios for recovered products, low recycling efficiency, and poor economic viability. Separation technologies should continue to be explored in terms of improving efficiency, reducing pollution, and lowering energy consumption. Achieving high-efficiency, complete separation can broaden application scenarios. (2) Integrated technology and equipment need improvement. Currently, China has not yet developed a complete set of technologies and equipment for PV module frame disassembly, component separation, metal recovery, reprocessing, and pollution control, and the level of intelligence and automation also needs to be enhanced. (3) Industry-leading effects have not yet formed. The large-scale decommissioning period for PV modules has not yet arrived, and most large companies utilizing resources have not yet entered the market. Small enterprises or workshops, due to their limitations, have led to initial rough and unregulated recycling practices, which are detrimental to long-term development. There is a need to further regulate industry market entry conditions.

Acknowledgments
This paper was supported by CHN Energy Investment Group (CSIEKJ220700493).

4. References

[1] Research Center of Shudao Youfang. Market Status of the Photovoltaic Industry in 2023 and Analysis of Trends and Prospects for 2024. Available from: https://www.doc88.com/p-77339579203549.html.
[2] Zhou Z, Sun K W, Jiang L X, et al. 2020 Research Progress on Recycling Technologies for End-of-Life Photovoltaic Modules. *Journal of Central South University (Science and Technology)* **51**(12) 3279-3288.
[3] Granata G, Pagnanelli F, Moscardini E, et al. 2014 *Solar Energy Materials and Solar Cells* **123** 239-248.
[4] Pagnanelli F, Moscardini E, Granata G, et al. 2017 *Waste Management* **59** 422-431.
[5] Tokoro C, Nishi M, Tsunazawa Y. 2021 *Adv Powder Technol* **32**(3) 841-849.
[6] Duan C L, Diao Z J, Zhao Y M, et al. 2015 *Minerals Engineering* **70** 170-177.
[7] Akimoto Y, Iizuka A, Shibata E. 2018 *Miner Eng* **125** 1-9.
[8] Azeumo M F, Germana C, Ippolito N M, et al. 2019 *Solar Energy Materials and Solar Cells* **193** 314-319.
[9] Song B P, Zhang M Y, Fan Y, et al. 2020 *Waste Management* **101** 180-187.
[10] Zhao Y, Zhang B, Duan C, et al. 2021 *Powder Technol* **269** 219–226.
[11] Yuta A, Atsushi I, Etsuro S. 2018 *Minerals Engineering* **125** 1–9.
[12] Wei X X, Wang S, Gao D D, et al. 2024 *Journal of Qinghai University* **42**(1) 17-23.
[13] Zhang X F, Yan Y, Wang D, et al. 2021 *Mining and Metallurgy* **30**(3) 19-23, 39.
[14] Wang S Y, Wang Z Y, Zhou H L, et al. 2012 A Method and Device for Decomposition and Recycling of Photovoltaic Modules: China, CN102544239A. 2012-03-07.
[15] Zhang Z S, Cui Z, Yang J, et al. 2017 *China Environmental Science* **37**(8) 3048-3055.
[16] Li D, Zhou X Y, Liu J Y, et al. 2020 *Environmental Pollution & Control* **42**(10) 1211-1215.
[17] DOI T, Tsuda I, Unagida H, et al. 2001 *Solar Energy Materials and Solar Cells* **67**(1/4) 397-403.
[18] Li D, Feng J Y, Liu J Y, et al. 2020 *Environmental Pollution & Control* **42**(6) 678-681, 689.
[19] Jiao Y T, Chen Q, Li D X, et al. 2021 *Chemical Engineering and Environmental Protection* **41**(2) 179-183.
[20] Tao M, Fthenakis V, Ebin B, et al. 2020 *Progress in Photovoltaics: Research and Applications* **28**(10) 1077-1088.
[21] Wang T Y, Hsiao J C, Du C H. 2012 *2012 38th IEEE Photovoltaic Specialists Conference* (IEEE) p002355-002358.
[22] Fiandra V, Sannino L, Andreozzi C, et al. 2019 *Waste Management* **84** 91-101.
[23] Tammaro M, Rimauro J, Fiandra V, et al. 2015 *Renewable Energy* **81** 103-112.
[24] Li D, Zhou X Y, Liu J Y, et al. 2020 *Environmental Pollution & Control* **42**(10) 1211-1215.
[25] Zhang Z, Liu M, Wang L, et al. 2021 *Solar Energy Materials and Solar Cells* **230** 111218.
[26] Zhang G, He Y, Wang H, et al. 2017 *Waste Management* **64** 228-235.
[27] Farrell C, Osman A I, Zhang X L, et al. 2019 *Scientific Reports* **9** 5267.
[28] Feng Y, He Y Q, Zhang G W, et al. 2023 *Solar Energy Materials and Solar Cells* **262** 112553.
[29] Doni A, Dughiero F. 2012 *2012 38th IEEE Photovoltaic Specialists Conference* (Austin: IEEE) p757-762.
[30] Ito M, Doi T. 2018 *2018 IEEE 7th World Conference on Photovoltaic Energy Conversion*

(WCPEC) (Waikoloa: IEEE) p709-711.
[31] Ercole P, Ramon L, Agnoletti O. 2016 Method and Apparatus for Detaching Glass from a Mono- or Polycrystalline Silicon-Based Photovoltaic Panel: EU, 15185389.2. 2016-03-23.
[32] Tammaro M, Rimauro J, Fiandra V, et al. 2015 *Renewable Energy* **81** 103-112.
[33] Ardente F, Latunussa C.E.L., Blengini G.A. 2019 *Waste Management* **91** 156-167.
[34] Farrell C.C., Osman A.I., Doherty R., et al. 2020 *Renewable and Sustainable Energy Reviews* **128** 109911.
[35] Kang S, Yoo S, Lee J, et al. 2012 *Renewable Energy* **47** 152-159.
[36] Doi T, Tsuda I, Unagida H, et al. 2001 *Solar Energy Materials and Solar Cells* **67** 397-403.
[37] Kang S, Yoo S, Lee J, et al. 2012 *Renewable Energy* **47** 152-159.
[38] Jiao Y T, Chen Q, Li D X, et al. 2021 *Chemical Engineering and Environmental Protection* **41**(2) 179-183.
[39] Pang S, Yan Y, Wang Z, et al. 2021 *Solar Energy Materials and Solar Cells* **230** 111213.
[40] Azeumo M F, Germana C, Ippolito N M, et al. 2019 *Solar Energy Materials and Solar Cells* **193** 314-319.
[41] Kim Y, Lee J. 2012 *Solar Energy Materials and Solar Cells* **98** 317-322.
[42] Pang S, Yan Y, Wang Z, et al. 2021 *Solar Energy Materials and Solar Cells* **230** 111213.
[43] Bruton T M, Scott R D W, Nagle J P. 1994 *Proceedings of the 12th EC-PVSEC* (Amsterdam: The Netherlands) p 303-304.
[44] Chang H R, Zhang Z, Yu S K 2018 A Method for Separating and Recycling Double-Glass Structure Modules: China, 201811425095. 2018-11-27.
[45] Xu Y. 2019 *Basic Research on the Separation and Recovery of Crystalline Silicon from Waste Photovoltaic Modules* (Shanghai: Shanghai Jiao Tong University) Master's Thesis.
[46] Axel B, Antoine L A, Olivier D, et al. 2022 *Journal of CO_2 Utilization* **5**(59) 101933.
[47] Li K, Wang Z, Liu C M, et al. 2022 *Solar Energy Materials and Solar Cells* **9**(245) 111870.
[48] Li K, Wang D, Hu K B, et al. 2023 *Clean Technology and Environmental Policy* 25 2203-2212.
[49] Min R, Li K, Wang D, et al. 2023 *Solar Energy* **253** 117-126.
[50] Li K, Wang Z, Liu C M, et al. 2022 *Solar Energy Materials and Solar Cells* **9**(245) 111870.

Research on Surface Roughness and Surface Topography of TC4 Titanium Alloy Processed by High-Speed Turning

Lianyao Tang* and Peiming Peng

School of Mechanical and Electrical Engineering, Guangdong University of Science and Technology, Dongguan 523083, Guangdong, China.
Email: *448795465@qq.com; 1418879376@qq.com

Abstract. Titanium alloy is a typical difficult to process material, in order to further improve the processing quality. In this study, the single factor experiment method was used to perform high-speed turning experiments on TC4 titanium alloy with TiAlSiN carbide coated blade. The experimental results show that the cutting speed (V_c), feed per tooth (f_z), and cutting depth (a_p) have certain effects on the surface roughness of TC4 titanium alloy. When the feed per tooth increases from 0.04 mm/r to 0.2 mm/r, the surface roughness value (Ra) increases by 7 times. In the cutting speed range of 40-200 m/min, the surface roughness value Ra is about 0.5-0.7 μm. In addition, the effect of depth of cutting (a_p) on surface roughness (Ra) is relatively small. The reliability of the experimental results was further verified by observing the surface morphology with an ultra depth field microscope. This study will provide a theoretical basis for the cutting of titanium alloys.
Keywords. TC4 titanium alloy; high speed cutting; surface roughness; surface morphology.

1. Introduction

Titanium alloys (TC4) are widely used in aerospace, weaponry, automotive, and shipbuilding industries due to their high specific strength, good oxidation resistance, corrosion resistance, heat resistance, and other advantages [1,2]. However, the unique properties of titanium alloys make them difficult to process, with challenges such as large cutting force per unit area, high cutting temperature, chilling phenomenon, and tool wear [3]. Extensive research has been conducted on high-speed cutting technology of titanium alloys by scholars worldwide. Venugopal [4] studied tool wear and service life of titanium alloy Ti-6Al-4V using uncoated carbide tool blades. Elmagrabi et al investigated the impact of feed rate on surface roughness when using PVD carbide cutting tools for processing Ti6Al4V, identifying adhesion and thermal cracking as main failure modes at high cutting speeds [5]. Zhao et al. performed high-speed milling experiments on Ti6A14V titanium alloy to analyze the effects of different cutting parameters on cutting force, while also studying surface roughness (Ra) through orthogonal testing [6]. In addition, Yao et al. explored high-speed milling of TB6 and TC11 titanium alloys [7,8], examining how various process parameters influence surface quality. Single factor experiments were utilized to investigate the influence of cutting parameters on surface roughness (Ra) and morphology of TC4 titanium alloy during high-speed milling. This ongoing research aims to improve the efficiency and quality of machining titanium alloys for various industrial applications.

2. TC4 Titanium Alloy High-Speed Turning Experiment Arrangement

2.1. Experimental Materials
This experiment uses a TC4 titanium alloy workpiece with a diameter of 50 mm, which is classified as an (α+β) type titanium alloy with a composition of Ti-6Al-4V. Its chemical composition iron content is less than 0.30%, carbon content is less than or equal to 0.10%, the nitrogen content is less than.05%, hydrogen content is less than 0.05%, oxygen content is less than 0.20%, its detailed chemical composition is iron (Fe) is less than 0.30%, carbon (C) is less than0.10%, alumin um content is between 5.5% to 6.8%, vanadium content is between 3.5% to 4.5%. TC4 titanium alloy exhibits good mechanical properties at room temperature with a tensile strength of ≥902 MPa, residual elongation stress of ≥824 MPa, elongation of ≥10%, and section shrinkage of ≥25% [9]. The tool used in this experiment is a TiAlSiN carbide coated blade, specifically the VNEG160404-NF blade type, mounted on a 100 mm milling cutter. The TiAlSiN coating, known for its high hardness and adhesion resistance [10] due to the inclusion of Si, is ideal for this application. Because of the strong chemical affinity of TC4 titanium alloy for tool materials, the TiAlSiN carbide coated blade produced by Zhuzhou Diamond Cutting Tool Co., Ltd. was selected for its resistance to adhesive wear. This blade model, VNEG160404-NF, is well-suited for processing TC4 titanium alloy.

2.2. Experimental Scheme
The turning experiment on the LBR-370 CNC lathe was conducted using the single factor experiment method. Figure 1 illustrates the setup of the equipment and tool. Combined with the machining performance characteristics of difficult-to-machine material titanium alloy and the technical parameters and performance characteristics of the equipment used, the speed of cutting (V_c) selected in this experiment is 40 m/min-200 m/min, the quantity of feed (f_z) selected is 0.04 mm/r-0.2 mm/r, and the depth of cutting (a_p) selected is 0.2 mm-0.6 mm. After the experiment was completed, the surface morphology was observed under the ultra-depth of field three-dimensional microscopic system.

Figure 1. Equipment installation and tool.

3. Experimental Results

In this experiment, the MarSurfM300 surface structure measuring instrument was usually used to detect the roughness date of the processed titanium alloy surface. To minimize the date of surface residue on the measurement results of roughness value (Ra), the workpiece was thoroughly cleaned with alcohol first, and then 4 points were measured along the direction perpendicular to the feed. After measuring each point three times, the average value was taken, and the measurement results of surface roughness value were finally obtained, as shown in table 1. In order to further observe the surface morphology after machining, the ultra-depth field 3D microscopic system of KEYENCEVHX-500FE was used to detect the finish of the surface grain.

Table 1. Presentation of experimental data.

No.	v_c (m/min)	f_z (mm/r)	a_p (mm)	Ra (μm)
1	40			0.61
2	80			0.535
3	120	0.12	0.4	0.577
4	160			0.497
5	200			0.611
6		0.04		0.172
7		0.08		0.42
8	120	0.12	0.4	0.752
9		0.16		1.063
10		0.2		1.286
11			0.2	0.602
12	120	0.12	0.4	0.577
13			0.6	0.493

4. Analysis of Experimental Results

According to the experimental data in table 1, origin software was used to draw a picture of the influence of the given cutting parameters on the date of surface roughness, as shown in figure 2. Figure 2(a) shows that the date of roughness value increases from 0.2 um to 1.2 um when the rate of feed increases from 0.04 mm/r to 0.2 mm/r. Figure 2(b) depicts the change in surface roughness as the cutting speed increases from 40 m/min to 200 m/min. Figure 2(c) depicts the change in surface roughness as the depth of cutting increases from 0.2 mm to 0.6 mm. Figures 3-5 display the surface topography resulting from variations in cutting parameters. These images reveal a distinct texture on the cutting surfaces, characterized by alternating high and low regions along the turning direction. Additionally, a periodic variation in surface texture profile is observed along the feed direction of the tool, with the displacement of evenly spaced convex ridges corresponding to the feed per tooth (f_z) during the cutting process. This phenomenon highlights the dynamic interaction between the tool and workpiece, accurately depicting the motion trajectory of the cutting edge.

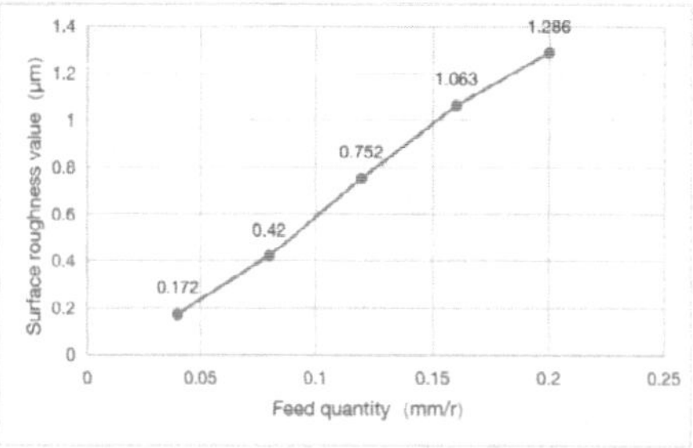

(a) v_{c}=120 m/min, a_{p}=0.4 mm

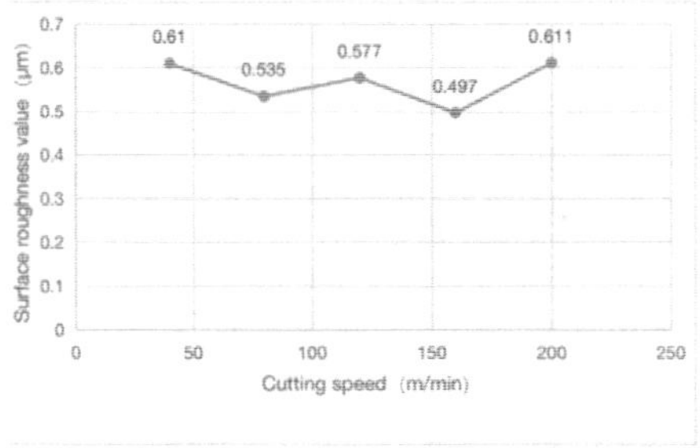

(b) f_z=0.12 mm/r, a_p=0.4 mm

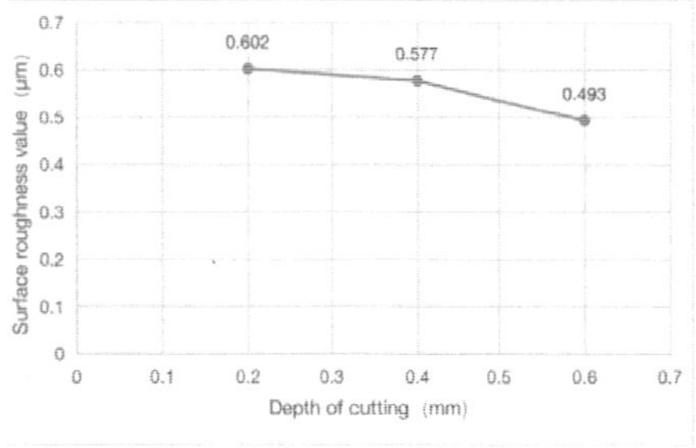

(c) v_c=120 m/min, f_z=0.12 mm/r

Figure 2. Influence of different cutting parameters on surface roughness value.

4.1. Analysis of the Relationship of the Feed Per Tooth (F_z) on Surface Roughness Value and Surface Morphology

According to the data in figure 3(a), as the feed amount in cutting increases, the surface roughness value (Ra) will also increase, and the surface quality will become worse and worse. Especially when the feed per tooth (f_z) increased from 0.04 mm/r to 0.2 mm/r, the surface roughness value (Ra) increased by 7 times. It seems that in order to obtain good surface quality, when cutting titanium alloy, it is not appropriate to choose too large a feed, which should be about 0.1 mm/r. As can be seen from the four figures in figure 3, all machined surfaces present regular machined textures, and the texture width is basically similar to the rate of feed per tooth. It can be observed from figure 3 that with the gradual increase of the rate of feed, the feature of "uplift" of the cutting mark texture in the feed direction becomes more and more obvious. This is due to the increase of the rate of feed (f_z), the amount of tool wear gradually increases, resulting in a bigger contact area between the tool surface and the workpiece machined surface, the workpiece processing position after extrusion, deformation, thermal coupling under the comprehensive action, the whole cutting process is more complex, the plastic flow becomes more and more obvious in this case, the date of roughness (Ra) controllability becomes worse. The absolute value of surface roughness increases.

(a) v_c=120 m/min, f_z=0.04 mm/r, a_p=0.4 mm

(b) v_c=120 m/min, f_z=0.08 mm/r, a_p=0.4 mm

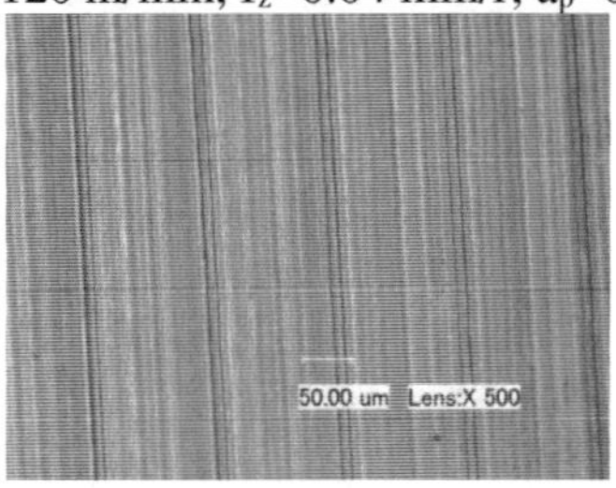

(c) v_c=120 m/min, f_z=0.12 mm/r, a_p=0.4 mm

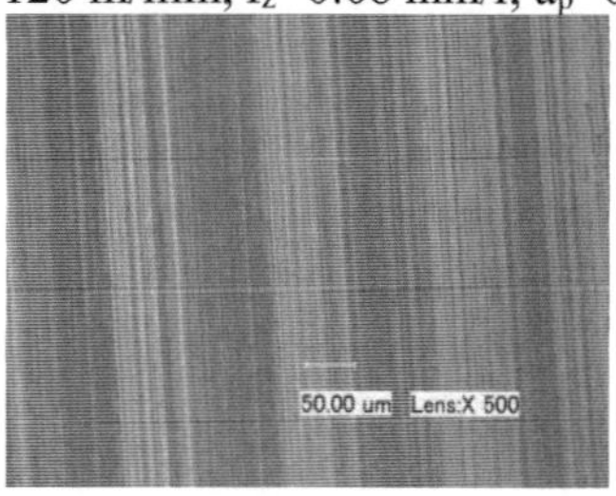

(d) v_c=120 m/min, f_z=0.16 mm/r, a_p=0.4 mm

Figure 3. Surface morphology of milling with different feed rates.

Journal of Physics: Conference Series **2941** (2025) 012012 doi:10.1088/1742-6596/2941/1/012012

4.2. Analysis of the Relationship of Cutting Speed (Vc) on Surface Roughness Value and Surface Morphology

As can be seen from figure 2(b) mentioned above, the cutting speed is in the range of 40 m/min-200 m/min, and the surface roughness value is between 0.5-0.7 μm. According to the cutting mechanism, we know that when the speed is small, the force is larger, which causes larger vibration, and therefore the surface quality is relatively poor; As the cutting speed increases, the energy consumption becomes larger and larger, and the cutting system temperature increases, the titanium alloy will become softened, and the cutting force required is small, so when the cutting speed reaches 160 m/min, the surface roughness value is minimum; however, with the further increase of the cutting speed and the further increase of the temperature, the tool will be worn very badly, making the processing quality decline. Therefore, when processing TC titanium alloy, choosing a moderate cutting speed is the key to ensuring good surface quality. According to the data in figure 4, it indicates that the surface quality is poor and the adhesion phenomenon may occur at a low speed. However, with the increase in speed, the surface quality is improved until it reaches 160 m/min.

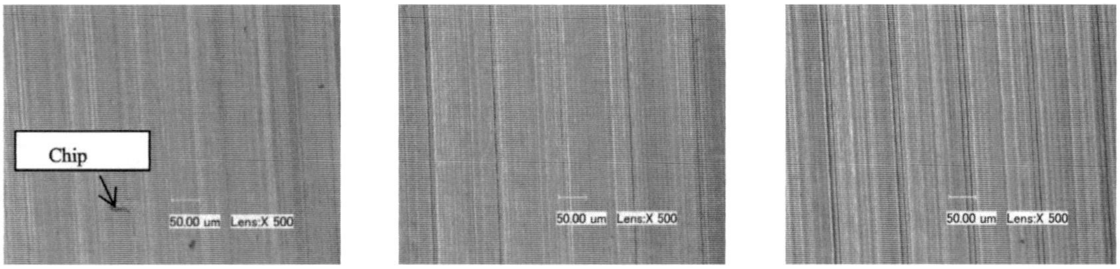

(a) v_c=40 m/min, f_z=0.12 mm/r, a_p=0.4 mm (b) v_c=80 m/min, f_z=0.12 mm/r, a_p=0.4 mm (c) v_c=120 m/min, f_z=0.12 mm/r, a_p=0.4 mm

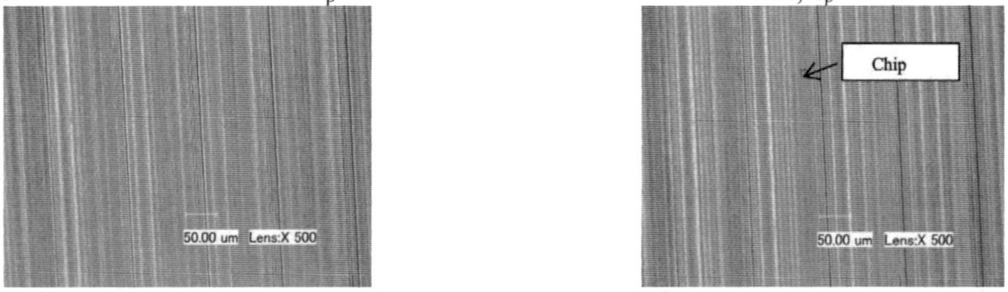

(d) v_c=160 m/min, f_z=0.12 mm/r, a_p=0.4 mm (e) v_c=200 m/min, f_z=0.12 mm/r, a_p=0.4 mm

Figure 4. Surface morphology of milling at different milling speeds.

4.3. Analysis of the Influence of the Depth of Cutting on the Date of Roughness and Surface Morphology

Figure 2(c) shows that the depth of cutting (a_p) has little influence on the date of roughness (Ra). When the cutting depth is between 0.2 mm and 0.6 mm, the surface roughness value is between 0.5 um-0.6 μm, and there is little change. It can also be seen from the surface topography in figure 5 that there is little difference in surface topography at different cutting depths. Because the cutting tool excitation frequency is far away from the natural frequency of the processing system during high-speed turning processing, although the cutting depth increases and the cutting force increases, the processing system is still relatively stable and can ensure better processing quality.

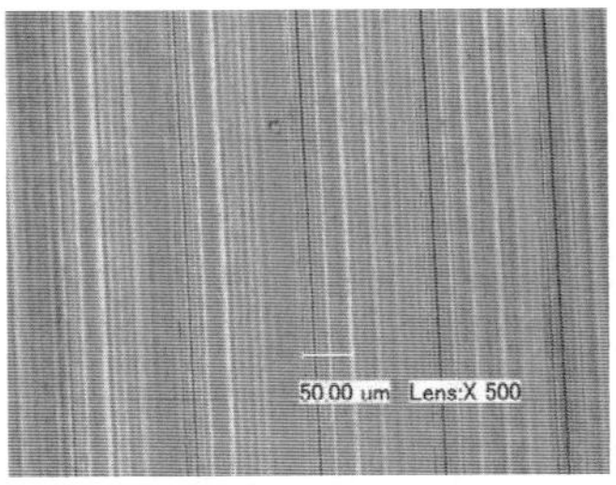

(a) v_c=120 m/min, f_z=0.12 mm/r, a_p=0.2 mm

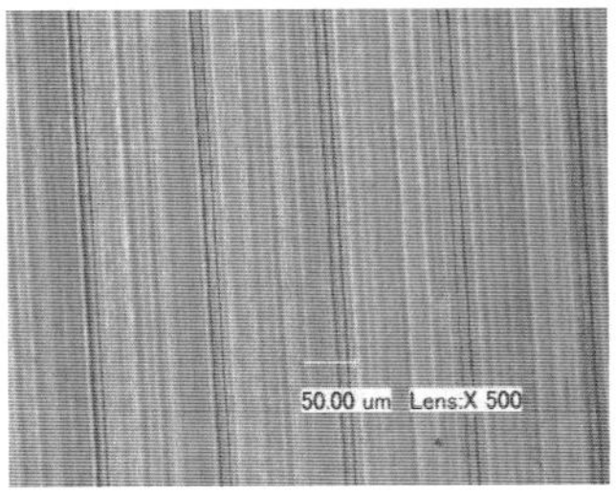

(b) v_c=120 m/min, f_z=0.12 mm/r, a_p=0.4 mm

(c) v_c=120 m/min, f_z=0.12 mm/r, a_p=0.6 mm

Figure 5. Surface morphology during milling at different milling depths.

5. Conclusions

Through the high-speed turning experiment of TC4 titanium alloy, the surface roughness value was collected, and the surface morphology was observed under the microscope. The following conclusions were drawn:

(1) Feed speed has a great impact on the date of the roughness of TC4 titanium alloy during high-speed cutting. When the feed of the tooth is less than 0.12 mm/r, the surface roughness value (Ra) can be maintained within 0.8 microns.

(2) When the cutting parameter is set to f_z=0.12 mm/rPM, a_p=0.4 mm, and the turning speed is 160 m/min, the surface roughness obtained is 0.497 microns. Experiments show that this set of parameters is the best combination.

(3) The experimental results show that the choice of the depth of cutting (a_p) has little influence on the date of roughness (Ra) of high speed turning Ti alloy, and has limited influence on the results.

Acknowledgment

This work is funded by Guangdong University of Science and Technology key project (GKY-2024KYZDK-9).

6. References

[1] Wang M H, Zhou D L, Liu N, Zhao Z Y and Shao C F 2017 Research on wear mechanism of titanium alloy TA15 milling tool *Manufacturing Automation* **39(12)** 1.

[2] Yuan G S, Chong L P and Han Y Y 2017 Application progress of titanium alloy *Hot Working Technology* **46(4)** 13.

[3] Wang Z H and Zhang C F 2022 Research on cutting force prediction in titanium alloy milling based on multi-layer fusion network *Manufacturing Technology and Machine Bed* **9** 90.

[4] Dalke A P, Karanjkar V A and Wankhede P 2024 Multi response optimization of process parameters in titanium alloy minimum quantity lubrication drilling operation *Engineering Research Express* **(4)** 045412.

[5] Burkov A A, Nikolenko V S, Krutikova O V, et al. 2024 Electrospark deposition of Ti-Ta coatings on Ti6Al4V Titanium alloy: oxidation resistance and wear properties *Physical Mesomechanics* **27(5)**618.

[6] Jiang Z H, Dong J C and Kong F Y 2021 Research status and development tendency of titanium alloy milling technology *New Technology &New Process* **9** 54.

[7] Sha Y D, Guan S X and Zhao H 2023 Prediction of cutting force of inner spiral groove by Support vector regression machine optimized by simulated annealing algorithm *Work with Technology* **57(01)** 100.

[8] Ye W J and Qian W 2022 Simulation and prediction of micro milling force of M2 high speed steel based on advantedge machining *Light Industrial Machinery* **40(05)** 81.

[9] Melkote S N, Grzesik W and Outeiro J 2017 Advances in material and friction data for

modelling of metal machining *CIRP Annals-Manufacturing Technology* **66(2)** 731.
[10] Chen Q, Zhang E G and Zhang S H 2017 Preparation and machinability of TiAIN, TiAISiN coatings *Surface Technology* **1** 118.

Simulation Study on the Operational Performance of Continuous Mixers Based on the Discrete Element Method (DEM)

Yaomeng Xiao[1], Zhaoyun Zhang[1], Zhongxiang Gao[1,*], Daixiang Zhang[1] and Guanguo Ma[1,2]

[1] Yankuang Energy Group Co., Ltd., Jining 272101, Shandong, China.
[2] College of Safety and Environmental Engineering, Shandong University of Science and Technology, Qingdao 266590, Shandong, China.
*Corresponding Author
Email: 1277643368@qq.com; 8388_000@163.com; 16118027289@qq.com; 2237782315@qq.com; ma156154682@126.com

Abstract. This study aims to systematically analyze the effects of various operational parameters on the continuous mixing process of concrete using the Discrete Element Method (DEM) combined with a wet mixing model, with the objective of optimizing mixing uniformity and material flow to improve concrete mixing quality and efficiency in practical engineering applications. The results indicate that an appropriate water addition rate (e.g., 10 m/s) and a moderate moisture content (e.g., 2:9) can significantly enhance the uniformity of concrete mixing. Additionally, a moderate rotational speed (e.g., 200 rpm) achieved good mixing effects in a short time, avoiding issues such as oversaturation and uneven mixing. This study provides theoretical support and a data foundation for optimizing concrete mixing processes, contributing to improvements in mixing quality and efficiency in engineering practice.
Keywords. Discrete Element Method (DEM), wet mixing model, concrete mixing, water addition rate, rotation speed.

1. Introduction

With the expansion of project scales and advancements in construction technology, the impact of concrete mixing processes on project quality has garnered increasing attention. The improvement of concrete uniformity is crucial for the durability and strength of building structures[1]. Continuous mixing processes offer the advantage of providing stable mixing quality in large-scale projects, and thus have received widespread attention in recent years[2].

The Discrete Element Method (DEM), as a numerical approach for simulating the dynamic behavior of particulate materials, has been widely applied in concrete mixing research. Krenzer et al.[3] and Schmelzle et al. [4] used DEM to study the effects of liquid bridge forces on particle separation during the wet mixing process, demonstrating that liquid bridges can effectively improve mixing uniformity. In recent years, Moreno-Juez et al. [5] and Gong et al. [6] analyzed the wear behavior of recycled concrete aggregates during mixing through experiments and simulations, verifying the accuracy and applicability of DEM simulations.Regarding the effects of water addition rate and moisture content on mixing performance, studies have shown that a moderate moisture content can significantly improve concrete uniformity, while excessive moisture can lead to localized oversaturation [7-9]. Additionally, the influence of particle shape on mixing performance has drawn attention; Saeed et al. [10] found that particles with complex shapes exhibited poorer mixing results. Three-dimensional simulation studies have also shown that rotational speed significantly impacts mixing uniformity [11]. In the latest research, Dai et al. [12] and Wang et al. [13] explored mixing performance under different rotational speeds and particle shapes, suggesting that an optimal rotational speed and suitable particle characteristics are essential for improving mixing results.

In summary, DEM shows great potential in optimizing concrete mixing processes, yet further research is needed to comprehensively study the effects of various operational parameters on the mixing process to achieve more efficient engineering applications.

2. Model Establishment

2.1. Wet Mixing Model

The wet mixing model, an extension of the Discrete Element Method (DEM), is designed to simulate particulate systems containing a liquid phase by accounting for liquid bridge forces between particles. This model enables a more accurate representation of wet particle behavior, which is essential in studying concrete mixing processes influenced by moisture. By simulating the formation and rupture of liquid bridges—primarily governed by capillary forces and liquid film pressure—the model captures particle interactions under different humidity conditions.

In this model, liquid bridge parameters such as surface tension and viscosity are defined, and numerical integration methods are applied to track particle motion and liquid forces dynamically. This approach allows for a detailed analysis of how moisture content and water addition rate affect mixing outcomes, providing valuable insights for optimizing concrete mixing processes.

2.2. Simulation Model

As shown in figure 1, the geometric model of the continuous mixer includes the feeding port, water inlet, mixing shaft, and discharge port. The mixing shaft is equipped with uniformly distributed mixing blades used for stirring the concrete. Regarding material properties, the physical characteristics of concrete particles, such as density, elastic modulus, and Poisson's ratio, are set based on experimental data. Particle parameters, including diameter and shape, are also defined to ensure the accuracy of the simulation model. The feeding port is used for material input, the water inlet controls the water addition rate, and the mixing shaft adjusts the rotational speed to achieve different mixing effects. Finally, the uniformly mixed concrete is discharged through the discharge port. This model allows for a systematic analysis of the impact of various operational parameters on the mixing process.

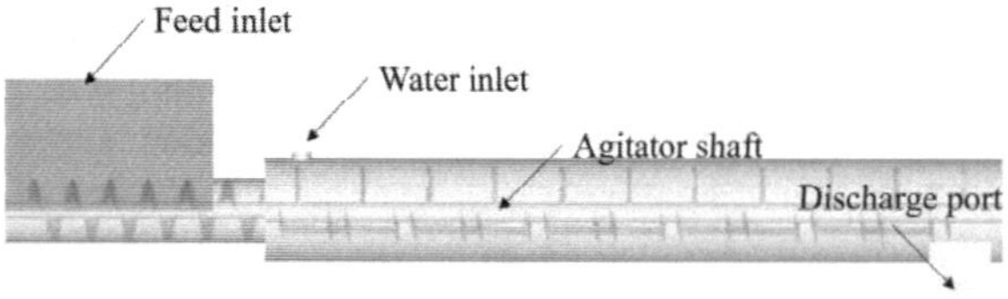

Figure 1. Geometric model establishment.

3. Discussion and Analysis

As shown in figure 2, the trajectories of concrete particles and the distribution of water saturation ratio inside the continuous mixer at different time points are displayed, with a rotational speed of 200 rpm, water addition rate of 1 m/s, and water-to-material ratio of 1:9. The figure illustrates the particle distribution and changes in water saturation at 2 s, 5 s, 10 s, and 15 s during the mixing process. Initially, dry particles at the feeding port gradually enter the mixer. Over time, water from the water inlet progressively enters and fully mixes with the particles. It can be observed that the rotation of the mixing shaft causes the particles to move axially along the mixer, showing different levels of mixing uniformity at various time points. At 2 s and 5 s, water is mainly concentrated in the inlet area of the mixer, indicating a lower level of mixing. As time progresses, at 10 s, the mixing of particles and water becomes more uniform, with a more even distribution of water saturation inside the mixer. Finally, at 15 s, the particles and water achieve a high level of mixing uniformity. This demonstrates that the rotation of the mixing shaft and the control of the water addition rate have a significant impact on the concrete mixing process. The dynamic analysis clearly shows the changes in particle trajectories and mixing uniformity inside the mixer under different operational parameters. By adjusting the rotational speed of the mixer and the water addition rate, the concrete mixing process can be optimized, improving mixing uniformity and construction quality. These results provide important theoretical support and data for further optimization of mixer design and operational parameters.

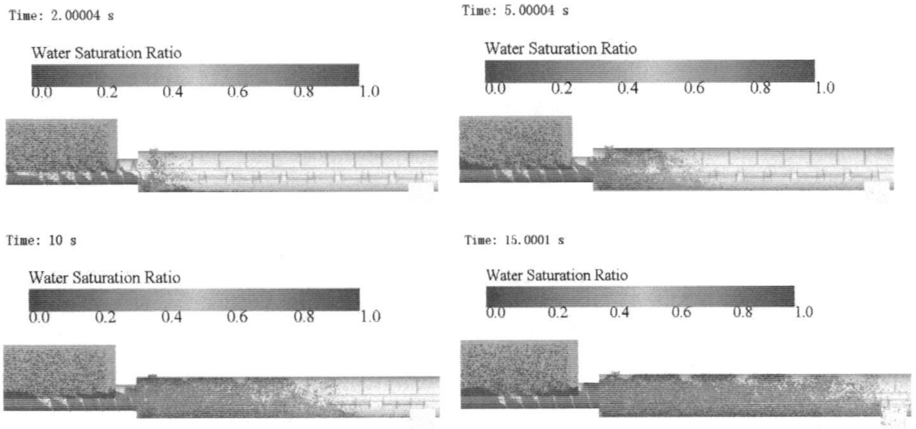

Figure 2. Dynamic analysis of the stirring process.

As shown in figure 3, the trajectories of concrete particles and the distribution of the water saturation ratio inside the continuous mixer at different water addition rates (1 m/s, 10 m/s, 20 m/s) are displayed. At a water addition rate of 1 m/s, the water distribution inside the mixer is relatively uniform, but the overall saturation is low. This indicates that a lower water addition rate may cause moisture to penetrate the concrete material slowly, thereby affecting the uniformity and efficiency of mixing. When the water addition rate is increased to 10 m/s, the water distribution inside the mixer significantly improves, with a higher overall saturation. At this point, the uniformity of mixing between concrete particles and moisture is significantly enhanced, suggesting that a moderate water addition rate can effectively promote the even distribution of moisture within the material, improving the mixing effect. However, when the water addition rate is further increased to 20 m/s, although moisture rapidly penetrates all parts of the mixer, localized oversaturation occurs at the inlet. This may be due to the excessively high water addition rate causing moisture to not mix thoroughly with the material promptly, leading to accumulation in the inlet area and affecting the overall mixing uniformity.

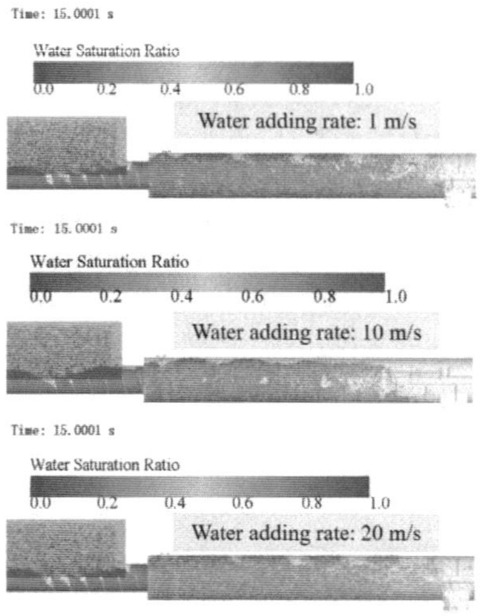

Figure 3. Stirring process at different water addition rates.

As shown in figure 4, the distribution of the movement trajectory and water saturation ratio of concrete particles in the continuous mixer under different moisture contents (1:9, 2:9, 1:3) is shown. When the moisture content is 1:9, that is, when each unit of material corresponds to 1 unit of moisture, the figure shows that the water saturation is lower than the overall ratio, especially in the middle and end of the mixer. This shows that lower moisture content is not enough to achieve a sufficient wet state of concrete particles, resulting in poor mixing uniformity. When the moisture content is increased to 2:9, the water saturation ratio in the mixer is significantly improved, and the moisture distribution in the material is more uniform. At this time, the mixing degree of concrete particles and moisture is significantly improved, indicating that moderate moisture content can effectively improve the mixing uniformity and mixing effect. However, when the moisture content was further increased to 1:3, water supersaturation occurred at the inlet of the mixer and in local areas. This may be due to the excessively high moisture content, which prevents the moisture from quickly and fully mixing with the particles. Instead, it forms accumulation in local areas, affecting the overall mixing effect and material fluidity.

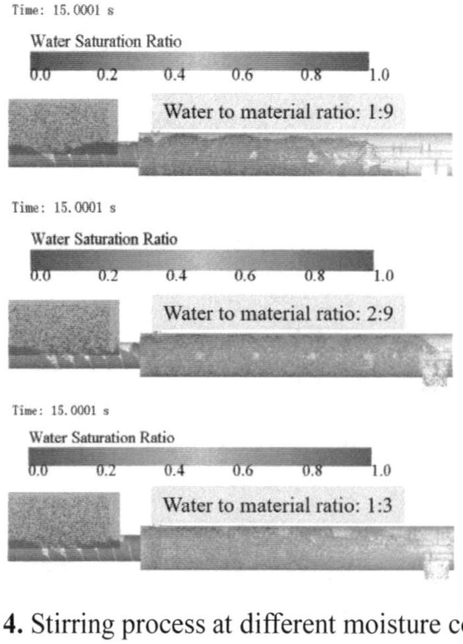

Figure 4. Stirring process at different moisture contents.

Journal of Physics: Conference Series **2941** (2025) 012013 doi:10.1088/1742-6596/2941/1/012013

As shown in figure 5, the trajectories of concrete particles and the distribution of the water saturation ratio inside the continuous mixer at different rotational speeds (150 rpm, 200 rpm, 250 rpm) are displayed. At a rotational speed of 150 rpm, the water saturation ratio inside the mixer is generally low, especially at the outlet, with moisture mainly concentrated in the inlet area. This indicates that a lower rotational speed results in insufficient mixing of particles and moisture, leading to poor mixing uniformity and an inability to effectively distribute moisture throughout the concrete material. When the rotational speed is increased to 200 rpm, the water saturation ratio inside the mixer significantly improves, with a more uniform distribution of moisture within the material. At this point, the mixing degree between concrete particles and moisture is significantly enhanced, suggesting that a moderate rotational speed can effectively promote the uniform mixing of particles and moisture, thereby improving the mixing effect. However, when the rotational speed is further increased to 250 rpm, although the water saturation ratio in most areas approaches 0.5, localized oversaturation occurs at the inlet and certain regions. This may be due to the excessively high rotational speed causing particles to move too quickly, preventing moisture from thoroughly mixing with the particles in a timely manner, leading to accumulation in localized areas and affecting the overall mixing uniformity and material flow.

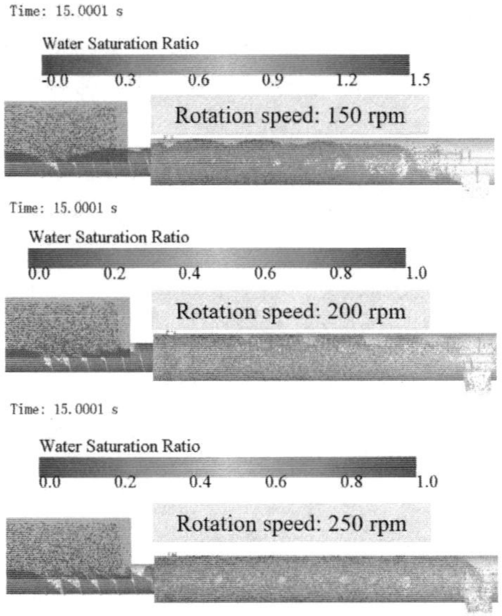

Figure 5. Stirring process at different speeds.

4. Conclusions

This study systematically analyzed the impact of different operational parameters on the continuous mixing process of concrete using the Discrete Element Method (DEM) and the wet mixing model. The results indicate that a moderate water addition rate (e.g., 10 m/s) and a medium moisture content (e.g., 2:9) can effectively enhance the mixing uniformity of concrete. Additionally, a moderate rotational speed (e.g., 200 rpm) achieved good mixing effects within a short time, avoiding issues of oversaturation and uneven mixing. These findings provide important theoretical support and data basis for optimizing concrete mixing processes, contributing to the improvement of mixing quality and efficiency in practical engineering applications.

5. References

[1] Mechtcherine V, Palzer U. "Simulating mixing processes of fresh concrete using the discrete element method (DEM) under consideration of water addition and changes in moisture distribution." *Cement and Concrete Research*, 2019, **115**, 274-282.

[2] Schmelzle S, Radel B. "Modelling of partially wet particles in DEM simulations of a solid mixing process." *Powder Technology*, 2018, **338**, 354-364.

[3] Moreno-Juez J, Tavares L M, Artoni R, et al. "Simulation of the attrition of recycled concrete aggregates during concrete mixing." *Materials,* 2021, **14(11),** 3007.
[4] Saeed M K, Siraj M S. "Mixing study of non-spherical particles using DEM." Powder Technology, 2019, **344,** 617-627.
[5] Gong S, Zuo Z, Xie G, et al. "Numerical simulation of wet particle flows in an intensive mixer." *Powder Technology*, 2019, **346**, 301-315.
[6] Krenzer K, Mechtcherine V. "Experimental and numerical study on the dispersion of fresh concrete in mixing processes using DEM." *Journal of Building Engineering*, 2020, **28**, 101071.
[7] Dai Q, Wang Y. "Effect of water-to-powder ratio on the dispersion of cement paste during mixing." *Construction and Building Materials*, 2022, **353**, 129052.
[8] Wang Z, Li Y, Chen S. "Impact of rotational speed on mixing homogeneity and efficiency of concrete mixers: A DEM study." *Journal of Materials Processing Technology*, 2023, **299**, 117363.
[9] Feng W, Gao H. "Enhanced DEM modeling for wet mixing processes in particulate materials." *Advanced Powder Technology,* 2020, **31**, 1457-1465.
[10] Li Z, Huang T. "Effects of particle shape on mixing performance in concrete mixers: A DEM study." *Powder Technology*, 2023, **418**, 118333.
[11] Zhang W, Liu Y. "DEM simulation of wet concrete mixing: Effects of rotational speed and particle shape." *Journal of Cleaner Production,* 2022, **356**, 131765.
[12] Dai Q, Lin X. "Influence of mixing parameters on concrete homogeneity using DEM analysis." *Construction and Building Materials*, 2022, **357**, 129322.
[13] Wang Y, Chen L. "Investigating the effects of particle properties on the mixing process of concrete using DEM." *Powder Technology*, 2021, **381**, 87-95.

Electrochemical Deposition of PEDOT/MXene Composite Films for Supercapacitor Electrodes

Shuhao Zheng, Jiaqiang Sun, Wenzhu Hu, Kaiqi Chen, Siwen Yang, Mengqian Li* and Haoran An*

School of Materials Science and Engineering, Hebei University of Science and Technology, Shijiazhuang 050018, Hebei, China.
Email: 734326990@qq.com; 2962582315@qq.com; 3319123805@qq.com; ckq0405@163.com; y18264577273@163.com; *2668575061@qq.com; *anhr@hebust.edu.cn

Abstract. PEDOT is a kind of electrode material for supercapacitors due to its excellent thermal stability, good film-forming properties, and electrical conductivity. However, PEDOT generates large volume changes during charging and discharging, resulting in structural collapse and hampering electrochemical performance. MXene, as a two-dimensional layered nanomaterial, is able to act as a support for PEDOT and reduce its volume change. In this study, PEDOT/$Ti_3C_2T_x$ composite films were prepared on substrates by electrochemical deposition using cyclic voltammetry method and used as electrode materials for supercapacitors. The effects of different molar ratios of EDOT to PSSNa and the number of CV cycles of growth on the performance of PEDOT were investigated. It was found that the best performance of the prepared PEDOT was achieved when the number of CV cycles was 40 and the ratio of EDOT: PSSNa was 1:3. Under optimum experimental conditions, the PEDOT/$Ti_3C_2T_x$ was obtained by in-situ polymerization and its specific capacity was up to 41.55 mF cm^{-2} at a current density of 1 mA cm^{-2}.

Keywords: MXene; PEDOT; Electrochemical deposition; Supercapacitor

1. Introduction

Supercapacitors have received a lot of attention from researchers because they have higher specific capacitance and power density than traditional batteries [1-3]. There are two mechanisms of supercapacitors, one is the double-layer capacitor and the other is the redox supercapacitor, where a rapid Faraday charge transfer reaction occurs on the electrode material [4].

Conductive polymer is a typical pseudocapacitive material. It has good electrical conductivity, excellent specific capacitance, and low cost [5,6]. Typical conducting polymers such as polypyrrole [7], polyaniline [8], polythiophene [9] etc. Poly (3,4-ethylenedioxythiophene) (PEDOT) is widely used in supercapacitors because of its excellent conductivity, wide potential range, and good thermal and chemical stability [10,11]. Patil et al. prepared PEDOT-PSS as an electrode material for supercapacitors, which can reach a specific capacity of 149 F g^{-1} at a current density of 0.5 mA cm^{-2} [12]. However, compared to other conductive polymers, PEDOT is insoluble and refractory, which requires the use of dispersant complexation such as poly (sodium p-styrene sulfonate) (PSSNa) to improve the solubility and dispersion of PEDOT.

MXene is considered an ideal electrode material for supercapacitors due to its unique two-dimensional layered structure, metal-like conductivity, and large specific surface area[13,14]. Lukatskaya et al. prepared porous MXenes and applied them in aqueous supercapacitors, achieving volume specific capacities of up to 1500 F cm^{-3} [15]. Compounding MXene with PEDOT reduces the effects of volume expansion of PEDOT through the supporting effect of MXene. Li et al. prepared $Ti_3C_2T_x$ /PEDOT:PSS hybrid films with a volume capacitance [16].

Therefore, in this paper, we explored the optimal process for the preparation of PEDOT by varying the ratio of PEDOT to PSS and the electrochemical parameters in cyclic voltammetry, and then synthesized PEDOT/$Ti_3C_2T_x$ by in-situ electrochemical polymerization. By analyzing the composition, morphology, and electrochemical data, we explored the effects of the cycle number of cyclic voltammetry in electrochemical deposition as well as the ratio of PEDOT to PSS on the structure and properties of PEDOT/$Ti_3C_2T_x$.

2. Experimental Section

2.1. Preparation of MXene

2 g of Ti_3AlC_2 powder was added to 20 mL of HF solution and reacted at 35°C for 24 h. The residual HF and soluble impurities were separated by centrifugation, and then dried to obtain the multi-layered $Ti_3C_2T_x$. The multi-layered $Ti_3C_2T_x$ was added to a hydrothermal reactor containing 30 mL of acetone and 1 g of ascorbic acid and heated at 120°C for 18 h. Afterward, the product was mixed with DMSO at 1:12, then mixed with deionized water and sonicated and centrifuged to obtain the supernatant which is the few-layer $Ti_3C_2T_x$.

2.2. Preparation of PEDOT

Disperse EDOT and PSSNa into deionized water according to different molar ratios of EDOT to PSSNa (2:1, 1:1, 1:2, 1:3, 1:4) and disperse well. The polymerization parameters were set: potential interval -0.5V-1.2V, scan rate of 50 mV·s^{-1} (10-50 cycles). The PEDOTs prepared with 10, 20, 30, 40 and 50 cycles were named CV-1, CV-2, CV-3, CV-4 and CV-5. The electrode sheets obtained with the ratios of EDOT to PSSNa were 2:1, 1:1, 1:2, 1:3, and 1:4 and were named EP-1, EP-2, EP-3, EP-4, and EP-5, respectively.

2.3. Page Numbers Preparation of PEDOT/Ti3C2Tx

A homogeneously dispersed mixed solution was obtained by sonication of PSSNa, EDOT, and few-layer $Ti_3C_2T_x$ proportionally with 20 mL of deionized water. The electrochemical parameters were set as pure PEDOT for polymerization.

2.4. Characterization

SEM was tested using S-4800 from HITACHI, Japan. The prepared electrode (size: 1*1 cm^2) was used as the working electrode, and the platinum sheet electrode and saturated Ag/AgCl electrode were used as the counter electrode and reference electrode, respectively. The electrochemical tests were carried out in 1 M H_2SO_4 electrolyte using a CHI660E electrochemical workstation. The voltage range of the CV test was -0.2-0.4 V with a scanning rate of 10 mV s^{-1}-1000 mV s^{-1}. The voltage range of the galvanostatic charge-discharge test was -0.2-0.4 V, and the current density was 1 A g^{-1}-100 A g^{-1}.

3. Results and Discussion

In order to investigate the effect of the number of scanning turns on the surface morphology of the PEDOT films, the films were viewed using SEM, figures 1(a) and 1(b) show the PEDOT films prepared at a CV cycle of 10 and 40 times, respectively. The PEDOT films prepared under 40 cycles had a more uniform surface and were denser than those prepared under 10 cycles. It indicates that the higher the number of cycles, the greater the PEDOT film loading. Figure 1(c) shows the SEM image of MXene. In $Ti_3C_2T_x$ the Al layer is selectively etched by hydrofluoric acid, and its microstructure shows an accordion-like layered structure with well-defined delamination and large, loose interlayer spacing. Figure 1(d) shows PEDOT/$Ti_3C_2T_x$ prepared with EDOT:PSSNa of 1:3. Compared to the dispersed structure of PEDOT, PEDOT/$Ti_3C_2T_x$ showed a continuous structure with a more uniform distribution. PEDOT was attached to the surface of $Ti_3C_2T_x$ with good contact, which was favorable for synergistic interactions and electron transport and provided more active sites.

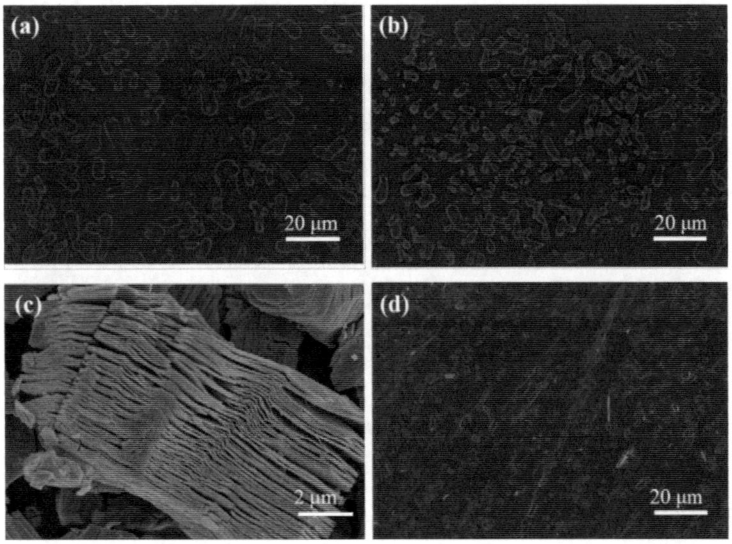

Figure 1. SEM images of PEDOT prepared by CV cycling for (a) 10 times, (b) 40 times, (c) SEM image of $Ti_3C_2T_x$, (d) SEM image of PEDOT/$Ti_3C_2T_x$.

Figure 2 shows the CV plot of CV-1, CV-2, CV-3, CV-4 and CV-5. The area of the CV curve of PEDOT increases gradually with the increase of the number of CV cycles and the circulation rate. The area of the curve reaches an extreme value at 40 times, indicating that the electrochemical performance of PEDOT is the best.

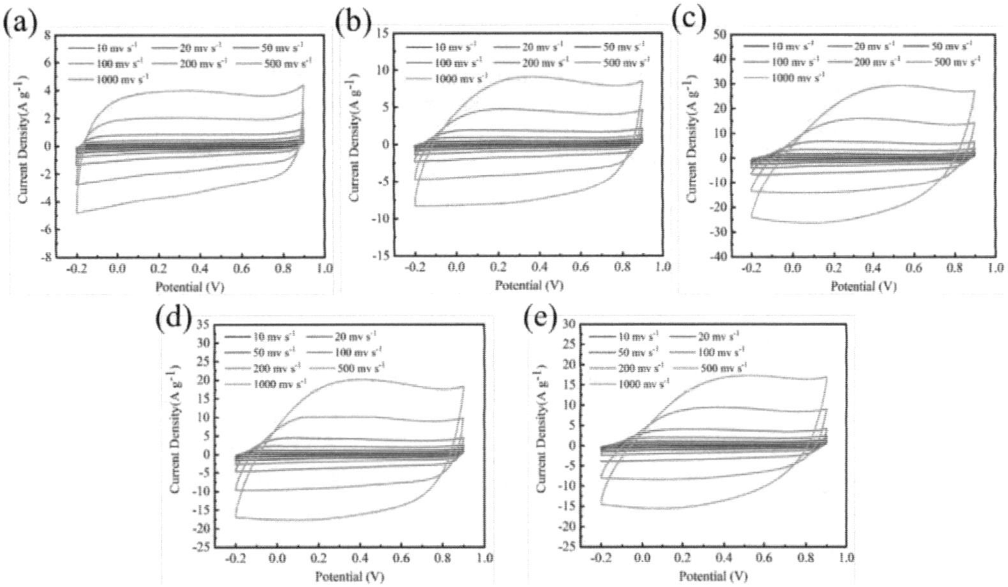

Figure 2. CV curves of (a) CV-1, (b) CV-2, (c) CV-3, (d) CV-4 and (e) CV-5.

Figure 3 shows the GCD curves of CV-1, CV-2, CV-3, CV-4, and CV-5 at current densities of 1, 2, 5, 10, 20, 50, and 100 mA cm^{-2}. The charge-discharge curves of all the samples were isosceles triangle-shaped, in which CV-4 had the longest discharge time. The capacities of CV-1, CV-2, CV-3, CV-4, and CV-5 are 3.85 mF cm^{-2}, 10.36 mF cm^{-2}, 15.825 mF cm^{-2}, 20.095 mF cm^{-2} and 18.195 mF cm^{-2}, respectively. This is due to that when the number of CV cycles is small, it has the least amount of PEDOT polymerized on the substrate surface, so the specific capacity is lower. As the number of CV cycles increases, more PEDOT is polymerized on the substrate surface, where the positively charged PEDOT polymerizes with the negatively charged PSS to form a homogeneous and dense film, thus increasing the charge storage capacity of the film. The specific capacity is maximized at 40 CV cycles. When the number of CV cycles is excessive, a large amount of PEDOT polymerizes on the substrate surface, reducing the electron mobility and thus the electrochemical properties of the film.

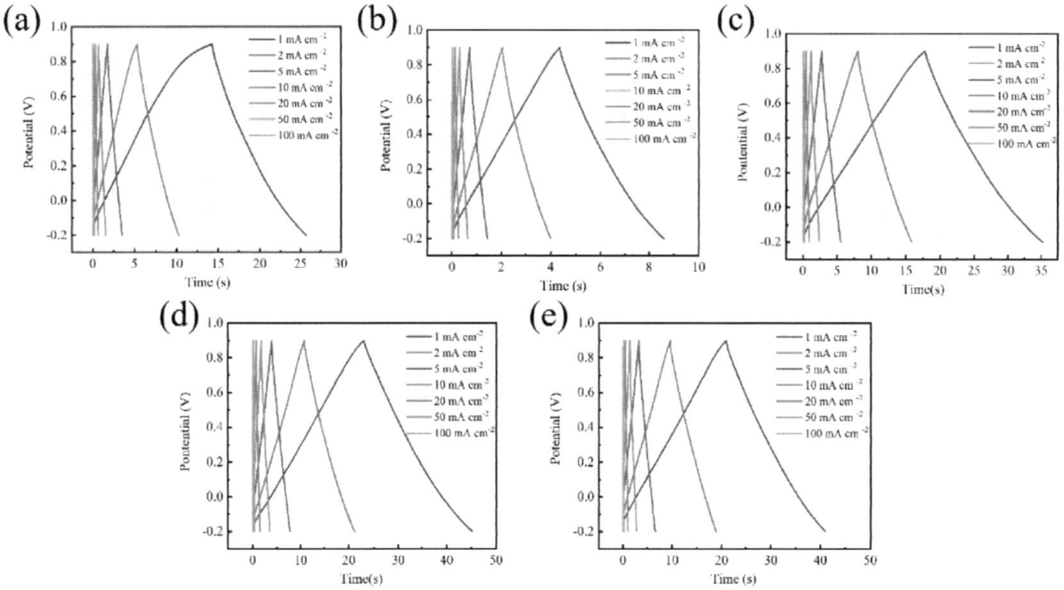

Figure 3. GCD plots of (a) CV-1, (b) CV-2, (c) CV-3, (d) CV-4 and (e) CV-5.

Figure 4 shows the CV curves of EP-1, EP-2, EP-3, EP-4, and EP-5 at different scan rates. It can be observed that the area of the CV curves increases as the scan rate increases. The largest CV area in EP-4 can be clearly observed in the figure.

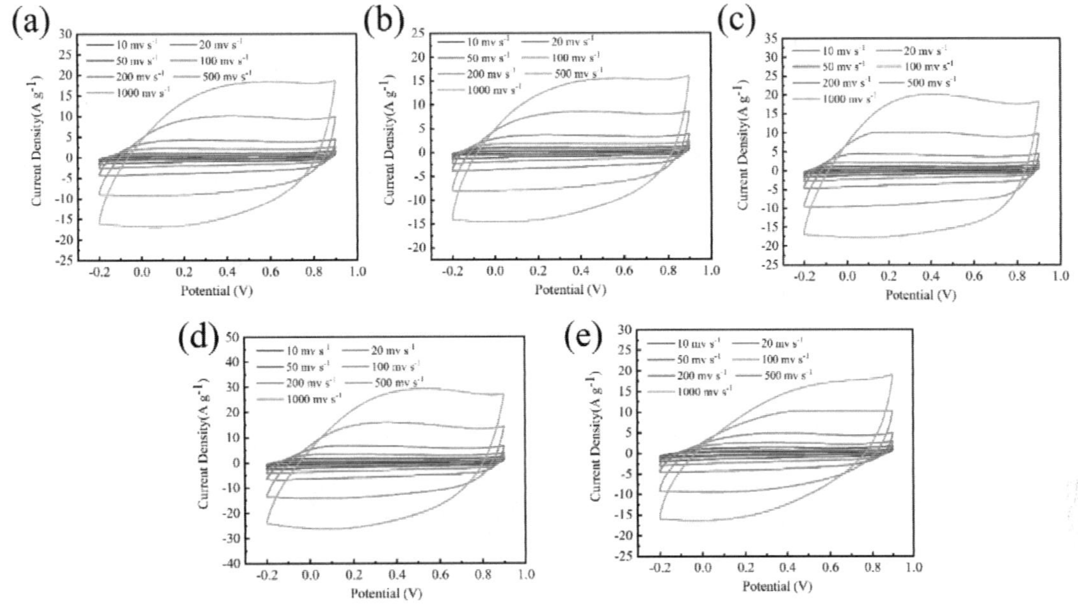

Figure 4. CV curves of (a) EP-1, (b) EP-2, (c) EP-3, (d) EP-4 and (e) EP-5.

Figure 5 shows the GCD curves of PEDOT prepared with different ratios of EDOT to PSSNa conditions at current densities of 1, 2, 5, 10, 20, 50, and 100 mA cm^{-2}. The specific capacitances of EP-1, EP-2, EP-3, EP-4, and EP-5 are 19.82 mF cm^{-2}, 16.91 mF cm^{-2}, 20.09 mF cm^{-2}, 37.37 mF cm^{-2} and 24.64 mF cm^{-2}, respectively. EP-4 has the longest discharge time and the GCD plots are isosceles triangular, this result is consistent with the results of the CV curve. The optimum process parameters for electrochemical deposition of PEDOT by cyclic voltammetry were derived from the different number of cycles and the ratio of EDOT to PSSNa was as follows: the number of CV curves was 40 times and the ratio of EDOT:PSSNa was 1:3.

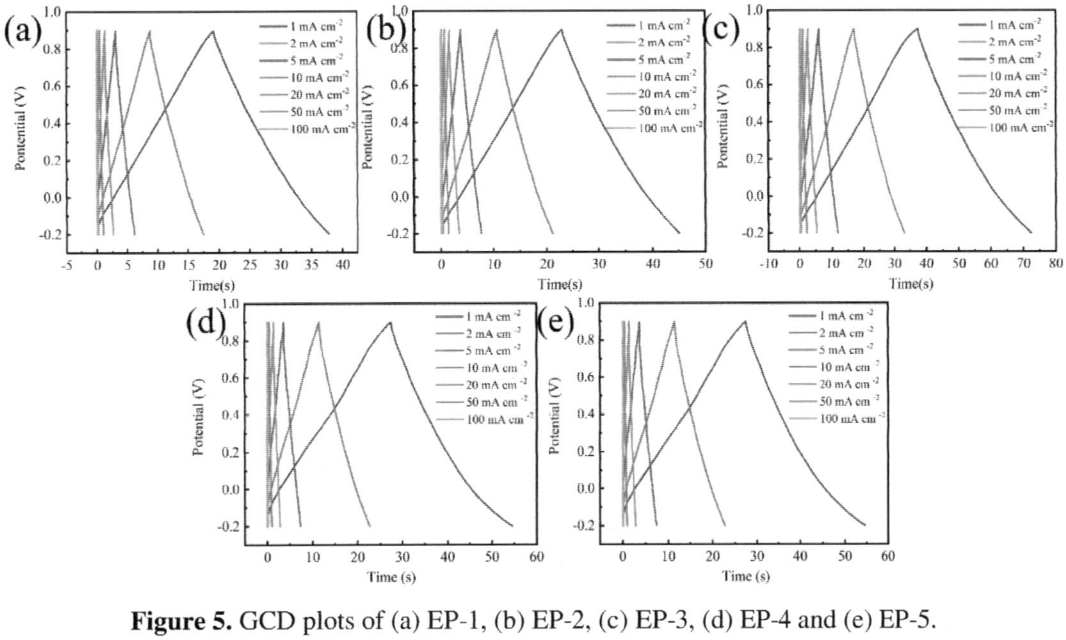

Figure 5. GCD plots of (a) EP-1, (b) EP-2, (c) EP-3, (d) EP-4 and (e) EP-5.

Figure 6(a) shows the GCD curves of PEDOT and PEDOT/$Ti_3C_2T_x$ at 1 mA cm^{-2} current density, and it can be clearly observed that the specific capacitance of PEDOT/$Ti_3C_2T_x$ is obviously higher than that of PEDOT. Figure 6(b) shows the CV curves of PEDOT/$Ti_3C_2T_x$, the curves are still rectangular at higher rates, which indicates that the composites have good rate properties. Figure 6(c) shows the GCD curves of PEDOT/$Ti_3C_2T_x$ at different current densities. The specific capacity is 41.55 mF cm^{-2} at a current density of 1 mA cm^{-2}. In this paper, PEDOT was polymerized in situ onto $Ti_3C_2T_x$ lamellae by cyclic voltammetry electrochemical deposition, and the $Ti_3C_2T_x$ not only acted as a linkage and support between PEDOT molecules, which exposed more active sites for PEDOT, but also as a conductive medium, which formed a multi-pathway electron transport channel, resulting in an enhancement of its electrochemical performance.

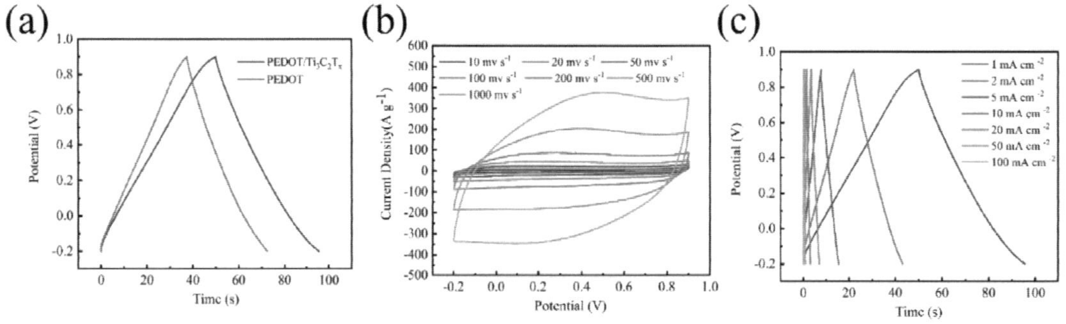

Figure 6. (a) GCD curve of PEDOT and PEDOT/$Ti_3C_2T_x$ at 1 A g^{-1} (b) CV curve of PEDOT/$Ti_3C_2T_x$ (c) GCD curve of PEDOT/$Ti_3C_2T_x$.

4. Conclusion

Herein, PEDOT was prepared by cyclic voltammetry electrochemical deposition, and the optimum reaction conditions for the preparation of PEDOT were discussed: the number of CV curves was 40 times, and EDOT:PSSNa was 1:3. Subsequently, PEDOT/$Ti_3C_2T_x$ composite films were prepared by in situ cyclic voltammetry electrochemical deposition with a specific capacity of 41.55 mF cm^{-2} at a current density of 1 mA cm^{-2}. The superior performance is attributed to the in situ polymerization of $Ti_3C_2T_x$ onto PEDOT, which acts as a support for PEDOT, exposing more active sites and also acts as a conductive medium, forming more ions transport channels, resulting in enhanced electrochemical performance.

5. References

[1] Jiang Q, Lei Y, Liang H, Xi K, Xia C and Alshareef N H 2020 Review of MXene electrochemical micro supercapacitors *Energy. Storage. Mater* **27** 78-95.

[2] Augustyn V, Simon P and Dunn B 2014 Pseudocapacitive oxide materials for high-rate electrochemical energy storage *Energ. Environ. Sci.* 1597-1614.

[3] Kumar N A and Baek J B 2014 Electrochemical supercapacitors from conducting polyaniline-graphene platforms *Chem, Commun.* **50** 6298-6308.

[4] Lota K, Khomenko V and Frackowiak E 2004 Capacitance properties of poly(3,4-ethylenedioxythiophene)/carbon nanotubes composites *J. Phys. Chem. Solids.* **65** 295-301.

[5] Ferraris J P, Eissa M M, Brotherston I D and Loveday D C 1998 Performance evaluation of poly(3-phenylthiophene) derivatives as active materials for electrochemical capacitors applications *Chem. Mater.* **10** 3528-3535.

[6] Tao C, Feng W, Yizhou Z, Lang L, Siya G, Xuanli Y, Shi W, Pengfei C and Wenyong L 2022 3D printable conductive polymer hydrogels with ultra-high conductivity and superior stretchability for free-standing elastic all-gel supercapacitors *Chem Eng J.* **450** 138311.

[7] Wu W, Wei D, Zhu J, Niu D, Wang F, Wang L, Yang L, Yang P and Wang C 2019 Enhanced

electrochemical performances of organ-like Ti_3C_2 MXenes/polypyrrole composites as supercapacitors electrode materials *Ceram. Int.* **45** 7328-7337.

[8] Wu W, Wang C, Zhao C, Wei D, Zhu J and Xu Y 2001 Facile strategy of hollow polyaniline nanotubes supported on Ti_3C_2-MXene nanosheets for High-performance symmetric supercapacitors *J. Colloid. Interf. Sci.* **580** 601-613.

[9] Goud G S, Park J H, Harpalsinh R, Kota M, Kim J H, Kim T, Gogotsi Y and Park H S 2019 MXene/polymer hybrid materials for flexible AC-filtering electrochemical capacitors *Joule.* **3** 164-176.

[10] Weikuan L, Juan C, Junjun Z, Jianrong Z and Junjie Z 2005 Application of ultrasonic irradiation in preparing conducting polymer as active materials for supercapacitor *Mater. Lett.* **59** 800-803.

[11] Chen J, Liu Y, Minett A I, Lynam C, Wang J and Wallace G 2007 Flexible, aligned carbon nanotube/conducting polymer electrodes for a lithium-ion battery *Chem. Mater.* **19** 3595-3597.

[12] Patil D S, Pawar S A, Patil P S, Kim J H and Shin J C 2016 Silver nanoparticles incorporated PEDOT-PSS electrodes for electrochemical supercapacitor *J Nanosci Nanotechno.* **16** 10625-10629.

[13] Yingyi M, Kaizheng X, Xiaoting L, Shun Y, Xinzheng L, Yu S and Xinran Li 2024 3D porous PEDOT/MXene scaffold toward high-performance supercapacitors *Chem Eng J.* **496** 154348-154348.

[14] Tao C, Xuan-Li Y, Sheng Y, Lang L, Zhong-Ting L, Jie Q, Chao-Fu M, Xiang-Chun L, Yi-Zhou Z and Wen-Yong L 2022 Flexible transparent bifunctional capacitive sensors with superior areal capacitance and sensing capability based on PEDOT:PSS/MXene/Ag grid hybrid electrodes *Adv Funct Mater.* **33** 10997.

[15] Lukatskaya M R, Kota S, Lin Z, Zhao M Q, Shipgel N, Levi M D, Halim J, Taberna P, Barsoum M W, Simon P and Gogotsi Y 2023 Ultra-high-rate pseudocapacitive energy storage in two-dimensional transition metal carbides *Nat. Energy.* 723-743.

[16] Li L, Zhang N, Zhang M, Zhang X and Zhang Z 2019 Flexible $Ti_3C_2T_x$/PEDOT:PSS films with outstanding volumetric capacitance for asymmetric supercapacitors *Dalton Transactions.* **48** 1747-1756.

Study on the Corrosion Failure Mechanism and Protective Materials of AC Charging Piles in Tropical Island Environment

Songling Pang[1], Jun Wang[2], Zhi Li[2,*] and Meiyi Huo[1]

[1] Hainan Power Grid Co., Ltd, Haikou, Hainan, China.
[2] State Key Laboratory of Environmental Adaptability for Industrial Products, China National Electric Apparatus Research Institute Co., Ltd, Guangzhou, Guangdong, China.
Email: *lizhi@cei1958.com

Abstract. This paper first analyses the corrosion of AC charging piles in tropical island environments in detail through field investigation and laboratory simulation. The survey results show that the corrosion of the charging pile shell and internal connectors is the most serious, especially in the area close to the coastline, where the corrosion rate is more than 3 times higher than that in inland areas. The laboratory test uses an artificial seawater environment to simulate the actual island environment and study the corrosion resistance of different materials. Through the electrochemical impedance spectroscopy (EIS) test, this paper analyses the electrochemical behavior of different materials in the simulated environment. The results show that the corrosion rates of nano-silver-coated steel and stainless steel are 0.12 mm and 0.38 mm per year, respectively. This paper introduces finite element analysis (FEA) to simulate and analyze the stress distribution and fatigue life of different protective materials in a corrosive environment. The results show that after three years of service, the unprotected stainless-steel material has obvious cracks in the structural stress concentration area, while the stainless steel with nano-silver coating effectively extends the service life of the material, and the cracks are delayed until six years after service. In addition, FEA analysis found that the uniformity and adhesion of the surface coating are the key factors affecting the protective effect. If the coating thickness is uneven or there is a problem of insufficient adhesion, corrosion points are likely to form at the microscopic level, eventually leading to local failure of the material.
Keywords. Tropical island environment; AC charging pile corrosion; electrochemical impedance spectroscopy (EIS) modeling; failure mechanism; protection.

1. Introduction

In tropical island environments, the corrosion problem of AC charging piles has become an important factor restricting their long-term use. Charging piles are exposed to extreme environments such as high humidity, high salinity, and high temperature for a long time, which can easily lead to material corrosion, thereby affecting the safety and service life of charging piles. Studying the corrosion failure mechanisms of AC charging piles in tropical island environments and their protective materials is of considerable practical importance. Recently, both domestic and international researchers have extensively explored corrosion issues associated with power equipment. Reference [1] introduced a research approach focused on surface coatings for power equipment, addressing the issue of coating detachment in high-humidity conditions. Reference [2] investigated the corrosion behavior of metallic materials in marine environments and proposed an innovative alloy design to enhance material corrosion resistance, thereby significantly prolonging the service life of the equipment. Reference [3] developed a self-healing protective coating, effectively mitigating accelerated corrosion due to microcracks through the material's intrinsic repair capability. Reference [4] systematically analyzed the impact of various environmental factors on the corrosion rate of metallic materials, proposed relevant protection strategies, and achieved a substantial reduction in equipment maintenance costs.

Reference [5] proposed a method of using corrosion-resistant alloy materials in view of the working environment of charging piles, which solved the safety hazard caused by the corrosion of charging piles. Reference [6] studied the relationship between environmental parameters and the corrosion rate of charging piles based on the characteristics of the tropical island environment and optimized the protection scheme of charging piles by combining an approximate dynamic programming algorithm and an adaptive weighted sum algorithm. Reference [7] studied the performance of different protective materials in high-salinity environments through simulation and verified the effectiveness of new materials in extending the service life of charging piles.

While numerous studies have offered solutions to the corrosion challenges faced by power equipment and charging piles, there remains a significant gap in systematic research specifically addressing the corrosion failure mechanisms and protective materials of AC charging piles within tropical island environments. This research seeks to thoroughly investigate the corrosion mechanisms affecting AC charging piles under the unique conditions of tropical islands, to identify and analyze the primary factors contributing to corrosion failure, and to subsequently evaluate and refine protective materials, aiming to enhance the longevity and durability of charging piles in these environments.

This study will use an approximate dynamic programming algorithm combined with an adaptive weighted sum algorithm to evaluate and optimize the performance of different protective materials through system simulation [8]. First, this paper will analyze in detail the corrosion mechanism of AC charging piles in tropical island environments and identify key corrosion factors. Secondly, by establishing a system simulation model, this paper will explore the corrosion behavior of different protective materials under complex environmental conditions, and optimize material selection and protection strategies in combination with an approximate dynamic programming algorithm. Finally, this paper will experimentally verify the optimized protective materials in order to provide a theoretical basis and practical guidance for the corrosion-resistant design of AC charging piles.

2. Research Theory and Algorithm Selection

In tropical island environments, AC charging piles face severe corrosion challenges. In order to effectively analyze and predict the performance of protective materials in such extreme environments, it is crucial to select appropriate research theories and algorithms [9]. This section will discuss the main theories and models used to analyze corrosion failures and introduce how to use these methods to simulate and predict the performance of different protective materials.

2.1. Electrochemical Impedance Spectroscopy (EIS)

Electrochemical impedance spectroscopy (EIS) is a highly effective method for analyzing the electrochemical properties of materials. By evaluating the system's impedance response across a range of frequencies, EIS provides critical insights into the charge transfer processes occurring at the material's surface and the kinetics of the associated electrochemical reactions [10]. This technique is particularly valuable in corrosion studies, where it facilitates the determination of the material's corrosion rate, the stability of surface films, and the integrity of protective coatings. Equation (1) illustrates the commonly employed equivalent circuit model in EIS, which typically consists of a combination of resistance (R), capacitance (C), and inductance (L) elements.

$$Z(\omega) = R_s + \frac{R_p}{1+j\omega R_p C_p} + j\omega L \tag{1}$$

$Z(\omega)$ is the impedance of the system, R_s is the solution resistance, R_p is the polarization resistance, C_p is the electric double layer capacitance, L is the inductance, and ω is the angular frequency. By fitting the experimental data to the model, key parameters can be obtained and the corrosion behavior of the material can be further analyzed.

2.2. Finite Element Analysis (FEA)

Finite Element Analysis (FEA) is a numerical computational technique extensively utilized for the analysis of stress and strain within the engineering domain. In the context of corrosion studies, FEA proves invaluable for simulating the stress distribution and evaluating the mechanical properties of protective materials when subjected to specific environmental conditions. Furthermore, it enables the prediction of the structural integrity of materials experiencing corrosion, providing insights into their durability and potential failure points under corrosive influences. By establishing a physical model of the charging pile and the environment, FEA can accurately simulate the performance of different protective materials under high temperature, high humidity, high salinity, and other island environments. Equation (2) shows the common discretization equations in FEA:

$$\mathrm{Ku} = \mathrm{f} \tag{2}$$

Where K is the stiffness matrix, u is the displacement vector, and f is the external load vector. By solving the above equations, the deformation and stress distribution of the material under the corrosive environment can be obtained, which provides a basis for the selection of protective materials.

3. Experimental Design and Methods

3.1. Experimental Device and Test Conditions

In order to study the corrosion failure mechanism of AC charging piles and the performance of protective materials in tropical island environments, this paper designs a set of experimental devices simulating island environments. The device includes a salt spray chamber, a high humidity constant temperature and humidity chamber, and a temperature cycle chamber to simulate the corrosion conditions experienced by charging piles in long-term exposure to island environments [11]. The experimental temperature is set to 35°C, the humidity is 95%, the salt spray concentration is 5%, and the test cycle is 60 days. In terms of sample preparation, this paper selects several commonly used protective materials on the market, including bare nanosilver particles and nanosilver particles coated with alumina. These materials are pretreated to ensure their surface cleanliness and are evenly covered on the surface of the metal substrate by spraying. The samples are then installed in the experimental device for long-term exposure tests to simulate the corrosion conditions in real island environments.

3.2. Experimental Steps of Electrochemical Impedance Spectroscopy (EIS)

The Electrochemical Impedance Spectroscopy (EIS) test is a crucial method for assessing the corrosion resistance of materials. In this investigation, a three-electrode configuration was employed, where the working electrode consisted of the protective material sample under evaluation, the reference electrode was a saturated calomel electrode, and the counter electrode was made of platinum [12-13]. The experiment was conducted under ambient conditions, with the frequency spectrum ranging from 100 kHz to 0.01 Hz, and the applied voltage amplitude set at 10 mV.

The experimental steps are as follows: 1) Soak the sample in 3.5% NaCl solution to simulate the seawater environment. 2) Connect the working electrode, reference electrode, and auxiliary electrode. 3) Gradually reduce the frequency, starting from high frequency and gradually scanning to low frequency. 4) Record impedance data to obtain Nyquist diagrams and Bode diagrams. In terms of data processing, this paper adopts the classic equivalent circuit model to obtain corrosion impedance parameters by fitting experimental data.

$$Z(\omega) = R_s + \frac{R_p}{1+\left(j\omega C_p\right)^n} \tag{3}$$

R_s is the solution resistance, R_p is the polarization resistance, C_p is the capacitance, n is the phase angle factor, and ω is the angular frequency.

3.3. Data Analysis Method and Calculation

This paper uses the nonlinear least squares method for data fitting and realizes automatic data processing through programming. This paper also uses the Fourier transform (FFT) to perform frequency domain analysis on time domain signals to enhance the resolution of signals [14-15]. In addition, this paper uses the finite element analysis (FEA) method to simulate the stress distribution of materials during corrosion. The calculation formula is as follows:

$$\sigma = \frac{F}{A} + \alpha\Delta T \tag{4}$$

σ is stress, F is external force, A is cross-sectional area, α is thermal expansion coefficient, andΔT is temperature change.

This paper also uses an adaptive filtering algorithm to remove noise from experimental data to improve data accuracy. Its basic expression is as follows:

$$y(t) = x(t) + n(t) \tag{5}$$

$y(t)$ is the observed signal, $x(t)$ is the real signal, and $n(t)$ is the noise.

4. Results and Discussion

4.1. Experimental Data and Analysis

This paper conducted corrosion tests on a variety of protective materials in a simulated tropical island environment to evaluate their corrosion resistance in harsh environments. The following experimental data were obtained through salt spray tests, high humidity exposure, and temperature cycle tests (table 1).

Table 1. Mass loss of different protective materials in salt spray tests (mg).

Materials	24 hours	48 hours	72 hours	Average corrosion rate (mg/hour)
Nanosilver particles coated with aluminum oxide	2.5	5.1	7.8	0.108
Materials	1.9	3.8	5.6	0.078

From the data in table 1, it can be seen that the nanosilver particles coated with aluminum oxide show lower mass loss in the salt spray test, indicating that they have stronger corrosion resistance. In addition, the electrochemical impedance spectroscopy (EIS) test results also show that the corrosion rate of nanosilver particles coated with aluminum oxide is significantly lower than that of bare nanosilver particles.

4.2. Finite Element Analysis and Material Failure

To gain deeper insights into the impact of corrosion on the mechanical properties of materials, this study utilizes Finite Element Analysis (FEA) to simulate the stress distribution under corrosive conditions. The simulation outcomes indicate that corrosion-induced mechanical stress tends to concentrate in specific critical regions of the material. Moreover, the stress in these regions progressively intensifies over time, which could eventually result in material failure.

Table 2. Maximum stress values (MPa) of different protective materials under corrosion conditions.

Materials	Initial state	30 days later	60 days later	Stress growth rate (%)
Nanosilver particles coated with aluminum oxide	150	185	220	46.7
Materials	140	170	195	39.3

The results of finite element analysis show that the stress concentration area of bare nanosilver particles is large, and the stress value increases significantly with the increase of corrosion time, indicating that its corrosion resistance is poor, as shown in Table 2. Simulation curve 1 shows the stress distribution of bare nanosilver particles and nanosilver particles coated with alumina after corrosion, which further verifies this conclusion (figure 1). In contrast, nano silver particles coated with alumina exhibit better corrosion resistance under the same corrosion conditions. Finite element analysis shows that the stress concentration of nanosilver particles coated with alumina is low, and the stress growth rate over time is also relatively mild. After 30 days of corrosion, the stress value of nanosilver particles coated with alumina increased from the initial 140 MPa to 170 MPa, and the stress value was 195 MPa after 60 days of corrosion. In contrast, the stress increase of nanosilver particles coated with alumina is significantly smaller than that of bare nanosilver particles, indicating that the alumina coating has an effective protective effect on nanosilver particles and slows down the effect of corrosion on the mechanical properties of the material.

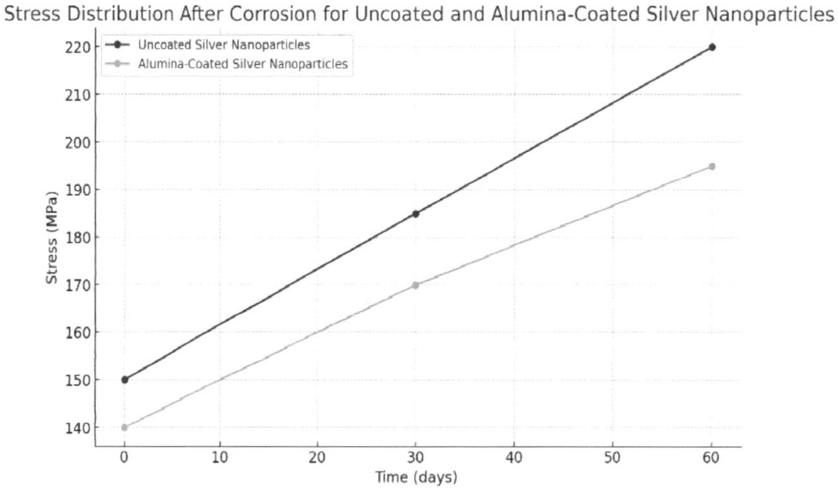

Figure 1. Stress distribution of bare silver nanoparticles and silver nanoparticles coated with alumina after corrosion.

In addition, through fatigue tests on materials after corrosion, this paper found that the failure cycle of bare silver nanoparticles was significantly shorter than that of silver nanoparticles coated with alumina, which is consistent with their higher stress concentration. Figure 2 shows the fatigue life comparison of different materials after corrosion. The fatigue life of bare silver nanoparticles in a corrosive environment is significantly shortened and can only withstand a small number of fatigue cycles, which is closely related to their higher stress concentration and rapid stress growth. The fatigue life of silver nanoparticles coated with alumina is significantly extended and can withstand more fatigue cycles, further proving its superiority in corrosion resistance and delaying fatigue failure.

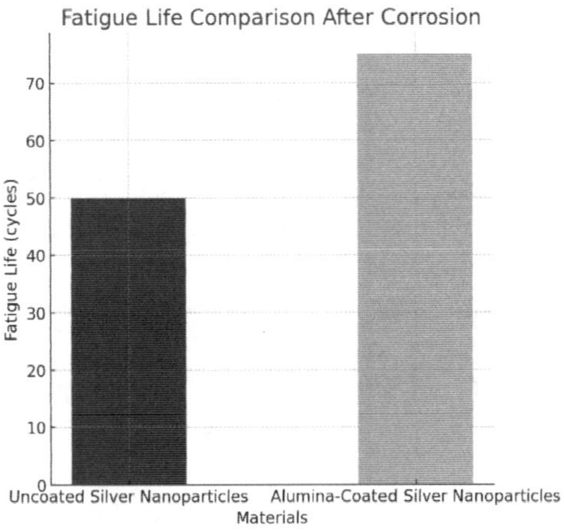

Figure 2. Comparison of fatigue life of different materials after corrosion.

4.3. Performance of Protective Materials in Island Environment

Combining experimental data and finite element analysis results, the performance of alumina-coated nanosilver particles is better than that of bare nanosilver particles in a simulated tropical island environment. It has less mass loss in a corrosive environment and lower mechanical stress concentration, showing better durability and corrosion resistance. This paper conducted a long-term exposure test and measured the changes in the mechanical properties of the material at different time points. Figure 3 shows the stress changes of alumina-coated nanosilver particles under long-term exposure. Alumina-coated nanosilver particles show better corrosion resistance in long-term exposure tests. Specifically, under long-term exposure, the stress growth of alumina-coated nanosilver particles is relatively slow, and there are fewer mechanical stress concentration areas, indicating that it has better durability in tropical island environments. In contrast, the stress value of bare nanosilver particles under the same conditions grows more rapidly, indicating that they are prone to failure in such corrosive environments.

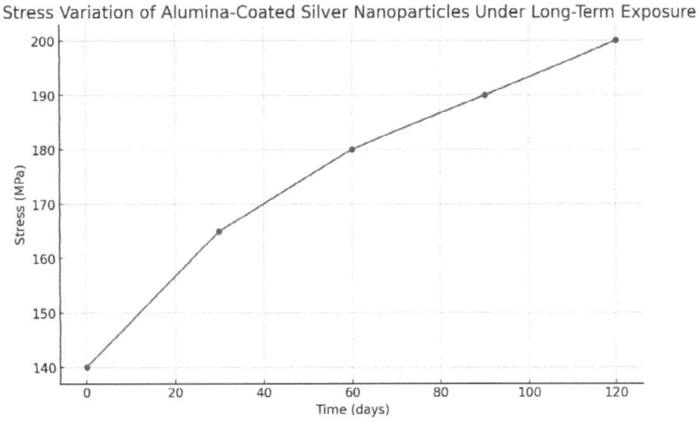

Figure 3. Stress changes of alumina-coated nanosilver particles under long-term exposure.

The performance of alumina-coated nanosilver particles in a corrosive environment is significantly better than that of bare nanosilver particles. It not only effectively slows down the expansion of stress concentration areas, but also significantly improves the fatigue life and overall corrosion resistance of the material. These research results provide an important reference for material selection under harsh environmental conditions and further verify the key role of surface coatings in improving the performance of nanomaterials.

5. Conclusion

This study simulates the tropical island environment and offers a comprehensive analysis of the corrosion failure mechanisms affecting AC charging piles under these challenging conditions. The investigation revealed that factors such as salt spray, high humidity, and temperature cycling exert a substantial influence on the corrosion behavior of the materials used in charging piles. Through the application of electrochemical impedance spectroscopy (EIS) technology, a detailed corrosion model of the material was established, enabling a deeper examination of its failure mechanisms. The experimental data demonstrated significant variations in corrosion resistance among different protective materials in the island environment. Notably, alumina-coated nanosilver particles exhibited a lower corrosion rate in the EIS tests and demonstrated superior mechanical stability during long-term exposure trials. Furthermore, the results from finite element analysis confirmed that stress concentration in alumina-coated nanosilver particles was minimal during the corrosion process, indicating their strong corrosion resistance. In the context of tropical island environments, selecting appropriate protective materials is essential for extending the operational lifespan of AC charging piles. Due to their outstanding protective characteristics, alumina-coated nanosilver particles are recommended as the optimal material choice for AC charging piles in such environments. This research provides a scientific foundation for the selection of materials and protective designs for charging piles and establishes a basis for future investigations.

Acknowledgments

This work was supported by the fund: Hainan Province Science and Technology Special Fund (ZDYF2021GXJS013).

6. References

[1] Yao Y C, Chen S X and Yang J 2022 Research and analysis on key technologies of high-speed railway in tropical marine environment *Journal of Railway Engineering Society* **39(4)** 11-16.

[2] Guo Q 2023 Corrosion law, protection design and control measures of vehicle equipment in tropical island and reef atmospheric environment *Equipment Environmental Engineering* **20(11)** 121-132.

[3] Luo L, Ye G Y, Guan Z, et al. 2022 Changes in vegetation coverage and landscape pattern in Jinchuan mining area under ecological restoration *Journal of Subtropical Resources and Environment* **17(1)** 8-34.

[4] Liu C, Li S B, Li X, et al. 2023 Research on monitoring of atmospheric corrosion environmental factors and application model of environmental data for a certain product *Equipment Environmental Engineering* **20(8)** 136-141.

[5] Jia J H, Liu M, Luo C, et al. 2023 Corrosion aging behavior of typical protective coating of 2A97 aluminum-lithium alloy in tropical marine atmospheric environment *Journal of Corrosion and Protection* **43(1)** 143-151.

[6] Hu J Z, Lan W J, Deng P C, et al. 2023 Study on the initial corrosion behavior of E690 steel in tropical marine atmosphere *Journal of the Chinese Society for Corrosion and Protection* **43(5)** 1140-1144.

[7] Li Y, Ruan X and Zhang S J 2023 Microscopic study on carbonization of bridge concrete anchors in subtropical hot and humid environment *Journal of Xi'an University of Architecture and Technology: Natural Science Edition* **55(4)** 504-514.

[8] Liu Q, Yu D Z, Qiao S R, et al. 2022 Protective performance of corrosion inhibitors on printed circuit boards in tropical marine atmosphere *Journal of Naval Aviation University* **37(6)** 495-500.

[9] Tong C H, Wang H, Tan S, et al. 2022 Effect of fruit plantation on the stability of red soil aggregates in subtropical hilly areas *Chinese Journal of Applied Ecology* **33(4)** 1012-1020.

[10] Yang H F, Yuan Z Z, Li J, et al. 2023 Effect of Ni content on corrosion behavior of copper-aged easy-to-weld steel in simulated tropical marine atmosphere *Journal of Corrosion and Protection* **43(5)** 1022-1030.

[11] Yang L H, Xu W C, Wang X T, et al. 2023 Application of oxidative polymerization coating technology (OTC) in bolts in tropical marine atmosphere *Equipment Environmental Engineering* **20(3)** 62-68.

[12] Yang Q Q and Du Z H 2023 Ecological stoichiometric characteristics of leaves and soils of tree species from different origins in tropical coastal shelterbelts *Bulletin of Soil and Water Conservation* **43(2)** 156-163.

[13] Ogawa F, Itoh T, Shimomura H, et al. 2023 Estimation of Fatigue Limit and Strength Degradation Mechanism of a High Strength Steel in Petrol Environment by Means of a Small Sized Specimen with Radial Notch. *Journal of the Society of Materials Science*, **23**.

[14] Altunbas A, Saygili O, Kelesoglu M K, et al. 2023 Numerical analysis of failure mechanism observed in backfills supported by masonry walls. *Journal of Corrosion and Protection* **43(3)**.

[15] Fojt J, Alferi D, Hybasek V, et al. 2023 Corrosion failure of nitinol stents in the upper gastrointestinal tract: The role of surface finishes and the importance of an appropriate test environment. *Materials Chemistry and Physics*, **309**.

Structural Diagnostic Study of Electrode Materials for Power and Energy Storage Lithium-Ion Batteries

Lu Li[1], Jianquan Liang[1,*], Changsong Dai[2], Lei Wang[1], Jian Zhang[1], Hang Zhang[1], Kexin Zhang[1], Yue Wang[1] and Zhongyuan Li[1]

[1] State Grid Heilongjiang Electric Power Co., Ltd., Electric Power Research Institute, Harbin, Heilongjiang, China.
[2] School of Chemical Engineering and Chemistry, Harbin Institute of Technology, Harbin, Heilongjiang, China.
Email: *ljq_hit@163.com

Abstract. As technology advances, the demand for electric vehicles and energy storage continues to expand, and the production of lithium-ion batteries has grown rapidly in recent years. When the battery enters the end-of-life stage, the recycling of its rare metals such as lithium, cobalt, nickel, and manganese can also alleviate the crisis of rare mineral resources and create huge economic value. The current recycling process for used batteries is based on destructive methods, which are costly and harmful to the environment. Therefore, in recent years, research has focused more on the short-range recovery of active substances in an attempt to repair the compositional and structural defects of the substance particles. This paper carries out research on lithium iron phosphate and ternary cathode materials for lithium-ion batteries with different cycle lifetimes, establishes the correspondence law between the cycle life of the battery and lithium loss, and explores the influence of lithium loss on the structure of the material, the electronic and ionic conductivity, and the structural stability of the law. The results show that the increase in the number of cycles of lithium loss aggravated, which in turn led to the deterioration of the structural stability of the material, which is the main reason for the decay of the crystal structure and performance of lithium iron phosphate and ternary materials decay, to realize the short-range and effective recycling and reuse of electrode materials to provide technical support.
Keywords. Lithium-ion battery; lithium iron phosphate anode materials; ternary anode materials; structural diagnostics.

1. Introduction

The importance of lithium-ion batteries (LIB) is being recognized as “green” power generation technologies, such as solar and wind, that face problems such as varying power quality and difficulties in connecting to the grid. Lithium-ion batteries offer outstanding advantages over other similar technologies, such as high energy density, long cycle life, and environmental friendliness. In addition, lithium-ion batteries have shown unlimited potential for electrochemical energy storage and conversion, driving a new revolution in portable electronic devices (e.g., cell phones, laptops, iPads, etc.), and contributing to the rapid development of new energy vehicles, where the annual value of lithium-ion batteries for electric vehicles is expected to be in the tens of billions to hundreds of billions of dollars, with a promising market outlook [1]. Among the developed materials for lithium-ion batteries, ternary cathode materials ($LiNi_xCo_yMn_{1-x-y}O_2$) have the advantages of low cost, large discharge capacity, and long cycle life, and have become one of the most promising cathode materials for lithium-ion batteries [2]. In addition to this, lithium iron phosphate cathode material ($LiFePO_4$) is considered to be the safest cathode material for lithium-ion batteries at present, and it is also gradually becoming an important development direction for lithium-ion power batteries due to its cost advantage [3].

Typically, the lifespan of power and energy storage lithium-ion batteries is 8-10 years. Since these batteries contain a large amount of Co, Ni, Al, Cu, Li, and other metals [4] and fluorine-containing electrolytes, when the life of these lithium-ion batteries is terminated, if they are not able to be utilized in a stepwise manner and are not recycled and reused, millions of tons of used lithium-ion batteries will be produced, which will inevitably cause serious pollution of the environment, as well as a huge waste of resources [5,6]. The recycling of used power and energy storage lithium-ion batteries can not only alleviate the shortage of the aforementioned metal resources but also bring considerable economic benefits.

The main recycling and reuse methods for used lithium-ion batteries are direct recycling [7], pyrometallurgy, and hydrometallurgy [8]. Direct recycling, is the method of physically separating the positive electrode material from the collector, and then repairing the compositional and structural defects of the material particles through processes such as lithium replenishment and heat treatment, so as to prepare a new electrode material. Pyrometallurgy is to recover battery materials by means of heat treatment, its process is relatively simple, but there are problems such as low recovery rate and easy cause of environmental pollution as well as reuse still need to be dissolved and precipitated process. In contrast, hydrometallurgy generally requires low-temperature leaching, purification, and separation to achieve recycling of battery materials.

In contrast, the restoration method is characterized by simplicity and convenience. Most of the repair techniques for lithium iron phosphate cathode materials involve replacing the missing lithium elements proportionally to the material to be repaired and subsequently repairing the composition, structure, and properties of the material using high-temperature solid-phase sintering [9]. Additionally, some studies have employed a combination of hydrothermal and sintering methods to repair the materials [10]. Kim et al. [11] proposed a simple solid-phase reaction for regenerating $LiFePO_4$ cathode material from spent cathode materials. Most of the studies on the repair of ternary cathode materials have been similar to the repair process of $LiFePO_4$. In addition to the above commonly used restoration methods, Chen et al. [12] used Li+ eutectic molten salt to recrystallize the used $LiNi_{0.5}Co_{0.2}Mn_{0.3}O_2$ cathode material under atmospheric pressure. The discharge capacity of the repaired material at 1C (C=150 mA/g) is 149.3 mAh/g, and the capacity after 100 cycles is 134.6 mAh/g. The results show that the discharge capacity and cycle stability of the NCM cathode material repaired by this method can be restored to the original level. It can be seen that the structure and properties of waste LFP and NCM materials can be restored through repair, but not all materials can be regenerated through repair, and the number of material cycles has a great impact on the repair effect. It is urgent to establish the evaluation criteria for lithium iron phosphate and ternary materials that can be regenerated through repair.

Therefore, exploring the influence law of cycling on the structural stability of cathode materials and carrying out structural diagnosis and compositional analysis of used LFP and NCM batteries with different lifespans are the primary measures to establish an effective recycling method for their short-range.

2. Structural Diagnosis of Waste Lithium Iron Phosphate Electrode Material

During the battery cycling process, lithium ions in the positive electrode material go back and forth between the positive and negative electrodes, while the charging and discharging process proceeds to form an SEI film at the negative electrode, which leads to the lack of lithium in the positive electrode material. In this paper, electron-coupled plasma atomic emission spectrometry (hereinafter referred to as ICP) tests were performed on lithium iron phosphate cathode materials with different cycle times. The elemental ratios of the battery materials obtained for three different cycle times are shown in table 1.

Table 1. ICP test results of recycled lithium iron phosphate materials with different cycles.

Type of Material	Li (mmol/L)	Fe (mmol/L)	P (mmol/L)	Lithium Deficiency
Lithium iron phosphate recycled material with 500 cycles	0.19507	0.24548	0.24908	20.5%
Lithium iron phosphate recycled material with 1000 cycles	0.21697	0.29329	0.29037	26.0%
Lithium iron phosphate recycled material with 1500 cycles	0.21510	0.27645	0.28037	22.2%

As shown in figure 1, the XRD spectra of the three materials also show a high degree of consistency, but after cycling the lithium iron phosphate material appears to have some impurities. The impurity comes from the partial incomplete conversion of $FePO_4$ due to insufficient active lithium and $Li_3Fe_2(PO_4)_3$ formed upon oxidation of Fe. Besides, impurity components in the raw material are not completely removed. These can lead to defects in the recycled material that directly affect the repair material. Among them, the peaks at 18°, 23.8°, 30.3° and 30.9° of the XRD pattern are attributed to the (200), (210), (211) and (020) crystal planes of FePO4, whereas the impurity peak at 26.5° is attributed to $Li_3Fe_2(PO_4)_3$. Comparing the XRD results of the recycled materials with different number of cycles, it can be found that compared with the peak positions of the standard pdf card of $Li_3Fe_2(PO_4)_3$, the peak positions of the plots of the recycled materials are all blueshifted to varying degrees, which suggests that the cellular parameter of the recycled materials becomes smaller.

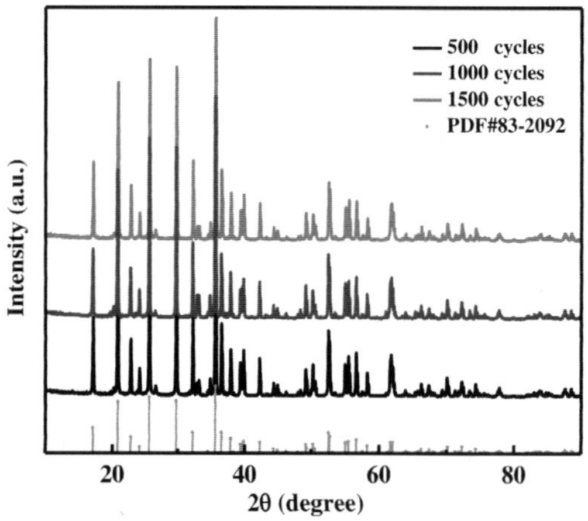

Figure 1. XRD patterns of recycled materials with different cycles.

3. Structural Diagnosis of Waste Ternary Electrode Materials

Similarly, ICP tests were done on the recycled materials, and the elemental ratios of the three ternary battery materials with different cycle times were obtained as in table 2. It was found that the lithium-ion content in the material gradually decreased with the increase in cycling times, while the content of nickel, cobalt, and manganese did not change significantly. This indicates that the cycling caused serious lithium loss in the ternary cathode material, which is presumed to be the direct cause of its structural and performance decay.

Table 2. ICP test results of recycled ternary materials with different cycles.

Sample	Li (mmol/L)	Ni (mmol/L)	Co (mmol/L)	Mn (mmol/L)
Primary ternary material	0.9684	0.50822	0.1990	0.3002
Ternary recycled material with 500 cycles	0.8374	0.50083	0.1989	0.3017
Ternary recycled material with 2000 cycles	0.7785	0.50912	0.1973	0.2938

The crystal structure of the waste NCM materials with different numbers of cycles was further analyzed and the results of the XRD tests are shown in figure 2. Compared to the pristine NCM diffraction peaks, the waste NCM material exhibits a slight angular shift, and the (003) diffraction peaks are shifted at a larger angle with the increase in the number of cycles, which may be attributed to the disruption of the crystal structure during the cycling process. The calculation results of the cell parameters are shown in table 3, and the decrease of I(003) and I(104) and the increase of [I(006) + I(102)]/I(101) also indicate that after cycling, the laminar structure of the material becomes disordered, the hexagonal structure is destroyed, and the lithium and nickel mixed rows gradually become heavier.

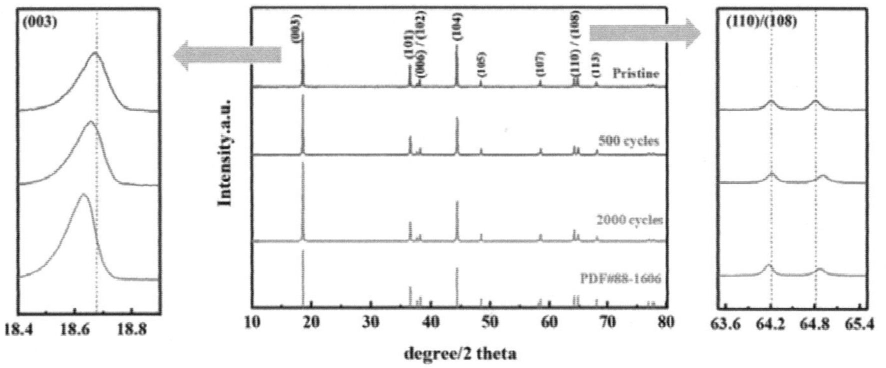

Figure 2. XRD pattern of waste materials with 500 and 2000 cycles for commercial materials (angle-peak strength).

Table 3. The calculation results of the cell parameters of the recycled ternary material with different cycles.

Sample	a (Å)	c (Å)	c/a	R (%)	I(003) /I(104)	[I(006)+ I(102)] /I(101)
Primary ternary material	2.8716	14.2754	4.9713	5.56%	1.5251	0.4598
Ternary recycled material with 500 cycles	2.8703	14.2650	4.9698	5.58%	1.5751	0.4701
Ternary recycled material with 2000 cycles	2.8703	14.2623	4.9689	5.05%	1.8123	0.5214

Next, this paper establishes the theoretical model of ternary cathode material on the basis of experimental research and applies the first nature principle [13] to its theoretical calculation to obtain the theoretical energy and energy band structure in the presence of different defects, as shown in figure 3. It was found that the energy of the ternary cathode material to generate antisite defects during the cycling process was -1351.4 eV. The energy is larger when both Li defects and antisite defects exist, which indicates that the cycling process is easy to form mixed rows, and the structural stability of the material deteriorates after multiple defects are generated, which is the main reason for the material failure, structural collapse, and performance decay.

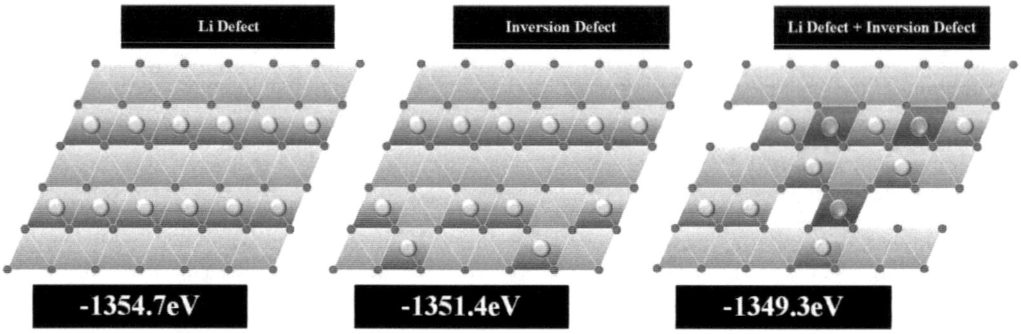

Figure 3. Structure optimization results of ternary cathode materials.

4. Conclusions

For lithium-ion battery lithium iron phosphate and ternary cathode materials with different cycle lifetimes, XRD and ICP were used to study the structural changes and lithium loss, and it was found that the lithium loss and structural disorganization were aggravated with the increase in the number of cycles. A theoretical model of ternary materials has been established, and the configurations, electronic structures, and energy band structures of different defect structures have been calculated using the first nature principle, it is found that the formation of microscopic defects inside the crystals reduces their structural stability. From the structural diagnostic results, it was found that cycling negatively affects the crystal structure and lithium content of the material. Initially, it was hypothesized that the evaluation criterion of whether the battery material can be repaired is the structural damage of the material. Thereafter, the electrode materials will be analyzed in depth by means of finer characterization to explore their restoration techniques, and the restoration evaluation criteria will be further studied and demonstrated in depth.

Acknowledgments

This work was financially supported by State Grid Heilongjiang Electric Power Co., LTD. Science and technology project 'Development of Lithium Iron Phosphate and Lithium Vanadium Phosphate Composite Materials and Batteries for Energy Storage in High Cold Areas' (522437240009) fund.

5. References

[1] Dehghanisanij A R, Tharumalingam E, Dusseault M B, et al. 2019 Study of energy storage systems and environmental challenges of batteries *J. Renewable and Sustainable Energy Review* **104** pp. 192-208.

[2] Harper G, Sommerville R, Kendrick E, et al. 2019 Recycling lithium-ion batteries from electric vehicles *J. Nature* **575(7781)** pp. 75-86.

[3] Hu M, Wang H and Chen Q 2020 Development status and trend of lithium-ion power battery for electric vehicle *J. Automotive Practical Technology* **9** pp. 8-10.

[4] Zheng X, Zhu Z, Lin X, et al. 2018 Research progress on recovery of valuable metals from waste lithium-ion batteries *J. Engineering* **4(03)** pp. 138-158.

[5] Chen Y, Liu N, Jie Y, et al. 2019 Toxicity identification and evolution mechanism of thermolysis-driven gas emissions from cathodes of spent lithium-ion batteries *J. ACS Sustainable Chemistry & Engineering* **7(22)** pp. 18228−18235.

[6] Schmidt T S, Beuse M, Zhang X, et al. 2019 Additional emissions and cost from storing electricity in stationary battery systems *J. Environmental Science & Technology* **53(7)** pp. 3379-3390.

[7] Zhu J, Tian J, Wu K, et al. 2024 Solid-phase direct repair and regeneration of waste lithium nickel-cobalt-manganese oxide cathode materials *J. Power Supply Technology* **48(04)** pp. 568-583.

Journal of Physics: Conference Series **2941** (2025) 012016 doi:10.1088/1742-6596/2941/1/012016

[8] Atia T A., Elia G A, Hahn R, et al. 2019 Closed-loop hydrometallurgical treatment of end-of-life lithium-ion batteries: Towards zero-waste process and metal recycling in advanced batteries *J. Journal of Energy Chemistry* **35** pp. 220-227.

[9] Wang L, Li J, Zhou H, et al. 2018 Regeneration cathode material mixture from spent lithium iron phosphate batteries *Journal of Materials Science: Materials in Electronics* **29(11)** pp. 9283-9290.

[10] Li X, Zhang J, Song D, et al. 2017 Direct regeneration of recycled cathode material mixture from scrapped $LiFePO_4$ batteries *Journal of Power Sources* **345** pp. 78-84.

[11] Kim H S and Shin E J 2013 Re-synthesis and electrochemical characteristics of $LiFePO_4$ cathode materials recycled from scrap electrodes *J. Bulletin of The Korean Chemical Society* **34(3)** pp. 851-855.

[12] Shi Y, Zhang M, Meng Y S, et al. 2019 Ambient-pressure lithiation of degraded $Li_xNi_{0.5}Co_{0.2}Mn_{0.3}O_2$ ($0<x<1$) via eutectic solutions for direct regeneration of lithium-ion battery cathodes *J. Advanced Energy Materials* **9(20)** pp. 1900454.

[13] Zhou X Y, 2022 *First-principles Computational Study on Doping Modification of High Nickel Ternary Cathode Materials for Lithium-Ion Batteries* (Central South University).

Research on the Pavement Performance of Rubber Powder /Anti-Rutting Agent Composite Modified Asphalt Mixture

Xue Liu, Shaowei Jiang* and Shoukun Zhang

China First Highway Engineering Co., Ltd., Beijing, China.
Email: 572757131@qq.com; *348268467@qq.com; 1037461275@qq.com

Abstract. Rutting is a prevalent issue in road engineering that adversely impacts driving safety and comfort. To improve the rutting resistance of pavement, this study formulated two groups of modified asphalt mixtures by incorporating varying amounts of rubber powder and anti-rutting agents as environmentally friendly modifiers. The performance of these mixtures was assessed using Rutting tests, Marshall stability tests, and low-temperature Three-point bending beam tests. The findings demonstrate that the two composite modified asphalt mixtures, 11% rubber powder + 0.25% anti-rutting agent (MA-A) and 15% rubber powder + 0.3% anti-rutting agent (MA-B), exhibited outstanding performance in terms of high and low-temperature resistance as well as water damage resistance. Notably, MA-B delivered superior overall performance. Compared to SBS-modified and base asphalt mixtures, the composite mixtures with anti-rutting agents significantly enhanced high-temperature resistance, surpassing that of SBS-modified asphalt. Although the low-temperature performance was markedly better than that of the base asphalt mixture, it was slightly inferior to that of SBS-modified asphalt. The water damage resistance of the composite modified mixtures was comparable to that of SBS-modified asphalt. In summary, the multi-phase composite modified asphalt mixtures significantly improved high-temperature, low-temperature, and water stability in comparison to both base and SBS-modified asphalt.
Keywords. Rubber powder; anti-rutting agent; composite modification; asphalt mixture.

1. Introduction

Asphalt mixtures serve as the predominant paving material for asphalt pavements, constituting over 95% of highway pavement types in China [1]. Pavements are directly exposed to wheel loads and various natural factors, such as sunlight, temperature, air, and rain, which can cause changes in their performance and condition, ultimately affecting the pavement's usability and service life [2]. During vehicle operation, asphalt pavements are subjected to cyclic loads, resulting in cumulative deformation and rutting, a common road defect in asphalt pavements. Incorporating anti-rutting agents and other composite modifiers into modified asphalt mixtures effectively mitigates rut formation, thereby improving the high-temperature stability of asphalt pavements. This enhancement occurs without sacrificing low-temperature stability or water resistance [3,4] This reduces or delays the occurrence of rutting. The addition of modifiers ensures that asphalt pavements maintain good performance over the long term. A significant body of research, both domestic and international, has been conducted on this topic. Hong [5] used a composite of two anti-rutting agents to prepare modified asphalt mixtures, finding that the composite anti-rutting agent had little effect on enhancing the water stability of the asphalt mixture but considerably improved its performance at both high and low temperatures. Hadiwardoyo [6] studied the physical parameters of asphalt mixtures with varying amounts of anti-rutting agents and found that at the optimal dosage, the asphalt concrete's high-temperature stability, low-temperature performance, and impermeability have improved. Shen[7] Investigated the performance of rubber powder modified asphalt catalysed by olefin metathesis, revealing that although high-temperature performance remained largely unchanged, low-temperature crack resistance improved significantly, alongside marked enhancements in fatigue resistance, aging resistance, and water damage resistance. Jia [8] investigated the performance of high-content rubber powder/SBS composite modified asphalt and mixtures, demonstrating superior high-temperature rutting resistance, low-temperature crack resistance, and water stability.

Considering the beneficial impact of anti-rutting agents and rubber powder modifications on improving the performance of asphalt mixtures, the integration of these two modifiers in a composite system holds significant potential for further improving pavement durability and resilience [9-11]. However, the optimal combination and dosage of rubber powder and anti-rutting agents, as well as their synergistic effects on the comprehensive pavement performance, remain to be fully elucidated [12-15]. Therefore, this study aims to systematically examine the effects of different amounts of rubber powder and anti-rutting agents on the high-temperature stability, low-temperature flexibility, and water damage resistance of asphalt mixtures. The goal is to develop a cost-effective and environmentally friendly multi-phase composite modified asphalt mixture that outperforms traditional SBS modified asphalt, thereby enhancing the safety, comfort, and durability of asphalt pavements.

2. Experimental Materials and Mix Design

2.1. Experimental Materials

2.1.1. Asphalt. The rubber asphalt used in the experiment was self-prepared. It is produced by adding a certain amount of waste rubber powder into molten base asphalt, stirring it evenly, and then placing the mixture in an oven for thermal insulation. The mixture is then heated in an electric furnace and subjected to high-speed shearing to produce rubber asphalt. The physical performance indicators of the rubber asphalt are presented in table 1.

Table 1. Rubber asphalt test index.

Test item	Unit	Test result	Technical requirement	Test procedure
Rotational Viscosity at 180°C	Pa·s	3.0	2.0~5.0	T0625
Penetration	0.1mm	55.9	30~60	T0604
Softening Point	°C	61	60	T0606
Ductility (5°C)	cm	6.2	5	T0605

2.1.2. Aggregates. The aggregates utilized comprise three grades of limestone, with 10-20 mm and 5-10 mm fractions serving as coarse aggregates, while the 0-5 mm fraction is designated as fine aggregate. Mineral powder is employed as the filler. The physical performance indicators of the aggregates are presented in table 2.

Table 2. Aggregate performance test index.

Test item		Test result	Technical requirement	Test procedure
Crushing Value, %		19.3	≤28	T0316
Los Angeles Abrasion Value, %		15.4	≤30	T0317
Flaky and Elongated Particle Content, %	4.75-9.5mm	15.7	≤20	T0312
	≥9.5mm	7.3	≤15	
Adhesion Class with Asphalt		4	≥4	T0616

Weak Particle Content, %	1.7	≤5	T0320

2.1.3. Additives. The additives include an anti-rutting agent and rubber powder, which together form a composite modifier. Table 3 presents the physical performance indicators of the anti-rutting agent.

Table 3. Test index of physical properties of anti-rutting agent.

Test item	Unit	Technical requirement	Test procedure	Test result
Appearance	/	Granular, uniform, well-shaped and non-agglomerated.	/	satisfy
Individual Particle Mass	g	≤0.03	JT/T860.1	0.023
Density	g/cm^3	≤1.0	GB/T 1033	0.946
Melt Flow Index	g/10min	≥1.0	GB/T 3682	1.3
Ash Content	%	≤5	JTG E20	3.3

2.2. Mix Design

2.2.1. Gradation Design. The gradation was based on the AC-20 type asphalt mixture, adhering to the JTG F40-2004. Coarse, medium, and fine gradations were initially adjusted and selected for the aggregate structure. The mix ratios for the three adjusted gradations are presented in table 4. The asphalt content was established at 4.1%, with the anti-rutting agent added at 0.3%. The asphalt mixture was subsequently prepared for testing. The gradation curves of the aggregates and the volumetric indicators of the composite gradation are presented in figure 1 and table 4.

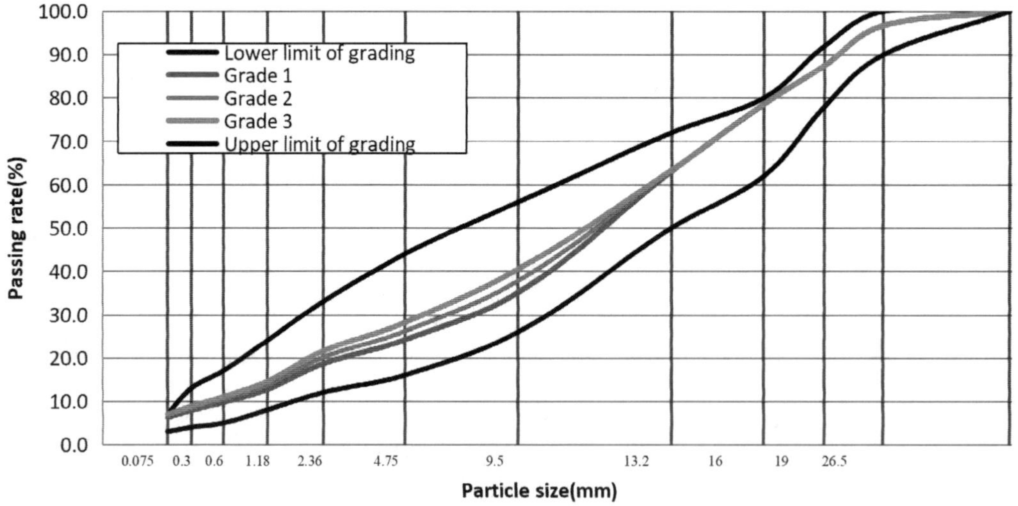

Figure 1. Gradation curves.

Table 4. Mix ratio composition.

Gradation type	Limestone 10-20mm	Limestone 5-10mm	Limestone 0-5mm	Mineral powder
Gradation 1	45	21	31	3
Gradation 2	45	18	34	3
Gradation 3	45	15	37	3

Table 5. Volume index table for the three synthetic gradations.

Gradation type	Gradation 1	Gradation 2	Gradation 3	Technical requirement
Asphalt content, %	4.1			/
The relative density of specimen gross volume, g/cm^3	2.405	2.419	2.442	Actual measurement
Maximum relative theoretical density, g/cm^3	2.523	2.530	2.536	Actual measurement
Void ratio AV, %	5.0	4.4	3.1	3-6
Ore clearance rate VMA, %	14.2	13.6	12.9	≥13.5
Asphalt saturation VFA, %	63.7	67.9	73.8	65-75
Stability, KN	11.65	12.6	10.32	≥8
Flow value, mm	2.5	3.3	3.7	1.5-4

The experimental data in table 5 indicate that Gradation 2 conforms to the specification requirements for AV, VMA, VFA, stability, and flow values. and it is closer to the design requirements. Therefore, Gradation 2 was selected as the design gradation.

2.2.2. Optimal Asphalt Content Selection. Marshall tests were conducted with five asphalt contents: 3.1%, 3.6%, 4.1%, 4.6%, and 5.1%. Table 6 provides the experimental results. The calculated optimal asphalt content was 4.11%, which was rounded to 4.1%.

Table 6. Summary of AC-20 Marshall test results.

Asphalt content	3.1	3.6	4.1	4.6	5.1	Technical requirement
Maximum theoretical density	2.568	2.549	2.530	2.511	2.493	/
Void ratio, %	6.3	5.3	4.4	3.4	2.5	3-6
Ore clearance rate, %	13.23	13.50	13.62	13.81	14.11	≥13.5%
Saturation, %	52.1	60.3	67.9	75.7	82.4	65-75
Stability, kN	10.1	11.9	12.6	12.6	11.5	≥8.0
Flow value (mm)	2.2	2.8	3.3	3.7	4.1	1.5-4

3. Performance Study of Asphalt Mixture

Following the JTG F40-2004, the asphalt mixture underwent evaluation for high-temperature performance, low-temperature performance, and water stability. Marshall specimens were created with an optimal asphalt content of 4.1%. The preparation of the specimens is largely similar to the preparation of other common asphalt mixture specimens, with the primary difference being an additional 15-second dry-mixing step to add the anti-rutting agent after the aggregate mixing is completed. After specimen preparation, they were cooled and allowed to rest for 24 hours.

Subsequently, the Marshall stability and dynamic stability tests were carried out on the asphalt mixture specimens, with the dynamic stability test specifically aimed at evaluating rutting resistance. Additionally, immersion residual stability and freeze-thaw splitting strength tests were conducted to assess the water stability of the mixture [16,17]. To assess the low-temperature stability of the mixture, splitting strength tests and three-point bending beam tests were also performed.

Comprehensive performance tests were conducted on the anti-rutting modified asphalt mixture, comparing the high- and low-temperature characteristics and water stability of the two multi-phase composite modified asphalt mixtures, MA-A and MA-B, with base asphalt (BA) and SBS modified asphalt (SBSMA).

3.1. High-Temperature Stability

3.1.1. Marshall Stability. Marshall stability tests were performed on the MA-A and MA-B under high-temperature conditions (60°C). Initially, the stability of the base asphalt and SBS modified asphalt mixture specimens was assessed as reference values. These results were then compared with those of the multi-phase composite modified asphalt mixtures to evaluate their high-temperature performance. The results of the tests are illustrated in figure 2.

3.1.2. Dynamic Stability. Currently, the rutting test serves as a crucial method for assessing the high-temperature performance of asphalt mixtures. The rutting test is by a certain load of the wheel at a limited temperature and a certain time in the mixture specimens repeatedly crushed on the same track to form deep and shallow rut, through the depth of the rut and the number of times each 1mm rut depth needs to be crushed that is, the dynamic stability of asphalt mixtures, assessed at a specified temperature, reflects their capacity to resist plastic flow deformation. The dynamic stability is calculated by the equation [18]:

$$DS = \frac{N(t_2 - t_1)}{d_2 - d_1} C_1 C_2 \tag{1}$$

In the formula: N represents the round-trip speed of the test wheel, set to 42 cycles/min; d_1 and d_2 are the deformation values (in mm) at times t_1 (45 minutes) and t_2 (60 minutes), respectively; C_1 is the correction factor; C_2 is the specimen factor.

Dynamic stability is a key indicator of the high-temperature resilience of asphalt mixtures. A rutting test was performed to assess the development of rut depth in asphalt mixture slabs under repeated rolling action at high temperatures (60°C), functioning as a benchmark for assessing the asphalt mixture's ability to withstand high-temperature deformation. The dynamic stability of base asphalt and SBS modified asphalt mixture specimens served as reference values. The dynamic stability of the two anti-rutting modified asphalt mixtures, MA-A and MA-B, was assessed using Marshall specimens at a high temperature of 60°C. The experimental procedures followed the guidelines outlined in JTG E20-2011. The displacement curves and dynamic stability values for the four asphalt mixture groups are shown in figures 3 and 4.

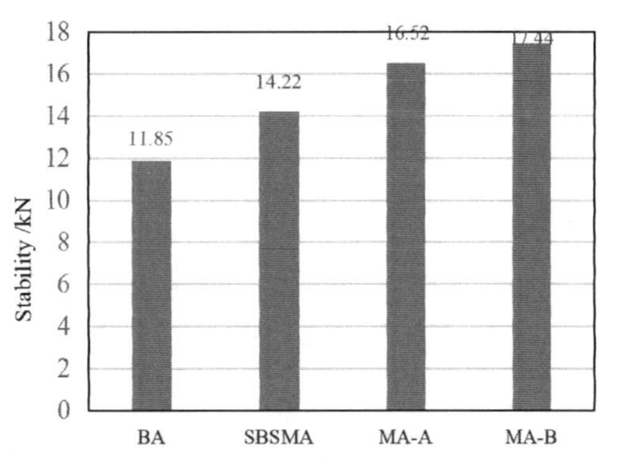

Figure2. Stability of multiphase composite modified asphalt mixture.

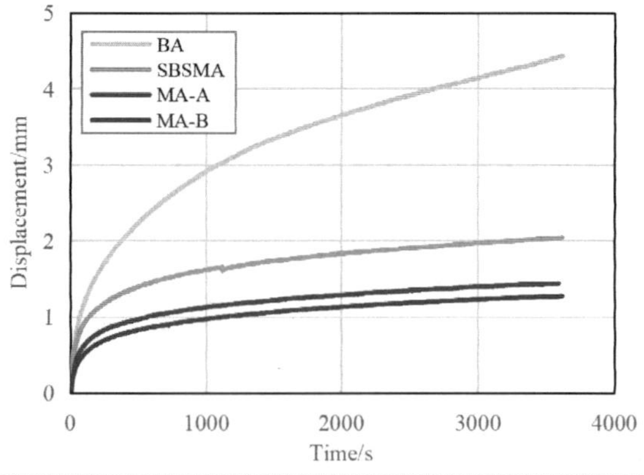

Figure 3. Displacement curve.

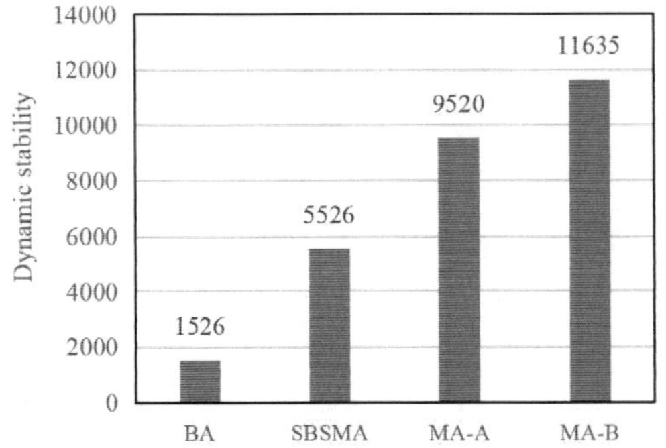

Figure 4. Dynamic stability diagram.

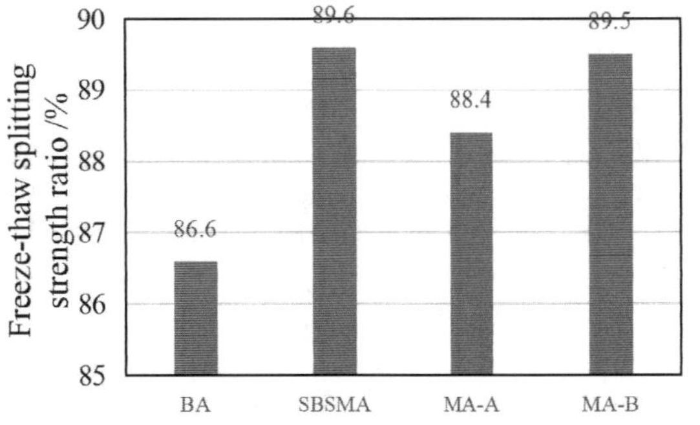

Figure 5. Immersion residual stability diagram.

Figure 4 demonstrates that, following testing, the rutting depth of the composite modified asphalt mixture was reduced by half compared to the base asphalt. Notably, the dynamic stability of MA-B reached a maximum of 11,635, while the dynamic stability of the SBS modified asphalt mixture was 5,526, which is three times greater than the dynamic stability of the base asphalt at 1,526. Furthermore, the rutting depth of MA-B was lower than that of the other groups. This indicates that the composite modified asphalt mixtures exhibit significant advantages in high-temperature rutting resistance, with MA-B showing particularly outstanding performance. This further substantiates the superior capability of composite modified asphalt in resisting high-temperature rutting.

3.2. Water Stability

3.2.1. Immersion Residual Stability. The immersion Marshall stability test was conducted to assess the resistance of MA-A and MA-B to stripping under water damage conditions. This test is essential for *evaluating* the viability of the mix design. To further examine the modified asphalt mixture's water damage resistance, the immersion stability of base asphalt and SBS modified asphalt mixture specimens was tested as reference values and compared with the composite modified asphalt mixtures. The formula for calculating immersion Marshall residual stability (MS_0) is provided in equation (2) [19].

$$MS_0 = \frac{MS_1}{MS} \tag{2}$$

Where: *MS*, MS_1 is the stability of the specimen after immersion in water for 35min and 48h respectively (kN).

The specific experimental procedure follows the JTG E20-2011 T0709 guidelines. The test results are illustrated in figure 5. From figure 5, it is evident that the immersion residual stability of the two composite modified asphalt mixtures is 92.3% and 92.7%, respectively, showing significant improvement compared to the matrix asphalt mixtures. When compared to the SBS modified asphalt mixtures, the values are lower but relatively close. This indicates a marked enhancement in the composite modified asphalt mixtures' ability to resist spalling under water damage. Additionally, MA-B exhibits a 0.4% higher immersion residual stability than MA-A, suggesting that MA-B has superior water stability performance.

3.2.2. Freeze-Thaw Splitting Strength. The freeze-thaw splitting test is conducted to evaluate the resistance of asphalt mixtures to water damage, featuring more stringent conditions than those used for standard water immersion stability tests. Under specified conditions, freeze-thaw cycles are performed, and the split strength of specimens from matrix asphalt and SBS modified asphalt mixtures serve as reference values for assessing the water damage resistance of the two rutting modified asphalt mixtures, MA-A and MA-B. The freeze-thaw split tensile strength ratio (TSR) of the specimens is utilized as the evaluation index, and its calculation formula is shown in equation (3) [20].

$$TSR = \frac{RT_2}{RT_1} \times 100 \quad (3)$$

Where: RT_1 represents the splitting tensile strength (MPa) of specimens without freeze-thaw cycles, while RT_2 indicates those subjected to freeze-thaw cycles.

To evaluate the water stability of the asphalt mixture, the strength ratio of split damage of the specimens before and after water exposure was determined. The specific experimental procedure follows the JTG E20-2011 T0709 guidelines. The test results are presented in figure 6 below.

From the chart can observed that both groups of anti-rutting mixtures show a significant improvement in splitting strength compared to the base asphalt. The combination of 15% rubber powder + 0.3% anti-rutting agent is comparable to SBS modified asphalt mixtures, while the combination of 11% rubber powder + 0.25% anti-rutting agent also shows a notable improvement over base asphalt, but the effect is less pronounced than that of MA-B. This indicates that the MA-B has a significantly improved resistance to water damage under extreme conditions, approaching that of SBS modified asphalt mixtures.

3.3. Low-Temperature Crack Resistance

3.3.1. Splitting Strength. Splitting strength is utilized to evaluate the mechanical properties of asphalt mixtures at a defined temperature and loading rate, relevant during both splitting failure and the elastic phase. This measure aids in selecting mechanical design parameters for asphalt pavement structures while also assessing the low-temperature crack resistance of the mixtures. Testing was performed at 15°C with a loading rate of 50 mm/min, using the splitting strength of base asphalt and SBS modified asphalt mixture specimens as reference values. The low-temperature crack resistance of the Marshall specimens for MA-A and MA-B was assessed under these conditions. The results are illustrated in figure 7.

The figure indicates that the splitting strength of these two rutting-resistant mixtures is improved by factors of 1.48 and 1.76 compared to the base asphalt, achieving performance comparable to that of SBS-modified asphalt mixtures. This demonstrates that the low-temperature performance of the rutting -resistant mixtures has significantly enhanced through composite modification.

3.3.2. Three-Point Bending Low-Temperature Beam Test. The use of a universal testing machine to test the asphalt mixture beam under low-temperature conditions (-10°C) of the bending and tensile strain and creep modulus of strength, through a comparison with the SBS-modified asphalt mixtures, evaluation of multi-phase composite asphalt mixtures of low-temperature cracking resistance. Specific experimental methods refer to the JTG E20-2011.

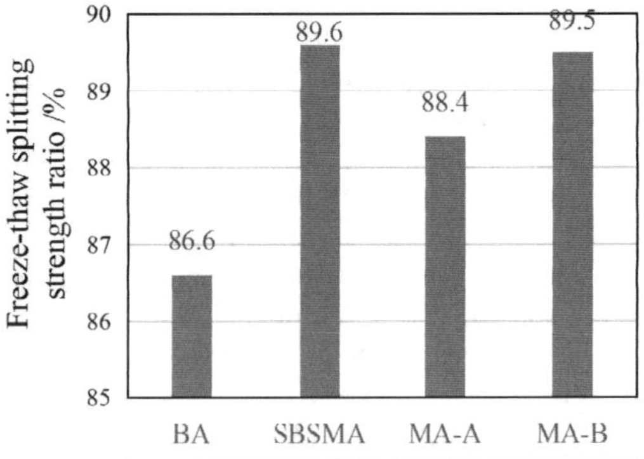

Figure 6. Tensile strength ratio of freeze-thaw

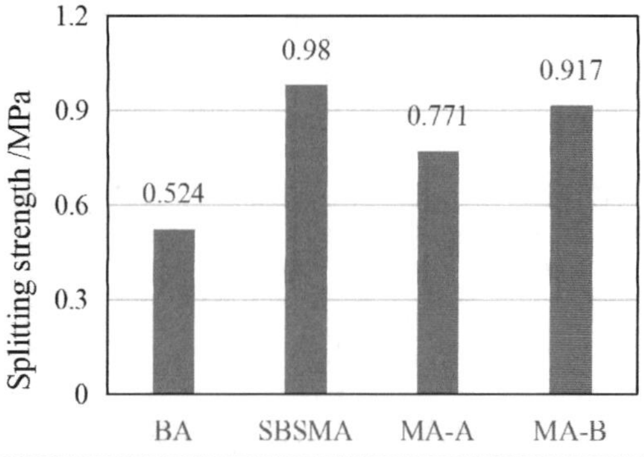

Figure 7. Splitting strength of different types of

splitting of different types of mixtures.

mixtures.

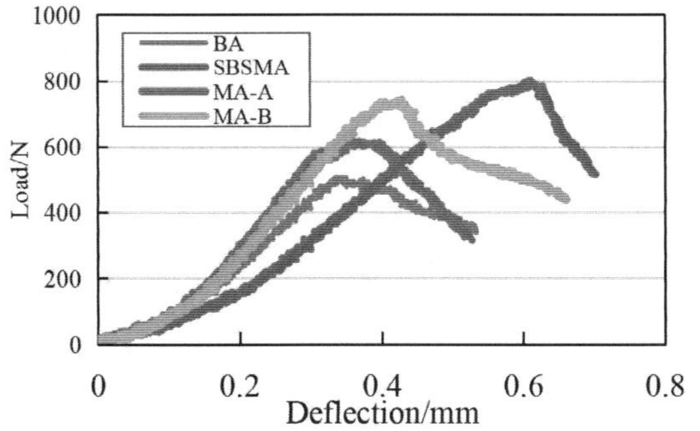

Figure 8. Low temperature trabecular test load-deflection curve.

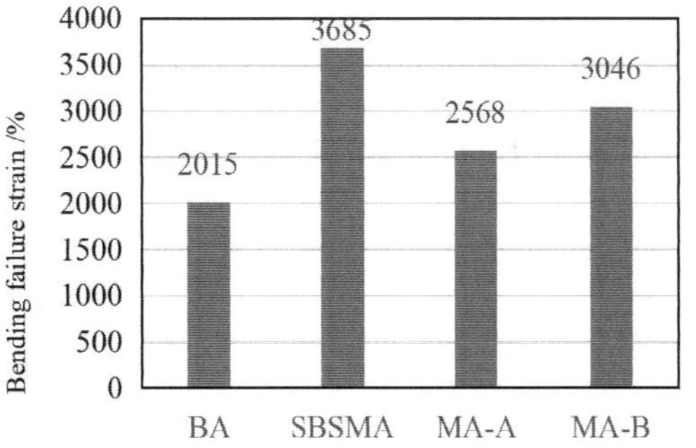

Figure 9. Low temperature trabecular test bending failure strain.

We extend the tangent line of the curve in figure 8 and the intersection line of the horizontal coordinates for the beginning of the deflection curve, the maximum value of the load corresponding to the deflection coordinates for the most termination value, the termination value and the difference from the initial value represent the magnitude of the deflection. According to the deflection and the height of the specimen, the span can be obtained from the maximum bending and tensile strain formula as shown in (4).

$$\varepsilon(t) = \frac{6hd(t)}{L^2} \tag{4}$$

Where $\varepsilon(t)$ is strain; h is the height at the span interruption, m; L is the span diameter, m; and $d(t)$ is the load, N;

The larger the curve deflection, the better the low-temperature cracking resistance. Figures 8 and 9 show that the SBS modified asphalt mixture achieves the highest load deflection, indicating superior low-temperature cracking resistance. Additionally, the deflection of MA-B is greater than that of the matrix asphalt group, demonstrating significant modification effects. A comparison of the maximum load values reveals that MA-B and the SBS modified asphalt mixture have similar maximum loads, which represent a substantial improvement over the matrix asphalt, indicating a greatly enhanced ability to withstand low-temperature cracking.

4. Conclusion

(1) Based on the comprehensive experimental research conducted on rutting-resistant pavement materials, a thorough evaluation of the high-temperature, low-temperature, and water damage resistance performance indicators of the two modified asphalt mixtures, MA-A and MA-B, was performed. The results indicate that MA-B demonstrates excellent overall road performance.

(2) The composite modified asphalt mixture containing rutting agents exhibits significantly improved high-temperature performance compared to both the SBS modified asphalt mixture and the matrix asphalt mixture, even surpassing the latter.

(3) In terms of low-temperature performance, the composite modified asphalt mixture shows greater enhancement over the matrix asphalt mixture, although it is slightly inferior to the SBS modified asphalt mixture. Additionally, its resistance to water damage is comparable to that of the SBS modified asphalt mixture.

5. References

[1] Li H, Zhang C, Zou X, Li W, Zhao G, Ma Q, et al. 2022 Analysis of the rheological

performance of composite asphalt rubber and the influence of gradation types on its mixture prepared under the optimal specifications *International Journal of Pavement Research and Technology* **15(3)** 659-675.

[2] Aiping Q U 2023 Research on planted performance of green ecological asphalt mixture *Journal of Environmental Protection and Ecology* **24(3)** 859-865.

[3] Ye Y, Hao Y, Zhuang C, Shu S and Lv F 2022 Evaluation of improvement effect of different anti-stripping agents on pavement performance of granite-asphalt mixture *Materials* **15**.

[4] Meng Y, Yan T, Muhammad Y, Li J, Qin P and Ling L, et al. 2022 Study on the performance and sustainability of modified waste crumb rubber and steel slag powder/SBS composite modified asphalt mastic *Journal of Cleaner Production* **(Mar.1)** 338.

[5] Hong Z, Yan K, Wang M, Ge D and Chen J 2023 Laboratory performance evaluation of waste tire rubber (WTR) and ethylene-vinyl acetate (EVA)-modified asphalt mixtures *The International Journal of Pavement Engineering* **24(1)**.

[6] Hadiwardoyo S P, Sumabrata R J, Nissa A C, Muhammad F A, Hia M, Iskandar D, Lumingkewas R H 2023 Improvement of button rock asphalt performance by adding nano-crumb rubber and waste engine oil *International Journal of Pavement Research and Technology* **16(5)** 1181-1195.

[7] Shen Q, Fan L, Yang Y H, Wei H, Xu H & Cao A, et al. 2023 Study on high temperature performance of coal based modified asphalt mixture *Advances in Engineering Technology Research.*

[8] Jia H, Sheng Y, Guo P, Underwood S, Chen H and Kim Y R, et al. 2023 Effect of synthetic fibers on the mechanical performance of asphalt mixture: a review *Journal of Traffic and Transportation Engineering (English Edition)* **(3)**.

[9] Musbah A, Ali A S B, Albeddal A M, Musbah M G and Ali H 2022 Effect of long-term aging on the behavior of batu pahat soft clay-modified asphalt mixture *Advanced Engineering Forum,* **47** 69 - 78.

[10] Khadim H M and Al-Mosawe H M 2024 Crumb rubber modification for enhanced rutting resistance in asphalt mixtures *Open Engineering* **14(1)** 103238-93.

[11] Ekmekci M, Sinanmis R and Woods L 2024 Predictive modeling for highway pavement rutting: a comparative analysis of auto-machine learning and structural equation models *Transportation Research Record* **2678(6)** 724-737.

[12] Jia J, Zandi Y, Agdas A S, Al Tamimi H M, Banaeipour A and Assilzadeh H, et al. 2022 Artificial intelligent modeling and optimization of the application of aluminum oxide nanoparticles modified cement asphalt toward non-polluted sustainable environment *International Journal of Energy Research* **46** 20939 - 20949.

[13] Gupta L and Kumar R 2022 Experimental investigation of reclaimed asphalt foamed bituminous mix: a cold mix technique *World Journal of Engineering.*

[14] Xue Y, Liu C, Qu J, Lv S, Ju Z and Ding S, et al. 2023 Research on pavement performance of recycled asphalt mixture based on separation technology of asphalt and aggregate in rap *Construction & Building Materials* **(Aug. 22)** 393.

[15] Ernas O, Vaitkus A and Kulteck J 2023 Performance of crumb rubber bitumen and asphalt modified in the wet process alone and in combination with SBS polymer *Road Materials and Pavement Design 24* **(Sup.1)** 107-123.

[16] Jin D, Yin L, Nedrich S, Boateng K A and You Z 2024 Resurface of rubber modified asphalt mixture with stress absorbing membrane interlayer: from laboratory to field application. *Construction and Building Materials* **441**.

[17] He J 2024 Investigation on anti-fuel erosion performance of Sasobit/SBS-modified asphalt and its mixtures *Materials* **17**.

[18] Liu F, Pan B, Bian J and Zhou C 2023 Experimental investigation on the performance of the asphalt mixture with ceramic fiber *Journal of cleaner production* **(Jan.15)** 384.

[19] Davoodi A, Esfahani M A, Bayat M, Mohammadyan-Yasouj S E and Rahman A 2022 Influence

of nano-silica modified rubber mortar and EVA modified porous asphalt on the performance improvement of modified semi-flexible pavement *Construction and Building Materials* **(Jun.27)** 337.

[20] Guo R, Zhang H and Tan Y 2022 Influence of salt dissolution on durable performance of asphalt and self-ice-melting asphalt mixture *Construction and Building Materials.*

Experimental Study on the Directional and Graded Conversion of Ferrous Oxalate and Ferric Hydroxide from Red Mud

Xingying Zhang[1], Wei Chen[2], Shiyi Fu[2] and Hongpan Liu[2,*]

[1] Ecological Environment Monitoring Station of Shuangqiao Economic and Technological Development Zones, Chongqing, China.
[2] College of Chemistry and Environmental Engineering, Chongqing, Key Laboratory for Resource Utilization of Heavy Metal Wastewater, Chongqing University of Arts and Sciences, Yongchuan, Chongqing, China.
Email: 3128738316@qq.com; 599396702@qq.com; 3128738316@qq.com; *983785769@qq.com

Abstract. Red mud is a highly alkaline industrial solid waste generated by the alumina industry. Its stacking and disposal not only occupy land resources, but also cause environmental pollution. It is of great significance to separate and purify target compounds from complex mixtures of red mud based on their resource utilization properties. Ferrous oxalate and ferric hydroxide were prepared by oriented conversion based on redox reaction and chemical precipitation in this research. The products prepared were characterized by XRD, SEM and IR under different conditions (iron powder, ascorbic acid and photocatalysis). The results show that the two-stage acid leaching process can improve the leaching rate of iron from red mud significantly, and the experimental of oriented conversion of ferrous oxalate and ferric hydroxide from red mud with reproducibility. The sample prepared with ascorbic acid as reducing agent has the best effect. The development of this experiment provides a new idea for the directional classification and purification of ferrous oxalate and iron hydroxide from red mud.
Keywords: Red mud; fractionation; oriented conversion; oxalic acid; ferrous oxalate.

1. Introduction

Red mud is an industrial solid waste produced in the process of producing alumina from bauxite by alkali dissolution or alkali roasting. It has the characteristics of high alkali and high salt. A large number of storages causes great hidden dangers and environmental hazards to soil, water and atmosphere [1]. In the 14th Five Year Plan, the National Development and Reform Commission emphasized the need to significantly improve the comprehensive utilization rate of bulk solid waste such as red mud. It clearly stated that “by 2025, the comprehensive utilization rate of newly increased bulk solid waste will reach 60%, and the stock of bulk solid waste will be orderly reduced.” It also encouraged the recovery of iron and aluminum oxide from red mud to improve the efficiency of mineral resource utilization [2].

Red mud contains a large number of iron-bearing minerals, which are small in size and cemented with each other [3]. It is difficult to effectively separate, resulting in poor resource utilization. At present, the technology of iron recovery from red mud is mainly divided into physical methods (magnetic separation, gravity separation, etc), chemical methods (acid leaching, roasting, etc.) and biological method. The physical iron extraction process is simple and cost-effective, but the recovery rate is not high and the product quality is poor. The fire reduction process in chemical methods is relatively mature and has a high recovery rate, but the process is complex and prone to secondary pollution. Wet acid leaching has low energy consumption, but there are problems such as coexistence of impurity ions. Biological method is environmentally friendly, but it takes a long time [3]· In view of the problems of large consumption of inorganic acid, unfriendly environment and coexistence of impurity ions in the current acid leaching process. Oxalic acid was introduced as the acid leaching medium in this experiment, and hydrochloric acid-oxalic acid two-stage grading acid leaching process was used to separate red mud. Compared with inorganic acids, oxalic acid is less toxic and dangerous. It is a biodegradable and environmentally friendly acid, which has been applied in metal leaching.

Combined with the principle of redox reaction and chemical precipitation equilibrium in this experiment, the low concentration hydrochloric acid-oxalic acid two-stage acid leaching process was used for leaching treatment, and the iron-containing solution was subjected to fractional separation and directional conversion to prepare ferrous oxalate and iron hydroxide colloids [4]. This experiment provides a reference for the reduction, harmlessness and resource application of red mud, which meets the inevitable requirements of sustainable development of the country and the alumina industry.

2. Experimental Part

2.1. Experimental Principle

Red mud is rich in various valuable metals, which can be separated and purified for resource recovery and utilization [5]. The experimental process is shown in figure 1. After being pretreated with hydrochloric acid and dealkalized, red mud is added with oxalic acid solution to allow iron to enter the solution in the form of ions. A reducing agent is added to precipitate through oxidation-reduction and chemical reactions, causing oxalate ions to combine with ferrous ions to produce yellow oxalate ferrous precipitation [6]. $K_{sp}(FeC_2O_4)=3.2\times10^{-7}$, $K_{sp}(Fe(OH)_3)=4.0\times10^{-38}$ are known.

$$Fe_2O_3+3H_2C_2O_4\cdot2H_2O=Fe_2(C_2O_4)_3+9H_2O$$

$$Fe_2(C_2O_4)_3\leftrightharpoons Fe^{3+}+Fe(C_2O_4)_3^{3-}$$

$$Fe^{3+}+\text{ reducing agent}\rightarrow Fe^{2+}$$

$$Fe^{2+}+C_2O_4^{2-}\leftrightharpoons FeC_2O_4\downarrow$$

$$Fe^{3+}+3(OH)^-=Fe(OH)_3\downarrow$$

Determination of Fe content in solution [7-9]:

(1) Phenanthroline spectrophotometric method

A section of hydrochloric acid dealkalization solution was determined by phenanthroline spectrophotometry. Transfer 1mL to 50mL volumetric flask, dilute with distilled water to 15mL, add 10mL of mixed solution (2.5mL hydroxylamine hydrochloride+2.5mL phenanthroline+5.0mL acetic acid amine buffer), add distilled water to the mark, and shake well. After being left for 15 minutes, use a 1cm colorimetric dish and measure at a distance of 510mm on the spectrophotometer. At this time, the leaching rate of Fe is calculated according to the following formula (1).

$$\eta_{Fe}=\frac{m_{Fe}\times V\times 1000}{V_1\times m_{RM}\times w_{Fe}}\times 100\% \quad (1)$$

In the formula: m_{Fe} is the mass of Fe in the sample solution checked from the standard curve, μg; V_1 is the volume of the removed solution, mL; V is the constant volume of the leaching solution, L; m_{RM} is the quality of the red mud used, g; $w_{(Fe)}$ is the content of Fe in red mud, %.

(2) Atomic absorption spectrophotometry

After diluting the two-stage oxalic acid leaching solution to a suitable multiple, the content is determined by atomic absorption spectrophotometer. At this time, the leaching rate of Fe is calculated according to the following formula (2).

$$\eta_{Fe} = \frac{V \times C_1 \times 1000}{m \times w(Fe)} \times 100\% \quad (2)$$

In the equation: C_1 is the content of Fe, mg/L; V is the constant volume of the leaching solution, L; m is the mass of red mud used, g; $w_{(Fe)}$ is the mass fraction of Fe in red mud, %.

Yield of ferrous oxalate:

$$y_{FeC2O4} = \frac{m_1}{\frac{m \times \eta_{Fe} \times w(Fe)}{M_{Fe}} \times M_{FeC_2O_4 \cdot 2H_2O}} \times 100\% \quad (3)$$

In the formula: m is the mass of red mud, g; η_{Fe} is the leaching rate of atomic absorption spectrophotometry; $w_{(Fe)}$ is the content of Fe in red mud, %; m_1 is the quality of the preparation of ferrous oxalate, g; M_{Fe}=55.845 g/mol; $M_{FeC2O4} \cdot 2H_2O$==179.9 g/mol.

Yield of iron hydroxide:

$$m_i = (\frac{m \times \eta_{Fe} \times w(Fe)}{M_{Fe}} - \frac{m_1}{M_{FeC_2O_4 \cdot 2H_2O}}) \times M_{Fe(OH)3} \times 100\% \quad (4)$$

$$y_{FeOH} = \frac{m_2}{m_i} \times 100\% \quad (5)$$

In the formula: m is the mass of red mud, g; η_{Fe} is the leaching rate of atomic absorption spectrophotometry; m_1 is the quality of the preparation of ferrous oxalate, g; m_2 is the quality of the preparation of iron hydroxide, g; $w_{(Fe)}$ is the content of Fe in red mud, %; M_{Fe}=55.845 g/mol; $M_{FeC2O4} \cdot 2H_2O$=179.9 g/mol; $M_{Fe(OH)3}$=106.87 g/mol.

2.2. Experiment Reagent

The experimental red mud comes from an alumina plant in Chongqing. The reagents used in the experiment are shown in table 1.

Table 1. Experimental reagents.

Name	Specifications	Production company
Hydrochloric acid	AR	Chongqing Chuandong Chemical (Group) Co., Ltd.
Oxalic acid	AR	Chongqing Chuandong Chemical (Group) Co., Ltd.
Iron powder	AR	Chongqing Chuandong Chemical (Group) Co., Ltd
Sodium hydroxide	AR	Chongqing Chuandong Chemical (Group) Co., Ltd
Ammonium acetate	AR	Chongqing Chuandong Chemical (Group) Co., Ltd
Ammonium ferrous sulfate	AR	Chongqing Chuandong Chemical (Group) Co., Ltd
Sulphuric acid	AR	Chongqing Chuandong Chemical (Group) Co., Ltd
Potassium Permanganate	AR	Chongqing Chuanbeibei Chemical Reagent Factory
Phenanthroline nitrogen	AR	Chongqing Chuandong Chemical (Group) Co., Ltd
Hydroxylamine hydrochloride	AR	Chongqing Chuandong Chemical (Group) Co., Ltd
Ascorbic acid	AR	Chemical Reagent Co., Ltd.

2.3. Instruments and Characterization Methods

2.3.1. Laboratory Apparatus. The experimental and testing instruments are shown in table 2.

Table 2. Experimental instruments.

Name	Model	Production company
Electronic balance (1/100)	YP2102	Shanghai Guangzheng Medical Instrument Co., Ltd.
Analytical balance	PTX-FA220	Huazhi (Fujian) Electronic Technology Co.,

		Ltd.
Circulating water vacuum pump	SHZ-DSIII	Gongyi Yuhua Instrument Co., Ltd.
Electric blast drying oven	101-2AB	Tianjin Taisite Instrument Co., Ltd.
Water bath thermostatic oscillator	SHA-B	Jiangsu Jintan Medical Device Factory
UV-visible spectrophotometer	UV-5500	Shanghai Yuanxi Instrument Co., Ltd.
Atomic absorption spectrophotometer	PE-AA900T	U.S. platinum Elmer
X-ray diffractometer	D8 Advance	Brook, Germany
Field emission scanning electron microscope	Quanta 400 FEG	FEI, USA
X-ray fluorescence spectrometer	S8 tiger	Brook, Germany
Infrared Spectrometer	Spectrum Two	U.S. platinum Elmer

2.3.2. Product Characterization.
(1) X-ray powder diffraction

(2) The sample to be tested was ground into powder, pressed into pieces on a glass plate, and placed in an X-ray diffractometer at a scanning speed of 10°/min, a step width angle of 0.08°, and a scanning angle of 5 -90°.

(3) Infrared spectroscopy

(4) The KBr tabletting method was used and placed in an infrared spectrometer with a measurement wavenumber range of 400 -4000cm^{-1}.

(5) Thermal field emission scanning electron microscope

(6) The microstructure of the samples was observed by Regulus 8100 field emission scanning electron microscope.

2.4. Experimental Steps

2.4.1. Dealkalization Pretreatment of Red Mud. (1) The experiment was carried out after the red mud was placed in the oven for drying, grinding, and passing through a 40-mesh sieve.

(2) 5.00 g red mud and 30 mL 1mol·L^{-1} HCl solution were added into100mL beaker, and the mixture was stirred in a magnetic stirrer for1h and then filtered. The residue was dried in an electric blast oven for1h to obtain the dealkalized red mud.

2.4.2. Two-Stage Oxalic Acid Leaching. Dealkalized red mud (2.4.1) and oxalic acid solution (21.00 g oxalic acid solid dissolved in 70.00 mL distilled water) were added to a 100 mL beaker and placed in a water bath thermostatic oscillator. The water bath was heated to 363 K for 2h, and the filtrate was obtained by hot filtration.

2.5. Preparation of Ferrous Oxalate
(1) Iron powder reduction

0.50 g iron powder was added to the second-stage oxalic acid leaching solution (2.4.2) and heated to 323K in a water bath. Slowly add 1 mol·L^{-1} NaOH with a dropper, stir while adding, adjust to a pH of 3-4, stand still, cool, the crystal is completely precipitated and filtered to obtain ferrous oxalate. Drying, weighing and calculating the yield.

(2) Ascorbic acid reduction

The enediol in the molecular structure of ascorbic acid has strong reducibility. 0.50 g of ascorbic acid was added to the second-stage oxalic acid leaching solution (2.4.2) and heated to 323 K in a water bath. After the crystal is completely precipitated, the ferrous oxalate is obtained by filtration. Drying, weighing and calculating the yield.

(3) Photocatalytic reduction

In the case of light, it is easy to cause the electron transfer of the oxalic acid complex to promote the reduction of Fe^{3+} to Fe^{2+} [5]. The two-stage oxalic acid leaching solution (2.4.2) was placed under a photocatalytic device (as shown in figure 1), and the crystal was completely precipitated and filtered to obtain ferrous oxalate. We dry, weigh and calculate the yield.

Figure 1. Photocatalytic device diagram.

2.6. Preparation of Iron Hydroxide Colloid

In the 363 K constant temperature water bath, the lower filtrate after the separation of ferrous oxalate was added, and 1 mol·L^{-1} NaOH was slowly added with a dropper, stirring while adding until no reddish brown precipitate was produced. The filtrate was filtered to obtain ferric hydroxide. Drying, weighing and calculating the yield.

3. Results and Discussion

3.1. Analysis of Original Red Mud

The experimental raw materials were from an alumina plant in Chongqing, which was a reddish-brown solid (figure 2). The moisture content of the red mud sample was determined by the "Determination of Solid Waste, Moisture and Dry Matter Content Gravimetric Method (HJ 1222-2021)," and the result was 22.21%. In order to determine the types and contents of various elements in red mud, its chemical composition and mineral phase were analysed, as shown in table 3. The results show that red mud contains a large amount of Fe, Al, Si and other elements, and the content of Fe element is high (wt_{Fe2O3}=54.87%), which can be classified as high iron red mud (wt_{Fe2O3}≥30%). The XRD pattern analysis shows that the main phase in the red mud is Fe_2O_3 (JCPDF NO.87-1166), as shown in figure 3. The SEM results of figure 4 show that the red mud is composed of particles of different sizes and is distributed in layered mosaic, which confirms the research results that the mineral particle size is small and mutually cemented and inlaid, and it is difficult to separate effectively [10,11].

Figure 2. Red mud original sample.

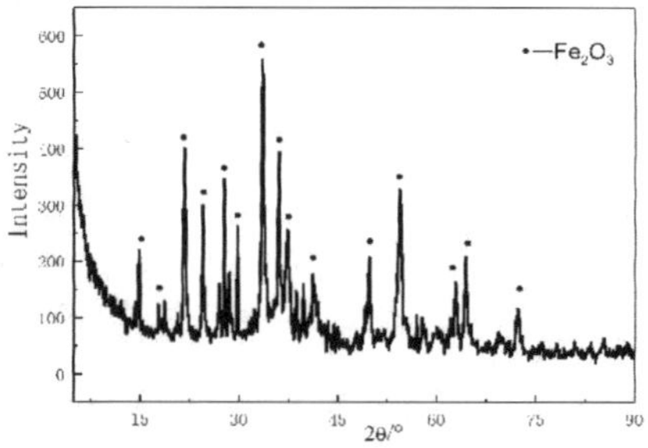

Figure 3. XRD pattern of red mud.

Figure 4. SEM of red mud.

Table 3. Main oxide components in red mud.

Components	Fe_2O_3	Al_2O_3	CaO	Na_2O	SiO_2	TiO_2
Content (wt%)	54.87	24.12	3.21	4.81	6.18	4.94

3.2. Analysis of Fe Leaching in Hydrochloric Acid-Oxalic Acid Two-Stage Acid Leaching Process

It was found that the leaching amount of Fe in the first-stage dealkalization process of hydrochloric acid was less through the pre-experiment, so the concentration of Fe in the first-stage dealkalization solution of red mud hydrochloric acid was determined by visible light spectrophotometer. The content of Fe in the solution during the two-stage acid leaching of oxalic acid was determined by atomic absorption spectrophotometer, as shown in table 4. In the first stage of hydrochloric acid dealkalization, the low concentration of hydrochloric acid in the first stage of hydrochloric acid dealkalization has less iron leaching in red mud, which mainly plays the role of adjusting pH, so that the alkalinity of red mud with pH=11 is adjusted to about 7; the second stage of oxalic acid greatly improved the leaching of iron in red mud, and the leaching rate reached 74.31%. The XRD phase analysis of the filter residue after two times of acid leaching showed that the characteristic peak of Fe_2O_3 basically disappeared. By comparing the XRD standard cards of ferrous oxalate and iron hydroxide, it was not found that there were ferrous oxalate and iron hydroxide crystals in the filter residue, as shown in figures 5 and 6.

Table 4. Iron leaching amount in different acid leaching process (mg/L).

Leaching stage	1	2	3	Mean value	Leaching rate (%)
A section of hydrochloric acid	3.035	3.101	3.062	3.066	0.0047
Oxalic acid second stage	4759	4753	4756	4756	74.31

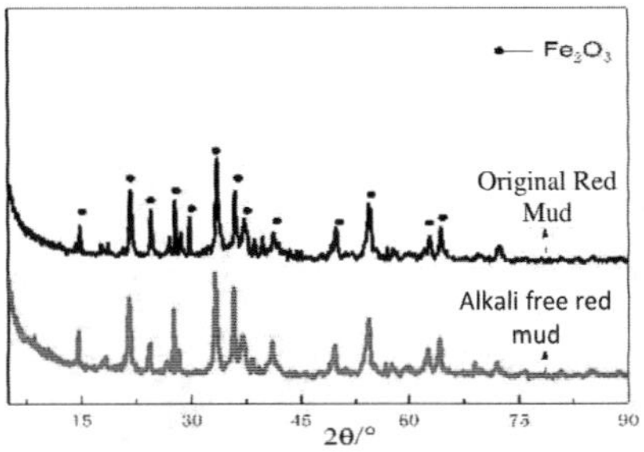

Figure 5. XRD of red mud before and after alkali removal.

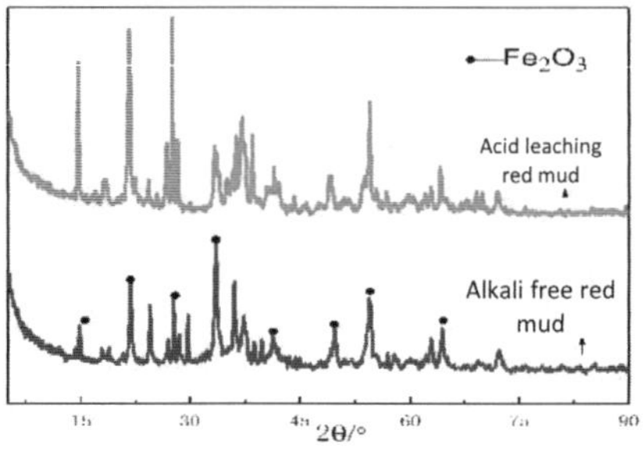

Figure 6. XRD of red mud and two-stage oxalic acid leaching red mud slag.

3.3. Effect of Different Reduction Conditions on the Preparation of Ferrous Oxalate

Ferrous oxalate can be used as a raw material for the new battery material lithium ferrous phosphate, and its application value is higher. Therefore, in the experimental design, priority is given to the directional conversion of iron in the solution into ferrous oxalate. The previous experimental study found that the red mud produced less ferrous oxalate after hydrochloric acid-oxalic acid two-stage acid leaching, which may be due to the fact that Fe^{2+} is easily oxidized to Fe^{3+} in the air. Therefore, three reduction conditions (iron powder Fe, ascorbic acid VC and photocatalytic reduction hv) were designed to reduce ferrous oxalate.

Through the analysis and characterization of appearance morphology, XRD, SEM and IR, it is found that ferrous oxalate is a yellow solid (figures 7 and 8). its microscopic morphology shows a regular rod-like structure and the crystal is complete. It can be seen from the infrared spectrum that the broad peak at 3300 cm^{-1} is the absorption peak of crystal water, the strong peak at 1619 cm^{-1} is due to the vibration of carboxyl C=O, the spectral peak at 820 cm-1 is the bending vibration absorption peak of O-C-O, and the two peaks below 700 cm^{-1} are the tensile vibration absorption peaks of Fe-O. The angle and intensity of the diffraction peak in the XRD pattern are consistent with the standard card of ferrous oxalate, and ferrous oxalate can be prepared by the three reducing agents. From the change of iron content in the solution under different reduction conditions, it can be seen that the iron ion content in the solution decreases rapidly and reaches equilibrium under the condition of ascorbic acid as reducing agent, and the factors such as the ease of operation of the experiment are comprehensively considered (figures 9-11). The reduction system formed by the addition of ascorbic acid has a better effect on the preparation of ferrous oxalate, and the time of producing ferrous oxalate is shorter and the yield is higher, as shown in table 5.

Figure 7. Ferrous oxalate solids.

Figure 8. SEM image of ferrous oxalate.

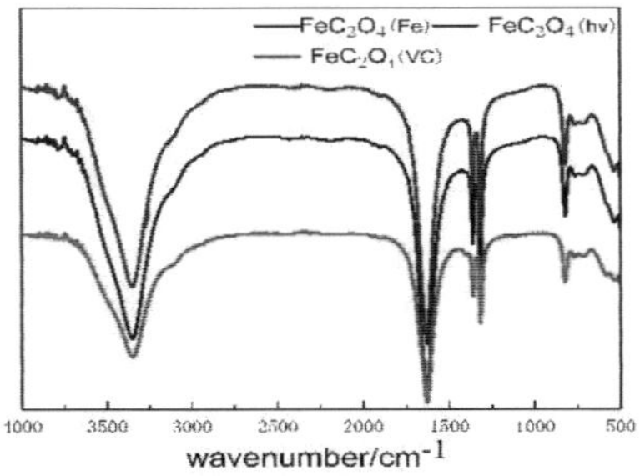

Figure 9. Infrared spectra of ferrous oxalate prepared by different reducing agents.

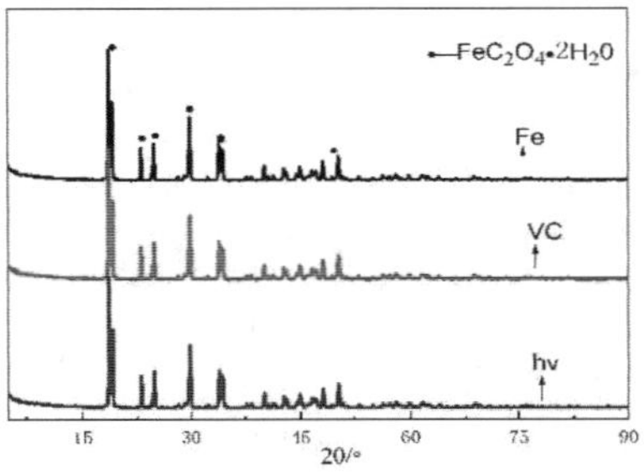

Figure 10. XRD patterns of ferrous oxalate prepared by different reducing agents.

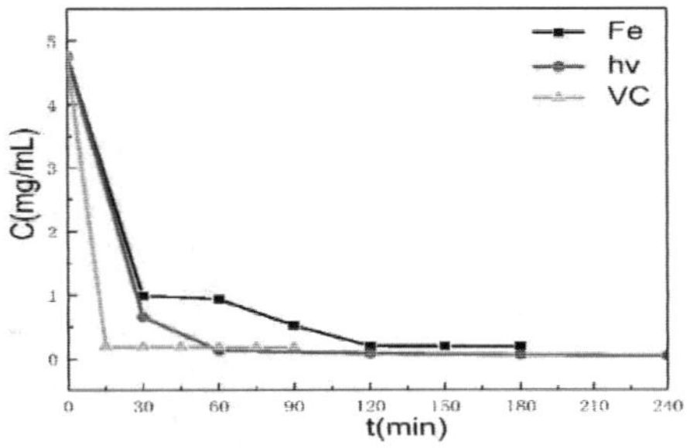

Figure 11. Variation of iron content in solution under different reduction conditions.

Table 5. The yield of ferrous oxalate prepared by different reducing agents (g).

Reducing agent	1	2	3	Mean value	Yield (%)
Ascorbic acid	6.5250	6.5245	6.5249	6.5248	85.04
Iron powder	5.4188	5.4183	5.4184	5.4185	70.62
Photochemical catalysis	6.0910	6.0801	6.0731	6.0814	79.26

3.4. Characterization of Iron Hydroxide

The filtrate obtained by filtration after the preparation of ferrous oxalate was reddish brown, indicating that it contained incompletely treated Fe^{3+}. If it is discharged directly, it is easy to feel uncomfortable and cause waste of resources. Therefore, according to the principle of precipitation balance, it is graded and recycled. The reddish-brown iron hydroxide (figure 12) recovered by three reducing agents was characterized by XRD, IR and SEM (figures 13-15). It was found that the angle and intensity of the diffraction peaks in the XRD pattern were consistent with the standard card of iron hydroxide (JCPDF NO.81-2022). Through IR analysis, it can be seen that the valence bond resonance frequency of the iron hydroxide samples recovered by the three reducing agents is in the same position; the morphology of iron hydroxide in SEM is spherical and different in size. It can be seen from table 6 that the yield of iron hydroxide under the condition of ascorbic acid as a reducing agent is lower than that under the other two reduction conditions. The reason may be that the addition of ascorbic acid promotes more iron in the solution to be converted into ferrous oxalate crystals.

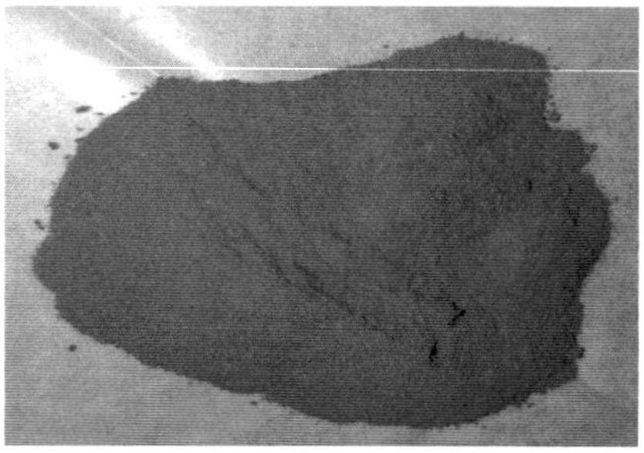

Figure 12. Iron hydroxide solid.

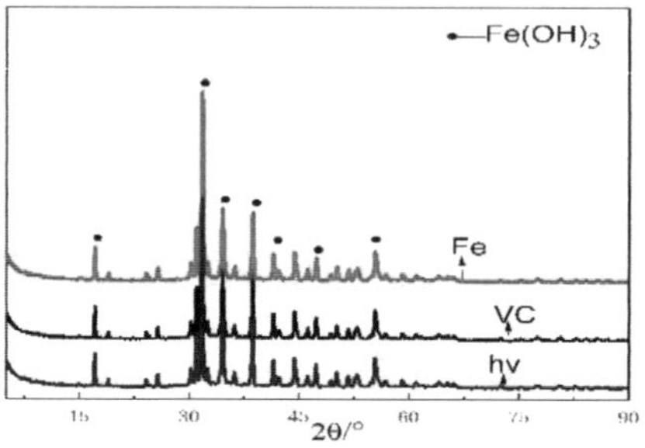

Figure 13. XRD of iron hydroxide prepared by different reducing agents.

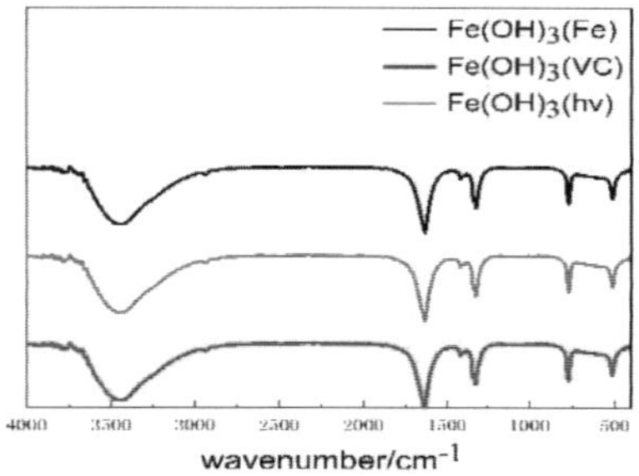

Figure 14. Infrared spectra of iron hydroxide prepared.

Figure 15. SEM image of iron hydroxide by different reducing agents.

Table 6. Production of iron hydroxide prepared by three reducing agents (g).

Reducing agent	1	2	3	Mean value	Yield (%)
Ascorbic acid	0.5814	0.5716	0.5865	0.5798	85.04
Iron powder	1.1615	1.1612	1.1609	1.1612	86.72
Photochemical catalysis	0.8522	0.8530	0.8532	0.8528	90.22

4. Conclusion

Oxalic acid is a widely existing organic acid. Its acidity is stronger than acetic acid, which is also a dibasic acid. It is a strong acid in organic acids. Compared with inorganic acids, its toxicity and danger are lower. Ferrous oxalate is the battery grade of lithium iron phosphate. Its preparation has great economic benefits. In this experiment, hydrochloric acid-oxalic acid two-stage grading acid leaching process was used to leach red mud. The sample prepared with ascorbic acid as a reducing agent had the best effect, which provided a new idea for the resource utilization of red mud.

Acknowledgments

This work was financially supported by science and technology research project of Chongqing Municipal Commission of Education (Grant No. KJQN202301336) and Natural Science Foundation of Yongchuan District (2023yc-jckx20075) fund.

5. References

[1] Zhang L X, Gao Y Q, Huang J H, et al. 2020 Research progress on resource comprehensive utilization of red mud *Bulletin of the Chinese Ceramic Society* **39** 144-149.

[2] Gu H N, Guo T F, Ma S C, et al. 2018 Review on separation, recovery, extraction and comprehensive utilization of iron from red mud *Chemical Industry and Engineering Progress* **37** 3599-3608.

[3] Liu X, Rong R, Dai M, Bian H, Peng C 2023 Preparation of red mud-based zero-valent iron materials by biomass pyrolysis reduction: Reduction mechanism and application study. *Science of The Total Environment* **864** 160907.

[4] Liu J J, Chen W, Zhou K G, et al. 2021 Research progress of iron recovery from red mud *Conservation and Utilization of Mineral Resources* **41** 70-75.

[5] Beuria P C, Biswal S K, Mishra B K, Roy G G 2017 Study on kinetics of thermal decomposition of low LOI goethetic hematite iron ore *International Journal of Mining Science and Technology* **27(6)** 1031-1036.

[6] He K, Zheng Z, Chen Z, Chen H, Hao W 2021 Kinetics of hydrogen reduction of Brazilian hematite in a micro-fluidized bed *International Journal of Hydrogen Energy* **46(5)** 4592-4605.

[7] Zhang B, Liu S, Zhu Y Y, Zhang J Y, Wu M, Liu C M 2022 Improvement and Optimization of Stoichiometric Ratio and Stability Constant Determination of Fe(phen)32+Complex *University Chemistry* **37(7)** 2110066-0.

[8] Zhang J L, Huang G J, Huang K R, Zhou L M, Cao M Y, Tang C Y 2016 Spectrophotometric determination of iron in 4 varieties of mangoes *Physical Testing and Chemical Analysis Part B: Chemical Analysis* **52(10)**.

[9] Huang Q L, Huang K, Lu Y H, et al. 2021 Recovery of iron and aluminum from red mud by oxalic acid leaching and solar photocatalysis *Environmental Engineering* **39** 199-205.

[10] Li H, Deng J, Jia Q, Zhu L, Huang L Z 2024 Enhanced $Fe(OH)_2$-driven reductive Dechlorination via shortened Fe-O bonds and colloidal medium *Water Research* **226** 118719.

[11] Ran G F, Ma H Z 2008 Simultaneous determination of trace Fe^{2+} and Fe^{3+} in rock salt by spectrophotometry *Journal of Salt Lake Research* **(1)** 22-26.

Identification Characteristics of Natural and Synthetic Cinnabar

Xihan Yang*, Wei Huang, Jiani Shen, Miao Li, Jing Wu, Shiyun Luo, Junfa Deng and Zhihao Xiao

National Jewelry Testing Center (Chongqing), Chongqing Academy of Metrology and Quality Inspection, Chongqing, China.
Email: *445088891@qq.com; 315683002@qq.com; 350003127@qq.com; 366279783@qq.com; 505273715@qq.com; 253933182@qq.com; 584104593@qq.com; 283582221@qq.com

Abstract. Using gemstone microscope, infrared spectrometer, micro laser Raman spectrometer, and laser ablation inductively coupled plasma mass spectrometer (LA-ICP-MS) to test and analyze synthetic and natural cinnabar on the market. The results showed that when synthetic cinnabar was in block form, obvious needle like straight longitudinal stripes were visible, which had significant differences in morphology from natural cinnabar. The aggregation of natural cinnabar often contains white calcite associated minerals, which appear as white clumps or veins on the surface. The luster differs greatly from the diamond luster of the main cinnabar. At the same time, the Se element content of natural cinnabar is much higher than that of synthetic cinnabar, and the Se content can be used as a reference to distinguish the two.
Keywords. Natural cinnabar; synthetic cinnabar; reducing sand; associated minerals; LA-ICP-MS.

1. Introduction

Mineral cinnabar was often used as a pigment and in pharmaceuticals in ancient times. Cinnabar powder, as a natural mineral pigment, has remained unchanged for thousands of years [1,2]. When used for medicinal purposes, it has the effects of calming and calming the nerves. Its bright red color and the traditional Chinese use and mysterious color of cinnabar make it unique in the jewelry industry and very popular among consumers.

The main chemical component of cinnabar is mercury sulfide (HgS), with crystals in plate-like or rhombohedral shapes, and aggregates in irregular granular, dense blocky, or earthy shapes. Single crystals are relatively rare. Cinnabar is a typical low-temperature hydrothermal mineral, often coexisting with calcite, quartz, and crystal, with numerous associated minerals. The main production areas of cinnabar in China are Tongren Wanshan in Guizhou and Xinhuang in Hunan. And its corresponding artificial compound, synthetic cinnabar, also known as 'Huan Yuan Sha' in the market, is a mercury sulfide crystal obtained by processing sulfur yellow powder and mercury. There is a lack of information on synthetic cinnabar in the field of gemology, but it has been recorded in traditional Chinese medicine for a long time. Tang Shenwei of the Song Dynasty first recorded it in the "Compendium of Materia Medica", and according to the refining method recorded, "Use one or two pieces of mercury and six baht of sulfur, finely grind it, first stir up the green sand head, then put it into a furnace of water and fire, and draw it like a bundle of needle patterns to achieve it." Its synthetic product is artificially synthesized cinnabar [3], and its characteristics are also recorded as needle shaped.

Natural cinnabar is relatively precious, especially when large crystals are extremely rare. Synthetic cinnabar is cheaper than natural cinnabar, only about one tenth of its price, and there may be residual free mercury in artificially synthesized cinnabar, which is not beneficial to the body. But in the market, there is a mixed bag of cinnabar products, with synthetic cinnabar replacing natural cinnabar for sale, and some even selling without distinguishing between them. Consumers are confused about how to distinguish between naturally produced cinnabar and synthetic cinnabar.

To study the identification characteristics of natural and synthetic cinnabar, six representative samples were collected from merchants for analysis, in order to further understand the relationship between natural and synthetic cinnabar and provide relevant scientific basis for consumers and jewelry testing departments.

2. Samples and Testing Methods

2.1. Sample Feature Description

The sample colors are all dark red to bright red, with some samples polished into the form of jewelry, while others maintain their original form. As shown in figure 1, CSN-1 and CSN-2 are natural cinnabar, while CSS-1, CSS-2, CSS-3, and CSS-4 are synthetic cinnabar, also known as 'Huan Yuan Sha'.

Figure 1. Samples photo (from left to right are CSN-1 CSN-2, CSS-1, CSS-2, CSS-3, CSS-4).

2.2. Test Method

This article conducted research and analysis using a gemstone microscope, infrared spectrometer, micro laser Raman spectrometer, and laser ablation inductively coupled plasma mass spectrometer (LA-ICP-MS). The main instruments and testing conditions are as follows:

2.2.1. Infrared Spectrometer. The infrared spectrometer used is Nicolet is5 Fourier Transform Infrared Spectrometer, and all tests were conducted at room temperature using the reflection method on the sample. The scanning frequency was 15 times, the testing range was 4000~400cm^{-1}, and the resolution was 4cm^{-1}.

2.2.2. Micro Laser Raman Spectrometer. Using Renishaw inVia laser confocal micro Raman spectrometer. Test conditions: 532nm and 785nm argon ion lasers are used to directly observe the sample through an eyepiece or camera, and separate the testing of cinnabar and cementitious materials. The testing range is 100-4000cm^{-1}.

2.2.3. LA-ICP-MS. Laser ablation inductively coupled plasma mass spectrometer (LA-ICP-MS) was used with iCAP RQ inductively coupled plasma mass spectrometer (Thermo Fisher) and New Wave 213 Nd: YAG laser ablation system (New Wave Research). The sample aerosol generated after laser ablation is transported by carrier gas helium and mixed with auxiliary gas argon before ICP. Experimental conditions: point erosion, laser *wavelength* of 213nm, energy density of 4.5J/cm^2, erosion spot size of 75 μm, pulse frequency of 10Hz; Standard samples: NIST 610 NIST612, BHVO-2G, BIR-1G, GSC-2G; The working conditions of the instrument shall be in accordance with GB/T 43589-2023 and T/CAQI 134-2020 [4,5].

3. Test and Analysis

3.1. Gemological Characteristics

CSN-1 is natural cinnabar, commonly known as 'Jing Ti Sha', which is a single crystal particle of cinnabar with a bright red color, a length of 11mm, and a width of 4.5mm. It exhibits the optical characteristics of a single crystal under a polarizing microscope, and its crystal is transparent and has developed cleavage. CSN-2 is natural cinnabar, commonly known as "raw ore cinnabar". It is a mineral aggregate of cinnabar, with a length of 22.2mm and a dense block shape. It is opaque and has a dark red color.

CSS-1 synthetic cinnabar is in the form of coarse and short columnar aggregates, with a length of 20.6mm. One end is flat while the other is rough, dark red, opaque, with a diamond like luster on the crystal surface, and fragile. CSS-2, CSS-3, and CSS-4 are synthetic cinnabar samples ground into jewelry, with a smooth, opaque surface and a dark red color; CSS-2 is a safety buckle with a diameter of 14.6mm and a thickness of 7.1mm. CSS-3 and CSS-4 are bead shaped with a diameter of 8.5mm. CSS-3 has a broken fracture surface where mineral particles can be seen.

3.2. Microscopic Magnification Observation

The natural cinnabar sample CSN-2 shows obvious white lumps or veins of other minerals under microscopic magnification of 10 times (figure 2).

The particle texture of the synthesized cinnabar samples is obvious, with clearly oriented needle shaped crystal particles visible; The crystal particles of CSS-1 are needle shaped, with vertical needle shaped crystals on the side, strong striped luster, and clear reflective crystalline particles arranged in a fence-like or clustered manner (figure 3); The CSS-2 reflected light shows the gaps formed by the boundaries of crystal particles on the surface of the visible sample, and the visible crystal particles are in the shape of long columns (figure 4); The cross-section of CSS-3 shows a clear columnar morphology of internal crystals (figure 5); The crystal morphology observed under magnification conforms to the record in 'Chinese Materia Medica': artificially synthesized cinnabar appears as flat blocks, and when viewed from the side, the grains usually form a neat fence like structure [3].

Figure 2. The white minerals in CSN-2.

Figure 3. The crystal particles of CSS-1.

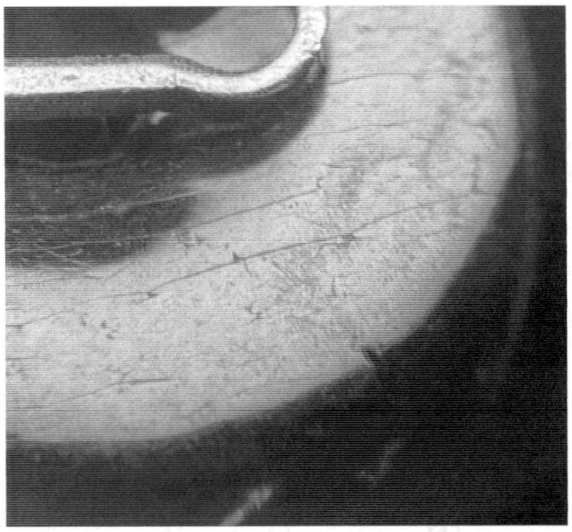

Figure 4. The gaps formed by the boundaries of crystal particles of CSS-2.

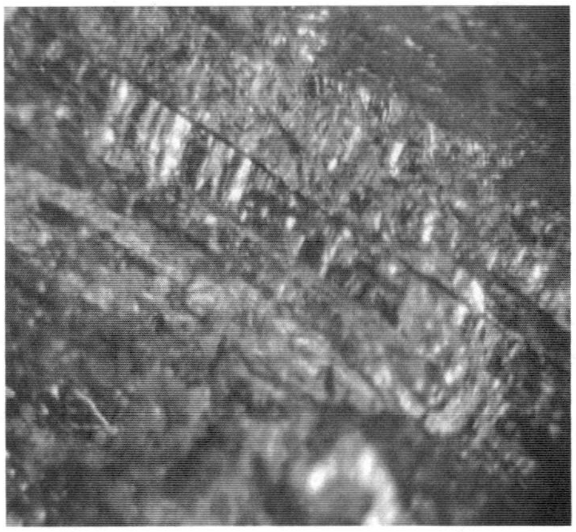

Figure 5. The crystal particles of CSS-3.

3.3. Infrared Spectrum

The white minerals in the CSN-2 sample were analyzed, and the characteristic spectral peaks were close to the infrared characteristic peak of calcite [6]. The clearly visible white opaque calcite in natural cinnabar aggregates is also one of the distinguishing features of natural cinnabar.

At the same time, under the same environment and conditions, infrared spectroscopy tests showed that the characteristic peaks of artificial resin, 3053cm^{-1}, 3034cm^{-1}, 2961cm^{-1}, 2927 cm^{-1}and 2870cm^{-1}, were measured in the infrared range of 2000-400cm^{-1} for CSS-2, CSS-3, and CSS-4 samples (figure 6), which overlapped with the infrared absorption peak of epoxy resin and matched the absorption band temperature caused by the coupling shift of benzene ring and - CH2-CH3 with benzene ring structure (i.e. artificial resin filling) [7]. It is inferred that CSS-2, CSS-3, and CSS-4 samples all contain a certain amount of artificial resin.

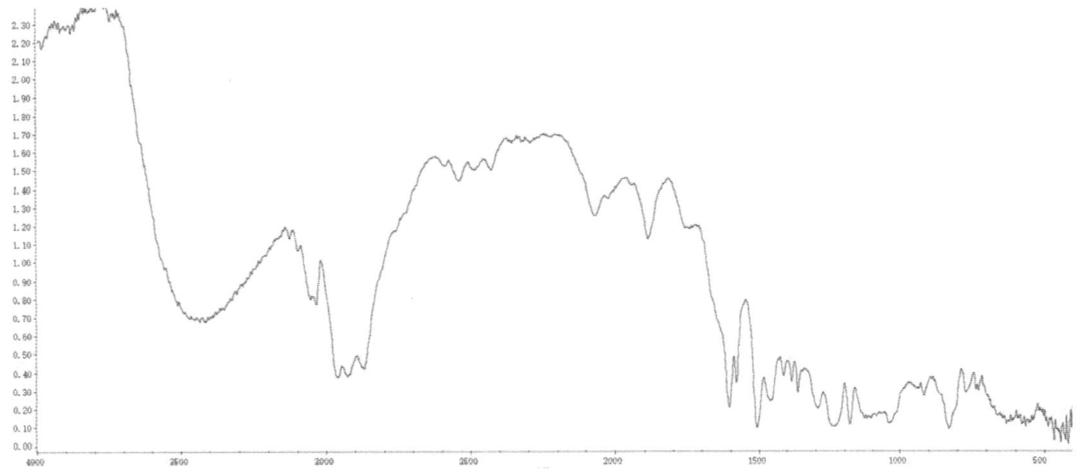

Figure 6. Infrared spectroscopy tests of artificial resin.

3.4. Micro Laser Raman Spectrometer

The Raman spectroscopy test results (figure 7) show that the six samples exhibit the same Raman shift, which is the characteristic Raman shift of cinnabar located at 254cm^{-1}, 287cm^{-1}, and 343cm^{-1}, respectively. 254cm^{-1} corresponds to the stretching vibration of Hg-S(A_1), while 287cm^{-1} and 343cm^{-1} correspond to the stretching vibration of Hg-S(E') [8]. However, the Raman spectra of synthesized cinnabar crystals and natural cinnabar are similar and cannot be distinguished from each other.

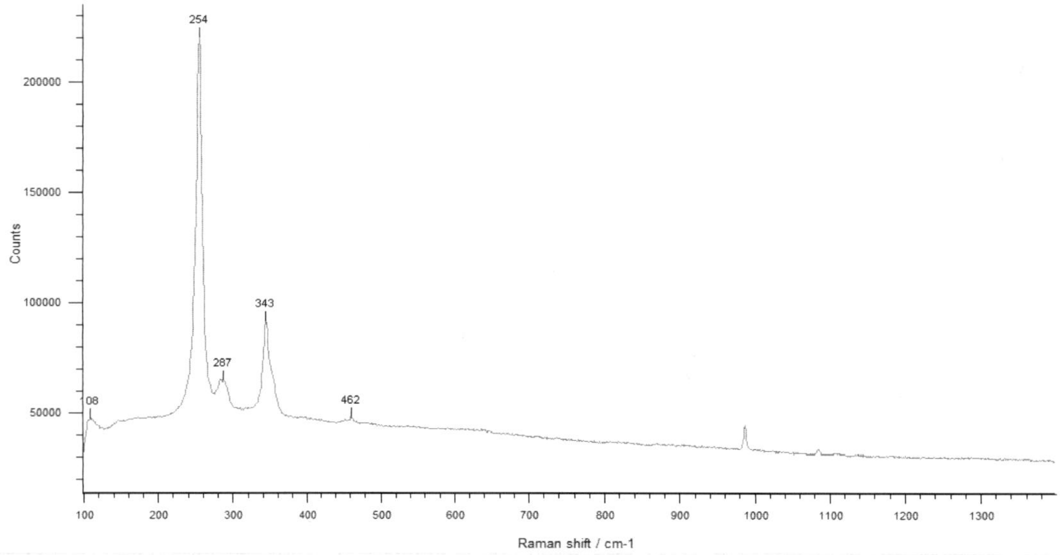

Figure 7. Raman spectra of cinnabar in the sample.

In addition, the gaps between crystal particles of samples CSS-2, CSS-3, and CSS-4 were tested (figure 8), and their Raman spectra exhibited characteristic Raman shifts of 1111cm^{-1}, 1184cm^{-1}, and 1607cm^{-1} for epoxy resin. The Raman shifts at 1607cm^{-1} and 1111cm^{-1} are caused by the stretching vibration of the carbon covalent bond of the phenyl group in the epoxy resin, while the Raman shift at 1184cm^{-1} is caused by the in-plane bending vibration of the carbon-hydrogen bond of the benzene ring [9]. It is inferred that the artificial resin of the sample mainly exists in the interstitial spaces between crystal particles.

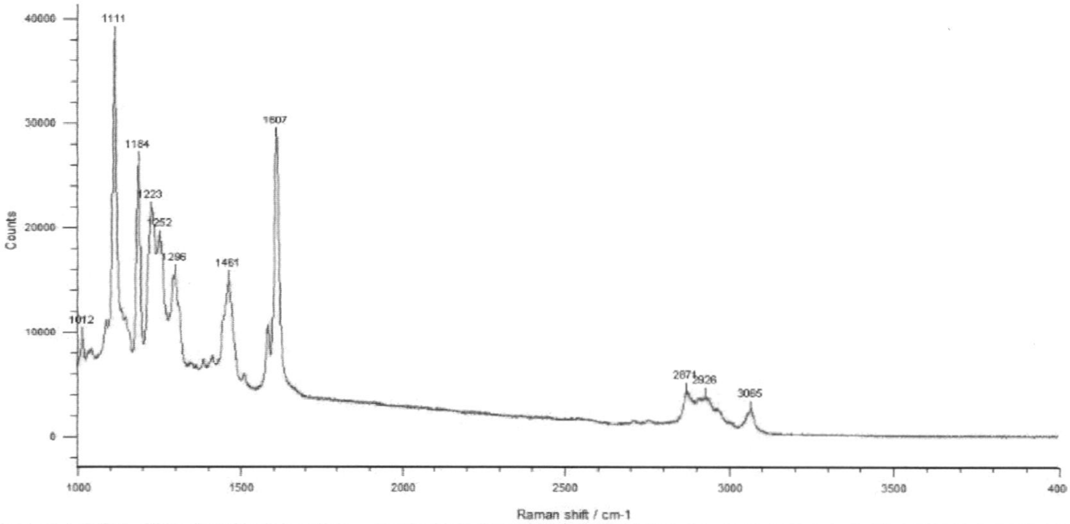

Figure 8. Raman spectra of artificial resin in the gaps between sample particles.

3.5. LA-ICP-MS

LA-ICP-MS was used for testing, and its accurate, in-situ, and micro area trace element detection method is suitable for analyzing only the cinnabar crystal particles in the sample. The results are shown in table 1.

Table 1. Trace elements in the sample.

	Ti	V	Ni	Cu	Zn	Se	Sr	Sb	Ba	Pb
CSN-1-1	4.062	0.013	0.035	2.193	43.321	153.99	-	33.993	-	80.208
CSN-1-2	4.729	0.007	0.278	16.744	2,751.64	181.19	0.729	61.956	0.275	255.147
CSN-1-3	4.006	0.021	0.035	0.997	35.704	161.42	0.033	46.255	-	158.322
CSN-2-1	5.375	0.006	0.035	0.797	26.659	115.67	0.298	1,699.86	1.374	1,135.36
CSN-2-2	5.451	0.006	-	4.784	12.853	106.88	0.331	1,213.68	1.099	503.725
CSN-2-3	5.189	0.002	0.556	24.12	49.51	122.2	1.26	1,440.60	4.946	508.071
CSS-1-1	6.123	0.019	0.278	25.515	87.119	1.56	1.127	91.997	1.099	190.067
CSS-1-2	5.968	0.013	0.382	25.116	52.366	1.418	0.53	67.211	1.649	171.369
CSS-1-3	5.388	0.027	0.452	37.475	99.496	1.277	1.657	93.618	1.649	226.422
CSS-2-1	5.728	0.012	0.07	3.189	21.422	0.142	0.232	22.316	0.275	154.325
CSS-2-2	5.843	0.018	-	6.179	18.09	0.615	0.961	19.202	-	121.212
CSS-2-3	6.612	0.019	0.104	9.369	63.792	0.52	0.265	23.613	-	139.084
CSS-3-1	6.317	0.012	0.139	14.95	17.138	1.229	0.033	9.86	-	55.647
CSS-3-2	6.133	0.014	0.417	13.156	158.054	0.804	0.133	11.352	0.275	91.06
CSS-3-3	5.324	0.013	0.348	8.97	20.47	1.04	0.265	12.326	-	77.827
CSS-4-1	7.195	0.015	0.382	31.097	1,739.61	1.466	0.365	3.957	1.099	80.084
CSS-4-2	7.093	0.016	0.348	30.1	444.191	1.797	0.994	4.217	3.297	202.99
CSS-4-3	6.716	0.013	0.104	18.937	168.528	2.553	0.265	3.698	2.473	29.44

[a] The unit is mg/kg.
[b] '-' represents below the detection limit.

Through comparative analysis of various elements, it was found that there is a clear distinction and significant differences in Se elements between natural and synthetic cinnabar crystals. The Se element content in natural samples is above 100mg/kg, while the Se element content in synthetic cinnabar crystals is below 3mg/kg, forming a clear distinction. Among the six samples, the Se element content in natural cinnabar particles is 40-300 times that of synthetic cinnabar particles.

According to relevant literature, natural cinnabar often contains a small amount of Se [10,11]. The common impurity elements in cinnabar can exist as microcrystalline solid inclusions in cinnabar, or undergo isomorphic substitution with cinnabar. Due to the similar crystal chemical properties of Se and S, they can form isomorphic substitution [12]. Therefore, during the production process of natural cinnabar, due to the substitution of isomorphic elements, its crystals contain a certain amount of Se, while synthetic cinnabar is made from mercury and sulfur as raw materials, making it difficult to contain Se elements. The purification of cinnabar is an ancient technology. If naturally produced cinnabar is artificially purified, it often does not contain or contains less impurity minerals, and cannot be distinguished from synthetic cinnabar by associated minerals and their corresponding elements. Its detection and discrimination have always been a difficult problem in the industry [13,14]. However, the results of this trace element test can effectively distinguish between the two through Se content.

4. Conclusion

After testing and analyzing the natural and synthetic cinnabar available in the market, it was found that:

- Synthetic cinnabar and natural cinnabar have significant differences in morphology. When synthetic cinnabar is in block form, obvious needle like straight longitudinal stripes can be seen, which are more obvious when observed from the fracture or side.
- The aggregate of natural cinnabar contains accompanying minerals of calcite, which appear as white clumps or veins on the surface. The luster differs greatly from the diamond luster of the main cinnabar, ranging from glassy luster
- to waxy luster. Infrared spectroscopy also shows obvious characteristic peaks of accompanying minerals, while synthetic cinnabar is usually prepared by chemical synthesis methods and is relatively pure.
- The Se element content of natural cinnabar is 40 to 300 times that of synthetic cinnabar, which can effectively distinguish.

Acknowledgments

This work was financially supported by the scientific research plan of the Chongqing Academy of Metrology and Quality Inspection (2023(YAN)-07) and (2023(YAN)-15).

5. References

[1] Lecroy B 2022 Gems on Canvas: Pigments Historically Sourced from Gem Materials *Gems and Gemology* **58(3)** 318-337

[2] Fang L and Bao L 2024 A Preliminary Study on the Function of Cinnabar in the Pre Qin Period *Identification And Appreciation To Cultural Relics* **282(15)** 148-153

[3] *State Administration of Traditional Chinese Medicine* 1999 Chinese Materia Medica (Shanghai Science and Technology Publishing House) p410

[4] Gems-Determination of Trace Elements-Laser Ablation-Inductively Coupled Plasma Mass Spectrometry T/CAQI 134-2020

[5] Gold Alloys Adornment-Determination of Multi-Element Contents-Laser Ablation-Inductively Coupled Plasma Mass Spectrometry GB/T 43589-2023

[6] Yang N, Kuang S Y and Yue Y H 2015 Infrared Spectra Analysis of Several Common Anhydrous Carbonate Minerals *Mineralogy and Petrology* **35(4)** 6

[7] Wang L, Deng L, Yang X, He L, Zhu C Y 2024 Ultraviolet Fluorescence and Infrared Spectrum Characteristic of "Traceless Repair" Treated Feicui (Jadeite Jade) *Journal of Gems &. Gemmology* **26(2)** 43-47

[8] Botticelli M, Maras A and Candeias A 2019 μ-Raman as a fundamental tool in the origin of natural or synthetic cinnabar: Preliminary data *Journal of Raman Spectroscopy* **51(9)** 1470-1479

[9] Zu E D, Chen D P and Zhang P X 2003 Identification of B Jade by Raman Spectroscopy *Spectroscopy and Spectral Analysis* **23(1)** 3

[10] Li S G, Xu H and Shen J F 2008 *Crystallography and Mineralogy* (Geological Publishing House) p170

[11] Gotoshia S V and Gotoshia L V 2008 Laser Raman and resonance Raman spectroscopies of natural semiconductor mineral cinnabar, α-HgS, from various mines *Journal of Physics D: Applied Physics* **(11)** 41

[12] Chen D F and Sun S Q 1991 Metacinnabar and Tiemannite in Hunan-Guizhou (Xiangqian) Mercury Metallogenic Belt *Acta Petrologica Et Mineralogica* **010(001)** 58-62.

[13] Shang J C, Zhou Z Y, Zhang Z W, et al. 2021 Gemological and Spectroscopy Identification Characteristics of Compacted Cinnabar *China Gems & Jades* **165(3)** 13-17

[14] Duan W W 2023 *Study on Identification Characteristics of Synthetic and Natural Cinnabar* (Beijing: China Academy of Chinese Medical Sciences) Master's Thesis p11

Impact of Various Warm Mixing Agents on the Performance of Asphalt Mixture

Fei Wang[1], Xinyong Zhang[1], Si Peng[1], Jia Wang[2,*] and Jie Yu[1]

[1] Tai Hang Urban and Rural Construction Group Co., Ltd., Shijiazhuang 050200, Hebei, China.
[2] Hebei Provincial Communications Planning Design and Research Institute Co., Ltd., Shijiazhuang 050091, Hebei, China.
Email: 782164963@qq.com; 3454252697@qq.com; duyinyu@126.com; *wjlvdyy@126.com; 99634740@qq.com

Abstract. Rubber asphalt mixtures facilitate the recycling of waste materials and enhance road performance. Nevertheless, they present challenges, including elevated construction temperatures and compaction difficulties. Additives effectively mitigate these issues by lowering the mixing temperature and improving compaction at reduced temperatures. In this study, we evaluated the influence of three distinct additives on rubber asphalt mixture properties. These additives uniformly decreased the mixing temperature by approximately 20°C without compromising porosity. The resultant warm mix rubber asphalt composites satisfy standard pavement performance criteria. While the additives bolstered high-temperature stability and moisture resistance, they slightly impaired the low-temperature crack resistance of the mixtures.
Keywords. Rubber asphalt; warm mixing agent; void ratio; compaction temperature; dynamic stability; flexural strain; freeze-thaw cracking.

1. Introduction

Rubber asphalt and its mixtures can not only achieve the recycling of waste materials and environmental protection but also effectively improve the road performance of materials, offering significant economic and social value [1-5]. However, with the increased incorporation of rubber powder, the viscosity of the asphalt becomes higher, leading to several issues during construction: a significant reliance on the construction season. In lower environmental temperatures, the temperature of the mixture drops too quickly, making it difficult to ensure the compaction and void ratio of the Rubber asphalt mixture; this results in lower compaction, affecting the construction quality and durability of the road surface, especially for surface layer Rubber asphalt mixtures, which are more prone to insufficient compaction. Both production and construction phases are completed in high-temperature environments, consuming enormous amounts of energy and emitting large quantities of toxic and greenhouse gases. This accelerates the aging of rubber asphalt, impacting its service life.

The warm mix agent effectively reduces the mixing temperature of asphalt mixtures, decreases emissions, and addresses issues related to paving and compaction in low-temperature environments. This extension of the construction season improves the overall asphalt pavement construction environment [6-9]. Given the increasing emphasis on energy conservation and environmental protection, warm mix asphalt represents the future direction of mixture technology development. The introduction of warm mix technology has lowered the production and construction temperatures at all stages of rubber asphalt mixtures. Despite the reduced mixing temperature, rubber asphalt mixtures still exhibit good workability at lower paving and compaction temperatures [10-13]. Consequently, this reduction in temperature enhances the paving and compaction efficiency of rubber asphalt mixtures. The author selected three types of warm mix agents, determined their optimal ratios, and added them to rubber asphalt containing 30% rubber powder content. The performance indicators of the asphalt were tested and compared, and the impact of different types of warm mix agents on the road performance of asphalt mixtures was analysed and evaluated [14,15].

2. Experimental Materials and Methods

2.1. Experimental Materials

The asphalt employed in this study is rubberized asphalt, characterized by a 30% rubber powder content. Its primary technical specifications are delineated in table 1.

Table 1. Technical indicators of asphalt rubber.

Test	Values	Experimental method
Penetration (100 g, 5 s, 25°C) /0.1 mm	54	ASTM D5
Softening Point (R&B) /°C	72.5	ASTM D36
Ductility (15°C, 5 cm/min), cm	15.9	ASTM D113
Flexible recovery (%)	87.5	
Viscosity (180°C)/Pa·s	3.1	

The chosen warm mix asphalt additives comprise EW04 (in liquid form), 1208 (in powdered form), and EC120 (in granular form). Both the coarse and fine aggregates employed are basalt, which have been evaluated in accordance with the "Highway Engineering Aggregate Test Specifications." All technical parameters satisfy the prescribed standards, and the adopted asphalt mixture gradation is ARHM-13.

2.2. Test Method Asphalt Performance Experiment

This study investigates the relationship between the volumetric parameters of asphalt mixtures and compaction temperature to determine optimal compaction temperatures for various warm-mix asphalt mixtures, following the experimental methods outlined in the "Test Regulations for Asphalt and Asphalt Mixture in Highway Engineering" and the technical indicators provided in the "Technical Specifications for Highway Asphalt Pavement Construction." Through a series of performance tests, the study verifies whether the high-temperature performance, water stability, low-temperature performance, and fatigue resistance of different warm-mix asphalt mixtures at their optimal compaction temperatures meet the required standards. Additionally, the study examines the patterns of change in relevant performance characteristics of various warm-mix asphalt mixtures.

3. The Impact of Warm-Mix Agents on the Performance of Asphalt Mixtures

3.1. Determining the Optimal Dosage of Warm Mix Agents

Viscosity, a critical parameter that characterizes the fluidity of asphalt, plays a pivotal role in the mixing and compaction processes of asphalt mixtures. This study initially examines the impact of varying dosages of warm mix agents on the viscosity of rubber powder-modified asphalt, aiming to establish the optimal blending ratios for different warm mix additives within this context. Table 2 illustrates how incremental additions of these agents influence viscosity. A discernible trend emerges: as the dosage of the warm mix agent escalates, there is a general decline in viscosity, albeit with varying magnitudes. To visually encapsulate this relationship, figure 1 presents a trend chart derived from table 2. Notably, the chart reveals a critical threshold for each of the three tested warm mix agents (EW04, 1028, and EC120) beyond which a substantial reduction in viscosity occurs. Specifically, these thresholds are set at 1.5%, 0.5%, and 1.0% for EW04, 1028, and EC120, respectively. Drawing from past construction practices, it is established that a modified asphalt viscosity approximating 2Pa·s optimizes the efficiency of mixing and compaction operations. Consequently, the ideal dosages for EW04, 1028, and EC120 warm mix agents are determined to be 1.5%, 1.0%, and 1.0% respectively, aligning with this operational sweet spot.

Table 2. Impact of warm mix agent content on the viscosity of rubber asphalt.

Dosage%	Viscosity Pa·s		
	RA+EW04	RA+1208	RA+EC120
0	3.1	3.1	3.1
0.5	2.82	2.22	2.92
1	2.72	2.09	2.18
1.5	1.75	1.78	1.9
2	1.57	1.63	1.67
2.5	1.42	1.44	1.52
3	1.29	1.28	1.39

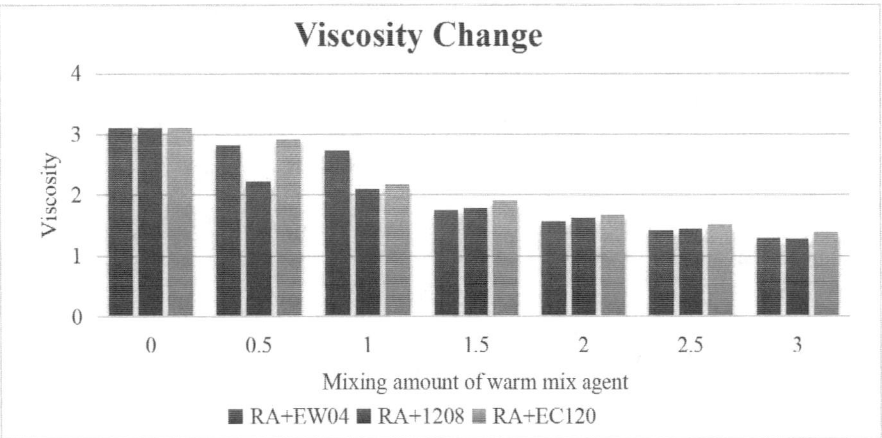

Figure 1. Impact of warm mix agent content on the viscosity of rubber asphalt.

3.2. Analysis of the Impact of Warm Mix Agents on the Performance of Asphalt Mixtures

The warm mix agent serves several key functions: (1) It acts as a surfactant between the asphalt and aggregate, improving the wettability of the asphalt and enhancing its ability to coat the aggregates. (2) It generates water droplets within the asphalt-water system, which reduces the surface tension between water and asphalt, thereby mitigating the impact of water on the asphalt-aggregate interface. (3) It prevents moisture penetration at the interface between asphalt and aggregate, thereby increasing the water stability of the asphalt mixture.

Asphalt mixtures prepared with crumb rubber-modified asphalt containing the optimal dosage of the warm-mixed agent exhibit a reduction in discharge temperature under the condition of ensuring proper air voids. Generally, they are about 20 degrees lower compared to those without the additive, as shown in table 3. This indicates that the addition of the warm-mixed agent allows mixing and compaction to be completed at lower temperatures, thereby achieving energy conservation and emission reduction goals.

Table 3. Impact of warm mix agent content on the viscosity of rubber asphalt.

Sample	Porosity	Discharge Temperature
30%RA	5.9	180
30%RA+EW04	6.0	161
30%RA+1208	5.9	162
30%RA+ EC120	5.4	162

3.3. The Effect of Warm Mix Additives on the High-Temperature Performance of Mixtures

Figure 2 presents the test results evaluating the high-temperature performance of warm mix rubber asphalt mixtures. Specifically, Hamburg wheel tracking data illustrates that after 20,000 load cycles, under combined water and temperature exposure, the rutting depth of the rubber asphalt mixture devoid of any warm mix additive was measured at 6.24 mm. In contrast, mixtures incorporating a warm mix agent exhibited reduced rutting depths, with the most significant reduction observed in the mixture containing the EC120 warm mix agent, which recorded a mere 3.15 mm—approximately a 50% decrease compared to the unadulterated mixture. The EW04 and 1208 warm mix agents also demonstrated efficacy, yielding reductions in rutting depth by 45% and 38%, respectively. These findings underscore the effectiveness of warm mix agents in enhancing both the rutting resistance and high-temperature stability of rubber asphalt mixtures.

Exceeding the melting point causes the warm-mix agent to liquefy and dissolve more effectively in asphalt, thereby enhancing its high-temperature stability. Furthermore, the integration of the warm-mix agent with asphalt amplifies its adhesive properties, leading to improved compaction performance. Conversely, when temperatures fall below the melting point, the agent crystallizes, encapsulating oil components within its structure. This crystallization process forms a network that substantially reduces asphalt's high-temperature flow deformation and bolsters its rutting resistance at elevated temperatures.

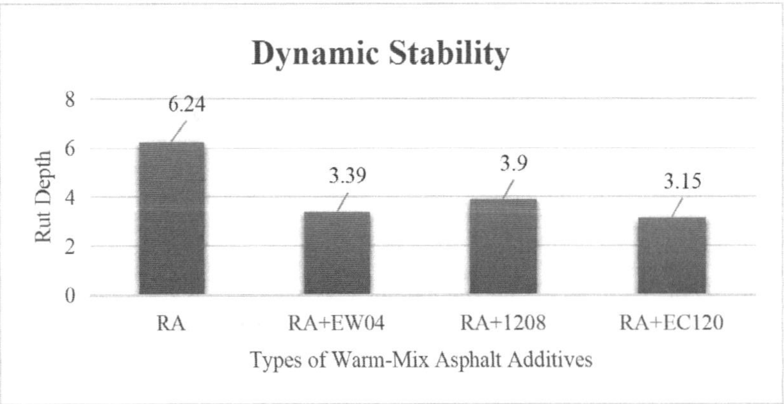

Figure 2. The effect of warm mix additives on the high-temperature performance of mixtures.

3.4. The Effect of Warm Mix Additives on the Low-Temperature Performance of Mixtures

Figure 3 presents the low-temperature bending test results for rubber asphalt mixtures before and after the incorporation of various warm-mixed agents. Notably, the maximum bending strain for the rubber asphalt mixture treated with the EC120 warm-mixed agent exhibited a substantial decline of 20%. This observation underscores the significant adverse effect that organic viscosity-reducing warm-mixed agents can have on the low-temperature crack resistance of the mixture. Conversely, while the addition of EW04 and 1208 warm-mixed agents also led to decreases in low-temperature bending strains compared to untreated mixtures, these reductions were marginal, amounting to 7% and 9%, respectively, thereby suggesting a lesser impact on the material's resistance to low-temperature cracking.

The incorporation of warm-mix agents diminishes the flexibility of rubber-modified asphalt mixtures, thereby reducing their stress-relieving capabilities. From the perspectives of flexural strength and load relaxation capacity, the low-temperature performance of these mixtures is not significantly influenced by warm-mix agents and may even be adversely affected. If the melting point of the warm-mix agent exceeds the ambient temperature, it precipitates in a crystalline form, creating a crystalline structure along with some saturated components that adsorb and dissolve. This can significantly reduce the crack resistance of asphalt mixtures once a certain number of cracks have formed, due to the stress concentration on the surface of the crystals caused by both external forces and temperature changes.

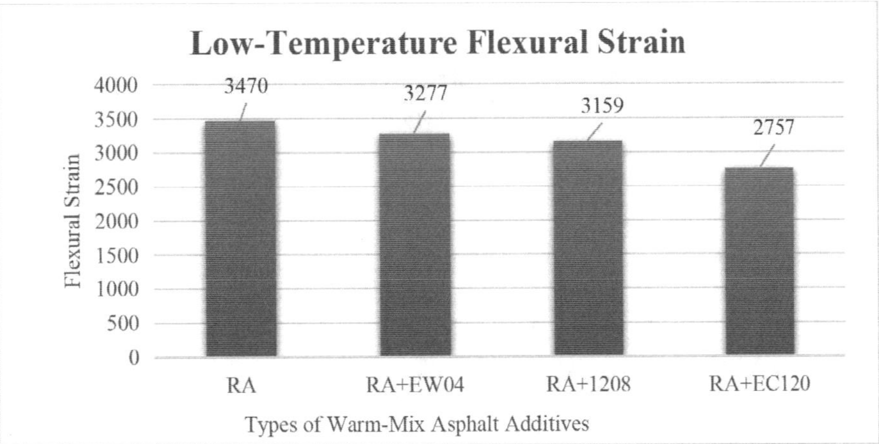

Figure 3. The effect of warm mix additives on the low-temperature performance of mixtures.

3.5. The Effect of Warm Mix Additives on the Water Stability Performance of Mixtures

The results of the freeze-thaw splitting tests for rubber asphalt mixtures with and without warm mix additives are presented in figure 4.

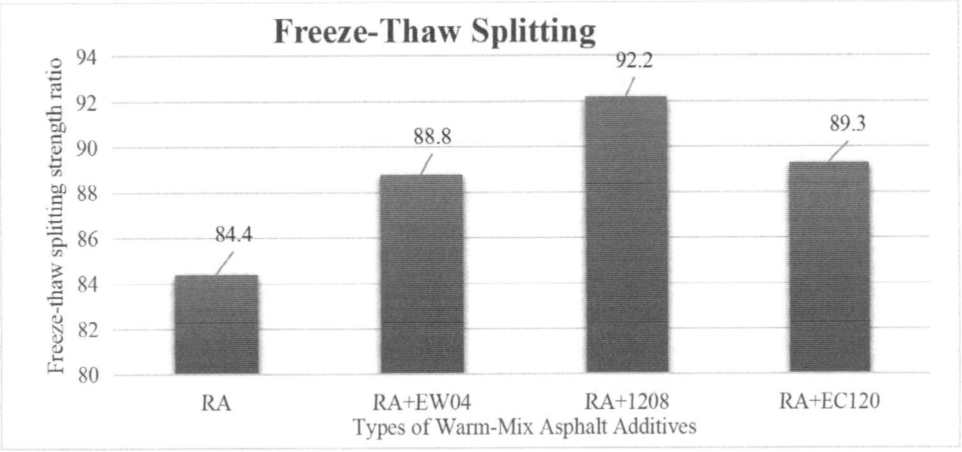

Figure 4. The effect of warm mix additives on the water stability performance of mixtures.

The figure shows that, compared to the mixture without a warm mix additive, the splitting strength ratios of the mixtures with the additive increased by 5%, 9%, and 6% respectively. This indicates that the addition of a warm mix additive effectively improves the anti-water damage capability of the mixture, thereby enhancing its water stability. The primary reason is that after adding the warm mix additive to the asphalt, the effective thickness of the asphalt film increases during the mixing process, which enhances the mutual adhesion between the asphalt and the aggregates. This prevents direct contact between the mineral materials and water, reducing the erosive effect of water on the mixture.

4. Conclusion

(1) Based on changes in viscosity indicators, the optimal dosages for the three warm mix agents—EW04, 1028, and EC120—were determined to be 1.5%, 1.0%, and 1.0%, respectively. Utilizing these dosages to prepare rubber powder-modified asphalt mixtures can reduce the discharging temperature by approximately 20 degrees Celsius while ensuring porosity, compared to asphalt mixtures without additives.

(2) The rut depth of asphalt mixtures incorporating warm mix agents decreased by 45%, 38%, and 50% respectively, compared to those without warm mix agents. This reduction indicates that the addition of warm mix agents effectively enhances the rutting resistance of the mixture, thereby improving the high-temperature stability of rubber asphalt mixtures.

(3) Compared to asphalt mixtures without warm-mix agents, those with added agents exhibited a reduction in low-temperature flexural strain by 7%, 9%, and 20%, respectively. This suggests that warm-mix agents negatively impact the crack resistance of the mixture, thereby reducing its low-temperature stability.

(4) The freeze-thaw splitting strength ratios of asphalt mixtures incorporating a warm mix agent increased by 5%, 9%, and 6% respectively, compared to those without the agent. This indicates that the addition of the warm mix agent effectively enhances the moisture damage resistance and improves the water stability of rubber asphalt mixtures.

5. References

[1] Wang H H and Dong Z 2021 Effect of surface-active warm mixing agent on the performance of SBS modified asphalt *Petroleum Asphalt* **34(4)** 43-46.

[2] Liu S Y 2021 Study on the influence of warm mixing agent on rheological property of warm mixing rubber asphalt *Hunan Communication Science and Technology* **47(3)** 71-74.

[3] Kanitpong K, Sonthong S, Nam K, et al. 2007 Laboratory study on warm-mix asphalt additives *Transportation Research Board 86th Annual Meeting*.

[4] Jones D, Wu R, Tsai B, et al. 2008 Warm-mix asphalt study: test track construction and first-level analysis of phase 1 HVS and laboratory testing *Asphalt Concrete* 1-140.

[5] Akisetty C, Lee K, et al. 2009 High temperature properties of rubberized binders containing warm asphalt additives *Construction & Building Materials* **23(1)** 565-573.

[6] Brown D 2010 The warm mix fix for rubberized asphalt *HMAT Hot Mix Asphalt Technology* **26 -27** 30-31.

[7] Li K N, Yan X L and Wang Y J 2023 Study on Rheological Properties and Microscopic Mechanism of Warm Mix SBS modified asphalt *Applied Chemical Industry* **52(12)** 3309-3315.

[8] Wang Y L, Lu J, Liang P F, et al. 2023 Influence of different warm mix agents on rheological and microscopic properties of high-viscosity asphalt *Materials Reports* **37(26)** 1-6.

[9] RodrÍGuez-Alloza A M, Gallego J and PÉRez I 2013 Study of the effect of four warm mix asphalt additives on bitumen modified with 15% crumb rubber *Construction & Building Materials* **43(3)** 300-308.

[10] Jones D, Farshidi F and Harvey J T 2013 Warm-mix asphalt study: summary report on rubberized warm-mix asphalt research *Pavement Performance* 1-17.

[11] Kim H H, Jeong K D, Lee M S, et al. 2015 Effect of FT paraffin wax contents on performance properties of crumb rubber-modified asphalt binders *American Society of Civil Engineers* **37** 107044.

[12] RodrÍGuez-Alloza A M and Gallego J 2017 Mechanical performance of asphalt rubber mixtures

with warm mix asphalt additives *Materials & Structures* **50(2)** 147.
[13] Ma X, Zhao Q and Wang H H 2023 Effect of surface active warm mix on high temperature performance of SBS modified asphalt *Shan Dong transportation technology* **24(5)** 31-34.
[14] Wang Q, Liu S, Wu X Y, Wei S F and Ou C F 2016 Weather aging resistance of different rubber modified asphalts *Construction and Building Materials* **106** 443-448.
[15] Li X, Chen J and Shi K J 2015 Thermal oxidative aging mechanism of crumb rubber/SBS composite modified asphalt *Construction and Building Materials* **5** 169-175.

Influence of Rubber Powder Content on the Rheological Properties of Rubber Asphalt

Long Wang[1], Kai Wang[1], Han Lu[1], Chao Sun[2,*], Huilin Meng[2] and Yang Luan[2]

[1] Shijiazhuang Jiaojian Expressway Construction Management Co., Ltd, Shiheng Branch, Shijiazhuang 050000, Hebei, China.
[2] Hebei Provincial Communications Planning, Design and Research Institute Co., Ltd, Shijiazhuang 050091, Hebei, China.
Email: sc1517@163.com; sc1517@126.com; tgd11517@163.com; *1207589155@qq.com; sctg1517@163.com; 861351921@qq.com

Abstract. Rubber asphalt has excellent durability and can maintain excellent performance under harsh service conditions. In order to further promote the application of rubber asphalt, the dynamic rheological shear test was used to test the complex shear modulus G *, rutting factor G */sin δ, fatigue dissipation factor G * sin δ of rubber asphalt under different test temperatures and loading frequencies. The conclusions are as follows: (1) For the dynamic rheological properties of rubber asphalt, the content of rubber powder is an important factor. When the rubber powder content is 30%, the rheological properties of rubber asphalt are optimal. (2) For the rutting factor and fatigue dissipation factor of rubber asphalt, temperature and frequency are the primary influencing factors. The rutting factor and fatigue dissipation factor of rubber asphalt samples in each group decrease with increasing temperature and increases with increasing frequency. (3) Among the three types of rubber powder-mixed rubber asphalt, the 30% rubber asphalt has higher high-temperature performance and fatigue performance, which can meet the performance requirements of the pavement surface layer.
Keywords. Dynamic rheological properties; complex shear modulus; rutting factor; fatigue dissipation factor.

1. Introduction

According to data from the Ministry of Commerce, the annual production of waste rubber tires in China is about 330 million, equivalent to more than 10 million tons in weight. Waste tires are extremely difficult to degrade, and traditional treatment methods such as stacking, burning, and burying can cause serious pollution [1,2]. The recycling and utilization of waste tires have always been a worldwide technical problem.

In order to effectively recycle waste rubber tires, scholars from various countries have made rubber powder from waste rubber tires in recent years and used it as a modifier for asphalt in road engineering [3-5]. As a typical viscoelastic and temperature-sensitive material, rubber asphalt has a complex mechanical response under temperature-load coupling conditions during its service life [6]; Generally, the rutting test and four-point bending test are often used to evaluate the high-temperature performance and fatigue performance of asphalt mixture [7]. Both processes of the two tests are complex, and the mechanical response of the mixture under complex stress conditions cannot be accurately described by the test results [8-10]; In order to compare the performance of rubber asphalt with different rubber powder content, rubber asphalt with 20%, 30% and 40% rubber powder content was prepared. The rheological properties of rubber asphalt with different rubber powder content were tested by the DSR test. According to the experimental results, the influence of different rubber powder content on the rheological properties of rubber asphalt was analyzed.

2. Preparation of Rubber Asphalt

The addition of rubber powder to asphalt will cause the fluidity of asphalt to deteriorate and some properties to decrease. The performance of rubber asphalt is usually improved by adding some admixtures. Admixtures mainly include (1) aromatic hydrocarbon softeners: mainly composed of naphthenes, paraffins, and nitrogen-containing organic bases, with a density of 0.92-1.11g/m3; (2) Activator: mainly composed of styrene-acrylic emulsion polymer and calcium carbonate-based substances, with a density of 0.97-1.12g/m3; (3) SBS modifier: relative density 0.92-0.95.

There are three key parameters in the process of rubber asphalt preparation: swelling time, shear time, and asphalt temperature. The preparation process of rubber asphalt with different rubber powder contents is as follows: heat 70# base asphalt to 175°C, add three additives, and stir for 10 minutes; Then add rubber powder to the asphalt, heat it to 190°C, and stir for 30 minutes to fully swell the rubber powder particles; Finally, use a shearing machine to shear the asphalt at a shear rate of 7000rpm for 45 minutes. After shearing, the rubber asphalt content is cooled to room temperature.

Table 1. Basic properties of rubber asphalt.

Project		20% glue to asphalt	30% rubber to asphalt	40% rubber to asphalt
Rotational viscosity at 180°C (Pa·s)		2.2	2.7	3.3
Penetration at 25°C (0.1mm)		43	64	68
Softening point (°C)		71	74	70
5°C ductility (cm)		13	20	17
Residue after TFOT	Mass loss (%)	-0.34	-0.34	-0.31
	Penetration ratio at 25°C (%)	84	77	76

As shown in table 1, the penetration of the three rubber asphalts is greater than 30 (0.1mm), indicating that rubber asphalt has high crack resistance; The softening point of different rubber powder contents is above 70°C, which means that the three rubber asphalts have strong resistance to deformation. Among them, the rubber asphalt with a 30% rubber powder content has the highest softening point and the strongest resistance to deformation. The rotational viscosity of the three rubber asphalts at 180°C is less than 3.5Pa·s, indicating that the Sion rubber asphalt has good fluidity and workability.

3. Dynamic Rheological Shear Test

The dynamic rheological test was conducted on different amounts of rubber-modified asphalt using a TA dynamic shear rheometer (DSR) from the United States. A parallel plate with a diameter of 25mm is selected, and the plate gap is set to 1mm. The loading frequency is 10rad/s±0.1rad/s, and the load type is sinusoidal oscillating load, with temperatures of 60°C, 66°C, 72°C, 78°C, 84°C, 90°C, and 96°C. The rheological test parameters such as G*, phase angle δ, and G*/sinδ were measured.

*3.1. Analysis of the Results of the Complex Shear Modulus Test G**

The complex shear modulus characterizes the strain response exhibited by asphalt under the application of shear stress, and it is frequently employed as a key parameter for evaluating the viscoelastic properties of asphalt materials.

Dynamic rheological shear testing was conducted to measure the complex shear modulus (G*) of rubber asphalt samples with varying rubber powder concentrations. The outcomes of this testing are illustrated in figure 1.

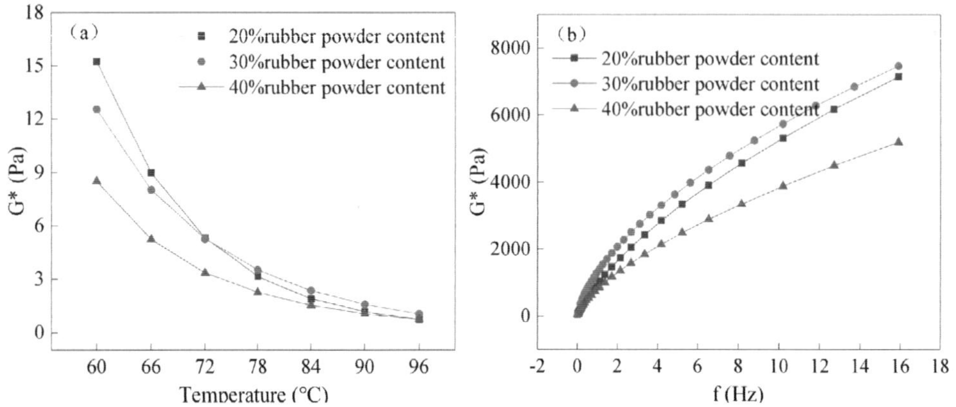

Figure 1. (a) Complex shear modulus of rubber asphalt under different temperature conditions; (b) Complex shear modulus of rubber asphalt at 90°C.

As shown in figure 1, Temperature and frequency serve as the primary influencing factors that determine the complex shear modulus of rubber asphalt.

The complex shear modulus of rubber asphalt, featuring three distinct rubber powder contents, exhibits a consistent trend of gradually decreasing as the temperature rises and gradually increasing as the frequency increases. When the temperature is the same, G* of rubber asphalt with a 30% rubber powder content under different loading frequencies is higher than that of the other two rubber asphalts, indicating that rubber asphalt with a 30% rubber powder content has excellent resistance to deformation; For rubber with a 20% rubber powder content, when the loading frequency is 10Hz and the test temperature is increased from 60°C to 90°C, G* of the asphalt sample decreases from 15.244Pa to 1.165Pa, a decrease of 92.36%. The frequency has a pronounced impact on the G* of rubber asphalt, demonstrating a clear positive correlation between frequency and the complex shear modulus for all three types of rubber asphalt samples. For rubber asphalt with a 30% rubber powder content, when the test temperature is 90°C and the loading frequency is increased from 1Hz to 15Hz, the complex shear modulus increases from 371.812Pa to 2273.481Pa, an increase of approximately 511.46%. The addition of rubber powder has a significant impact on the complex shear modulus of rubber asphalt. When the addition of rubber powder is 30%, the dynamic rheological properties of rubber asphalt are optimal.

The reason for the above test results can be understood as follows: rubber asphalt is a viscoelastic material. When the test temperature is constant, under the action of the test load, there is a certain amount of hysteretic deformation, i.e., viscoelastic deformation, in the asphalt. When the frequency increases, the hysteretic deformation gradually recovers with the increase of frequency, and the accumulated energy in the asphalt gradually releases, which is manifested macroscopically as an increase in the complex shear modulus of rubber asphalt; When the test temperature is low, the mechanical response of asphalt is close to elasticity, that is, the deformation immediately recovers after unloading. At this time, the complex shear modulus of rubber asphalt is high. When the test temperature is high, the mechanical response of rubber asphalt is viscoelastic. When the specimen is subjected to load, the stress, strain, and deformation have a certain hysteresis, and the deformation cannot be immediately restored. The macroscopic manifestation is a decrease in the complex shear modulus of rubber asphalt.

3.2. Analysis of Rutting Factor G/Sinδ Results*

The rutting factor G*/sinδ is commonly used to evaluate the high-temperature stability of asphalt. Figure 2(a) shows the rutting factor of rubber asphalt at different temperatures, and figure 2(b) shows the rutting factor of rubber asphalt at 90°C.

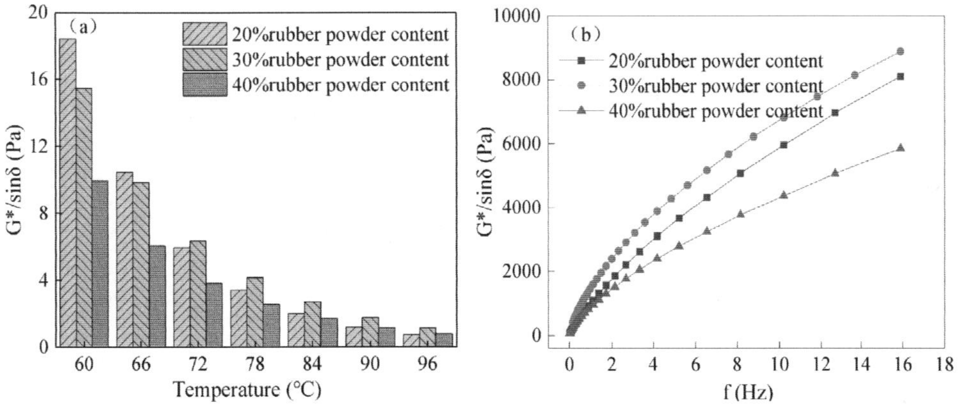

Figure 2. (a) Rubber asphalt rutting factor under different temperature conditions; (b) Rubber asphalt rutting factor at 90°C.

As can be seen from figure 2, (1) temperature and frequency are pivotal factors influencing the rutting factor of rubber asphalt. The variation pattern of the rutting factor (G/sinδ) across various rubber asphalt samples within each group, in relation to temperature and loading frequency, is fundamentally similar: it decreases with escalating temperatures and increases with rising frequencies. This signifies that higher frequencies and lower temperatures enhance the shear deformation resistance of rubber asphalt. At a constant loading frequency, as the test temperature rises from 60°C to 90°C, the rutting factor (G/sinδ) of rubber asphalt with 20% rubber powder content diminishes substantially, from 18.415Pa to 1.213Pa, a decrement of 93.42%. Conversely, at a fixed temperature of 90°C, when the frequency augments from 1Hz to 15Hz, the rutting factor (G*/sinδ) of rubber asphalt containing 30% rubber powder intensifies, escalating from 2914.1Pa to 8889.9Pa. (3) Under identical conditions, the rubber powder content exerts a notable influence on the rutting resistance of rubber asphalt. Specifically, rubber asphalt with 30% rubber powder demonstrates superior shear deformation resistance compared to those with 20% and 40% rubber powder content, indicating that the rutting resistance of 30% rubber asphalt outperforms that of the other two compositions.

The reason for the above test results can be understood as follows: rubber asphalt is a typical viscoelastic material. When the temperature is low, rubber asphalt mainly exhibits elastic characteristics. After being subjected to load, the deformation of rubber asphalt can be quickly restored. When the temperature is high, rubber asphalt mainly exhibits viscous characteristics. When the asphalt pavement is subjected to vehicle load, the deformation recovery is delayed. Therefore, when the temperature is low, the anti-rutting factor is larger, and vice versa when the temperature is high.

*3.3. Analysis of the Results of Fatigue Dissipation Factor G*Sinδ*

The fatigue dissipation factor G*sinδ is commonly used to evaluate the fatigue resistance of asphalt. The larger the fatigue dissipation factor, the more rapidly the fatigue damage accumulates in the asphalt sample, and the worse the fatigue resistance. Figure 3(a) shows the fatigue dissipation factor of rubber asphalt under different temperature conditions, and figure3(b) shows the fatigue dissipation factor of rubber asphalt at a temperature of 90°C.

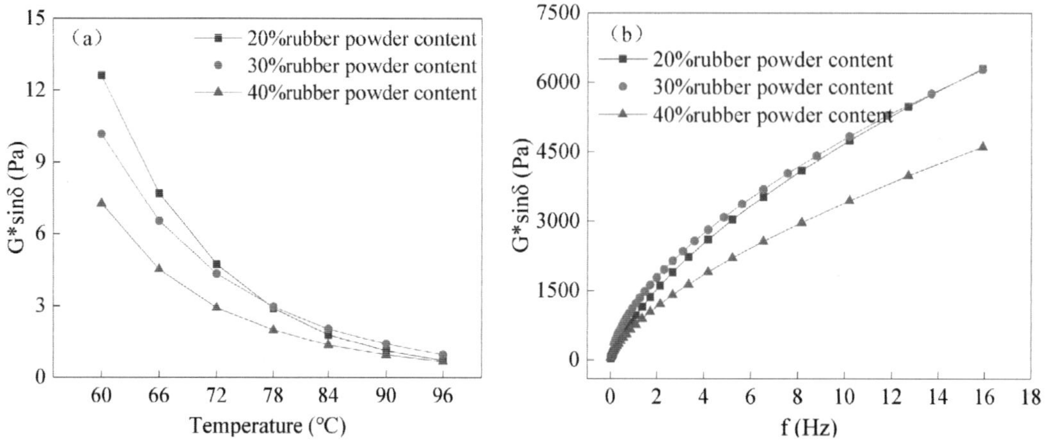

Figure 3. (a) Complex shear modulus of rubber asphalt under different temperature conditions; (b) Complex shear modulus of rubber asphalt at 90°C.

In figure 3, (1) temperature and frequency serve as the primary variables that govern the fatigue dissipation factor of rubber asphalt, exerting a substantial influence on its performance. The variation trend of the fatigue dissipation factor of rubber asphalt with temperature and frequency is consistent for the three types of rubber powder content, that is, the higher the temperature, the smaller the fatigue dissipation factor, and the higher the frequency, the larger the fatigue dissipation factor. When the loading frequency is consistent, the test temperature increases from 60°C to 90°C, and the fatigue dissipation factor G*sinδ of 20% rubber powder content decreases from 12.619 Pa to 1.120 Pa, a decrease of 91.12%. When the temperature is 90°C and the loading frequency is increased from 1Hz to 15Hz, the fatigue dissipation factor G*sinδ of rubber asphalt with 30% rubber powder content increases from 2143.622Pa to 6276.247Pa.

4. Conclusion

(1) The inclusion of rubber powder significantly influences the dynamic rheological properties of rubber asphalt, with an optimal performance observed for the blend containing 30% rubber powder content. The complex shear modulus (G*) of rubber asphalt samples with varying rubber powder percentages displays a consistent trend of diminishing with rising temperature and augmenting with increasing frequency.

(2) Temperature and frequency are crucial factors that significantly influence the rutting resistance and fatigue dissipation characteristics of rubber asphalt. Specifically, as the temperature rises, both the rutting factor and fatigue dissipation factor of each group of rubber asphalt samples undergo a reduction, whereas they exhibit an enhancement with an increase in frequency.

(3) Among the three types of rubber powder-containing rubber asphalts, the 30% rubber asphalt has higher high-temperature performance and fatigue performance, which can meet the performance requirements of the pavement surface layer.

5. References

[1] Wang X C, Jiang B Y and Nie X M 2022 Prospects and suggestions for the recycling of waste tires in China *Rubber & Plastics Technology and Equipment* **48(08)** 5-8.

[2] Gao J N, Li G f and Gu L T 2022 The current situation and research of resource recovery and harmless technology of waste tires *Ageing and Application of Synthetic Materials* **51(02)** 117 -118+14.

[3] Zhang Z Q, Fang Y and Li Z L 2020 Experimental study on the performance of desulfurized rubber asphalt and ordinary rubber asphalt *Journal of Chongqing Jiaotong University (Natural Science Edition)* **39(04)** 74-79+91.

[4] Wei W F, Zhang Y P, Wei Z, et al. 2024 Study on the rheological and fatigue properties of rubber high-viscosity asphalt based on DSR/BBR *Applied Chemical Engineering* **08(07)** 1-9.
[5] Lv L, Zhao C, Li R, et al. 2023 Preparation process and performance evaluation of olefin metathesis catalyzed rubber powder modified asphalt C*hina Journal of Highway and Transport* **36(12)** 143-154.
[6] Wang G Q, Cao D W, Wang Z B, et al. 2022 Comparative study on the rheological properties of high-volume rubber powder modified asphalt cement *Journal of Highway and Transportation Research and Development* **39(07)** 7-14+22.
[7] Wang L, Wang T, Wang J. 2024 Rheological properties of graphene oxide/polyphosphate composite modified asphalt *Contemporary Chemical Industry* **53 (09)** 2061-2066.
[8] Shi F Z, Wang Y Y, Long D, et al 2024 Preparation and rheological properties of high viscosity rubber modified asphalt *Aging and Application of Synthetic Materials* **53 (04)**: 62-65.
[9] Chen L H, 2024 Rheological and microscopic properties of surface modified diatomite modified bio asphalt *Western Transportation Technology* **20 (10)** 20-22+161.
[10] Zhang Y, Zhang X Q, Xu M, et al 2024 Review on coupling aging of asphalt binder *Highway transportation technology*, **41 (07)**: 1-13+74.

Research on the Performance of PTFE Gasket Materials under High-Temperature and High-Pressure Environments

Mingyu Bao[1,*], Yan He[2], Chang Liu[1], Qian Ma[3], Xiaoyan Zhang[3] and Shuliang Wang[3,#]

[1] Safety, Environment & Technology Supervision Research Institute, PetroChina Southwest Oil & Gasfield Company, Chengdu 610041, Sichuan, China.
[2] Sichuan Changning Natural Gas Development Co., Ltd., Chengdu 610000, Sichuan, China.
[3] School of New energy and materials, Southwest Petroleum University, Chengdu 610500, Sichuan, China.
Email: *mingyukpau@petrochina.com.cn; hey@petrochina.com.cn; liuchang02@petrochina.com.cn; 2941683587@qq.com; 702091960@qq.com; #wsliang1465@126.com

Abstract. To study the performance changes of polytetrafluoroethylene (PTFE) under high temperature and high-pressure environment, PTFE gasket material was selected to soak in the high-temperature and high-pressure reactor (6 MPa) at different temperatures (60, 90°C) for different times (4, 8, 12, 16 days) for aging failure test. The microstructure, hardness, functional group, and phase structure of the material were measured and analyzed by scanning electron microscope, hardness meter, infrared spectrometer, and X-ray diffractometer respectively. The results showed that with the extension of soaking time, defects and cracks appeared on the surface, and the mechanical properties of the PTFE gasket material were weakened. After soaking at 6 MPa and 90°C for 16 days, no reaction occurred in the PTFE gasket material, but the spiral twisted chain structure of the molecule was destroyed, which affected the performance of the material. Therefore, PTFE is suitable for long-term use in environments below 6 MPa and 60°C, but not suitable for long-term use in environments above 6 MPa and 90°C.
Keywords. PTFE gasket material; high-temperature and high-pressure; soak; mechanical properties.

1. Introduction

Non-metallic pipelines are widely used in the transportation of oil, natural gas, and water because of their light weight, good corrosion resistance, and coil bending during transportation and installation. The sealing material is very important for the sealing life and reliability of the pipeline. Research on the performance of the sealing material, especially the material characteristics under the application condition, has important guiding significance for the analysis of the failure mode and cause of the sealing. Early gasket materials used with non-metallic pipes are nitrile rubber, fluorine rubber, PTFE, and so on. Nitrile rubber with poor heat resistance is cheap and Fluorine rubber with high cost has good high temperature resistance. At the same time, Nitrile rubber and fluorine rubber have poor wear resistance, which affects the service life. PTFE has excellent strength, stiffness, elastic modulus, and tensile strength, with multi-channel sealing and a certain self-tightening function, which can partially eliminate the influence of pressure, temperature changes, and mechanical vibration [1]. At the same time, the PTFE temperature range (about -200°C-260°C) is wide, and can better adapt to different working environments [2,3].

Although PTFE has many good properties, its damage behavior is very different from metal materials due to the specific inherent properties of polymer materials. In recent years, the academic and engineering circles have done a lot of research on mechanical behavior under different loading conditions.

Liu et al. [4] studied the uniaxial ratchet, creep, ratchet creep interaction, and uniaxial torsion tests of cold-pressed PTFE round rod samples at different temperatures, and found that temperature had a great influence on the deformation behavior of PTFE. Zheng et al. [5] tested the cumulative deformation of polytetrafluoroethylene (PTFE) gaskets considering temperature and stress rate under cyclic stress-controlled compression load and found that accumulated strain of PTFE gaskets is temperature sensitive due to material softening and creep over time, especially when the temperature exceeds 100°C. Qi Huimin [6] made a comparative study on the tribological properties of PTFE and aluminum oxide/PTFE composites in the range of -100-100°Cand found that the friction coefficient of PTFE gradually decreased and the wear rate gradually increased during the heating process due to creep. The tribological behavior of PTFE is significantly affected by the introduction of aluminum oxide filler. The friction film formed during sliding is a key factor determining the tribological behavior, which has important guiding significance for the design of polymer composite materials under extreme working conditions.

In this paper, the PTFE gasket material was immersed in the simulated water environment of a high-temperature and high-pressure reactor. After soaking at different temperatures and different times, the microstructure and mechanical properties of the samples were analyzed, and the influence of high-temperature and high-pressure environments on the structure and properties of PTFE gasket material was studied.

2. Experiment

2.1. Main Raw Materials

PTFE gasket (commercially available).

2.2. Main Instruments and Equipment

High-temperature and high-pressure reactor, PARR 4584, United States PARR Company; Hardness tester, HXD-2000TM/LCD, Shanghai Shuangxu Electronics Co., LTD; Scanning Electron Microscope, 0ZEISS EV0 MA15, Zeiss, Germany; X-ray diffractometer, DX2700B, Dandong Tongda Technology Co., LTD.

2.3. Performance Test and Structural Characterization

(1) The on-site water quality environment and the ion concentrations are respectively 87 mg/L calcium, 39.4 mg/L potassium, and 1.46×10^3 mg/L sodium. The aqueous solution is configured according to the following ratio: water of 1 L, NaCl of 1.106 g, $NaHCO_3$ of 0.193 g, $CaCl_2$ of 0.241 g, $MgCl_2$•$6H_2O$ of 0.401 g and KCl of 0.073 g.

(2) PTFE gasket materials were soaked at high pressure of 6 MPa for 4 days, 8 days, 12 days, and 16 days at 60°C and 90°C, respectively, to simulate the aging failure process and analyze the effects of soaking temperature and time under high pressure on the microstructure, organizational structure and mechanical properties of gasket materials.

3. Results and Discussion

3.1. SEM Analysis

In order to analyze the influence of high temperature and high-pressure conditions on the microstructure of materials, scanning electron microscopy (SEM) was used to test the surface morphology of PTFE samples at different temperatures and soaking times. As shown in figures 1 and 2, with the aging process under high temperature and high pressure, small defects and cracks began to appear on the surface of materials, and the soaking time increased at any time. With the aging process under high temperature and high pressure, defects and cracks began to appear on the surface of the material, and as the immersion time increased, the number of defects and voids increased. Ultimately, the molecular chains on the surface of the material were twisted and deformed, leading to fracture. Compared with the failure process of the material at 60°C shown in figure 1, the material surface showed more severe defects and cracks at 90°C, as shown in figure 2, indicating that more rapid structural failure would occur at higher temperatures, leading to failure.

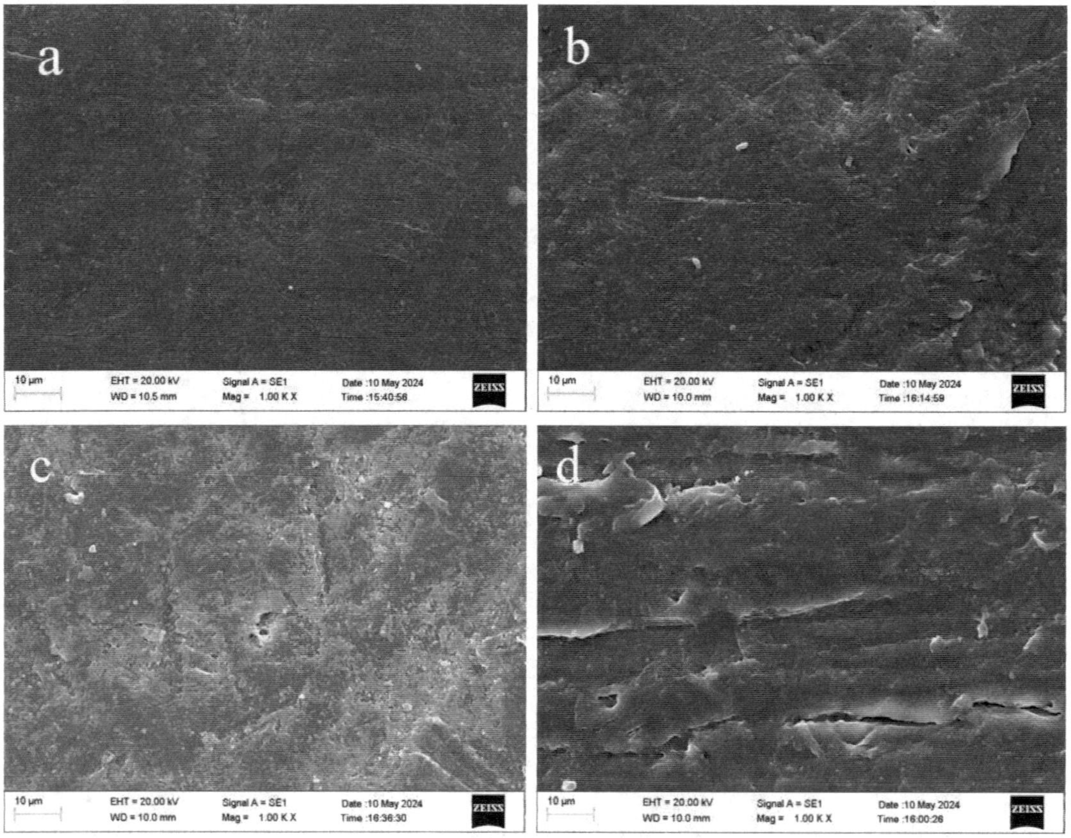

Figure 1. SEM images of 6 MPa, 60°C of PTFE gasket material soaked for different periods: (a) 4 days, (b) 8 days, (c) 12 days, (d) 16 days.

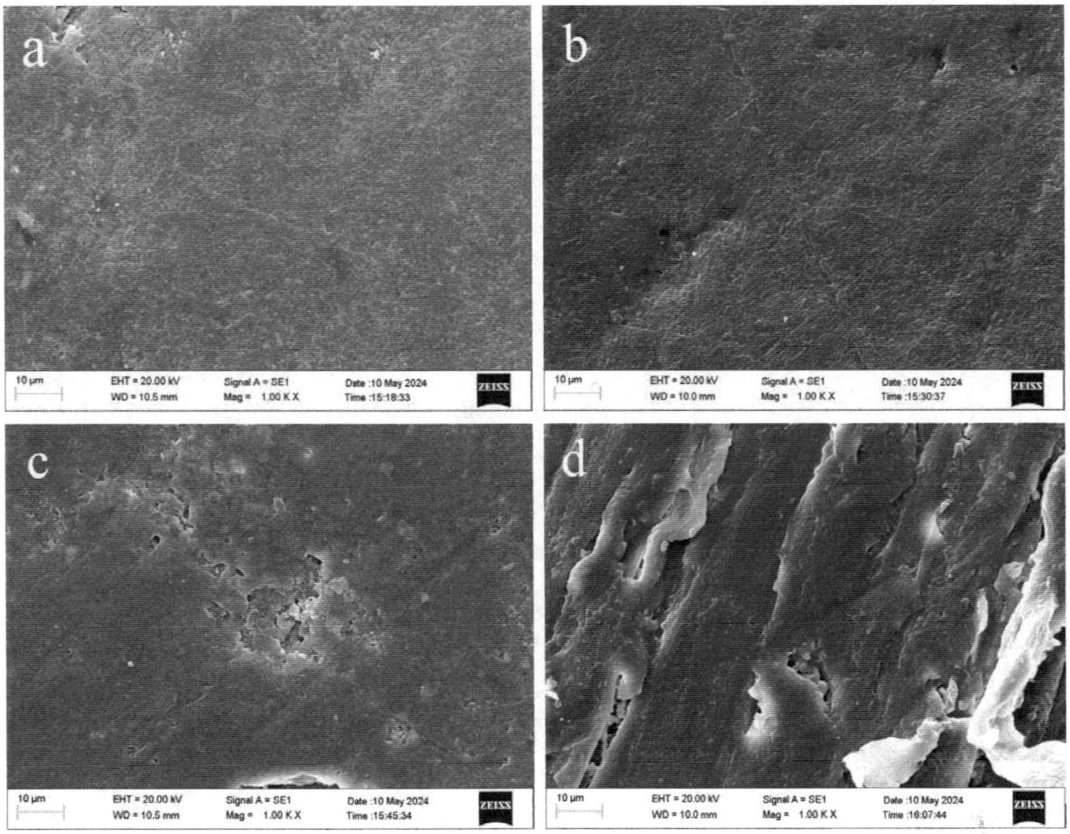

Figure 2. SEM images of 6MPa, 90°C of PTFE gasket material soaked for different periods: (a) 4 days, (b) 8 days, (c) 12 days, (d) 16 days.

3.2. Hardness Analysis

Figure 3 shows the hardness change curves of PTFE gaskets soaked at 6 MPa, 60°C, and 6 MPa, 90°C for different times respectively. As can be seen from the figure, with the increase of soaking time, the hardness of the material decreases, which is because the hardness of the material is related to the chemical bond force of its main chain and the force between molecular chains. When the solvent molecules penetrate into the interior of the material under a high-pressure soaking environment, they gradually enter the space between molecules and destroy the intermolecular force. At the same time, the extension of soaking time can also lead to the formation of some holes in the material, which can also lead to a reduction in hardness as a structural defect. It is worth noting that the hardness of the material soaked at 90°C is slower than that of the material at 60°C, which is because the crystallinity of PTFE has increased at a higher temperature, making the orderly arrangement of molecular chains increase, which slows down the reduction of hardness to a certain extent.

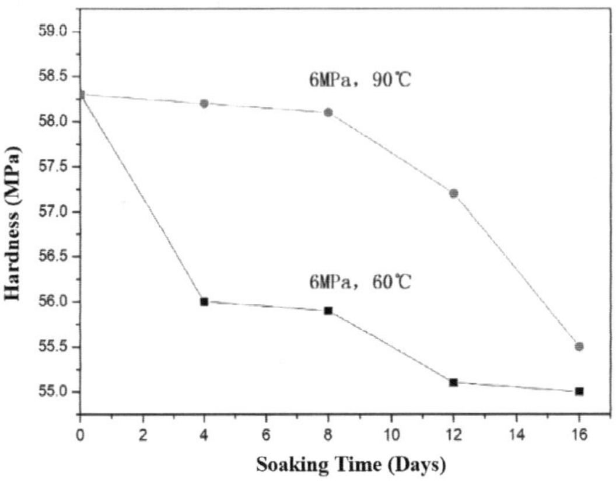

Figure 3. Changes in the hardness of PTFE during soaking at different temperatures.

3.3. FTIR Analysis

The infrared spectra of PTFE gasket material soaked at 6 MPa at 60°C and 90°C for different times are shown in figure 4. The basic component unit of the PTFE molecular chain is CF2(F-C-F), and the strong absorption peaks corresponding to 1199 cm^{-1} and 1145 cm^{-1} in the infrared spectrum are the antisymmetric stretching vibration and symmetric stretching vibration of CF group, CF2(F-C-F) is the basic unit of the molecular chain [7,8]. The weak absorption peak at 637 cm^{-1} is attributed to the bending vibration of CF in PTFE. As can be seen from figure 4, with the increase of the soaked time, the infrared spectrum of PTFE did not appear obvious new vibration peaks, and the peak change was not large, indicating that the material has excellent heat resistance and oxygen resistance, which is due to the high C-F bond energy in the PTFE molecular bond, stable performance, and not easily replaced by oxygen in the air [9,10]. At 90°C, with the increase of soaking time, the absorption peak frequency at 1199 and 1145 cm^{-1} showed a redshift, mainly because the helically twisted chain in the molecular structure of PTFE was highly sensitive to the degree of aging, and aging would destroy the helically twisted chain structure of PTFE molecules, resulting in a redshift phenomenon. With the extension of aging time, this molecular structure will be further damaged, and the redshift will be intensified [11].

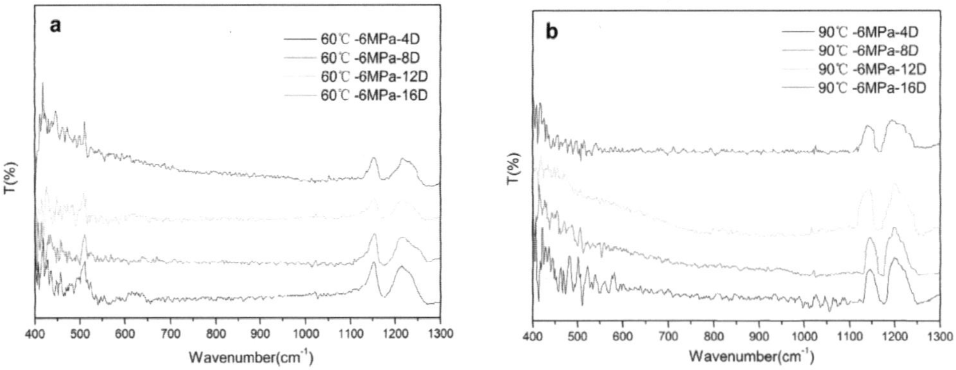

(a) 60°C (b) 90°C

Figure 4. Infrared spectra of PTFE gasket material soaked at different temperatures for different times.

3.4. XRD Analysis

Figure 5 illustrates the X-ray diffraction pattern of the soaked PTFE gasket material. It can be seen that at higher temperatures, the crystallinity of the material increases with the increase of soaking time, which increases the ordered arrangement of molecular chains and the intensity of diffraction peaks. However, no new diffraction peak appeared, indicating that the phase composition and crystal structure of PTFE did not change significantly, so it can be inferred that no new oxidation products were generated during the soaking process, which is consistent with the infrared spectrum analysis results.

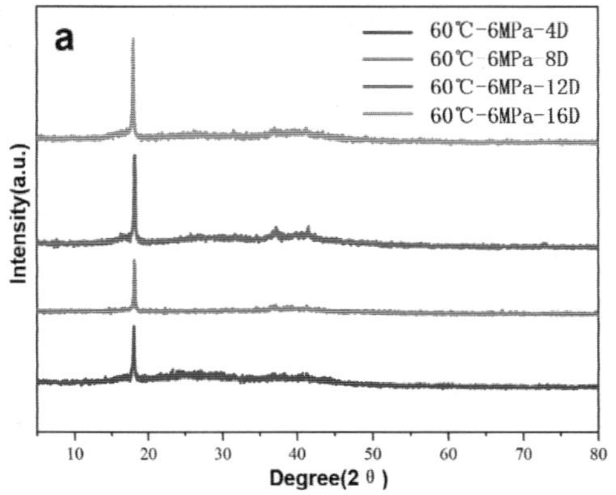

(a) 60°C

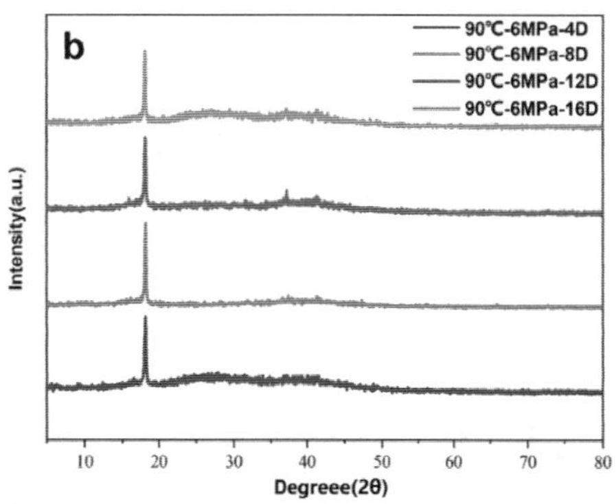

(b) 90°C

Figure 5. XRD patterns of PTFE gasket material soaked at different temperatures for different times.

4. Conclusions

(1) Under the conditions of high pressure (6MPa) and 60, 90°C, with the increase of soaking time, the PTFE gasket material undergoes surface defects and cracks due to aging, especially at 90°C where surface damage is more significant.

(2) As the soaking time of high temperature and high-pressure increases, solvent molecules penetrate into the interior of the material and enter the gaps between molecules, breaking the intermolecular forces and causing a decrease in the hardness of the PTFE gasket material.

(3) During the high-temperature and high-pressure soaking process, the PTFE gasket material did not form new phase structures, which indicates it has good heat and oxygen resistance properties. At 90°C, the helical twisted chain structure of the molecule is disrupted, which affects its performance and use.

5. Reference

[1] Dhanumalayan E and Joshi G M 2018 Performance properties and applications of polytetrafluoroethylene (PTFE)-a review *Advanced Composites and Hybrid Materials* **1** 247-268.
[2] Feng S, Zhong Z, Wang Y, et al. 2018 Progress and perspectives in PTFE membrane: Preparation, modification, and applications *Journal of Membrane Science* **549** 332-349.
[3] Luo X Y, Wang Y, Peng B and Fan Q 2010 Research on the sealing properties of PTFE oil seal *Lubrication Engineering* **35(3)** 112-115.
[4] Liu H Y, Zhang L, Lu K and Gao B J 2023 Study of cyclic plasticity and creep ratchet behavior of PTFE *Applied Sciences* **13(18)** 10039.
[5] Zheng X T, Zhang X H, Ma L W, Lin W, Yu J Y and Xu J M 2018 Rate-dependent mechanical characteristics of polytetrafluoroethylene (PTFE) gaskets under cyclic pulsating compression *Polymer Bulletin* **75(10)** 4783-4796.
[6] Qi H M, Hu C, Li J, Huang C Y, Zhou Y, Cui J N, Yu J X and Zhang G 2022 Tribological performance of PTFE and its composite in wide temperature range *Tribology* **42(1)** 65-73.
[7] Shen D Y 1982 *Application of Infrared Spectroscopy in Polymer Research* (Beijing: Science Press) p 149-152.
[8] Feng J M 2010 *Application of Infrared Spectroscopy in Trace Evidence Analysis* (Beijing: Chemical Industry Press) p 95-99.

Journal of Physics: Conference Series **2941** (2025) 012022 doi:10.1088/1742-6596/2941/1/012022

[9] Mohammadian-Kohol M, Asgari M and Shakur H R 2018 Effect of gamma irradiation on the structural, mechanical and optical properties of polytetrafluoroethylene sheet *Radiation Physics and Chemistry* **145** 11-18.
[10] Ye J F and Pei Y J 2015 Study on radiation degradation effect of PTFE by electron beam *Nuclear Techniques* **38(1)** 010201-1-010201-5.
[11] Wu K, Fan Y Q, Shen Q, Miao Z and Zhou C 2022 Study on aging state and life prediction of PTFE seals for nuclear power stations *China Plastics* **36(10)** 98-103.

Research on the Improvement and Application of State Estimation Algorithm for Lithium Iron Phosphate Batteries in Low-Temperature Environments

Qiwen Wang*, Qiang Sheng and Wei Liu

School of New Energy Engineering and Automotive, Huzhou Vocational and Technical College, Huzhou, Zhejiang, China.
Email: *wangqiwenyy@163.com; 376041718@qq.com; liu_wei_cjlu@163.com

Abstract. Accurately predicting the state of charge (SOC) and state of health (SOH) of lithium iron phosphate batteries is crucial for the safe operation of electric vehicle battery systems. However, the Unscented Kalman Filter (UKF) algorithm, commonly used for battery state estimation, exhibits significant errors under low-temperature conditions due to changes in ion mobility and increased internal resistance. To address this, a UKF algorithm incorporating a temperature compensation model is proposed. First, a SOC-OCV functional relationship is established through analysis of experimental data, achieving a high fitting degree of 0.999. Then, a temperature compensation model is integrated to refine the UKF estimation by accounting for the increased electrolyte viscosity and reduced reaction kinetics at low temperatures. Simulation results demonstrate that the proposed UKF with temperature compensation exhibits lower mean squared errors in SOC and SOH estimation, ensuring robust and accurate performance, even as ambient temperatures decrease.
Keywords. Lithium iron phosphate battery; state estimation; temperature compensation model; state of charge (SOC); state of health (SOH).

1. Introduction

In electric vehicles and energy storage systems, ensuring the safety and reliability of battery management is crucial, and it generally requires the use of a Battery Management System (BMS) to monitor and regulate battery performance [1]. The BMS system can effectively prevent failures and optimize battery operation under various conditions, with key metrics including State of Charge (SOC) and State of Health (SOH) [2]. Currently, research on SOC estimation for power batteries primarily focuses on room temperature environments, while in low-temperature conditions, the reduced charge and discharge capacity make overcharge and over-discharge control a technical challenge. The literature has explored the aggregation of lithium polysulfides in lithium-sulfur batteries at low temperatures and proposed an effective mechanism by which lithium nitrate can promote the conversion of lithium polysulfides [3]. Additionally, studies have shown that microalgae can not only remove heavy metals at low temperatures but also promote lipid synthesis, with glycine betaine specifically enhancing the efficiency of heavy metal removal and the antioxidant capacity of microalgae. However, the removal mechanisms for different heavy metals require further investigation, and the long-term application effects in real-world environments remain to be verified [4]. In SOC estimation, novel architectures such as the FuFi-CNN-Bi-LSTM model have performed well under different temperatures and driving cycles, but its generalization ability and computational resource requirements for more complex and long-term applications still need further validation [5]. The bidirectional pulse current heating method at low temperatures optimizes frequency, amplitude, and the charge/discharge pulse duration ratio, significantly improving the heating efficiency of lithium-ion batteries. However, the generalization of this method for different battery types and sizes still requires further study [6]. Experimental studies on SOC estimation of lithium-ion batteries at different temperatures using fractional-order models and BP neural network algorithms showed that errors could be controlled within 10%, meeting basic requirements for battery charge state estimation. However, for scenarios requiring high accuracy, a 10% error may still be insufficient [7]. Another study proposed a SOC evaluation method based on robust Kalman filtering, suitable for accurately assessing battery SOC under complex noise environments. However, its adaptability to various environmental factors and noise types, especially long-term stability in practical applications, still requires further validation [8].

In low-temperature environments, the prediction of power battery life is highly dependent on temperature, and there is currently a lack of mature strategies for SOH estimation under low-temperature conditions. Based on experimental data from lithium iron phosphate batteries, this paper establishes the SOC-OCV functional relationship under low-temperature conditions and optimizes the Unscented Kalman Filter algorithm through a temperature compensation model, accurately estimating the SOC and SOH of the battery, thereby enhancing battery management performance under low temperatures.

2. Analysis of Lithium Iron Phosphate Battery Performance in Low-Temperature Environments

2.1. Data of OCV-SOC

The experimental OCV-SOC data of the lithium iron phosphate battery is shown in table 1 (from 100% SOC to 5% SOC).

Table 1. OCV-SOC data of lithium iron phosphate battery.

SOC (%)	OCV (V)	SOC (%)	OCV (V)
100	3.65	50	3.40
95	3.62	45	3.38
90	3.60	40	3.35
85	3.58	35	3.32
80	3.55	30	3.30
75	3.53	25	3.28
70	3.50	20	3.25
65	3.48	15	3.20
60	3.45	10	3.15
55	3.42	5	3.10

The OCV-SOC relationship is typically nonlinear, f_{OCV}^{-1} representing the inverse function of OCV and SOC. The equation following this explanation will likely describe how SOC is derived as a function of OCV or how the relationship can be expressed mathematically [9].

$$SOC = f_{OCV}^{-1}\left(OCV\right) \tag{1}$$

Polynomial, exponential, and Sigmoid functions were used for fitting, and their MSE and R2 values were calculated. The specific values are shown in table 2, and the fitting curves are illustrated in figure 1.

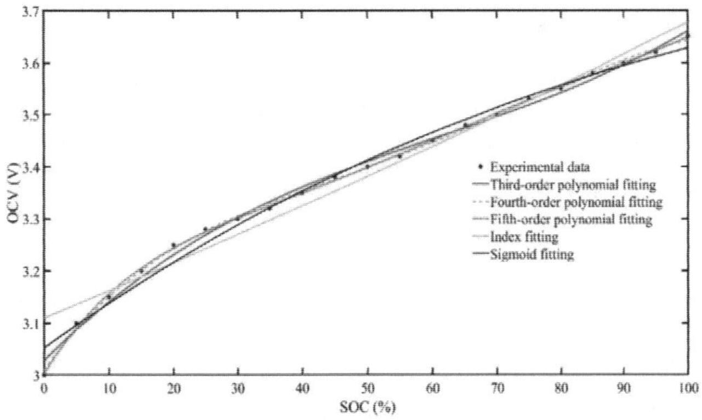

Figure 1. Fitting curves.

From figure 2, it can be observed that the fourth-order polynomial fitting effectively balances fitting accuracy and curve smoothness. It captures the nonlinear changes in the OCV-SOC curve while avoiding overfitting.

Table 2. Fitting curves of OCV-SOC.

Fitting method	MSE	R^2	Smoothness
Third-order polynomial	0.0015	0.980	Relatively Smooth
Fourth-order polynomial	0.00009	0.999	Very Smooth
Fifth-order polynomial	0.00012	0.987	Fluctuation Exists
Index	0.00016	0.991	Fluctuation Exists
Sigmoid	0.00014	0.986	Relatively Smooth

From table 2, it can be observed that the MSE of the fourth-order polynomial fitting value is low, and the R2 value is close to 1, indicating very high fitting accuracy. The OCV-SOC relationship of the battery is experimentally determined to be a polynomial relationship, which can be expressed as follows:

$$OCV = a_0 + a_1 \cdot SOC + a_2 \cdot SOC^2 + a_3 \cdot SOC^3 + \cdots + a_n \cdot SOC^n \tag{2}$$

The fourth-order polynomial coefficients solved using MATLAB are:

$$P = \left[-1.5647\times10^{-5}, 5.1264\times10^{-3}, -6.7238\times10^{-1}, 3.8\right] \tag{3}$$

Hence, the fitted fourth-order polynomial expression is:

$$OCV(SOC) = -1.5647\times10^{-5}\cdot SOC^4 + 5.1264\times10^{-3}\cdot SOC^3 + (-6.7238\times10^{-1})\cdot SOC^2 + 3.8\cdot SOC + 3.0 \tag{4}$$

2.2. The Impact of Low Temperature on Battery Internal Resistance, Capacity, and Voltage

Battery internal resistance R_{int} is an important parameter that describes the conductivity inside the battery. In low-temperature environments, the internal resistance of the battery increases significantly, primarily because low temperatures increase electrolyte viscosity, reduce ion diffusion rates, and increase charge transfer resistance. In low-temperature environments, the effective capacity of the battery is greatly reduced, especially under extremely low temperatures (such as -20°C), where the discharge performance of the battery declines severely, with some capacity even being irreversibly lost.

Table 3 presents the HPPC experimental test data for the Tesla 21700 battery.

Table 3. HPPC experimental test data.

Temperature (°C)	Internal Resistance (Ω)	Capacity (%)	OCV (v)
25	1178	100	3.70
15	1176	94	3.72
5	2727	88	3.74
-5	4349	82	3.71
-15	7191	76	3.70
-25	7238	70	3.69

From the data in table 3, it can be observed that as the temperature decreases, the internal resistance rapidly increases from 1178Ω at 25°C to 7238Ω at -25°C. According to the Arrhenius equation (Equation 5), in low-temperature environments, the relative increase in activation energy hinders electrochemical reactions, slows down ion migration rates, and ultimately results in a sharp increase in internal resistance [10].

$$R_{\text{int}} = R_0 \exp\left(\frac{E_a}{RT}\right) \tag{5}$$

Where, R_{int} is the internal resistance of the battery, R_0 is the internal resistance at room temperature (at 25°C), E_a is the activation energy, representing the energy required to overcome the electrochemical reaction, R is the gas constant (8.314 J/mol·K), and T is the absolute temperature (in Kelvin, K).

2.3. Temperature Compensation Model

In low-temperature environments, the internal electrochemical characteristics of the battery change, and the relationship between OCV and SOC is also affected [11]. Specifically, at low temperatures, the OCV value of the battery decreases compared to room temperature, especially in high SOC regions where the difference is more pronounced. Therefore, a temperature-dependent correction factor $\Delta OCV(T)$ is introduced to describe the impact of temperature on OCV. Typically, the open-circuit voltage at low temperatures can be expressed as the open-circuit voltage at room temperature plus a temperature compensation term:

$$OCV(T) = OCV_{ref} + \Delta OCV(T) \tag{6}$$

$OCV(T)$ is the open-circuit voltage at the current temperature, OCV_{ref} is the open-circuit voltage at the reference temperature (25°C), and $\Delta OCV(T)$ is the temperature compensation term, expressed as:

$$\Delta OCV(T) = k \cdot (T - T_{ref}) \tag{7}$$

k is the temperature compensation coefficient related to the battery materials and chemical properties. In this study, for lithium iron phosphate batteries, k is -3.5mV/°C, and T_{ref} is the reference temperature [12].

At low temperatures, due to changes in open-circuit voltage, SOC estimation errors also increase, making it necessary to apply temperature compensation for SOC estimation. SOC can be corrected using the modified OCV equation.

$$SOC(T) = f_{OCV}^{-1}\left(OCV_{ref} + k \cdot \left(T - T_{ref}\right)\right) \tag{8}$$

$$SOC(T) \approx SOC_{ref} + \frac{dSOC}{dOCV} \cdot \Delta OCV(T) \tag{9}$$

$$OCV(T) = a_1 \cdot SOC + a_2 \cdot SOC^2 + a_3 \cdot SOC^3 + \cdots + a_n \cdot SOC^n + k \cdot \left(T - T_{ref}\right) \tag{10}$$

3. Improvement of Unscented Kalman Filter (UKF) Algorithm

3.1. Unscented Kalman Filter (UKF) Algorithm

The Unscented Kalman Filter (UKF) is an improvement over the classical Kalman Filter (KF) and the Extended Kalman Filter (EKF). UKF uses the Unscented Transformation (UT) to estimate the nonlinear state of the system, avoiding the linearization error problem found in EKF and achieving higher accuracy in state estimation for nonlinear systems [13].

3.1.1. Fundamentals of Kalman Filtering. Classical Kalman filtering (KF) is used for linear systems, with its basic state estimation equations as follows [14]:

State Equation

$$X_{k+1} = AX_k + Bu_k + W_k \tag{11}$$

Where X_k is the system state vector at time k, A is the state transition matrix, B is the control matrix, u_k is the control input, and W_k is the process noise, typically assumed to follow a Gaussian distribution with zero mean.

Observation Equation

$$y_k = CX_k + V_k \tag{12}$$

y_k is the observation vector, C is the observation matrix, and V_k is the measurement noise, also typically assumed to follow a Gaussian distribution with zero mean.

3.1.2 Extended Kalman Filtering (EKF). For nonlinear systems, the Extended Kalman Filter (EKF) linearizes the nonlinear functions using Taylor expansion.

State Equation

$$X_{k+1} = f\left(X_k, u_k\right) + W_k \tag{13}$$

Observation Equation

$$y_k = h\left(x_k\right) + V_k \tag{14}$$

3.1.3 Unscented Kalman Filtering (UKF). UKF is an improvement over EKF, directly addressing nonlinear problems through the unscented transformation, thereby avoiding linearization errors. UKF uses a set of sigma points to approximate the probability distribution of the state and propagates the state and covariance updates through these sigma points [15]. The specific process is as follows:

(1) Generation of sigma points

Given the mean $\hat{X}_k$ and covariance of the system state P_k, UKF generates a set of $2L+1$ sigma points using the following formula:

$$X_K^0 = \hat{X}_k \tag{15}$$

$$X_K^i = \hat{X}_k + \left(\sqrt{(L+\lambda)P_k}\right), i = 1, \cdots, L \tag{16}$$

$$X_k^{L+i} = \hat{X}_k - \left(\sqrt{(L+\lambda)P_k}\right), i = 1, \cdots, L \tag{17}$$

Where L is the dimension of the state vector, and λ is the scaling parameter for the unscented transformation, typically defined as $\lambda = \alpha^2(L+k) - L$, where α and k are control parameters.

(2) Propagation of sigma points

The generated sigma points are propagated through the nonlinear state transition function $f(\bullet)$ to the next time step:

$$X_{k+1}^i = f\left(x_k^i, u_k\right) \tag{18}$$

The predicted mean and covariance of the state are then calculated:

$$\hat{X}_{k+1} = \sum_{i=0}^{2L} W_i^{(m)} x_{k+1}^i \tag{19}$$

$$P_{k+1} = \sum_{i=0}^{2L} W_i^{(c)} \left(x_{k+1}^i - \hat{X}_{k+1}\right)^{T+1} + Q_k \tag{20}$$

Where $W_i^{(m)}$ and $W_i^{(c)}$ are the weights for the mean and covariance of the sigma points, and Q_k is the process noise covariance

(3) Observation update

Similarly, the sigma points are propagated into the observation space using the observation function $h(\bullet)$:

$$y_k^i = h\left(x_k^i\right) \tag{21}$$

The predicted mean and covariance of the observation are then calculated:

$$\hat{y}_k = \sum_{i=0}^{2L} W_i^{(m)} y_k^i \tag{22}$$

$$P_{yy} = \sum_{i=0}^{2L} W_i^{(c)} \left(y_k^i - \hat{y}_k \right)^{T+1} + R_k \tag{23}$$

R_k is the observation noise covariance.

(4) Kalman gain and state update

The cross-covariance between the state and observation is calculated:

$$P_{xy} = \sum_{i=0}^{2L} W_i^{(c)} \left(x_k^i - \hat{X}_k \right)\left(y_k^i - \hat{y}_k \right)^T \tag{24}$$

The Kalman gain is calculated as:

$$K_k = P_{xy} P_{yy}^{-1} \tag{25}$$

The state and covariance are updated as follows:

$$\hat{X}_k = \hat{X}_k + K_k \left(y_k - \hat{y}_k \right) \tag{26}$$

$$P_k = P_k - K_k P_{yy} K_k^T \tag{27}$$

3.2. UKF Improvement Method Based on Temperature Compensation

In low-temperature environments, the open-circuit voltage (OCV) of the battery decreases, and a temperature compensation model can correct it using a compensation coefficient K. Similarly, the battery's internal resistance R varies with temperature, with the temperature and resistance compensation formulas as follows:

$$OCV(T) = OCV_{ref}(SOC) + K\left(T - T_{ref}\right) \tag{28}$$

$$R(T) = R_{ref} + \beta\left(T - T_{ref}\right) \tag{29}$$

In this study on lithium iron phosphate batteries, α is set to -0.35mV/°C, and β is set to 1.5mΩ/°C, with the reference temperature being 25°C.

The temperature-compensated OCV is used for voltage observation:

$$V_k = OCV\left(SOC_k, T_k\right) - I_k \bullet R\left(T_k\right) + v_k \tag{30}$$

After correcting with the temperature compensation model, an accurate terminal voltage observation of the battery can be obtained. The sigma points are propagated through the observation equation, and the corrected OCV model is used to update the voltage observations [16]:

$$V_k^i = OCV\left(SOC_k^i, T_k\right) - I_k \bullet R\left(T_k\right) \tag{31}$$

4. Simulation Verification and Analysis

To verify the superiority of the Unscented Kalman Filtering (UKF) combined with a temperature compensation model in estimating the State of Charge (SOC) of lithium batteries under low-temperature conditions, a simulation experiment was conducted using experimentally obtained OCV-SOC data. SOC and SOH estimates were made, and the accuracy and reliability of the model were analyzed based on these estimates, as shown in figure 2.

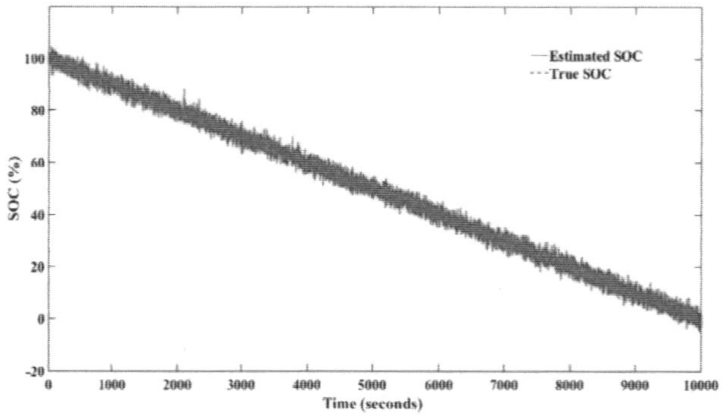

Figure 2. SOCs of estimated and true.

After introducing temperature compensation, the reduction in OCV is corrected by the model. The temperature compensation model effectively reduces the impact of low temperatures on OCV, making the SOC estimation closer to the actual value. The curve error decreases, particularly in the low SOC region, resulting in a more stable estimation. Without temperature compensation, the SOH estimation fails to account for the effects of low temperatures on the battery's electrochemical reactions and the increase in internal resistance, as shown in figure 3.

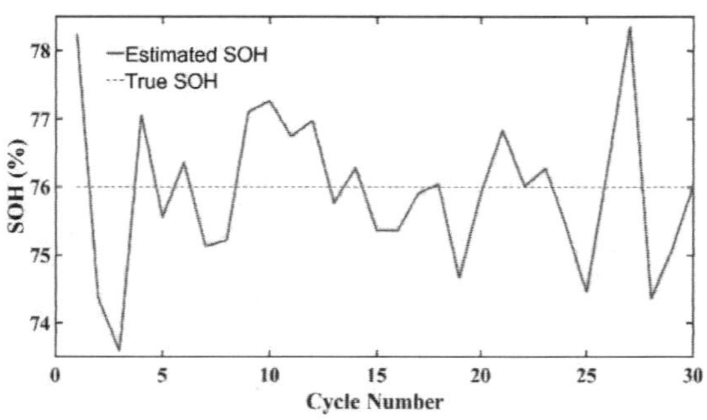

Figure 3. Estimated SOH without temperature compensation.

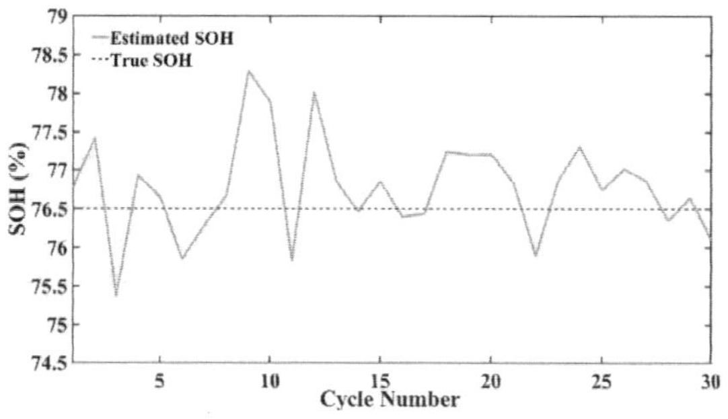

Figure 4. Estimated SOH with temperature compensation.

Low temperatures increase uncertainty in SOH estimation, and as the battery undergoes more cycles, the estimation error tends to accumulate. However, incorporating temperature compensation results in a smoother SOH curve, exhibiting higher robustness, as shown in figure 4. The temperature compensation model effectively mitigates the effects of low temperatures on the battery's internal resistance and electrochemical reaction rate, thereby enhancing the stability of SOH estimation. As the number of cycles increases, the SOH estimation curve becomes more consistent, accurately reflecting the degradation trend of the battery's health state with significantly reduced fluctuations. Figures 3 and 4 illustrate the challenges in accurately estimating SOH andSOC of lithium-ion batteries under low-temperature conditions. Low temperatures significantly impact the electrochemical behavior of lithium-ion batteries in the following ways:

(1) **Ion Transport Rate:** At low temperatures, the electrolyte viscosity increases significantly, directly reducing the mobility of lithium ions. This increased viscosity is due to reduced thermal motion

within the electrolyte molecules. As a result, lithium-ion migration from the cathode to the anode is hindered, leading to increased internal resistance.

(2) **Electrode/Electrolyte Interface Reactions:** The electrochemical reactions occurring at the electrode/electrolyte interface also become sluggish at low temperatures. The kinetics of the Solid Electrolyte Interphase layer are notably affected, becoming denser and less permeable at low temperatures, which impedes lithium-ion insertion and extraction and increases polarization. These effects reduce instantaneous power output and contribute to deviations in battery state estimation, as indicated by the increased SOC and SOH errors without temperature compensation, as shown in table 4.

(3) **Lithium Plating on the Anode:** Under extremely low temperatures, lithium plating may occur on the anode surface instead of the typical intercalation process. Lithium plating can result in dendrite formation, increasing internal resistance and posing safety risks. These irreversible processes lead to capacity loss, which impacts SOH estimation. This effect is evident in figure 3, where the uncorrected SOH shows large fluctuations and deviates significantly from the true value. During repeated cycling at low temperatures, both lithium plating and accelerated structural degradation of electrode materials can occur. For cathode materials, particularly $LiFePO_4$, reduced reaction kinetics at low temperatures lead to the accumulation of lattice stress, which can eventually cause microcrack formation. Similarly, for anode materials such as graphite, lithium plating can irreversibly degrade the material structure. The fluctuations observed in the SOH curve under low-temperature conditions, particularly in the absence of temperature compensation (figures 3 and 4), can be explained by the uneven degradation of material structures. However, the temperature compensation model adjusts dynamically to account for electrochemical behavior, thereby reducing these fluctuations. The temperature compensation model is not just a mathematical adjustment; it is also closely tied to the physical properties of battery materials. Different material systems demonstrate varying temperature dependencies in their electrochemical performance. For example, Nickel Manganese Cobalt Oxide and $LiFePO_4$ display different internal resistance growth and OCV behavior under low-temperature conditions. Thus, the parameters of the temperature compensation model must be calibrated and optimized according to the specific material systems in use. One key advantage of the temperature compensation model is its universality and adaptability across different material systems.

Table 4. Estimation errors.

Temperature (°C)	Without temperature compensation		With temperature compensation	
	SOC	SOH	SOC	SOH
-25	4.50	1.05	3.90	0.79
-15	3.93	0.78	3.78	0.75
0	2.80	0.65	2.60	0.59
25	1.50	0.50	1.40	0.45

Without temperature compensation, the error accumulates as the ambient temperature decreases, making it unsuitable for state estimation, as shown in table 2. With temperature compensation, as the ambient temperature decreases, the error in the UKF algorithm remains almost unchanged, allowing for better estimation of SOC and SOH.

5. Conclusion

This paper addresses the accurate prediction of SOC and SOH for lithium iron phosphate batteries in low-temperature environments, proposing a state estimation method based on the temperature-compensated UKF. Using SOC test data under low-temperature conditions, a fourth-order polynomial was used for parameter identification, achieving a fitting degree of 0.999. Simulation results show that compared to the traditional UKF algorithm, the temperature-compensated UKF algorithm has a lower mean squared error in SOC and SOH estimation, demonstrating stronger robustness and accuracy. This study only conducted state estimation on individual battery cells, but battery pack management is more complex and requires further research in combination with the state of the entire battery pack.

Acknowledgments
This project is supported by the General Scientific Research Project of Zhejiang Provincial Department of Education (Y202455633).

6. References

[1] Nascimento M, Ferreira S M and Pinto L J 2017 Real time thermal monitoring of lithium batteries with fiber sensors and thermocouples: A comparative study *Measurement* **111** 260-263.
[2] S H and M A K 2016 Investigation on the functionality of battery management system by using application-specific integrated circuit for lithium-ion phosphate battery *Journal of Industrial Technology* **24(1)** 87-100.
[3] Zhang H J, Fan Z X and Zhou H X 2024 Charging lithium polysulfides by cationic lithium nitrate species for low-temperature lithium-sulfur batteries *Energy Storage Materials* **73** 103786-103799.
[4] Song X, Kong F and Liu F B 2024 Lipidomics analysis of microalgal lipid production and heavy metal adsorption under glycine betaine-mediated alleviation of low-temperature stress *Journal of Hazardous Materials* **480** 135831-135846.
[5] Zafar H M, Khan M N and Houran A M 2024 A novel hybrid deep learning model for accurate state of charge estimation of Li-Ion batteries for electric vehicles under high and low temperature *Energy* **292** 130584-130597.
[6] Huang R, Wei G and Wang X 2024 A non-destructive heating method for lithium-ion batteries at low temperatures *Renewable and Sustainable Energy Reviews* **205** 114868-114883.
[7] Yuhang C, Hongjun X and Yi G 2022 SOC Estimation of Lithium-ion Batteries at Different Temperatures *Journal of Physics: Conference Series* **2401(1)** 55-79.
[8] Jian Z, Qiang W and Fan Jie M 2022 New Energy Vehicle Battery SOC Evaluation Method based on Robust Extended Kalman Filter *Journal of Physics: Conference Series* **2196(1)** 80-92.
[9] Ahmadzadeh F and Lundberg J 2014 Remaining useful life estimation: review *International Journal of System Assurance Engineering and Management* **5(4)** 461-474.
[10] Li F and Xu J P 2015 A new prognostics method for state of health estimation of lithium-ion batteries based on a mixture of Gaussian process models and particle filter *Microelectronics Reliability* **55(7)** 1035-1045.
[11] Wang Y J, Liu C and Pan R 2017 Experimental data of lithium-ion battery and ultracapacitor under DST and UDDS profiles at room temperature *Data Brief* **12** 161-173.
[12] Chen L, Wang Z and Lyu Z Q 2018 A novel state-of-charge estimation method of lithium-ion batteries combining the grey model and genetic algorithms *IEEE Transactions on Power Electronics* **33(10)** 8797-8817.
[13] Ma Y, Chen Y and Zhou X 2019 Remaining useful life prediction of lithium-ion battery based on Gauss-Hermite particle filter *IEEE Transactions on Control Systems Technology* **27(4)** 1788-1795.
[14] Huang C, Yin X and Huang H 2020 An enhanced deep learning-based fusion prognostic method for RUL prediction *IEEE Transactions on Reliability* **69(3)** 1097-1109.
[15] Sun Y, Hao X and Pecht M 2018 Remaining useful life prediction for lithium-ion batteries based on an integrated health indicator *Microelectronics Reliability* **88** 1189-1194.
[16] Pang X, Liu X and Jia J 2021 A lithium-ion battery remaining useful life prediction method based on the incremental capacity analysis and Gaussian process regression *Microelectronics Reliability* **127** 114405.

Experimental Study on Milling BK7 Optical Glass Assisted by Longitudinal-Torsional Ultrasonic Vibration

Zhe Liu[1*], Jinglin Tong[2] and Bo Zhao[2]

[1] School of Mechatronics and Automotive Engineering, Qingyuan Polytechnic, Qingyuan, Guangdong, China.
[2] School of Mechanical and Power Engineering, Henan Polytechnic University, Jiaozuo, Henan, China.
Email: *liuzhe201326@126.com; tongjinglin@hpu.edu.cn; zhaob@hpu.edu.cn

Abstract. The purpose of this paper is to fully explore the cutting performance of BK7 optical glass. Two milling methods ordinary milling and ultrasonic vibration (UV) milling are used to conduct a cutting test on BK7 optical glass on a self-made test platform for milling assisted by longitudinal-torsional two-dimensional ultrasonic vibration and research influence law of milling speed, feed rate per tooth and milling depth on milling force (MF) and surface quality for purpose of fully exploring cutting properties of BK7 optical glass. Test results show that the MF in the process of milling assisted by longitudinal-torsional two-dimensional ultrasonic vibration is less than ordinary milling and the surface quality of Longitudinal-Torsional Ultrasonic Vibration (LTUV) milling is superior to ordinary milling under the same conditions.
Keywords. Ultrasonic vibration; milling; longitudinal-torsional; optical glass; test research.

1. Introduction

Optical glass has lots of superior properties, such as high transparency, accurate optical constant, high chemical stability, good thermal stability, and so on, and accordingly, it is widely used in large-scale integrated circuits, high-power semiconductor devices, high-performance laser systems, lithography system, laser gyro, and other fields [1-5]. However, as a hard and brittle material, the high hardness and brittleness of optical glass cause great difficulties in its machining, for this reason, domestic and foreign scholars have conducted extensive research. At present, the machining methods of optical glass in industry can be mainly divided into contact machining and non-contact machining. Contact machining mainly includes super-precision cutting, super-precision grinding, and polishing with abrasives [6-11]. Non-contact machining mainly includes ion beam machining, plasma machining, etc. [12-15]. The efficiency of contact machining is higher, but the surface quality, sub-surface quality, and shape accuracy are relatively poor, and the reverse is true of non-contact machining. It has always been an eternal subject of how to obtain high-quality machining surfaces efficiently in the field of optical glass machining.

In recent years, there have been some results achieved in the field of optical glass machining since scholars introduced UV. Zhu Zijun and others optimized machining parameters by introducing UV into the grinding process of optical glass, test results showed that surface roughness of ordinary grinding was 0.1μm but surface roughness of UV grinding was 0.04μm under the same conditions, so the latter greatly improved surface quality [16]. Lu Mingming and others studied the subsurface damage of BK7 optical glass during cutting assisted by UV and established a subsurface damage depth prediction model [17]. Zhou Ming and others conducted test research on the subsurface damage of optical glass during grinding assisted by UV, according to the test results, showed that UV was introduced, which improved machining conditions during grinding, helping to improve the machining surface and subsurface quality [18]. Hong Jing and others machined optical glass by combining UV with milling machining, according to research results, showed that the introduction of UV could greatly improve the quality of the machining surface while reducing MF and tool wear [19]. Huang Cheng and others studied the UV grinding force and machining damage of BK7 optical glass, according to the research results, showed that it made the critical cutting depth of optical glass brittle-plastic transition increased from 0.54μm to 1.85μm due to the introduction of UV, meanwhile, and tangential loads were reduced by 42.0% and 38.5% respectively, and the optical glass surface roughness decreased by 36.75% and maximum subsurface crack damage depth by 44.27% [20]. In summary, there is a lack of in-depth research, especially in multidimensional ultrasonic vibration-assisted machining there are few reports although some achievements have been made in the field of optical glass machining by means of ultrasonic vibration-assisted machining technology.

In this paper, it is intended to take BK7 optical glass as the machining object to research the MF and surface quality during LTUV-assisted milling of optical glass by combining longitudinal-torsional two-dimensional ultrasonic vibration with milling machining for the purpose of helping to reveal the mechanism of optical glass milling assisted by multidimensional UV.

2. Test Platform and Test Method

2.1. Test Platform and Equipment

The drawing of the test platform is shown in figure 1 and the image of the test platform is shown in figure 2. One end of the amplitude transformer is attached to the milling cutter, the other end is attached to the transducer, and it is connected with the spindle of the machining center through the amplitude transformer flange, rotating with the spindle. The longitudinal-torsional vibration of the cutting tool is achieved by opening a chute on the amplitude transformer. The workpiece is mounted on the dynamometer which is mounted on the machine tool workbench so that the MF can be collected during milling. The dynamometer used in the test is a 9257B piezoelectric three-way dynamometer manufactured by Kistle, and the surface morphology is examined by a VHX-2000 microscope manufactured by KEYENCE. The workpiece is BK7 optical glass, and the cutting tool is a PCD two-edge milling cutter.

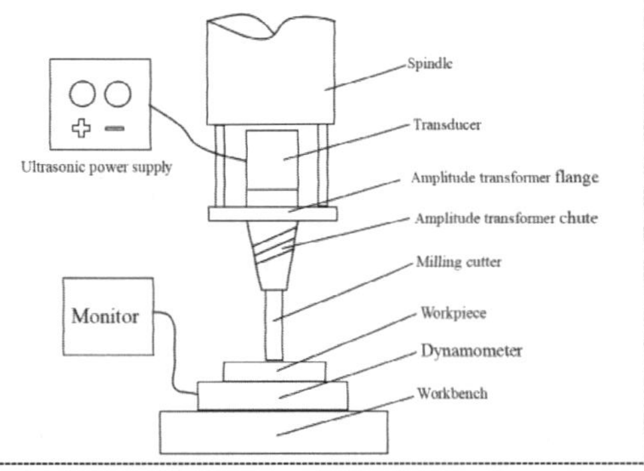

Figure 1. The drawing of the test platform.

Figure 2. The image of the test platform.

2.2. Test Method

The test is performed by using single factor method, with spindle speeds 2000 r/min, 3000 r/min, 4000 r/min, and 5000 r/min respectively, milling depths 0.1mm, 0.2mm, 0.3mm and 0.4mm respectively, feed rates per tooth 0.005 mm/tooth, 0.01 mm/tooth, 0.015 mm/tooth and 0.02 mm/tooth respectively. LTUV milling and ordinary milling are conducted respectively under the same conditions for the purpose of comparing the effect of UV with the machining procedure. In this paper, in view of the indefinite dynamic change of MF in the cutting process, the root mean square of MF is adopted to measure MF.

3. Test Results and Analysis

3.1. Test Results and Analysis of MF

3.1.1. Influence of Milling Speed on MF. The influence of milling speed on cutting force is demonstrated in figures 3 and 4. It is shown that no matter whether it is ordinary milling or ultrasonic milling, with milling speed increases, MF exhibits an upward trend, but slows down in a later stage. The MF during milling assisted by LTUV is lower than ordinary milling under the same conditions. MF also increases with milling speed, since milling speed increases, which increases the material strain rate, thus increasing the material strain rate, consuming more energy and increasing the MF. The increase of MF shows a gradual slowing trend because the high temperature leads to the material having a softening influence due to large cutting heat as the milling speed increases, reducing cutting resistance. MF during milling assisted by UV is lower than ordinary milling, on the one hand, high-frequency UV generates lubrication between the cutting tool and material to reduce frictional force, thereby reducing MF; on the other hand, it causes fatigue damage to the surrounding materials to a certain extent owing to high-frequency UV of milling cutter, thereby reducing the difficulty in subsequent cutting.

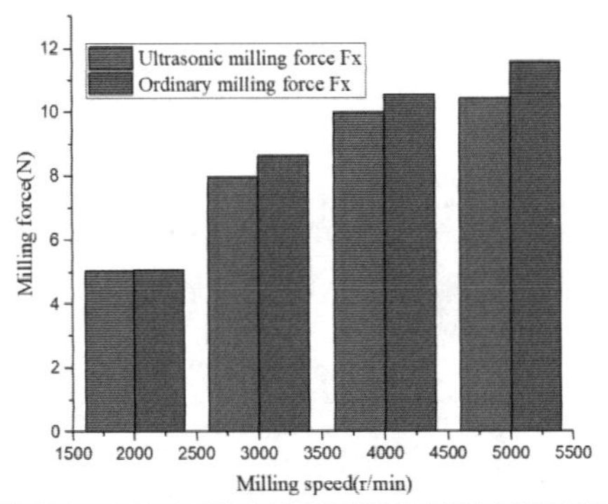

Figure 3. Fluence of milling speed on MF Fx.

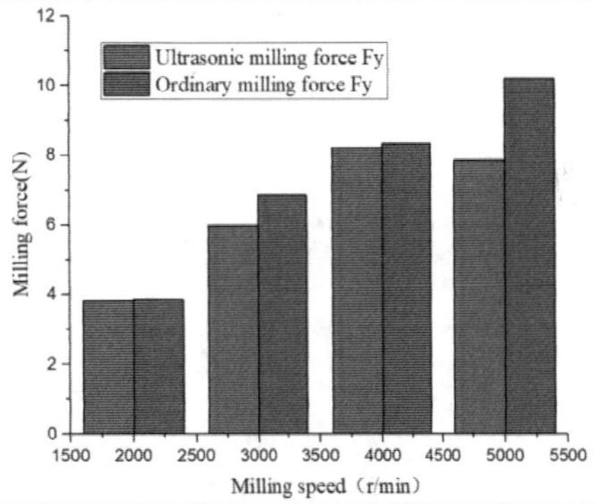

Figure 4. Influence of milling speed on MF Fy.

3.1.2. Influence of Feed Rate Per Tooth on MF. In figures 5 and 6, the MF varies with the feed rate per tooth. It can be found in the figures that there is a trend that ordinary milling and ultrasonic vibration MFs both increase first and then decrease as the increase of feed rate per tooth, MF milling assisted by UV is less than ordinary milling in X and Y directions. When the feed rate per tooth is small, as the feed rate per tooth increases, chip thickness gradually thickens, causing the MF to increase constantly, at this point, the material is mainly removed through fine particles or even plastic manner. However, to some extent the chip thickness is large after an increase in feed rate per tooth, at this point, the material is removed in the form of large particles or chunks, and material removal requires less energy, therefore, MF decreases.

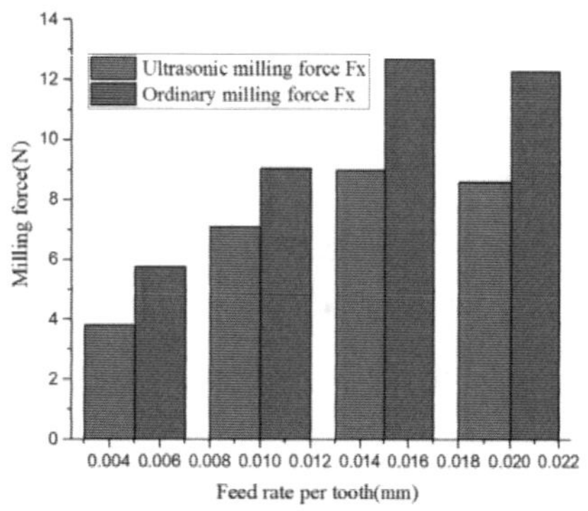

Figure 5. Influence of feed rate per tooth on MF Fx.

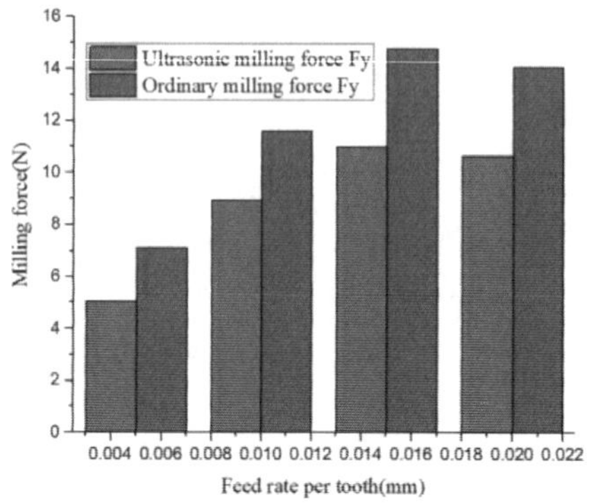

Figure 6. Influence of feed rate per tooth on MF Fy.

3.1.3. Influence of Milling Depth on MF. In figures 7 and 8, the MF varies with milling depth. The figures indicate the MF's significant increase with milling depth in the early stage, but the MF tends to flatten out or even shows a downward trend after the milling depth reaches 0.3mm. The MF during milling assisted by UV is lower than in ordinary milling. Contact length between the cutting edge and workpiece grows when milling depth increases and frictional force grows, thus causing an increase of MF; the difficulty of chip removal increases, and the cutting heat at the contact point between the cutting tool and workpiece increases when milling depth reaches a certain value, meanwhile, high temperature causes the material to have a softening influence, reducing cutting difficulty, consequently, cutting force decreases in a later stage.

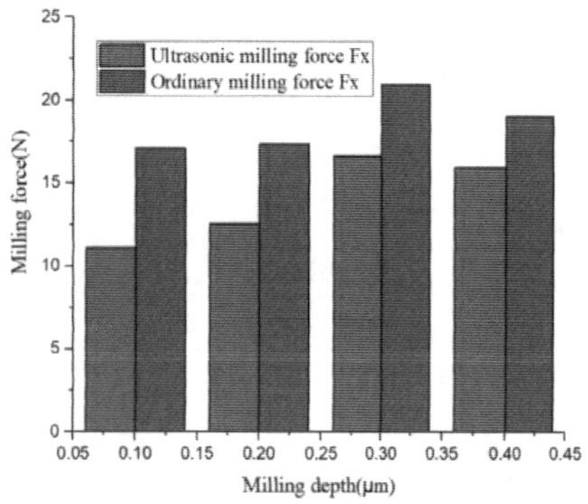

Figure 7. Influence of milling depth on MF Fx.

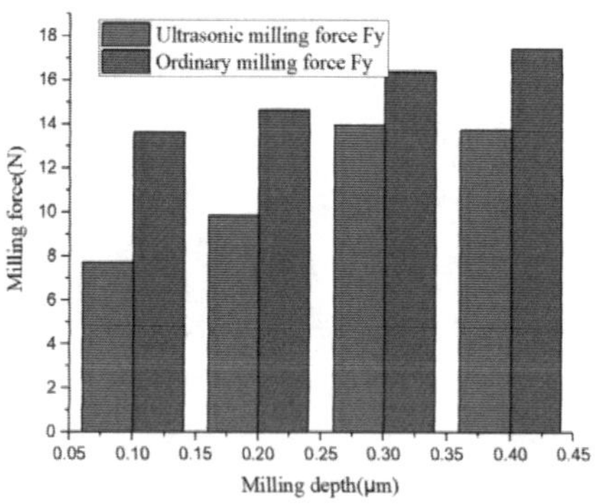

Figure 8. Influence of milling depth on MF Fy.

3.2. Surface Quality Test Results and Analysis

3.2.1. Influence of Milling Speed on Surface Roughness. In figure 9, surface roughness varies with milling speed. Surface roughness in the figure during ordinary milling and milling assisted by LTUV decreases with an increase in milling speed, and surface roughness during milling assisted by LTUV is less than in ordinary milling. On the premise that other cutting parameters remain unchanged, the increase of milling speed will not only bring the chips out of the machining area in time so as to avoid excessive accumulation and blockage of chips but also reduce the thickness of the cutting layer, thus improving the quality of machining surface. Compared with ordinary milling, it will have a hammering and ironing influence on the workpiece surface as the cutting tool is attached with longitudinal and torsional high-frequency UV during milling assisted by LTUV, thus reducing surface roughness.

3.2.2. Influence of Feed Rate Per Tooth on Surface Roughness. Figure 10 shows surface roughness varies with feed rate per tooth. It can be found that surface roughness grows constantly as the feed rate per tooth grows. The reason for this is that the thickness of the chip layer grows significantly as the feed rate per tooth grows, meanwhile, it is difficult to remove chips, which causes the cutting force and heat to increase, thus reducing surface quality. It can be found in figures 11 and 12 that the surface pits during ordinary milling and milling assisted by UV are relatively fine and compact when the feed rate per tooth is 0.01mm, however, the pits become large and sparse when the feed rate per tooth is 0.02mm, which shows that the material is removed through large particles as feed rate per tooth grows.

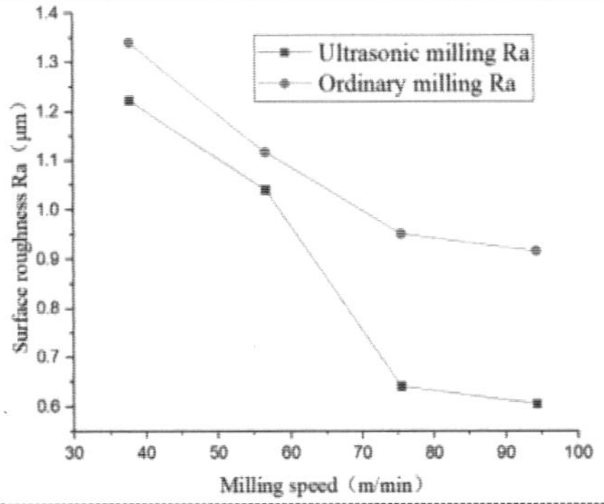

Figure 9. Influence of milling speed on surface roughness.

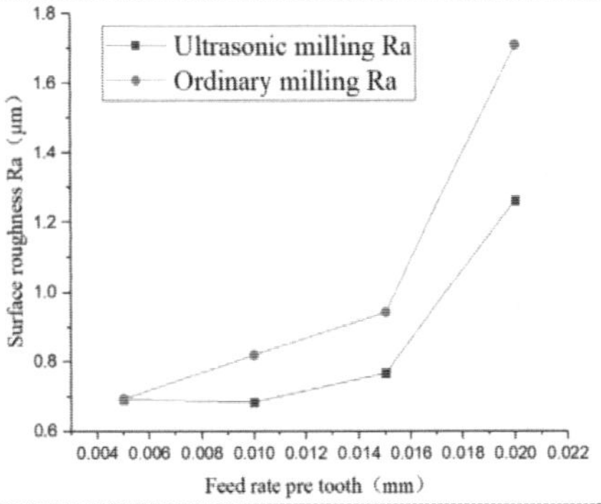

Figure 10. Influence of feed rate per tooth on surface roughness.

(a) Feed rate per tooth 0.01mm (b) Feed rate per tooth 0.02mm

Figure 11. Surface morphology of ordinary milling.

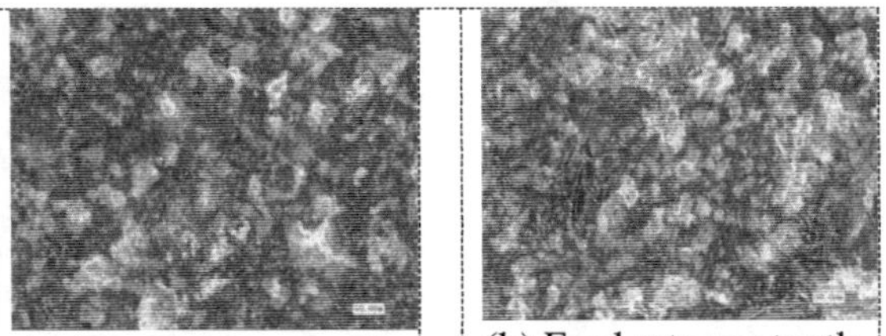

(a) Feed rate per tooth 0.01mm (b) Feed rate per tooth 0.02mm

Figure 12. Surface morphology of ultrasonic vibration-assisted milling.

3.2.3. Influence of Milling Depth on Surface Roughness. Figure 13 demonstrates the roughness of the surface varies with milling depth. It can be found in the figure that the roughness of the surface grows on the whole as the cutting depth grows since the contact area of the cutting tool and workpiece enlarges and the volume of removed materials grows as the cutting depth grows, which leads to increased chip removal difficulty, increased cutting force and cutting heat, resulting in deterioration of cutting conditions, thereby reducing surface quality. It can also be found in the figure that the roughness of the surface shows a downward trend when the milling depth reaches 0.2mm since it changes the material removal mode from plastic removal to brittle removal as the depth of milling increases, meanwhile, it consumes less energy during brittle removal, which optimizes machining conditions to some extent, reducing the roughness of the surface, however, it leaves large pits on the surface when the material is removed through large particles as the cutting depth further increases, thus increasing the roughness of surface additionally.

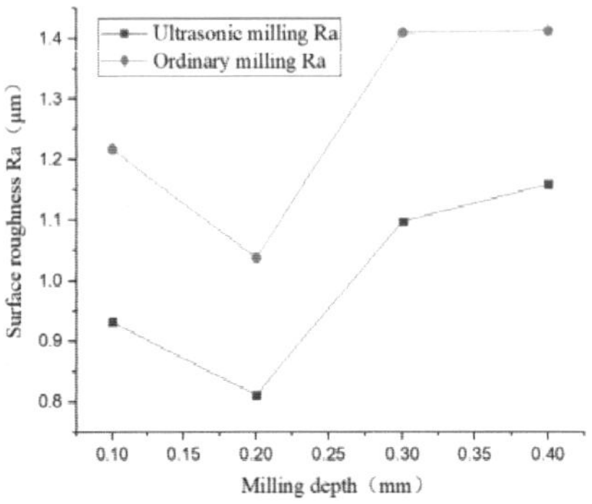

Figure 13. Influence of milling depth on surface roughness.

4. Conclusions

This paper draws the following conclusions through the test research on ordinary milling and milling of BK7 optical glass assisted by LTUV:

(1) MF of LTUV milling is lower than ordinary milling under the same conditions, on one hand, the lubrication influence of UV abates frictional force between the cutting tool and workpiece, on the other hand, high-frequency UV can lead to fatigue damage of workpiece material, reducing difficulty of cutting; no matter it is ultrasonic milling or ordinary milling, there is a trend that the MF grows greatly in early stage and flattens out in the late stage as milling speed, feed rate per tooth and milling depth increase;

(2) Quality of surface during milling assisted by LTUV is superior to ordinary milling under the same conditions, and roughness of the surface diminishes as the milling speed grows; on the whole, as feed rate per tooth and milling depth increases, surface roughness increases.

Acknowledgments

The authors would like to thank Professor Cui Feng of Henan Polytechnic University for his hard work on the experiment of this paper. At the same time, the authors greatly appreciate the Natural science project of Qingyuan Polytechnic (No. ZK23007) and the National Natural Science Foundation of China (No. 51475148) for support.

5. References

[1] Wang Z Z Shi C C and Zhang P F 2021 Recent progress of advanced optical manufacturing technology *Journal of Mechanical Engineering* **57(08)** 23-56.

[2] Yi K Cheng J J and Zhu K F 2020 Design and realization of a novel poly-silicon light-emitting device based on standard CMOS technology *Chinese Journal of Lasers* **47(07)** 308-314.

[3] Yng C Li F K and Ren T 2020 Fast and high quality composite processing method for silicon carbide wafers *Acta Optica Sinica* **40(13)** 147-152.

[4] Jin S P Fu Y G and Jin Y H 2019 High-precision transmission-wavefront-processing technology of large aperture ti: sapphire crystal at full spatial frequency *Chinese Journal of Lasers* **46(09)** 137-146.

[5] Li M K Xiang X S and Zhou C H 2019 Two-dimensional grating fabrication based on ultra-precision laser direct writing system *Acta Optica Sinica* **39(09)** 38-46.

[6] Bu Y B Wan Z P and Cheng Y 2024 Experimental study on grinding of BK7 glass by orderly arrangement grinding wheel *Journal of Hunan University (Natural Sciences)* **51(04)** 54-63.

[7] Chen F J Yin S H and Hu T 2013 Single-point inclined-axis error compensation grinding for small aspheric optical molds *Journal of Mechanical Engineering* **49(17)** 59-64.

[8] Lu Z K Guo H R and Wu Y B 2023 Magnetic compound fluid polishing of concave optical glass with small curvature *Optics and Precision Engineering* **31(09)** 1314-1324.

[9] Geng Q D and Li C Y 2018 Experimental study on magnetic abrasive finishing K9 optical glass *Surface Technology* **47(07)** 112-118.
[10] Zhong B, et al. 2017 The effect of tool wear on the removal characteristics in high-efficiency bonnet polishing *The International Journal of Advanced Manufacturing Technology* **91(2017)** 3653-3662.
[11] Hashimoto F, Yamaguchi H and Krajnik P 2016 Abrasive fine-finishing technology *CIRP annals* **65(2)** 597-620.
[12] Tang W Deng W J and Li R G 2015 Correction of removal function of ion beam figuring highly steep off-axis asphere *Optics and Precision Engineering* **23(06)** 1572-1579.
[13] Shi B L 2016 Study on key techniques in arc-enhanced plasma machining for silicon carbide mirrors *National University of Defense Technology* PhD dissertation.
[14] Zeuner M and Kiontke S 2012 Ion beam figuring technology in optics manufacturing:An established alternative for commercial applications *Optik & Photonik* **7(2)** 56-58.
[15] Castelli M, Jourdain R and Morantz P 2012 Rapid optical surface figuring using reactive atom plasma *Precision Engineering* **36(3)** 467-476.
[16] Zhu Z J Liu S and Han B 2020 Study on hybrid technology of ultrasonic vibration assisted abrasive lapping K9 optical glass *Surface Technology* **49(04)** 74-80.
[17] Lu M M Ma Y H and Du Y S 2023 Study on prediction of subsurface damage depth of non-resonant elliptical vibration cutting BK7 optical glass *Tool Engineering* **57(09)** 51-57.
[18] Zhou M Huang C and Zhao P Y 2017 Experimental investigation on subsurface damages in ultrasonic assisted grinding of optical glass *Tool Engineering* **51(07)** 15-19.
[19] Hong J 2011 *Research of Processing Technology in Ultrasonic Vibration Milling Optical Glass Materials* (Jiaozuo: Henan Polytechnic University) Doctor's Thesis.
[20] Huang C 2022 *Research on Grinding Forces and Machining Damages in Ultrasonic Vibration Grinding of BK7 Optical Glass* (Harbin: Harbin Institute of Technology) Doctor's Thesis.

Effect of Oblique Laser Shock Peening on Fretting Fatigue of FGH95 Powder Superalloy

Xiangming Qu[1,2,3,*], Wenjun Wang[4], Yanyun Zhao[4], Jun Cao[4], Haiyan Zhu[4], Xiaojun Guo[4], Xu Wang[4] and Yongkang Zhang[2]

[1] College of Mechanical Engineering, Hunan Institute of Science and Technology, Yueyang, Hunan, China.
[2] Guangdong Provincial Key Laboratory of Advanced Manufacturing Technology of Marine Energy Equipment, Guangdong University of Technology, Guangzhou, Guangdong, China.
[3] Foshan Yuantian Mattress Machinery Co., Ltd, Foshan, Guangdong, China.
[4] AECC Hunan Aviation Powerplant Research Institute, Zhuzhou, Hunan, China.
Email: *quxm1990@163.com; mifeng_2005@163.com; 15073361032@163.com; Tech_zz01@163.com; hyzhu2016@163.com; gxj608@163.com; liushch88@163.com; zykseu@163.com.

Abstract. Due to the high-frequency vibration contact of blade tenon, the contact surface of the mortise surface of the turbine disk is prone to fretting wear and fatigue fracture. The effects of shock times of Oblique Laser Shock Peening (LSP) and temperature on the fretting fatigue properties were studied. The fretting fatigue life of Oblique LSP treated samples at room temperature could be increased by 103.0%; At high temperatures, the fretting fatigue life could be increased by 68.5%. Through the analysis of fracture morphology, the fracture developing zone of fretting fatigue and tensile fatigue samples were compared. It is revealed that the wear in the transverse contact pads would promote the initiation of fatigue cracks; Due to the compressive force of the transverse contact pads, the degree of plastic deformation of the fretting specimens would increase, the quasi-cleavage facets appeared, and the tear ridge characteristics were obvious. The results of this paper provide a reference for the Oblique LSP of the FGH95 powder superalloy turbine disk and other aero-engine parts.
Keywords. Oblique LSP; FGH95 powder superalloy; fretting fatigue; fracture morphology.

1. Introduction

Due to the high alloying degree and volume fraction of γ' phase, the FGH95 alloy has excellent processability, and its yield strength reaches the maximum at about 600°C [1]. FGH95 alloy is a nickel -base powder superalloy, which can be used to manufacture turbine disks, a high-temperature structure of aero-engine. The turbine disk is the key component that affects the service life of the engine. The rotor of the turbine disk is connected with the blades through a mechanical structure of dovetail mortise-tenon joints.

During the operation of the engine, the mortise structure of a turbine disk as shown in figure 1 will bear the thermal stress caused by high temperature and the fluctuating centrifugal tensile stress caused by high-speed rotation. In the high-frequency cyclic rotation, the tooth root (red area in figure 1) of the mortise structure is prone to form microcracks and then break. The vibration phenomenon of slight relative displacement occurs in the contact area (yellow area in figure 1) between the mortise structure of a turbine disk and the blade tenon, which eventually leads to fretting wear on the contact surface of the mortise structure [2].

Surface treatment technology can improve the surface quality of vulnerable areas, thus improving their fatigue performance and ensuring the operation reliability of the engine. Surface coating [3] and deposition [4] can be used to improve the fatigue resistance of metal materials. In addition, the microstructure can be changed by surface mechanical treatment, and the residual compressive stress field can be generated near the treated surface to inhibit crack initiation and propagation. These treatment methods include LSP [5], Low Plastic Burnishing (LPB) [6], Deep Rolling (DR) [7] and Shot Peening (SP) [8]. Compared with other surface treatment technologies, LSP technology has attracted more attention due to its advantages of strong operability, excellent strengthening effect and high surface smoothness[9-13].

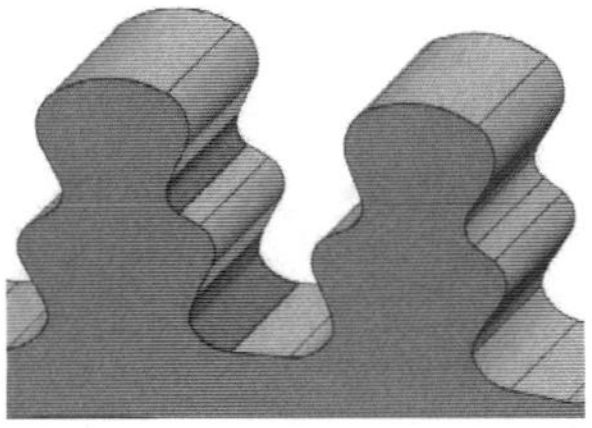

Figure 1. The mortise structure of a turbine disk.

As an advanced surface treatment method, the contact surface treated by LSP has a deep compressive residual stress, which can significantly improve the fretting fatigue life. Srinivasan et al. [14] used LSP-treated Ti-6Al-4V alloys for fretting fatigue experiments. The lifetimes of the LSP-ed samples increased by more than 5 times compared to the untreated samples. Through Metallography and fractographic analysis of the contact area, it was shown that LSP increased the fretting fatigue life mainly by suppressing the fretting crack propagation through the compressive residual stress induced. Dharmesh et al. [15] carried out the effect of the fluence of LSP on the fretting behavior of Ti-6Al-4V. Compared to the untreated substrate, the optimized laser fluence could obtain the compressive residual stress of -604 MPa, increase the hardness by 35%, and significantly reduce the fretting wear. Yang et al. [16] investigated the effect of LSP on the fretting fatigue property of TC11 titanium alloy. The results showed that the LSP significantly improved the fretting fatigue lives of the fretting specimens, and the fretting fatigue life increased most when the laser power density was 4.8 GW/cm2. It was also found that with the increase of the laser power density, the fatigue crack initiation location tended to move from the surface to the interior of the specimen. At present, most of the literature on fretting fatigue of LSP on mortise and tenon structure is focused on the titanium alloy of a turbine blade, while the nickel-base powder superalloy of turbine disk in the aero-engine hot-end components is less concerned. However, there are few reports on the fretting fatigue behavior of nickel-based alloy after surface strengthening, and the literature on the fretting fatigue behavior of FGH95 powder superalloy after LSP has not been published.

In order to understand the influence of LSP on tensile and fretting fatigue properties of FGH95 powder superalloy, it is necessary to be familiar with the strengthening effect of LSP technology on FGH95 turbine disk and carry out relevant research on LSP of FGH95 powder superalloy. Due to the narrow structure of the mortise area of the turbine disk, LSP could only be carried out at oblique incidence, that is, the angle between the laser beam and the normal line on the surface of the site to be impacted was greater than 0°. The range of oblique impact angle of each area was determined by using MotoSim EG CRV offline programming software with a 2 mm beam [5]. The results show that when the angle of 60° was selected as the oblique impact angle, there was no interference in the region to strengthen the mortise structure.

2. Materials and Methods

2.1. Experimental Materials

The experimental materials used in this study were FGH95 powder superalloys with the specific chemical compositions shown in table 1.

Table 1. Chemical compositions of the FGH95 superalloy.

Element	Mass Fraction/%	Element	Mass Fraction/%	Element	Mass Fraction/%
Cr	13	W	3.4	C	0.02
Al	3.6	Ti	2.65	Ni	Margin
Co	3.5	Nb	1.8		
Mo	3.5	B	0.1		

2.2. Treatment and Arrangement

The dog-bone samples with a length of 130 mm and a thickness of 4 mm were obtained by wire cutting [17]. The middle area of the specimens was 36 mm×12 mm and were treated by LSP, as shown in figure 2. The burrs and scratches on the surface of the samples should be removed successively through the metallographic sandpaper of 180# to 3000# then polished with diamond polish and wool felt polishing cloth. Finally, the samples were cleaned by an ultrasonic cleaning machine equipped with an anhydrous ethanol solution, and the surface of the samples was blown dry by a hair dryer.

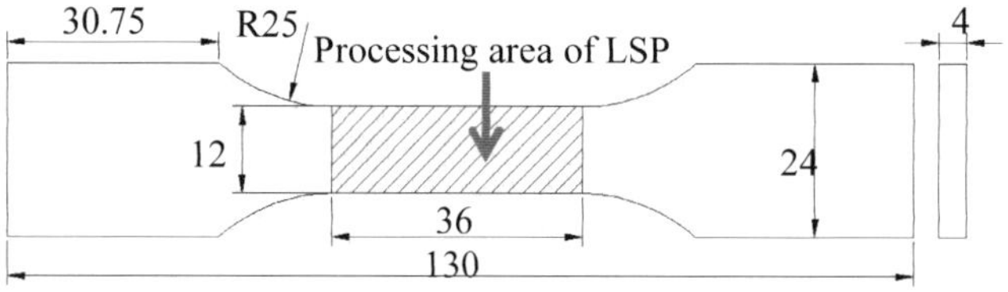

Figure 2. Size and impact area of the dog-bone sample.

Table 2. Experimental parameters of LSP and heat treatment

Number	Shock times	Temperature and time
1	Untreated	Untreated
2	1	Untreated
3	2	Untreated
4	3	Untreated
5	Untreated	600°C/1h
6	1	600°C/1h
7	2	600°C/1h
8	3	600°C/1h

The LSP equipment used in this paper was PROCUDO®200, a third-generation single-longitudinal mode laser peen system produced by LSP Technologies in the United States, which employs a diode-pumped pulsed YLF laser. Parameters of LSP: laser energy 8 J, circular spot 2 mm, pulse width 20 ns, spot lap rate 30 %, frequency 5 Hz, incidence angle 60°, restraint layer water, absorption layer 3M

Vinyl Tape Black Tape. The processing path was an "S"-shaped path, that is, from top to bottom, while impacting from left to right. Two groups of dog-bone fretting fatigue test samples after LSP treatment were obtained by impact times of 1, 2, and 3. Then, one group of the test samples were subjected to heat treatment at 600°C for 1 hour through the 1300°C high-temperature energy-saving box-type electric furnace produced by Tianjin Zhonghuan Electric Furnace Co., Ltd. The specific experimental parameters are shown in table 2.

2.3. Fretting Fatigue Test

The experiment scheme of fretting fatigue was designed according to the stress characteristics that the contact surface of the mortise-tenon structure bears the cyclic tensile force in the vertical direction and the friction force in the parallel direction at the same time, as shown in figure 3. The MTS Landmark 370.10 fatigue test system was used to apply a cyclic tensile load in the vertical direction to the dog-bone sample, while the frictional forces for fretting wear on both sided surfaces of the dog-bone sample were applied through the pressure loads of the contact pads. The contact strength of the tenon structure of the turbine blade ranges from 10-15 MPa [18]. Because the contact area of the contact pad in this experiment was 180 mm2, the pressure load was 2178.36 N, and the contact strength was 12.1 MPa.

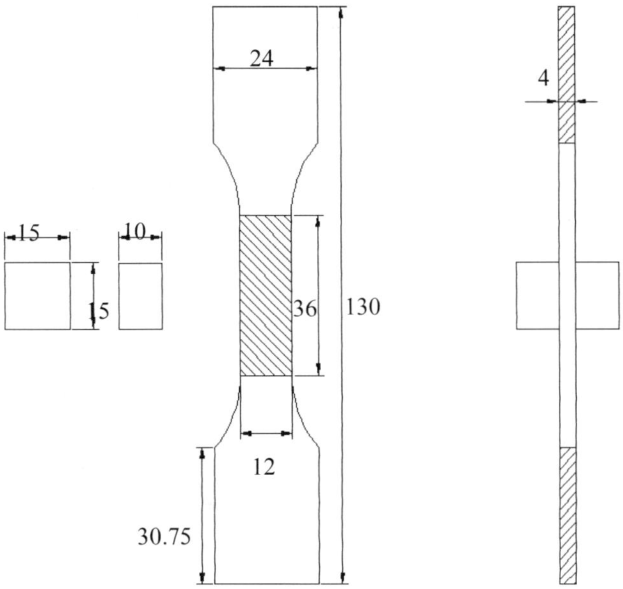

Figure 3. Schematic of geometry and load direction of the dog-bone specimen and contact pad.

2.4. Fracture Morphology Observation

After the fretting fatigue experiments, the fracture of the samples was obtained by a wire cutting machine, then the fractures were cleaned by an ultrasonic cleaning machine filled with absolute ethyl alcohol, and the surfaces were dried by a hair dryer. Finally, the surface morphology of the fractures was obtained by SU8220 field emission scanning electron microscopy produced by Hitachi, Japan.

3. Results

3.1. Data Analysis of Fretting Fatigue Life

As can be seen from the data in figure 4, at 25°C or 600°C, the number of cycles of samples treated with LSP was higher than those of untreated samples, which showed that LSP treatment can obviously strengthen the fretting fatigue property of FGH95 powder alloy.

At 25°C, the cycle number of untreated samples was 14021, while after 1, 2, and 3 LSP treatments, the cycle numbers were 21208, 24385, and 28467, respectively, and the fretting fatigue properties were increased by 51.3%, 73.9%, and 103.0%, respectively. At 600°C, the cycle number of the untreated sample was 15263, while the cycle number of the sample after one, two, and three LSP treatments were 18660, 23173, and 25716, respectively, and the fretting fatigue properties were increased by 22.2%, 51.8%, and 68.5%, respectively. Meanwhile, compared with 1 shock, the increase of fretting fatigue property of 2 shocks increased by 24.2% and that of 3 shocks increased by 11.0%. This indicates that the fretting fatigue property increases with the increase of the number of shocks, and the increase of fretting fatigue property slows down with the increase of the number of shocks. The fatigue performance of the increase is mainly due to the FGH95 sample after dealing with the LSP forming residual stress field, grain refinement and dislocation density increases, and the comprehensive influence factors such as nanocrystalline increase. With the increase in the number of shocks, the larger the plastic deformation is, the larger the dislocation density is, and the fatigue property is constantly strengthened, but the deformation resistance increases and the plastic deformation increment decreases, thus showing a trend of decreasing the increase of fretting fatigue property [19,20].

The cycle numbers of different temperatures with the same LSP processing parameters were compared. The cycle numbers of the untreated samples treated at 600°C were slightly higher than those at 25°C. This is mainly due to the fact that during the heat treatment process, various carbides in nickel-base powder superalloy are precipitated as particles along the grain boundaries, and these particles will show a nail-like pinning effect during the Test process of fretting fatigue, which will impede the dislocation structure and grain boundary movement, thus improving the fretting fatigue property of powder alloy [21]. In addition, the cycle number of samples treated with the same time of LSP at 600°C was slightly less than those at 25°C, indicating that the fatigue-strengthening effect of LSP treated at 600°C was stable. The residual compressive stress induced by LSP was mostly relaxed after heat treatment at 600°C, but the grain refinement, dislocation density, and nanocrystalline induced by LSP still had good thermal stability, and the strengthening effect of LSP on FGH95 could still be well presented [22].

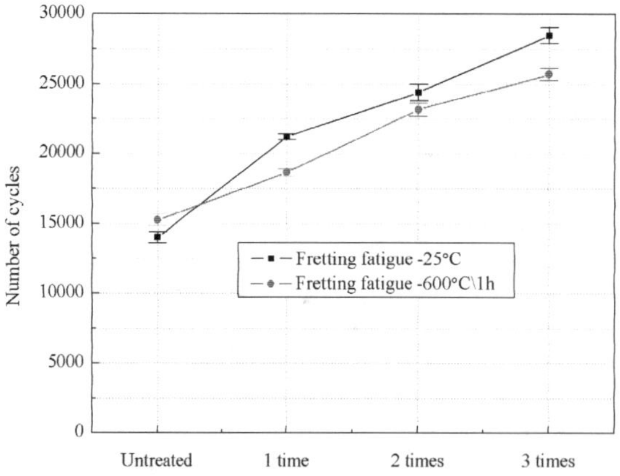

Figure 4. Cycles number of fretting fatigue under different treatment processes.

3.2. Fracture Morphology Analysis

3.2.1. Fracture Source Area

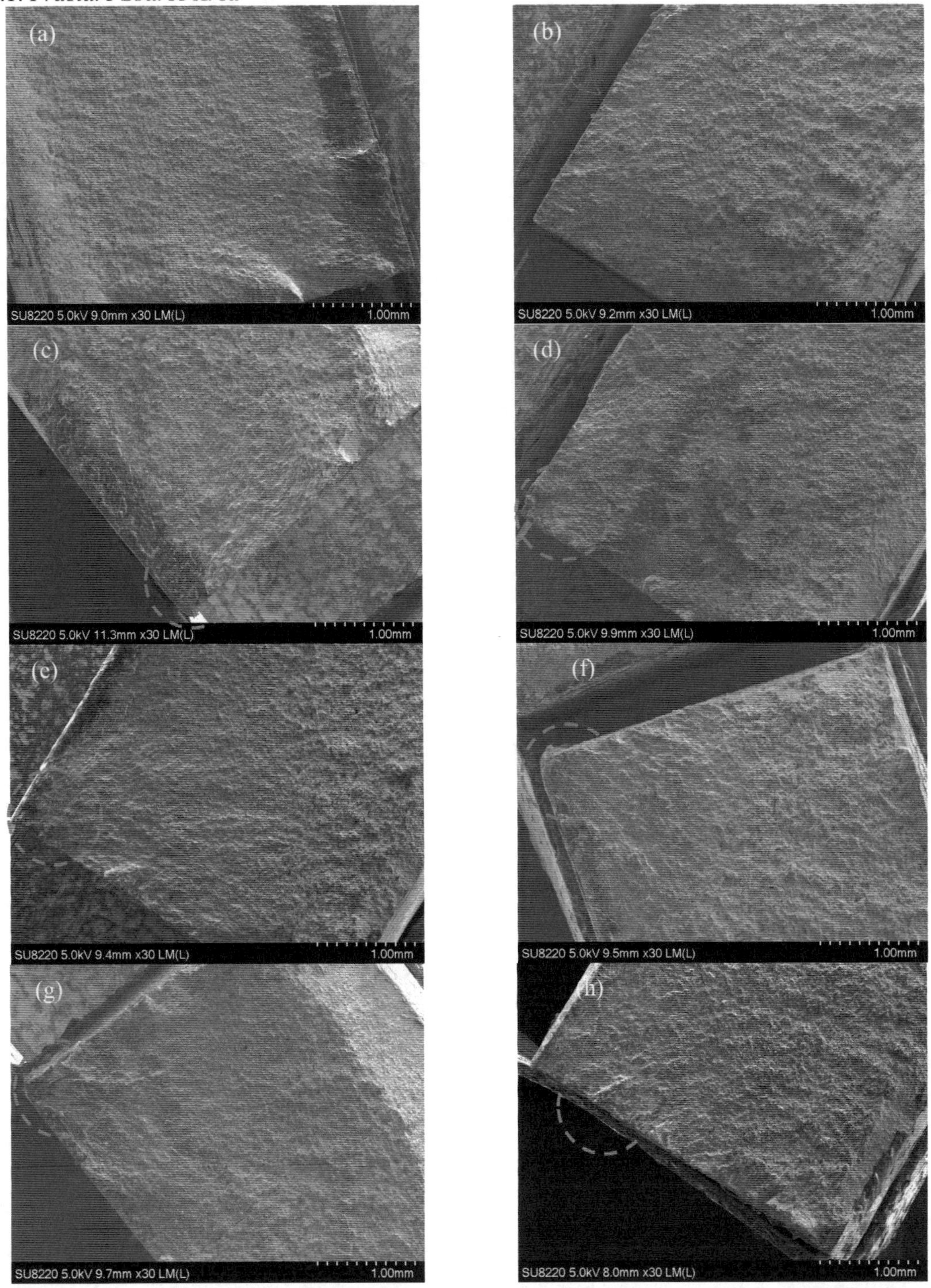

Figure 5. Fracture source zone of fretting fatigue fracture under different treatment methods: (a)Untreated-25°C; (b)Untreated-600°C; (c)1 shock-25°C; (d)1 shock-600°C; (e)2 shocks-25°C; (f)2 shocks-600°C; (g)3 shocks-25°C; (h)3 shocks-600°C.

Figure 5 shows the source zones of fretting fatigue fracture under different treatments. Figures 5(a) and 5(b) show the untreated samples and the fracture source zone was sprouted on the surface area of the samples, presenting a multi-source phenomenon. After high temperature treatment at 600°C, the direction of crack propagation changed, and the fluctuation of the fracture source zone was intensified, due to the blocking effect of carbide precipitation. As shown in figures 5(c) to (h), the surface area of the samples had plastic deformation after LSP treatment. Due to the refinement of the surface microstructure and residual compressive stress, the fracture source area shifted towards the area with unreinforced and stress concentration. At 25°C, there was a certain depth of the affected layer with fretting wear in the fracture source zone after fretting on the surface layer of the strengthened sample. The fretting affected layer refers to the fact that the wear particles formed by fretting wear were oxidized and diffused into the material cracks due to the extrusion of fretting blocks [23]. The depth of the fretting affected layer decreased with the increase of shock times, which was caused by the slowing down of the crack growth rate in the fracture source zone. After high temperature treatment at 600°C, there was a certain thickness of oxide layer on the surface of the sample, which prevented the diffusion of wear particles, and the presence of a fretting influence layer was not obvious.

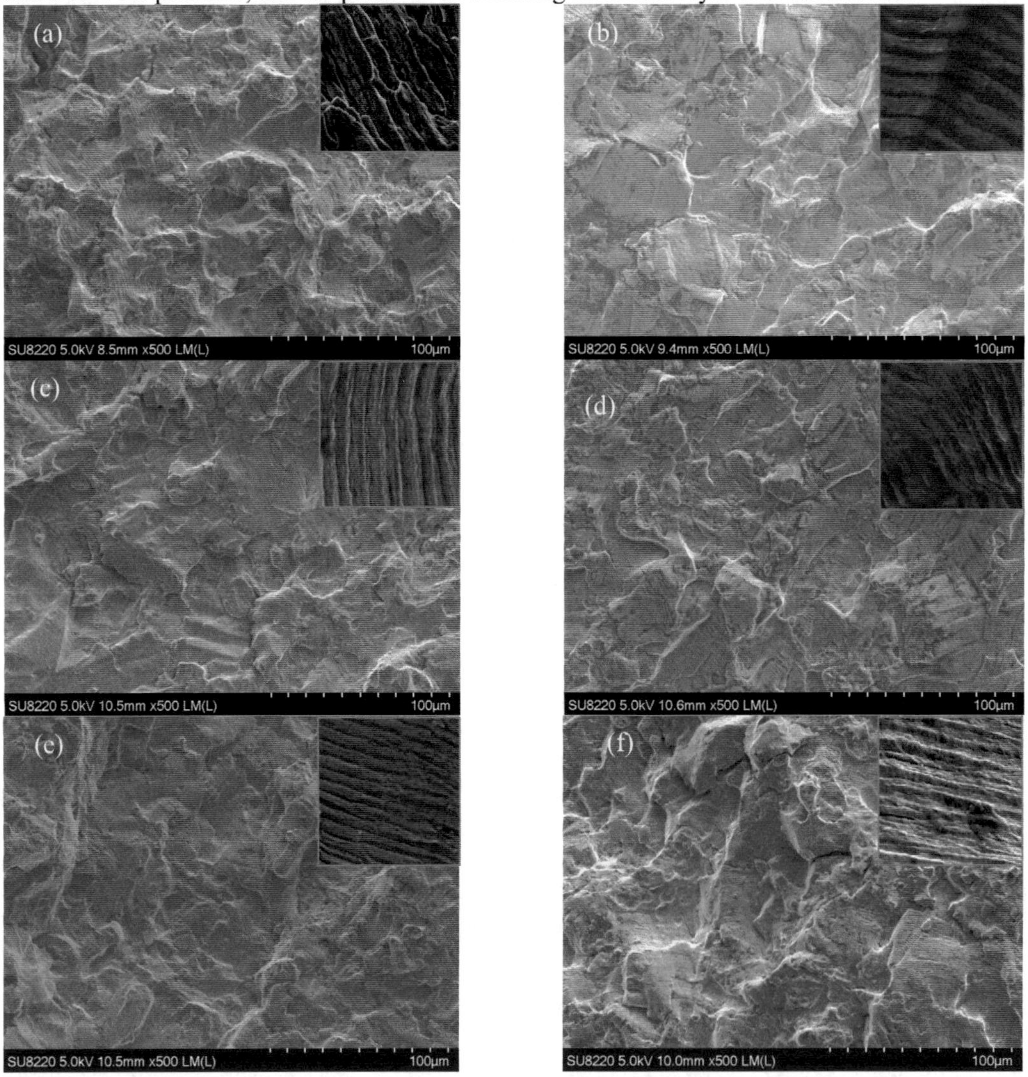

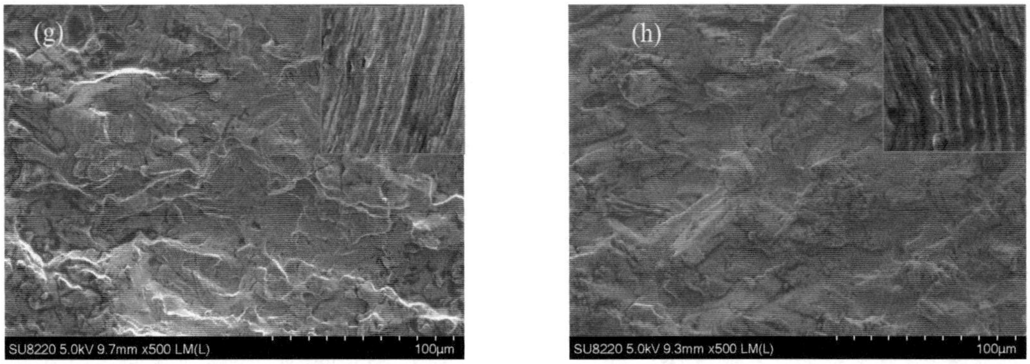

Figure 6. Fatigue propagation zone of fretting fatigue under different treatment methods: (a)Untreated -25°C; (b)Untreated-600°C; (c)1 shock-25°C; (d)1 shock-600°C; (e)2 shocks-25°C; (f)2 shocks-600°C; (g)3 shocks-25°C; (h)3 shocks-600°C.

3.2.2. Fracture Propagation Area.

The propagation zones of fretting fatigue fracture under different treatments are shown in figure 6. It can be seen from the figure that there were some approximate flat areas in the fracture propagation zone of fatigue specimens under all treatment methods, and these flat areas contained a large number of nearly parallel striations. Each striation indicates that the sample had undergone a fatigue cycle. During each fatigue cycle of the tensile specimen, the cracks in the specimen would be subjected to tensile/compressive stress, and the cracks would extend under tensile stress and close under compressive stress, and the extension was greater than the closure, thus forming permanent plastic deformation. The crack extension and closure continued on the propagation path, thus forming fatigue striations. Meanwhile, the fatigue strip spacing was different under different surface treatments. According to the measurement of image analysis software, the fatigue striations spacing of samples with no shock, 1 shock, 2 shocks, and 3 shocks was 0.84μm, 0.70μm, 0.55μm and 0.48μm, respectively, at 25°C. After being treated at 600°C for 1/h, the corresponding fatigue striation spacing was 0.81μm, 0.73μm, 0.62μm and 0.52μm, respectively. At both 25°C and 600°C, the fatigue striation spacing decreased with the increase in shock times. After LSP treatment, a certain depth of residual compressive stress layer, microstructure of smaller size and high density of dislocation structure were formed, which hindered the crack growth and slowed down the crack propagation rate. As the number of shocks increased, the residual compressive stress layer became deeper, the microstructure size became smaller, and the density of the dislocation structure increased, which further slowed down the crack propagation rate and made the fatigue fringe spacing smaller. Due to the carbide precipitated at high temperature, the crack propagation was hindered and the spacing between fatigue stripes was reduced in the sample without shock treatment. The fatigue strip spacing of LSP-ed samples increased after high temperature treatment, but it was still smaller than those of non-shock-treated samples. This is mainly due to the relaxation of residual compressive stress induced by LSP after high temperature treatment, but the microstructure and dislocation structure is more stable at high temperatures, and still could better reflect the benefits of LSP treatment.

3.2.3. Final Fracture Area.

Figure 7 shows the final fracture zone of fretting fatigue under different treatments. It can be seen from the figure that there were some dimples of different sizes and depths in the final fracture zone of fretting fatigue samples under all treatments, which indicates that the fracture feature of the samples was a plastic fracture. The mechanism of dimple formation is void aggregation, that is, microvoids nucleate, grow, and aggregate until fracture [24]. Dimples were different in size and depth under different treatments. At 25°C, the dimples of non-shock treated samples were smaller and shallower equiaxial dimples. After 1 shock, the dimples widened and deepened. As the number of shocks

Journal of Physics: Conference Series **2941** (2025) 012025 doi:10.1088/1742-6596/2941/1/012025

continues to increase, the size of dimples further increases. In 2 shocks, the characteristics of a tearing ridge appeared in the final fracture zone. There were even some slip stripes at 3 shocks. This indicates that the plasticity of the final fracture zone of shock-treated specimens increased, and the plasticity increased with the increase of shock number. After treatment at 600°C, the morphology of the dimple changed due to carbide precipitation. The dimple of non-shock specimens increased obviously and their depth increased slightly. Due to the release of compressive stress, the dimple size of treated samples was reduced and shallower. Due to the compressive force of the transverse contact pads, the degree of plastic deformation of the fretting specimens would increase, the quasi-cleavage facets appeared, and the tear ridge characteristics were obvious.

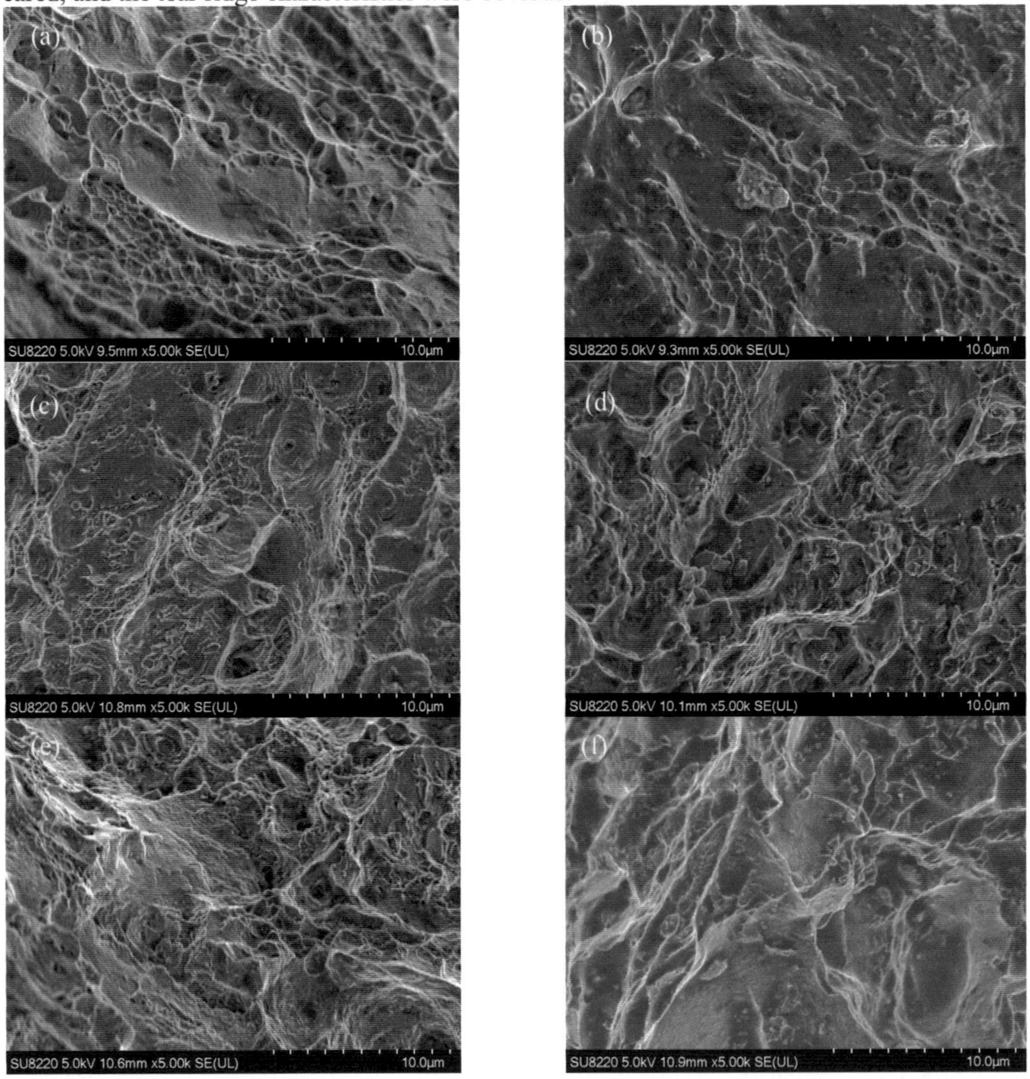

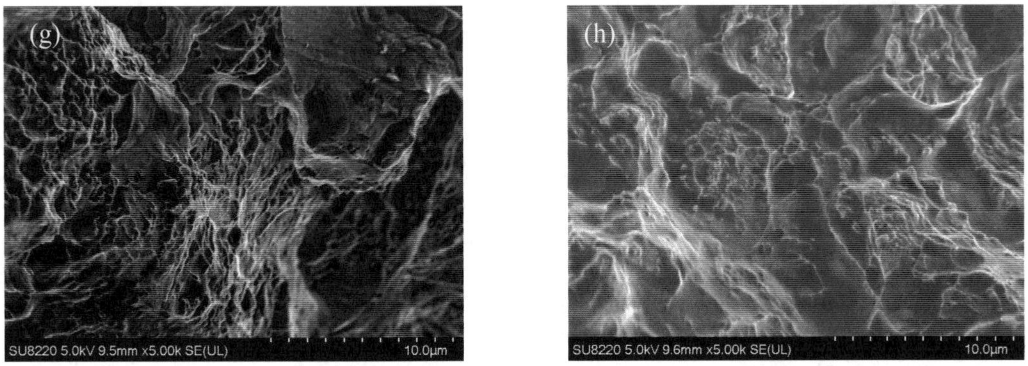

Figure 7. Final fracture zone of fretting fatigue under different treatment methods: (a)Untreated-25°C; (b)Untreated-600°C; (c)1 shock-25°C; (d)1 shock-600°C; (e)2 shocks-25°C; (f)2 shocks-600°C; (g)3 shocks-25°C; (h)3 shocks-600°C.

4. Conclusions

In this paper, the influences of LSP on fretting fatigue properties of FGH95 powder superalloy were studied. The following conclusions were obtained:

(1) LSP treatment had a significant strengthening effect on the fretting fatigue property of FGH95 powder alloy. The cycle number of LSP-treated samples was higher than those of untreated samples at 25°C and 600°C. At 25°C, the cycle number of untreated samples was 14021, while after 1, 2 and 3 LSP treatments, the fretting fatigue properties were increased by 51.3%, 73.9% and 103.0%, respectively. At 600°C, the cycle number of untreated samples was 15263, while the fretting fatigue properties of samples after one, two and three LSP treatments were increased by 22.2%, 51.8% and 68.5%, respectively. The residual compressive stress induced by LSP was mostly relaxed after heat treatment at 600°C, the cycle number of samples treated with the same time of LSP at 600°C was slightly less than those at 25°C. Due to the precipitation of various carbides in nickel-based powder superalloys after high-temperature treatment, The cycle number of untreated samples treated at 600°C was slightly higher than those at 25°C.

(2) The fracture source zone germinated in the surface area of untreated samples and presented a multi-source phenomenon, the crack propagation directions were different after 600°C treatment. After LSP treatment, the fracture source zone was transferred to the area without reinforcement and stress concentration. With the increased shock number, the fluctuation of the fracture source zone intensified. At 25°C, there was a certain depth of the affected layer with fretting wear in the fracture source zone after fretting on the surface layer of the strengthened sample, the depth of the fretting affected layer decreased with the increase of shock times. After high temperature treatment at 600°C, Due to the obstacle of the oxide layer, the presence of the fretting influence layer was not obvious. There were some approximate flat areas in the fracture propagation zone of fatigue specimens under all treatment methods, and these flat areas contained a large number of nearly parallel striations. The fatigue striations spacing of samples with no shock, 1 shock, 2 shocks and 3 shocks was 0.84μm, 0.70μm, 0.55μm and 0.48μm, respectively, at 25°C. After being treated at 600°C, the corresponding fatigue striation spacing was 0.81μm, 0.73μm, 0.62μm and 0.52μm, respectively. There were some dimples of different sizes and depths in the final fracture zone of fretting fatigue samples under all treatments, which indicates that the fracture feature of the samples was a plastic fracture. After LSP treatment, the size and depth of dimples increased with the increase of shock number, and tear ridges and slip stripes appeared. After treatment at 600°C, there were carbide particles in the dimples. The dimple of non-shock specimens increased obviously and their depth increased slightly. The dimple size of treated samples was reduced and shallower. Due to the compressive force of the transverse contact pads, the degree of plastic deformation of the fretting specimens would increase, the quasi-cleavage facets appeared, and the tear ridge characteristics were obvious.

Acknowledgments
This work was supported by the General Project of Hunan Education Department (No.23C0213); and the Science and Technology Innovation Project of Foshan City of China (No. 2018IT100112).

5. References

[1] Qin M, Chen C, Xuan H, Liu Y and Huang B, 2025. A modified cyclic elastoplastic constitutive model considering the variational dynamic recovery term of back stress for FGH95 under asymmetrical cyclic loading. *International Journal of Fatigue*, **190**, pp.108601.

[2] Vantadori S, Fortese G, Ronchei C and Scorza D, 2017. A stress gradient approach for fretting fatigue assessment of metallic structural components. *International Journal of Fatigue* **101** pp.1-8.

[3] Baradaran S, Zalnezhad E, Basirun W, Hamouda A, Sookhakian M, Sarhan A. and Alias Y, 2014. Statistical optimization and fretting fatigue study of Zr/ZrO2 nanotubular array coating on Ti-6Al-4V. *Surface and Coatings Technology*, **258**, pp.979-990.

[4] Du D, Liu D, Ye Z, Zhang X, Li F, Zhou Z and Yu L, 2014. Fretting wear and fretting fatigue behaviors of diamond-like carbon and graphite-like carbon films deposited on Ti-6Al-4V alloy. *Applied Surface Science*, **313**, pp.462-469.

[5] Qu X, Zhang Y, and Liu J, 2020. Numerical Simulation on Residual Stress Field of Flat-Topped Laser Oblique Shocking of Ni-Based Alloy GH4169. *Advances in Materials Science and Engineering*, 8824824.

[6] Prevéy P, Jayaraman N, Ravindranath R and Shepard M, 2010. Mitigation of fretting fatigue damage in blade and disk pressure faces with low plasticity burnishing. *Journal of engineering for gas turbines and power*, **132(8)**, 082105.

[7] Bertini L, Santus C, 2015. Fretting fatigue tests on shrink-fit specimens and investigations into the strength enhancement induced by deep rolling. *International Journal of Fatigue*, **81**, pp.179-190.

[8] Yang Q, Zhou W, Zhong Y, Zhang X, Fu X, Chen G and Li Z, 2018. Effect of shot-peening on the fretting wear and crack initiation behavior of Ti-6Al-4V dovetail joint specimens. *International Journal of Fatigue*, **107**, pp.83-95.

[9] Ye Y, Zhang Y, Huang T, Zou S, Dong Y, Ding H, Vasudevan V and Ye C, 2023. A critical review of laser shock peening of aircraft engine components. *Advanced Engineering Materials*, **25(16)**, pp.2201451.

[10] Pan X, Zhou L, Wang C, Yu K, Zhu Y, Yi M, Wang L, Wen S, He W and Liang X, 2023. Microstructure and residual stress modulation of 7075 aluminum alloy for improving fatigue performance by laser shock peening. *International Journal of Machine Tools and Manufacture*, **184**, pp.103979..

[11] Maleki E, Unal O, Guagliano M and Bagherifard S, 2021. The effects of shot peening, laser shock peening and ultrasonic nanocrystal surface modification on the fatigue strength of Inconel 718. *Materials Science and Engineering: A*, **810**, 141029.

[12] Gao Y, Zhong Z, Lei L, 2016. Influence of Laser Peening and Shot Peening on Fatigue Properties of FGH97 Superalloy. *Rare Metal Materials and Engineering*, **45(05)**, pp.1230-1234.

[13] Li T, Ding W, 2020. Characterization of Ni-Based FGH96 Superalloy Surface Modified by Ageing and Laser Shock Peening. *Chinese Journal of Vacuum Science and Technology*, **40(04)**, pp.342-346.

[14] Srinivasan S, Garcia D, Gean M, Murthy H and Farris T, 2009. *Fretting fatigue of laser shock peened Ti-6Al-4V*. Tribology International, **42(9)**, pp.1324-1329.

[15] Kumar D, Akhtar S, Patel A, Ramkumar J. and Balani K., 2015. Tribological performance of laser peened Ti-6Al-4V. *Wear*, **322**, pp.203-217.

[16] Yang X, Zhang H, Cui H and Wen C, 2020. Effect of laser shock peening on fretting fatigue life of TC11 titanium alloy. *Materials*, **13(21)**, pp.4711.

[17] Li K, Fu X, Li R, Gai P, Li Z, Zhou W.L. and Chen G.Q., 2016. Fretting fatigue characteristic of Ti-6Al-4V strengthened by wet peening. *International Journal of Fatigue*, **85**, pp.65-69.
[18] Wu G, Liu X, Li H, Sha W. and Huang L.J., 2015. Effect of contact pressure on fretting fatigue behavior of Ti-1023. *Wear*, **326**, pp.20-27.
[19] Yang J, Zhou W, Yang T, 2017. Nanocrystallization of Ti-6Al-4V alloy by multiple laser shock processing. *Laser Technology*, **41(5)**, pp.754-758.
[20] Li S, Zhang C, Li X, 2020. Effect of Laser Shock Peening on Fatigue Life of 2524 Aluminum Alloy.*Surface Technology*, **49(05)**, pp.207-213.
[21] Shu D, Tian S, Tian N, Xie J, Su Y, 2017. Thermodynamic analysis of carbide precipitation and effect of its configuration on creep properties of FGH95 powder nickel-based superalloy. *Materials Science and Engineering: A*, **700**, pp.152-161.
[22] Zhou L, Long C, He W, Tian L and Jia W, 2018. Improvement of high-temperature fatigue performance in the nickel-based alloy by LSP-induced surface nanocrystallization. *Journal of Alloys and Compounds*, **744**, pp.156-164.
[23] Wei S, Qiang C, Shurong Y, Yanni H and Pengfei J, 2020. Fretting wear behavior of TC4 alloy in different environmental media. *Rare Metal Materials And Engineering*, **49(07)**, pp.2393-2399.
[24] Lin C, Yu L, Zeng J, Wu H, Guo X, Liu J and Zhang Y, 2021. Experimental study on FGH95 superalloy turbine disk joint material by oblique laser shock processing. *Metals*, **11(11)**, 1770.

Optimization on the Preparation Conditions of Fe-N/BC Catalyst and its Pollution Removal Efficiency

Ting Zhou[1], Yang Yang[1], Tianxing Du[1], Wenyi Dong[1,2,3] and Hongjie Wang[1,2,3,*]

[1] School of Civil and Environmental Engineering, Harbin Institute of Technology Shenzhen, Shenzhen 518055, Guangdong, China.
[2] Shenzhen Key Laboratory of Water Resource Utilization and Environmental Pollution Control, Shenzhen 518055, Guangdong, China.
[3] State Key Laboratory of Urban Water Resource and Environment, School of Environment, Harbin Institute of Technology, Harbin 150090, Heilongjiang, China.
Email: 1226133395@qq.com; 284212494@qq.com; 1169777374@qq.com; dwy1967@qq.com; whj1533@qq.com; *whj1533@qq.com

Abstract. This study was aimed to optimize the preparation conditions of Fe-N/BC catalyst. Citric acid iron and melamine was used to co-doped modified selected biomass, optimized factors included biomass type, acid washing time, iron-nitrogen doping amount and pyrolysis temperature, with nitrobenzene as the targeted degradation object. The optimal preparation conditions for Fe-N/BC iron-nitrogen co-doped biochar catalyst were determined. The results showed that using onion as the biomass, after 8 hours of acid washing, immersed in an ultrasonic solution of citric acid iron with a concentration of 8 g/L for 30 minutes, followed by adding melamine at a doping ratio of 5 g/g biomass and another 30 minutes of ultrasonication, centrifuged, water washed, and dried, then pyrolyzed at 900°C under N_2 atmosphere for 1 hour. The obtained Fe-N/BC iron-nitrogen co-doped biochar catalyst exhibited the best removal efficiency for nitrobenzene in a catalytic reduction system, with a removal rate of 96.6% after 15 minutes of reaction.
Keywords. Iron and nitrogen-doped biochar; preparation optimization; nitrobenzene; catalytic reduction.

1. Introduction

The biochar material contains graphite carbon, persistent free radicals, and other catalytically active structures. By doping with Fe and N, zero-valent iron with catalytic reduction properties and Fe-Nx active sites with catalytic oxidation properties can be introduced into the biochar, enabling the catalytic oxidation removal of nitrobenzene. In this study, catalysts (Fe-N/BC) were prepared by co-doping biochar with citric acid iron and melamine, and a catalytic reduction-catalytic oxidation system was constructed using sodium borohydride ($NaBH_4$) as reducing agents to investigate the removal efficiency of nitrobenzene.

2. Catalyst Preparation Method

2 g of biomass powder was weighed into a beaker and 10 mL 1 mol/L hydrochloric acid was added. Then, the beaker was placed on a magnetic stirrer and stirred at a speed of 350 r/min (0, 2, 4, 6, 8, 12 h). The residue was filtered and washed with pure water three times. The filter cake was transferred to a beaker containing 25 mL citric acid iron solution with different concentrations (0.8, 1.6, 4, 6, 8, 10, 12, 16 g/L), and sonicated at a frequency of 40 kHz for 30 minutes. Varying masses (2, 4, 6, 8, 10, 12 g) of melamine were added, stirred well, and sonicated again at the same frequency for 30 minutes. After that, the mixture was transferred to a 50 mL centrifuge tube and centrifuged at a speed of 6000 r/min for 5 minutes. The obtained solid was washed after centrifugation with pure water, and was dried in a blast drying oven at 60°C until completely dry. The solid was then removed and grinded evenly,

and placed in a tube furnace for heating at a rate of 5°C/min in a nitrogen atmosphere to different temperatures (500, 700, 800, 900, 1000, 1100°C) for high-temperature pyrolysis (0, 0.5, 1, 2, 3 h). At last, the biochar was grinded and sieved through a 100-mesh sieve to obtain the Fe-N/BC catalyst.

3. Nitrobenzene Degradation

3.1. Degradation Experiment

100 mL 10 mg/L nitrobenzene solution was added to a beaker, followed by a certain amount of $NaBH_4$ solution. Then 1 mg of Fe-N/BC was added. The speed was set to 350 r/min and the reaction was carried out at room temperature. At timed intervals (0, 1, 3, 5, 7, 10, 15 min), 1 mL reaction solution was taken and filtered through a 0.22 μm membrane into a small vial pre-loaded with 10 μL sodium thiosulfate solution. The UPLC was used to determine the concentration of nitrobenzene.

3.2. Determination of Pollutant Concentration

The concentration of nitrobenzene in the solution was measured using a high-performance liquid chromatograph. Separation was carried out using a C18 liquid chromatography column (ACQUITY UPLC BEH C18 1.7μm). The mobile phase consisted of acetonitrile to water in a ratio of 45:55, with a flow rate of 0.25 mL/min. Detection was performed using a PDA detector at a wavelength of 250 nm, with a retention time of 3 min and a peak time of 1.9 min for nitrobenzene. The removal efficiency of pollutants was calculated using formula (1).

$$\eta = \frac{C_0 - C}{C_0} \times 100\% \quad (1)$$

Where C_0 was the initial concentration (mg/L); C was the remaining concentration (mg/L); η was the removal efficiency (%).

4. Results and Discussion

4.1. Types of Biomass

The structures of different types of biomass exhibit significant differences [1-3], such as cellulose being the main component of plant biomass and chitin being the main component of crustacean biomass. These structural variances often have a profound impact on the performance of biochar catalysts. Biomass including shrimp shells, spirulina, onions, straw, bamboo powder, pine needles, etc., were acid-washed for 12 h with 2 grams of melamine per 1 gram of biomass, impregnated with a 2 g/L citric acid iron solution, and pyrolyzed at 800 °C for 1 h to produce biochar catalysts. In degradation experiments on nitrobenzene with an initial concentration of 10 mg/L, using a catalyst dosage of 0.2 g/L, and adding 0.5 g/L each of $NaBH_4$ and PMS, the results were shown in figure 1.

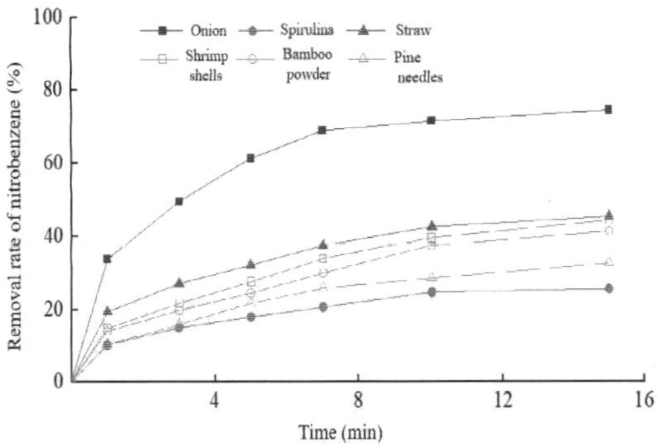

Figure 1. The influence of biomass types on the removal rate of nitrobenzene.

As shown in figure 1, when onion powder was used as the biomass, the catalytic system showed the best removal of nitrobenzene, with a removal rate of 74.4% after 15 min. Different biomass components led to differences in the thermal decomposition into char, resulting in varying specific surface areas of char, char structures, and sites that could provide coordination structures, ultimately causing significant differences in the catalytic performance of catalysts made from different biomass types.

4.2. Acid Washing Time

Biomass feedstocks often contain a lot of impurities such as ash and inorganic salts, which can block pores, affect the loading of miscellaneous atoms, and lead to a decrease in catalyst activity. Relevant studies [2-6] have shown that acid washing biomass with hydrochloric acid could effectively remove impurities, improve biomass pore structure, and have a certain help in increasing Fe and N loadings. Using hydrochloric acid (1 mol/L) for acid washing, immersing biomass in a 1.6 g/L citric acid iron solution with a melamine doping amount of 2 g/g biomass, pyrolysis temperature of 800°C, and pyrolysis time of 1 h, the effects of different acid washing times (2, 4, 6, 8, 12 h) on the catalytic performance of Fe-N/BC catalyst prepared were investigated for nitrobenzene removal. Nitrobenzene degradation experiments were conducted under conditions of catalyst addition of 0.2 g/L, $NaBH_4$ addition of 0.5 g/L, and initial nitrobenzene concentration of 10 mg/L, as shown in figure 2.

As shown in figure 2, with the increase in acid washing time from 2 to 8 h, the removal rate of nitrobenzene by the catalytic reduction system increased from 89.5% to nearly 100% after 15 min. Further extending the acid washing time to 12 h, the removal rate of nitrobenzene decreased to 74.4%, indicating that the catalyst had the best activation effect on $NaBH_4$ at an acid washing time of 8 h. Acid washing the catalyst for a certain period of time could produce more defects in biomass [7,8], allowing better doping of active components such as Fe and N. If the acid washing time was too long, the carbon skeleton of the biomass would collapse, leading to a decline in catalyst performance. Therefore, the optimal acid washing time for onion biomass was 8 h.

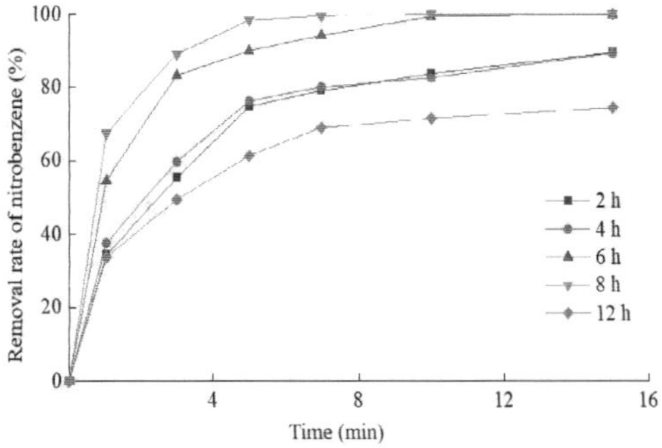

Figure 2. The effect of pickling time on the removal rate of nitrobenzene.

4.3. Iron Impregnation Solution Concentration

Citric acid iron was the main source of Fe component in the catalyst, and the citric acid iron impregnation concentration would affect the loading amount of Fe in the Fe-N/BC biochar catalyst, as well as the composition of Fe elements in the char, ultimately affecting the number and types of active sites in the catalyst. To study the effect of citric acid iron loading amount on the treatment effect of the system, under the conditions of 8 h acid washing time, melamine doping amount of 2 g/g biomass, 800°C pyrolysis temperature, and 1 h pyrolysis time, the effect of citric acid iron impregnation solution (0, 0.8, 1.6, 4, 6, 8, 10, 12, 16 g/L) on the catalytic reduction system for reducing nitrobenzene was studied. Nitrobenzene reduction experiments were conducted under the conditions of 0.1 g/L catalyst addition and 0.1 g/L $NaBH_4$ addition, and the results were shown in figure 3.

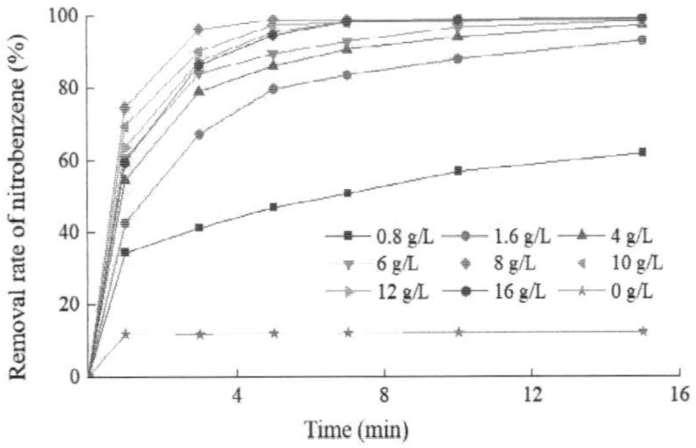

Figure 3. Effect of citric acid iron impregnation solution concentration on nitrobenzene removal rate.

As shown in figure 3, when no citric acid iron impregnation was carried out, the removal rate of nitrobenzene by the catalytic reduction system was only 12.3% after 15 min. When the citric acid iron impregnation solution concentration was increased from 0.8 to 8 g/L, the removal rate of nitrobenzene after 15 min increased from 61.7% to 99.2%. Further increasing the concentration of citric acid iron impregnation solution didn't significantly affect the removal rate of nitrobenzene [4,5]. Therefore, the optimal concentration of citric acid iron impregnation solution was 8 g/L.

4.4. Nitrogen Doping Amount

Under the conditions of 8 h biological acid washing, 8 g/L citric acid iron impregnation solution concentration, and 1 h pyrolysis at 800°C, the influence of melamine doping amount (0, 1, 2, 3, 4, 5, 6 g/g biomass) on the catalytic performance was investigated. With a 10 mg/L catalyst addition and 0.1 g/L $NaBH_4$ addition, the nitrobenzene removal rate was shown in figure 4.

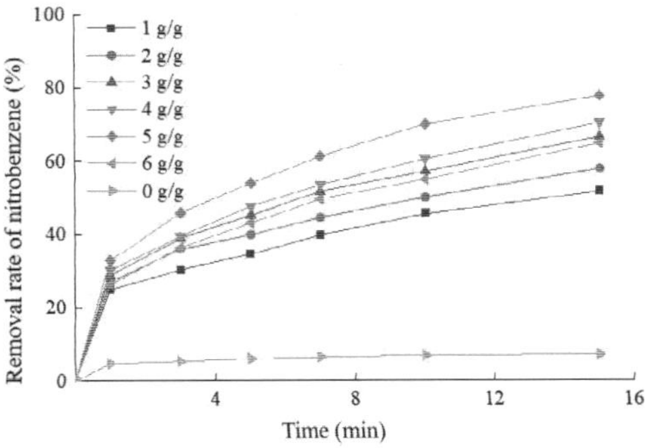

Figure 4. Influence of nitrogen doping amount on nitrobenzene removal rate.

As shown in figure 4, with the increase of melamine doping amount, the efficiency of the catalytic reduction system was significantly improved. When the melamine doping amount increased from 0 to 5 g/g biomass, the nitrobenzene removal rate increased from 7.1% to 77.6% after 15 min. However, when the melamine doping amount was further increased to 6 g/g biomass, the nitrobenzene removal rate decreased to 64.8%.

4.5. Pyrolysis Temperature

Under the conditions of biomass acid washing for 8 h, melamine doping at 5 g/g biomass, citric acid iron impregnation solution at 8 g/L, and pyrolysis time of 1 h, the influence of pyrolysis temperature (500, 700, 800, 900, 1000, 1100°C) on the catalyst efficiency was investigated, as shown in figure 5.

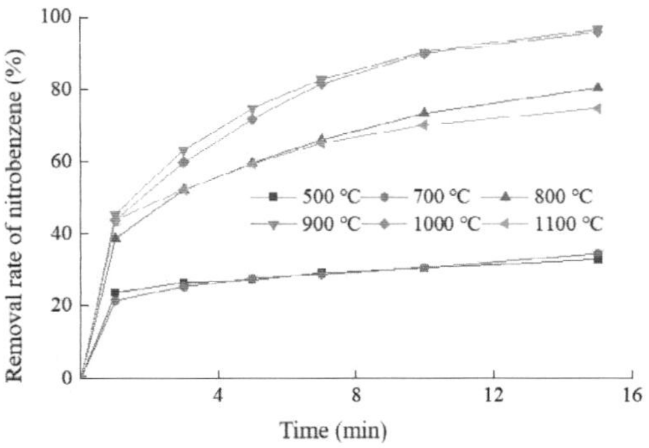

Figure 5. Influence of pyrolysis temperature on nitrobenzene removal rate.

It can be seen from figure 5 that with the increase of pyrolysis temperature, the removal rate of nitrobenzene by the catalytic reduction system showed an initial increase followed by a decrease trend. When the pyrolysis temperature was raised from 500 to 900°C, the removal rate of nitrobenzene increased from 32.9% to 96.6% after 15 min. Continuing to increase the pyrolysis temperature to 1100°C, the removal rate of nitrobenzene decreased to 74.6%. An appropriate pyrolysis temperature could help to enhance the content of pyridine nitrogen and graphite nitrogen and regulate the existence form of Fe in the catalyst [5,8]. Considering the comprehensive removal rate of nitrobenzene, 900°C was chosen as the optimal pyrolysis temperature.

5. Conclusion

The main conclusions of this study were as follows:

The optimal preparation conditions and process for Fe-N/BC catalyst were as follows: onion was selected as the biomass, and after 8 h acid washing, it was immersed in a citric acid iron solution with a concentration of 8 g/L for 30 min under ultrasonic treatment, followed by doping with melamine at a ratio of 5 g/g biomass, then ultrasonic treated again for 30 min. After centrifugation, water washing, and drying, the biochar catalyst co-doped with iron and nitrogen was pyrolyzed at 900°C under an N_2 atmosphere for 1 h. When applied to the catalytic reduction system, the Fe-N/BC catalyst showed the best removal effect on nitrobenzene, with a removal rate of nitrobenzene as high as 96.6% after 15 min reactions.

Acknowledgment

This work was supported by the National Natural Science Foundation of China (Grant No. 42177044).

6. References

[1] Gao F, Zhang M, Zhang W, et al. 2023 Synthesis of carboxymethyl cellulose stabilized sulfidated nanoscale zero-valent iron (CMC-S-nZVI) for enhanced reduction of nitrobenzene *Separation and Purification Technology* **315**.

[2] Wang R Z, Huang D L, Liu Y G, et al. 2019 Recent advances in biochar-based catalysts: properties, applications and mechanisms for pollution remediation *Chemical Engineering Journal* **371** 380-403.

[3] Zaeni J R J, Lim J W, Wang Z, et al. 2020 In situ nitrogen functionalization of biochar via one-pot synthesis for catalytic peroxymonosulfate activation: characteristics and performance studies *Separation and Purification Technology* **241**.

[4] Roy D, Neogi S and De S 2021 Highly Efficient reduction of P-nitrophenol by sodium borohydride over binary Zif-67/G-C3N4 heterojunction catalyst *Journal of Environmental Chemical Engineering* **9(6)**.

[5] Zeng C, Wang Y, Wan J, et al. 2023 Three-dimensional hollow catalysts with edge-defective Fe-Nx sites and enhanced peroxymonosulfate activation performance: preparation,

Journal of Physics: Conference Series **2941** (2025) 012026 doi:10.1088/1742-6596/2941/1/012026

applications, and mechanism *Separation and Purification Technology* **325**.
[6] Xu L, Fu B, Sun Y, et al. 2020 Degradation of organic pollutants by Fe/N co-doped biochar via peroxymonosulfate activation: synthesis, performance, mechanism and its potential for practical application *Chemical Engineering Journal* **400**.
[7] Sun M, Wang X, Li Y, et al. 2022 Bifunctional Pd-Ox center at the liquid–solid–gas triphase interface for H_2O_2 photosynthesis *ACS Catalysis* **12(4)** 2138-2149.
[8] Huang Y M, Li G, Li M, et al. 2021 Kelp-derived N-doped biochar activated peroxymonosulfate for ofloxacin degradation *Sci Total Environ* **754** 141999.

Degradation Kinetics of Nitrobenzene and Aniline by Fe-N/BC Redox System

Ting Zhou[1], Yang Yang[1], Tianxing Du[1], Wenyi Dong[1,2,3] and Hongjie Wang[1,2,3,*]

[1] School of Civil and Environmental Engineering, Harbin Institute of Technology Shenzhen, Shenzhen 518055, Guangdong, China.
[2] Shenzhen Key Laboratory of Water Resource Utilization and Environmental Pollution Control, Shenzhen 518055, Guangdong, China.
[3] State Key Laboratory of Urban Water Resource and Environment, School of Environment, Harbin Institute of Technology, Harbin 150090, Heilongjiang, China.
Email: 1226133395@qq.com; 284212494@qq.com; 1169777374@qq.com; dwy1967@qq.com; whj1533@qq.com; *whj1533@qq.com

Abstract. This study was aimed to analysis the degradation kinetics of nitrobenzene and aniline by the Fe-N/BC redox system. Factors such as the initial pH of the solution, PMS/$NaBH_4$ concentrations, and pollutant concentrations were investigated to explore the variations in pollutant degradation rates in the system. The results indicated that the Fe-N/BC-catalyzed $NaBH_4$ reduction of nitrobenzene followed first-order kinetics, while the Fe-N/BC-catalyzed PMS oxidation of aniline followed pseudo-first-order kinetics. When the pH was alkaline, the reaction was inhibited. Increasing the addition of $NaBH_4$ and PMS from 0.05 to 0.5 g/L resulted in the reaction rate constants for nitrobenzene and aniline increasing from 0.115, 0.120 min^{-1} to 0.743, 0.354 min^{-1}, respectively. As the pollutant concentration gradually rose from 2 to 50 mg/L, the reactions were inhibited, with the reaction kinetics for nitrobenzene and aniline decreasing from 0.348, 0.288 min^{-1} to 0.147, 0.091 min^{-1}, respectively.
Keywords. Fe-N/BC; redox system; nitrobenzene; aniline; reaction kinetics.

1. Introduction

Nitrobenzene is generally removed by the reduction method, but its reduction product is aniline, which still has toxicity and needs to be completely removed by combining with the oxidation method. Existing catalysts often only have a single function of catalyzing reduction or oxidation. To achieve the complete removal of nitrobenzene and its products, a combination of two catalysts is needed. Biochar material contains catalytically active structures such as graphitic carbon and persistent free radicals, and by doping with Fe and N, zero-valent iron with catalytic reduction properties and Fe-Nx active sites with catalytic oxidation properties can be introduced into biochar [1-4]. In theory, the same catalyst can first achieve the reduction of nitrobenzene and then achieve the oxidation removal of its reduction product aniline [2-6]. Therefore, the research group prepared a catalyst (Fe-N/BC) by co-doping modified biochar with citric acid iron and melamine in advance, and used sodium borohydride ($NaBH_4$) and peroxymonosulfate (PMS) as reducing agent and oxidizing agent to construct a catalytic reduction-catalytic oxidation system. The Fe and N doping form multiple active sites in biochar, enabling Fe-N/BC to catalyze both the reduction of nitrobenzene by $NaBH_4$ and the oxidation of aniline by PMS [3-8].

Based on the previous research on catalyst preparation and pollutant degradation efficiency, this study further investigates the reaction kinetics of the Fe-N/BC reduction-oxidation system in degrading nitrobenzene and aniline. It explores the changes in pollutant degradation rate in the system under different initial solution pH values (3-7), PMS/$NaBH_4$ concentrations (0.05-0.5 g/L), and pollutant concentrations (2-50 mg/L), to examine the removal efficiency of nitrobenzene and aniline.

2. Preparation Conditions and Process of Fe-N/BC

Based on previous research, onion was selected as the biomass in this study. After 8 hours of acid washing, it was immersed in an 8 g/L citric acid iron solution for 30 minutes with ultrasound. Then, 5 g/g biomass of melamine was doped and ultrasonicated for another 30 minutes. After centrifugation, washing, and drying, the Fe-N/BC iron and nitrogen co-doped biochar catalyst was pyrolyzed at 900°C under an N_2 atmosphere for 1 hour. When applied to catalytic reduction and oxidation systems, it showed the best removal effect on nitrobenzene and aniline, with removal rates of 96.6% and 99.3% respectively after 15 minutes of reaction.

3. Degradation and Detection of Nitrobenzene and Aniline

100 mL 10 mg/L nitrobenzene (aniline) solution was added to a beaker, followed by the addition of a certain amount of $NaBH_4$ (PMS) solution. Then, 1 mg Fe-N/BC was added. The stirring speed was set at 350 r/min, and the reaction was carried out at room temperature. At timed intervals (0, 1, 3, 5, 7, 10, and 15 min), the reaction solution of 1 mL was taken and filtered through a 0.22 μm membrane into a vial pre-filled with 10 μL sodium thiosulfate solution. The concentration of nitrobenzene (aniline) was determined using UPLC (ACQUITY UPLC BEH C18 1.7μm) [3].

4. Results and Discussion

4.1. Initial pH

The removal efficiency of nitrobenzene and aniline under different pH conditions (3-9) was fitted with first-order and second-order reaction kinetics, obtaining the reaction rate constants k and correlation coefficients R^2, as shown in figures 1 and 2 respectively.

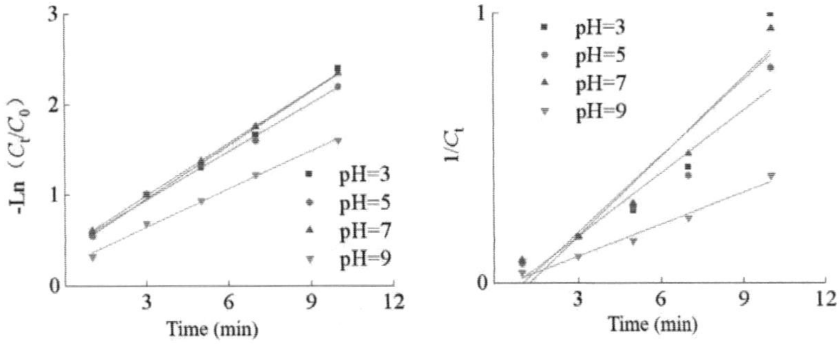

(a) First-order kinetics (b) Second-order kinetics

Figure 1. Kinetic fitting of nitrobenzene removal rate as a function of pH.

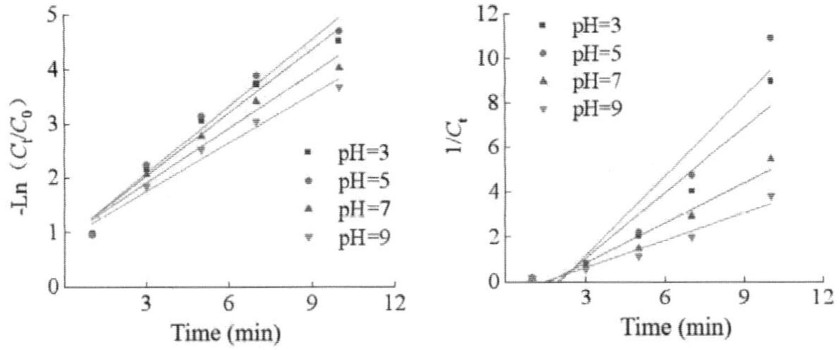

(a) First-order kinetics (b) Second-order kinetics

Figure 2. Kinetic fitting of aniline removal rate as a function of pH.

Journal of Physics: Conference Series **2941** (2025) 012027 doi:10.1088/1742-6596/2941/1/012027

From figure 1, it can be seen that the first-order reaction kinetics R^2 of nitrobenzene reduction reaction at different pH values were higher than the second-order reaction kinetics, indicating that the removal of nitrobenzene at different pH values was carried out through first-order reaction kinetics. At pH values of 3, 5, and 7, the reaction rate constant of nitrobenzene showed no significant change, remaining around 0.2. However, when the pH value increased to 9, the reaction rate constant of nitrobenzene decreased to 0.1445. It can be observed that in acidic environments with pH values from 3 to 7, pH was not the primary limiting factor for the reaction, but when the pH was alkaline, the reaction was inhibited.

From figure 2, it can be observed that for the reaction of aniline removal, the first-order reaction kinetics R^2 at different pH values were higher than the second-order reaction kinetics, indicating that the removal of aniline at different pH values occurs through first-order reaction kinetics. The rate constant k was highest at pH=5 and lower at pH=9, suggesting that in alkaline environments, the oxidation of aniline was slightly inhibited.

4.2. PMS/NaBH4 Concentration

First-order and second-order reaction kinetics fitting for the removal of nitrobenzene and aniline under different $NaBH_4$ and PMS dosages (0.05-0.5 g/L) were conducted, obtaining reaction rate constants k and correlation coefficients R^2, as shown in figures 3 and 4, respectively.

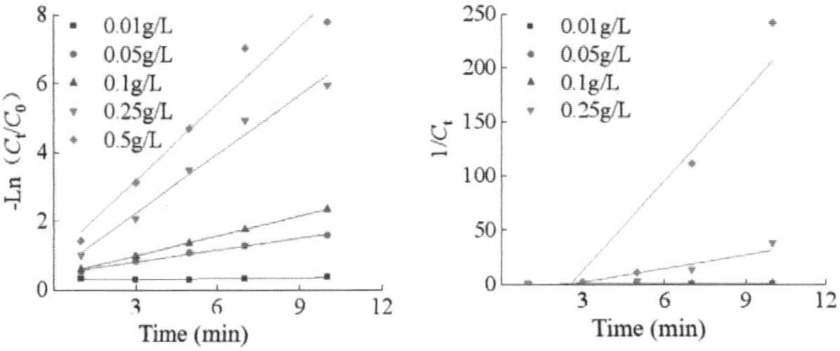

(a) First-order kinetics (b) Second-order kinetics

Figure 3. Kinetic fitting of nitrobenzene removal rate as a function of $NaBH_4$ dosage.

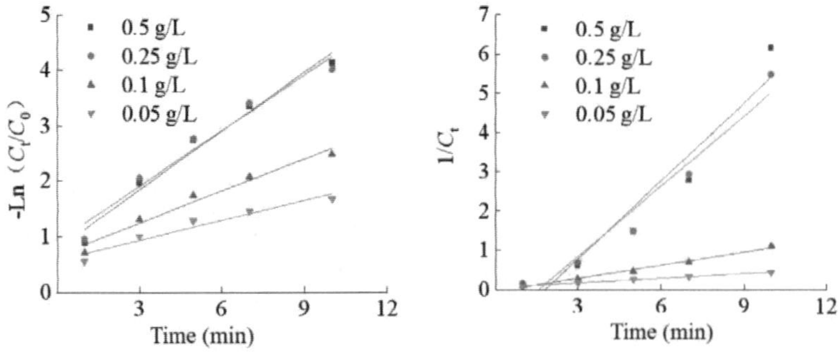

(a) First-order kinetics (b) Second-order kinetics

Figure 4. Kinetic fitting of aniline removal rate as a function of PMS dosage.

From figure 3, it can be seen that for the removal reaction of nitrobenzene, when the $NaBH_4$ addition amounts were 0.05, 0.1, 0.25, and 0.5 g/L, the first-order reaction kinetics R^2 were all higher than the second-order reaction kinetics, while when the $NaBH_4$ addition amount was 0.01 g/L, neither kinetics could fit well. This indicated that the removal pattern of nitrobenzene at different $NaBH_4$ addition amounts followed first-order reaction kinetics. As the $NaBH_4$ addition amount increased from

0.05 to 0.5 g/L, the reaction rate constant increased from 0.115 to 0.743, indicating that $NaBH_4$ addition was the main factor affecting the removal rate of nitrobenzene.

From figure 4, it can be observed that for the removal reaction of aniline, when the PMS addition amounts were 0.5, 0.25 g/L, the first-order reaction kinetics R^2 were higher than the second-order reaction kinetics R^2, while when the PMS addition amounts were 0.1, 0.05 g/L, the second-order reaction kinetics R^2 were higher than the first-order reaction kinetics R^2. This suggested that the removal reaction of aniline cannot be well described by a first-order reaction under different PMS addition amounts. Relevant studies [4-10] have shown that in the removal of aniline by PMS, intermediate products such as quinone, azobenzene, and dodecane were generated, and quinone and azobenzene can further oxidize to oxalic acid and other small organic acids in the catalytic oxidation system. When the PMS addition amount was low, the intermediate products generated from the oxidation of aniline would compete with the active species in the reaction system, whereas when the PMS addition amount was high, the intermediate products generated from the oxidation of aniline would be rapidly further oxidized. Therefore, the removal reaction of aniline followed pseudo-first-order kinetics.

4.3. Pollutant Concentration

The removal efficiency of nitrobenzene and aniline with different initial pollutant concentrations (2-50 mg/L) was fitted with first-order and second-order reaction kinetics, obtaining the reaction rate constants k and correlation coefficients R^2, as shown in figures 5 and 6, respectively.

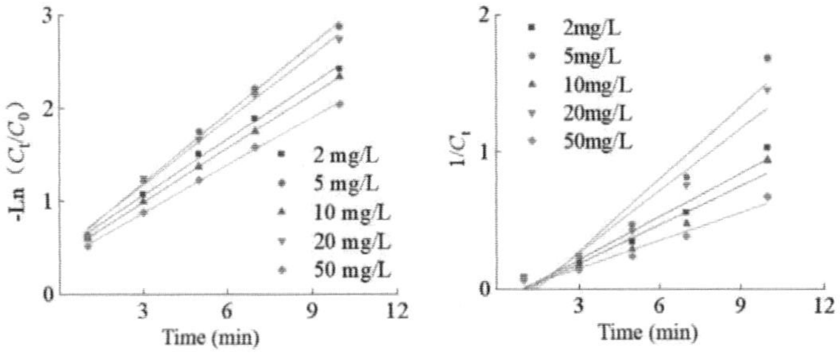

(a) First-order kinetics (b) Second-order kinetics

Figure 5. Kinetics fitting of nitrobenzene removal rate as a function of nitrobenzene concentration.

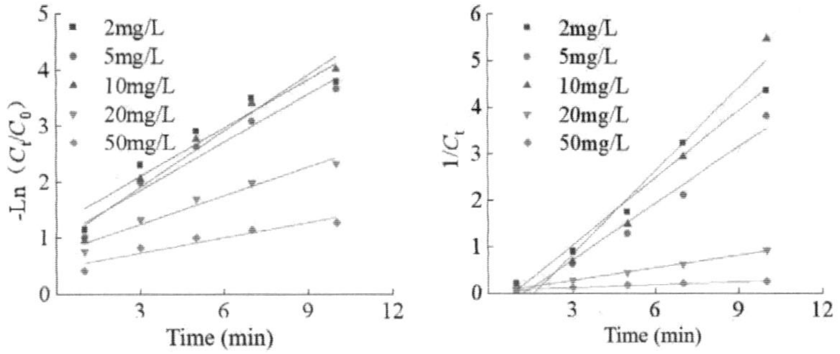

(a) First-order kinetics (b) Second-order kinetics

Figure 6. Kinetics fitting of aniline removal rate as a function of aniline concentration.

From figure 5, it can be seen that for the removal reaction of nitrobenzene, at different initial concentrations of nitrobenzene, the first-order reaction kinetics R^2 were higher than the second-order reaction kinetics, and the fitting effect of the first-order reaction kinetics were better (R^2>0.99), indicating that the removal reaction of nitrobenzene at different initial concentrations were carried out with first-order reaction kinetics.

From figure 6, it can be observed that for the removal reaction of aniline, when the initial concentration of aniline was 10 mg/L, the first-order reaction kinetics R^2 was slightly higher than the second-order reaction kinetics R^2, while when the initial concentration of aniline were 2, 5, 20, 50 mg/L, the second-order reaction kinetics R^2 were higher than the first-order reaction kinetics R^2, further indicating that the intermediate products generated after the oxidation of aniline would compete with the active species in the aniline oxidation system [6-8]. When the initial concentration of aniline was increased from 10 to 50 mg/L, the reaction rate constant decreased from 0.596 to 0.022, indicating that the initial concentration of aniline significantly affected the oxidation reaction of aniline.

5. Conclusion

The main conclusions of this study were as follows:

(1) The Fe-N/BC catalyzed reduction of nitrobenzene by $NaBH_4$ was a first-order reaction, while the Fe-N/BC catalyzed oxidation of aniline by PMS was a pseudo-first-order reaction.

(2) The reaction was inhibited under alkaline pH conditions.

(3) When the addition of $NaBH_4$ and PMS increased from 0.05 to 0.5 g/L, the reaction rate constants for nitrobenzene and aniline respectively increased from 0.115, 0.120 min^{-1} to 0.743, 0.354 min^{-1}.

(4) As the pollutant concentration gradually increased from 2 to 50 mg/L, the reactions were inhibited, and the reaction kinetics of nitrobenzene and aniline decreased from 0.348, 0.288 min^{-1} to 0.147, 0.091 min^{-1}.

Acknowledgments

This work was supported by the National Natural Science Foundation of China (Grant No. 42177044).

6. References

[1] Liang G, Yang Z, Wang Z, et al. 2021 Relying on the non-radical pathways for selective degradation organic pollutants in Fe and Cu Co-Doped biochar/peroxymonosulfate system: the roles of Cu, Fe, Defect sites and ketonic group *Separation and Purification Technology* **279**.

[2] Fang C, Yan J, Wang Y, et al. 2024 Facile synthesis of n-doped carbon nanorods for antibiotics degradation via Pms activation: mechanism insight and biotoxicity assessment *Separation and Purification Technology* **340**.

[3] Ghanbari S, Fatehizadeh A, Taheri E, et al. 2022 Degradation of 4-chlorophenol using MnOOH and Gamma-MnOOH nanomaterials as porous catalyst: performance, synergistic mechanism, and effect of co-existing anions *Environ Res* **215(Pt 2)** 114316.

[4] Yan Z, Gu Y, Wang X, et al. 2021 Degradation of aniline by ferrous ions activated persulfate: impacts, mechanisms, and by-products *Chemosphere* **268** 129237.

[5] Xu Q, Zhou F, Yu Q, et al. 2023 Aniline degradation and AS(III) oxidation and immobilization by thermally activated persulfate *Chemosphere* **338** 139573.

[6] Hu J, Zhou S, Liao S, et al. 2023 Effect of oxidized pyrite activating persulfate on the degradation of aniline aerofloat: mechanism and degradation pathway *Minerals Engineering* **201**.

[7] Peng X, Fan Z, Wang Q, et al. 2024 Recent advances on the role of high-valent metals formed during persulfate activation *Separation and Purification Technology* **345**.

[8] Wang X, Zheng M, Qian Y, et al. 2023 Synergizing redox of zerovalent iron and singlet oxygen to remove aniline, chromium and antimony in printing and dyeing wastewater synchronously: multifunctional effect of sludge derived biochar *Chemical Engineering Journal* **476**.

[9] Song M, Nguyen Q B, Kim C, et al. 2023 Sustained activation of persulfate by slow release of Fe(II) from silica-coated nanosized zero-valent iron for in situ chemical Oxidation *Water Res*

246 120715.

[10] Ke Y, Zhang X, Ren Y, et al. 2024 Remediation of Polycyclic aromatic hydrocarbons polluted soil by biochar loaded humic acid activating persulfate: performance, process and mechanisms *Bioresour Technol* **399** 130633.

Fracture Analysis of Cylinder Support Bolt of Natural Gas Compressor

Shuai Wang[1,2,*], You Qi[3], Chao An[3], Jun Xian[3], Zhenjie Qin[3], Yan Fang[3] and Xinmin Xiong[3]

[1] Tubular Goods Research Institute, China National Petroleum Corporation & State Key Laboratory of Oil and Gas Equipment, Xi'an 710065, Shaanxi, China.
[2] College of Chemistry & Environmental Engineering, Yangtze University, Jingzhou 434023, Hubei, China.
[3] Tarim Oilfield Company, PetroChina Company Limited, Korla 841000, Xinjiang Uighur Autonomous Region, China.
Email: *574916060@qq.com; 13619950901@139.com; 18699658079@139.com; xianjun-tlm@petrochina.com.cn; 13999620609@139.com; 282837568@qq.com; 13999028150@qq.com

Abstract. Failure of accessories of high-pressure natural gas compressors is harmful. In this paper, taking the fracture accident of the cylinder support bolt of a natural gas compressor in an oil field as an example, the causes of the fracture failure of the bolt are studied, as the chemical composition, hardness, metallographic structure, and inclusions of the bolt are tested and analyzed, the macro morphology of the fracture is observed and analyzed, the micromorphology of the fracture is observed by scanning electron microscope and energy spectrum analysis, and the installation structure and operating conditions of the bolt are studied. The results show that the fatigue fracture of the bolt at the stress concentration at the root of the thread under the vibration alternating load of the compressor is the reason for the fracture failure of the bolt. It is suggested to strengthen compressor vibration monitoring and take appropriate vibration reduction measures to reduce the impact of compressor vibration.
Keywords. Compressor; bolt; fracture; fatigue.

1. Introduction

As a gas pressurizing and conveying device, the compressor is widely used in modern industry, not only involving civil industries such as automobile and household appliance manufacturing but also being the "heart equipment" of heavy industries such as the petrochemical industry and mineral mining. There are various types of compressors, which can be divided into volume type and speed type according to their different working principles. A reciprocating compressor is a typical volumetric compressor, which has the advantages of a wide range of pressure regulation, high thermal efficiency, mature technology, and convenient maintenance. It is currently a dominant model in the application of compressors in the whole industry. Although the reciprocating compressor has many advantages and is widely used, it also has a prominent and inevitable disadvantage, that is, the airflow pulsation caused by its working principle itself, especially the pressure pulsation of airflow. Excessive pressure pulsation will cause vibration of the reciprocating compressor and its pipeline and other accessories, and cause a series of serious harm to production activities. The common ones are: fatigue failure of pipeline structure; The pipeline and its connecting parts are loose or broken; Causing pipeline leakage, explosion, combustion, and other malignant accidents; Damage to key parts such as valve plate; Failure or destruction of the test instrument installed on the pipeline.

In addition, equipment vibration problems often directly lead to huge economic losses. As early as the 1990s, the annual losses caused by equipment vibration in the United States reached 10 billion dollars, and the proportion of damage events caused by vibration is only second to mechanical failures. Therefore, it is of great significance to study the failure mechanism of pipes or connecting bolts and other accessories caused by compressor vibration and the corresponding vibration reduction measures to reduce the accident risk caused by vibration [1,2].

In December 2021, two fastening bolts connecting the secondary cylinder of the compressor and the support seat were found to be broken after an on-site inspection of the compressor in a gas field natural gas processing station. The failure location is shown in figure 1.

(a) Position of connecting bolt between cylinder block support and cylinder block

(b) The connecting bolt broken and dropped.

Figure 1. Failure location diagram.

The natural gas compressor will be put into operation in March 2021. The model is DYT1600. The vibration speed during operation is 7.1 mm/s. The accumulated operation time of a single unit is about 4000 hours. The failed bolt is a hexagon head bolt with the specification of M20 × 95mm, the strength grade is 8.8, the specific material is not clear, and the manufacturing standard is GB/T 3098.1-2010. In order to prevent the reoccurrence of this kind of accident, the author analyzed the causes of the fracture of the joint.

2. Experiment

In order to analyze the cause of the fracture of the support bolt of the compressor, a macro analysis of the fracture surface of the failed bolt was carried out. Mark the two broken hexagon bolts as No. 1 and No. 2 samples respectively. In order to confirm whether there are other surface cracks in the failed bolt, the fluorescent magnetic powder was used to detect the defects on the outer wall of No. 1 and No. 2 studs respectively. Take the chemical composition analysis sample on No. 1 bolt, use ARL 4460 direct reading spectrometer to analyze the chemical composition of the sample, respectively take the thread cross-section and bolt surface hardness samples on No. 1 bolt, use KB30BVZ-FA Vickers hardness tester to test the Vickers hardness (HV10) of the bolt cross-section and surface, take the metallographic analysis sample from the bolt, and use OLS4100 laser confocal microscope to analyze the microstructure The grain size and non-metallic inclusion were tested and analyzed. The fracture surface of the No. 1 bolt was observed by scanning electron microscope and analyzed by energy dispersive X-ray spectroscopy (EDS).

3. Experiment Results

3.1. Visual Inspection

The macro morphology of bolts is shown in figure 2. It can be seen from figure 2 that the bolt of sample 1 remains intact and breaks into two parts. The total length of the screw is 95mm. Sample 2 is the tail of the screw after the bolt breaks, with a length of 15mm. The two bolts break at the same location, which is the end face of the threaded hole where the screw is screwed into the cylinder head of the compressor. There are many scratches on the fracture surface at the tail side of 1 # and 2 # bolts, and the morphology has been damaged, as shown in figures 2(b) and 2(c). Figure 2(d) shows the macroscopic picture of the fracture at the head side of 1 # bolt. It can be seen from figure 2(d) that the fracture is flush, the section is perpendicular to the bolt axis, and no obvious necking is found on the fracture. The section is divided into three areas: crack source region, propagation region, and final fracture region. The crack originates from the bottom of the bolt thread, with the deepest surface color. The crack propagation area is large, and the surface of the transient fracture area is rough, showing a bright metal color. The whole fracture surface is characterized by brittle fracture.

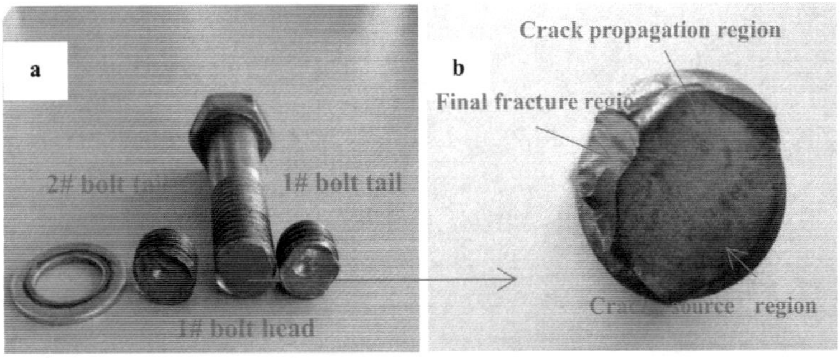

Figure 2. Macro photos of failed bolts: (a) fractured bolt morphology (b) macro morphology of No. 1 bolt fracture.

3.2. Non-Destructive Test

The magnetic particle testing results show that no other defects are found on # 1 and # 2 bolts.

3.3. Chemical Composition

Chemical composition analysis results show that the bolt is made of 45# steel and its chemical composition meets the requirements of GB/T 3098.1-2010 for Grade 8.8 high-strength bolts.

3.4. Hardness Test

The measurement location of the bolt cross-section and surface hardness is shown in figure 3, and the measurement results are shown in table 1. It can be seen from table 1 that the hardness within 1/2 radius of the bolt cross section is 228-258 HV10, which does not conform to the provisions of the standard GB/T 3098.1-2010, and the surface hardness conforms to the provisions of GB/T 3098.1-2010.

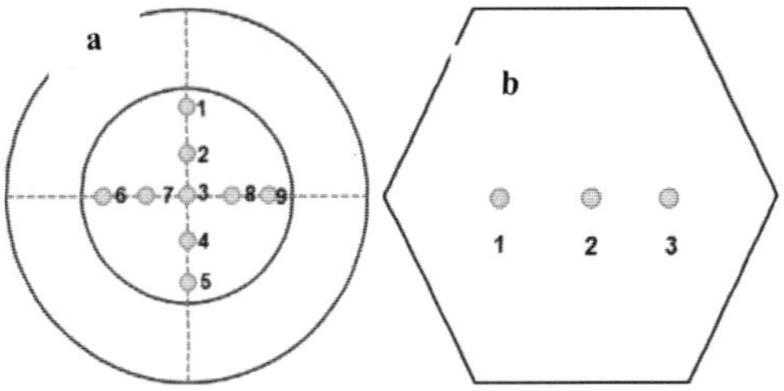

Figure 3. Schematic diagram of hardness point test location: (a) cross section (b) surface.

Table 1. Vickers hardness test results (HV10).

Test position	1	2	3	4	5	6	7	8	9
Cross section hardness value	258	251	228	240	252	254	247	249	255
Surface hardness value	288, 299, 302								
GB/T 3098.1-2010requirement	255-335HV10								

Journal of Physics: Conference Series **2941** (2025) 012028 doi:10.1088/1742-6596/2941/1/012028

3.5. Metallographic Analysis

The inspection and analysis results of bolt microstructure, grain size and non-metallic inclusion are shown in figures 4-7. It can be seen from figures 4-7 that the metallographic structure of the bolt surface is tempered sorbite with a grain size of 8.5, the metallographic structure of the core is ferrite+pearlite+bainite+a small amount of widmanstatten with a grain size of 9.5, and the non-metallic inclusions are A0.5, B0.5, D1.0.

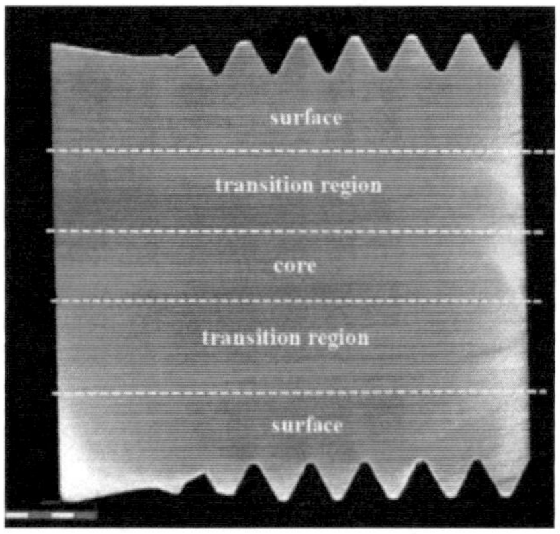

Figure 4. Macro metallographic structure.

Figure 5. Metallographic structure of bolt surface.

Figure 6. Metallographic structure of transition zone.

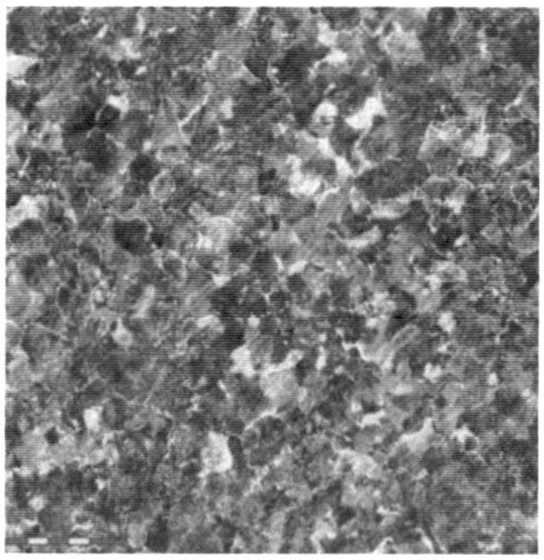

Figure 7. Metallographic structure of core.

3.6. Scanning Electron Microscope Analysis of Fracture

Figure 8 shows the scanning electron microscope (SEM) photos of different areas of the fracture. It can be seen from figure 8 that the fracture surface is flat, and the crack originates from the root step of the thread root (figure 8(b)) and extends outwards; High magnification observation shows that there are extrusion wear marks on the near source area section (figure 8(c)), obvious fatigue striations (figures 8(d) and 8(e)) are visible in the expansion area, no dimple morphology is found in the transient fracture area, and tear edges are visible in some

Journal of Physics: Conference Series **2941** (2025) 012028 doi:10.1088/1742-6596/2941/1/012028

areas (figure 8(f)). In addition, X-ray energy spectrum (EDS) analysis was carried out on different areas of the fracture surface (figure 8(a)), and the results are shown in table 2. It can be seen from table 2 that the chemical composition of the section is mainly C, O and Fe, and the content of O near the fracture source area is the highest, mainly because the area is the first to crack, the largest contact time with air, and the highest content of oxygen corrosion products.

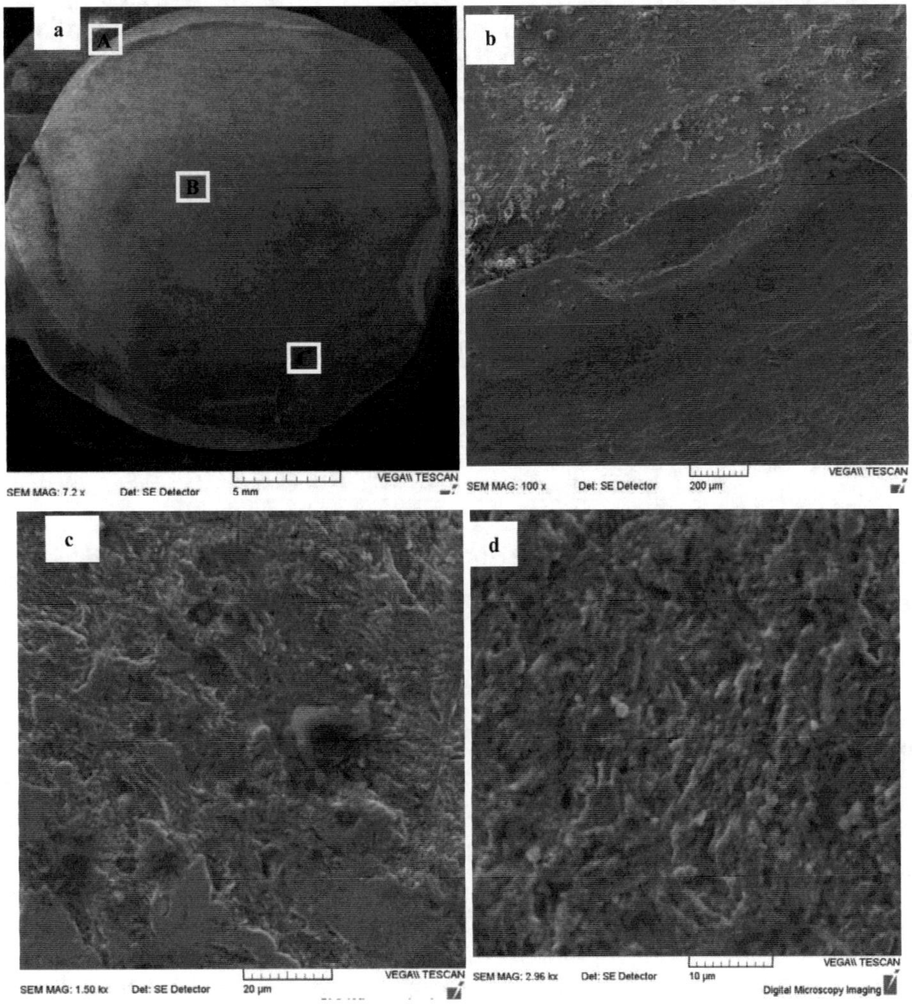

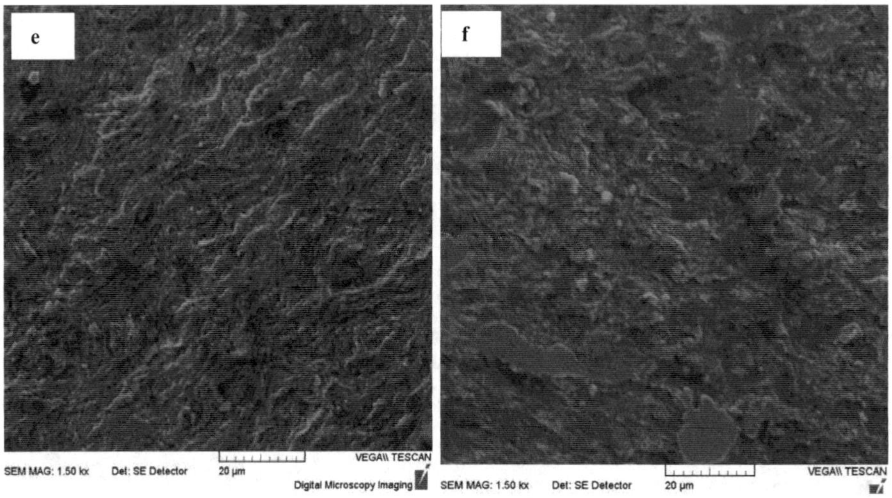

Figure 8. SEM photos of different areas of fracture: (a) At low magnification, (b) and (c) are the magnified morphologies in area A, (d) and (e) are the magnified morphologies in area B, (f) are the magnified morphology of C region.

Table 2. Energy spectrum analysis results in different regions (wt.%).

Element	Area A	Area B	Area C
C	38.45	19.88	16.80
O	22.20	14.21	
Fe	34.42	62.04	82.30
Si	1.48	2.54	0.90
Ca	1.22		

4. Comprehensive Analysis

From the macroscopic appearance of the bolt fracture, the fracture is bright and flat without obvious necking. The crack originates from the bottom of the bolt thread, and the entire fracture is flat, showing brittle fracture characteristics [3-6]. The microscopic morphology of the fracture surface shows obvious fatigue striation, indicating that the bolt fracture type is fatigue fracture. According to the field working conditions, the vibration speed of the compressor during operation is 7.1 mm/s, and the vibration direction is perpendicular to the bolt. Based on this, it can be judged that the failure mode of the bolt is vibration fatigue. The reasons for the fracture of bolts are analyzed in terms of material properties, operating conditions and installation forces.

Firstly, the performance of the bolt material is analyzed. The bolt is a grade 8.8 high-strength bolt. According to the requirements of GB/T 3098.1-2010, the bolts with grade 8.8 and thread diameter exceeding 20mm must be subject to quenching and tempering heat treatment (quenching+high temperature tempering). The microstructure of the steel after quenching and tempering treatment is tempered sorbite. It is also required that the materials used for this grade should have sufficient hardenability to ensure that the core of the thread section of the fastener can obtain martensite with a volume fraction of about 90% in the hardening state and before tempering, so as to obtain higher strength, plasticity and toughness. It is specified that the difference of hardness values measured in the half radius area of the thread cross section should be less than 30HV to verify that the martensite in the material has reached the requirement of 90%. According to the metallographic structure analysis results of the fractured bolt, the metallographic structure of the bolt core is ferrite+pearlite+bainite+a small amount of widmanstatten, which is different from the surface metallographic structure (tempered sorbite). The inconsistency between the internal and external structures is caused by an improper heat treatment process, which is often said to be not quenched. In view of the inconsistency of the internal and external structure of the bolt, the internal and external hardness of the bolt was tested. The cross section hardness test results show that the hardness of the bolt core is low, which is due to the compact structure of the external tempered sorbite with a high hardness value, while the core structure, especially the ferrite structure, is relatively soft, reducing the strength level of the core area. The hardness difference at the test location in the half radius area of the cross section of the thread is 30HV, which also indicates that the bolt core is not hardened and is unqualified during quenching heat treatment. Relevant research shows that heat treatment has a great impact on the fatigue performance of bolt materials. The bolt core is not hardened and the distribution of metallographic structure is uneven, which will lead to the reduction of its fatigue strength [7-12].

Secondly, the operating conditions of bolts are analyzed. In general, the installation of mechanical equipment should consider the impact of vibration on their normal operation, especially for compressors and other equipment that will generate vibration. In the international standard ISO 2372, the vibration speed in the range of 0.11 mm/s to 71 mm/s is divided into 15 levels, and the operating state of mechanical equipment is divided into four levels: good, allowable, tolerable and not permissible, and the mechanical equipment is divided into four categories according to its type and installation conditions. The vibration speed of the compressor is 7.1 mm/s, and the power is 1600 kW. According to the standard, the vibration level of the compressor belongs to level C. At this level, the operation status of the equipment is tolerable, and there is a certain risk of failure.

Finally, the installation force of bolts is analyzed. According to the field investigation, the two bolts are tightened according to the torque on site. The hand-held vibration tester is used on site to detect the vibration of the support seat, and the historical vibration parameters are obtained for analysis. The maximum horizontal vibration speed of the compressor is 7.1 mm/s, and the vertical vibration speed is 3.0 mm/s. Therefore, the bolt bears both horizontal and vertical vibration loads. Through the observation of the external surface morphology of the bolt thread, after a long period of impact and friction with the inner wall of the supporting circular hole on one side of the bolt fracture source area, the outer side of the thread is smooth and bright, the friction trace is very obvious, the coating has been worn off, and there is no obvious wear trace on the thread on one side of the transient fracture area. Therefore, it is judged that the bolt is close to or has a small gap with the supporting circular hole on the fracture side, and the gap on the other side is large. Because the compressor vibrates in the horizontal direction, the bolt contacts the inner wall of the circular hole of the support seat, so the bolt on the inner wall of the circular hole of the support seat bears a one-way bending fatigue load, and finally, the bolt breaks at the stress concentration of the root of the thread on the upper end face of the circular hole of the support seat.

To sum up, the core of the bolt was not hardened during heat treatment, and the distribution of the metallographic structure was uneven, leading to the reduction of its fatigue strength. Under the action of the alternating vibration load of the compressor, fatigue fracture occurred at the stress concentration at the root of the thread, leading to the failure of the bolt.

5. Conclusions and Recommendations

(1) The chemical composition of the failed bolt meets the requirements of GB/T 3098.1-2010 for Grade 8.8 high-strength bolt; The bolt surface structure is tempered sorbite with a grain size of 8.5. The core metallographic structure is ferrite+pearlite+bainite+a small amount of widmanstatten with a grain size of 9.5. The non-metallic inclusions are A0.5, B0.5, and D1.0; The core hardness does not meet the requirements of GB/T 3098.1-2010.

(2) The fatigue strength of the bolt was reduced due to the nonhardening of the core during heat treatment. Under the action of the alternating vibration load of the compressor, the fatigue fracture at the stress concentration at the root of the thread was the cause of the fracture failure of the bolt.

(3) It is suggested to strengthen the quality acceptance of bolts and conduct performance evaluation on the same batch of materials by sampling inspection. Strengthen the vibration monitoring of the compressor, and take appropriate vibration reduction measures to reduce the impact of compressor vibration.

Acknowledgment

The authors are grateful to the fund support of China's national key R&D plan "Research on Safety Assessment and Risk Assessment and Early Warning of Crude Oil and Natural Gas Storage Tanks, Auxiliary Pipelines and Auxiliary Facilities" (2017YFC0805804).

6. References

[1] Liu E B, et al. 2021 Analysis and research on pipeline vibration of a natural gas compressor station and vibration reduction measures *Energy Fuels* **35** 479-492.
[2] Liu E B, et al. 2022 Research on abnormal vibration and vibration reduction measures of a natural gas compressor station: a case study of the JYG *Energy Fuels* **36** 897-909.
[3] Yang J Y, et al. 2019 Analysis of fracture reason of high strength bolt *Hot Working Technology* **48** 173-176.
[4] Li W P, et al. 2008 Analysis on failure of supporting bolt for compressor cylinder *Refining and Chemical Industry* **19** 39-40.
[5] Han S S, et al. 2020 Causes analysis and study on the fracture of M12×1.25-10.9 bolt under vibration *Construction Machinery and Equipment* **51** 44-47.
[6] Jia F L, et al. 2019 Analysis on cracking failure of bolt *MW Metal Forming* **10** 12-14.
[7] Yang X, et al. 2017 Constant amplitude fatigue test research on M20 high-strength bolts

Journal of Physics: Conference Series **2941** (2025) 012028 doi:10.1088/1742-6596/2941/1/012028

in grid structure with bolt-sphere joints *Advances in Structural Engineering* **20** 1466-1475.

[8] Hedayat A A, et al. 2017 Prediction of the bolt fracture in shear using finite *Element Method, Structures* **12** 188-210.

[9] Zhang Y J, et al. 2018 Experimental study on shear behavior of high strength bolt *Connection in Prefabricated Steel-concrete Composite Beam, Composites part B* **10** 007.

[10] Guo H F, et al. 2019 Failure analysis on 42CrMo steel bolt fracture *Hindawi* **10** 2382759.

[11] González P V, et al. 2020 Failure analysis study on a fractured bolt *Engineering Failure Analysis* **109** 104355.

[12] Fu C X, et al. 2020 Fracture failure analysis on a bolt *Journal of Physics: Conference Series* **1676** 012015.

[13] Giller M, et al. 2024 Fatigue dimples? -Fractured bulkhead ring bolts in a large gas turbine engine *Practical Metallography* **11** 879-888.

[14] Warren M, et al. 2022 A review of experimentation and computational modeling of dynamic bolt fracture *Journal of Constructional Steel Research* **194** 107293.

[15] Rudman A, et al. 2021 Conventional I-shape brace member bolted connections under seismic loading: Laboratory study *Journal of Constructional Steel Research* **184** 106795.

Preparation of Amphiphilic Biosilica for Oil Recovery Enhancement

Yanxin Wang[1], Hua Liu[2], Longxue Bi[1], Wenlin Liu[1] and Li Sheng[1,*]

[1] College of Materials Science and Engineering, Jilin Institute of Chemical Technology, Jilin 132022, Jilin, China.
[2] Shouguang Nuomeng Chemical Co., Ltd, Shouguang 262700, Shandong, China.
Email: wangyanxin1@jlict.edu.cn; yishuilh@163.com; bilongxue@jlict.edu.cn; liuwenlin@jlict.edu.cn; *shengli@jlict.edu.cn

Abstract. SiO_2 nanofluid is frequently employed to enhance the extraction of crude oil. The enhancement of the stability of nano-sized SiO_2 and its affinity for oil droplets is crucial to the advancement of this technology. By using ring-opening polymerization (ROP) and atom transfer radical polymerization (ATRP), a polypeptide-containing triblock copolymer of poly (ethylene glycol)-b-poly (L-lysine)-b-poly (styrene) (PEG-b-PLL-b-PS) was synthesized and characterized by ^1H NMR, FTIR, and GPC. After a mild biomimetic synthesis process, the amphiphilic biosilica particles were successfully prepared. The oil displacement efficiency of the resultant SiO_2 nanofluid can exceed 65%.
Keywords. Nanofluid; poly(L-lysine); biosilicification; amphiphilic biosilica; oil displacement efficiency.

1. Introduction
Nanofluid flooding technology has been demonstrated to be an effective approach for enhancing oil recovery in tight oil reservoirs within oil fields [1]. The fundamental principle of this technology involves the manipulation of the size, morphology, and surface modification characteristics of nanoparticles [2]. Especially, the nanofluids which utilize amphiphilic nanoparticles have the potential to streamline formulation design and enhance oil displacement efficiency. Current methods for the preparation of nanoparticles do not enable the production of amphiphilic particles in a one-step reaction [3]. Additionally, the "grafting to" [4] and "grafting from" [5] techniques for surface modification of organic macromolecules present challenges, including an uncontrollable grafting rate and complex procedural steps. Consequently, achieving precise regulation of morphology, structure, and composition of nanoparticles is of particular importance in the preparation and application of nanofluids.

Drawing inspiration from the natural process of biosilicification, a collection of cationic proteins derived from diatom cell walls, along with synthetic polypeptides, has been utilized to facilitate mild biosilicification [6]. In this instance, cationic poly(lysine) exhibited catalytic properties in the deposition of silicic acid, and its conformational changes during the biosilicification process were examined in detail. This approach offers a straightforward method for the production of silica nanostructures, allowing for precise control over their morphology. Therefore, Synthetic poly(lysine)-containing block copolymers receives extensive research interests for their novel and unexpected aggregate morphologies due to the unique secondary structures in self-assembly and biosilicification process. For example, Deming et al. firstly prepared nonionic, water-soluble, methylated mono- and diethyleneglycol functionalized poly(lysine) [7]. The group of Kataoka has extensively studied drug and gene delivery based on supramolecular assembly of poly (ethylene glycol)-b-poly (L-lysine) (PEG-b-PLL) hybrid block copolymers [8]. Later, the poly(styrene)-poly(L-lysine) block copolymers with narrow molecular weight distribution were synthesized via ring-opening polymerization [9]. It can be attributed to amphiphilic diblock copolymers after removing the protect group of polypeptide. According to this, Klok and co-workers studied the aqueous solution self-assembly of a series of poly(styrene)$_m$-b-poly(L-lysine)$_n$ linear block copolymers [10]. In addition to the self-assembly behavior, cationic PLL hybrid copolymers have been applied to direct the biomimetic synthesis of silica structures. Patwardhan et al. first found that new amorphous silica structures can be uniquely synthesized using synthetic PLL as organic templates [11]. More recently, our group has reported a facile method to prepare PEG grafted silica nanoplates using double hydrophilic PEG-b-PLL [12]. What the above findings have in common is that they all utilized the hydrophilic or hydrophobic of synthetic units to design the products unique structures.

Here we report the synthesis of PLL-containing triblock copolymer as a template molecule for biosilicification of amphiphilic silica organic-inorganic hybrid nanoparticles, and the oil displacement efficiency of the resulting fluids was examined. Bionic technology represents an effective approach to the design of hybrid materials. Furthermore, it has the potential to overcome the limitations associated with the application of biosilicification technology in the engineering field, thereby broadening its range of applications.

2. Experimental

2.1. Materials

Monomethoxy-poly (ethylene glycol)-amine (mPEG-NH_2) with a molecular weight of 1000 was obtained from JemKem Technology USA. Nε-Trifluoroacetyl-L-lysine (lys(TFA)) was purchased from GL Biochem (Shanghai) Ltd. Styrene (St) was supplied by Aldrich, and passed through an alumina column to remove inhibitors. Other reagents were purchased from Aldrich and used as received.

2.2. General Procedure for the Preparation of PEG-PLL-PS Template Molecule

The synthesis of lys (TFA) N-carboxyanhydride (lys (TFA)-NCA) was performed using established protocols [13]. Following this, PEG-b-PTLL-NH_2 was synthesized through the ring-opening polymerization (ROP) of lys (TFA)-NCA, with PEG-NH_2 serving as the initiator in dimethylformamide (DMF). After purification, PEG-b-PTLL-NH_2 was reacted with an excess of 2-bromopropionyl bromide in a dichloromethane/tetrahydrofuran (DCM/THF) solvent system, resulting in the formation of the corresponding atom transfer radical polymerization (ATRP) macroinitiator, PEG-b-PTLL-Br. This macroinitiator was subsequently utilized for the ATRP of styrene, employing Cu(I)Br/PMDETA as the catalyst in DMF. The resultant product was precipitated in excess of diethyl ether followed by removing any residual copper salts step. After deprotection [14], the PEG-b-PLL-b-PS was obtained readily.

2.3. General Procedure for the Biosilicification of Amphiphilic SiO_2 Nanoparticles

Typically, phosphate buffer and silicic acid precursors, derived from the fresh hydrolysis of tetramethoxysilane (TMOS) using hydrochloric acid, were added sequentially to the aqueous solution of the tripolymer. The reaction medium became turbid within a few minutes. Subsequently, the resulting product was washed three times with distilled water and ethanol, yielding a white solid.

2.4. Characterizations

^1H NMR spectra were measured at room temperature (20±1°C) by an AV-400 NMR spectrometer. GPC/LS was performed at 50°C using DMF as eluent. The flow rate was 1.0 mL/min. FTIR spectroscopy was conducted using a Bruker EQUINOX 55 spectrometer. For core flooding experiments, nanofluid was injected into the core at a flow rate of 2 mL/min, while oil was displaced at a constant temperature of 105°C until the water cut exceeded 98%. Subsequently, the oil displacement efficiency was determined.

3. Results and Discussion

3.1. Preparation of PEG-PLL-PS Template Molecule

Firstly, $PEG\text{-}NH_2$ was used as an initiator to carry out the ROP of TFA-Lys NCA to get a diblock copolymer $PEG\text{-}b\text{-}PTLL\text{-}NH_2$. The complete reaction was proved by FTIR spectroscopy (figure 1(A)). The carbonyl signal of anhydride around 1855 cm^{-1} and 1785 cm^{-1} had disappeared after the esterification and a new signal of PEG appeared at 1100 cm^{-1}. The absorption peak at 3292 cm^{-1} was attributed to υ_{NH} stretch vibration and υ_{CO} stretching vibrations of the amide group arose predominantly in the region 1600-1700 cm^{-1}, suggesting the formation of the polypeptide block. The second step involved the synthesis of PEG-b-PTLL-Br by reacting the primary amine of the PEG block with α-bromoisobutyryl bromide. The esterification reaction was carried out at 0°C in the presence of TEA and DMAP. Subsequently, the PEG-b-PTLL-Br was used as macroinitiator of ATRP at 90°C, the polymerization was carried out in DMF and CuBr/PMDETA was used as a catalytic system. ^1H NMR of PEG-b-PTLL-b-PS (figure 1(B)) reveals the appearance of resonance between 6.5 and 7.2 ppm, which is attributed to the aromatic protons from benzene ring of the PS, as the mixture solvent of DMSO-d6 and $CHCl_3$ was selected, the other chemical shifts of triblock copolymer unit was the same as corresponding diblock intermediate in DMSO-d6. Figure 1(C) shows the GPC results of PEG-b-PTLL-b-PS triblock copolymers, compared with the corresponding PEG-b-PTLL-Br macroinitiator, the triblock copolymers display a substantial decrease in elution volume. The copolymer sample showed no residual macroinitiator, suggesting a nearly quantitative chain extension from the diblock macroinitiator. After deprotection, the PEG-PLL-PS template molecule was successfully synthesized.

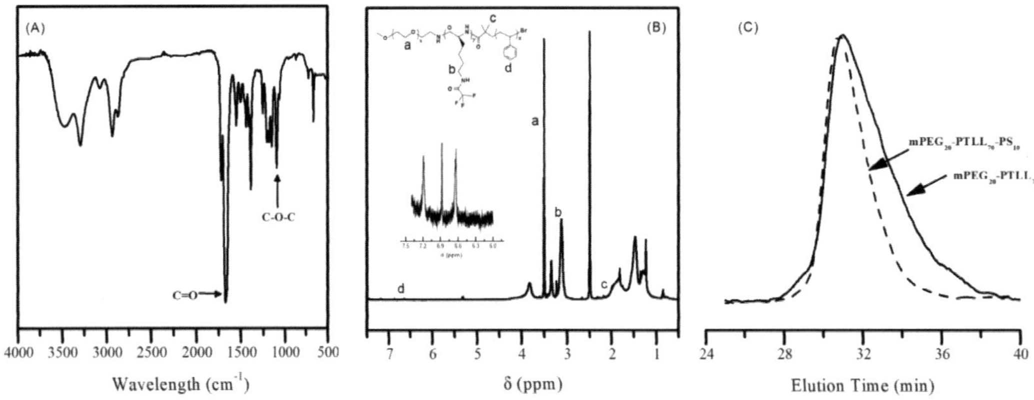

Figure 1. (A) FTIR spectra of $PEG_{20}\text{-}b\text{-}PTLL_{70}\text{-}NH_2$, (B) ^1H NMR spectra of $PEG_{20}\text{-}b\text{-}PTLL_{70}\text{-}PS_{10}$, (C) GPC traces of $PEG_{20}\text{-}b\text{-}PTLL_{70}\text{-}NH_2$ and $PEG_{20}\text{-}b\text{-}PTLL_{70}\text{-}PS_{10}$.

3.2. Preparation of Amphiphilic Biosilica

Based on previous research [12], it has been established that homo-PLL is capable of inducing silicate mineralization, resulting in the formation of regular nanosheets. The PEG-PLL block copolymers exhibit the ability to extend freely in aqueous environments. Notably, the PEG segment does not interfere with the secondary structural transformations or the assembly dynamics of the PLL segment during the mineralization process. Consequently, the PEG-PLL template molecule can be employed to produce regular nanosheets akin to those generated through homopolylysine mineralization. Additionally, the incorporation of water-soluble PEG onto the surface of the nanosheets facilitates the synthesis of water-soluble silica nanosheets in a single step. In the case of the biosilicification of PEG-PLL-PS triblock copolymer, the polymer aqueous solution exhibits a white coloration, indicating the presence of the hydrophobic segment polystyrene (PS) within the triblock copolymer, in contrast to the clear aqueous solution of the hydrophilic PEG-PLL copolymer. Upon the introduction of a phosphate buffer at pH 7.4, the white solution remained largely unchanged. However, the addition of freshly prepared silicic acid hydrolyzate resulted in a rapid transformation of the solution into an emulsion, accompanied by the formation of a visible white precipitate. This precipitate was subsequently subjected to centrifugation and washing, yielding a white solid. Scanning electron microscopy (SEM) analysis (figure 2(B)) revealed that the particle size of the resultant SiO_2 ranged from 400 to 500 nm. Drawing upon previously documented mineralization mechanisms involving PEG-PLL [12], we propose a speculative mineralization mechanism for PEG-PLL-PS, as illustrated in figure 2(A). Under neutral conditions, the amphipathic PEG-PLL-PS tripolymer is dispersed in aqueous solution, with the PLL segment carrying a positive charge. Following the addition of phosphate buffer, hydrogen bonding interactions with multivalent anions induce a degree of aggregation in the PLL chain, although no significant alteration in the secondary structure occurs. The introduction of silicic acid subsequently triggers a condensation reaction. At this juncture, the agglomeration effect of the hydrophobic PS segment is likely to influence the conformation of the macromolecule in aqueous solution, thereby hindering the transition of the connected PLL segment from random coils to α-helices. The resultant mineralization product consists of irregular SiO_2 nanoparticles that are modified on their surfaces with both PEG and PS (figure 2(B)). It is evident that the hydrophilic and hydrophobic characteristics of the SiO_2 nanoparticle surface can be further optimized by adjusting the polymerization degree of the PEG and PS blocks.

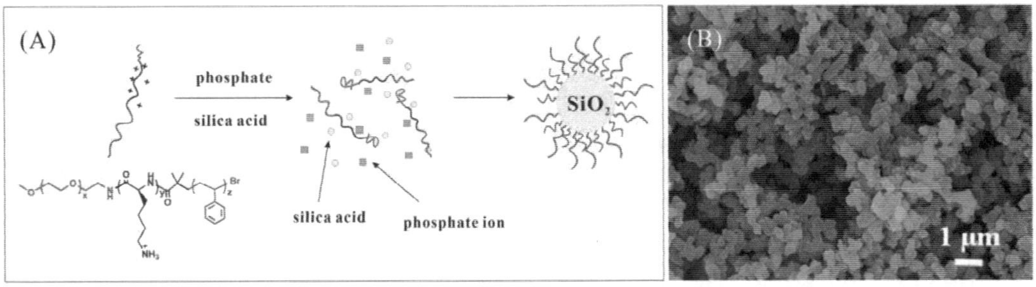

Figure 2. (A) Illustration of PEG-PLL-PS triblock templated amphiphilic biosilica formation mechanism, (B) SEM images of amphiphilic biosilica nanoparticles.

3.3. Characterization of Nanofluid Oil Displacement Effect

Nanofluid flooding represents a novel approach in the field of enhanced oil recovery. This technique employs an aqueous solution as the transport medium, which facilitates the formation of numerous small nanoparticles within the water. The nanoparticles possess a significant specific surface area and elevated surface energy, thereby promoting the detachment of crude oil into smaller droplets that can be effectively displaced by the flooding fluid. Specifically, SiO_2 nanoparticles have the capacity to temporarily occlude pore spaces, increase the swept volume, lower the oil-water interfacial tension, and adsorb onto rock surfaces, thereby influencing the wettability characteristics of the rock. Consequently, surface modification of these nanoparticles is of paramount importance. In this study, PEG molecules were utilized to modify the surface of the SiO_2 nanoparticles, enhancing the dispersibility and stability of the nanofluid. Additionally, PS molecules were incorporated to augment the lipophilicity of the nanoparticles. The synergistic effect of these modifications allows for the amphiphilic nature of the silica to effectively regulate wettability, significantly reduce oil-water interfacial tension, and enhance the efficiency of oil extraction processes. Adding amphiphilic biosilica nanoparticles to the experimental water in proportion to prepare a 1000 mg/L nanofluid, as illustrated in figure 3, there is a progressive increase in the oil displacement efficiency (ODE) corresponding to the continuous augmentation of the pore volume (PV) of injected nanofluid. Following the injection of 0.5 PV of nanofluid, the ODE exceeds 60%. Upon achieving a cumulative injection of 1.87 PV of nanofluid, the pressure stabilizes, and no oil is expelled from the tail end, thereby concluding the oil recovery experiment. The final ODE surpasses 65%, indicating that the synthesized nanofluid demonstrates effective oil recovery capabilities. Taking into account considerations such as oil recovery efficiency and associated costs, the optimal injection volume of nanofluid is determined to be 1.0 PV.

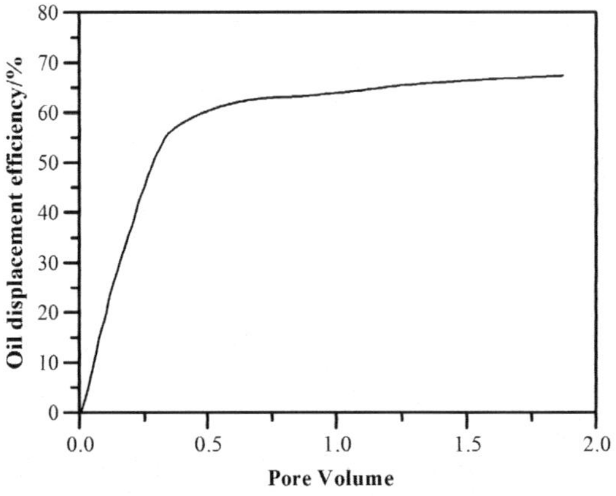

Figure 3. Variation of oil displacement efficiency with the injection volume in the process of injecting the nanofluid into the core.

4. Conclusion

In conclusion, we have prepared the unprecedented PEG-b-PTLL-Br macroinitiator to initiate the ATRP of styrene. ^{1}H NMR, FTIR, and GPC analysis manifested that new PEG-b-PTLL-b-PS triblock copolymers have been successfully synthesized. After the deprotection of the protecting group, the PEG-PLL-PS was applied for the direct biomimetic synthesis of amphiphilic biosilica nanoparticles. When flowing in porous media (105°C), the amphiphilic SiO_2-containing nanofluid exhibits good mobility control ability, and the ODE during core flooding is greater than 65%, thus expanding the application of bionic mineralization technology in the petrochemical field.

5. References

[1] Suleimanov B A Abbasov H FandIsmayilov R H 2023 Enhanced oil recovery with nanofluid injection *Pet Sci Technol* **41(18)** 1734-1751.
[2] Liu D Zhang X Tian F Liu X Yuan JandHuang B 2022 Review on nanoparticle-surfactant nanofluids: formula fabrication and applications in enhanced oil recovery *J Dispersion Sci Technol* **43(5)** 745-759.
[3] Hu N Zhang Z Wang Q Li Y Pei YandZhang Z 2024 Enhanced foam fractionation of perfluorooctane sulfonate (PFOS) from water using amphiphilic Janus SiO_2 nanoparticles *Chem Eng J* **499** 155829.
[4] Ling X Y Phang I Y Acikgoz C Yilmaz M D Hempenius M A Vancso G JandHuskens J 2009 Janus Particles with Controllable Patchiness and Their Chemical Functionalization and Supramolecular Assembly *Angew Chem Int Edit* **48(41)** 7677-7682.
[5] Berger S Synytska A Ionov L Eichhorn K-JandStamm M 2008 Stimuli-Responsive Bicomponent Polymer Janus Particles by "Grafting from"/"Grafting to" Approaches *Macromolecules* **41(24)** 9669-9676.
[6] Xia L Liu YandLi Z 2010 Biosilicification Templated by Amphiphilic Block Copolypeptide Assemblies *Macromol Biosci* **10(12)** 1566-1575.
[7] Yu M Nowak A P Deming T JandPochan D J 1999 Methylated mono- and diethyleneglycol functionalized polylysines: Nonionic, alpha-helical, water-soluble polypeptides *J Am Chem Soc* **121(51)** 12210-12211.
[8] Kataoka K Harada AandNagasaki Y 2001 Block copolymer micelles for drug delivery: design, characterization and biological significance *Adv Drug Deliver Rev* **47(1)** 113-131.

[9] Dimitrov IandSchlaad H 2003 Synthesis of nearly monodisperse polystyrene-polypeptide block copolymers via polymerisation of N-carboxyanhydrides *Chem Commun* **(23)** 2944-2945.
[10] Lubbert A Castelletto V Hamley I W Nuhn H Scholl M Bourdillon L Wandrey CandKlok H A 2005 Nonspherical assemblies generated from polystyrene-b-poly(L-lysine) polyelectrolyte block copolymers *Langmuir* **21(14)** 6582-6589.
[11] Patwardhan S V Mukherjee N Steinitz-Kannan MandClarson S J 2003 Bioinspired synthesis of new silica structures *Chem Commun* **(10)** 1122-1123.
[12] Chen H Xia L Fu W Yang ZandLi Z 2013 One-step synthesis of water dispersible silica nanoplates *Chem Commun* **49(13)** 1300-1302.
[13] Poché D S Moore M JandBowles J L 1999 An Unconventional Method for Purifying the N-carboxyanhydride Derivatives of γ-alkyl-L-glutamates *Synthetic Commun* **29(5)** 843-854.
[14] Rodríguez-Hernández J Gatti MandKlok H-A 2003 Highly Branched Poly(l-lysine) *Biomacromolecules* **4(2)** 249-258.

Study on Bending Performance of Basalt Fiber Ultra-High Performance Concrete

Fan Gao[1], Lihuan Wang[1], Yaning Ren[1], Chu Li[1], Jinyong Yang[2] and Zuohu Wang[2,*]

[1] State Grid Hebei Electric Power Co., Ltd. Economic and Technological Research Institute, Shijiazhuang, Hebei, China.
[2] Beijing University of Civil Engineering and Architecture, Beijing, China.
Email: gaofan0427@126.com; 461152972@qq.com; 840772394@qq.com; 3786327651@qq.com; 2339778066@qq.com; *wangzuohu@bucea.edu.cn

Abstract. In order to study the effect of basalt fiber on the flexural properties of ultra-high performance concrete (UHPC) slabs, basalt fiber UHPC containing coarse aggregate was prepared. The basic mechanical properties of basalt fiber UHPC were tested, and the bending test of four-sided simply supported two-way slabs was carried out. The effects of different fiber types, steel mesh, waterborne epoxy, and their mixture on the toughness and flexural strength of UHPC slabs were studied. The results show that the basalt fiber can significantly improve the basic mechanical properties of UHPC; Comparing the UHPC plates mixed with different types of basalt fiber, Compared to the plain UHPC plates, UHPC with higher tensile strength and shorter length exhibit higher bending resistance, The peak load increased by 154.7% and 196.6%, respectively, The maximum deflection increased by 38.1% and 326.2%, respectively; The ductility of the UHPC plate mixed with basalt fiber and steel wire mesh has been significantly improved, The maximum increase is 48.9%; Based on the assumption theory, the bidirectional theory is calculated and compared with the test value, and the results agree well.
Keywords. Ultra-high performance concrete; basalt fiber; bidirectional plate; bending performance and bending capacity.

1. Introduction

Compared with ordinary concrete (NC), ultra-high performance concrete stands out because of its high strength, excellent ductility and excellent durability. The improvement of these properties is due to the dense microstructure of UHPC and the bridging effect of internal fibers. Therefore, based on its excellent performance, UHPC is expected to be used as a sustainable building material and widely used in new buildings and building restoration. This can not only prolong the service life of the building but also effectively reduce the cost of its entire life cycle [1,2]. In the composition of ultra-high performance concrete, fiber reinforced materials are also one of the key factors that determine its mechanical properties, in addition to the design of mix ratio, curing method and the selection of mineral admixtures. Studies have shown that the volume content, type and geometry of fibers significantly affect the mechanical properties of UHPC, such as compressive strength, tensile strength and toughness [3,4]. At present, steel fiber [5] is widely used as a reinforcing material in UHPC systems. However, the lack of high temperature resistance and corrosion resistance of steel fiber limits its applicability in extreme environments. In addition, the high density and cost of steel fiber are also some of the main reasons that restrict its large-scale application.

Compared with steel fiber, basalt fiber (BF), as an inorganic fiber made of natural basalt ore by high temperature melting, has received extensive attention in recent years. Basalt fiber has high tensile strength, high elastic modulus, excellent impact resistance, high temperature resistance, and good chemical stability, and especially shows excellent corrosion resistance in acidic and alkaline environments [6-8]. These characteristics make it a potential steel fiber substitute material, which can improve the durability of UHPC in harsh environments. In addition, basalt fiber has good compatibility with cement-based materials and excellent dispersibility, which helps improve the overall mechanical properties of UHPC [9]. Therefore, the application of basalt fiber provides a new research idea and practical basis for improving the performance of UHPC in special projects. Scholars have introduced coarse aggregate and found that its high strength and elastic modulus can effectively reduce the autogenous shrinkage of UHPC and improve its stiffness. Studies have shown that the volume fraction, mineral properties, and particle size of coarse aggregate significantly affect the compressive, tensile, and flexural properties of UHPC. Optimizing the selection and ratio of coarse aggregate is very important to improve the performance of UHPC [10-12].

In addition, many scholars usually do not use coarse aggregate when studying the mechanical properties of UHPC, but only use expensive quartz sand, resulting in a significant increase in the preparation cost of UHPC, which is difficult to popularize to practical engineering applications on a large scale. At present, the research on UHPC containing coarse aggregate is limited, and there are few reports on the bending performance of UHPC plates reinforced by steel wire mesh. Therefore, it is of great significance to carry out research on the influence of basalt fiber and steel wire mesh on the mechanical properties of UHPC.

In this paper, basalt fiber UHPC containing coarse aggregate was first prepared, and the related basic mechanical properties were tested. The effects of fiber content on the slump, compressive strength, splitting tensile strength, and bending strength of UHPC were studied. Through the four-sided simply supported two-way slab test, the effects of fiber type, waterborne epoxy, and steel wire mesh on the bending performance of UHPC two-way slab were further discussed, which provided a reference for practical engineering applications.

2. Test Overview

2.1. Raw Materials and Mix Ratio Design

The Portland cement (PC) used in this study has a strength grade of 42.5. The silica fume(SF) is Elkem 951 grade silicon powder produced by Eken International Trade (Shanghai) Co., Ltd. The silica content is not less than 95%, and the loss on ignition at 950 °C is not more than 1%. S95 grade granulated blast furnace slag powder is used for slag powder. The medium-coarse sand with uniform and nearly circular shape of river sand particles is selected, and the mesh number is between 20 and 40. The particles of coarse aggregate (CA) are well graded and continuously graded, clean, with few needle-like particles, and the particle shape is close to the cube. The mechanical properties of basalt fiber (BF) from Zhejiang Shijin basalt fiber are shown in table 1. The test UHPC is shown in table 2. The steel wire mesh used in the test is provided by Hebei Anping County Wire Mesh Co., Ltd., with a diameter of 1mm and a pore size of 20mm × 20mm, the sample is shown in figures 1 and 2. The water reducing agent (SP) and defoaming agent (DA) are white powder, and the water reducing rate of water reducing agent is 35%. The waterborne epoxy used in this paper is composed of epoxy resin (EP) H258A, epoxy curing agent (ZDMA) H258B and water in a ratio of 1:1.5: 3.

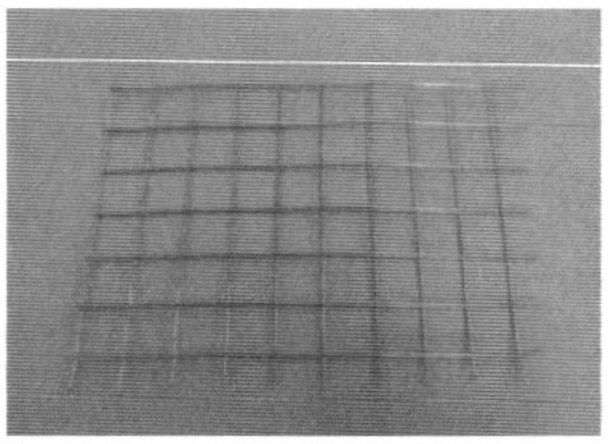

Figure 1. Wire mesh.

Figure 2. Basalt fiber.

Table 1. Basalt fiber performance parameters.

Sample name	Diameter /μm	Tensile strength /MPa	Density /(kg/m^3)	Modulus of elasticity /GPa	Length /mm	Break elongation rate /%
Basalt fiber 1	13	1050	2600	75	30	3.1
Basalt fiber 2	18	1700	2600	40	20	2.1
Steel mesh	/	582	7850	200	/	2.6

Table 2. Recipes of UHPC (kg/m^3).

Code	UHPC matrix						EP	ZDMA	BF	SP	DA
	PC	SF	Breeze	Sand	CA	Water					
UHPC-1	510	70	380	486	900	163.2	0	0	0	3.5	2
UHPC-2	510	70	380	486	900	163.2	0	0	15	3.5	2
UHPC-3	510	70	380	486	900	163.2	54.4	86.4	15	3.5	2

2.2. Test Piece Preparation

In order to test the basic mechanical properties of ultra-high performance concrete, two types of specimens were poured: 18 cube specimens of 100mm × 100mm × 100mm and 9 prism specimens of 100mm × 100mm × 400mm. The cube specimens were used to test the compressive strength (f_{cu}) and splitting tensile strength (f_{ct}) of UHPC, while the prism specimens were used to test the flexural strength (f_w). After the pouring is completed, the specimens are demoulded within 24 hours, and the basic mechanical properties are tested after 28 days of curing.

In this paper, a total of eight ultra-high performance concrete two-way slabs of 500mm × 500mm × 50mm were prepared. The effects of the following mixing methods on the flexural properties of UHPC two-way slabs were studied: no basalt fiber, single basalt fiber, basalt fiber mixed with steel wire mesh, steel wire mesh mixed with waterborne epoxy, and three mixed UHPC two-way slabs. Among them, two UHPC plates with single basalt fiber were prepared, and the specific information is shown in table 3.

For the two-way slab with a grid, the layered pouring method is adopted, and the protective layer is 3mm. The pouring process of the concrete slab is as follows: (1) The mold with a specification of 500mm × 500mm × 50mm was placed horizontally, and a 3mm-thick UHPC pad was placed around the bottom of the template. (2) The steel wire mesh is placed horizontally on the pad (the thickness of the steel wire mesh is 2mm, and the flatness of the steel wire mesh should be ensured when laying), and then 5mm thick UHPC is poured and oscillated to ensure that the UHPC at the bottom is uniform and just covers the steel wire mesh. (3) The remaining 45 mm thick UHPC is poured on it, slightly oscillating to ensure that the steel wire mesh does not float, and finally the surface is smoothed. In the table, the B number indicates whether the fiber and the type of fiber are incorporated, the S number indicates whether the steel wire mesh is incorporated, and the E number indicates whether the waterborne epoxy is incorporated. '√' indicates the incorporation of steel mesh or waterborne epoxy. For example, B1S1E1 indicates that the UHPC two-way slab is laid with a layer of steel wire mesh and mixed with waterborne epoxy and basalt fiber 1.

Table 3. The UHPC board numbering scheme.

Label	Fiber species	The UHPC matrix species	Steel mesh	Water-based epoxy
B0S0E0	0	UHPC-1	0	0
B1S0E0	BF1	UHPC-2	0	0
B1S0E0-1	BF1	UHPC-2	0	0
B1S1E0	BF1	UHPC-2	√	0
B1S1E1	BF1	UHPC-3	√	√
B1S0E1	BF1	UHPC-3	0	√
B2S0E0	BF2	UHPC-2	0	0
B2S0E0-1	BF2	UHPC-2	0	0

2.3. Test Method

The compressive strength test of concrete is a basic and important evaluation to ensure the quality of concrete. It determines the ability of the concrete specimen to resist the axial load or force applied along its axis without significant deformation. As shown in figures 3(a) and 3(b), the compressive strength and splitting tensile strength of UHPC were tested. The loading rates were 1.2 MPa/s and 1.0 MPa/s, respectively. The loading rate of the bending test is 0.2mm/min, as shown in figure 3(c).

Figure 3. Test loading device. (a) Compression test device; (b) Splitting tensile test device; (c) Four point bending device; (d) Double-way plate bending and loading device.

As shown in figure 3(d), a two-way plate is placed in the center of a square rigid base. The center of the slab top transfers the concentrated load through a prefabricated steel block of 80mm×80mm×100mm and a square steel plate of 150mm×150mm. Two displacement meters are installed on the rigid supports on both sides to measure the deflection of the plate. Two strain gauges are pasted on the UHPC surface in the center of the bottom plate and around the top plate to measure the strain change. The load is measured by the force sensor, and the force, deflection and strain data are collected synchronously by the data acquisition instrument. The whole loading process is controlled by the MTS electro-hydraulic servo testing machine, and the loading rate is 0.5 mm/min.

3. Basic Mechanical Properties of UHPC

3.1. Basic Mechanical Properties of UHPC

The results of UHPC are shown in table 4.

Table 4. Basic mechanical properties of UHPC.

Label	Slumps /mm	Compression strength f_{cu} /MPa	Cleavage tensile strength, f_{ct} /MPa	Flexural strength f_w /MPa
UHPC-1	78	104.4	9.8	8.65
UHPC-2	72	107.4	10.1	10.44
UHPC-3	75	106.4	10.5	10.62

As can be seen from table 4, when the compressive, splitting tensile strength and bending strength reach 107.4MPa, 10.5MPa, 10.62MPa, and the UHPC without fiber increases by 2.8%, 3%, 22.8% with 2.5%, and the slump decreases by 7.6%, indicating that although the basalt fiber improves the UHPC density, it will reduce the working performance. Water epoxy incorporation can also increase the basic mechanical properties of UHPC, but the increase effect is relatively poor.

3.2. Bending Process and Failure Pattern of UHPC Bidirectional Plate

3.2.1 Bending and Failure Process of the UHPC Plate. Based on the two-way plate bending test, the load-deflection curve diagram of the UHPC two-way plate under concentrated load is drawn. As shown in figure 4, the whole bending process can be divided into three stages. The first stage is the linear elastic stage (OA stage): at the initial stage of loading, due to the relatively small load, the concrete does not crack, and the load deflection curve is approximately a straight line. At this stage, the load is shared by the concrete and the fiber. However, as the load gradually increases, when the concrete cracking load is reached, the first main crack appears at the bottom of the square plate. The second stage is the internal force redistribution stage (AB stage): when the first crack appears, due to the irregular arrangement of fibers in the plate, the internal force begins to transfer along the fiber to the surroundings, so the second main crack also gradually extends from the first main crack to other directions irregularly. As the fiber at the first main crack is gradually broken, the load shows a downward trend, and the internal force and stress redistribution continue. However, due to the existence of the second main crack fiber, the specimen does not completely break and can continue to bear the load. With the increase in load, the load-deflection curve rebounds. The third stage is the failure stage (BC stage): At this stage, because the concrete has lost its bearing capacity, the load is mainly borne by the fiber. At this time, there will be no new cracks. The small cracks gradually extend outward and develop into the main cracks. As the deflection increases, the fibers at the cracks are continuously broken. Due to the brittle characteristics of basalt fiber, the load shows a gradual downward trend until the specimen completely loses its bearing capacity.

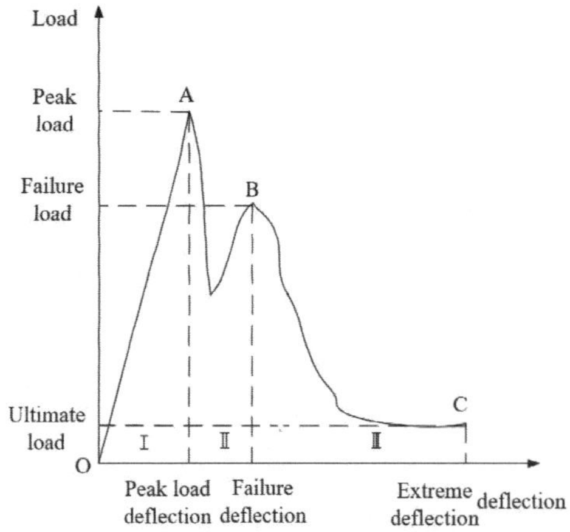

Figure 4. Load-deflection curve.

3.2.2 UHPC Plate Destruction Pattern. Figure 5 shows the failure modes of each UHPC slab. It can be seen from the figure that there is only one main crack along the diagonal line when the plain UHPC plate is destroyed, and the section is relatively flat. When the concrete is mixed with basalt fiber 1 or basalt fiber 2, the number of main cracks in the slab increases to 4 and presents a cross-shaped shape, while the three specimens without steel wire mesh show certain brittle failures when they are destroyed.

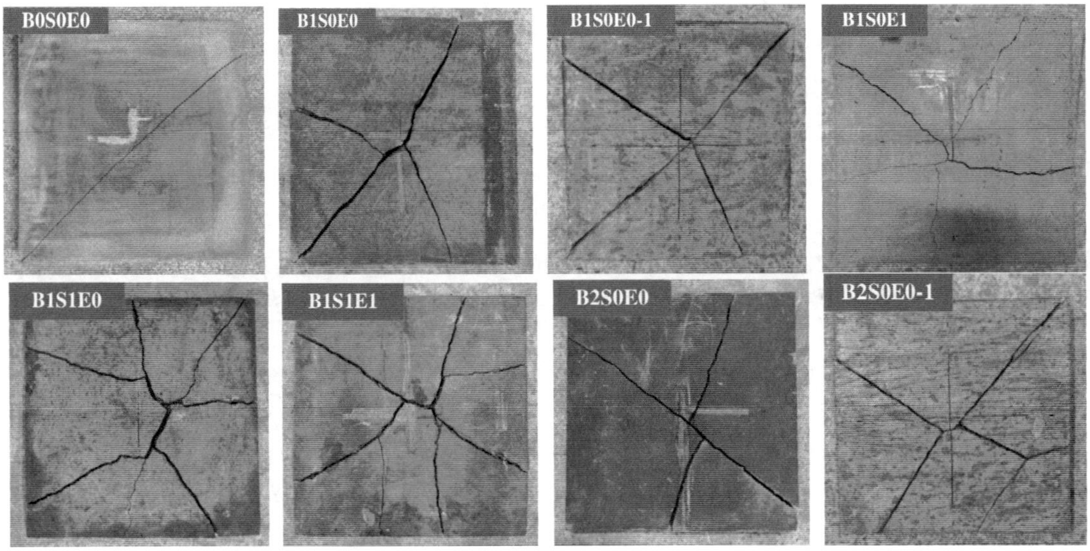

Figure 5. UHPC plate destruction morphology.

3.3. Main Factors Affecting the Bearing Capacity of UHPC Bidirectional Board

When basalt fiber 1 and waterborne epoxy are mixed, the number of main cracks in the specimen increases to 5, and the crack width is smaller than that of the specimen doped with basalt fiber alone, but the flatness of the section is poor. This is due to the random distribution of chopped basalt fiber in concrete and the different bonding forces of waterborne epoxy at different cross-section positions, which leads to the inconsistent cross-section position of UHPC under the same strength, so it shows irregularity when breaking. When basalt fiber 1 and steel wire mesh are mixed, the main cracks of the specimen increase to 6. Due to the good ductility of the steel wire mesh, the integrity of the specimen is better when UHPC is completely destroyed, and the ductility is significantly improved compared with the specimen with a single fiber. When the three are mixed, the main cracks increase to 7, and the UHPC plate has an unbroken steel wire mesh connection at the main crack when it is destroyed, and the integrity is better than that of the meshless plate.

3.3.1 Influence of Fiber Type on the Bearing Capacity of UHPC Plate. Figure 6 shows the load-deflection curves of plain UHPC slab and UHPC slab with single basalt fiber, while table 5 lists the peak load and corresponding deflection of each UHPC slab. It can be seen from figure 6 and table 5 that the peak loads of specimens B1S0E0 and B2S0E0 are 154.7% and 196.6% higher than those of specimens B0S0E0, respectively, and the displacement increases are 38.1% and 326.2%, respectively. This shows that basalt fiber significantly enhances the bearing capacity of UHPC two-way slabs.

In addition, the bearing capacity of UHPC slabs with single basalt fiber 2 is higher than that of single basalt fiber 1, mainly because the tensile strength of basalt fiber 2 is higher than that of fiber 1, and its length is shorter, which makes the number of fiber 2 more under the same quality, so the enhancement effect is more significant. All three show certain brittle characteristics during failure, but the UHPC plate with single basalt fiber shows a double peak phenomenon during failure, which further illustrates the influence of fiber on its mechanical properties.

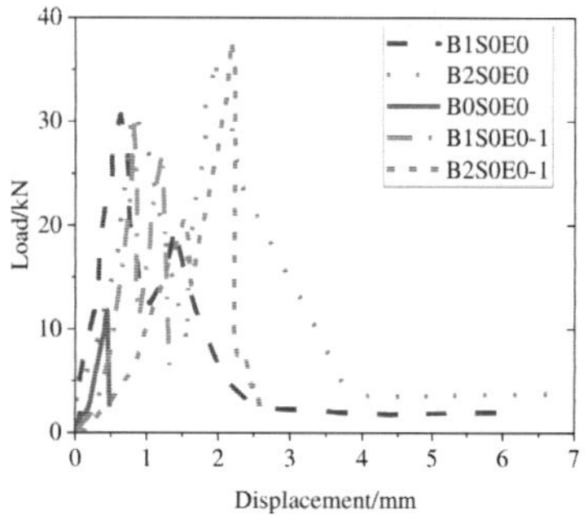

Figure 6. Load-deflection curve of UHPC board.

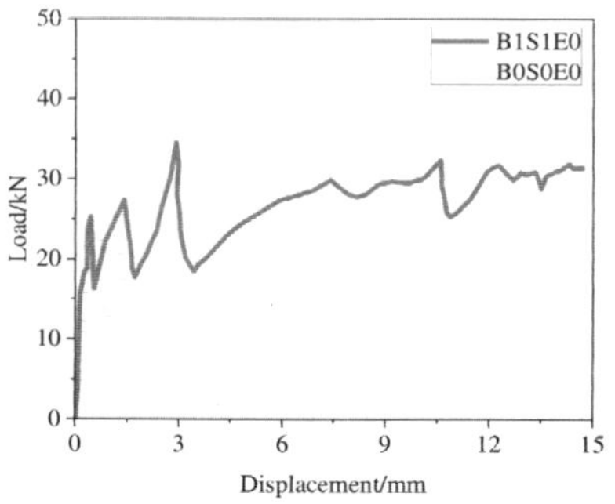

Figure 7. The load deflection curve of the UHPC plate.

Table 5. Peak load and deflection of UHPC plates.

Label	Peak load /kN	Peak load deflection /mm
B0S0E0	12.22	0.42
B1S0E0	31.12	0.58
B1S0E0-1	30.59	0.76
B1S1E0	34.89	2.81
B1S1E1	46.05	1.30
B1S0E1	38.82	1.85
B2S0E0	36.24	1.79
B2S0E0-1	38.18	2.01

3.3.2 Influence of Steel Wire Mesh on the Bearing Capacity of UHPC Plate. Figure 7 shows the comparison of load-deflection curves of mixed steel wire mesh with basalt fiberboard and single basalt fiberboard. It can be seen from figure 7 and table 5 that the peak load of specimen B1S1E0 is 10.9% higher than that of specimen B1S0E0, which indicates that the laying of steel wire mesh can increase the peak bearing capacity of UHPC slab.

The load-deflection curves of the two specimens fluctuate greatly because the basalt fiber has brittle characteristics. When the fiber at the crack is broken, the load will suddenly decrease. For the specimen without steel wire mesh, the crack has led to fracture failure; For the specimens with steel wire mesh, the load at the crack begins to be borne by the steel wire mesh, and the load can continue to rise. However, due to the smooth surface of the steel wire mesh and the small bonding force with the concrete matrix, it is easy to slip, so the curve after the peak load is wavy. At the same time, the steel wire mesh has good ductility, which leads to the strong bearing capacity of the plate after the peak load, and the curve decreases slowly, showing good ductility and bearing capacity.

3.3.3 Influence of Aqueous Epoxy on the Bearing Capacity of UHPC Plate. Figure 8 shows the comparison of load-deflection curves of mixed water-based epoxy with basalt fiberboard and single-doped basalt fiberboard. Combined with table 5, it can be seen that the peak load of specimen B1S0E1 is 24.7% higher than that of specimen B1S0E0, and the displacement is increased by 124.1%, which indicates that the mixing of waterborne epoxy and basalt fiber can significantly improve the bearing capacity of UHPC slab compared with single basalt fiber.

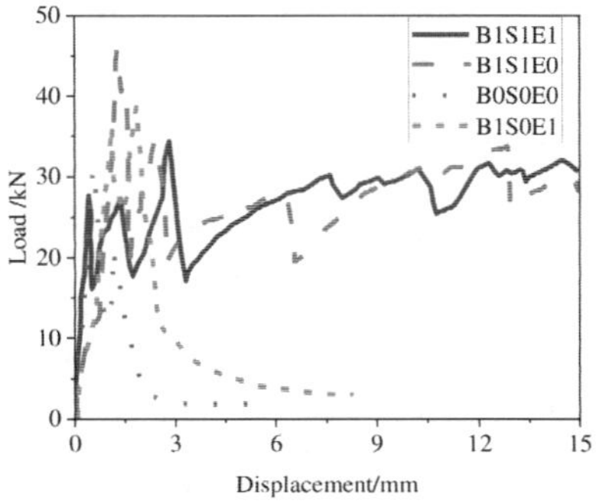

Figure 8. The load deflection curve of the UHPC plate.

The peak load of specimen B1S1E1 is 18.6%, 32.0% and 48.0% higher than that of specimens B1S0E1, B1S1E0, and B1S0E0, respectively. This shows that in the three mixed (basalt fiber, steel wire mesh, water-based epoxy) than only mixed with steel wire mesh and basalt fiber or single-doped basalt fiber, the bearing capacity of the UHPC plate is more significant. This is because the surface of steel wire mesh and fiber is smooth, and it is easy to slip when bearing load in the matrix. The incorporation of waterborne epoxy enhances the bonding force between fiber and matrix, fiber and grid, and grid and matrix, thus significantly enhancing the bearing capacity of the component, showing a significant positive hybrid effect. Since the load is mainly borne by the steel wire mesh in the later stage of loading, the load strength and curve change trend of the specimens B1S1E1 and B1S1E0 after the peak value is obtained.

4. Evaluation of the Bending Toughness of Basalt Fiber UHPC Bidirectional Plates

The bending toughness of the plate is evaluated by destroying the absorption energy in the bidirectional plate, as calculated by equation (1):

$$Q = \int_0^a F(x)dx \tag{1}$$

Where: Q is the energy absorption value; a is the central deflection of the plate; $F(x)$ is the load value when the deflection is x.

Table 6. Calculated values of energy absorption at different deflection of UHPC plates.

Label	Energy absorption value							
	Q_m/J	$Q_{0.3}$/J	$Q_{0.5}$/J	Q_1/J	Q_2/J	Q_5/J	Q_{10}/J	Q_{15}/J
B0S0E0	2.03	0.91	-	-	-	-	-	-
B1S0E0	9.24	2.64	6.85	16.86	29.31	37.36	-	-
B1S0E0-1	6.71	0.52	1.81	10.63	-	-	-	-
B1S1E0	61.78	3.89	7.93	18.17	40.48	110.42	251.47	404.04
B1S1E1	23.47	1.58	3.53	12.20	45.25	122.27	254.47	407.80
B1S0E1	40.55	1.62	3.47	14.62	44.70	74.17	-	-
B2S0E0	34.06	1.85	4.43	18.05	40.85	67.92	-	-
B2S0E0-1	25.84	0.38	0.99	4.66	25.44	-	-	-

Table 6 shows the calculated values of energy absorption at different deflection points of UHPC plates. Q_m indicates the energy absorbed when the load reaches the peak; $Q_{0.3}$, $Q_{0.5}$, Q_1, Q_2, Q_5, Q_{10}, and Q_{15} indicate the energy absorbed by the UHPC plates at deflection 0.3 mm, 0.5 mm, and 1 mm, 2 mm, 5 mm, 10 mm and 15 mm.

4.1. Effect of Fiber Type on the Bending Toughness of UHPC

Figure 9 is the energy absorption value comparison diagram of a plain UHPC plate and a UHPC plate with a single basalt fiber. Combined with table 6, it can be seen that the energy absorption values of specimen B1S0E0 and B2S0E0 at 0.3 mm deflection are 190.1% and 103.3% higher than that of specimen B0S0E0, respectively, indicating that the toughening effect of basalt fiber on UHPC plate is more significant when the deformation is small. When the deflection continues to increase, no fiber plate is damaged; when the deflection increases to 1mm, the energy absorption value of the plate mixed with basalt fiber 1 is greater than that of basalt fiber 2, and the toughening effect is better. This is because the elastic modulus of basalt fiber 1 is higher, the control effect of micro cracks is better, and more energy can be absorbed; when the deflection increases to 5 mm, the energy absorption value of the specimen B2S0E0 is 131.7% higher than that of B1S0E0. This is because the tensile strength of fiber 2 is significantly higher than that of fiber 1. When the crack is large, the fiber is not easy to break and can absorb more energy.

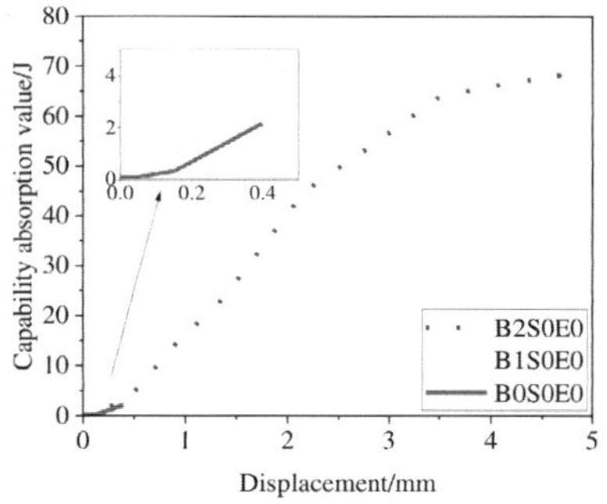

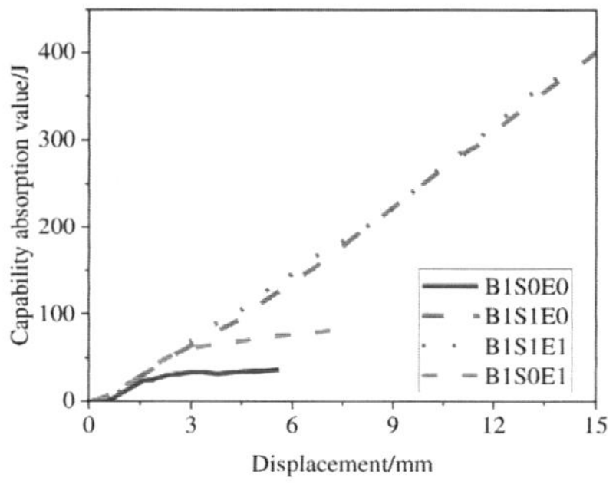

Figure 9. Energy absorption value of the plate. **Figure 10.** Energy absorption value of the plate.

4.2. Effect of Wire Mesh and Aqueous Epoxy on the Bending Properties of UHPC

Figure 10 is the energy absorption value of a UHPC plate mixed with steel mesh or waterborne epoxy. According to table 6, the energy absorption value of specimen B1S1E0 at deflection 5 mm is 195.6% higher than that of specimen B1S0E0, indicating that the incorporation of steel mesh can significantly increase the flexural toughness of UHPC plate, and with the increase of deflection, the toughening effect gradually increases. The energy absorption value of specimen B1S0E1 at the deflection of 5 mm is 98.5% higher than that of specimen B1S0E0, indicating that the addition of waterborne epoxy to the UHPC plate significantly increases the flexural toughness of the UHPC plate. However, due to the absence of steel wire mesh, the specimen will break when the deflection increases to 8.5 mm, and the flexural toughness will no longer increase. The energy absorption value at the deflection of 5 mm of specimen B1S1E0 is 48.9% higher than that of B1S0E1, indicating that the toughening effect of steel mesh on UHPC plate is stronger than that of waterborne epoxy. When the three are mixed, the energy absorption value of the specimen B1S1E1 at the deflection of 5 mm is 10.7 % and 64.9% higher than that of B1S1E0 and B1S0E1, respectively, and the deflection of 10 mm is 1.2% higher than that of the specimen B1S1E0, indicating that the three are mixed. The toughening effect of the UHPC plate is not significant.

5. Bending Bearing Capacity of UHPC Bidirectional Plate

When calculating the flexural capacity of flexural members, the stress pattern of concrete in the compression zone is generally in the form of a quadratic parabola plus rectangle, which is not conducive to the calculation of stress. Different from ordinary concrete, UHPC has excellent tensile properties, so it should be subjected to the action of concrete in the tension zone during calculation. When calculating the bearing capacity, the actual stress diagram is equivalently processed. Based on the resultant force point and the action point, the tensile and compressive stress curves of the section are equivalent to the rectangular stress diagram, as shown in figure 11.

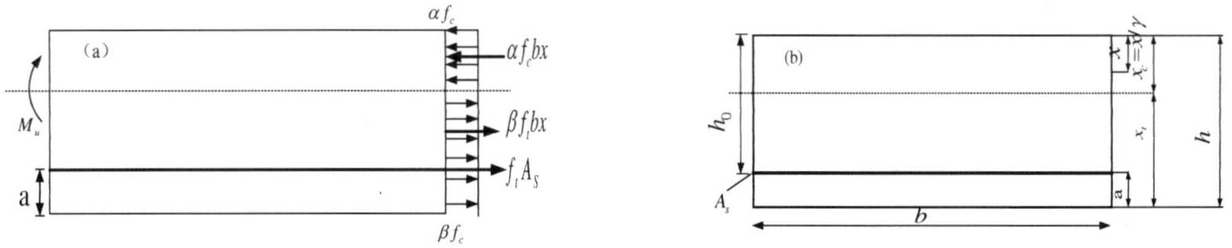

Figure 11. Effect diagram of UHPC bidirectional plate. (a) Distribution of rectangular section; (b) Equivalent rectangular stress block.

According to the axial force balance is available:

$$\alpha f_c bx = E_s \varepsilon_s A_s + \beta f_t bx_t \tag{2}$$

$$x_t = h - x_c = h - \frac{x}{\gamma} \tag{3}$$

$$\varepsilon_s = (6.7264 f_c + 2460.9) \times 10^{-6} \tag{4}$$

Where: f_c and f_t are the average of compression and tensile strength of the UHPC axis respectively; The f_s is the effective stress value of the wire mesh in the matrix; x_c and x_t are the height of UHPC compression and tensile area; The x is the equivalent rectangular stress height in the compression area of UHPC plate; α, β, γ The influence coefficient of the equal effect force value is 0.88,0.35 and 0.72 respectively; ε_s is the peak strain; h is the actual height of the UHPC board; b is the calculated width of the UHPC board; A_s is the cross-section area of the steel wire mesh; a is the distance between the edge of the bottom section of the plate and the joint force point of the wire mesh in the tensile area; The E_s is the elastic modulus of the wire mesh. The axial compression and tensile strength of UHPC and the tensile strength of the cube are as follows:

$$f_c = 0.89 f_{cu} \tag{5}$$

$$f_t = 2.31 f_{ct}^{0.56} \tag{6}$$

Based on the flat section assumption, the strain coordination relationship between concrete and wire mesh in the compression area is:

$$\frac{\varepsilon_c}{x_c} = \frac{\varepsilon_s}{h_0 - x_c} \tag{7}$$

Where: ε_c is the compressive strain in the concrete compression area; h_0 is the effective height of the UHPC plate; To obtain the bending bearing capacity equation of the UHPC plate:

$$M_u = \alpha f_c bx(h_0 - \frac{x}{2}) + \beta f_t bx_t(a - \frac{x_t}{2}) \tag{8}$$

The height of the concrete compression zone is obtained by the iterative method. The specific steps are as follows: Firstly, it is assumed that the height of the concrete compression zone is x_{c1}, and the tensile strain ε_s of the steel mesh is solved according to equation (7). Then x_1 is calculated by the axial force balance equation (2), and x_1 is substituted into equation (3) to calculate x_{c2}. The above steps are repeated with x_{c2} instead of x_{c1}. When the error between x_{c1} and x_{c2} is less than 1%, the assumption is valid, and x_{c2} is the height of the concrete compression zone.

Table 7. Calculation results of UHPC plate bearing capacity.

Label	f_c/MPa	f_t/MPa	ε_s	h_0/mm	M_u/(kN·m)	M_e/(kN·m)	M_e/M_u
B1S0E0	95.586	8.29	0.00310	50	3.65	4.06	1.11
B1S0E0-1	95.586	8.29	0.00310	50	3.65	4.12	1.12
B1S1E0	95.586	8.43	0.00310	46.5	4.12	4.67	1.13
B1S1E1	94.696	8.61	0.00309	46.5	5.71	6.19	1.08
B1S0E1	94.696	8.61	0.00309	50	4.59	5.09	1.10
B2S0E0	95.586	8.43	0.00310	50	4.15	4.92	1.19
B2S0E0-1	95.586	8.43	0.00310	50	4.15	5.12	1.23

It can be seen from table 7 that the theoretical flexural capacity of UHPC two-way slabs with single basalt fiber 1 and basalt fiber 2 is 3.65 kN·m and 4.15 kN·m, respectively, both of which are higher than that of plain UHPC slabs, but the improvement effect of basalt fiber 2 is greater. It is considered that the longer basalt fiber plays a 'bridging' role between the matrix and the aggregate during the failure process, which in turn increases the bearing capacity of the slab. Compared with the single steel mesh, the bearing capacity of the two-way slab mixed with the three is increased by 38.6%. This is because the waterborne epoxy increases the friction between the fiber and the steel mesh, making the concrete more dense, and thus enhancing the flexural toughness. Based on the plastic hinge line theory, the maximum bending moment Me calculated by the virtual work principle is compared with the theoretical bending moment, and the results are relatively consistent.

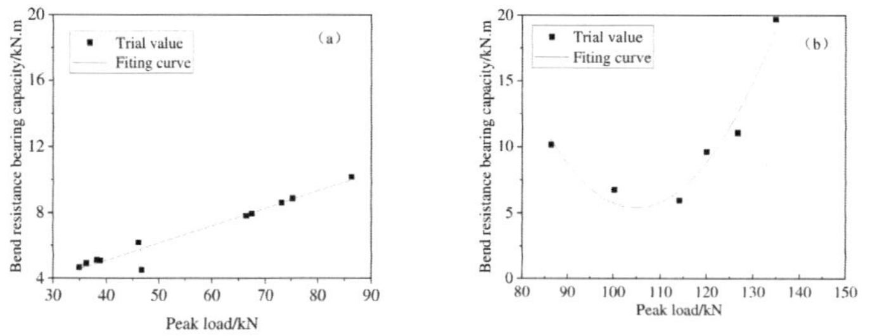

Figure 12. Relationship between bending bearing capacity and peak load of UHPC plate. (a) Peak load of 30–85 kN; (b) Peak load of 85–150 kN.

Based on the test data provided in this paper and literature, figure 12 draws the fitting curve of bidirectional plate bending capacity and peak load, and gives the relevant curve equation:

$$y=\begin{cases}0.105x+0.862 & 30\le x\le 85\\ 0.015x^2-3.153x+170.7 & 85\le x\le 150\end{cases} \tag{9}$$

From figure 12, when the peak load is 30-85 kN, the bearing capacity and load into linear distribution, when the peak load exceeds 85 kN, the bending bearing capacity and load curve into a secondary parabolic shape, and the bending bearing capacity calculated by equation (9) is 1.01, the mean variance is 0.015, which is consistent with the literature the bending bearing capacity of the bidirectional plate on the basis of a large number of tests and theoretical analysis.

6. Conclusion

(1) The increase in fiber incorporation will reduce UHPC and vulnerability but can enhance the compressive strength. Compared with UHPC, when the basalt incorporation is 15 kg/m^3, the compressive strength, split tensile strength, and bending strength of UHPC are increased by 2.8%, 3.0%, and 20.6%, respectively.

(2) Compared with non-fiber UHPC board, single mixed basalt fiber will increase the number of cracks of UHPC board. The mixing of steel wire mesh or water-based epoxy and basalt fiber 1 can improve the destruction form of the UHPC board and delay the initiation and expansion of cracks. The integrity of the UHPC board mixed with wire mesh remains intact.

(3) The load-deflection curve of single-doped basalt fiber UHPC board shows a double peak, and the ability of shorter basalt fiber 2 to enhance and toughen UHPC board is higher than that of basalt fiber 1.

(4) Basalt fiber, steel wire mesh, and waterborne epoxy can increase the bearing capacity and flexural toughness of UHPC two-way slabs. The ductility of UHPC slabs mixed with basalt fiber 1 and steel wire mesh is significantly improved. The UHPC slabs mixed with basalt fiber 1 and waterborne epoxy have a more significant enhancement effect. The mixing of the three has a significant increase in the bearing capacity of UHPC slabs and a small increase in toughness.

(5) The calculated value of bending bearing capacity of the bidirectional plate is well consistent with the test value results, the curve equation of bending bearing capacity and peak load is given.

Acknowledgments

The study was jointly carried out by the Economic and Technical Research Institute of State Grid Hebei Electric Power Co., LTD and Beijing University of Civil Engineering and Architecture, focusing on the key mechanical properties of basalt fiber ultra-high performance concrete (Project No. 00362023264). We would like to express our gratitude.

7. References

[1] Hung C C, El-Tawil S and Chao S H 2021 A review of developments and challenges for UHPC in structural engineering: Behavior, analysis, and design *Journal of Structural Engineering* **147(9)** 03121001.

[2] Wang D, Shi C, Wu Z, et al. 2015 A review on ultra high performance concrete: Part II. Hydration, microstructure and properties *Construction and Building Materials* **96** 368-377.

[3] Yoo D Y, Lee J H, Yoon Y S, et al. 2013 Effect of fiber content on mechanical and fracture properties of ultra high performance fiber reinforced cementitious composites *Composite Structures* **106** 742-753.

[4] Yu R, Spiesz P, Brouwers H J H, et al. 2014 Static properties and impact resistance of a green Ultra-High Performance Hybrid Fibre Reinforced Concrete (UHPHFRC): Experiments and modeling *Construction and Building Materials* **68** 158-171.

[5] Kang S T, Choi J I, Koh K T, et al. 2016 Hybrid effects of steel fiber and microfiber on the tensile behavior of ultra-high performance concrete *Composite Structures* **145** 37-42.

[6] Dhand V, Mittal G, Rhee K Y, et al. 2015 A short review on basalt fiber reinforced polymer composites *Composites Part B: Engineering* **73** 166-180.

[7] Bao X, Liu F, He J, et al. 2021 Mechanical properties and water-aging resistance of glass ionomer cements reinforced with 3-aminopropyltriethoxysilane treated basalt fibers *Journal of the Mechanical Behavior of Biomedical Materials* **116** 104369.

[8] Gao W Y, Hu K X, Dai J G, et al. 2018 Repair of fire-damaged RC slabs with basalt fabric-reinforced shotcrete *Construction and Building Materials* **185** 79-92.

[9] Khan M and Cao M 2021 Effect of hybrid basalt fibre length and content on properties of cementitious composites *Magazine of Concrete Research* **73(10)** 487-498.

[10] Jia P L and Hua X Z 2017 Effects of coarse aggregate and steel fiber on uniaxial tensile property of ultra-high performance concrete *Materials Review* **31(12)** 109-114.

Journal of Physics: Conference Series **2941** (2025) 012030 doi:10.1088/1742-6596/2941/1/012030

[11] Wu F, Xu L, Chi Y, et al. 2020 Compressive and flexural properties of ultra-high performance fiber-reinforced cementitious composite: The effect of coarse aggregate *Composite Structures* **236** 111810.
[12] Li P P, Yu Q L, Brouwers H J H, et al. 2018 Effect of coarse basalt aggregates on the properties of ultra-high performance concrete (UHPC) *Construction and Building Materials* **170** 649-659.

Study on the Properties and Strengthening Treatment of Recycled Aggregates

Ting Liu[1,2], Xiaohui Yan[1,*] and Bei Zhang[1]

[1] Department of Municipal and Ecological Engineering, Shanghai Urban Construction Vocational College, Shanghai, China.
[2] Prefabricated Building Research Institute of Anhui Province, Hefei, Anhui, China.
Email: liuting@succ.edu.cn; *406779136@163.com; zhangbei@succ.edu.cn

Abstract. Concrete material is the basic material of infrastructure construction, and its dosage is quite large. With the development of society, environmental protection, energy saving, resource recycling, and recycling have aroused people's great attention. Therefore, it is necessary to improve the recycling of building material resources. Based on the research status of recycled coarse aggregate concrete at home and abroad, this paper summarizes the quality evaluation and grade classification of recycled coarse aggregate, physical strengthening, chemical strengthening, microbial sedimentation and other methods to improve the performance of recycled coarse aggregate, and analyzes the existing problems and further research content.
Keywords. Recycled coarse aggregate; quality grade; physical strengthening; chemical strengthening; microbial deposition.

1. Introduction

With the rapid development of China's economy and the continuous advancement of urbanization, the importance of the civil engineering industry has become increasingly prominent, and new buildings have sprung up, but at the same time, it has also brought some adverse effects. On the one hand, in order to make full use of space resources, more and more old buildings are demolished; On the other hand, in order to meet the current situation of the increasing use of engineering materials, the exploitation of raw materials is rampant, and the problem of environmental deterioration is becoming more and more serious. From the perspective of sustainable development, the rational use of useful resources in these construction waste, through the processing and production of recycled aggregate, on this basis to develop recycled concrete and its structure, is of great significance for reducing the site required for landfill, reducing the waste of useful resources and the consumption of natural raw materials and protecting the ecological environment.

Recycled concrete aggregate is the aggregate formed by cleaning, crushing, screening, and matching waste concrete blocks in a certain proportion. According to the size of the recycled aggregate particle size, it can be divided into recycled coarse aggregate (particle size >4.75mm) and recycled fine aggregate (particle size <4.75mm). The concrete prepared by using recycled concrete aggregate to partially or completely replace natural aggregate such as sand and stone (mainly coarse aggregate) is called recycled concrete. Previous studies have found that, compared with natural aggregate concrete, the compressive strength of recycled aggregate concrete decreases significantly [1-3]. In order to improve the performance of recycled aggregate concrete, a variety of high-quality recycled aggregate production processes [4,5] and recycled concrete production processes [6-9] have been developed at home and abroad. This paper summarizes the current research status from the aspects of quality evaluation grade and performance improvement of recycled coarse aggregate and points out the existing problems and future research directions.

2. Regenerated Coarse Aggregate Quality Evaluation Grade

An important aspect of recycled concrete research is to classify and evaluate the quality grade of recycled concrete aggregate (RCA), which is reflected in: On the one hand, the classification and evaluation of RCA quality grade is very important for the correct evaluation of the quality of recycled concrete aggregate and reasonable preparation of recycled aggregate. On the other hand, the classification and evaluation of RCA quality grade is an important basis for mix ratio design and optimization design of recycled concrete. At the same time, the evaluation method is also the theoretical basis for formulating and revising the quality standard of recycled concrete aggregate in the future. Therefore, in order to realize the rational utilization of recycled concrete coarse aggregate, it is necessary to determine its quality grade. There are many indicators to evaluate the quality of recycled coarse aggregate abroad, and the representative ones are Japan, the United Kingdom and the United States. In the "Trial Regulations on the Quality of Recycled Concrete Materials" promulgated by Japan in 1994, water absorption and crushing indexes are used as evaluation indexes for the quality grade of recycled concrete coarse aggregate [10]. The British Standard (BS) [11] and the American Society for Testing Materials Standard (ASTM) [12,13] are the minimum apparent density, maximum water absorption, maximum needle flake particle content, maximum impact value, maximum chlorine content and maximum sulfate content corresponding to the extreme values of evaluation indexes for recycled coarse aggregates in different projects. China's research on recycled aggregate started late. In 2010, it promulgated the Standard of Recycled Aggregate for Coagulation (GB/T25177-2010), which stipulated 7 indicators as a quality evaluation system for recycled concrete coarse aggregate, and divided its quality into four grades: high quality, good quality, qualified quality and unqualified quality. The classification and limit values of quality grades were shown in table 1.

Table 1. Summary of quality grade division and limit value of recycled coarse aggregate.

Index	I	II	III	IV
Apparent density /kg•m^{-3}	>2450	(2350,2450]	(2250,2350]	≤2250
Voidage /%	<47	[47,50)	[50,53)	≥53
Crushing index /%	<12	[12,20)	[20,30)	≥30
Solidity /%	<5.0	[5.0,9.0)	[9.0,15.0)	≥15.0
Lump content /%	<0.5	[0.5,0.7)	[0.7,1.0)	≥1.0
Fine powder content /%	<1.0	[1.0,2.0)	[2.0,3.0)	≥3.0
Water absorption /%	<3.0	[3.0,5.0)	[5.0,7.0)	≥7.0

It can be seen from table 1 that the level of each indicator system can be determined, but the comprehensive index level cannot be known. Many Chinese and foreign scholars have carried out a large number of evaluation methods for recycled coarse aggregate. At present, there are many methods for dividing and evaluating the quality grade of concrete recycled aggregate, such as the fuzzy evaluation method [14], analytic hierarchy process [15], gray interval method [16], combined weight-extension theory [17], etc. All these methods can give the final evaluation results well. The fuzzy evaluation method and combinatorial weight-extension theory are calculated with the same set of experimental values, and the results are consistent. However, compared with these methods, the fuzzy comprehensive evaluation method has strong subjectivity, the analytic hierarchy process has less quantitative data, the gray theory method requires a large sample size, and it is difficult to determine the optimal value of some indicators. Moreover, all algorithms are studied based on the index value, and how to evaluate the comprehensive level according to the grade division of the index value needs to be further explored.

3. Physical Modification Method of Recycled Aggregate

The application of recycled aggregate in recycled concrete is limited due to its many corners, rough surface, micro-cracks, high porosity, high water absorption and high crushing value. Domestic and foreign scholars put forward the method of physical modification and strengthening of recycled aggregate. The physical modification method is to dry the recycled aggregate by mechanical grinding, heating grinding and so on. The particle shaping and strengthening method can remove the cement mortar attached to the surface and make the aggregate particles round. Compared with the coarse aggregate that is simply broken, the packing density and water absorption rate are improved, the crushing value is reduced, and its various properties are basically close to that of natural gravel [18]. Akbarnezhad et al. [19] strengthened the reclaimed coarse aggregate by microwave with high power (10 kW), indicating that microwaves do have a certain effect on removing the surface mortar of the reclaimed coarse aggregate. However, excessive power instantaneous heating will lead to the excessive temperature of the natural aggregate inside the recycled aggregate, resulting in high temperature damage to the aggregate itself. Xiao et al. [20] adopted the low-power microwave heating cycle method to remove the cement mortar attached to the reclaimed coarse aggregate, thus improving the quality of the reclaimed coarse aggregate. The modified reclaimed coarse aggregate mortar reduced its content by 10.8%, water absorption by 2.2%, and crushing index by nearly 30%. Su et al. [21] adopted the method of first heating and then removing the surface mortar of recycled coarse aggregate through the concrete shaking table, and found that the modified performance of coarse aggregate was the best when the heating temperature reached 500°C. Compared with unmodified coarse aggregate, the water absorption rate decreased by 34.9%, the apparent density increased by 6.5%, and the crushing index decreased by 42.9%. Katz [22] treated RAC with ultrasonic cleaning technology, which increased the compressive strength of recycled concrete by 7%.

The physical modification method is mainly strengthened by the mechanical method, which is complicated and has high energy consumption. Therefore, the strengthening method with a simple process and low energy consumption will become the mainstream research direction.

4. Chemical Modification Method of Recycled Aggregate

Chemical strengthening mainly takes the regenerated aggregate with organic or inorganic chemical solution impregnation, leaching or other means. Vivian et al. [23] used HCl, H_2SO_4 and H_3PO_4 solutions to pre-soak recycled coarse aggregate. Although the modification effect is obvious, soaking with an acidic solution is easy to damages the structure of aggregates containing limestone and other components, which has strong limitations, and its durability also needs to be studied. Kou and Poon et al. [24] modified the reclaimed coarse aggregate by soaking it in polyvinyl alcohol (PVA). After modification, the apparent density of the modified coarse aggregate was significantly higher than that without modification, and the water absorption was also significantly reduced. Yang et al. [25] sprayed PVA solution on the surface of the reclaimed aggregate and then wrapped a layer of cement paste on the surface of the PVA binding layer, and the water absorption and crushing value of the reclaimed coarse aggregate were lower than those of the unmodified coarse aggregate. Zhu et al. [26] used PVA and silicone solution with an effective concentration of 10% to soak the reclaimed coarse aggregate, and the water absorption rate of the coarse aggregate dropped the most.

Shan et al. [27] added nano-silica sol to achieve in-situ strengthening of recycled aggregate, which can effectively improve the strength loss caused by the replacement of recycled coarse aggregate. Enhanced by a high concentration of CO_2 gas, both the apparent density and bulk density of recycled coarse aggregate increased by 1.2%, the water absorption decreased by 27.3%, and the crushing index decreased by 10.5% [28]. Studies have shown that the porosity and water absorption of regenerated coarse aggregate decreased after being treated with permeable crystalline materials [29].

Other studies [30] showed that compared with unstrengthened aggregates, the crushing index of recycled aggregates strengthened by different strengthening methods decreased, and the apparent density increased. Only the water absorption of cement coated with PVA solution decreased, and the water absorption of other strengthening methods increased. As can be seen from table 2, the crushing index of aggregate strengthened by pure cement slurry, mineral powder mixed with cement slurry, diatomaceous earth mixed with cement slurry, silica powder mixed with cement slurry, and PVA solution coated cement is reduced by 20.9%, 21.6%, 27.6%, 23.5% and 29.5%, respectively, compared with unstrengthened aggregate. The apparent density increased by 4.5%, 4.0%, 2.1%, 2.6% and 1.2%, respectively. The water absorption rate increased by 2.4%, 10.9%, 5.3% and 9.7%, respectively, while the water absorption rate of cement coated PVA solution decreased by 6.4%.

Table 2. Properties of recycled aggregate before and after reinforcement [31].

Reinforcement method	Moisture content (%)	Water absorption (%)	Apparent density (kg/m^3)	Crushing index (%)
Not enhanced	2.58	6.77	2470	16.73
Pure cement paste	3.37	6.93	2580	13.24
Mineral powder in cement slurry	3.78	7.51	257.	13.12
Cement paste mixed with diatomaceous earth	2.71	7.13	2523	12.11
Silica powder in cement slurry	2.79	7.43	2534	12.80
Pva solution coated with cement	2.63	6.34	2500	11.80

5. Microbial Degradation

Inspired by the phenomenon of diagenesis, the researchers found that microorganisms can complete mineralization outside the cell to produce calcite. Therefore, microbial mineralization restoration technology has been applied in the field of civil engineering [31,32]. Gollapudi et al. [33] first tried to control building leakage by mineralized deposition of bacteria in 1995 and achieved success. Qian et al. [34,35] conducted an in-depth study on the mechanism of deposition of calcium carbonate by microbial mineralization and successfully used microbial mineralization technology to consolidate loose particles. Navarro et al. [36] found that the surface cracks of limestone can be repaired by microbial mineralization deposition. Hill et al. [37] filled the microbial mixture into the broken rocks, and the microbial mineralization sedimentation can effectively repair the gaps between the broken rocks and improve impermeability. Later, Dic et al. [38] selected suitable bacillus and deposited calcium carbonate on the surface of the degraded limestone, greatly reducing the capillary water absorption coefficient of the surface of the limestone. The microbially induced mineralization reaction produces calcium carbonate, which not only protects the environment but also has excellent compatibility with cement-based materials. Qiu J used different bacterial concentrations, PH values, temperatures, and CaCl solution quantities as variables to study regenerated coarse aggregate after surface treatment with microbial carbonate precipitation. The greater the carbonate content, the greater the aggregate weight, and the lower the water absorption [39], but under different bacterial concentrations and PH values, the shape of the settling amount is different, as shown in figure 1. In this paper, the author used a CaCl solution, but chloride ions have an erosion effect on concrete, and the performance of microbial decomposition on aggregate is still unclear, and further research is needed.

(a) (b) (c)

Figure 1. Microstructure of recycled aggregate under different bacterial concentrations and PH values. (a) untreated recycled aggregate; (b) Bacterial concentration of 106 cell/ml; PH=8.5; (c) Bacterial concentration of 108 cell/ml; PH=9.5.

6. Conclusion

(1) Most studies on recycled coarse aggregate focus on particle size below 40 mm. Compared with recycled coarse aggregate with conventional particle size, recycled large aggregate with particle size over 40 mm is easy to break, has low energy consumption, and is easy to obtain, while relevant studies are relatively few. Therefore, the research on recycled large aggregate and recycled concrete has certain innovations.

(2) There are two main physical methods and chemical methods for the modification and strengthening of recycled coarse aggregate, but both methods have shortcomings. For physical modification, more research can be done to reduce the process flow and reduce mechanical energy consumption. Chemical modification can be considered cheap, efficient, energy-saving saving, and environmental protection of high quality strengthening reagents. In addition, the current research scope only considers one-time recycling, if secondary recycling, its performance indicators need to be further studied.

(3) In recent years, the use of microbial mineralization technology to repair concrete cracks has been gradually developed in many countries and has shown good application prospects.

Acknowledgments

The authors acknowledge the financial support of the Prefabricated Building Research Institute of Anhui Province (No. AHZPY2022KF02).

7. Reference

[1] Feng C H and Huang Y H 2022 Research progress on the strengthening method of recycled aggregate in building construction *Materials Science and Technology* **36(21)** 88-95.

[2] Xiao J Z and Zhang H H 2023 The principle of recycling of waste concrete and basic issues of regenerated concrete *Science Bulletin* **68(05)** 510-523.

[3] Wang B and Yan L 2021 A comprehensive review on recycled aggregate and recycled aggregate *Concrete Resources Conservation and Recycling* **71(5)** 1-29.

[4] Tam V W Y and Tam C M 2007 Removal of cement mortar remains from recycled aggregate using pre-soaking approaches *Resources Conservation and Recycling* **50(1)** 82-101.

[5] Tateyashiki H and Shima H 2001 Properties of concrete with high quality recycled aggregate by heat and rubbing method *Proc JCI* **23(2)** 61-66.

[6] Zhai L and Zhang Z J 2019 Research status and development prospect of recycled concrete *Henan Building Materials* **(3)** 309-311.

[7] Yang K and Wang J J 2018 A review of the research status of recycled concrete at home and abroad *Shanxi Architecture* **44(5)** 107-108.

[8] Zhao W and Feng Z X 2006 Study on strengthening mechanism of recycled aggregate concrete by vibration mixing *Concrete* **(8)** 17-20.

[9] Kong D and Lei T 2010 Effect and mechanism of surface-coating pozzalanics materials around aggregate on properties and ITZ microstructure of recycled aggregate concrete *Construction and Building Materials* **24(5)** 701-708.

[10] Tomosawa F 1992 The recycling of concrete: the Japanese experience *Proceedings of the 4th CANMET/ACI/JCJ International Conference on Recent Advances in Concrete Technology (London)* 221-237.

[11] BS 882 1992 *Specification for Aggregates from Natural Sources for Concrete* (London UK British Standards Institution) 4.

[12] ASTM D 2940-03 2003 *Standard Specification for Graded Aggregate Material for Bases or Sub-bases for Highways or Airports* (USA: American Society for Testing and Materials) 16.

[13] ASTM D 448-03 2003 *Standard Classification for Sizes of Aggregate for Road and Bridge Construction* (USA: American Society for Testing and Materials) 3.

[14] Wang S Y and Zhu P H 2010 Fuzzy comprehensive evaluation method for coarse aggregate quality grade of recycled concrete *Concrete* **(7)** 87-89.
[15] Kong D Yand Lu X X 2023 Optimization evaluation of recycled aggregate production mode based on improved analytic hierarchy process *Journal of Zhejiang University of Technology* **51(5)** 504-508.
[16] Bao X Y and Wang Q C 2014 Application of grey clustering method in quality grade evaluation of recycled concrete coarse aggregate *Civil Architectural and Environmental Engineering* **36(3)** 112-117.
[17] Li F H and Jin H S 2019 Quality classification optimization evaluation of recycled concrete coarse aggregate *Journal of Southwest Jiaotong University* **54(4)** 793-800.
[18] Li Q Y and Li Y X 2006 Research on strengthening technology of recycled concrete aggregate *Concrete* **(1)** 74-77.
[19] Akbarnezhad T and W J 2011 Microwave-assisted beneficiation of recycled concrete aggregates *Construction and Building Materials* **25(8)** 3469-3479.
[20] Xiao J Z and Wu L 2012 Experimental study on modification of reclaimed coarse aggregate by microwave heating *Concrete* **(7)** 55-57.
[21] Su Y and Gu S 2015 Experimental study on coarse aggregate of recycled concrete modified by heating *Concrete and Cement Products* **(10)** 6-10.
[22] Katz A 2004 Treatments for the improvement of recycled aggregate *Journal of Materials in Civil Engineering* **16(6)** 597-603.
[23] Tama W Y and Tam C M 2007 Removal of cement mortar remains from recycled aggregate using pre-soaking approaches *Resources Conservation and Recycling* **(50)** 82-101.
[24] Kou S C and Poon C 2010 Properties of concrete prepared with PVA-impregnated recycled concrete aggregates *Cement and Concrete Composites* **(32)** 649-654.
[25] Yang N and Wang C 2011 Research on strengthening technology of recycled aggregate *New Building Materials* **(3)** 45-47.
[26] Zhu Y G and Xu P Z 2015 Experimental study on modification of coarse aggregate of recycled concrete by silane and PVA *Concrete* **(3)** 93-95.
[27] Shan Y C and Shen C 2016 Effect of in-situ strengthening of recycled aggregate on recycled concrete and its Structural properties *Building Structures* **(12)** 37-40.
[28] Ying J W and Meng Q J 2017 CO2 strengthening of recycled aggregate and its effect on compressive strength of concrete *Journal of Building Materials* **(2)** 277-282.
[29] Wang H C and Chen C 2017 Experimental research and analysis on strengthening mechanism of recycled aggregate *Concrete* **(12)** 95-98.
[30] Yang N and Wang C G 2011 Research on intensifying technique of recycled aggregate *New Building Materials* **38(3)** 45-47.
[31] Jonkers H M and Schlangen E 2008 *Development of a Bacteria-based Self-healing Concrete* (In: Joost C. Walraven, Dick Stoelhorst. Tailor Made Concrete Structures: New Solutions for our Society. London: CRC Press) 425-430.
[32] Muynck W D and Belie N D 2010 Microbial carbonate precipitation in construction materials: A review *Ecological Engineering* **36(2)** 118-136.
[33] Gollaudi U K and Knutson C L 1995 A new method for controlling leaching through permeable channels *Chemosphere* **30(4)** 695-705.
[34] Qian C X and Luo M 2013 Formation mechanism of microbial mineralized calcite in self-healing concrete *Journal of the Chinese Silicate* **(5)** 620-626.
[35] Qian C X and Ren L F 2015 Research progress of surface defects and crack repair technology of concrete based on microbial induced mineralization *Chinese Journal of Ceramics* **(5)** 619-631.
[36] Rodriguez-navarro C and Rodriguez-gallego M 2003 Conservation of ornamental stone by myxococcusxanthus induced carbonate biomineralization *ApplEnviron Microbiol* **69(4)** 2182-2193.
[37] Hill D D 2002 Sleep B E. Effects of biofilm growth on flow and transport through a glass parallel

plate fracture *J Contam Hydrol* **56(3/4)** 227-246.
[38] Dick J and De W W 2006 Bio-deposition of a calcium carbonate layer on degraded limestone by Bacillus species *Biodegradation* **(17)** 357-367.
[39] Qiu J and Tng D Q S 2014 Surface treatment of recycled concrete aggregates through microbial carbonate precipitation *Construction and Building Materials* **57** 144-150.

Sol-gel Method Preparation and First-Principles Calculations of $CaCu_3Ti_4O_{12}$ (CCTO)

Xiaofen Jiang and Wenjie Chen*

College of Chemical Engineering and Materials, Quanzhou Normal University, Quanzhou, Fujian, China.
Email: 17019@qztc.edu.cn; *chenwenjie@qztc.edu.cn

Abstract. Copper Calcium Titanate (CCTO) is widely used in high-dielectric applications, such as capacitors, where its exceptionally high dielectric constant enables the storage of large amounts of electrical energy in a small volume. In this paper, CCTO powder was successfully prepared using the sol-gel method. The morphology, composition, and crystalline state were characterized using X-ray diffraction analysis (XRD) and scanning electron microscopy (SEM). The X-ray diffraction analysis indicated that the prepared powder had good crystallization, high purity, and quality comparable to commercially purchased CCTO powder. Scanning electron microscopy results showed that the grain boundaries of the powder were clear, and the particles were uniformly complete. The relationship between the dielectric constant and capacitance density of CCTO with frequency reveals that as the frequency increases, the dielectric constant (K value) of the electrode decreases rapidly at first and then the rate of decrease slows down, indicating that the dielectric constant is higher at lower frequencies. The capacitance density of the electrode also decreases rapidly at first and then more slowly as the frequency increases. Moreover, first-principles calculations were performed to study the electronic structure characteristics of CCTO. The band structure diagram indicates that CCTO has an indirect band gap structure. The density of states (DOS) showed that the Cu-O and Ti-O bonds in CCTO form the band structure near the Fermi surface, and it was also found that the contribution of calcium atoms to the band structure is relatively small. The study will provide experimental foundations and theoretical basis for the study of CCTO as a high-dielectric-constant ceramic material.
Keywords. Sol-gel method; $CaCu_3Ti_4O_{12}$; first-principles calculations.

1. Introduction

Microelectronics has rapidly developed along with the rise of the electronics information sector. As people pursue more lightweight and convenient electronic products such as mobile phones and laptops, the demand for miniaturization and high integration of electronic components has increased significantly. Dielectric materials, as the raw materials for manufacturing microelectronic devices and storage devices, largely determine the size of electronic products and occupy a central position in this field [1-4].

Copper calcium titanate (CCTO) ceramic materials, due to their high dielectric properties, have received significant attention in the era of rapid development of microelectronics and solid-state electronics. For the CCTO crystal structure, there are eight oxygen octahedra with oxygen atoms and titanium atoms located at the vertices and the centers of the octahedra, respectively [5,6]. Additionally, copper atoms are bonded to oxygen atoms, forming CuO bonds, and are directly connected to calcium atoms, but calcium atoms do not form chemical bonds with oxygen atoms. Copper atoms form a planar square coordination structure with four adjacent oxygen atoms, leading to different orientations and varying degrees of tilting of the oxygen octahedra. As a result, positive and negative ions are closely arranged. The large ionic polarization in the CCTO crystal structure causes the displacement of positive and negative ions within the crystal when an external electric field is applied, resulting in a polarization phenomenon. This ionic polarization significantly increases the dielectric constant of the material, giving it high dielectric response performance, good dielectric stability over a wide range of frequencies and temperatures, and excellent non-linear I-V characteristics [7-10].

Subramanian et al. [11] discovered that CCTO ceramic materials possess giant dielectric properties and exhibit very good temperature stability. The term "giant dielectric" refers to the high dielectric constant of CCTO materials, which can reach up to 12,000 at a frequency of 10 kHz. The non-linear I-V characteristics of CCTO materials make them potential candidates for the next generation of varistor materials [12,13]. These characteristics are closely related to the internal microstructure of the material. The presence of insulating grain boundaries and semiconducting grains within the material results in the formation of barriers at the boundary interfaces when current flows, which hinders the movement of charge carriers.

Due to the giant dielectric behavior of CCTO materials, researchers have proposed many theoretical models to explain this characteristic of CCTO ceramic materials. One such model, the Internal Barrier Layer Capacitance (IBLC) model, was proposed by Sinclair et al. [14]. They conducted in-depth studies on CCTO materials, which consist of grain boundaries and grains. The different properties between the grains and grain boundaries cause a significant amount of charge to accumulate on both sides of the grain boundaries when current passes through the material. This phenomenon is caused by the electrostatic barrier effect of the grain boundaries. The accumulation of a large amount of charge at the grain-grain boundary interfaces forms many micro-capacitors (grain-grain boundary-grain), which are interconnected in series and parallel within the material, leading to the giant dielectric behavior of CCTO. Another model, the Surface Barrier Layer Capacitance (SBLC) model, was proposed by Lunkenheimer et al. [15]. They believed that the depletion layer between the CCTO ceramic sample and the electrodes exhibits Maxwell-Wagner type polarization, which results in high dielectric behavior in CCTO ceramic materials. Currently, all theories explaining the high dielectric properties of CCTO materials have certain limitations, and no single theory has been able to perfectly explain the giant dielectric behavior of CCTO materials.

To study the dielectric behavior of CCTO materials, CCTO powder was successfully prepared using the sol-gel method. The morphology, composition, and crystalline state were characterized using X-ray diffraction analysis (XRD) and scanning electron microscopy (SEM). The dielectric constant and capacitance density of CCTO was studied. The first-principles calculations were performed to study the electronic structure characteristics of CCTO.

2. Materials and Methods

The precursor of CCTO was synthesized via the sol-gel technique. For this process, high-quality reagents from Sinopharm Chemical Reagent Co., Ltd. (AR grade) were used. Specifically, $Ti(OC_4H_9)_4$ metal alkoxide served as the titanium source, $Cu(NO_3)_2{\cdot}3H_2O$ as the copper source, and $Ca(CH_3COO)_2{\cdot}H_2O$ as the calcium source. The ions of these elements were mixed in a molar ratio of 1:3:4 for calcium, copper, and titanium, respectively. To prepare the solution, $Ti(OC_4H_9)_4$ and $Cu(NO_3)_2{\cdot}3H_2O$, both prone to hydrolysis, were dissolved in anhydrous ethanol and stirred magnetically until a homogeneous solution was obtained. Separately, $Ca(CH_3COO)_2{\cdot}H_2O$ was dissolved in deionized water under magnetic stirring. To combine these solutions and ensure a stable mixture, CH_3COOH was added as a chelating agent. The three solutions were then mixed together and stirred magnetically to produce the CCTO precursor solution.

2.1. Preparation CCTO Powder

The obtained CCTO precursor was placed it in an oven at 80°C, and dry it for 52 hours to get a blue-green xerogel. The xerogel was grinded into a powder using an agate mortar and placed it on a watch glass, and then put it in an oven at 125°C for 72 hours to obtain green particles. These particles were grinded into fine powder and placed them in an alumina crucible that has been washed with anhydrous ethanol. The muffle furnace program was set to calcine at 350°C for 3 hours, , followed by calcination at 900°C for 10 hours.

2.2. Performance Characterization of CCTO

The characterization of CCTO powders by X-ray diffraction (XRD) was conducted using a Bruker D8 Advance diffractometer (manufactured by Carl Zeiss AG, Germany). This equipment featured a Cu Kα radiation source with a wavelength of 1.54 Å, a scanning speed of 0.02 degrees (2θ) per second, and a scanning range from 25 to 60 degrees. The objective of this analysis was to investigate the phase composition and crystal structure of the prepared samples. Additionally, the samples were analyzed using a ZEISS MERLIN Compact scanning electron microscope (Carl Zeiss AG, Germany) for SEM imaging. This step aimed to characterize the micro-morphology of the material and conduct micro-area composition analysis.

2.3. First-Principles Calculations of CCTO

First-principles calculations were performed to study the electronic structure characteristics of CCTO. The calculations were performed using the VASP software package, with the interaction between ions and valence electrons described by the Projector Augmented Wave (PAW) method, and the exchange-correlation function was approximated using the Perdew-Burke-Ernzerhof (PBE) Generalized Gradient Approximation (GGA). After convergence tests, the plane wave cutoff energy was set to 500 eV, and the k-points were chosen as $7\times7\times7$. The convergence criteria for the geometric optimization of the system were listed as follows. The total energy of the system in the self-consistent iteration process was below 1.0×10^{-5} eV/atom, and the inter-atomic forces were below 0.01 eV/Å.

3. Results and Discussion

3.1. Structural and Morphological Analysis of CCTO Powder

Figure 1(a) shows the XRD spectrum of CCTO prepared by the sol-gel method. All peaks in the XRD spectrum correspond to the cubic phase of CCTO (JCPDS card No.75-2188). After calcination at 900°C for 10 hours, it was shown that a pure CCTO phase was generated. However, the obtained powder still contained certain impurities. On the left side of the strongest peak (220), there were weaker impurity peaks of titanium dioxide and calcium titanate, and on the right side of the strongest peak, there were impurity peaks of copper oxide. Figure 1(b) The XRD characterization of the commercial purchase of CCTO powder (preparation by high temperature solid phase sintering) shows that it also has some impurities, with weak mixed peaks of calcium titanate and copper oxide and titanium dioxide. The comparison implied that the CCTO powder prepared by sol-gel method was similar to the commercial powder in purity.

Figures 1(c) and 1(d) show the scanning electron microscope images of CCTO powders prepared by the sol-gel method and commercial CCTO powders, respectively. From the images, it can be observed that both samples have clear grain boundaries, relatively uniform particles, and intact morphologies. The differences in morphological structure are almost similar.

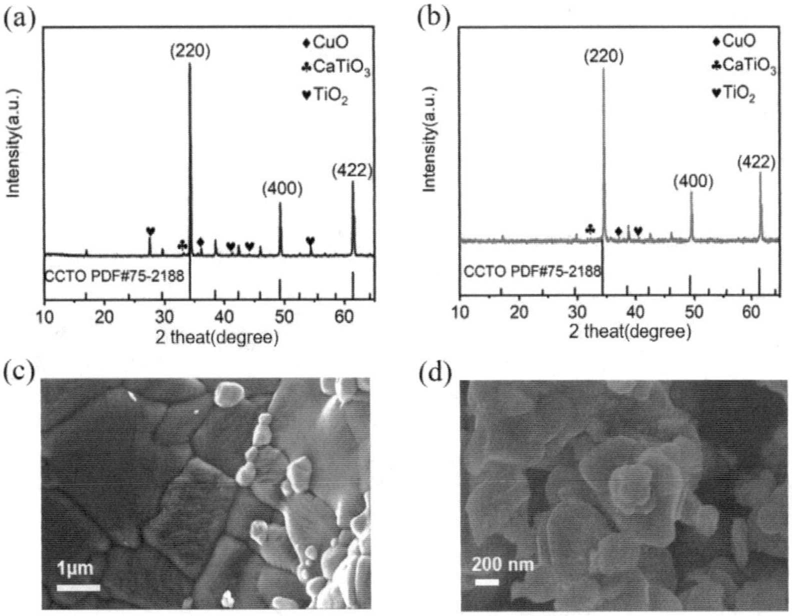

Figure 1. (a) and (c) are the XRD and SEM images of CCTO prepared by the sol-gel method;(b) and (d) are the XRD and SEM images of the commercial CCTO.

3.2. The Dielectric Constant and Capacitance Density of CCTO

The relationship between the dielectric constant and capacitance density of CCTO with frequency is shown in Figure 3. As the frequency increases, the dielectric constant (K value) of the electrode decreases rapidly at first and then the rate of decrease slows down, indicating that the dielectric constant is higher at lower frequencies, with a maximum value of 265.80 (Figure 2a). The capacitance density of the electrode also decreases rapidly at first and then more slowly as the frequency increases (Figure 2b). Higher capacitance density indicates better electrode performance, suggesting that lower frequencies are more favorable for the electrode, with a maximum capacitance density of 1.78522×10^{-6} F/m^2.

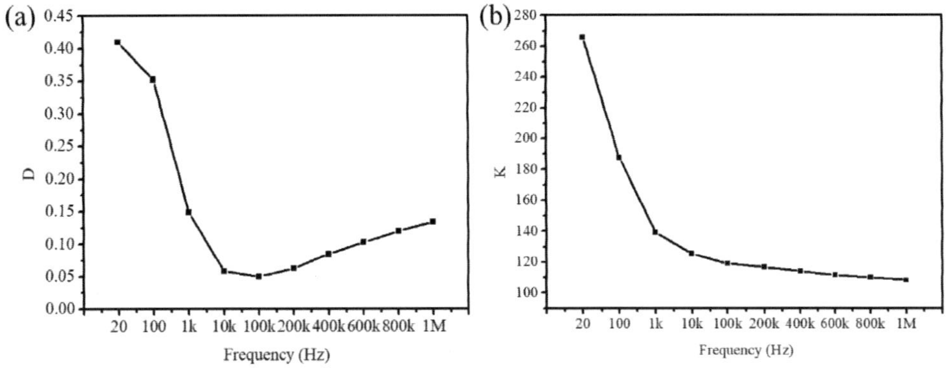

Figure 2. The relationship between the dielectric constant (a) and capacitance density (b) of CCTO with frequency.

3.3. The DOS Density of States Images of the CCTO Powder

To analyze the electronic structure characteristics of CCTO, this section employed first-principles (ab initio first-principles calculation) computational simulations to model the band structure and density of states (DOS) of the compound. The GGA-PBE exchange-correlation functional method can obtain the density of states of the system with less computational effort, hence this paper adopts the GGA-PBE method. Figure 3(a) shows the total density of states (TDOS) and partial density of states (PDOS) of CCTO with the Fermi level at E = 0eV, which is marked with a red dashed line in the figure. The Fermi level is near the valence band formed by the orbitals of copper, calcium, oxygen, and titanium atoms. It can be seen that the orbitals of copper and oxygen atoms contribute significantly to the formation of the valence band, while less contribution of the orbitals of titanium and calcium atoms. Moreover, around an energy of 2.4eV, the conduction band is mainly composed of the orbitals of titanium and oxygen atoms. The Cu-O and Ti-O bonds of CCTO constitute the band structure near the Fermi level, and it can also be observed that the contribution of calcium atoms to the band structure is relatively small.

In the band structure diagram of CCTO, the letters on the x-axis refer to the high-symmetry points in the crystal's reciprocal space, as shown in figure 3(b), which displays the band structure of CCTO along high-symmetry directions, with the conduction band minimum at the X point and the valence band maximum at the R point, indicating an indirect band gap.

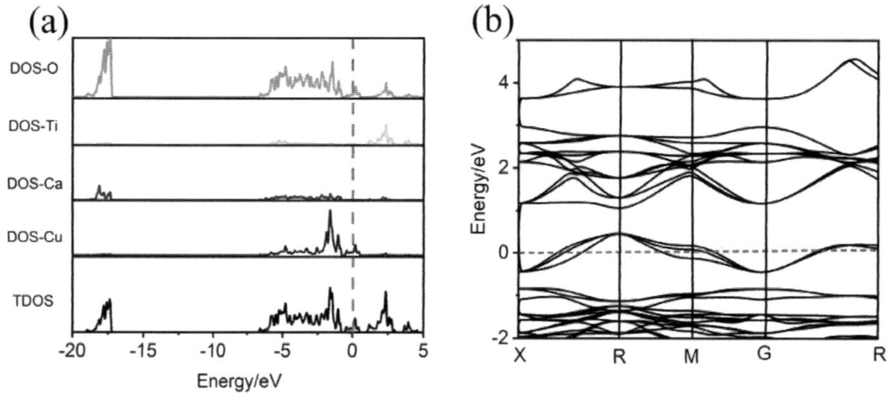

Figure 3. (a) The total density of states (TDOS) and partial density of states (PDOS) of CCTO. The position of E = 0 eV is the Fermi energy level; (b) Energy band structure of CCTO.

4. Conclusion

$CaCu_3Ti_4O_{12}$(CCTO) powder was successfully prepared using the sol-gel method. The morphology, composition, and crystalline state were characterized using X-ray diffraction analysis (XRD) and scanning electron microscopy (SEM). The X-ray diffraction analysis indicated that the prepared powder had good crystallization, high purity, and quality comparable to commercially purchased CCTO powder. Scanning electron microscopy results showed that the grain boundaries of the powder were clear, and the particles were uniformly complete. The relationship between the dielectric constant and capacitance density of CCTO with frequency reveals that as the frequency increases, the dielectric constant (K value) of the electrode decreases rapidly at first and then the rate of decrease slows down, indicating that the dielectric constant is higher at lower frequencies. The capacitance density of the electrode also decreases rapidly at first and then more slowly as the frequency increases. Moreover, first-principles calculations were performed to study the electronic structure characteristics of CCTO. The band structure diagram indicates that CCTO has an indirect band gap structure. The density of states (DOS) showed that the Cu-O and Ti-O bonds in CCTO form the band structure near the Fermi surface, and it was also found that the contribution of calcium atoms to the band structure is relatively small.

5. Acknowledgments

This work was supported by the Natural Science Foundation of Fujian Province (No. 2024J02021), the

Innovation and Entrepreneurship Projects for High-Level Talents of Quanzhou (No. 2024QZC003R), Educational Research Projects for Middle-Aged and Youth in Fujian Province (No. JAT220274).

6. References

[1] Wallace R M 2017 Dielectric Materials for Microelectronics In: S. Kasap, P. Capper, (eds.) Springer Handbook of Electronic and Photonic Materials. Springer Handbooks. S pringer, Cham. pp1.

[2] Infantiya S G, Aslinjensipriya A, Reena, R S, Deepapriya S, Rodney J D, Jerome Das S, Justin Raj C 2022 Calcium Copper Titanate a Perovskite Oxide Structure: Effect of Fabrication Techniques and Doping on Electrical Properties—a Review. *Journal Material Science: Material Electronics* **33** 15992-16028.

[3] Schmidt R and Sinclair D 2013 $CaCu_3Ti_4O_{12}$ (CCTO) Ceramics for Capacitor Applications. In: Kristofer N. M.(eds) Capacitors: Theory of Operation, Behavior and Safety Regulations. Nova Science Publishers pp1-33.

[4] Yang L, Wang H, Zhang F, Yang Y, Qu S, Leng D, Ma X 2023 Synthetic Technologies, Property Enhancements and Versatile Applications of Calcium Copper Titanate: A review, *Nano Energy* **114** 108609.

[5] Luo Liang K C, Liu H L, Yang H D, Mei W N, Ling D C 2008 Structural and Optical Studies of High Dielectric Constant $(Na_{0.5}A_{0.5})Cu_3Ti_4O_{12}$ (A = La and Bi). *Journal of Physics: Condensed Matter* **20** 275238.

[6] Chung S Y, Kim I D and Kang S J L 2004 Strong Nonlinear Current-voltage Behaviour in Perovskite-derivative Calcium Copper Titanate. *Nature Materials* **3** 774-778.

[7] Ou R B 2014 *Study on Doping Modification of Giant Dielectric Coefficient $CaCu_3Ti_4O_{12}$ Ceramic Materials* (Guangzhou: South China University of Technology) Master' s Thesis.

[8] Xiao M and Huang H 2016 Effects of Temperature and Ti-Non-Stoichiometry on Electric Properties of $CaCu_3Ti_4O_{12}$Thin Films *Journal of Materials Science Materials in Electronics* **27(12)** 12550-12556.

[9] Wang J Q 2020 *Study on the Photoelectric Properties of Titanate Compounds and Their Doping Systems Using First-Principles* (Jinan: Shandong University) Master' s Thesis.

[10] Gao Y, Zhao Q H, Shi J F, Duan Q Q, Gang L I and Wang K Y 2016 Preparation of $CaCu_3Ti_4O_{12}$ Thin Films by Sol-Gel Method and Study of Their Electrical Properties *Journal of Synthetic Crystals* **45(12)** 2774-2777.

[11] Subramanian M A and Sleight A W 2002 A $Cu_3Ti_4O_{12}$ and A $Cu_3Ru_4O_{12}$ Perovskites: High Dielectric constant and Valence Degeneracy. *Solid State Science* **4** 347-351.

[12] Mao P, Wang J, Liu S, Zhang L, Zhao Y, He L 2019 Grain Size Effect on the Dielectric and Non-Ohmic Properties of $CaCu_3Ti_4O_{12}$ Ceramics Prepared by the Sol-gel Process *Journal of Alloys and Compounds* **778** 625-632.

[13] Mao P, Sun J, Guo Y, Li W, Xiao P, Gerhard M S, Liu Z, Xie B, Zhang L 2024 Giant Permittivity Response and Enhanced Nonlinear Electrical Properties in a Novel Perovskite-like Ceramic with Multiple Elements *Ceramics International* **50(13)** 22501-22513.

[14] Sinclair D C, Adams T B, Morrison F D, West A R 2002 $CaCu_3Ti_4O_{12}$: One-step Interna l Barrier Layer Capacitor Appllied Physics Letter 80 2153-2155.

[15] Lunkenheimer P, Fichtl R, Ebbinghaus S G 2004 Non-intrinsic Origin of the Colossal Dielectric Constants in $CaCu_3Ti_4O_{12}$ P*hysical Review B: Condensed Matter* **70** 3352-3359.

Analysis of the Causes of Stripe Black Spot Defects in Hot-Dip Galvanized Strip of Thin Slab Continuous Casting and Rolling Production Line

Jiangong Wang*

State Key Laboratory of Metallurgical Intelligent Manufacturing System, Automation Research and Design Institute of Metallurgical Industry Co., Ltd., Beijing 100071, China.

Email: *wall_2021@qq.com

Abstract. After the thin slab continuous casting and rolling production line of a domestic steel plant was put into operation, with the vigorous development of hot-dip galvanized products, the strip blackening defect on the surface became more prominent. After hot-dip galvanizing, the coil had a certain spacing of strip black spot defects on the surface, which could not meet the needs of high-end users and restricted the improvement of the manufacturing capacity of the headless production line.Therefore, this article uses scanning electron microscopy (SEM), energy dispersive spectroscopy (EDS), three-dimensional profilometer, and optical microscope (OM) to study the microscopic characteristics and causes of black spot defects on the surface of hot-dip galvanized steel.There are a large number of oxidized black spots on the surface of the stripe black area. The zinc layer thickness of the cross-section at the blackened area is smaller than that of the normal area, and the secondary dendrites inside the zinc layer at the blackened area are not well-developed in the normal area. The surface roughness Ra and Rt of the blackened position in the three-dimensional morphology display show significant differences from the normal position. Electron microscopy observation shows that the grain size of the substrate at the blackened area is different from that at the normal area after removing the zinc layer, which is the direct cause of the final surface showing stripe black spot defects. By optimizing the heating temperature uniformity of the IH on the thin slab continuous casting and rolling production line, the stripe black spot defects were ultimately eliminated.

Keywords. Hot dip galvanizing; stripe black spot defects; cause analysis; temperature uniformity.

1. Introduction

In recent years, with the continuous development of the photovoltaic industry, the demand for hot-dip galvanized steel has gradually increased. The products are widely used in high-end products such as photovoltaic brackets, underground pipe galleries, light steel keels, and three-dimensional garages. With the increasingly stringent requirements for their surface quality in the market, domestic and foreign research on the surface quality control of hot-dip galvanized products has focused more on the galvanizing process [1-3]. For example, Liang et al. [4] studied the mechanism and influencing factors of black spots and horizontal stripes in hot-dip galvanized products, and the main control methods include the stability of strip steel operation, furnace process, zinc liquid composition, zinc liquid, and strip steel temperature, etc; Wang et al. [5] believe that slight mechanical scratches during the production process of hot-dip galvanized products are more likely to meet the nucleation conditions of the zinc layer at the defect site during galvanizing, resulting in a large number of localized nucleation defects such as centipede patterns; Xiong et al. [6] believe that the zinc removal defect is due to the long storage time of the acid washed plate, high content of surface C and O elements, and easy enrichment of Si and Mn elements on the surface of the acid washed plate. During the annealing process, the surface reduction is not sufficient, and the reaction between the Al element in the zinc solution and oxygen on the surface of the strip steel is not sufficient, resulting in excessive oxides between the substrate and the zinc layer, causing poor adhesion between the zinc layer.

The hot base raw materials studied above generally come from conventional hot rolling production lines, while there is relatively little research on the surface quality and defect control of hot-dip galvanized steel sheets produced by thin slab continuous casting and rolling production lines, especially in short process production lines under headless rolling mode. Compared with conventional hot rolling production lines, there are significant differences in the process path and strip heat history, as shown in figure 1 and 2. The oxidation influencing factors of the process are more complex and varied [7,8], which poses more risks to the inheritance of defects in downstream processes and affects the product quality of acid washing, hot-dip galvanizing, and other processes. The design of the third-generation thin slab continuous casting and rolling production line involves either casting the billet directly into the roughing mill after leaving the fan-shaped section or equipping it with a roller bottom tunnel furnace for uniform heating [9,10] before entering the roughing mill. This article takes the hot-dip galvanized steel plate produced by the third generation thin slab continuous casting and rolling production line of a certain factory as the research object, systematically studies the microscopic characteristics of the strip like a blackening defect on the surface after hot-dip galvanizing, explores the causes and control measures of blackening defects, improves the surface quality of hot-dip galvanizing, meets user requirements, and provides guidance for optimizing the variety structure and improving product quality of thin slab continuous casting and rolling production lines.

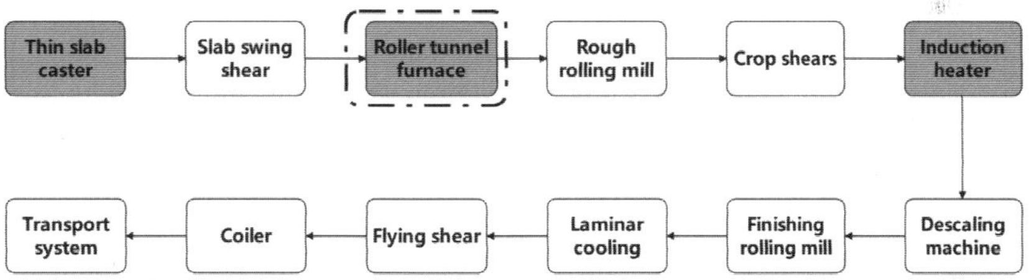

Figure 1. Typical process flow diagram of thin slab continuous casting and rolling production line.

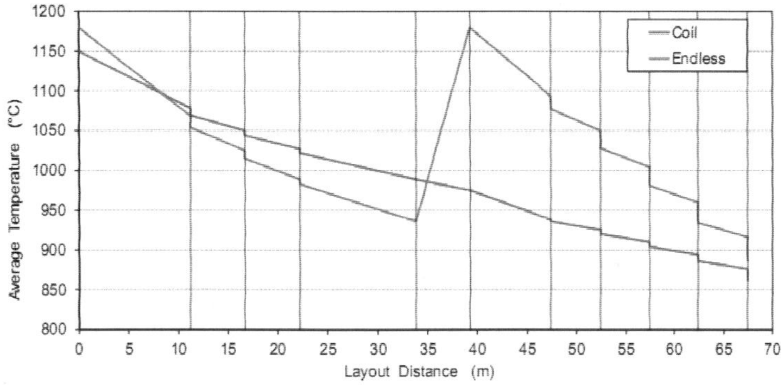

Figure 2. The difference in thermal history between headless rolling and conventional hot rolling.

2. Introduction to Stripe Black Spot Defects

The hot-dip galvanized steel sheet produced by a thin slab continuous casting and rolling production line has strip-like black spot defects on the surface, without tactile sensation, with certain spacing in the width direction and intermittent distribution in the length direction. There is a significant color difference between the blackened position and the normal position, which affects the end user's use and processing and does not meet the release standards. The sample was taken from a thin slab continuous casting and rolling production line of a steel plant, and the raw materials produced in the headless mode were hot-dip galvanized. The chemical composition is shown in table 1. The macroscopic morphology of the defects is shown in figure 3. The surface and cross-sectional morphology of the sample were observed using a JSM-6480LV scanning electron microscope, and the defect locations were analyzed for composition using a built-in energy spectrometer; Using the ST400 3D profilometer to observe the surface morphology, after stripping the zinc layer with 4% nitric acid, the metallographic structure of the sample substrate surface was observed using a ZEISS optical microscope.

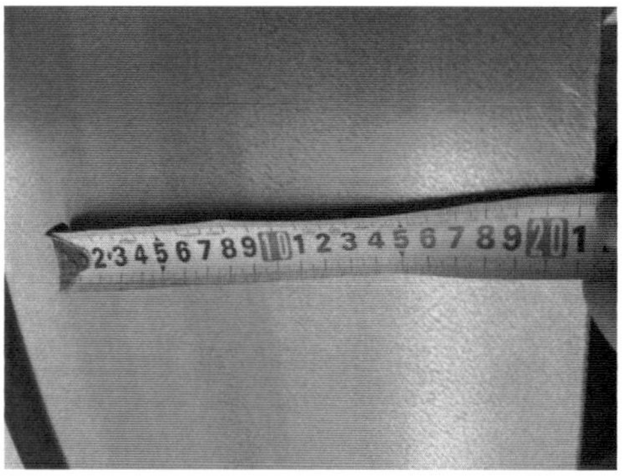

Figure 3. Macroscopic morphology of surface defects on hot-dip galvanized steel sheet.

Table 1. Chemical composition of hot-dip galvanized steel sheet (mass fraction) %.

C≤	Si≤	Mn≤	Nb≤	Alt≤	P≤	S≤
≤0.06	0.2	2.0	0.05	0.05	0.01	0.02

3. Analysis and Discussion of Defect Causes

3.1. Microscopic Analysis of Blackening Defects

Microscopic analysis was conducted on the blackened and normal positions of the galvanized sheet using scanning electron microscopy. The microstructure is shown in figure 4. At low magnification, there is a significant difference in color contrast on the surface of the blackened position. At high magnification, there is a large amount of black block-like material inside, with a small amount of black dot-like appearance around it, as shown in figures 4(a) and 4(b); The difference in color contrast between normal positions is significantly reduced and relatively uniform, and no black dot like morphology is found around it, as shown in 4(c) and 4(d) of figure 4. By comparison, it was found that the higher the density of black block distribution, the more severe the surface blackening.

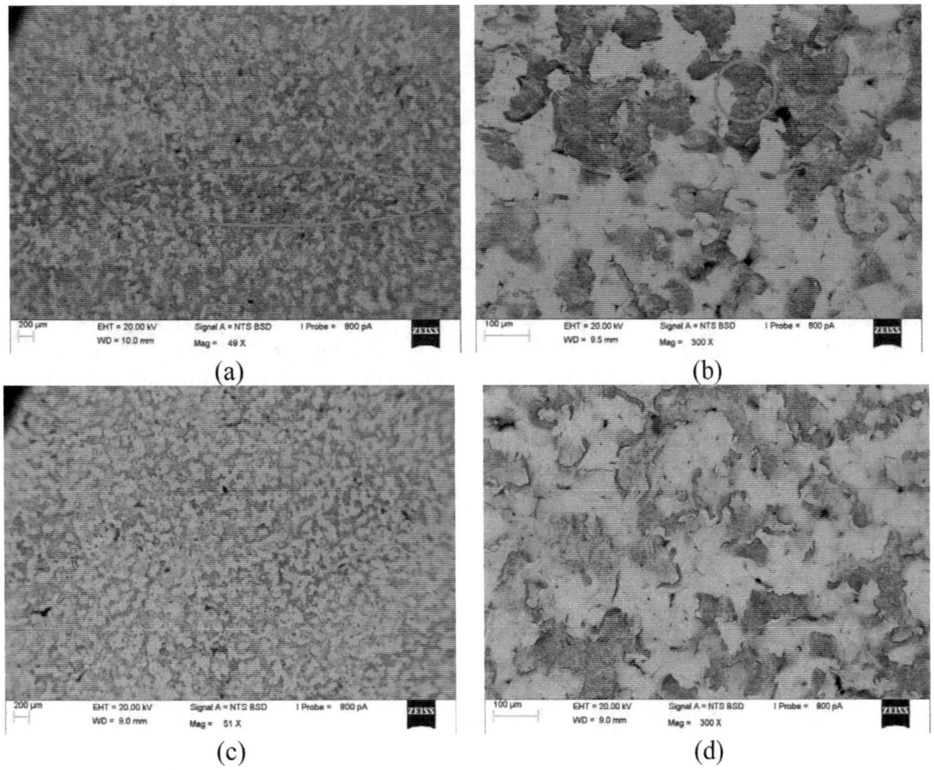

Figure 4. Microscopic morphology of blackening defects and normal positions.

Randomly analyzing the composition of areas with dense black spots in the blackened position using an energy spectrometer, it was found that the oxygen content was 7-14%, and there was no oxygen content in normal positions, indicating that the surface of the zinc layer in the blackened position cannot undergo varying degrees of local oxidation. As shown in figure 5 and the spectrum in table 2.

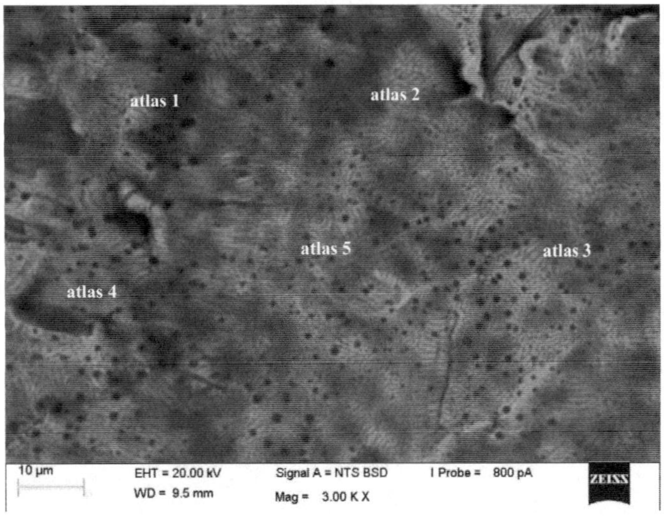

Figure 5. Microscopic morphology and spectral location of blackened surface areas.

Journal of Physics: Conference Series **2941** (2025) 012033 doi:10.1088/1742-6596/2941/1/012033

Table 2. Composition analysis of blackened surface areas (mass fraction) %.

atlas label	atlas 1	atlas 2	atlas 3	atlas 4	atlas 5
O	7.33	8.62	14.05	6.91	13.08
Mg	1.57	1.74	1.39	2.22	1.37
Al	1.88	1.29	1.65	2.32	3.51
Si	13.60	1.69	1.48	10.08	1.38
Zn	41.19	51.15	41.13	47.12	43.13

The macroscopic morphology of the galvanized sheet surface is different. Comparative analysis of its three-dimensional morphology reveals that the surface roughness at the blackened position is relatively large, with a Ra value of 1.5-1.8μm and a maximum of over 2.0μm, and a micro-roughness Rt of 19-22μm; The surface roughness Ra at the normal position is between 1.2 and 1.3μm, and the surface micro-roughness Rt is between 15 and 16μm. There is a significant difference from the blackened position, as shown in figure 6 for the three-dimensional morphology and parameter analysis. At the same time, the greater the difference between surface roughness and micro roughness, the more severe the color difference.

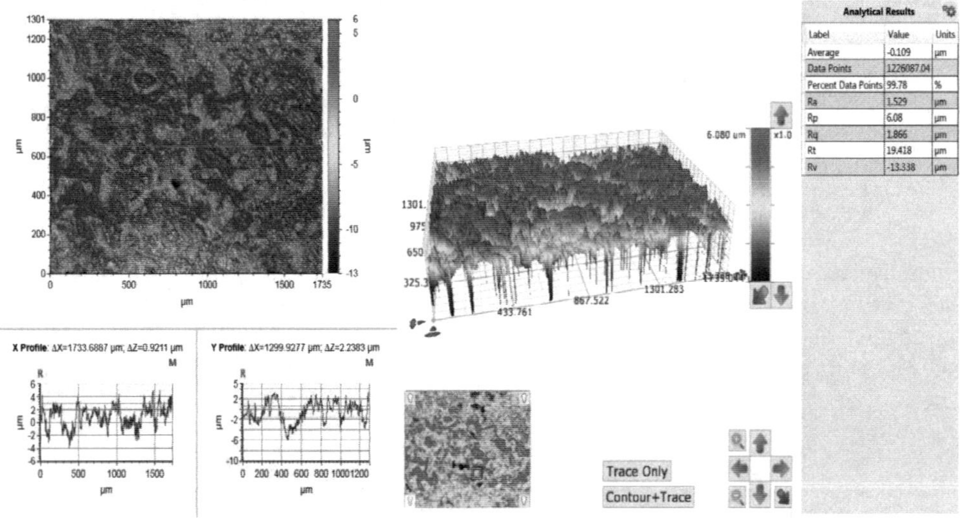

(a) Blackened areas

Journal of Physics: Conference Series **2941** (2025) 012033 doi:10.1088/1742-6596/2941/1/012033

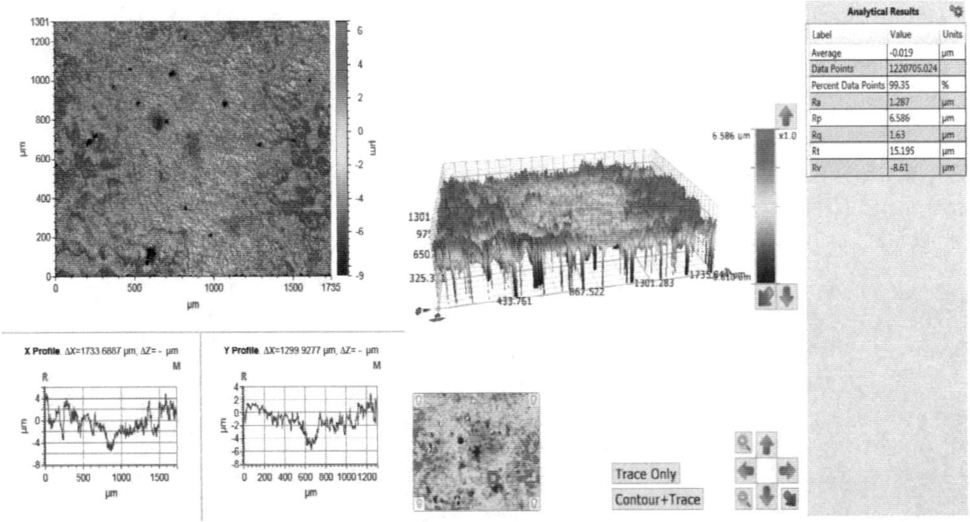

(b) Normal areas

Figure 6. Comparison of three-dimensional morphology of blackened (a) and normal areas (b).

By comparing the microstructure of the zinc layer thickness interface between the blackened and normal positions, it can be seen that the zinc layer thickness at the blackened position is significantly smaller than that at the normal position, and the dendrites at the blackened position are not as developed as those at the normal position. This indicates that there are significant differences in the growth process of the zinc layer, leading to uneven organization and thickness, as shown in figure 7. Therefore, through the above analysis, it can be concluded that the uneven thickness of the zinc layer and the differences in roughness and flatness lead to different colors and glossiness on the surface of the steel plate, resulting in color differences and macroscopic strip-shaped black spot defects, which also affect the corrosion resistance.

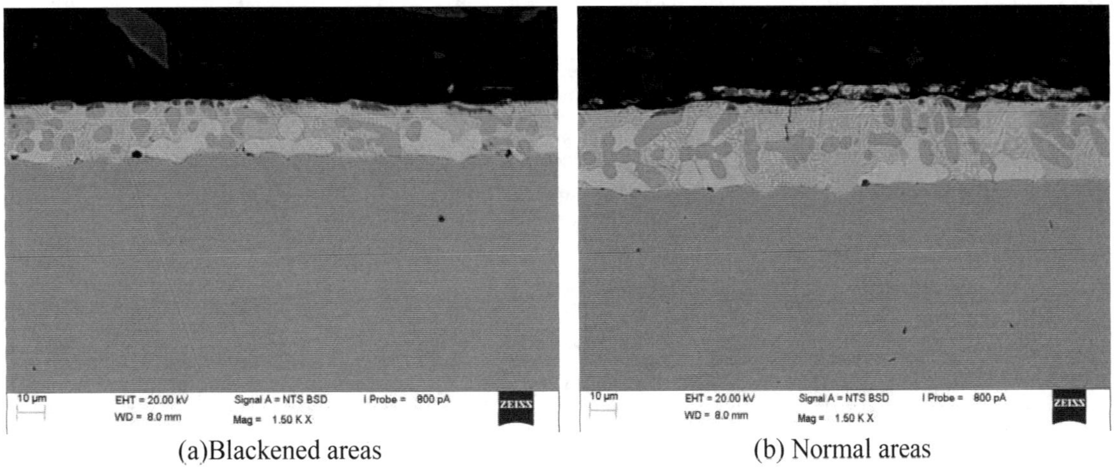

(a)Blackened areas (b) Normal areas

Figure 7. Comparison of cross-sectional zinc layer morphology of blackened (a) and normal areas (b).

3.2. Correspondence Analysis of Key Processes

Further, analyze the key processes for the uneven thickness and roughness of the zinc layer. According to previous research, factors such as the rapid cooling process in the galvanizing process, polishing equipment, and oxide inclusions in the substrate may all cause color difference defects after galvanizing [11-14]. However, the black spot defects in the strip discussed in this article have significant differences in their causes and characteristics of headless rolling production lines.

During the headless rolling production process on this production line, there are irregular and slight strip-like iron oxide scale defects on the surface of the hot-rolled plate. However, after acid washing, the strip-like iron oxide scale can be effectively removed, and there is no significant difference from other normal positions. The morphology of the surface inspection instrument on different production lines is shown in figure 8. Therefore, it can be concluded that there is no corresponding relationship between the surface strip-like blackening defects of hot-dip galvanizing and the actual slight iron oxide scale defects on the surface of hot-rolled substrates [15].

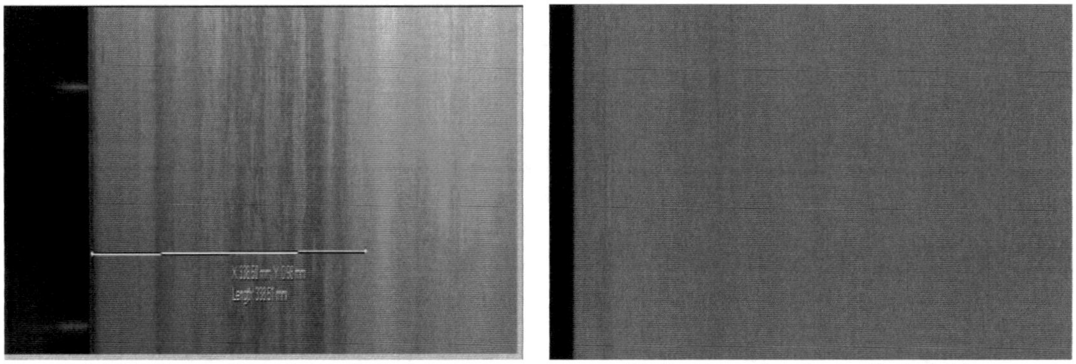

(a) Endless rolling hot rolled plate (b) Surface after hot-rolled pickling

Figure 8. The appearance of the hot-rolled plate (a) and pickled steel sheets (b) surface inspection instrument.

After further peeling off the zinc layer, the surface microstructure of the blackened and normal positions was observed using a light microscope. It was found that there was significant grain refinement in the blackened position, and the phenomenon of mixed crystals was more severe compared to the normal position, as shown in figure 9. The microstructure differences of the substrate result in differences in the nucleation conditions of the zinc layer during the subsequent galvanizing process. Small grains on the substrate surface promote the formation of crystal nuclei in the coating, generating zinc flowers with relatively small grain diameters. In areas with severe crystal mixing, local nucleation points change, and the size of zinc flowers produced is inconsistent, leading to significant fluctuations in surface roughness after galvanizing. At the same time, there is a difference in the thickness of the zinc layer, and the grain structure on the surface of the steel plate will affect the thickness of the hot-dip galvanized zinc layer, mainly reflected in the interaction between the steel and the zinc liquid. The coarse and uniform position of the grain allows more zinc liquid to precipitate on the substrate surface, thereby increasing the contact area between the steel and the zinc liquid. Therefore, the thickness of the zinc layer is relatively thick and uniform. However, the small grain size, especially in the presence of coarse grains, can affect the contact area and reaction activity with the zinc solution due to the unevenness of the coarse grain structure, resulting in uneven deposition of the zinc solution and overall thin coating thickness.

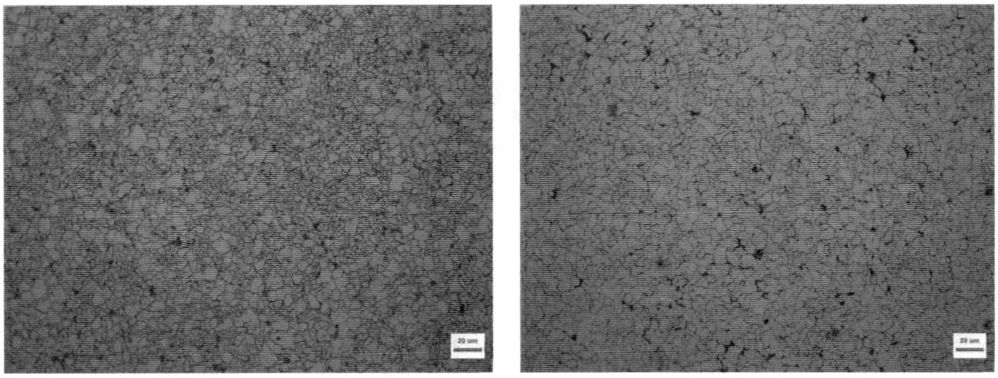

(a) Blackened areas (b) Normal areas

Figure 9. Microstructure of blackened position (a) and normal position (b) on the surface of galvanized sheet.

The uneven thickness of the zinc layer will result in different contact stresses with the smooth roller during subsequent cooling and finishing processes. That is, under the same flat reduction rate or rolling force conditions, different positions of the zinc layer will experience significant differences in finishing effects, resulting in black strip-shaped color difference defects macroscopically. However, actual comparative studies have shown that the surface roughness of the strip steel can be controlled by varying the flatness reduction rate. With a large flatness reduction rate, both the Ra and Rz values of the substrate roughness decrease, as shown in figure 10. The reduction rate of the large leveling machine and the flatness of the substrate will also be significantly improved, thereby improving the uniformity and adhesion of the zinc layer structure, which is conducive to improving the surface condition of the galvanized sheet [16].

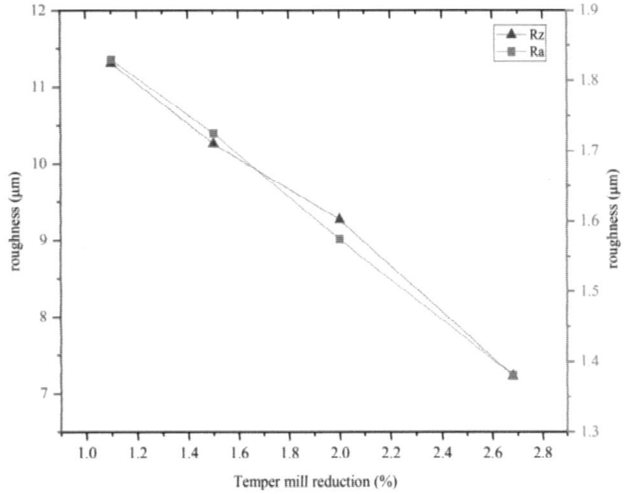

Figure 10. The influence of temper mill reduction rate on the roughness of galvanized substrate.

Further analysis of the differences in the microstructure of the substrate revealed a close correlation with the temperature distribution in the width direction of the substrate. Due to the use of induction heaters for secondary heating of the intermediate billet in the headless rolling mode, the lateral uniformity of the induction heating temperature is relatively poor, and there are high and low-temperature zones, as shown in figure 11. In the low-temperature zone, the grain size becomes small and uneven during the subsequent cooling process, forming a mixed crystal zone. Therefore, the focus is on optimizing the induction heater. On the one hand, by optimizing the lateral displacement of the heater, and on the other hand, by improving the structure of the heater coil and optimizing the magnetic circuit gap of the iron core, a smooth transition of lateral temperature can be achieved. The temperature uniformity of the induction heating plate width is significantly improved, and the quality of hot-dip galvanizing is also greatly improved. The temperature distribution of the induction heating plate width before and after improvement is shown in figure 12.

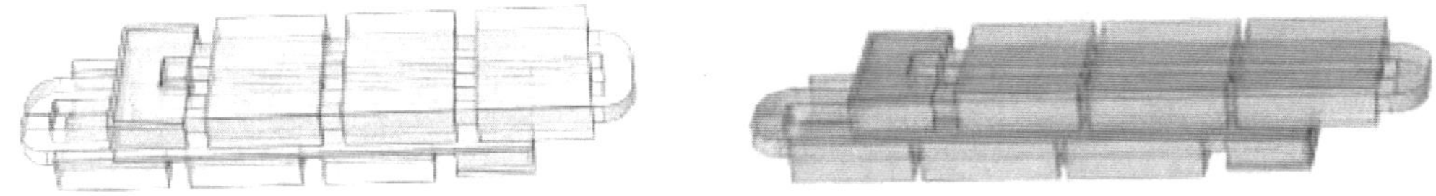

(a) The magnetic circuit gap before improvement (b) Improved magnetic circuit gap

Figure 11. Optimization of magnetic circuit gap of induction heater.

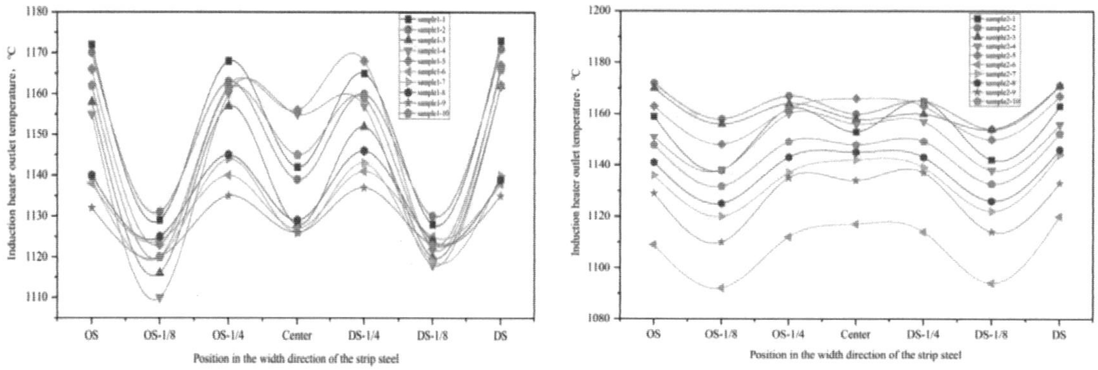

(a) Lateral temperature of width before improvement (b) Improved lateral temperature of width

Figure 12. Temperature distribution of the heating plate width after induction heater heating.

4. Conclusions

Compared with conventional hot rolling production lines, the process path and strip heat history of the third-generation thin slab continuous casting and rolling production line have significant differences. It is more difficult to control the oxidation and heating uniformity of the process, and the genetic risk of defects in downstream processes is also greater. This article analyzes the root cause of strip black spot defects caused by hot-dip galvanized substrates, which is of great significance for improving the ability of short-process production lines to produce high-quality products, obtain the following conclusion:

(1) The thickness of the zinc layer on the surface of the hot-dip galvanized sheet with black spots is abnormal, and the secondary dendrites inside the zinc layer at the blackened position are not developed in the normal position. After removing the zinc layer, the grain structure of the substrate at the blackened position is finer and more uneven.

(2) The uneven thickness of the zinc layer and the differences in roughness and flatness result in different colors and glossiness on the surface of the steel plate, leading to color differences and macroscopic strip-shaped black spot defects.

(3) The poor temperature uniformity of induction heaters is the fundamental reason for the difference in grain size. By optimizing and transforming the induction heater, the temperature uniformity of the plate width has been improved, and the quality of hot-dip galvanizing has also been greatly improved.

(4) The size of the surface roughness of the strip steel can be controlled by varying the flatness

reduction rate. With a large flatness reduction rate, the roughness Ra and Rz values of the substrate are reduced, which is beneficial for improving the polished surface state of the galvanized sheet.

5. References

[1] Li J L, Xu X F and Li S H 2011 *Q&A on Continuous Hot Dip Galvanizing Production of Strip Steel* (Beijing: Metallurgical Industry Press) p10-99.

[2] Song J 2000 Production and development of hot rolled galvanized sheet *Steel Rolling* **(03)** p38-41.

[3] Zhang L Y, Li J and Zuo L 2005 The current status of continuous hot-dip galvanizing process technology for strip steel *Steel Rolling* **(02)** p38-43.

[4] Liang Y Y, Xia X S and Zhang Z C 2021 The mechanism and control of black spots and transverse defects in hot-dip galvanized products *Hebei Metallurgical* **(06)** p78-81.

[5] Wang F, Zhang P and Wang X 2020 Analysis and control of centipede pattern defects on the surface of hot dip galvanized sheet *Hebei Metallurgical* **(05)** p34-36+69.

[6] Xiong Z L and Liu H Q 2016 Control mechanism of adhesion of zinc layer on hot-dip galvanized sheet *Hot Working Process* **45(14)** p110-113+116.

[7] Yan B, Wang X L and Jiao S 2009 Investigation on oxide scale and surface roughness of strip steel hot-rolled in CSP *Steel Rolling***26(01)** p10-11.

[8] Liu Z Z, Liu J and Liu W Q 2011 Analysis and control of surface defects on strip steel in CSP Soaking Furnace *Journal of Wuhan Engineering Institute* **23(03)** p15-16.

[9] Tian N Y 1998 *Thin Slab Continuous Casting and Rolling* (Beijing: Metallurgical Industry Press) p22-102.

[10] Wang S Z, Mao X P and Gao J H 2022 Development status and prospects of thin slab continuous casting and rolling technology *Journal of Engineering Science* **44(4)** p534-545.

[11] Chen R Y and Yuen W Y D 2003 Review of the high-temperature oxidation of iron and carbon steels in air or oxygen *Oxidation of Metals* **59(5,6)** p433-468.

[12] Zhu H J 2006 Design parameter analysis of push-pull pickling line *Steel Rolling* **23(1)** p37-40.

[13] Yu Y and Tang S 2006 Research on the formation mechanism and control strategy of iron oxide scale on hot rolled coil *Iron and Steel* **41** p50-52.

[14] Wang Y J, Mu H L and Dong H J 2006 Analysis and countermeasures of difficulties in acid pickling SPHC tropical iron oxide Scale *Steel Rolling* **23(4)** p51-54.

[15] Zhang L L, Lv B and Wang J G 2023 Analysis of causes of scale defects in low carbon steel by short process *Forging and Stamping Technolgy* **48(04)** p147-150.

[16] Huang Y P, Wang Z H and Men J L 2023 Mechanism and control strategy of black defect on galvanized pipe surface *Metal World* **(5)** P61-65.

Finite Element Analysis of the Mechanism of Transverse Crack Formation in 15CrMoG Water-cooled Walls

Ye Li[1,*], Yang Pei[2] and Yue Zhao[2]

[1] China Power Hua Chuang Electricity Technology Research Company Ltd, Suzhou, Jiangsu, China.
[2] Guoneng Boiler and Pressure Vessel Inspection Company Ltd, Beijing, China.
Email: *15951916426@163.com; wy163yang@163.com; 289736138@qq.com

Abstract. In recent years, during the maintenance of 660MW ultra-supercritical boiler water walls, it has been discovered that large-scale surface transverse cracks have developed on the fire-facing tubes within specific regions, posing considerable safety hazards to the boiler's operation. This paper conducts mathematical modeling of a water wall panel and specific tubes in a unit, simulating the following scenarios: the influence of the soot-blowing process on the stress distribution within the water wall; the stress response when significant temperature gradients exist within a specific panel; and the extent of stress peak changes when abnormal heat exchange occurs due to local blockage or coking of the water wall. The results indicate that while the soot-blowing process affects the stress distribution, the maximum stress values in the tube wall change relatively little. However, when abnormal heat transfer occurs in the tubes, the thermal stress peaks increase sharply, significantly impacting the formation of transverse cracks in the tube wall and potentially even causing short-term overheating and bursting of the tubes.
Keywords. Transverse cracks; FEM; thermal stress; water-cooled wall tube.

1. Introduction
The 660MW ultra-supercritical coal-fired boiler is one of the more advanced power generation units in China, with technology that is more mature compared to the 1000MW units. However, during the maintenance of multiple power plants after commissioning, large areas of surface transverse cracks have been found on the fire-facing side of the water-cooled wall tubes in certain specific areas [1]. These cracks pose a significant threat to the safe operation of the boiler. This paper conducts a general analysis of such issues to identify the causes and mechanisms behind the formation of transverse cracks in the water-cooled wall. The findings aim to provide valuable reference and guidance for resolving these issues.

2. Boiler Water-Cooled Wall Tube Specifications
The boiler under study is a type of ultra-supercritical parameter-variable direct current boiler, featuring a vertical rising membrane water-cooled wall in the furnace. The specifications of the water-cooled wall tubes are φ28.6mm×6mm, made from 15CrMoG steel, with a pitch of 44.5mm. The flat steel welded between the tubes is 15.9mm wide and 6mm thick, made from 15CrMo material.

During maintenance, statistical analysis of surface transverse cracks in a specific area of the front wall showed that approximately 30% of these cracks occur around the soot blower holes [2]. The remaining cracks are mostly concentrated around the fins and the apex of the fire-facing side of the

Journal of Physics: Conference Series **2941** (2025) 012034 doi:10.1088/1742-6596/2941/1/012034

tube, as shown in figure 1. Research was conducted to analyze the impact of soot blower operation on tube wall stress. Temperature measurement points were placed in areas with a higher concentration of cracks, and control data was used as a basis for finite element analysis. The goal was to identify the causes of crack formation from the perspective of stress distribution.

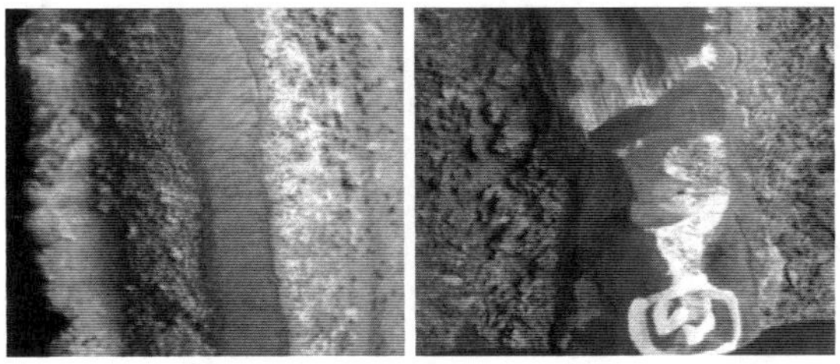

Figure 1. Schematic diagram of cracks on the outer wall of the water-cooled wall tube.

The water-cooled wall tubes are all made of 15CrMoG, and their physical properties are shown in table 1:

Table 1. Physical properties of 15CrMoG at different temperatures.

Temperature (°C)	100	150	200	250	300	350	400	450	500	550	600	650
Elastic Modulus (103 MPa)	210	/	204	/	197	/	187	/	177	/	/	/
Poisson's Ratio	0.295	/	0.300	/	0.301	/	0.304	/	0.308	/	/	/
Coefficient of Linear Expansion (10-6 /K)	13.37	/	13.38	/	13.70	/	14.06	/	14.36	/	14.63	/
Thermal Conductivity (W /(m·k))	40.6	40.4	40.1	39.5	38.7	37.8	36.8	35.8	34.8	33.9	32.8	31.6
Allowable Stress (MPa)	116	114	114	114	113	110	107	103	88.3	40.3	17.3	7.4
Yield Strength (MPa)	/	/	269	256	242	228	216	205	198	182	174	/
Temperature (°C)	100	150	200	250	300	350	400	450	500	550	600	650

3. Boundary Conditions for Thermal Analysis Calculations

In the heat transfer process of the water-cooled wall, the outer wall absorbs heat from the fire-facing side, while the inner wall absorbs heat through the flowing working fluid. The outer wall on the fire-back side is insulated to achieve thermal insulation [3]. Therefore, the boundaries are divided into the following three: the outer wall facing the fire, the outer wall facing away from the fire, and the inner wall.

First, determine the heat flux density boundary condition for the outer wall facing the fire. Due to the different angles and modes of contact with the flame, separate considerations are made for the fins and the tube wall.

The angle coefficient for the fins facing the fire is shown in figure 2.

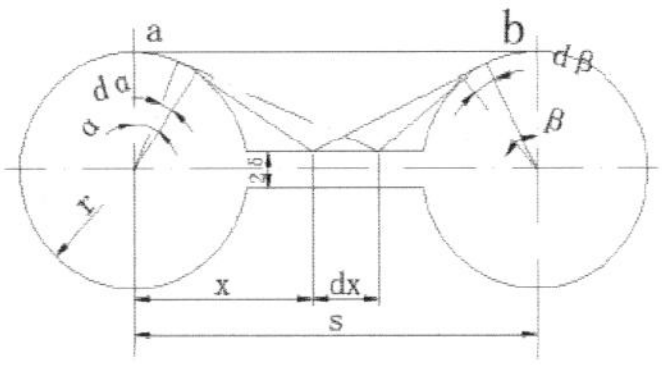

Figure 2. Schematic diagram of the calculation of the fins' angle coefficient.

As shown in figure 3, the amount of heat absorbed by an infinitesimal segment of length *dx* on the fin is $Qdx = q(x)\cdot dx$, where $q(x)$ is the heat flux density at position x on the fin. If the angle coefficient of *ab* with respect to dx is φ_x, then the heat flux density on the infinitesimal segment is:

$$q(x) = \frac{\varphi_x}{dx} \cdot q \cdot s \tag{1}$$

The angle coefficient for the tube wall facing the fire is shown in figure 3.

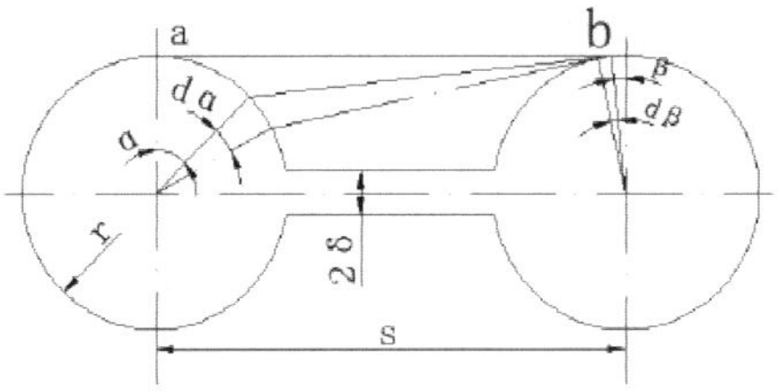

Figure 3. Schematic diagram of the calculation of the tube wall angle coefficient.

The angle coefficient for the tube wall facing the fire is:

$$\varphi_\alpha = \frac{1}{2S}[r \cdot (1 - \sin\alpha \cdot \csc\beta)d\alpha + r \cdot \cot\beta \cdot \csc\beta\,(\cos\beta - \cos\alpha) \cdot d\beta] \tag{2}$$

Substitute the angle coefficient into the following heat flux density calculation formula to obtain the final expression for the tube wall heat flux density.

$$q(\alpha) = \frac{\varphi_\alpha}{r \cdot d\alpha} \cdot q \cdot s \tag{3}$$

Normalize the distribution of heat flux density on the tube wall and fins facing the fire to obtain a unified heat flux density coefficient.

$$K_q = \frac{q(x)}{q} or \frac{q(\alpha)}{q} \tag{4}$$

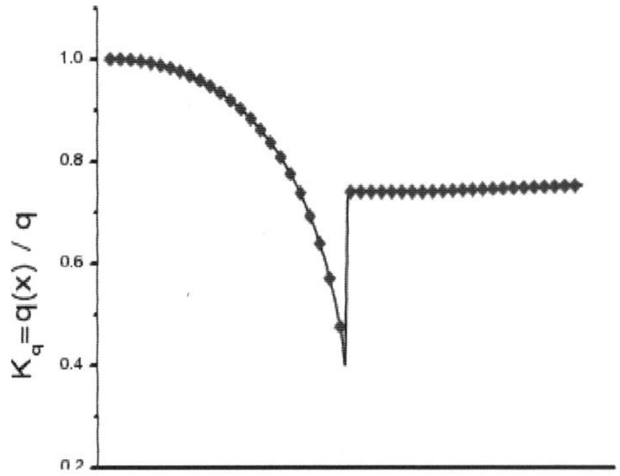

Figure 4. Normalized heat flux density distribution.

Figure 4 shows the normalized heat flux density function. The left-side curved portion represents the area from the apex of the tube wall facing the fire to the fin location, while the right-side horizontal portion represents the fin position.

To obtain the heat flux density value at a specific point on the fire-facing boundary, multiply the heat flux coefficient at that point by the average heat flux density data for that area. According to relevant research, the heat flux coefficient at the apex location is set to 380 kW/m² [4].

Next, the insulation layer on the fire-back side of the water-cooled wall is approximated as adiabatic for analysis purposes.

Finally, the heat exchange between the inner wall of the water-cooled wall and the working fluid is considered. Since the tube screen of the water-cooled wall is located in the upper part of the boiler furnace, and the steam inside the tube is superheated, the temperature difference between the inner wall and the working fluid does not exceed 10°C. The heat transfer coefficient variation around the tube's circumference also does not exceed 10%. Given that this study focuses on the thermal stress response within a small range of the water-cooled wall, and to improve the efficiency of the calculation model, the heat transfer coefficient for the inner tube wall is set to 50 kW/(m²·K) [5].

Based on the thermal conductivity characteristics at the above three locations, boundary conditions are set for the inner and outer walls of the water-cooled wall. Temperature field calculations are then performed, which will be used as conditions for structural field calculations.

4. Analysis of Finite Element Calculation Results for the Water-Cooled Wall**

4.1. Impact of Soot Blowing on Tube Screen Stress Distribution

A region of 1440mm × 3500mm (length × height) centered around a specific soot blower hole is selected for the study. The temperature and stress field distributions before and after soot blowing are compared.

The following settings are applied to the calculation model:

Turbulence Model: RNG k-ε model, with a hydraulic radius of 0.025m.

Operating Environment: Ambient pressure of 0.1 MPa.

Boundary Conditions: Medium flow rate of 40 kg/s.

Ambient Temperature: 1273 K.

The FLUENT module is used to calculate the impact of the soot blower on the temperature distribution of the water-cooled wall. The calculation results are shown in figure 5.

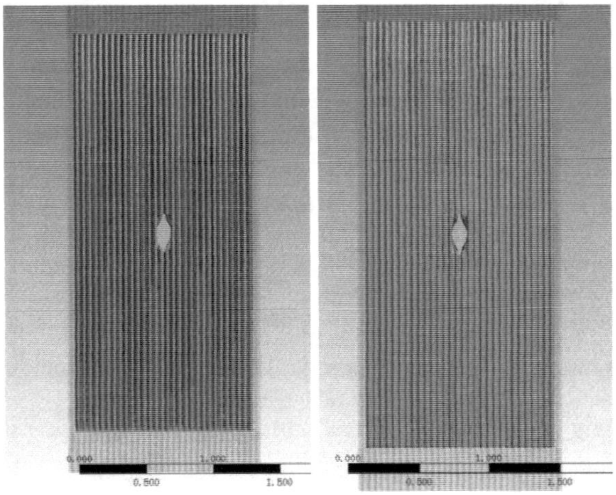

Figure 5. Temperature distribution contour maps of the tube screen before and after soot blower operation.

On the left side of figure 6, the temperature distribution before the soot blower operation is shown. The results indicate that the temperature on the fire-facing side is higher than that on the fire-back side, with the highest temperature located in the middle of the fins facing the fire, reaching up to approximately 470°C. On the tube cross-section, the temperature on the fire-facing side ranges from 435.8°C to 450.3°C (from bottom to top), while the temperature on the fire-back side ranges from 410.3°C to 420.8°C (from bottom to top). The temperature of the inner wall at the same height is the same as the temperature at the center point of the fire-back side outer wall.

The temperature field of the water-cooled wall before the soot blower operation is applied as a load, with corresponding constraints imposed in the selected area. Since the upper and lower sides of the selected region are within the constraint range of horizontal rigid beams, the upper part is set with Uy=0 and Ux=0; the lower part is set with Ux=0, while all other degrees of freedom are set as free. The calculation focuses on the vertical normal stress, which has the greatest impact on the formation of transverse cracks. The calculation results are shown in figure 6.

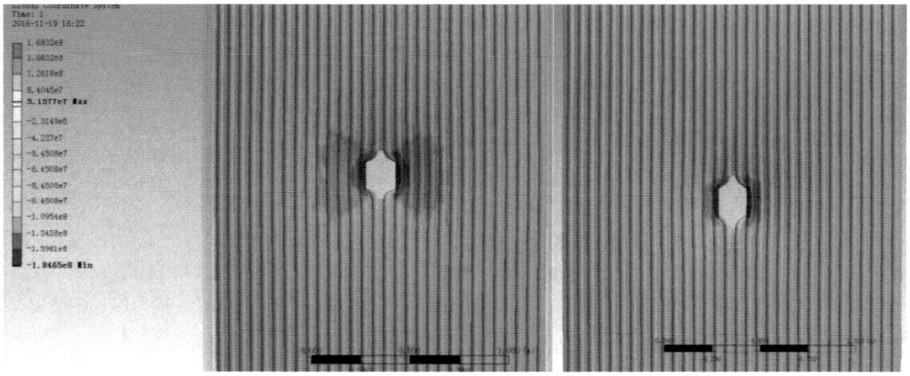

Figure 6. Stress distribution contour maps of the tube screen before and after the soot blower operation.

Before the soot blower operation, the vertical (Y-direction) normal stress in most areas of the tube screen ranged from -54 MPa to -84 MPa (where negative values indicate compressive stress). Compressive stress was higher on the two tubes on each side of the soot blower hole, reaching -184.65 MPa. Additionally, the tubes on the upper and lower sides of the soot blower hole exhibited a noticeable reduction in compressive stress within a certain range. This is because the presence of the soot blower hole alleviates some of the thermal expansion of the upper and lower-end tubes and causes the tubes on the sides to bear more expansion pressure.

After the soot blower operation, significant changes in stress were observed within the temperature fluctuation region. The maximum compressive stress on the two tubes on each side of the soot blower hole decreased to -162.94 MPa, with a fluctuation range of 20 MPa. There were also substantial stress variations in the Y-direction at fixed positions on the upper and lower sides.

In summary, the normal operation of the soot blower, which results in a temperature decrease around the soot blower hole, does not significantly impact the vertical stress in that area. Additionally, the low frequency of soot blower operation is insufficient to cause low-cycle fatigue in the water-cooled wall tubes. Therefore, the cause of the cracks in that area should be investigated from other perspectives.

4.2. Stress Distribution in Tube Rows with Temperature Deviations of the Working Fluid Within a Certain Range

Based on the temperature measurement points in the areas with a higher concentration of cracks, significant temperature deviations often occur in the tube wall within a certain time frame, with temperature differences of up to 70°C between adjacent 10 tubes.

Using the field-measured temperatures, a tube screen of 20 water-cooled wall tubes was selected and numbered from 1 to 20 from left to right. Three operating conditions were simulated for calculations:

(1) The working fluid temperature inside all 20 tubes is the same at 440°C.

(2) The working fluid temperature inside tubes 1 and 20 is 430°C, with the temperatures of the remaining adjacent tubes increasing towards the center, with tubes 10 and 11 having a temperature of 460°C.

(3) The working fluid temperature inside tubes 1 and 20 is 420°C, with the temperatures of the remaining adjacent tubes increasing towards the center, with tubes 10 and 11 having a temperature of

Journal of Physics: Conference Series **2941** (2025) 012034 doi:10.1088/1742-6596/2941/1/012034

490°C.

The calculation results shown in figure 7:

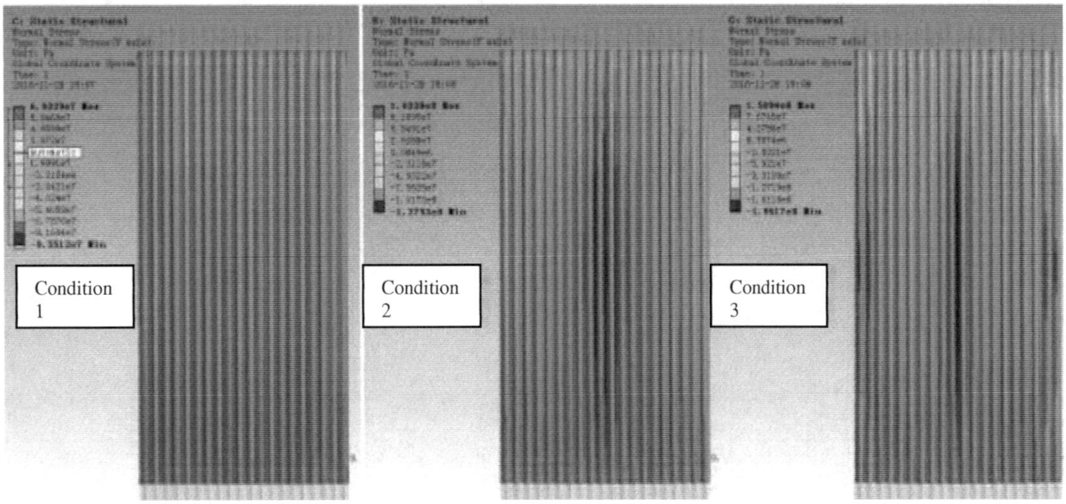

Figure 7. Stress distribution contour maps of the water-cooled wall tube screen under three operating conditions.

For Condition 1, the stress distribution on the tube screen is uniform, with the vertical (Y-direction) normal stress at the centerline of the fire-facing outer wall of each tube being 26 MPa (compressive stress).

In Condition 2, due to the temperature differences on the fire-facing outer wall of the tubes, the stress field redistributes. Specifically, tubes 5–16 still exhibit compressive stress at the centerline of the fire-facing outer wall, with values being smaller at the ends (5.8 MPa) and the maximum compressive stress in the center at tubes 10 and 11 (78.7 MPa). The stress at the positions of tubes 1–4 and 17–20 has shifted from compressive to tensile, with values as follows: for tubes 1–4, 98.7 MPa, 71.2 MPa, 53.8 MPa, and 31.8 MPa, respectively, and the opposite for tubes 17–20.

In Condition 3, the increased temperature differences on the fire-facing outer wall further amplify the redistribution of the stress field. Tubes 6–15 display compressive stress at the centerline of the fire-facing outer wall, with values being smaller at the ends (10.8 MPa) and the maximum compressive stress at tubes 10 and 11 (175.7 MPa). Tubes 1–5 and 16–20 exhibit tensile stress at the centerline of the fire-facing outer wall, with values, as follows: for tubes 1–5, 140.7 MPa, 115.9 MPa, 79.2 MPa, 49.3 MPa, and 22.6 MPa, respectively, and the opposite for tubes 17–20.

When there is a 60°C temperature difference between adjacent 9 tubes, the tensile stress in tube 1 can reach 140.7 MPa. At this point, the overall temperature of tube 1 is approximately 440°C to 460°C. For 15CrMoG at this temperature, the allowable stress of the material is around 103 MPa. This indicates that significant temperature differences within a small localized area of the water-cooled wall should be given sufficient attention.

4.3. Stress Analysis of a Single Tube with Localized Heat Transfer Degradation

Due to the operation of many domestic ultra-supercritical units in sliding pressure adjustment mode, the properties of the working fluid in the water-cooled wall vary significantly, often leading to localized heat transfer degradation and resulting in rapid temperature increases on the tube wall [6]. The large temperature gradients observed in the tube screen within a small area are likely due to heat transfer degradation within the tube, which affects the heat absorption capability of the working fluid [7,8].

Based on this, a single tube with a length of 2000 mm is selected for analysis. A 50 mm long and

14 mm wide area on the inner wall facing the fire is simulated to represent heat transfer degradation. In this region, the heat transfer coefficient is reduced to 2 kW/(m²·K) [9]. The resulting temperature and stress changes in the surrounding area are analyzed. The calculation results are shown in figure 8.

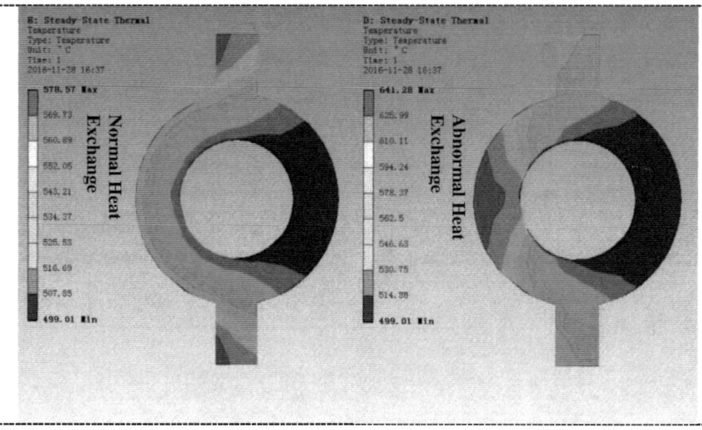

Figure 8. Comparison of cross-sectional temperature fields before and after heat transfer degradation.

Under normal heat transfer conditions, the temperature distribution in the water-cooled wall tube is consistent, with the highest temperature at the fin position being 520.5°C. When localized heat transfer degradation occurs, the overall temperature on the fire-facing side increases, and the highest temperature shifts to the fire-facing outer wall vertex of the tube at the degradation location, rising to 561.9°C. The stress distribution in the water-cooled wall tube under these conditions is shown in figure 9.

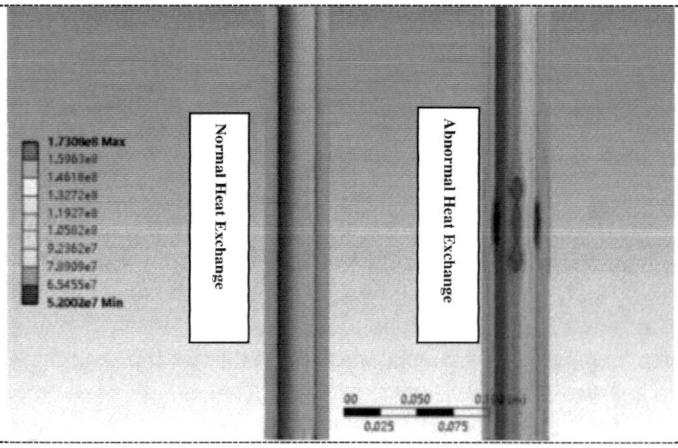

Figure 9. Equivalent stress distribution on the fire-facing outer wall before and after heat transfer degradation.

The figure above shows that in the heat transfer degradation area, the temperature at the fire-facing vertex of the tube increases sharply within a small range, causing significant and uneven expansion. This leads to a corresponding rise in equivalent stress, increasing from approximately 120 MPa to 179 MPa. According to table 1, this stress is close to the yield strength of the tube material at the corresponding temperature.

For an intuitive comparison, a curve is used to represent the changing relationship between temperature and equivalent stress at the vertex along the tube axis (with the center of heat transfer deterioration as the zero point). The result is shown in figure 10.

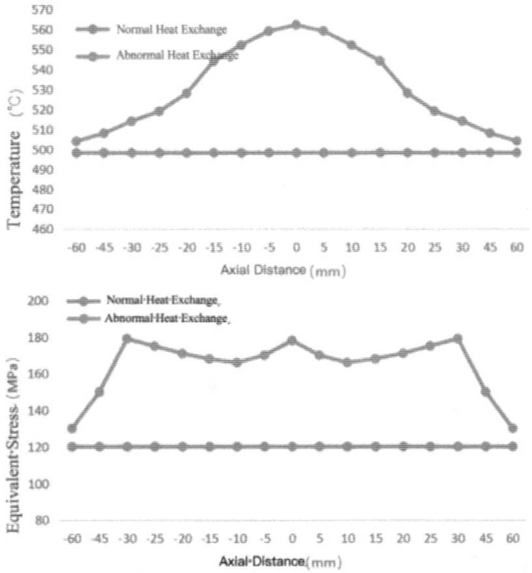

Figure 10. Temperature and equivalent stress variation along the tube axis at the fire-facing vertex before and after heat transfer degradation.

It is evident that the maximum temperature point is located at the outer wall in the center of the heat transfer deterioration area. The stress levels affected by heat transfer deterioration extend not only to the center of the affected region but also to the boundaries 25 to 30 mm above and below this region.

Additionally, for the equivalent stress value at the top of the central area affected by heat transfer deterioration, various operating conditions and different degrees of heat transfer deterioration were studied. Four sets of vertex heat flux density values were established, and based on the normalized distribution shown in figure 4, thermal convection loading was applied to the water-cooled wall tubes. For each operating condition, five different heat transfer coefficients for the deteriorated area of the tube wall were set. The specific loading values and calculation results are shown in table 2:

Table 2. Calculation results of temperature and equivalent stress at the fire-facing vertex under different heat loads and heat transfer deterioration levels.

Heat load (kW/m^2)	Heat transfer coefficient in the abnormal heat exchange area of the inner tube wall (kW/(m^2·K))									
	0.2		2		10		30		50 (Normal heat exchange)	
	Temperature (°C)	Equivalent stress (MPa)	Temperature (°C)	Equivalent stress (MPa)	Temperature (°C)	Equivalent stress (MPa)	Temperature (°C)	Equivalent stress (MPa)	Temperature (°C)	Equivalent stress (MPa)
250	621	181	512	120	476	101	470	82	461	68
300	659	176	534	148	501	116	505	99	485	80
400	698	165	562	178	538	120	528	113	520	120
450	722	172	588	172	545	143	537	120	525	131

As shown in table 2, the equivalent stress at the vertex under different levels of heat transfer deterioration tends to have a clear linear relationship with the heat load and temperature in most cases. However, when the heat transfer coefficient drops to 0.2 kW /(m²·K), this linear relationship no longer holds. It can be concluded that stress concentration caused by heat transfer deterioration is significantly affected by the rapid rise in temperature, as well as by the deformation misalignment induced by local complex temperature gradients.

In summary, when the heat transfer coefficient in the deteriorated area falls to 2 kW/(m²·K) or below, the stress level in the central region of deterioration approaches or exceeds the yield strength at the corresponding temperature, making it an important cause of localized transverse microcracks.

5. Full Conclusion

This paper establishes a temperature-stress calculation model for water-cooled walls and focuses on several key areas: the impact of soot blower operation on temperature and stress fields, stress distribution under temperature gradients, and the effect of localized heat transfer deterioration on stress distribution. The main conclusions are as follows:

(1) Impact of Soot Blower Operation:

The soot blower operation significantly affects the temperature field on the fire-facing side, with the influence concentrated in a circular area centered around the soot blower, with a diameter of approximately 500 mm. However, the frequency and magnitude of the stress changes induced by soot blower operation are relatively small and are not sufficient to be the primary cause of frequent transverse cracks in this area. A comprehensive analysis should also consider factors such as combustion flow distribution in the furnace, high-temperature flue gas erosion, and the presence of elbow pipes.

(2) Effect of Temperature Gradients:

Data collected from monitoring a region with a high density of transverse cracks indicates that this area frequently experiences large temperature gradients. Under these conditions, significant axial tensile stresses develop on the fire-facing surface of the water-cooled wall tubes, which can facilitate the initiation and propagation of cracks.

(3) Analysis of Heat Flux Density:

Analysis of the heat flux density on the fire-facing side shows substantial differences in thermal load between the fins and the adjacent tubes, leading to a significant temperature difference between the fins and the shoulder of the tube. The fins and their associated temperatures and stresses are elevated across various operating conditions of the water-cooled wall, making them high-risk areas for transverse cracks.

(4) Localized Heat Transfer Deterioration:

When localized heat transfer deterioration occurs, there is a notable increase in both temperature and thermal stress within a certain range of the tube wall. The most significant impact is observed at the fire-facing vertex. When the heat transfer coefficient in the deterioration area drops to 2 kW/(m²·K) or below, yielding deformation occurs at the fire-facing vertex under various thermal load conditions, which can lead to material damage.

6. References

[1] Liu C 2024 *Analysis, Research, and Countermeasures on the Failure Causes of Water-Cooled Walls in 600MW Subcritical Controlled Circulation Boilers* (Nanjing: Southeast University) Master's Thesis.

[2] Zhang Y, Zhang L, Li, G and Duan Z 2024 Analysis of Cracking Failures in Water-Cooled Wall Fins and Thermal Peaking Unit Mitigation Strategies *IOP Publishing Ltd.*

[3] Jin Y, Zhang Y, Li L and Li Y 2022 Effects of flow deviation on wall temperature in the water-cooled wall of power plant boiler *SSRN Electronic Journal*

[4] Hu Z, Chen T and Sun D 2001 Experimental study on the heat transfer characteristics of vertical plain tubes and internally rifled tubes in the near-critical and supercritical pressure regions *Journal of Thermal Science and Power Engineering* **2001(03)** 267-270+348.

[5] Li C 2010 *Numerical Calculation and Research on the Temperature of Water-Cooled Wall Tubes in Supercritical Boilers* (Hebei: North China Electric Power University) Master's Thesis.
[6] Wu H, Wang L and Wei P 2023 Study on the corrosion mechanism of water-cooled walls under over-temperature conditions during deep peak shaving of power station boilers *Journal of Xi'an Jiaotong University* **57(11)** 118-129.
[7] Hou Z, Huang X and Gao Y 2023 Experimental investigation on heat transfer of supercritical water flowing in a horizontal tube with axially non-uniform heating *International Journal of Heat and Mass Transfer*.
[8] Rezini D, Baki T 2023 Effect of thermal barriers on the martensitic transformation in a bar jominy *Strojnícky časopis - Journal of Mechanical Engineering* **73(2)** 133-144.
[9] Sako N, Noda K and Hayashi J 2023 Liquid film and heat transfer characteristics during superheated wall cooling via pulsed injection of a liquid jet *International Journal of Heat and Mass Transfer*.

Suspension Parameter Optimization and Simulation of Wheel-Edge Electric Vehicle

Peilei Shi* and Guanjun Xu

Anhui Technical College of Mechanical and Electrical Engineering, Wuhu 241000, Anhui, China.

Email: *674792327@qq.com; 0125000630@ahcme.edu.cn

Abstract. The wheel-edge drive mechanism is a forward-looking technology of new energy electric vehicles, but the wheel-edge drive mechanism increases the unsprung mass due to the addition of the hub drive motor and the planetary wheel reducer, which deteriorates vehicle smoothness and grounding. In light of the problems above, the worm gear two-stage reducer wheel edge drive mechanism is adopted. The 1/4 vehicle vibration model was established in MATLAB/Simulink and simulated for improving the ride comfort of the vehicle and the grounding of the wheel, based on the considerations of the influence of damping coefficient, suspension stiffness, and unsprung mass on smoothness and grounding. The stiffness and damping coefficient of the suspension was optimized in consideration of the calculation results and relevant evaluation indicators. The results indicate that the optimization of suspension system parameters effectively reduces the impact of vertical vibration on vehicle ride comfort.

Keywords. Wheel-Edge Drive; suspension; ride comfort; parameter optimization.

1. Introduction

Currently, the study of the comfort impact of wheel-drive electric vehicles has become a focus of research on the performance of new energy vehicles [1,2]. There are three measures to optimize the vertical performance of automobiles: firstly, material lightweight including engineering plastics and combined materials, reduces the unsprung mass, and improves the smoothness of automobiles. The second is the improvement of mechanical structure, by optimizing the design of the structure and changing the arrangement of the structure, making the ratio of spring mass to spring mass more reasonable. The third is to optimize the suspension parameters, by optimizing the stiffness and damping values of the suspension springs, to decrease the effects of the wheel drive structure on the vertical performance of the vehicle [3,4]. Parameter optimization is a method of achieving design goals by parameterizing the design objectives, using optimization methods, continuously adjusting design variables, and making the design results approach the parameterized target values. This article uses a genetic algorithm to optimize the suspension parameters of a newly designed wheel drive system structure and compares and simulates the vehicle vibration characteristics before and after the improvement of the wheel drive deceleration structure and the optimization of the suspension parameters.

2. Established 1/4 DOF Vertical Vibration Model

1/4 three DOF vertical vibration model was established according to the structure of the decelerating mechanism driven by the worm wheel edge, as shown in figure 1, and the differential equation of motion was derived. In the figure, the wheel mass is m_1, the body mass is m_2, and the integrated mass of the deceleration mechanism is m_3; The suspension damping is C_2, the damping between the reducer and the frame is C_3, and the damping between the reducer and the cantilever is C_4; The wheel vertical displacement is Z_1. The body's vertical displacement is Z_2. The vertical displacement of the reducing mechanism is Z_3. The tire stiffness is k_1. The suspension stiffness is k_2. The spring stiffness is k_3. The stiffness of the connecting element is k_4. Road surface excitation is denoted by q [5].

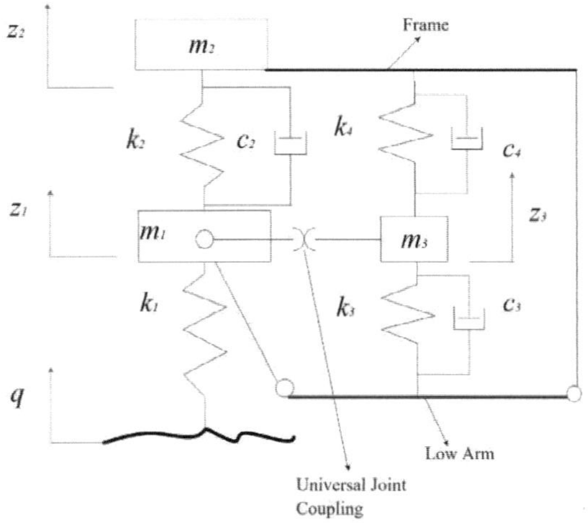

Figure 1. 1/4 DOF vehicle vertical vibration system model.

According to the 1/4 DOF vehicle vertical vibration model, the differential equation of its dynamic motion is established by the Lagrange equation:

$$\frac{d}{dt}\left(\frac{\partial T}{\partial \dot{q}_i}\right) - \frac{\partial T}{\partial q_i} + \frac{\partial U}{\partial q_i} + \frac{\partial D}{\partial \dot{q}_i} = Q_i \tag{1}$$

$$T = \frac{1}{2}m_1\dot{z}_1^2 + \frac{1}{2}m_2\dot{z}_2^2 + \frac{1}{2}m_3\dot{z}_3^2 \tag{2}$$

$$U = \frac{1}{2}k_1 z_1^2 + \frac{1}{2}k_2(z_2 - z_1)^2 + \frac{1}{2}k_3(z_3 - z_2)^2 + \frac{1}{2}k_4(z_3 - z_2)^2 \tag{3}$$

$$D = \frac{1}{2}c_2(\dot{z}_2 - \dot{z}_1)^2 + \frac{1}{2}c_3(\dot{z}_3 - \dot{z}_2)^2 + \frac{1}{2}c_4(\dot{z}_3 - \dot{z}_2)^2 \tag{4}$$

The above four equations can derive the dynamic differential equation of automotive vertical vibration system as follows:

$$\begin{cases} m_1\ddot{z}_1 + c_2(\dot{z}_1 - \dot{z}_2) + k_2(z_1 - z_2) + k_1 z_1 = k_1 q \\ m_2\ddot{z}_2 + c_2(\dot{z}_2 - \dot{z}_1) + c_3(\dot{z}_2 - \dot{z}_3) + c_4(\dot{z}_2 - \dot{z}_3) + k_2(z_2 - z_1) + k_3(z_2 - z_3) + k_4(z_2 - z_3) = 0 \\ m_3\ddot{z}_3 + c_3(\dot{z}_3 - \dot{z}_2) + c_4(\dot{z}_3 - \dot{z}_2) + k_3(z_3 - z_2) + k_3(z_3 - z_2) = 0 \end{cases} \tag{5}$$

The dynamic differential equation is obtained by Laplace transformation:

$$\begin{cases} z_1(m_1 s^2 + c_2 s + k_2 + k_1) = z_2(c_2 + k_2) + k_1 q \\ z_2(m_2 s^2 + c_2 s + c_3 s + c_4 s + k_2 + k_3 + k_4) = z_1(c_2 s + k_2) + z_3(c_3 s + c_4 s + k_3 + k_4) \\ z_3(m_3 s^2 + c_3 s + c_4 s + k_3 + k_4) = z_2(c_3 s + c_4 s + k_3 + k_4) \end{cases} \tag{6}$$

According to the above equation:

$$\frac{z_2}{z_1}=\frac{A_1A_3}{A_3A_4-{A_2}^2} \tag{7}$$

$$\frac{z_3}{z_2}=\frac{A_2}{A_3} \tag{8}$$

$$\frac{z_1}{q}=\frac{k_1(A_3A_4-{A_2}^2)}{A_5A_3A_4-A_5{A_2}^2-{A_2}^2A_3} \tag{9}$$

$$\frac{z_2}{q}=\frac{k_1A_1A_3}{A_5A_3A_4-A_5{A_2}^2-{A_2}^2A_3} \tag{10}$$

where, $A_1=c_2s+k_2, A_2=c_3s+c_4s+k_3+k_4, A_3=m_3s^2+c_3s+c_4s+k_3+k_4, A_4=m_2s^2+c_2s+c_3s+c_4s+k_2+k_3+k_4, A_5=m_1s^2+c_2s+k_2+k_1$

Transfer-Function $H_1(s)$ of vehicle acceleration$\ddot{z}_2$ to wheel velocity $\dot{q}$:

$$H_1(s)=\frac{\ddot{z}_2}{\dot{q}}=\frac{s^2z_2}{sq}=(s+0)\frac{z_2}{q}=(s+0)\frac{k_1A_1A_3}{A_5A_3A_4-A_5{A_2}^2-{A_2}^2A_3} \tag{11}$$

Transfer-Function $H_2(s)$ of wheel dynamic load F_d/G to q:

$$H_2(s)=\frac{z_1-q}{q}\left(\frac{k_1}{(m_1+m_2+m_3)g}\right)=\left[\frac{k_1(A_3A_4-{A_2}^2)}{A_5A_3A_4-A_5{A_2}^2-{A_2}^2A_3}-1\right]\left(\frac{k_1}{(m_1+m_2+m_3)g}\right) \tag{12}$$

Transfer-Function $H_3(s)$ of suspension dynamic deflection to $\dot{q}$:

$$H_3(s)=\frac{f_d}{\dot{q}}=\frac{z_2-z_1}{q}=\frac{z_2}{q}-\frac{z_1}{q}=\frac{k_1(A_1A_3-A_3A_4+{A_2}^2)}{A_5A_4A_3-A_5{A_2}^2-A_3{A_2}^2} \tag{13}$$

3. Optimization of Suspension Parameters

3.1. Established Optimization Objective Function

Generally, vehicle performance is evaluated by body acceleration, dynamic deflection of suspension spring, and tire dynamic load [6]. Therefore, when optimizing suspension parameters, the smoothness index of the vehicle is still used as the basis for reference. The parameters generally optimized for suspension parameter optimization are stiffness K and damping C. In this paper, stiffness K and damping C are taken as variables, and according to the indexes of vehicle ride comfort, an objective function is established based on the minimum sum of the root-mean-square value of the power spectral density of body acceleration, dynamic deflection of suspension spring and tire dynamic load. Then the objective function is as follows:

$$j_{min}=\sigma_{\ddot{z}_2}+\sigma_{f_d}+\sigma_{F_d/G} \tag{14}$$

Where, the root-mean-square value of the power spectral density of vehicle acceleration:

$$\sigma_{\ddot{z}_2}=\sqrt{4\pi G_q(n_0){n_0}^2u\int_0^\infty |H(j\omega)|^2{}_{\ddot{z}_2\sim\dot{q}}\,df} \tag{15}$$

The root-mean-square value of power spectral density of dynamic deflection of suspension spring:

$$\sigma_{f_d}=\sqrt{4\pi G_q(n_0){n_0}^2u\int_0^\infty |H(j\omega)|^2{}_{f_d\sim\dot{q}}\,df} \tag{16}$$

Root-mean-square e value of power spectral density of wheel dynamic load:

$$\sigma_{F_d/G}=\sqrt{4\pi G_q(n_0){n_0}^2u\int_0^\infty |H(j\omega)|^2{}_{F_d/G\sim\dot{q}}\,df} \tag{17}$$

Where, $G_q(n_0)$ is the road roughness coefficient, n_0 is the reference spatial frequency, n_0 is 0.1m^{-1}, u is speed, $|H(j\omega)|_{\ddot{z}_2\sim\dot{q}}$ is the amplitude-frequency characteristics of vehicle acceleration, $|H(j\omega)|_{f_d\sim\dot{q}}$ is the amplitude-frequency characteristics of suspension dynamic deflection, $|H(j\omega)|_{F_d/G\sim\dot{q}}$ is the amplitude-frequency characteristics of wheel dynamic load [7,8].

3.2. Optimization Variables and Constraints of the Objective Function

The optimization parameters in this chapter include the stiffness k_2 and damping C_2 of the suspension system, as well as the stiffness k_3 and k_4, damping C_3 and C_4 of the connection between the reducer and motor and the frame. Determined the constraint conditions based on the practical scope of inherent frequency f_0, static deflection f_s, limit travel $[f_d]$, and damping ratio ξ of the suspension system in automotive theory, as illustrated in table 1.

Table 1. Practical scope of the value of f_0, f_s, $[f_d]$, ξ.

Vehicle model	f_0/Hz	f_s/cm	$[f_d]$/cm	ξ
Car	1.2~1.1	15~30	7~9	0.2~0.4
Truck	2~1.5	6~11	6~9	
Bus	1.8~1.2	7~15	5~8	
SUV	2~1.3	6~13	7~13	

Determined the constraint conditions according to the value range of f_0, f_s, $[f_d]$andξ of the car in the table above, the constraint conditions are set as follows:

$$1.1 \leq f_0 \leq 1.2 \tag{18}$$

$$15 \leq f_s \leq 30 \tag{19}$$

$$7 \leq [f_d] \leq 9 \tag{20}$$

$$0.2 \leq \xi \leq 0.4 \tag{21}$$

Where, $f_0 = \frac{1}{2\pi}\sqrt{\frac{K}{m_2}}, \xi = \frac{C}{2\sqrt{m_2 K}}$.

The constraints for optimizing the objective function obtained by calculating equations (18)-(21) are as follows:

$$1.1 \leq \frac{1}{2\pi}\sqrt{\frac{k_2+k_3+k_4}{m_2+m_3}} \leq 1.2 \tag{22}$$

$$15 \leq \sigma_{f_d} \leq 30 \tag{23}$$

$$7 \leq \sigma_{F_d/G} \leq 9 \tag{24}$$

$$0.2 \leq \frac{c_2+c_3+c_4}{2\sqrt{(m_2+m_3)(k_2+k_3+k_4)}} \leq 0.4 \tag{25}$$

3.3. Optimization Parameter

Using the calculation rules of the genetic algorithm, the optimization calculation of the objective function is optimized, according to the constraint condition of the optimization objective function [9-11]. Before programming, set the population size to 100, set the probability of crossover in the population to 0.7, set the probability of mutation in the population to 0.3, and set the maximum allowed number of evolutions in the population to 100. The optimized parameter values for stiffness and damping calculated by programming in MATLAB are shown in table 2.

Table 2. Vehicle model parameters.

Parameter	Unoptimized parameter values	Optimized parameter values
$k_2\,N/m$	20000	18587
$k_3\,N/m$	10000	17365
$k_4\,N/m$	12000	0
$c_2\ N\cdot s/m$	2000	2134
$c_3\ N\cdot s/m$	1000	1862
$c_4\ N\cdot s/m$	1200	0

4. Simulation of Optimized Vibration Model

4.1. Establishment of Optimization Model

According to the optimized stiffness and damping values shown in table 1, the optimized vertical vibration dynamic model of 1/4 vehicle is established, as illustrated in figure 2.

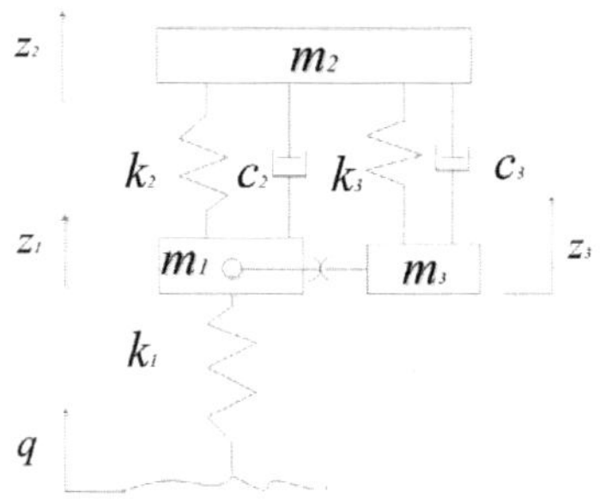

Figure 2. The optimized vertical vibration Dynamic Model.

According to figure 2, obtain the differential equation of dynamic vibration:

$$\begin{cases} m_2\ddot{z}_2 + c_2(\dot{z}_2 - \dot{z}_1) + c_3(\dot{z}_2 - \dot{z}_3) + k_2(z_2 - z_1) + k_3(z_2 - z_3) = 0 \\ m_1\ddot{z}_1 + c_2(\dot{z}_1 - \dot{z}_2) + k_2(z_1 - z_2) + k_1(z_1 - q) = 0 \\ m_3\ddot{z}_3 + c_3(\dot{z}_3 - \dot{z}_2) + k_3(z_3 - z_2) = 0 \end{cases} \tag{26}$$

Perform Laplace transform on equation (26):

$$\begin{cases} (m_2s^2 + c_2s + c_3s + k_2 + k_3)z_2 - (c_2s + k_2)z_1 - (c_3s + k_3)z_3 = 0 \\ (m_1s^2 + c_2s + k_1 + k_2)z_1 - (c_2s + k_2)z_2 = k_1q \\ (m_3s^2 + c_3s + k_3)z_3 = (c_3s + k_3)z_2 \end{cases} \tag{27}$$

Let s=jω, organized equation (27), we can obtain:

$$\frac{z_2}{z_1} = \frac{A_3A_4}{A_1A_3 - {A_5}^2} \tag{28}$$

$$\frac{z_3}{z_2} = \frac{A_5}{A_3} \tag{29}$$

$$\frac{z_1}{q} = \frac{k_1(A_1A_3 - {A_5}^2)}{A_1A_2A_3 - A_2{A_5}^2 - A_3{A_4}^2} \tag{30}$$

$$\frac{z_2}{q} = \frac{k_1A_3A_4}{A_1A_2A_3 - A_2{A_5}^2 - A_3{A_4}^2} \tag{31}$$

Where, $A_1 =- m_2\omega^2 + c_2j\omega + c_3j\omega + k_2 + k_3$, $A_2 = -m_1\omega^2 + c_2j\omega + k_1 + k_2$, $A_3 = -m_3\omega^2 + c_3j\omega + k_3$, $A_4 = c_2j\omega + k_2$, $A_5 = c_3j\omega + k_3$

4.2. Frequency Domain Simulation

Amplitude-frequency characteristics of vehicle acceleration $\ddot{z}_2$ to $\dot{q}$:

$$|H(j\omega)|_{\ddot{z}_2\sim\dot{q}} = \left|\frac{\ddot{z}_2}{\dot{q}}\right| = \omega\left|\frac{z_2}{q}\right| = \omega\left|\frac{k_1A_3A_4}{A_1A_2A_3-A_2A_5{}^2-A_3A_4{}^2}\right| \tag{32}$$

Amplitude-frequency characteristics of suspension dynamic deflection f_d to $\dot{q}$:

$$|H(j\omega)|_{f_d\sim\dot{q}} = \left|\frac{z_2-z_1}{q}\right| = \left|\frac{z_2}{q} - \frac{z_1}{q}\right| = \left|\frac{k_1(A_3A_4-A_1A_3+A_5{}^2)}{A_1A_2A_3-A_2A_5{}^2-A_3A_4{}^2}\right| \tag{33}$$

Amplitude-frequency characteristics of wheel dynamic load F_d/G to q:

$$|H(j\omega)|_{F_d/G\sim q} = \left|\frac{F_d}{Gq}\right| = \left|\frac{z_1-q}{q}\frac{k_1}{(m_1+m_2+m_3)g}\right| = \left|\left(\frac{k_1(A_1A_3-A_5{}^2)}{A_1A_2A_3-A_2A_5{}^2-A_3A_4{}^2} - 1\right)\frac{k_1}{(m_1+m_2+m_3)g}\right| \tag{34}$$

For the purpose of studying the effect of vertical vibration on comfort and comparing the effects of unoptimized and optimized suspension parameters, the Curve variation chart of the root mean square values of power spectral density of vehicle acceleration, suspension dynamic deflection, and relative dynamic load were simulated in MATLAB, as illustrated in figures 3-5. Model one in the graph is the existing traditional vibration model simulation curve, model two in the graph is the unoptimized worm and worm wheel system vibration model simulation curve, and model three in the graph is the optimized worm and worm wheel system vibration model simulation curve.

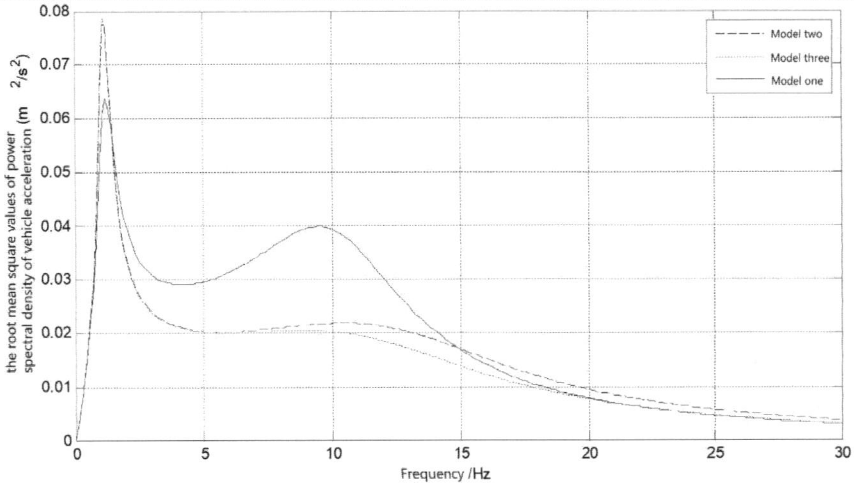

Figure 3. The curve variation chart of the root-mean-square values of power spectral density of vehicle acceleration.

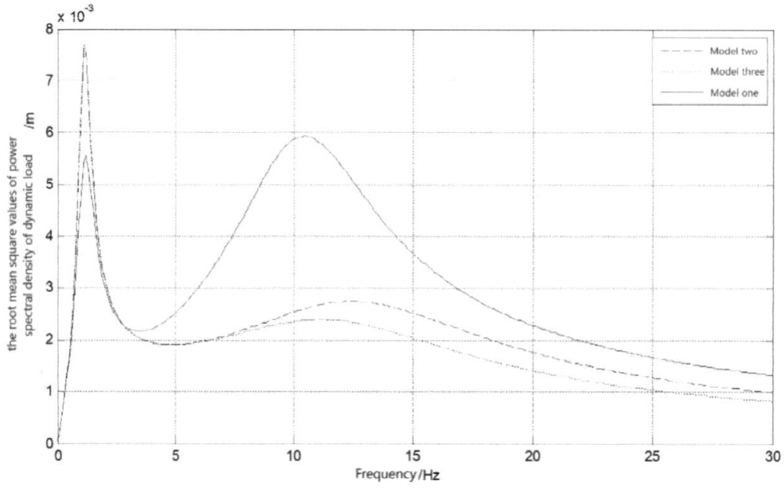

Figure 4. The curve variation chart of the root-mean-square values of power spectral density of relative dynamic load.

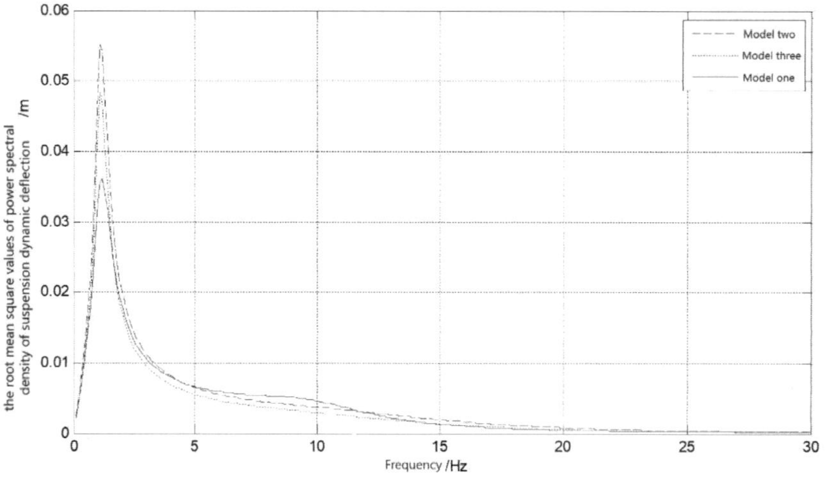

Figure 5. The curve variation chart of the root-mean-square values of power spectral density of suspension dynamic deflection.

From the analysis of the curve changes in figures 3-5, concluded that the worm wheel edge drive system electric vehicle has significantly reduced the main frequency vibration amplitude in terms of the body vertical acceleration, the tire dynamic load, and suspension dynamic deflection compared to the existing traditional Wheel-Edge drive system electric vehicle. In figure 3, the optimized vibration model has a smaller vibration amplitude than the unoptimized vibration model when the excitation frequency is around 7Hz, and its peak vibration frequency of 10Hz is lower than the peak vibration frequency of 11.6Hz of the unoptimized vibration model. In figure 4, the optimized vibration model shows a significant reduction in vibration amplitude compared to the unoptimized vibration model when the excitation frequency is around 7.4Hz. Its peak vibration frequency of 10.8Hz is lower than the peak vibration frequency of 12.5Hz of the unoptimized vibration model. In figure 5, the vibration amplitude and peak vibration frequency of the optimized vibration model are slightly lower than the unoptimized vibration model. According to the simulation results analysis, the vertical vibration performance of the suspension of the optimized parameter is better than that unoptimized, effectively improving ride comfort.

4.3. Time Domain Simulation

In order to better and more comprehensively analyze the influence of vertical vibration on vehicle ride comfort, a simulation model was built in Simulink according to the optimized vertical vibration dynamic equation of equation (26). Obtained the curve changes of wheel dynamic load *Ft*, body vertical acceleration $\ddot{z}_2$, body vertical velocity $\dot{z}_2$, wheel vertical acceleration $\ddot{z}_1$, and wheel vertical velocity $\dot{z}_1$ before and after optimization in the simulation software, as shown in figures 6-10. The black curve is the optimized simulation curve, the red curve is the unoptimized simulation curve, and the green curve is the existing simulation curve in the picture.

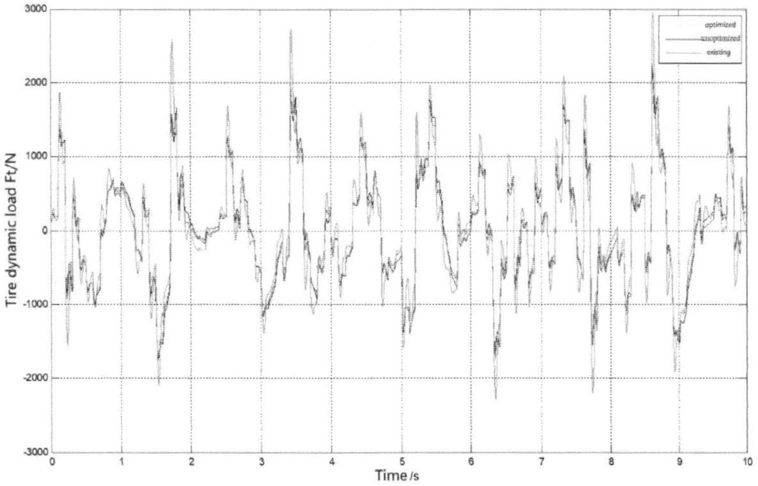

Figure 6. Tire dynamic load Ft.

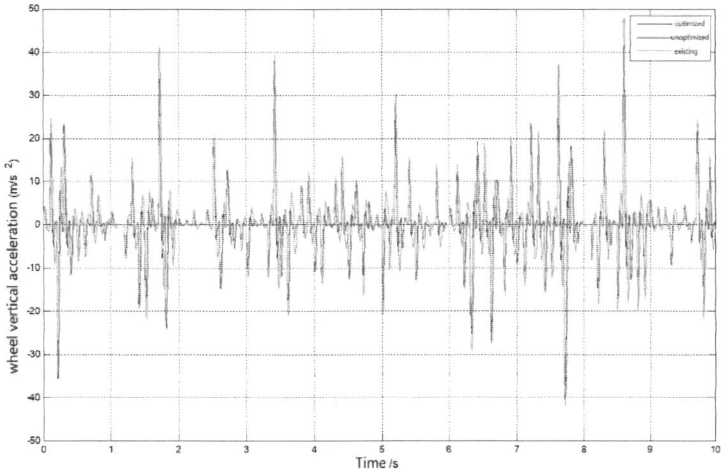

Figure 7. Wheel vertical acceleration.

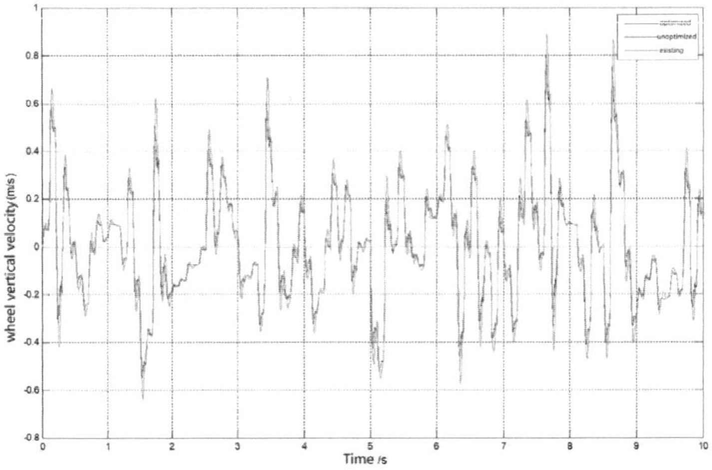

Figure 8. Wheel vertical velocity.

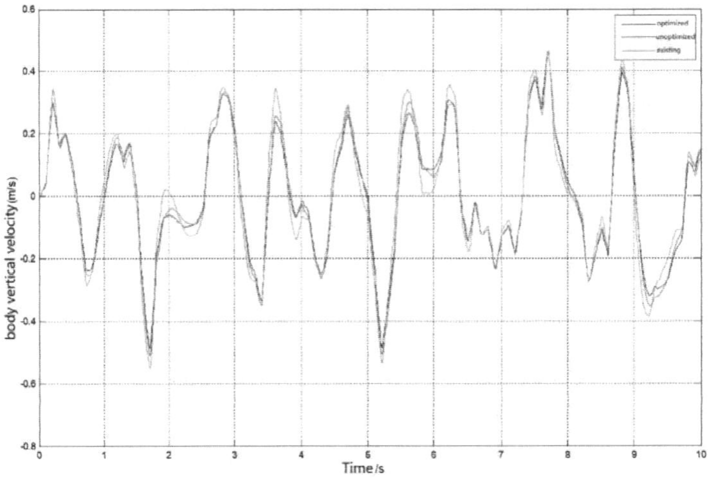

Figure 9. Body vertical velocity.

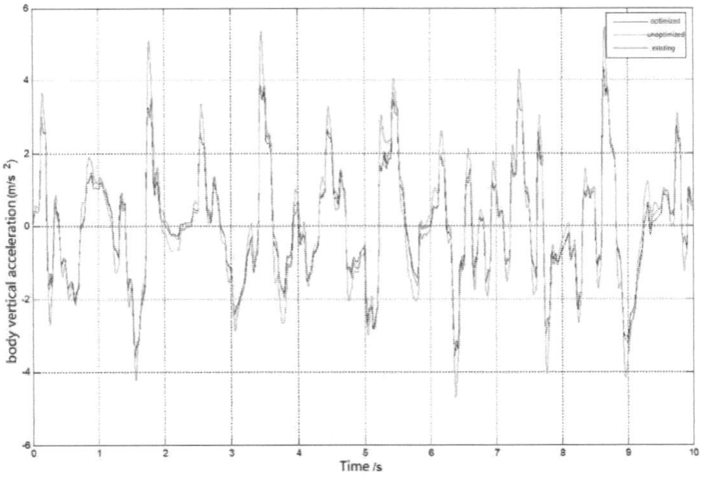

Figure 10. Body vertical acceleration.

From the simulation curve in figure 6, it has been proved that the peak value of the tire dynamic load of the wheel-edge drive electric vehicle of the optimized suspension parameter has decreased from 2254N to 2143N compared to the unoptimized. The overall change in the optimized tire dynamic load value is slightly smaller than unoptimized. The tire dynamic load values of the worm wheel edge drive structure before and after optimization are both lower than those of the existing wheel drive structure. From figures 7 and 8, it can be observed that wheel vertical acceleration and wheel velocity in the Wheel-Edge drive electric vehicle after suspension parameter optimization remain relatively unchanged compared to before optimization. However, the values of wheel acceleration and wheel velocity in the worm wheel edge drive structure before and after optimization are lower than those in the existing wheel drive structure. From this, it can be seen that optimizing suspension parameters has little effect on the vibration generated by road excitation on the wheels. From figures 9 and 10, it can be concluded that the peak vertical velocity of the Wheel-Edge drive electric vehicle after optimizing the suspension parameters has decreased from 0.523 to 0.496 compared to the peak vertical velocity before optimization, with a reduction rate of approximately 5.2%; The peak vertical acceleration of Wheel-Edge drive electric vehicle after optimizing the suspension parameters decreased from 4.2 to 3.8 compared to the peak vertical acceleration before optimization, with a reduction rate of about 9.5%. The vertical acceleration and velocity values of the car-body of the worm wheel edge drive structure before and after optimization are both lower than those of the existing wheel drive structure. From this, it has been proved that the optimized suspension vibration isolation effect is significantly improved. The above simulation results indicate that the optimization of suspension parameters effectively reduces the vertical vibration of the car, which is beneficial for improving the vehicle ride comfort.

5. Conclusion

According to the simulation comparison of vibration models before and after optimizing suspension parameters in frequency and time domains. The results show that in the frequency domain, when the excitation frequency is around 7Hz, the root-mean-square values of the power spectral density of body acceleration, dynamic deflection of suspension spring, and relative dynamic load of the optimized wheel edge electric vehicle are significantly reduced. In the time domain, the tire dynamic load, vertical acceleration, and vertical velocity of the Wheel-Edge electric vehicle are significantly reduced after optimizing the suspension parameters. In summary, the optimization of suspension system parameters effectively reduces the impact of vertical vibration on vehicle smoothness.

Journal of Physics: Conference Series **2941** (2025) 012035 doi:10.1088/1742-6596/2941/1/012035

Acknowledgment
The authors would like to thank the financial support of Anhui University Natural Science Research Project (Grant No. KJ2021A1516), Anhui University Research Project (Grant No. 2024AH050209).

6. References

[1] Chen X B, Kong A J and Wang 2024 W In-wheel drive system design with dynamic vibration absorption based on the adaptive transmission mechanism *Proceedings of the Institution of Mechanical Engineers* **238(9)** 2685-2695.
[2] Lireza H, Akbar R, Ahmed A, et al. 2024 Calculating torque, back-EMF, inductance, and unbalanced magnetic force for a hybrid electrical vehicle by in-wheel drive application *Scientific Reports* **14(1)** 12912-12912.
[3] Fadel T, Reine T and Moustapha D 2023 Multi-objective control architecture for an autonomous in-wheel driven electric vehicle *IFAC PapersOnLine* **56(2)** 11470-11476.
[4] Chang J B, Rae H C, Saurabh Y, et al. 2022 Cooling Effect of water channel with vortex generators on in-wheel driving motors in electric vehicles *Energies* **15(3)** 722-722.
[5] Shi P L and Xu L 2023 Design and simulation of two stage reduction wheel-edge drive system *Journal of Physics: Conference Series* **2463(1)**.
[6] Park J 2024 Design method for suspension parameters of high-speed railway vehicles considering uncertainties *Advances in Mechanical Engineering* **16(6)**.
[7] Chen X W, Shen L J, Hu X Y, Li G and Yao Y 2024 Suspension parameter optimal design to enhance stability and wheel wear in high-speed trains *Vehicle System Dynamics* **62(5)** 1230-1252.
[8] Park J 2023 Suspension parameters for low-damped carbody oscillation of high-speed railway vehicle *Advances in Mechanical Engineering* **15(12)**.
[9] Das P, Nayak K P and Kesavan K R 2024 Artificial neural networks (ANN)-genetic algorithm (GA) optimization on thermosonicated achocha juice: kinetic and thermodynamic perspectives of retained phytocompounds *Preparative Biochemistry & Biotechnology* 11-16.
[10] Mechouche S M, Merouane F, Addad A, et al. 2024 Enhanced biosynthesis of coated silver nanoparticles using isolated bacteria from heavy metal soils and their photothermal-based antibacterial activity: integrating Response Surface Methodology (RSM) Hybrid Artificial Neural Network (ANN)-Genetic Algorithm (GA) strategies *World Journal of Microbiology & Biotechnology* **40(8)** 252-252.
[11] İbrahim H T, Beytullah E, Oğuz K, et al. 2024 Dynamic viscosity prediction of nanofluids using artificial neural network (ANN) and genetic algorithm (GA) *Journal of the Brazilian Society of Mechanical Sciences and Engineering* **46(7)**.

Degradation of Sulfadiazine in Water Using Bismuth Molybdate Under Visible Light

Ling Xiao, Yu Li*, Jie Xin and Xuan Wang

Department of Basic Course, Naval University of Engineering, Wuhan 430033, Hubei, China.

Email: 359086326@qq.com; *liyuyuli720@163.com;35623973@qq.com; 1407051020@nue.edu.cn

Abstract. As a clean and efficient means of pollutant degradation, photocatalysis has received widespread attention and application in the degradation of wastewater pollutants and residual antibiotics, with promising prospects. In this paper, bismuth molybdate was prepared through the hydrothermal method under mild conditions, and the products were characterized. This study explored the key factors affecting the photocatalytic degradation of dimethoxy-sulfadiazine by bismuth molybdate, aiming to provide insights into the photocatalytic degradation of residual antibiotics in wastewater. The findings indicate that bismuth molybdate, when used as a photocatalyst with hydrogen peroxide as the oxidant, effectively degrades sulfonamide in water.
Keywords. Bismuth molybdate; sulfadiazine dimethyl oxide; photocatalysis; antibiotic degradation.

1. Introduction

In recent years, the presence of antibiotic residues in aquatic environments has garnered increasing concern. The residues of various antibiotics, their fate, and their impact on the environment have become significant areas of research. Sulfonamides, broad-spectrum antibiotics synthesized for use in clinical, veterinary, and aquaculture applications, often enter the environment in their original form or as metabolites, leading to their detection in natural waters and soil sediments [1]. These residues harm the ecosystem and pose health risks through bioaccumulation in the food chain. Managing environmental antibiotic contamination, especially of widely used agents like sulfadimethoxine (SDM), is crucial. Current water treatment strategies for sulfonamides combine physical, chemical, and biological methods [2], but these approaches often fail to degrade sulfonamides completely and efficiently. Photocatalytic oxidation, a technology that breaks down organic pollutants into carbon dioxide and water, offers a cost-effective, efficient, and environmentally friendly solution, showing promise for sulfonamide removal [3]. Yet, the application of bismuth molybdate, a novel visible light-responsive photocatalyst, in sulfonamide degradation has not been extensively researched.

Before the development of visible light-responsive bismuth-series catalysts, traditional photocatalytic materials like titanium dioxide, which respond to ultraviolet (UV) light with low quantum efficiency, limited their utility [4,5]. Bismuth-containing composite oxides, responsive to visible light, have emerged as potential photocatalytic materials [6].

Traditional water treatment methods for sulfonamides, including chemical degradation, physical adsorption and sedimentation, and biological decomposition [7], face limitations that prevent the complete and efficient degradation of sulfonamides. Conventional sewage treatment technologies involve microbial flocculation [8], adhesion [9], sedimentation [10], and microbial oxidation and decomposition.

The continuous industrial advancement has led to an increase in industrial wastewater discharge, containing complex organic chemicals that challenge standard treatment methods to meet discharge standards. The development of photocatalysts, notably titanium dioxide, has introduced an effective approach for degrading organic pollutants in wastewater [11-15].

2. Preparation of Bismuth Molybdate

2.1. Test Instruments and Reagents

The main instruments used in the study are shown in table 1, and the main reagents are shown in table 2.

Table 1. Main instruments.

Name	Model	Manufacturer
Tabletop low-speed automatic balance centrifuge	L420	Hunan Xiangyi Laboratory Instrument Development Co., Ltd
Ultrasonic cleaner	KQ-100	Kunshan Ultrasound Instruments Co., Ltd
Electric constant-temperature blast drying oven	DHG-9145A	Shanghai Heng Science Instruments Co., Ltd
Reaction kettle	LC-KH-100	Shanghai Precision Scientific Instruments Co., Ltd
Xenon lamp controller	350W	Nanjing Xujiang Electromechanical Plant
UV spectrophotometer	UV-2450	Shanghai Metash Instruments Co., Ltd
Electric furnace temperature controller	KSW	Beijing Zhongxingweiye Instruments Co., Ltd
X-ray diffraction analyzer	DX-2700BH	Dandong Haoyuan Instrument Co., Ltd
3D super depth-of-field microscope	VHX-5000	KEYENCE Corporation, Japan
Specific surface area tester	TriStar3000	Micromeritics, USA
UV-visible spectrophotometer	Shimadzu UV-3600	Henan Fude Technology Co., Ltd

Table 2. Main reagents.

Name	Specification	Manufacturer
Bismuth nitrate pentahydrate	Analytical reagent (AR)	Tianjin Kemiou Chemical Reagent Co., Ltd
Ammonium molybdate tetrahydrate	AR	Kaida Plant of the Fourth Tianjin Chemical Reagent Factory
Anhydrous ethanol	AR	Tianjin Tianli Chemical Reagent Co., Ltd
Ethylene glycol	AR	Tianjin Baishi Chemical Industry Co., Ltd
30% hydrogen peroxide	AR	Tianjin Tianli Chemical Reagent Co., Ltd
SDM	AR	Shanghai Jingchun Reagent Co., Ltd

2.2. Synthesis of Bismuth Molybdate

Using the hydrothermal synthesis method, 5 mmol of $Bi(NO_3)_3{\cdot}5H_2O$ and 0.358 mmol of $(NH_4)_6Mo_7O_{24}{\cdot}4H_2O$ were mixed in 20 mL of ethylene glycol and sonicated for 10 min. After dissolving all reagents, a solution of 60 mL anhydrous ethanol containing 0.6067 g of urea was incorporated. Following thorough stirring, this mixture was transferred to a 100 mL reaction kettle, which was then sealed and placed in an oven for constant-temperature reactions at 160 °C for 12 h. The reactor was subsequently allowed to cool to room temperature naturally, and the precipitate was collected and washed with deionized water and ethanol via centrifugation until the wash solution reached a neutral pH. Afterward, the product was dried in an oven at 60 °C for 24 h.

3. Characterization of Bismuth Molybdate

3.1. X-Ray Diffraction (XRD) Analysis

The XRD pattern of the resultant yellow powder, as shown in figure 1, revealed crystal cell parameters a = 5.502, b = 16.210, and c = 5.483, aligning with JCPDS standard card 21-0102. This confirmed that the yellow powder was bismuth molybdate Bi_2MoO_6.

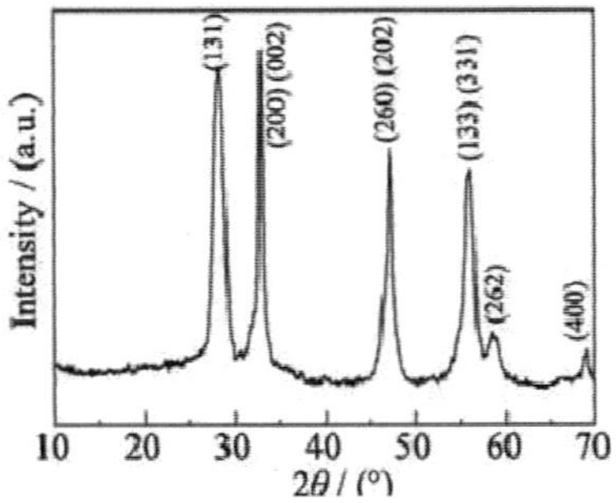

Figure 1. XRD diffraction pattern of the sample.

3.2. UV-Visible Diffuse Reflectance Spectroscopy

UV-visible (UV-vis) diffuse reflectance spectroscopy detected the light absorption capability of the yellow powder sample, as illustrated in figure 2. The analysis demonstrated that the bismuth molybdate had effective absorption in both UV and visible light spectrums, indicating strong photocatalytic activity. Reflectivity increased with wavelength, mitigating the interference from the xenon lamp in the experiment, which emitted light at wavelengths above 800 nm.

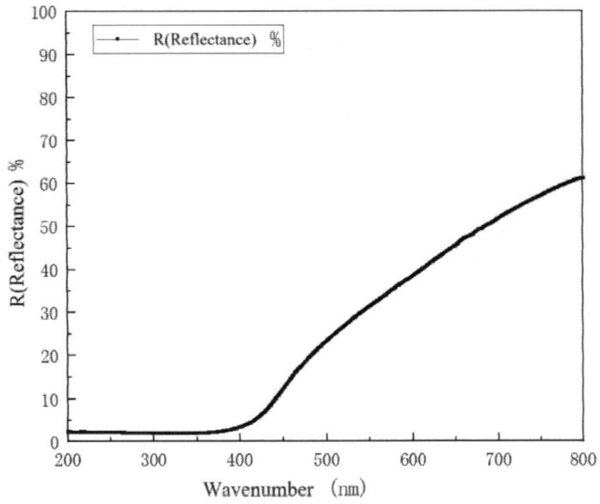

Figure 2. UV-vis diffuse reflectance curve of the bismuth molybdate sample powder.

3.3. Determination of Specific Surface Area

The specific surface area of the yellow powder was measured using the Brunauer–Emmett–Teller (BET) method, with results presented in figure 3, indicating a specific surface area of 41.9342 m^2/g.

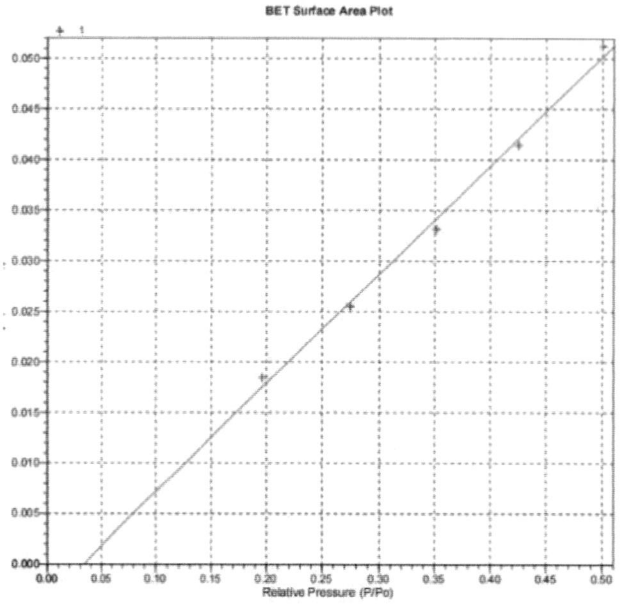

Figure 3. BET specific surface area chart.

3.4. 3D Super Depth-of-Field Microscope Observation

3D super depth-of-field microscope, at 5,000 times magnification, revealed the microstructure of the bismuth molybdate surface, shown in figure 4. The material appeared as irregular granules, with an average particle size of 3.3 μm, ranging from 1.8 to 4.7 μm.

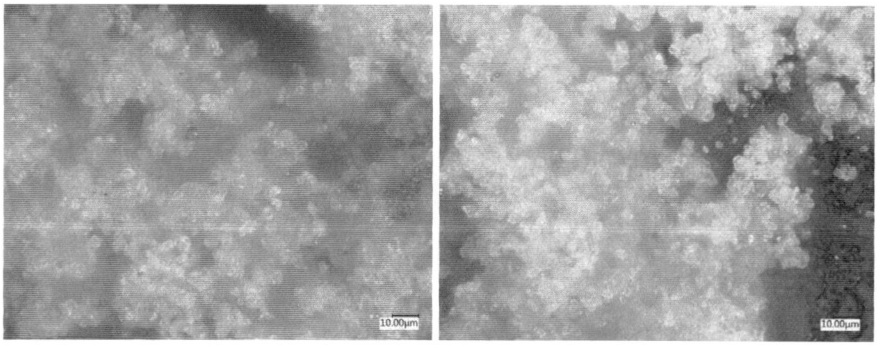

Figure 4. 3D microscopic observation images of the bismuth molybdate sample powder.

4. Research on Photocatalysis

4.1. Catalytic Degradation of SDM by Bismuth Molybdate under No Lighting

UV absorption spectra of various bismuth molybdate-SDM systems were recorded under dark conditions, and the SDM concentration at each time point was calculated. The concentration (C) versus time (T) curves, depicted in figure 5, showed minimal fluctuation in SDM concentration over time under dark conditions, indicating that bismuth molybdate does not catalyze the degradation of the antibiotic SDM without lighting.

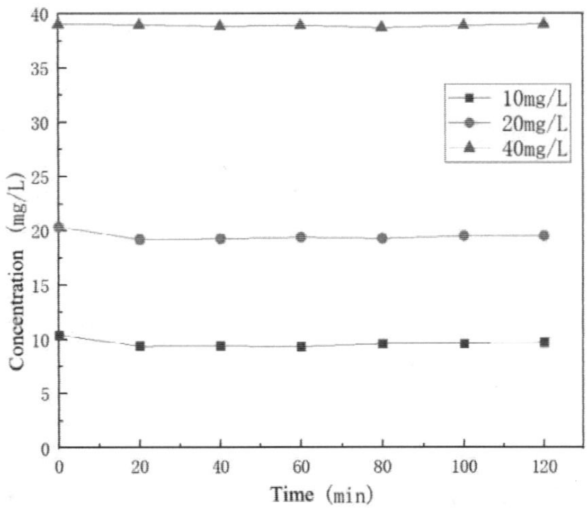

Figure 5. Degradation of SDM by bismuth molybdate under no lighting.

4.2. Catalytic Degradation of SDM by Bismuth Molybdate under Simulated Natural Light Conditions

Under simulated natural light, UV absorption spectra for SDM in various catalytic systems (including bismuth molybdate, titanium dioxide, hydrogen peroxide, and a blank group) were measured. The resulting concentrations were calculated and plotted over time in figure 6. Except for the blank group, all systems demonstrated some degree of SDM degradation, with the bismuth molybdate system showing the highest efficiency. However, the degradation rate was relatively low, reaching only up to 10% within 2h.

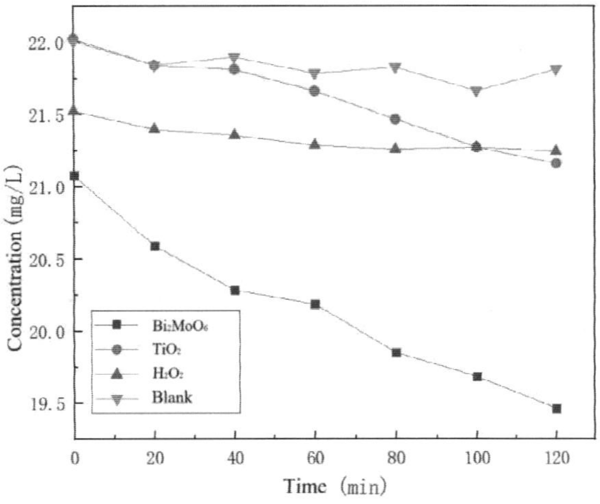

Figure 6. Effect of the catalytic system on SDM degradation efficiency.

Further experiments under simulated natural light showed that the concentration of SDM decreased over time across different bismuth molybdate-SDM systems, as illustrated in figure 6. With increasing initial concentrations of SDM, the degradation efficiency of bismuth molybdate initially increased, peaking when the SDM concentration was 20 mg/L, then slightly decreased as the SDM concentration continued to rise.

Journal of Physics: Conference Series **2941** (2025) 012036 doi:10.1088/1742-6596/2941/1/012036

5. Conclusion

(1) Bismuth molybdate was synthesized via hydrothermal method and characterized using XRD and UV-vis diffuse reflectance spectroscopy.

(2) In the absence of lighting, bismuth molybdate was ineffective in catalyzing the degradation of the antibiotic SDM.

(3) Under simulated natural light conditions, bismuth molybdate demonstrated a significant catalytic effect on SDM degradation.

6. References

[1] Li H, Zhou L, Li T, et al. 2012 Advanced treatment of printing and dyeing wastewater by ozone oxidation *Chemical Environmental Protection* **32(1)** 5.

[2] Chen F Z, Fu J M, Sheng G Y, et al. 1999 Research progress of organic matter degradation in water by electrocatalytic oxidation *China Water Supply and Drainage* **15(3)** 3.

[3] Zhang Y, Wang G R, Li Z H, et al. 2000 Treatment of reactive dye solution by photocatalytic oxidation *Fine Chemical Industry* **17(2)** 3.

[4] Li C C, Shen W H and Chen X Q 2009 Mechanism of photocatalytic oxidation reaction and its application in the treatment of papermaking wastewater *China Papermaking* **2009(08)** 191-197.

[5] Wei P Y, Yang Q L and Guo L 2009 Photocatalyst for bismuth oxyhalide compounds *Advances in Chemistry* **21(9)** 1734-1741.

[6] Zou Y 2016 *Preparation and Photocatalytic Performance of Photocatalysts Based on Bismuth Ferrite and Bismuth Tungstate* (Huainan: Anhui University of Science and Technology) Master's Thesis.

[7] Huang M S, Sun P and Zhu Li 2000 Preparation of microbial flocculants and their flocculation conditions *Environmental Science* **021(001)** 23-26.

[8] Wu J, Li Q B, Deng X, et al. 1998 Research progress of bioabsorption of heavy metals *Ion Exchange and Adsorption* **14(2)** 8.

[9] Feng Q X, Xie J and Liu J 2003 Biological oxidation of sediment and treatment of urban black and odorous river swell *China Water Resources* **23(006)** 2-44.

[10] Gao S and Li A C 2000 Research prospect of modern shallow sea sedimentation *Marine Science* **02(001)** 1-3.

[11] Tao Q, Lu T, et al. 2021 *Energy Chem.* **60** 351-359.

[12] Bi Y and Westerhoff P 2019 *Chemosphere* **223** 275-284.

[13] Zhao Y X, Gao J Y, et al. 2024 *Green Energy &Environment* **9** 1-6.

[14] Halling-Serensen B, Nors-Nielsen S, Lanzky P F, et al. 2013 Occurrence, fate and effects of pharmaceutical substances in the environment-a review *Chemosphere* **36** 357-393.

[15] Sivanandham V, Krishnan M, et al. 2012 *Pollution Bulletin* **64(4)** 790-800.

Synthesis and Characterization of Nitrogen-Doped Porous Carbon Materials for Enhanced Adsorption Performance

ShaSha Wang[1,*], Zhi Tian[2] and Yuelong Xu[1]

[1] Institute of Energy Resources, Hebei Academy of Sciences, Shijiazhuang 050081, Hebei, China.
[2] Hebei Baoli Engineering Equipment Corporation Limited, Hengshui 053000, Hebei, China.
Email: *517263436@qq.com; 441976122@qq.com; 408237993@qq.com

Abstract. Carbon materials are highly effective for CO_2 adsorption due to their large surface area and porosity. In this study, we developed a carbon-based material (C/N-0) featuring a high specific surface area of 1095 cm^2/g and a pore volume of 0.67 cm^3/g by utilizing a synthesis process involving polypotassium aspartic acid, chitosan, and melamine. The resulting C/N-0 material exhibits significant potential for CO_2 capture applications, demonstrating a high adsorption capacity of 3.09 mmol/g at 273 K, along with a low initial adsorption heat, making it an efficient candidate for practical CO_2 sequestration.
Keywords. Carbon materials; CO_2; adsorption capacity.

1. Introduction

The use of fossil fuels has significantly advanced the development of human society. However, the resulting substantial CO_2 emissions have far outpaced the rate of the natural carbon cycle, leading to an enhanced greenhouse effect that poses a major threat to human survival [1]. Therefore, the development of green and efficient carbon dioxide adsorbents is both urgent and crucial for addressing global warming [2].

Porous solid adsorbent materials (SCM) are required to be environmentally friendly, cost-effective, have high adsorption capacity, exhibit excellent selectivity, and require low energy consumption [3,4]. At present, extensive research is focused on developing high-performance porous solid adsorbents that meet these criteria. Porous carbon materials, due to their low cost, abundant availability, tunable structure and composition, ease of surface functionalization, high chemical and thermal stability, and robust regeneration capabilities, are particularly well-suited for CO_2 adsorption and separation, attracting significant attention from researchers [5].

Optimizing the microstructure to enhance specific surface area and refine internal pore architecture is a key strategy for improving material performance. Furthermore, studies have shown that incorporating heteroatoms or functional groups into carbon materials can substantially enhance their properties, including electrochemical performance as well as gas adsorption and separation efficiency [6,7]. The elemental composition of carbon materials can be modified through doping or surface functionalization, which introduces active sites either on the surface or within the material [8], improves surface hydrophilicity or hydrophobicity [9], and thereby enhances overall performance [10].

2. Methods

Five grams of potassium polyaspartic acid, 5 grams of chitosan, 3 grams of zinc chloride, and 4 grams of melamine were weighed and placed in a high-energy ball mill. The rotation speed was set to 600 rpm, and the mixture was milled for 1 hour to obtain the intermediate product. The intermediate product was then heated under a nitrogen atmosphere at 900°C for 1 hour, resulting in a black solid. This black solid was washed in a 1 M hydrochloric acid solution until it reached a neutral pH and then dried to produce a nitrogen-doped porous carbon material, designated as N/C. The material prepared without melamine was designated as N/C-0.

3. Results and Discussion

We performed N_2 adsorption-desorption experiments on the SCM samples, as shown in figure 1. The BET surface area and pore volume for N/C (1095 cm²/g and 1.00 cm³/g) and N/C-0 (887 cm²/g and 0.67 cm³/g) are summarized in table 1. The results indicate that all isotherms are of type IV, with a significant increase in adsorption observed in the range of $p/p_0 = 0.2$-1.0, suggesting the presence of mesopores. The overlapping curves in figure 1(a) further suggest a narrow pore size distribution within the mesoporous range [11]. These findings are corroborated by the pore size distribution data.

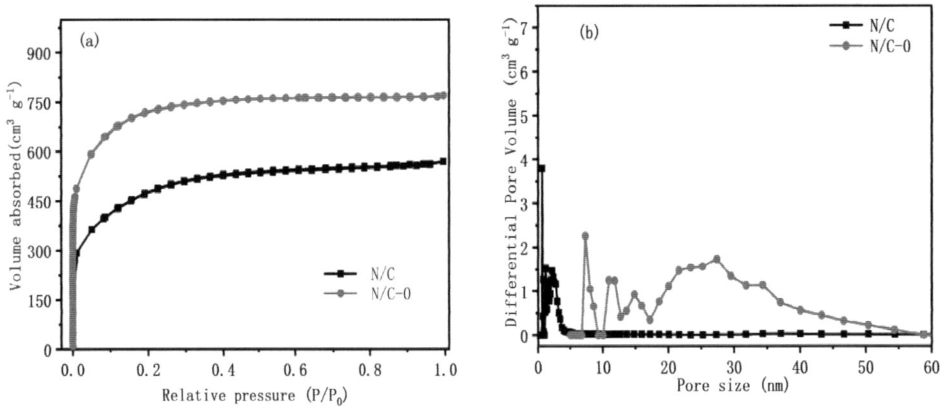

Figure 1. N_2 adsorption-desorption curves.

Table 1. Porous properties of N/C-0 and N/C.

	S_{BET} (m²/g)	S_{micro} (m²/g)	V_{total} (cm³/g)	V_{micro} (cm³/g)
N/C-0	887	215	0.67	0.18
N/C	1095	338	1.00	0.32

The SEM and TEM images in figures 2 and 3 reveal a typical three-dimensional nanonetwork structure, where nanoparticles are cross-linked and stacked to form an interconnected network. This unique architecture is characteristic of both N/C-0 and N/C. The EDS analysis confirmed the presence of O and N elements in N/C, as shown in figure 2.

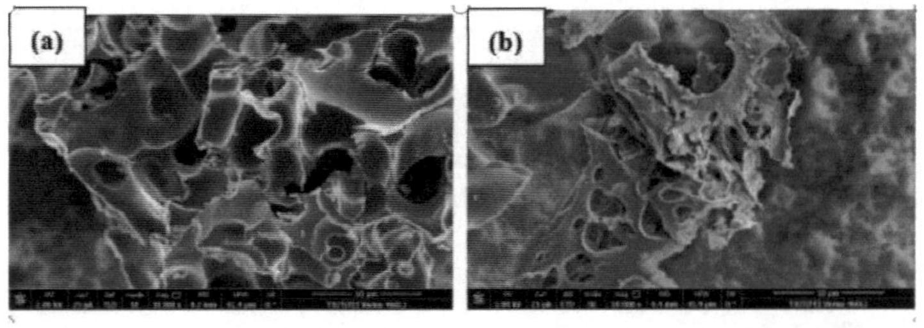

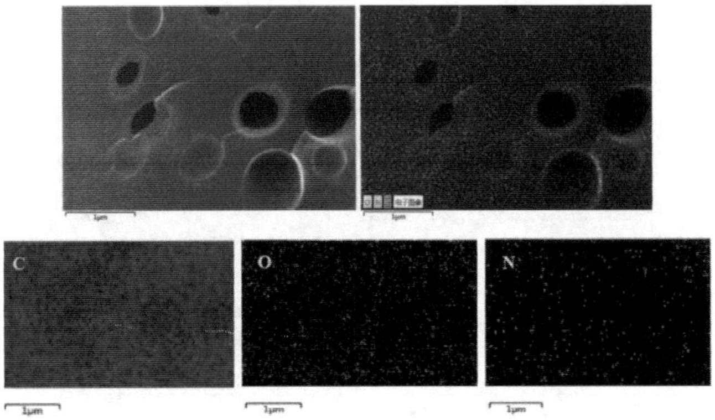

Figure 2. SEM images and mapping.

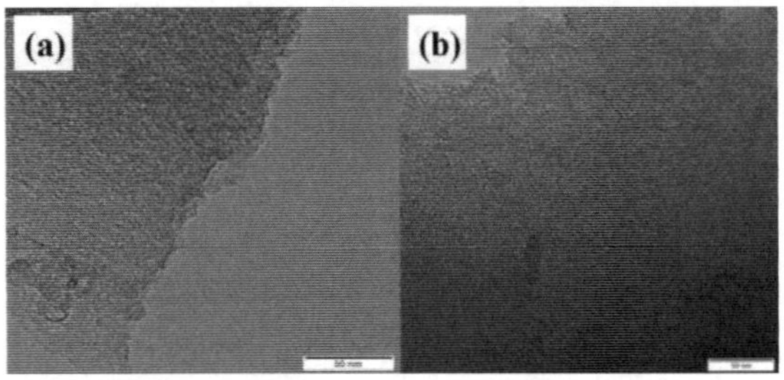

Figure 3. TEM images and mapping.

The XRD spectrum (figure 4) displays two distinct peaks at approximately 42.8° and 23.9°, corresponding to the (101) and (002) diffraction peaks of graphitic carbon, respectively [12]. N/C-0 and N/C consist of symmetrical C-C covalent bonds, allowing even slight structural changes in these materials to be detected [13]. The Raman spectra (figure 4) show characteristic peaks at 1350 cm^{-1} and 1529 cm^{-1}. The I_D/I_G ratio is presented in figure 4.

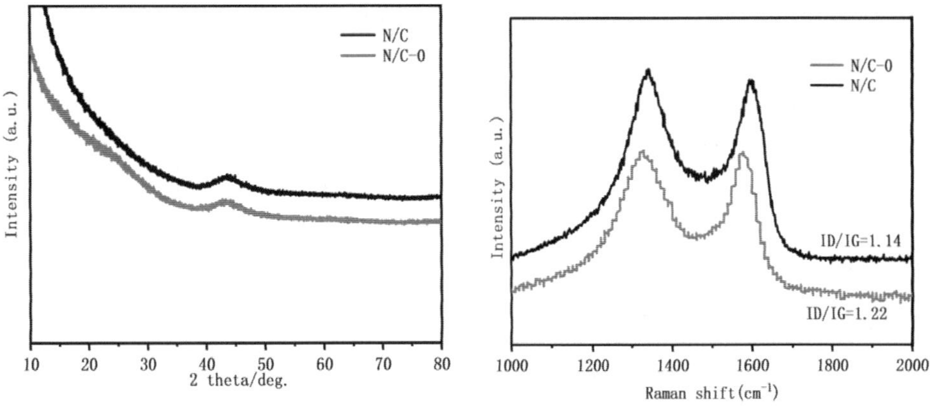

Figure 4. XRD and Raman spectra.

Figures 5(b)-5(d) display the magnified C1s, N1s, and O1s spectra, respectively. As shown in figure 5(b), the C1s spectrum of CM can be deconvoluted into four distinct component peaks [14]. Similarly, the N1s spectrum can be resolved into four peaks [15]. The O1s binding energies are also categorized into four peaks. Table 2 presents the atomic content of C, N, and O in the samples.

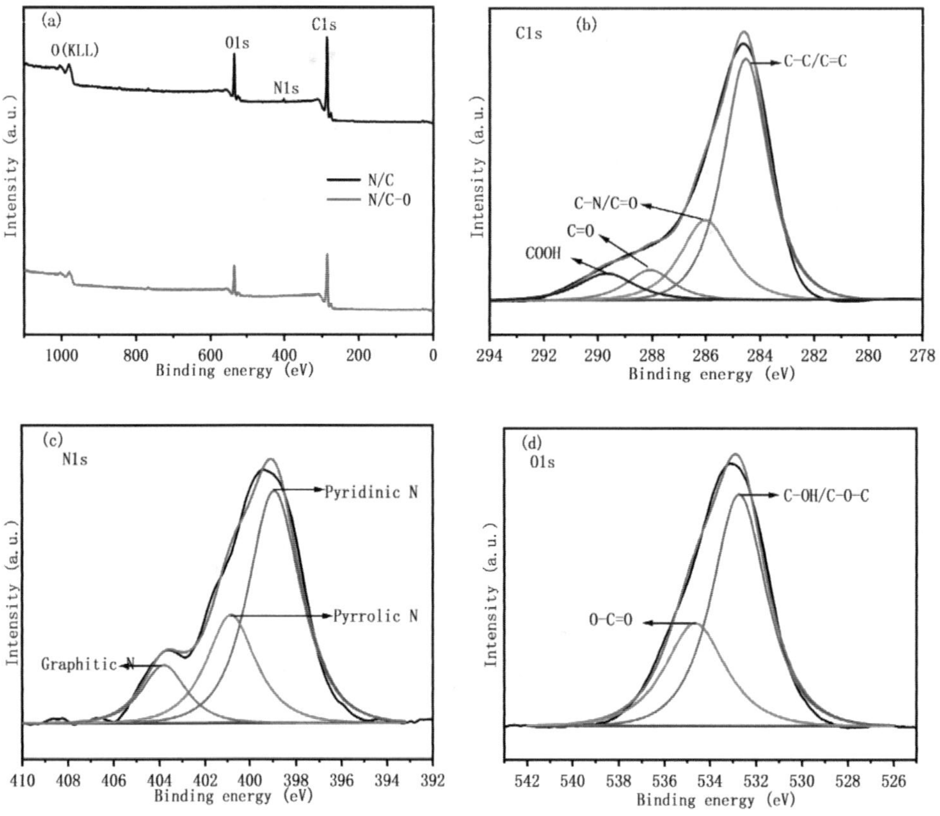

Figure 5. XPS spectrum.

Table 2. Elemental composition of C, N, and O in the samples.

	C (at%)	N (at%)	O (at%)
N/C-0	83.83	0.82	15.35
N/C	85.75	1.05	13.20

The CO_2 adsorption performance of N/C-0 and N/C is illustrated in figure 6. The absence of hysteresis in the isotherms indicates that these materials can readily release adsorbed CO_2 during the desorption process. The CO_2 adsorption isotherms exhibit a marked increase, particularly at lower pressures (0-0.2 P/P^{o}), due to the excellent polarizability and quadrupole moment of CO_2, which interact strongly with N/C-0 and N/C [16]. Figure 6 shows that the CO_2 adsorption capacities of the materials are 3.09 mmol/g at 273 K and 2.88 mmol/g at 298 K.

Figure 6 also depicts the relationship between CO_2 adsorption capacity and adsorption heat. The initial adsorption heat values for both materials range from 9 to 15 kJ/mol, significantly lower than the typical values for chemical adsorption (>60 kJ/mol) [17,18]. This suggests that N/C-0 and N/C primarily adsorb CO_2 through physical processes, offering advantages such as rapid adsorption and desorption, as well as reduced regeneration costs when used for gas adsorption.

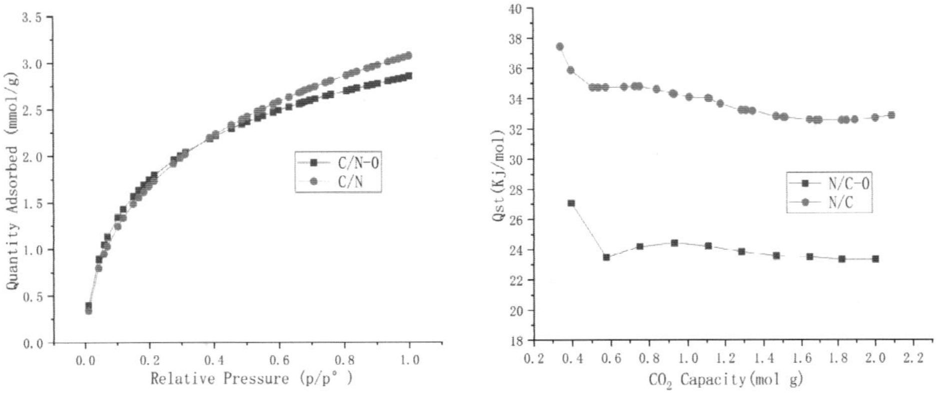

Figure 6. CO_2 adsorption performance and adsorption heat.

4. Conclusion

Using potassium aspartic acid and chitosan as raw materials, and melamine as an activator, N/C materials with a high BET surface area (1095 cm^2/g) and pore volume (1.00 cm^3/g) were produced, predominantly consisting of mesopores. Nitrogen and oxygen were doped into both N/C and N/C-0. The CO_2 adsorption capacities of N/C and N/C-0 at 273 K are 3.09 and 2.88 mmol/g, respectively. The initial adsorption heat values for both materials are significantly lower than the typical values associated with chemical adsorption.

Acknowledgments

This work was financially supported by the Hebei Academy of Sciences high-level talent training and funding project (2024G03).

5. References

[1] Bo G 2006 Adsorption of carbon dioxide on activated carbon *Journal of Natural Gas Chemistry* 223-229.
[2] Biniak S 2021 Enrichment of the surface functional groups of activated carbon by modification method *Surfaces and Interfaces* **22** 100873-100878.
[3] Kochervinskii V V 2022 The application of organic graphene-based electrodes for studies of electrophysical properties of polymer dielectrics and ferroelectrics *Ferroelectrics* **600** 59-72.
[4] Lashina E A 2023 Mathematical modeling of the homogeneous-heterogeneous non-oxidative CH_4 conversion: the role of gas-phase H or CH_3 *Reaction Kinetics and Catalysis Letters*.
[5] Huang X L 2021 Graphite felt modified-graphene oxide/graphene as novel vanadium battery electrode material *Ferroelectrics* **579** 12-22.
[6] Zhu L 2017 A carbon aerogels the removal of oils *Micro Meso Mater* **241** 285-292.
[7] Zulamita Z B 2013 Cu- and Ag-doped carbon aerogels *Mater. Chem. Phys.* **138** 870-876.
[8] Lui A P 2017 In situ catalytic fast pyrolysis of crude and torrefied *Energy* **128** 701-712.
[9] Chen C S 2020 Multi-walled carbon nanotubes enhanced the property of epoxy asphalt composite *Ferroelectrics* **565** 108-115.
[10] Liang C D 2008 Mesoporous CM: synthesis and modification *Angew. Chem. Int. Ed.* **47** 3696-3717.
[11] Hu Y J 2016 Electrode and CO_2 adsorption *RSC Adv.* **6** 15788-15795.
[12] Paul A W 2014 Adsorption technology for CO_2 separation and capture: a perspective *Adsorption* **20** 225-231.
[13] Yang J S 2013 Benzimidazole grafted polybenzimidazoles *J. Mater. Chem.* A. **4** 4768-4775.
[14] Liu D F 2012 Experimental and molecular simulation studies of CO_2 adsorption on zeolitic

imidazolate frameworks: ZIF-8 and amine-modified ZIF-8 *Adsorption* **19** 2-37.

[15] Cheng H Y 2012 A review of CO_2 capture by absorption and adsorption *Aerosol and Air Quality Research* **12** 745-769.

[16] Rashidi N A 2014 Experimental and modelling studies of carbon dioxide adsorption by porous biomass derived activated carbon *Clean Technologies and Environmental Policy* **16** 1353-1361.

[17] Ye H 2024 Construction of mannose-grafting Fe-doped carbon dots for CpG delivery and photothermal ablation of tumor *Journal of Nanoparticle Research* **26** 25-30.

[18] Mu D H 2024 Quercetin-crosslinked chitosan nanoparticles: a potential treatment for allergic rhinitis *Scientific Reports* **14** 4021-4028.

Thermodynamics Analysis of CrMnFeCoNi High-Entropy Alloys with Weak Local Segregation

Daohang Wang[1], Lanjia Yi[1], Siyu Lv[1], Xida Yu[1], Yi Wang[2], Yudong Wang[1,*]

[1] Faculty of Public Security and Emergency Management, Kunming University of Science and Technology, Kunming 650093, Yunnan, China.
[2] Zhengzhou Nonferrous Metals Research Institute Co., Ltd., Zhengzhou, Henan, China.
Email: 17468161@qq.com; 1169592198@qq.com; 3251248324@qq.com; 1398485463@qq.com; 12607199@qq.com; 584772219@qq.com; *wyd@kust.edu.cn

Abstract. To study CrMnFeCoNi High-Entropy Alloys with Weak Local Segregation, here we mixed Back Scattered Electron (BSE) and Second Electron (SE) images (BSE+SE) to find the weak local segregations, which commonly exist in CrMnFeCoNi high-entropy alloy (HEA) as-cast ingots, but the contrast of the segregations is too low to draw attention before. These segregations are located mostly in the grain interiors, indicative of homogeneous nucleation. Microstructure is investigated using Scan Electron Microscopy and Energy Dispersive X-ray Spectroscopy too. Further, the thermodynamic analysis is performed around the melting point by differential scanning calorimetry (DSC). It was observed that the melting point did not vary significantly with changes in the specimen mass and heating rate except for whether weak local segregation exists or not. The melting points of CrMnFeCoNi HEA without segregation was 1409.45±4.15°Cwhich was around 74°C higher than that with segregation, and the melting enthalpy (ΔHM) of CrMnFeCoNi HEA without segregation was 17.84±2.09 kJ/mol.
Keywords: CrMnFeCoNi high-entropy alloys; scanning electron microscopy (SEM); thermodynamics; differential scanning calorimetry (DSC); casting.

1. Introduction

The high-entropy alloys have high configurational entropies and might favor the formation of solid solutions over the precipitation of intermetallic compounds [1,2]. The equiatomic CrMnFeCoNi HEA was melted by arc [2-6] or induction [7] melting, hot rolling [3], room temperature rolling [2,4,5], room temperature swaging [7], cryogenic rolling [8], even high-pressure torsion (HPT) at room temperature [9] were processed to produces single-phase equiaxed microstructures homogeneously. Several studies have discussed the melting points of HEA ingots annealed at 800-1100°C for 1-6 h for complete recrystallization [7,9]. However, there have been few studies on these alloys in the as-cast ingot form.

Further, there have been several studies on equimolar CrMnFeCoNi HEAs, which have been confirmed to be monophasic using X-ray diffraction analysis [2], scanning electron microscopy, transmission electron microscopy [11], electron probe microanalyzer measurements [10], and energy-dispersive X-ray spectroscopy [12]. However, there have been few attempts to use DSC to show that the solid solution of these alloys is monophasic, and there have been no thermodynamic characterization studies on equimolar CrMnFeCoNi alloys, even the enthalpy of crystallization is one of the most important parameters with respect to alloy casting.

As Otto [12] pointed out the entropic stabilization of the solid solution state was limited in vulnerable temperature ranges, the stability of other HEAs currently thought to be solid solutions readily exhibit fault of weak local segregation in high temperatures. Thus, in this study, we measured the freezing temperature and enthalpy of a CrMnFeCoNi HEA, and further, confirmed the common presence of a monophase and weak local segregation in the HEA using DSC.

2. Experimental

Raw Co, Cr, Fe, Mn, and Ni metal grains with purities exceeding 99.9 wt% were used in this study. To remove the surface oxides, chemical etching with dilute nitric acid was performed on the raw metal grains before the melting process. The HEA was prepared by high-frequency electromagnetic induction melting in a water-cooled copper crucible (cold crucible melting) in the air; the melting temperature was up to 1600°C. Then, casting was performed to shape the ingots of the CrMnFeCoNi HEA. The surfaces of the ingots contained oxides and inhomogeneous components were peeled off using a lathe. After cooling, samples with a diameter of 10 mm and a thickness of approximately 1 mm were obtained by copper-based diamond saw blade.

Two points were cut on each sample at randomly chosen locations, and the obtained fragments were analyzed by X-ray fluorescence (XRF) measurements. The results were averaged. Next, the sample fragments were mixed homogeneously and subjected to chemical analysis.

Table 1. The samples and the DSC parameters.

Sample name	With/without segregation	Heating rate (°C/min)	Specimens mass (mg)
50-10gn	without	50	10.3
50-20gn	without	50	19.9
50-40gn	without	50	40.2
50-50gn	without	50	49.7
50-80gn	without	50	79.7
5-80g	with	5	80.3
20-80g	with	20	80.5
50-40g	with	50	39.6
50-50g	with	50	50.6
50-80g	with	50	79.6

The melting point and enthalpy were determined according to the ISO 11357-3 standard [13]. DSC was performed using a NETZSCH STA 449F3 thermal analyzer. First, a sample with a mass of 10 to 80 mg was placed in an alumina crucible, which was heated up to 1600°C at the rates of 10, 20, 40, or 50°C/min in an Ar purge flow. The sample was then cooled to room temperature at the same rate. All the samples and the DSC parameters are summarized in table 1. The alumina crucibles used were made by Shanghai Kai Zheng Instrument Co. Ltd. When the apparatus had cooled to the ambient temperature, the crucible was removed to confirm whether the specimen had melted completely. In order to measure the enthalpy quickly, an 80 mg specimen and a heating rate of 50°C/min were used. It should be noted that the melting transition is characterized by endothermic peaks in the DSC curve, while the crystallization process is represented by exothermic peaks. Next, heating rates of 5 and 50°C/min were used, in order to confirm the reliability of the measurement results. The area under the curves was calculated using the software program NETZSCH Proteus.

3. Results

3.1. Microstructure Analysis of Weak Local Segregations
The chemical composition of the alloy was determined by XRF measurements as well as the results of a chemical analysis performed on the alloy after the induction melting of the as-cast sample are shown in table 1. All the data indicated that the HEA samples' chemical composition is homogenous in general.

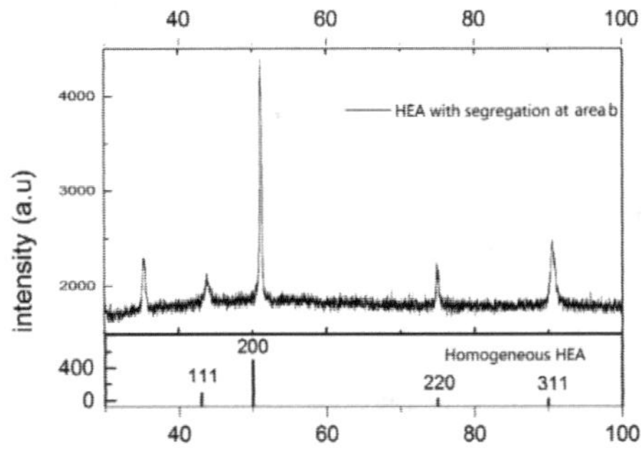

Figure 1. XRD patterns of as-cast FeCoNiCrMn HEA with and without segregation.

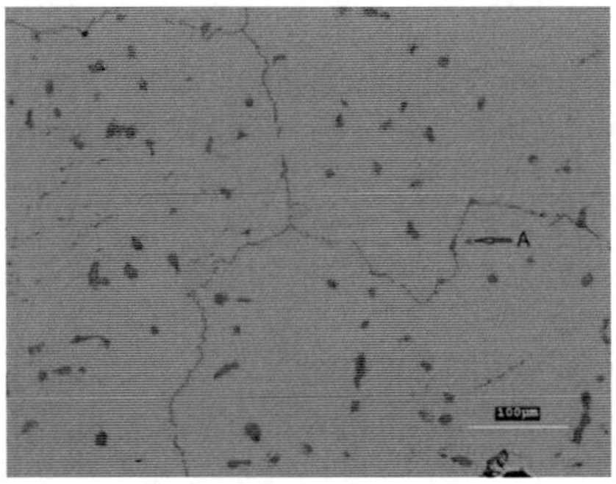

Figure 2. Microstructure of the as-cast CrMnFeCoNi HEA with weak local segregation.

X-ray diffraction patterns of the as-cast samples are shown in figure 1 The result of as-cast CrMnFeCoNi HEA measured [5] was also included for comparison. All patterns revealed an FCC single phase with some unknown object.

Figure 2 shows a representative low-magnification SEM image of the CrMnFeCoNi HEA with weak local segregations by a mixture of BSE and SE signals (BSE+SE). Many inclusions are clearly visible within the grains which have never been reported before, since the contrast of inclusions is too low to draw attention before. Indeed, the EDX point analysis, line scan, and elemental maps cannot distinguish the difference between these inclusions and the background materials, showing that the constituent elements of the HEA are homogenously distributed other than the very weak local segregations. However, the EDX zone analysis (figure 3) provided some evidence of this weak segregation (figure 4). Figure 3 represents a BSE + SE image indicating the presence of an accumulation area of spot segregations. EDX analysis was conducted at the locations marked by squares 1 and 2 (segregation and matrix, respectively). Figure 4 represents the corresponding EDX spectra. The EDX zone analysis depicted in figure 4 reveals a minor enrichment of Mn in the segregation and a slight depletion of Fe and Co (relative to their concentrations in the HEA). Cr and Ni remain unchanged.

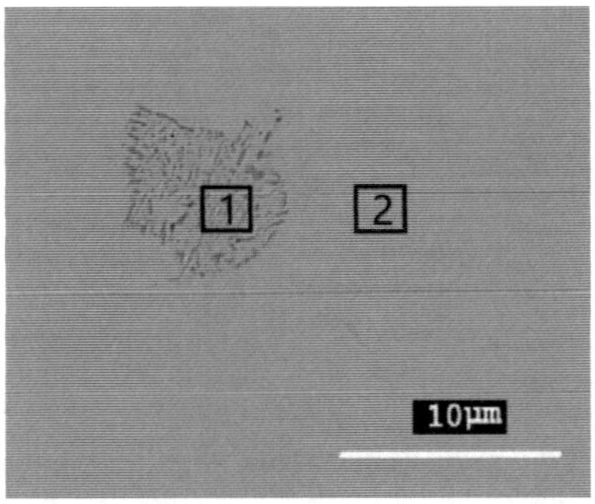

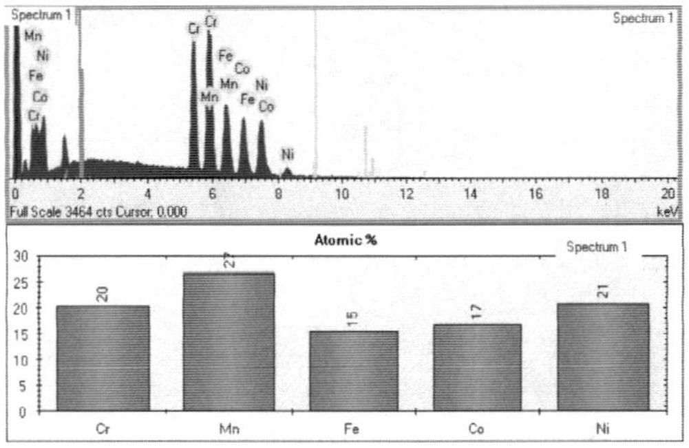

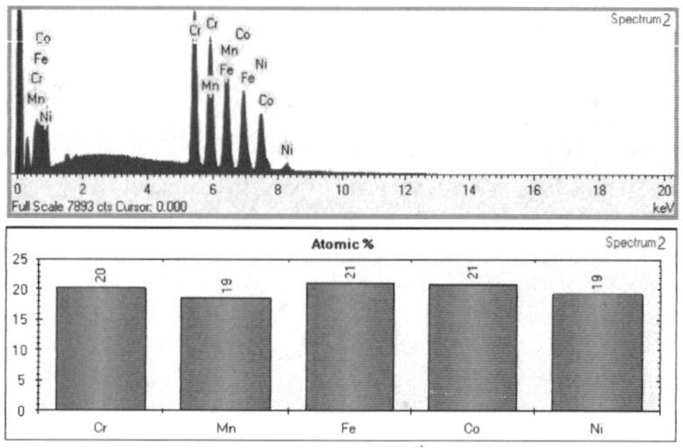

Figure 3. Typical BSE + SE image.

Figure 4. EDX zone analysis of this weak segregation

After solidification, the CrMnFeCoNi HEA grain boundaries are also visible, and coarse grains exceed 1mm in size. There is not any casting porosity, holes, or intact particles [3,5,6,9], which have been previously found in this alloy [12], indicative of high-quality casting.

The CrMnFeCoNi HEA was also characterized by DSC. This is the first instance where a single-phase structure with and without some weak local segregation, in which the constituent elements are lightly enriched or depleted, was observed as a HEA at the bulk scale. Our present results are consistent with previous EDX results at similar length scales [6] and APT results at much finer scales [9,10] which showed that the CrMnFeCoNi HEA was a single-phase solid solution.

3.2. Thermodynamic Analysis

The ISO 11357 standard suggests using samples with a mass of 5 to 10 mg for the DSC measurements. Ideally, a sample with a low mass should be used for DSC measurements, as the lower the mass, the more accurate the measurements. However, in this study, we used samples with greater masses in order to detect the tiny phase-transition-related DSC peaks, owing to the presence of small inhomogeneous.

The thermodynamic analysis includes measurements of liquidus, solidus, melting point, and melting enthalpy, figure 5 illustrates the method of measurement referred to by The ISO 11357 standard. In order to measure the melting point and enthalpy quickly, a heating rate at 50°C/min was performed up to 1500°C, the solidus (fusion onset) of CrMnFeCoNi HEA without segregation was 1387.7°C, liquidus (fusion end) was 1421.7°C, melting point (peak temperature) was 1413.6°C and melting enthalpy was 19.93kJ/mol. The DSC curves of both CrMnFeCoNi samples did not contain any individual peaks between 30 and 1387.7°C, confirming that no phase transition occurred before the melting of the alloy. Further, all the components formed a homogeneous monophasic solid solution at the bulk scale.

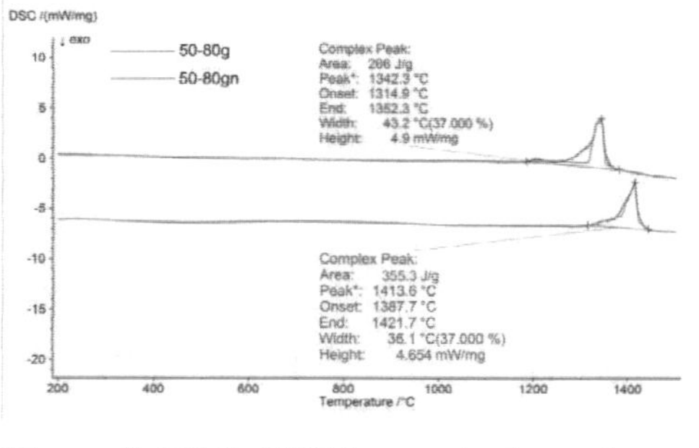

Figure 5. DSC of HEA re-melted samples V.S. that of as-cast sample.

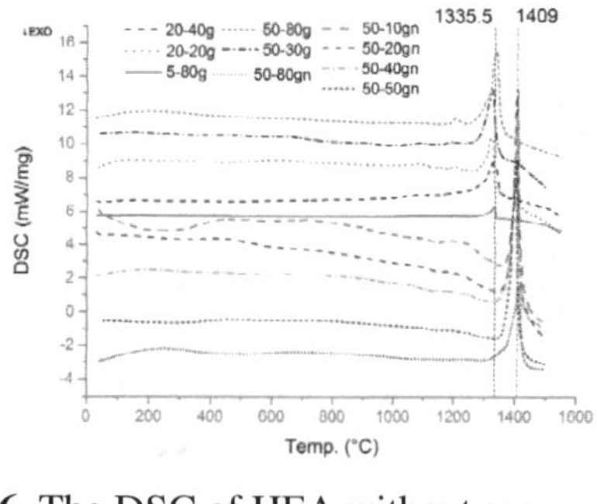

Figure 6. The DSC of HEA without segregation V.S. that of segregation.

Melting peaks were analyzed via extrapolated onset/end temperature and integrated area. The DSC of HEA re-melted samples was contrasted with that of the as-cast sample. Both scan rates were 50°C/min and the specimen mass was approximately 80mg (figure 5). Figure 6 presents the Heating DSC of HEA samples without segregation in comparison with those of samples with segregation, where the heating rate was 50°C/min. Figure 6 shows the DSC curves of the 10, 20, 40, 50, and 80 mg specimen mass, which were tested under the same conditions. All melting peaks largely at the same point of 1409.45±4.15°C which was around 74°C higher than that with segregation, and the melting enthalpy (ΔHM) of CrMnFeCoNi HEA without segregation was 17.84±2.09 kJ/mol, that proved the melting point and enthalpy of CrMnFeCoNi HEA without segregation are not influenced significantly by specimen mass.

In order to ensure that the ideal static state was reached, so that the melting-related peaks could be observed, 5 and 20°C/min heating rates were chosen to test the CrMnFeCoNi HEA with segregations. Figure 6 shows the DSC heating curves of the CrMnFeCoNi HEA with or without segregations at different heating rates and different specimen masses. All melting DSC curves showed all melting peaks of HEA with segregation largely around the same point of 1335.5°C and melting enthalpies in the same range of area 14.00 kJ/mol (table 2), which also proved the melting temperature and enthalpy of as-cast CrMnFeCoNi HEA are not influenced significantly by heating rate, because the melting activation energy of CrMnFeCoNi HEA is too small to influence the melting point significantly. As the above discussion, the DSC curve of the 80-mg sample at heating rates 50°C/min is more sensitive to small details and thus the most suitable for determining the melting properties of as-cast CrMnFeCoNi HEA.

Table 2. Compared the HEA sample with segregations to that without segregations at different specimen masses, the heating rate was 50°C/min, and purge Ar flow was 30 ml/min.

Sample name	Enthalpy of melting (J/g)	Enthalpy of melting (kJ/mol)	Fusion peak (°C)	Fusion onset (°C)	Fusion end (°C)	Width (37%) (°C)	Fusion height (mW/mg)
50-40g	270.4	15.16	1330.5	1292.9	1349.3	47.8	1.903
50-50g	227.3	12.75	1329.6	1285.7	1338.9	44.0	3.9
50-80g	266.0	14.92	1342.3	1314.9	1352.3	43.2	4.90
20-80g	257.6	14.45	1328.7	1301	1332.7	32.7	2.3
5-80g	233.0	13.07	1331.6	1316.1	1339.1	33.5	0.6
50-10gn	280.9	15.75	1410.7	1407.3	1415.2	19.4	12.01
50-20gn	326.5	18.31	1408.0	1391.5	1413.7	18.6	13.1
50-40gn	346.6	19.44	1408.6	1390.0	1417.1	23.3	11.68
50-50gn	329	18.45	1405.3	1384.3	1410.2	22.9	11.32
50-80gn	355.3	19.93	1413.6	1387.7	1421.7	36.1	4.65

4. Discussion

As above mentioned, the inducing as-cast ingots of CrMnFeCoNi HEA readily exhibit faults such as weak composition inhomogeneous and segregation which is difficult to find by common SEM analysis except for BSE+SE. In order to find a feasible and practical analysis method to evaluate weak composition segregation, the as-cast CrMnFeCoNi HEA samples with or without weak segregation were investigated by DSC. All melting peaks of HEA with segregations are distinctly different from those without segregation. The peak temperature of the HEA with segregations was substantially decreased, by 74°C. This meant that the weak local segregation could decrease the melting point significantly.

For the melting process of HEA without segregation, the fusion height and width of the melting peak are usually not equal to that with segregation, the melting peak shape is sharper than that with segregation which is more like the combination of two or multiple peaks. Stantially the melting peak of as-cast samples without segregation may be analogous to the melting peak of annealed HEA sample studded [10]. The annealing process caused a fusion temperature increase resulting from homogenizing the weak local segregation.

The consistency and repeatability of the melting process were tested using specimens with masses of 10, 20, 40, 50, and 80 mg (figure 6). In contrast to the case for the melting of HEA with segregations, the difference between the fusion peak temperatures, onset temperature, ending temperature, and width of DSC peak of all specimens mass within 0.8% (table 2), indicated that the melting points were not influenced by specimen mass and heating rates (figure 6) as the properties of HEA without segregations. On the other hand, the difference in melting enthalpies of all specimen mass is within 4% except for the 10-mg specimen which is 12% less than others. Thus, the specimen mass below 20-mg sample is too little to measure the melting enthalpy and unsuitable for determining the thermodynamic properties of CrMnFeCoNi HEA.

The homogeneous solid solution without segregations formed when the proper re-melting and annealing technique was used. The average melting enthalpy was 17.84 kJ/molt, this is greater than the enthalpy of HEA with segregation and 13 kJ/mol, which is the melting enthalpy [10]. Thus, the melting enthalpy of the homogeneous CrMnFeCoNi HEA is higher than the melting enthalpy with segregation but is more affected by specimen mass than the melting point. The weak local segregation can be detected successfully by comparing the peak temperature of HEA with segregation to that of HEA without segregation.

The results of the thermal analysis indicated that the enthalpy of melting of the face-centered-cubic single-phase solid solution was substantially different from that determined from the DSC curves of conventional alloys, for which the melting point is influenced significantly by heating rate because the melting activation energy is higher than that of HEA. The critical role of a high configuration entropy determines the multiple-element solid solution and unusual thermodynamic behaviors are expected. Zhang et al. [14] reported that the presence of atoms with different diameters results in significant distortion of the lattice compared to the case of conventional alloys, in which the concentration of the alloying elements is only a few atm%. The investigated HEA exhibited unique thermodynamic performance, especially with regard to melting behavior. Further, a high degree of distortion of the lattice means that less activation energy is needed for melting and it is more affected by segregation.

5. Conclusion

An equimolar CrMnFeCoNi alloy was produced by cold-crucible induction melting in the air. The obtained alloy was characterized by X-ray diffraction, the mixture of BSE and SE signals (BSE+SE), EDX zone analysis, and differential scanning calorimetry (DSC).

The main results are the following:

- For the very first time, a CrMnFeCoNi HEA alloy was characterized by BSE+SE and weak local segregation down to the grain scale was evidenced.
- The differential scanning calorimetry (DSC) analysis indicated that the melting point of CrMnFeCoNi HEA did not vary significantly with changes in the specimen mass and heating rate except for whether weak local segregation exists or not.
- The melting point and the melting enthalpy (ΔHM) of CrMnFeCoNi HEA without segregation were measured to be 1409.45±4.15°C and 17.84±2.09 kJ/mol respectively.
- In the future, a combination of research works, i.e. thermodynamic analysis and characterization of HEAs with various compositions will be performed to evaluate the composition and temperature range of stability of the CrMnFeCoNi system.

6. References

[1] Tang Y, et al. 2023 A review on the dynamic-mechanical behaviors of high-entropy alloys *Progress in Materials Science* **135**.

[2] Mehranpour M S, Shahmir H and Nili-ahmadabadi M 2021 Precipitation kinetics in heavily deformed CoCrFeNiMn high entropy alloy *Materials Letters* **288**.

[3] Hachet D, Gorsse S and Godet S 2021 Microstructure study of cold rolled Al0. 32CoCrFeMnNi high-entropy alloy: interactions between recrystallization and precipitation *Materials Science and*

Journal of Physics: Conference Series **2941** (2025) 012038 doi:10.1088/1742-6596/2941/1/012038

Engineering: A **802** 140452.
[4] Mehranpour M S, Shahmir H, Derakhshandeh A, Nili-Ahmadabadi M 2021 Significance of Ti addition on precipitation in CoCrFeNiMn high-entropy alloy *Journal of Alloys and Compounds* **888** 161530.
[5] Gao P, et al. 2022 Influence of Plasma Arc Current on the Friction and Wear Properties of CoCrFeNiMn High Entropy Alloy Coatings prepared on CGI through plasma transfer arc cladding *Coatings* **12(5)** 633.
[6] Wang Z, et al. 2024 Microstructure evolution, wear and corrosion behavior of WC reinforced CoCrFeNiMn high-entropy alloy composite coatings by induction cladding *Surface and Coatings Technology* **48** 130938.
[7] Hu Y, et al. 2022 Friction stir welding of CoCrNi medium-entropy alloy: recrystallization behaviour and strengthening mechanism *Materials Science and Engineering: A* **848** 143361.
[8] Lu Y, et al. 2021 Influence of carbon on the mechanical behavior and microstructure evolution of CoCrFeMnNi processed by high pressure torsion *Materialia* **16** 101059.
[9] Jiang Y, Chen Y, Li S, Lan S and Wang J T 2024 The impact of precipitation on recrystallization in a severely plastically deformed CoCrFeMnNi multi–principal element alloy *Materials Characterization* **211** 113910.
[10] Campari E G and Casagrande A 2022 Microstructural study of CrNiCoFeMn high entropy alloy obtained by selective laser melting *Materials* **15(16)** 5544.
[11] Wang Z, Du J, Su G, Sun Y, Zhang C and Kong X 2022 Microstructure, preparation and properties of TiC-Fe/FeCoCrNiMn cermet with a core-rim structure *Vacuum* **200** 110984.
[12] Otto F, Dlouhý A and Pradeep K G 2016 Decomposition of the single-phase high-entropy alloy CrMnFeCoNi after prolonged anneals at intermediate temperatures *Acta Materialia* **112** 40-52.
[13] ISO 11357-3, plastics-differential scanning calorimetry Part 3 2011
[14] Zhang Y, Zhou Y J, Lin J P, Chen G L and Liaw P K 2008 Solidsolution phase formation rules for multi-component alloys *Adv. Eng. Mater.* 10534-538.

The Application of SEM Test Technology in Full-graded Concrete's Long Age Alkali Reactivity Test Research in A Hydropower Station

Xiaomei Shao*, **Liang Zhang and Lu Zhou**

Yangtze River Scientific Research Institute, Wuhan 430010, Hubei, China.
Email: *409500448@qq.com; 59945766@qq.com; 2216256464@qq.com

Abstract. This paper is based on the study of the sandstone aggregate alkali reactivity in a large hydropower station in China, with full consideration to the engineering practice creatively carries out the long age simulation experiment of full-graded and mass volume concrete Alkali aggregate's reaction, using SEM test technology and combining with existing alkali aggregate reaction assessment system, studies aggregate alkali activity inhibition effect of long-term effectiveness. The results of the study include that using the SEM test technology in the long aged tests of full-graded concrete alkali aggregate reaction in a hydropower station can obviously observe the characteristics of reaction products that react in the active alkali-aggregate reaction aggregate interface, learn about its active alkali-aggregate reaction in semi-quantitative degree, and predict trends of concrete deterioration.
Keywords. SEM; Dams; concrete aggregate-grading; concrete; reactivity.

1. Introduction

Since the alkali-aggregate reaction problem was discovered at the beginning of the 1940s, the aggregate alkali activity test has been highly valued in the engineering community. For more than half a century, China has constructed many concrete high arch dams, and people have been concerned whether these works have caused damage to alkali-reactive aggregate reactions for a long time. According to the investigation of the 32 completed high dams and more than 40 water gate's concrete durability and aging disease state in the mid of the 1980s, summarizing the effect of hydraulic concrete durability of six main types of diseases (crack, leakage, frozen thawing damage, abrasion cavitation, water erosion carbonation, and steel corrosion damage), the investigation has not found the engineering damage example which due to alkali-reactive aggregate reaction. The main reason is that China pays early attention to the prevention of the deterioration problem in hydraulic concrete by alkali-reactive aggregate reaction. Therefore, the construction of water conservancy and hydropower engineering always pay high attention to the problem of alkali-aggregate reaction [1]. On the construction of projects, especially the national key projects, when exploring the geology, and selecting the material field, the requirements of the aggregate alkali activity detection should be carried out. At present, a system that has the feasibility of the concrete alkali active aggregate reaction risk prediction and evaluation system has been formed, and at the same time, through a large number of engineering experiences sum up the effective method to determine the activity of alkali aggregate, and try exploratory research test to the new method. Due to the restriction by regional geological conditions, some projects have to use reactive aggregate and take corresponding suppression measures, Therefore, how to evaluate the effect of measures to restrain concrete alkali aggregate reaction is a problem that researchers must face.

In recent years, as the SEM(SEM--scanning electron microscope) test technology developed, people have begun to use SEM to observe the products of alkali aggregate reactions in the concrete and carry out the meso analysis of the reaction product's characteristics [2]. SEM test method, which is used in the research and application of the alkali aggregate reaction and expansion mechanism in concrete, has a very important meaning in further understanding the research of alkali aggregate reaction product, deterioration mechanism of concrete subjected to alkali silica reaction, and initiation process of expansion source. This article is on the basis of a sandstone aggregate alkali activity test in a large China hydropower station, paying full consideration to the need for engineering practice, creatively carried out the long age period simulation test of alkali aggregate reaction in full-graded and mass volume concrete, using SEM technology combined with the existing alkali aggregate reaction evaluation system to doing long-term effective study of Aggregate alkali activity inhibition's reception, in order to comprehensively grasp the characteristics of the expansion and deterioration of the actual engineering concrete from the macroscopic and microscopic system. The main research experiments are: SEM analysis of sandstone rock and sandstone in alkali solution and the SEM analysis of the reaction products of full-graded long-age concrete aggregate alkali.

2. Test Materials and Methods

2.1. Raw Materials for Testing

The cement is E mei heat 42.5 cement; the powdered coal is Xuanwei I grade powdered coal; the aggregate is sandstone coarse aggregate which has ASR reaction activity and marble fine aggregate for a hydropower station.

2.2. Test Scheme

2.2.1. The SEM Test on Sandstone. Using SEM to analyze the sandstone aggregates taken from a hydropower station. The specific steps are as follows: Processing the rock samples into particles whose size are less than 5mm, respectively, put into each stainless steel tube, adding 1mol/L NaOH 10ml solution in each test tube, tightening the lid and put the tube into constant temperature water bath at 80 centigrade; Taking out the sample particles after maintaining it to the age of 3d, 7d, washing with anhydrous alcohol then put into the crucible and dry them; after drying, the samples are observed by JSM-6610 Japanese scanning electron microscope (SEM).

2.2.2. SEM Testing of Long Age Test Specimens of Full-graded Mass Volume Concrete. Two kinds of aggregate that adopt E mei heat 42.5 cement and Xuanwei I grade powdered coal (ash content is 35%) are combined (Sandstone coarse aggregate+ artificial sand, sandstone coarse aggregate + marble artificial sand). The proportion selects actual engineering mix proportion formed concrete specimen. The size of specimens is 450mm*450mm*1350mm, and the curing conditions are standard curing room with a temperature of 20 centigrade and outdoor natural moisture curing. These specimens are taken long age observation of the expansion ratio of concrete. The mix proportion of the test is shown in table 1, the total alkali content of concrete is referred to as the total alkali content of the actual engineering ratio to control. Select the all-graded concrete combined with two aggregates, under the condition of a wet natural-outdoor environment and the 20°C standard maintenance, the relatively smooth and integrated aggregate, and the surface of set cement on the premise of reducing the breaking vibration as much as possible [3-4]. Then gold-plate them, vacuum them for standby, and use SEM to test them [5].

3. Test Results and Discussion

3.1. The SEM Analysis of Sandstone

SEM tests were carried out on sandstone rock and sandstone which was soaked in alkali solution.

Journal of Physics: Conference Series **2941** (2025) 012039 doi:10.1088/1742-6596/2941/1/012039

Figures 1-3 are 3-day and 7-day photos of the SEM in which the original sandstone and the sandstone were soaked in NaOH 1mol/L solution.

Table 1. Long age test full-graded concrete mix proportion.

Water-binder ratio	Aggregate combination	Dosage of concrete(kg/m^3)							
		Water	Cement	Fly ash	Sand	Extra-large stone	Large stone	medium stone	small stone
0.43	sandstone + sandstone	82	124	67	438	617	441	353	353
0.43	sandstone + marble	93	141	76	470	587	420	336	336

Note: the dosage of ZB-1 high-efficiency water-reducing agent is 0.7%, and the dosage of AEA202 is 1.7/million.

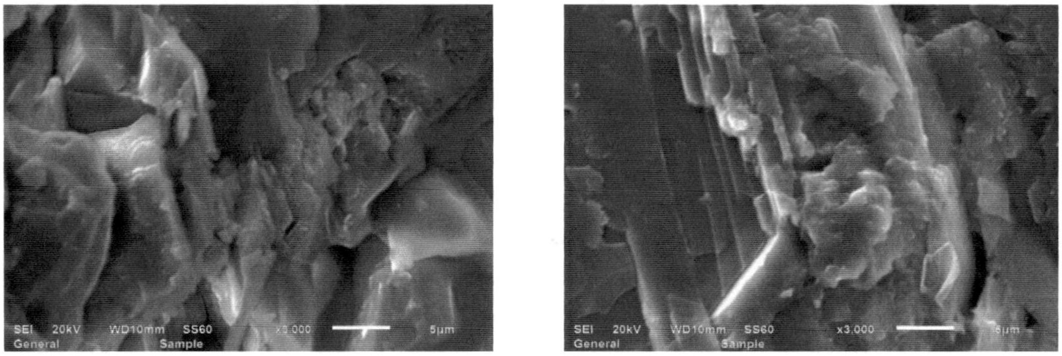

Figure 1. SEM photograph of sandstone original sample.

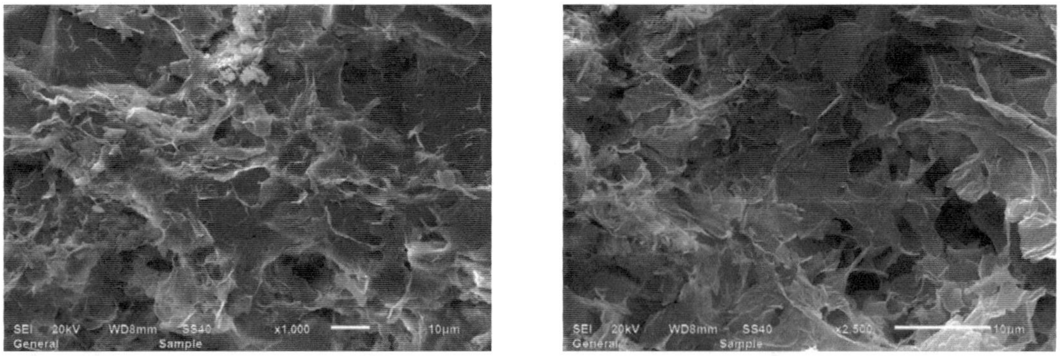

Figure 2. SEM photograph of sandstone soaked for 3 days.

Figure 3. SEM photograph of sandstone soaked for 7 days.

As the SEM photos show in figures 1-3: the surface structure of the sandstone is dense and the joint is clear; the surface of the sample which has soaked 3 days obviously has a petal shaped material distribution, and its structure is soft and uniformly-distributed; after soaking for 7days, the sample surface of flower shape substances increased, which was lush and luxuriant. Compared with the original one, alkali-silica reaction had occurred between the active components in sandstone sample which had been soaked and alkali, and with the increase of the age of reaction, the morphology of the reaction products of the sandstone surface changed, and the reaction products increased obviously. Through careful observation, it is further verified that the sandstone has ASR reaction activity [6].

3.2. SEM Analysis of Full-graded Long-age Concrete Specimens

Four sets of specimens were tested according to the proportion of full-graded concrete tests, which are divided into two groups respectively. As the specimen photos show in figures 4 and 5, the specimens have been nearly 4 years of age from the formation of observation. SEM test was carried out on two aggregate concrete specimens(coarse sandstone aggregate+ sandstone artificial sand, coarse sandstone aggregate+ marble artificial sand)taken from 20 degrees centigrade standard curing and outdoor natural moisture conditions. Figures 6-9 are SEM photos of 20°C standard curing sandstone coarse aggregate +sandstone grit, 20°C standard curing sandstone coarse aggregate +marble sand grit, outdoor natural curing sandstone coarse aggregate and marble grit full-graded concrete specimens for 4 years of age [7-8].

Figure 4. 20°Cstandard curing full-graded concrete prismoid specimens.

Figure 5. Outdoor natural curing full-graded concrete prismoid specimen.

Figure 6. SEM photograph of 20°C standard curing sandstone coarse aggregate & sandstone grit full-graded concrete specimens for 4 years age.

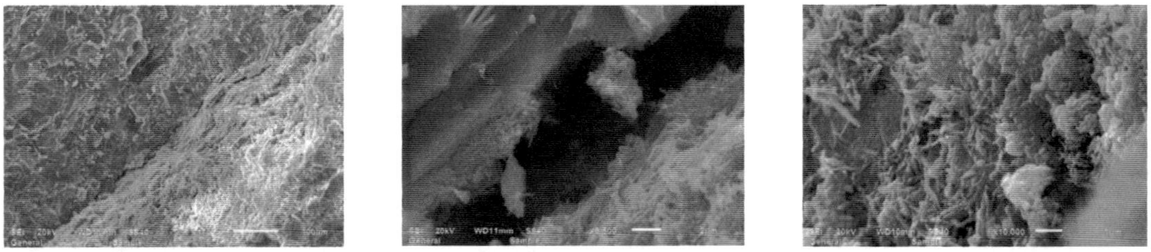

Figure 7. SEM photograph of outdoor natural curing sandstone coarse aggregate & sandstone grit full-graded concrete specimens for 4 years age.

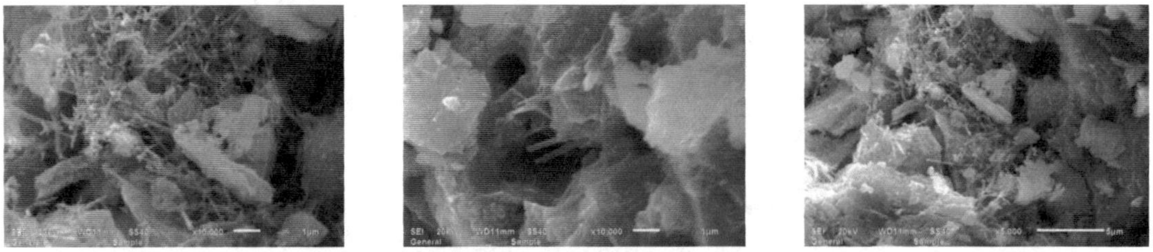

Figure 8. SEM photograph of 20°C standard curing sandstone coarse aggregate & marble grit full-graded concrete specimens for 4 years age.

Figure 9. SEM photograph of outdoor natural curing sandstone coarse aggregate & marble grit full-graded concrete specimens for 4 years age.

We can see from figures 6-9 in micro topography that: (1) The sandstone aggregate long age concrete in 20°C standard curing and outdoor natural curing conditions can visibly see small shells and cotton-like soft concrete alkali reactive product; (2) marble and sandstone combined aggregate in standard curing condition and natural curing conditions can be partially saw active alkali reaction products but more was long column shape or fine columnar cement hydration product, which was tightly packed. That shows that the marble sandstone effectively slows down the sandstone alkali aggregate reaction expansion; (3) Under the standard curing condition, the amount of these two kinds of combined aggregate alkali active reaction products is more than in natural curing conditions, which is also verified that the speed and degree of alkali aggregate reaction is closely related to the temperature and humidity of concretes [9-11].

4. Conclusion

(1) Applying the SEM test technique to the study of the alkali activity of concrete, which can clearly observe the morphology of the typical reaction product of aggregate interface and understand the degree of alkali reactivity to predict the deterioration trend of concrete.

(2) In the test of full-graded long-age concrete aggregate alkali reactivity in a hydropower station, the active aggregate reaction conditions in SEM images show that marble sand can slow down alkali activity expansion of the sandstone aggregate; If this mix proportion is applied to the project, it is necessary to combine with other methods to make a comprehensive decision.

(3) Due to the limitation of experimental conditions, in order to better observe the full-graded long-age concrete alkali reactivity aggregate from the microscopic and try to find out the expansion process of long-age concrete alkali reactivity, we will take SEM observation to different-age all-graded concrete alkali reactivity in the next step.

Acknowledgments

The paper is funded by the "Research and Application of Key Technologies for Fine Composite Grouting Treatment of Hydraulic High-Pressure Tunnels", a major scientific and technological project of the Ministry of Water Resources (Project No.: SKS-2022106), and "Demonstration and Promotion of High-Pressure Chemical Grouting under All Working Conditions and Its Supporting Equipment", a water conservancy technology demonstration project (Project No.: SF-202301).

5. References

[1] Liu C X and Wen Z Y 1995 *Monographon Dam-Engineering Concrete (1): Concrete Alkali Aggregate Reaction* (Guangzhou: South China University of Technology Press) p199-p227.

[2] Wu C H, Hung Y H and Hong C W 2012 On-line supercapacitor dynamic models for energy conversion and management *Energy Conversion and Management* **(1)** 337-345.

[3] Cai Y B, Ji G J, Ding J T, Yang H Q, Li G W, Li W W and Yang S 2014 *Code for Testing Aggregates of Hydraulic Concrete (DL/T5151-2014)* (Beijing: China Electric Power Press) p1-p80.

[4] Li Z, Xiao L and Zhang R W 2010 40 Microscopic tests and analysis of concrete core sample in a 40-year-old dam *China Water Power & Electrification* **(5)** 46-50.

[5] Li Z, Xiao L and Chen L 2012 The microstructure characteristics research of alkali-silica reaction products *Journal of Yangtze River Scientific Research Institute* **29(3)** 62-66.
[6] Li Z, Xiao L and Wang Z Q 2011 Microscopic test and analysis of concrete core samples in Danjiangkou Dam *Journal of Yangtze River Scientific Research Institute* **28(5)** 59-62.
[7] 2019 *Standard Guide for Petrographic Examination of Aggregates for Concrete (ASTM C295/C295M-19)* (Pennsylvania: American Society for Testing Materials) p1-p9.
[8] He X F, Miao C W, Zhang Y S and Hong J X 2009 Review on the analytical methods for the structure and properties of the interfacial transition zone in cement-based materials *Concrete* **(10)** 19-23.
[9] Fan Q 2023 Research based on the alkali activity test of concrete aggregate *Henan Water Resources & South-to-North Water Diversion* **52(7)** 109-110.
[10] Gao X P, Zhang A M and Zhang L Y 2019 Development and application of scanning electron microscopy technology and characterization technology *Science and Technology Innovation Herald* **16(19)** 99-103.
[11] Garboczi E J 1993 Computational materials science of cement-based materials *Materials and Structures* **(4)** 191-195.

Research Progress on Surface Modification of Biomedical Magnesium Alloys

Jie Song*, Yongbiao Chen and Kexin Wang

School of Mechanical and Electrical Engineering, Qingdao Institute of Technology, Qingdao, Shandong, China.

Email: 419796673@qq.com; 459942082@qq.com; 478392197@qq.com

Abstract. Biomedical magnesium alloys have good biocompatibility, degradability, and mechanical properties, which have become an important medical implant material. However, the corrosion resistance of magnesium alloys is poor, and their degradation rate in vivo is very fast, making it a difficult point for their clinical application. The surface treatment technology of magnesium alloys was studied in this paper, and the current research status of commonly used surface treatment technologies for magnesium alloys was discussed. The development trends of surface treatment technologies for magnesium alloys were explored, and prospects for their future research and development directions were also put forward.

Keywords. Magnesium alloy; surface modification; degradation.

1. Introduction

The medical metal materials currently used in clinical practice mainly include pure Titanium, Titanium alloys, Co-Cr alloys, and 316L stainless steel. These metal materials for clinical applications still have some shortcomings such as the possibility of material corrosion and degradation during long-term use in the physiological environment of the human body, resulting in the release of toxic metal ions [1,2]. At the same time, the elastic modulus of metal materials is usually much higher than that of human bones, which can easily cause stress shielding after implantation, preventing normal muscles and bones from receiving sufficient external stimulation, leading to atrophy, and even implant failure. Moreover, when metal medical materials are used as temporary implants such as vascular stents, bone plates, and bone nails, secondary surgery is usually required to remove them, which increases the economic burden and physical pain of patients.

Magnesium alloy has the advantages such as low density, high specific strength, low elastic modulus, good biocompatibility, and biodegradability, and has attracted the attention of medical material researchers. Research and development of biodegradable magnesium alloys that meet clinical requirements can replace traditional inert metal materials, accelerate the repair of bone injury sites, reduce discomfort during surgery and rehabilitation, reduce patient pain caused by secondary surgery, and lower medical costs.

The corrosion resistance of magnesium alloys has always been an important reason limiting their clinical application. The corrosion rate of magnesium is usually faster than the regeneration, healing, and repair rate of bone tissue, making it difficult to achieve the goal of promoting bone growth after implantation. Moreover, local corrosion occurs during the degradation process of magnesium alloys, leading to the accumulation of magnesium ions, hydrogen gas, and local PH values, and corrosion precipitates in the body, resulting in adverse reactions and treatment failure. At the same time, the rapid degradation and corrosion of magnesium alloys damage their structural integrity, reduce their mechanical properties, and fail to achieve the expected therapeutic goals [3].

In order to improve the corrosion resistance of medical magnesium alloys and meet the clinical requirements of the adjustable degradation rate of magnesium alloys, many researchers have adopted alloying with added alloying elements [4] and material surface modification methods for research. Currently, various surface modification technologies have been developed in the field of medical magnesium alloys. Surface modification is a low-cost and simplest effective method to improve the properties of degradable metals. This article reviews the major surface modification technologies for magnesium alloys in recent years.

2. Surface Modification Technology for Magnesium Alloys

2.1. Surface Nano-Modification

Surface nano-modification is a specific plastic deformation treatment on the surface of magnesium alloy without introducing other substances, forming a structural layer different from the original surface, thereby improving the hardness, fatigue resistance, and corrosion resistance of the material. Xie achieved surface nanoscaling of magnesium alloys through continuous hammering combined with annealing treatment [5]. Researchers analyzed and studied the corrosion performance of magnesium alloys subjected to hammering and annealing using an electrochemical comprehensive tester and electron microscopy. The results showed that the corrosion resistance of the nanosized magnesium alloy was significantly improved. The continuous hammering and annealing process for achieving surface nanosizing of magnesium alloys is relatively simple and cost-effective, but it has certain requirements for material size and structure, and further processing is needed in the future.

2.2. Anodic Oxidation

Anodizing is a method of generating oxides using electrochemical principles. Under specific electrolyte and voltage process conditions, a firmly bonded oxide film is formed on the surface of magnesium and magnesium alloys, improving their corrosion resistance. The anodic oxide film of magnesium alloys has a strong bonding ability with the substrate, and its corrosion resistance is significantly higher. The structure of the anodized film layer is porous and usually requires sealing treatment to improve corrosion resistance. The porous structure on the surface of magnesium alloy formed by anodizing can promote rapid tissue growth and promote bone growth when implanted into living organisms. The porous structure formed by anodizing can also provide a substrate for the formation of composite coatings [6].

In order to improve the corrosion resistance of medical AZ31 magnesium alloy, Li comprehensively applied anodizing and chemical conversion processes to prepare a composite film layer on its surface [7]. The film layer prepared by the anodizing process is rough and uneven, mainly composed of Mg (OH) 2 and Al2O3. The composite film obtained after chemical conversion is dense and smooth. The analysis of the potentiodynamic polarization curve of the chemical bonding between the anodic oxide film and the chemical conversion film shows that the composite film has a significant protective effect on AZ31 magnesium alloy.

2.3. Micro Arc Oxidation

Micro arc oxidation (MAO) is a kind of technique for the in-situ growth of ceramic layers on metal surfaces It involves placing metal in an electrolyte solution and generating a ceramic layer on the surface through the combined action of thermochemistry, plasma chemistry, and electrochemistry, resulting in spark discharge. Micro arc oxidation technology can significantly improve the corrosion resistance, wear resistance, and hardness of materials, and is a promising surface biological modification technology that has been successfully applied on the surface of titanium implants. Micro arc oxidation treatment has better corrosion resistance and wear resistance. Closely bound to the substrate and not easily detached; It can form a rough surface on the metal surface, which is conducive to the attachment of osteoblasts. Compared with the usual anodizing treatment, the reduction in fatigue strength of the material is minimal. As an effective measure to slow down the degradation rate of magnesium alloys, it has the advantages of a simple process, high film formation efficiency, and excellent overall performance of the film layer, achieving dual functions of controllable degradation rate and improved biocompatibility [8].

The disadvantage of micro-arc oxidation is high production costs, requiring the development of pollution-free electrolytes, suppression of arc energy consumption, and reduction of production costs, which are difficult for ordinary enterprises to carry out. In addition, the micro-arc oxidation film layer is mostly a single film layer, which makes it difficult to meet the clinical application needs of magnesium alloys.

2.4. Surface Rare Earth Conversion Coating

Conversion film technology is an effective method for surface anti-corrosion treatment of magnesium alloys. The conversion film itself not only has good corrosion resistance, which can prevent the erosion of the substrate by the medium but also can be used as the bottom for subsequent processing. The chemical conversion treatment method is simple, the process is easy to control, and it is widely used. Due to the presence of highly toxic and carcinogenic elements in the chemical conversion treatment solution of magnesium alloys, their application should be strictly restricted.

Yan [9] used orthogonal experiments to study the treatment process of rare earth conversion coatings on magnesium alloys with CeCl3 as the main salt and prepared a relatively dense rare earth conversion coating on the surface of AZ31B magnesium alloy. The results show that the AZ31B magnesium alloy treated by this process has good corrosion resistance in both physiological saline and Hank's solution; The surface rare earth conversion film has good anticoagulant properties. The film-forming mechanism of the conversion film was also discussed.

2.5. Surface Laser Peening Strengthening

Laser shot peening, also known as laser shock peening, is a technology that uses a strong pulse laser beam to irradiate the metal surface. The metal surface absorption layer absorbs the laser energy to generate high-pressure shock waves that interact with the material to modify the material surface and improve its fatigue resistance, wear and corrosion resistance, and other properties. Zhang used a laser with a wavelength of 1064nm, a half peak width of 15 ns, a spot size of Φ3mm, a pulse energy of 10J, and a frequency of 5Hz to perform surface shot peening treatment on AZ31B magnesium alloy thin plates [10]. The results showed that after laser peening, nanocrystals with a grain size of around 24nm were obtained on the surface of AZ31B magnesium alloy, and the surface roughness of the sample decreased from 2.36μm to 1.14 μm. At the same time, the corrosion potential of the sample increased by 51mV and the corrosion current decreased by 92.1%. The grain refinement and surface roughness reduction induced by laser shot peening have improved the corrosion resistance of AZ31B magnesium alloy.

2.6. Laser Surface Cladding

Laser surface melting is a laser surface treatment technology that melts the alloy powder that needs to be prepared and designed by laser to form a cladding layer. The cladding layer melts with a thin layer of the base metal and forms a metallurgical bond. The bioceramic layer generated by laser surface cladding technology has a fine and dense structure with a certain degree of preferred orientation. Gao Yali used laser technology to prepare hydroxyapatite (HA) coatings on the surface of magnesium alloys [11]. The research mainly adopts the following two methods to prepare HA coatings: laser cladding method and plasma spraying+laser remelting method. The research results indicate that due to the significant physical and chemical differences between magnesium alloy and HA, the coating prepared by the laser cladding method is discontinuous teardrop-shaped, which is very difficult to form. The magnesium dilution in the coating is large, seriously affecting the corrosion resistance and biocompatibility of the coating. By using plasma spraying and laser remelting treatment, it is easier to prepare HA coating on the surface of magnesium alloy.

3. Composite Coating on Magnesium Alloy Surface

Single-layer coatings on magnesium alloy surfaces often have problems such as surface quality, difficult control of degradation rate, and biocompatibility after implantation. Therefore, many scholars have conducted research on composite coatings on magnesium alloy surfaces.

3.1. Anodizing /Organic Composite Coating

Li conducted research on the preparation process and corrosion resistance of magnesium alloy surface anodizing/tea polyphenol composite conversion layer [12]. The researchers first used anodizing to construct an anodized layer on the surface of magnesium alloy and then prepared a polyphenol conversion layer on the anodized layer by soaking method to obtain a composite conversion layer. The conversion layer on the outer layer of the composite coating effectively fills the porous structure generated by the anodic oxidation layer, improving the corrosion resistance of the magnesium substrate. The selected composite coating raw materials are inexpensive and have multiple effects such as anti-oxidation and anti-inflammatory, providing a new solution for surface modification of medical magnesium alloys.

3.2. Micro Arc Oxidation/Organic Composite Coating

The presence of micropores on the surface of a single coating of micro-arc oxidation (MAO) makes it easy for corrosive media to penetrate and damage the internal dense layer and substrate. Li used three different concentrations of polylactic acid to seal the pores of a single porous micro-arc oxidation coating [13]. The results showed that the micropores were completely sealed by polylactic acid, and the roughness and corrosion resistance of the coating was increased. Among them, the micro-arc oxidation/polylactic acid composite coating with a concentration of 6% had the best corrosion resistance.

Polylactic acid (PLA) has a good sealing effect on micro-arc oxidation, and the surface of the composite coating after sealing is uniform, dense, smooth, and flat. The mechanical interlocking between the micro pores and cracks on the outer layer of micro-arc oxidation and polylactic acid can achieve a strong bond at the interface of the composite coating. The MAO/PLA composite coating overcomes the physical bonding defects between a single PLA coating and the substrate and has excellent processing performance, excellent biocompatibility, and acid-base neutralization ability of degradation products [14]. At present, research on MAO/PLA composite coatings mainly focuses on the sealing effect of PLA on MAO and the bonding mechanism between coating interfaces.

MAO/polycaprolactone (PCL) composite coating is also a current research hotspot. Li prepared MAO/PCL composite coatings on pure magnesium surfaces using the immersion pulling method. The PCL coating effectively filled the micropores and cracks in the MAO coating. HBSS (Hank's balanced salt solution) corrosion test showed that the MAO/PCL composite coating effectively reduced the corrosion rate of magnesium, and the composite coating was tightly bonded to the substrate without pitting or peeling during the corrosion process [15].

4. Conclusion and Prospect

As a biodegradable medical material, magnesium alloy has broad application prospects, but its rapid degradation rate limits its clinical application. Although current surface modification techniques can enhance the degradation performance of magnesium alloys, there is still no coating that can meet clinical applications. At present, most of the research on surface-modified coatings of magnesium alloys is carried out under in vitro conditions. Considering the extremely complex biological environment, a single coating is difficult to meet the requirements of the human environment. Many studies now tend to focus on the preparation of composite coatings, and multifunctional composite coatings will have greater development prospects in the future.

Acknowledgments

The authors would like to thank Youth Natural Science Research Project of Shandong Province

(24ZRK010), The fundamental science (natural science) research project of Shandong higher education institutions(24GJZ10030), and the Research Project of Qingdao Institute of Technology (Preparation and Corrosion Degradation Study of Controllable Zn/PLA Composite Coating on Biomedical Magnesium Alloy, 2023JY067).

5. Reference

[1] Hassan D, Amirhossein J M and Hamdy I 2023 Corrosion and corrosion-fatigue behavior of magnesium metal matrix composites for bio-implant applications: A review *Journal of Magnesium and Alloys* **2023(04)** 1125-1161.

[2] Violeta T, Christu T and Florentina M C 2021 Biodegradable Mg alloys for orthopedic implants. A review *Journal of Magnesium and Alloys* **2021(06)**.

[3] Wang G, Li G and Liu H F 2024 Research progress in application and surface modification of medical degradable magnesium alloys *Surface Technology* **53(07)** 15-30.

[4] Li H f, Zheng Y X and Wang L N 2021 Research progress on functional surface modification of biodegradable metals *Materials Reports* **35(1).**

[5] Wang C, Cui Y J and Liu H Y 2015 Research progress on the alloying of biodegradable magnesium alloys *Materials Reports* **29(11)** 55-60.

[6] Xie A, Tu F Q and Xu R B 2022 Study on surface nano crystallization and corrosion resistance of medical magnesium alloys *Hot Working Technology* **51(6).**

[7] Huo H W, Zheng Z G and Li M S 2005 Investigation on anodization of magnesium alloys *Rare Metal Materials and Engineering* **34(9).**

[8] Li S, Dong Y S and Sheng X B 2010 Preparation and characterization of a composite coating on AZ31 for biomedical application *Surface Technology* **9(1)**.

[9] Cheng C, Xiao H and Jing B 2014 Surface modification of biomedical magnesium alloy wires by micro-arc oxidation *Transactions of Nonferrous Metals Society of China* **24(04)** 1058-1064.

[10] Yan T T, Tan L L and Xiong D S 2009 Preparation and properties studies of rare earth conversion coating on bio-medical AZ31B magnesium alloy *Rare Metal Materials and Engineering* **38(5)** 918-923.

[11] Zhang Y, Ge M Z and Su S 2017 Effect of laser peening on electrochemical properties of AZ31B Mg alloy *Mechanicai Engineering & Automation*, **2017(1)** 146-148.

[12] Gao Y, Xiong D S and Wang C S 2009 Laser cladding hydroxyapatite coating on magnesium alloy for biomaterials application *Applied Laser* **29(6)** 526-530.

[13] Li H, Zhang H and Shen X L 2023 Corrosion resistance of anodic oxidation/polyphenol conversion composite coating on biomedical magnesium alloy *Surface Technology* **52(1)** 196-205.

[14] Li W 2020 *Preparation and Performance Study of Micro arc Oxidation and Polylactic Acid Sealed Coating on Pure Magnesium Surface* (Chongqing: Chongqing University). Master's Thesis

[15] Ji S Y, Chang C and Chang S B 2023 Research status and evolution direction of micro-arc oxidation /organic composite coating on medical magnesium alloy surface *Surface Technology* **52(12)** 315-334.

[16] Tian P, Xu D M and Liu X Y 2016 Mussel-inspired functionalization of PEO/PCL composite coating on a biodegradable AZ31 magnesium alloy *Colloids and Surfaces B: Biointerfaces* **141** 327-337.

Research on the Isothermal Forging Process for TC6 Titanium Alloy Blades with Thin-Walled and Variable Cross-Section Characteristics

Youhong Chen*, **Xing Sun and Bo Lan**

AECC Beijing Institute of Aeronautical Materials, Beijing 100095, China.
Email: *cyh198243@163.com; 18515470036@163.com; 89620411@qq.com

Abstract. TC6 titanium alloy blades with thin-walled and variable cross-section characteristics are key components of aero-engine compressors. Conventional die forging methods for such parts have issues with low dimensional accuracy and poor microstructural uniformity. This study employs the DEFORM software for the simulation and optimization of preform design and investigates the effects of isothermal forging and heat treatment processes on the microstructure and properties of TC6 titanium alloy blade forgings. The process parameters were used to manufacture TC6 titanium alloy blades, and the microstructure and mechanical properties were characterized. The results indicate that an optimized preform is conducive to obtaining blade forgings with high dimensional precision. The forging temperature should be controlled at 45-75°C below the β phase transformation temperature, with a holding time of 60-120 minutes, a deformation speed of 0.01-0.1 mm/s, and a deformation degree of 25-50%. The heat treatment method and parameters significantly affect the content and morphology of primary α and secondary α in the blade forgings. The duplex annealing parameters of "850°C for 1 hour, air cooling + 600°C for 2 hours, air cooling" can achieve the optimal balance of strength and plasticity in the blade forgings.
Keywords. TC6 Titanium Alloy; Isothermal Forging; β phase transformation temperature; Duplex annealing.

1. Introduction

TC6 titanium alloy is a martensitic α+β titanium alloy, equivalent to the Russian BT3-1 titanium alloy. It contains α-stabilizing element Al, isomorphous β element Mo, and eutectic β elements Cr, Fe, and Si [1]. This alloy possesses an ideal combination of mechanical properties, allowing it to work at temperatures between 400°C and 450°C for extended periods. It has been widely applied in aviation engines, primarily for the fabrication of compressor disks and blades [2-4].

TC6 titanium alloy blades, used in the fans and compressors of aviation engines, are key components for energy conversion. Their quality directly affects the engine's thrust-to-weight ratio (power-to-weight ratio), safety, and reliability. Titanium alloys are sensitive to process parameters during forming, with high deformation resistance, making them difficult to form [5,6]. Particularly for blade forgings with thin-walled and variable cross-section characteristics, as shown in figure 1, conventional die forging techniques can lead to poor shape filling and significant variations in microstructure and properties. Isothermal forging technology offers the advantages of a uniform and stable temperature field, low deformation rates, high forming accuracy, and uniform and stable microstructure and properties [7-9]. This paper focuses on the TC6 titanium alloy blades with thin-walled and variable cross-section characteristics, conducting DEFORM-optimized preform design, and investigating the effects of isothermal forging and heat treatment processes on the microstructure and properties of the forgings. It provides experimental evidence for the development of TC6 titanium alloy blade components characterized by thin walls and varying cross-sections, ensuring high dimensional precision, and meeting the criteria for balanced strength and ductility, as well as exhibiting a uniform microstructure.

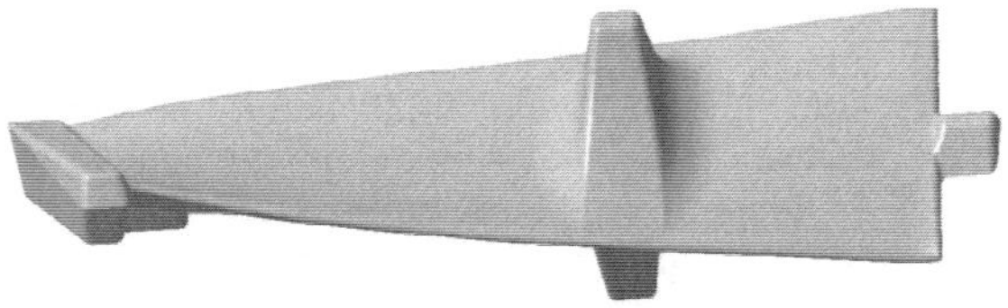

Figure 1. TC6 titanium alloy blade with thin-walled and variable cross-section characteristics.

2. Experimental Materials and Methods

The experimental material was a TC6 titanium alloy bar, Φ100mm in diameter, with its chemical composition detailed in table 1. The β phase transformation temperature of the alloy, as determined by metallographic methods, was 1005°C. The microstructure of the bar exhibited a typical α+β phase, with coarse equiaxed primary α uniformly distributed, accounting for approximately 50%, and secondary α in the form of short rods, distributed among the primary α, as shown in figure 2.

Table 1. Chemical composition of TC6 titanium alloy (wt%).

Element	Ti	Al	Mo	Cr	Si	Fe	O	H
Content	Bal.	6.41	2.53	1.50	0.24	0.46	0.13	0.0047

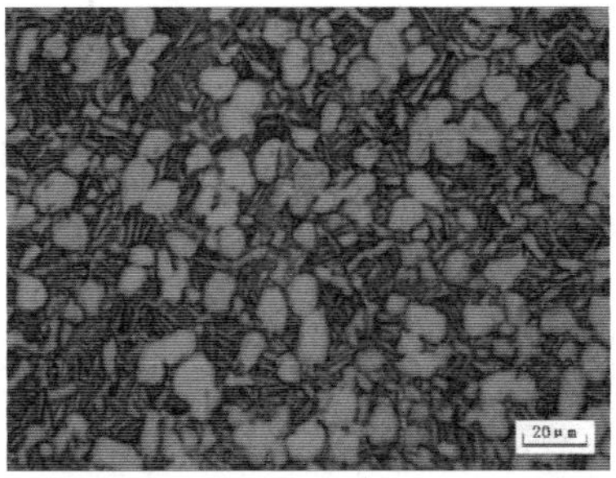

Figure 2. Microstructure of TC6 titanium alloy bar.

Adopting the inverse design method, the preform of the blade forging was initially designed based on the volume distribution results. Utilizing the DEFORM software, the forming process of the blade forging was simulated to determine the optimal preform design.

Specimens were machined from the bar to study the effects of forging temperature and holding time on the microstructure. The forging temperatures included 850, 900, 920, 940, 950, and 960°C, with holding times of 60 minutes and 120 minutes. Under constant forging temperature, the influence of forging speed and deformation degree on the alloy microstructure was investigated. The forging speeds (v) selected were 0.05, 0.01, and 0.5 mm/s, and the deformation degrees (ε) were designed to be 25% and 50%.

Specimens were extracted from the same location of TC6 titanium alloy blade forgings to study the effects of heat treatment processes on the microstructure and properties. Initially, the optimal heat treatment method was determined from four typical methods for titanium alloys [10]: (1) Conventional annealing: 800°C for 1 to 2 hours, followed by air cooling (AC); (2) Isothermal Annealing: 870°C for 1 to 2 hours, followed by furnace cooling (FC) to 550-650°C or transferring to another furnace at 550-

650°C and holding for 2 hours, then air cooling (AC); (3) Duplex annealing: 880°C for 1 to 2 hours, followed by AC, then 550-650°C for 2 to 5 hours, followed by AC; (4) Solution Treatment and Aging: 850°C for 1 hour, water cooling (WC), followed by 550°C for 1 to 4 hours, AC. Subsequently, under the optimal method, the influence of various temperature regimes on the microstructure and properties of blade forgings were analysed to establish the best heat treatment regime.

The forging process was conducted using a 2000T isothermal forging hydraulic press, and the heat treatment was carried out in an RJX-13 type high-temperature electric furnace. For the microstructural analysis of the specimens, an etching solution composed of HF, HNO3, and H_2O in a volumetric ratio of 1:2:5 was used to reveal the microstructure, which was then examined under a LEICA DMI5000 M optical microscope (OM). Room temperature tensile property tests were performed in accordance with the GB/T 228.1.

3. Experimental Results and Discussion

3.1. Numerical Simulation of Forging and Optimization Design

Figure 3 illustrates the volume distribution across different cross-sections of the forging. Utilizing the inverse design method, a preliminary preform was designed based on the results of the volume distribution.

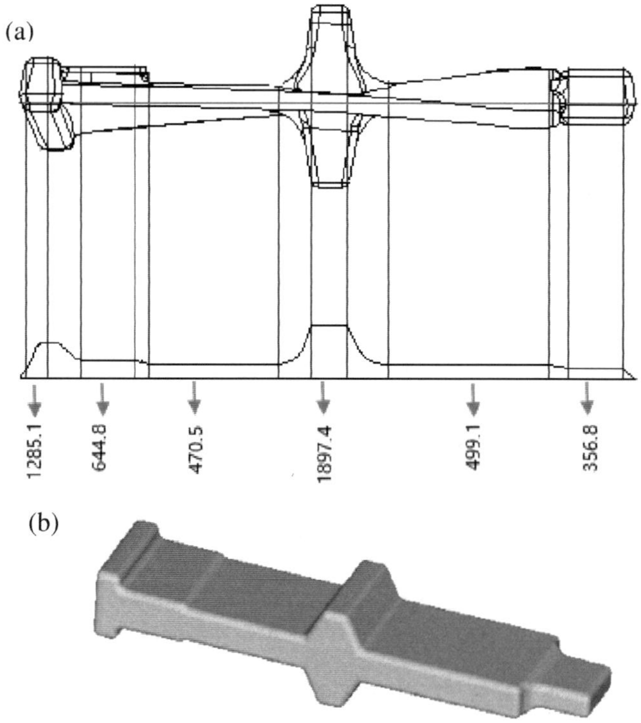

Figure 3. Design of the preform for TC6 titanium alloy blade forgings:(a) Blade forging volume partitioning calculation; (b) 3D model of the preform in the initial design

Incorporating a preforming stage is essential in metal forging processes, as it significantly mitigates the risk of forging defects [11,12]. Historically, industrial production has been heavily reliant on the designer's experience and a trial-and-error approach, which is inherently costly. In recent years, finite element modeling software, such as Deform-3D, has been increasingly employed to assess and optimize conventional metal forming processes [13]. In this paper, Deform-3D was employed to conduct isothermal forging simulation and emulation for the blade forging. With the objectives of achieving rational metal flow, complete filling of the forging, and a reasonable distribution of stress and strain within the forging, the preform structure dimensions were optimized. Figure 4 illustrates the optimized preform and the final forging schematic of the blade. Simulation results indicate that the preform can meet the requirements for complete filling of the blade root, dampening platform, and blade body. After the forging is fully formed, the internal stress distribution within the forging is relatively uniform.

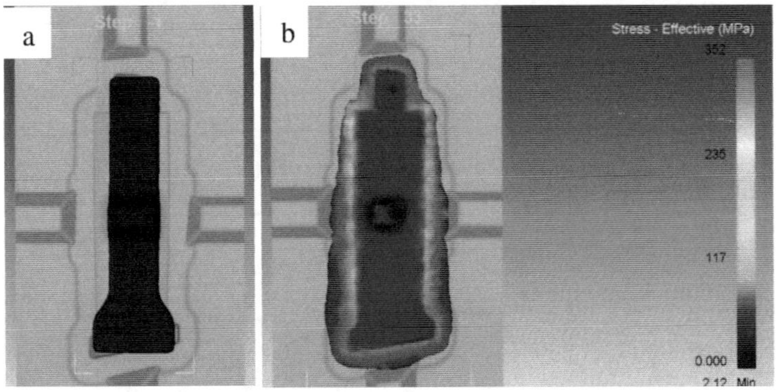

Figure 4. Schematic diagram of the isothermal forging preform (a) and final forging simulation for TC6 titanium alloy blade (b).

3.2. The Influence of Forging Temperature and Holding Time on the Microstructure of the Alloy
Figure 5 displays the microstructure of the alloy under various forging temperatures and holding times. Figure 5 reveals that as the heating temperature increases, the content of the primary α phase in the billet decreases, falling far below 30 % at 960°C; when the holding time is extended from 60 minutes to 2 hours, the content of the primary α phase slightly increases, with the size remaining essentially unchanged. Considering that higher forging temperatures facilitate easier forming in actual production, the forging temperature should be selected 45-75°C below the phase transformation point, provided that the microstructure requirements of the forgings are met. From an energy-saving and emission-reduction perspective, the holding time for the billet should not be excessively long; it should be arranged to maximize on-site efficiency.

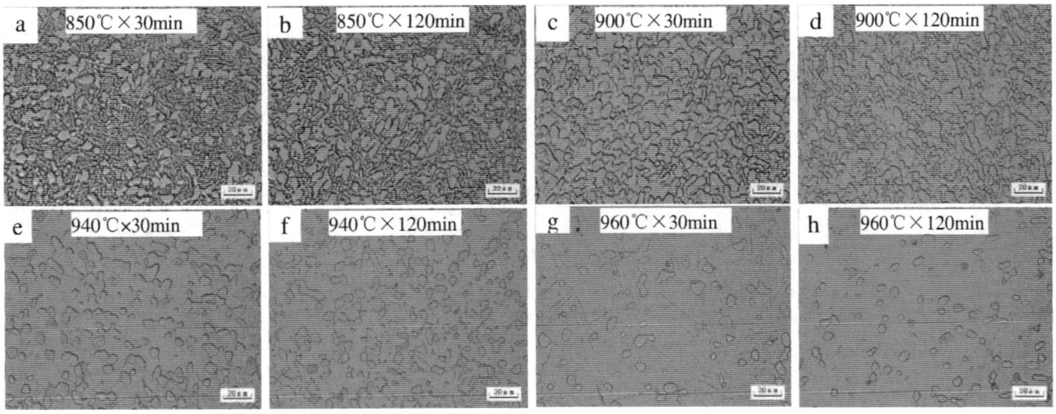

Figure 5. The influence of forging heating temperature and holding time on microstructure.

3.3. The Influence of Heat Treatment Methods on the Properties of Blade Forgings
Table 2 presents the room temperature tensile properties of TC6 titanium alloy samples after heat treatment according to four typical methods. There are significant differences in the tensile properties of the alloy after heat treatment by these four methods. Samples treated by solution treatment and aging

and duplex annealing exhibit high strength but low plasticity; those treated by conventional annealing and isothermal annealing show low strength but high plasticity. Based on the project's required performance specifications for blade forgings, namely 1030 MPa$\leq\sigma_b\leq$1230 MPa and $\delta\geq 9\%$, the duplex annealing method is superior.

Table 2. Typical heat treatment regimes and properties of tc6 titanium alloy.

Type		$\sigma_{0.2}$, MPa	σ_b, MPa	δ, %	ψ, %
Conventional annealing	800°C, 2h, ac	1036	1057	18.0	51.6
		1035	1055	18.0	50.4
Isothermal annealing	870°C, 1h, furnace cooling to 650°C, 2h, ac	1028	1094	19.0	47.4
		1022	1073	19.6	50.5
Duplex annealing	880°C, 1h, ac+600℃, 2h, ac	1197	1243	15.4	40.6
		1199	1245	13.9	40.1
Solution treatment and aging	850°C, 1h, wc+550℃, 4h, ac	1238	1309	11.9	36.3
		1227	1291	13.8	41.3

To further determine the optimal heat treatment parameters, six duplex annealing regimes were designed as shown in table 3. The primary annealing temperature for the duplex annealing was set at 830 and 850°C, with a holding time of 1 hour and AC; the secondary annealing temperatures were selected as 550, 600, and 620°C, with a holding time of 2 hours and AC.

Table 3. Duplex annealing of blade forgings.

Number	Heat Treatment Regime
A1	830°C×1h, AC + 550°C×2h , AC
A2	830°C×1h, AC + 600°C×2h, AC
A3	830°C×1h, AC + 620°C×2h, AC
A4	850°C×1h, AC + 600°C×2h, AC
A5	850°C×1h, AC + 620°C×2h, AC
A6	850°C×1h, AC + 650°C×2h, AC

Figure 6 illustrates the room temperature tensile properties under six different duplex annealing regimes. After duplex annealing heat treatment of blade forgings, as the secondary annealing temperature increases from 550°C to 620°C, the tensile strength generally exhibits a decreasing trend, while the elongation rate gradually increases. Under conditions of the same secondary annealing temperature, as the primary annealing temperature increases from 830°C to 850°C, the tensile strength, of the blade forgings increases, while the elongation rate decreases.

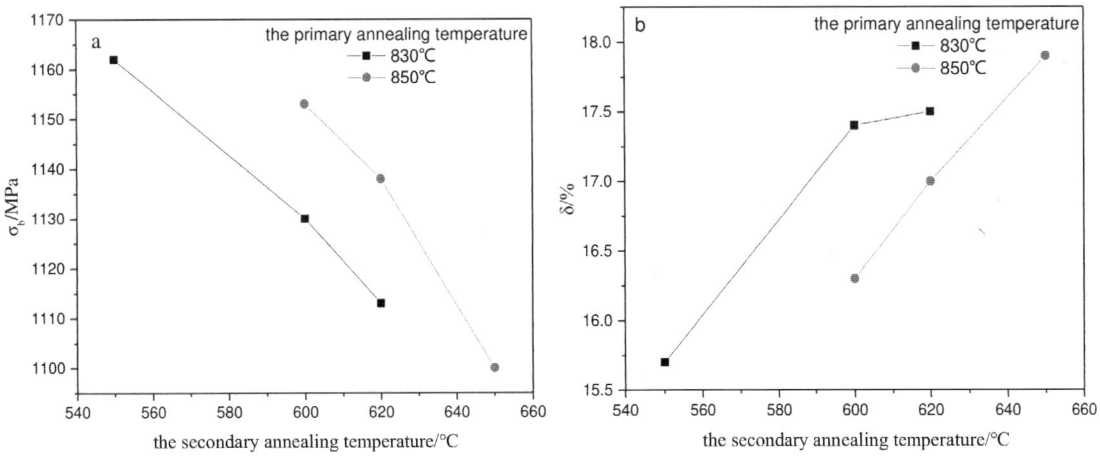

Figure 6. Mechanical property of blade forgings under different duplex annealing processes: (a) tensile strength; (b) elongation rate.

Figure 7 displays the microstructures of the blade forgings under six different duplex annealing regimes. Under the same primary annealing temperature, the content and morphology of the primary α phase remain consistent, but the short rod-shaped secondary α phase becomes finer. Under the same secondary annealing temperature conditions, the content of the equiaxed primary α phase decreases, while the content of the short rod-shaped secondary α phase increases without a change in morphology. In conjunction with the performance variation patterns observed in figure 6, it is evident that within the experimental temperature range, the primary annealing temperature primarily affects the content of the primary α phase. As the primary annealing temperature increases, the content of the primary α phase decreases, and the content of the secondary α phase increases, leading to increased strength but reduced plasticity. The secondary annealing temperature mainly influences the morphology of the secondary α phase. As the secondary annealing temperature increases, the secondary α phase becomes finer, resulting in decreased strength but increased plasticity. Therefore, the appropriate matching of the primary and secondary annealing temperatures plays a decisive role in achieving the desired balance of strength and plasticity in TC6 titanium alloy blade forgings. Considering the microstructural and property requirements of TC6 titanium alloy blade forgings, within the experimental temperature range, the heat treatment regime of "850°C × 1 h, AC + 600°C × 2 h, AC" provides the optimal balance of strength and plasticity.

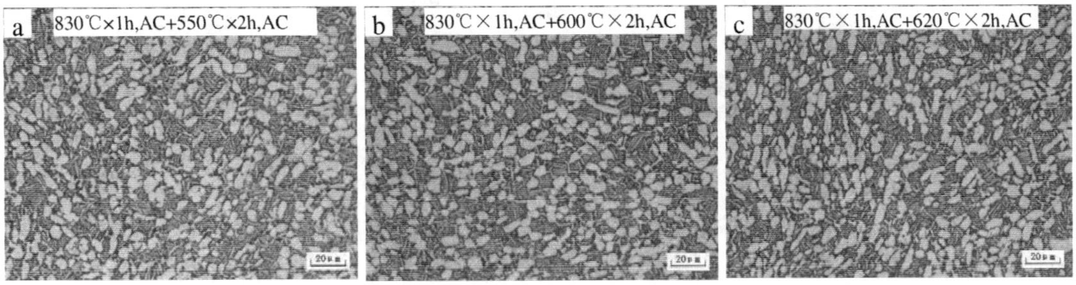

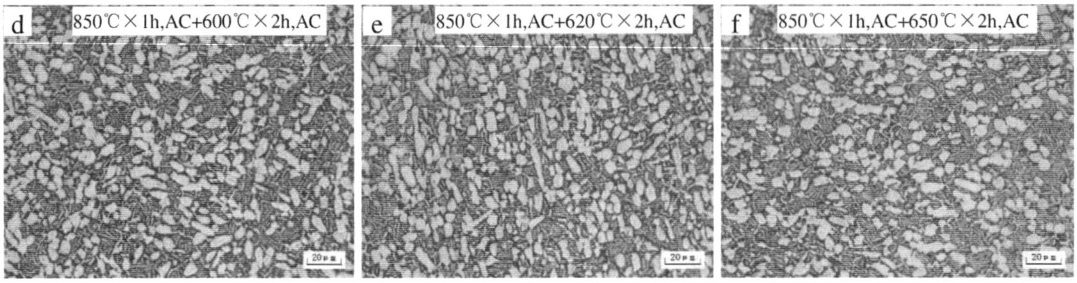

Figure 7. The microstructure of TC6 titanium alloy blade forgings after different duplex annealing.

3.4. The Influence of Isothermal Forging Speed and Deformation on the Microstructure and Properties of the Blade Forging

Figure 8 shows the microstructure of the blade forging after forging at a constant temperature of 940°C with different deformation speeds and degrees, followed by the same heat treatment. It can be observed that the α phase size of the alloy is refined after deformation. Comparing figures 8(a) and 8(b), when the deformation speed is constant, the α phase size slightly increases with increasing deformation degree. Comparing figures 8(b)-8(d), when the deformation degree is constant, the α phase size slightly increases as the deformation speed decreases. The slight increase in α phase size is due to the fact that whether it is an increase in deformation degree at the same deformation speed or a decrease in deformation speed at the same deformation degree, the overall result is an increase in deformation time, allowing the α phase more time to grow.

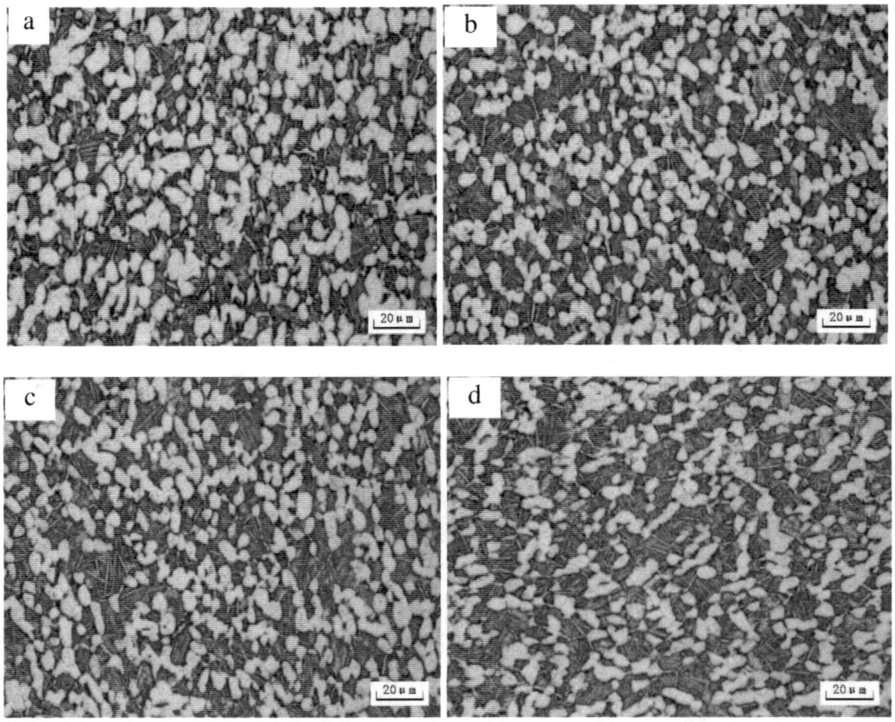

Figure 8. Microstructures under different deformation conditions:(a) ε=50%, v=0.1 mm/s;(b) ε=25%, v=0.1 mm/s;(c) ε=25%, v= 0.05 mm/s;(d) ε=25%, v= 0.01 mm/s.

The analysis of the mechanical properties under the four different forging regimes, as shown in figure 9, reveals that the strength and plasticity are essentially consistent with no significant differences observed. In conjunction with the microstructural photographs from figure 8, there is a slight variation in the size of the α phase between deformations of 25% to 50% and deformation speeds from 0.1 mm/s to 0.01 mm/s. Different speeds and deformation degrees have little impact on the mechanical properties.

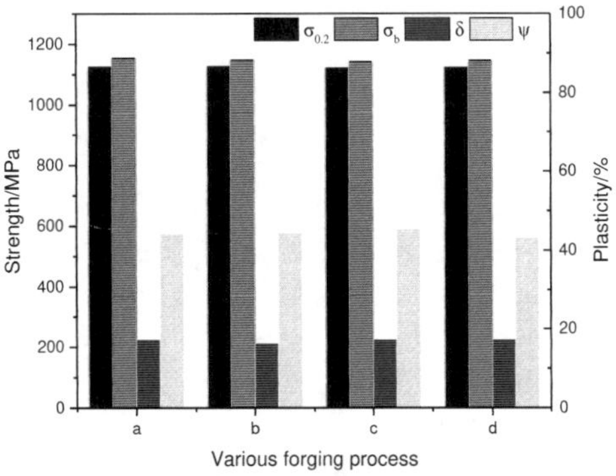

Figure 9. Room temperature mechanical properties forged under different deformation speeds and degrees:(a) ε=50%, v=0.1 mm/s;(b) ε=25%, v=0.1 mm/s;(c) ε=25%, v= 0.05 mm/s;(d) ε=25%, v= 0.01 mm/s.

3.5. Verification of Isothermal Forging Forming Process for Blade Forgings

Isothermal forming experiments for TC6 titanium alloy blade forgings were conducted on a 2000T hydraulic press. The forging process parameters were selected within the ranges determined above: the billet and mold heating temperature were 955°C (50°C below the β phase transformation temperature), the holding time was 60 minutes, the forging speed was 0.1 mm/s, and the deformation degree was between 30% to 50%. The heat treatment process was "850°C × 1 h, AC + 600°C × 2 h, AC". Figure 10 shows the TC6 titanium alloy blade forging and its microstructure. The blade forging is well-filled, with high dimensional precision, and the microstructure exhibits a uniform distribution of α phase at a content of 63%. The room temperature tensile results were σ_b = 1171 MPa, $\sigma_{0.2}$ = 1113 MPa, δ = 16.9%, and Ψ = 43.1%, achieving an optimal balance of strength and plasticity.

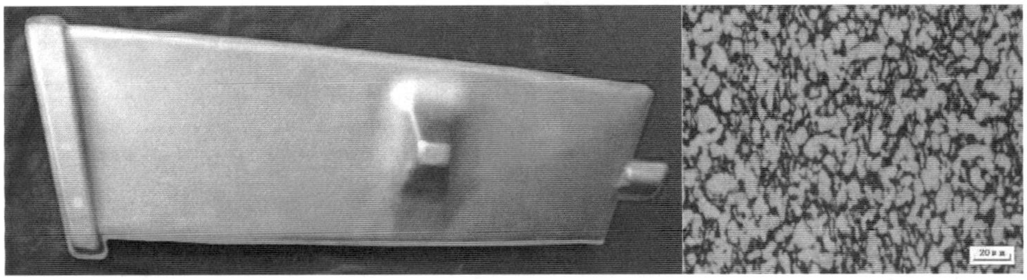

Figure 10. TC6 Titanium alloy isothermal forging blade forgings and microstructure.

4. Conclusions

(1) By analysing the structural characteristics of the thin-walled variable cross-section blade forgings, the preform was optimized, which is conducive to forging blade forgings with high dimensional precision and full filling.

(2) There are significant differences in the properties of TC6 titanium alloy blade forgings under four typical heat treatment methods. Materials treated by solution treatment and aging and duplex annealing exhibit the highest strength but the lowest plasticity; those treated by conventional annealing and isothermal annealing show reduced strength but higher plasticity. In conjunction with the project's required properties for blade forgings, duplex annealing is superior.

(3) Based on the results of experimental studies, reasonable parameters were established, and through

actual production trials, blade forgings with smooth surfaces, full filling, and free from defects such as folding were prepared. Additionally, the best balance of strength and plasticity and a uniform microstructure distribution were achieved.

5. References

[1] Yan M G, Liu B C, et al. 2002 *China Aeronautical Materials handbook (Second Edition)* **Volume 4** (Beijing: Standards Press of China) p 132.

[2] Zhu L, Xu Y, Chen L P, Hu S S and Zhou Q 2022 Deformation behavior of TC6 titanium alloy at high temperature and low strain rate *Special Casting &Nonferrous Alloys* **42(2)** 226-229.

[3] Xiong A M, Chen S H, Huang W C, Lin H and Li M Q 2003 Thermal deformation behavior and microstructure evolution for TC6 titanium alloy *Rare Metal Materials and Engineering* **32(6)** 447-450.

[4] Shen Z, Wu R, Yuan C and Wei J 2021Comparative study of metamodeling methods for modeling the constitutive relationships of the TC6 titanium alloy *Journal of Materials Research and Technology* **10** 188-204.

[5] Guo L F, Zhu Y C, Kong H X, Yuan M N and Wu F J 2012 Isothermal forging process of titanium alloy complex components *Chinese Journal of Rare Metals* **36(2)** 357-362.

[6] Williams J C and Boyer R R 2020 Opportunities and issues in the application of titanium alloys for aerospace components *Metals* **10(6)** 705.

[7] Pang K C 2002 Isothermal forging technology for precision forging of aerospace material *Acta Metallurgica Sinica* **38** 356-359.

[8] Zhang L J, Chang H and Xue X Y 2010 Isothermal forging technology and its application in aviation industry *Hot Working Technology* **39(21)** 21-24.

[9] Jiang H Y and Chen F 2017 Experimental study on isothermal forging technology for a complex-shaped titanium alloy wing *Rare Metal Materials and Engineering* **46(11)** 3182-3187.

[10] Zhu Z S, Wang Q R, Zhou Y and Wang X N 2004 Study of heat treatment processings of TC6 titanium alloy bars *Journal of aeronautical materials* **24(5)** 15-20.

[11] Pérez C L, Pérez D S and Arbizu I P 2014 Design and mechanical property analysis of ultrafine grained gears from AA5083 previously processed by equal channel angular pressing and isothermal forging *Materials and Design* **63** 126-135.

[12] Le T K and Bui T A 2021 Cold forging effect on the microstructure of motorbike shock absorbers fabricated by tube forming in a closed die *Applied Sciences* **11** 2142.

[13] Obiko J O, Mwema F M and Bodunrin M O 2019 Finite element simulation of X20CrMoV121 steel billet forging process using the Deform 3D software *SN Applied Sciences* **1** 1-10.

Preparation and Properties of Functionally Graded Lightweight High Insulation Materials

Xiaoxing Wei[1], Wenhao Lu[1], Xingdou Li[2], Zhixiong Huang[1], Hechen Liu[2], Jinyin Zhang[1], Xiang Xiao[1] and Le Li[2,*]

[1] Electric Power Research Institute of EHV Transmission Company, CSG, Guangzhou, Guangdong, China
[2] Department of Electrical Engineering, NCEPU, Baoding, Hebei, China
Email: weixiaoxing@ehv.csg.cn; luwenhao@ehv.csg.cn; 2468529829@qq.com; huangzhixiong@ehv.csg.cn; hc.liu@ncepu.edu.cn; zhangjinyin@ehv.csg.cn; xiaoxiang@ehv.csg.cn; *lile@ncepu.edu.cn

Abstract. In this study, functional gradient epoxy resin /high-performance composite foam was prepared, and its properties and structural characteristics were studied. By changing the ratio of epoxy resin and high-performance composite materials, the layered control of material density and structure was achieved during the preparation process, thereby achieving optimization of material performance. The results show that the prepared lightweight composite foam based on functionally graded epoxy resin has high strength, high modulus, low density, and excellent energy absorption properties. Through the analysis of the water diffusion leakage current experiment and partial discharge experiment, it is proved that the material has excellent structural characteristics, such as a layered structure. In addition, the effects of functional gradient changes of different types and contents on material properties and structures were also studied, and it was found that appropriate changes can significantly improve material properties and structures. This study may provide a theoretical and practical basis for the research and development of more excellent lightweight composite materials.
Keywords. Insulation material, gradient epoxy resin, composite foam, lightweight composite material.

1. Introduction

Epoxy resin based lightweight high insulation material is an important class of electrical insulation materials, because of its electrical insulation properties, physical properties, mechanical properties, chemical stability and other aspects of good characteristics, widely used in power, electronics, communications, aviation, automotive and other fields [1]. With the continuous development of science and technology, the types of epoxy resin based lightweight high insulation materials continue to increase, such as epoxy foam, epoxy molding materials, epoxy glass fiber reinforced plastics, epoxy ceramics and so on [2]. With the continuous improvement of production technology, the performance of epoxy resin based lightweight high insulation material products has been significantly improved, including material refinement, and processing technology improvement. Application of reinforcement modification technology: Through the reinforcement modification of epoxy resin based lightweight high insulation materials, it can improve its mechanical properties, heat resistance, anti-aging properties, etc. For example, by adding filling materials, adding fibers, etc., the material properties can be improved [3,4]. In order to develop the demand for high-performance applications: the demand for performance of electrical equipment and the increasing trend of environmental protection, epoxy resin based lightweight high insulation materials are also constantly improving their performance and application fields [5,6], such as power transformers, high-voltage cables, precision electronic equipment, spacecraft and other fields. In short, the development of epoxy resin based lightweight high insulation materials continues to grow and improve and will have broader application and development prospects in the future.

As important equipment in substations and transmission lines, pillar composite insulators play an important role in the mechanical support and electrical insulation of isolation switches, tubular busbars and flexible conductors [7,8]. At present, pillar composite insulators are widely used at home and abroad, and their internal core is the most important structural component of composite insulators. It not only plays the role of electrical internal insulation but also plays the role of bearing external mechanical loads [9]. At present, the foam-filled insulation core can realize the preparation of a large-size core system while ensuring its own lightweight and high insulation and has the prospect of becoming the mainstream insulation core [10]. Syntactic Foam (SF) with hollow microspheres as filler and epoxy resin as the matrix has good performance in water absorption, water diffusion leakage current, breakdown strength, etc., which better improves the poor water permeability of traditional polyurethane foam materials. However, there are still some problems in the practical application of pillar composite insulators, such as the poor resistance of high and low voltage end materials to partial discharge, mainly because there is distortion in the electric field at both ends of the pillar, which leads to the increase of voltage at both ends and causes partial discharge [11,12]. The above defects will have a great impact on the insulation properties of composite foam materials [13].

So we introduce the concept of functional gradient. Functionally gradient materials (Functionally Graded Material, FGM) refers to according to the specific requirements, choosing two or more than two kinds of materials through continuous change the composition of the material and structure, as a result, this project proposes a gradient control based on the composition and organization of new heterogeneous composite material [14]. By controlling the composition and organization of the material, the functional requirements for different parts of the same material or device are realized.

2. Experimental

2.1. Material

Both NMP and DMP30 were supplied by McLean, and bisphenol A type epoxy (E51) and organic microspheres were purchased from Aladdin.

2.2. Sample Preparation

Using the above materials, combined with the preparation method of FGM materials, functional gradient lightweight composite materials can be prepared. The preparation process is as follows:

(1) Mix the E51 resin with the methylhexa accelerator and heat to a certain temperature until the mixture is completely dissolved.

(2) Add the catalyst to the mixture and mix evenly.

(3) Add the organic microspheres into the mixture and mix evenly.

(4) Pour the mixture into a cylindrical mold.

(5) Using a heater or oven, control the temperature gradient to gradually cure the mixture. When the mixture is cured at the bottom of the mold, the mold is moved so that the height of the mixture in the mold gradually increases, thus forming a material composition gradient.

(6) After the material is fully cured, remove the material from the mold and cut it to the desired size.

By the above preparation method, the successful energy gradient material can be prepared, and the proportion and arrangement of different materials can be adjusted to further control the performance gradient of the material. This method can control the density, strength and thermal stability of the material, and is widely used in the manufacture of lightweight materials and high-performance composite materials.

2.3. Preparation of Leakage Current Test Materials

2.3.1. Sample Preparation. First, open the vacuum blast drying oven for preheating, preset 80°C, then use the electronic balance through the beaker to accurately weigh the epoxy resin, methyl-six curing agent, DMP-30 accelerator, and add 1.5%, 2%, 2.5% organic microspheres in batches (Note: Conditions allow as far as possible placed in the bellows to complete, the mass of the microsphere is too easy to float and harmful to human absorption in the lungs), according to the proportion of preparation completed several (Note: In order to prevent too much resin sticking to the beaker and causing too much loss, the input sequence can be appropriately changed), seal the beaker mouth with plastic wrap, place it in the planetary mixer for vacuum treatment and stir for 3-5min, put it into the preheated vacuum drying oven for 40min, take it out and stir it again in the mixer for 2-3min, and then pour it into the rubber mold. It is stored in a preheated 100°C drying oven and left for 2h to wait for solvent curing. At the end of the process, the mold is naturally cooled to room temperature, and then the product is removed from the mold. According to this process, more than ten materials of 1.5%, 2% and 2.5% batches were prepared respectively, and the experiments in the following stage were carried out, as shown in Figure 1.

Figure 1. Finished material with a single microsphere concentration.

2.3.2. Segmentation and Polishing of Materials. The cured material is cut into small pieces with a thickness of about 1cm and a diameter of not less than 3cm by a cutting machine, which is divided into the prepared containers with 1.5%, 2% and 2.5%, and then further processed by a grinding machine to polish off the smooth surface and ensure the smooth cutting surface, which is conducive to the splicing in the following experiments, as shown in Figure 2. (Note: Since the next test bonding is placed in the rubber mold in turn, unnecessary bubbles may be generated, so the crescent shaped notch can be polished during polishing so that the gas can be removed during the second polishing).

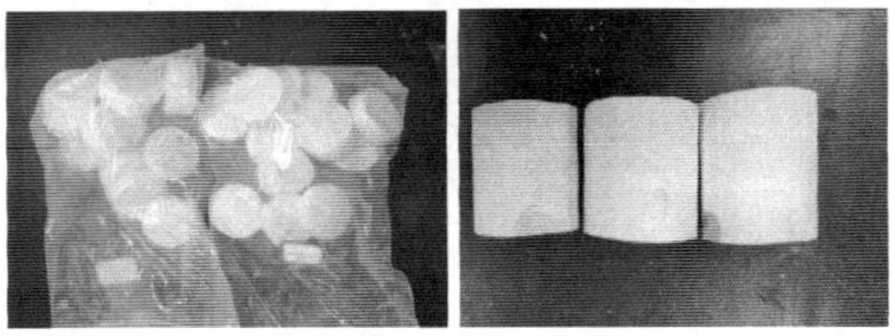

Figure 2. Material cut finished product (left) and samples before splicing and trimming of organic microspheres of different concentrations (right).

2.3.3. Glue and Trim. The reagent with 2% concentration is prepared again as the adhesive, and the materials with good specifications and sizes are splice according to the principle that the middle concentration is greater than the two-side concentration, and two types of materials are splice respectively (2% on the middle 2.5% side and 1.5% on the middle 2% side), a total of 12 groups of materials are polished and shaped again by the sander, and then packaged and marked for use, as shown in Figure 3.

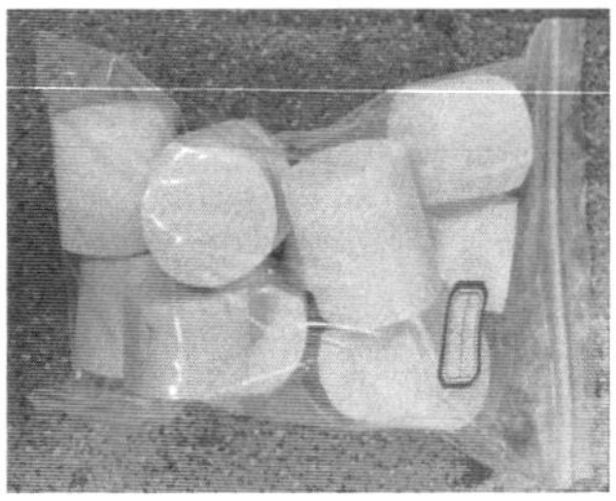

Figure 3. Different concentrations of microspheres spliced finished products.

2.3.4. Water Bath. A number of sodium chloride solutions with 0.1% solubility are configured, the spliced materials are placed in a mesh bag to make a distinction mark, a water bath is prepared, and it is sunk into the configured 0.1% sodium chloride solution with a heavy weight, and then boiled for 100h.

2.3.5. Outcome Measurement. At the end of the water bath, the sample was dried and a voltage of 12 kV was applied to the plate electrode, and then the leakage current was determined by Rigio-DM3068 with an accuracy of ±0.1 microamps.

2.4. Preparation of Partial Discharge Test Materials
Firstly, the mold was pretreated, a layer of release oil was evenly applied on the surface of the mold, and then it was put into a drying oven for heat treatment, and the vacuum drying oven was opened to preheat at 80°C, and the E51 epoxy resin, methyl-6 curing agent and DMP-30 catalyst of the established scheme were weighed with an electronic balance. Add 1.5%, 2% and 2.5% organic microspheres into them respectively, and place them in a planetary mixer for vacuum stirring for about 3-5 minutes, and place them in a preheated vacuum drying oven for 40min for pre-treatment. During this period, the molds in the drying oven can be taken out for needle embedding operation, and after solvent pretreatment is completed, the molds can be placed in a planetary mixer for stirring for 2-3 minutes and taken out for use, as shown in Figure 4.

Figure 4. Partial discharge test sample mold.

We Install the isolation plate on the mold using a pre-prepared clip so that the solvent in different microsphere concentrations is poured in stages (note: During the dumping should be as far as possible to lean on the same position, with the solvent's own liquidity to cover the mold, so as not to produce bubbles, have an uncontrollable impact on the test results), the first section of 3mm, the second section of 5mm required spacing, still in accordance with the principle that the middle concentration is greater than the concentration on both sides, remove the isolation plate and buckle after the solvent is installed, and rely on the solvent's own fluidity to connect and fuse. In this way, several parts of materials with different concentrations and ratios are sufficient for packaging and marking, as shown in Figure 5 and Figure 6.

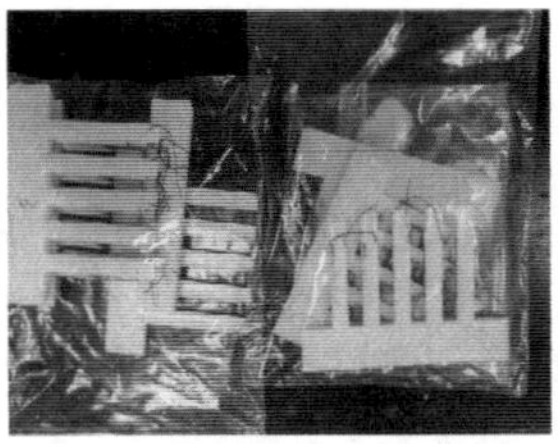

Figure 5. Partial discharge material finished.

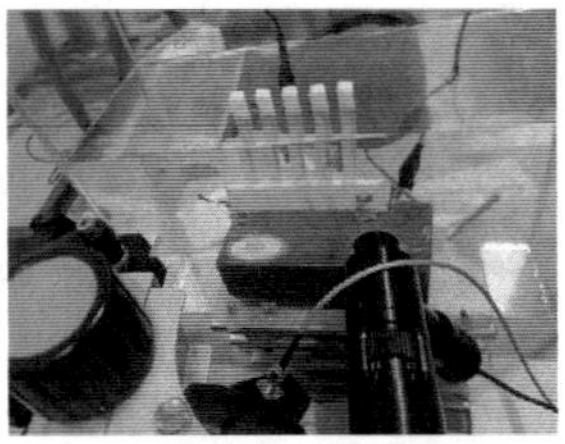

Figure 6. Partial discharge field experiment connection diagram.

3. Results and Discussion

3.1. Leakage Current Test

Leakage current test of functional gradient light composite material after a water bath can not only reflect the insulation resistance of the material itself but also detect the microscopic interface and defects of the filling material after a long time of high-temperature boiling. As shown in the bar chart, it can be seen that the splicing effect between different concentrations of organic microsphere composites is different. According to the data, the performance of the combination of 1.5% and 2% is superior, and the leakage current is low. On the contrary, the combination of 2% and 2.5% not only has high leakage current but also has extremely high instability. This indicates that in the environment of sodium chloride solution boiling at high temperature, sodium chloride molecules will enter the epoxy resin matrix composite foam material, and the composite foam material with different concentrations of organic microspheres will cause different effects. The combination of organic microspheres with greater solubility will increase the space between polymer molecules, and sodium chloride solution is easy to penetrate, resulting in increased leakage current and tends to be unstable, as shown in Figure 7.

Table 1. Leakage current experimental date.

Double-sided 2% clamp 2.5%	25.1	24.83	24.45	24.93	25.75	25.25
Double-sided 1.5% clamp 2%	24.45	24.22	24.24	24.35	24.33	24.27

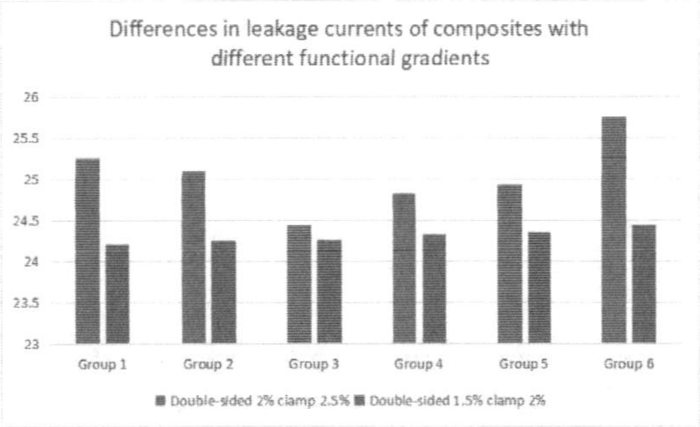

Figure 7. Comparison of the discharge current of different functional gradient combination samples.

3.2. Partial Discharge Experiment

A partial discharge test is one of the important methods to measure the electric strength and insulation level of insulating materials. The analysis of partial discharge test data can help to understand the nature and state of insulation materials and detect the type, location, and severity of partial discharge defects, as shown in Figure 8 and Figure 9. The data obtained from the combination data of pure 2% organic microsphere epoxy resin matrix composite foam material and other five groups of functionally graded materials with different concentrations were averaged as a control experiment. It can be seen from the data graph of local discharge amplitude and discharge time that functionally graded materials have lower amplitude and tend to be stable, as shown in Figure 10 and Figure 11. However, the amplitude of the pure 2% concentration organic microsphere epoxy resin matrix composite foam is high and has great fluctuation, and the relationship between the discharge energy and phase of the two groups of data does not have a great impact, and the discharge gradient of the functional gradient material is slightly better than that of the functional gradient material, so the functional gradient material can be used to meet the actual demand and have better amplitude effect. The performance is naturally better than that of the pure concentration material, so it can be seen that the functional gradient composite foam has a better development prospect and stable comparison effect than the pure concentration organic microsphere epoxy resin matrix composite foam, and can better serve the expansion and development of the functional gradient material experiment, as shown in Figure 12 and Figure 13.

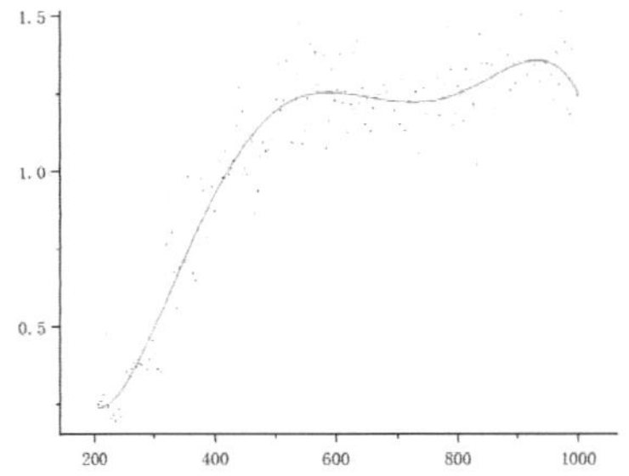

Figure 8. Relation between the amplitude of local discharge and discharge time of pure 2% microsphere concentration.

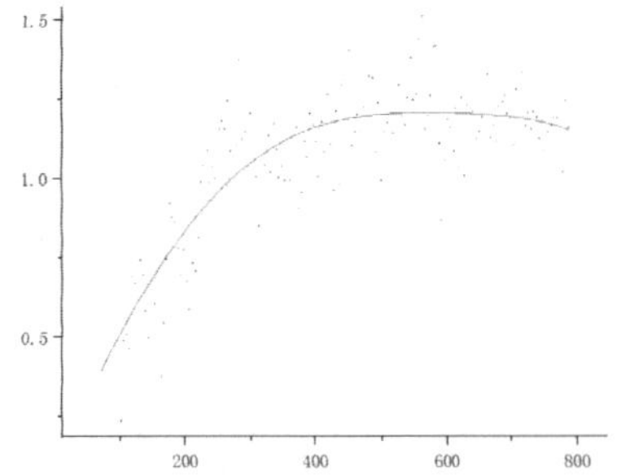

Figure 9. The relationship between the amplitude of local discharge and discharge time is taken for each point of different groups of functionally graded materials.

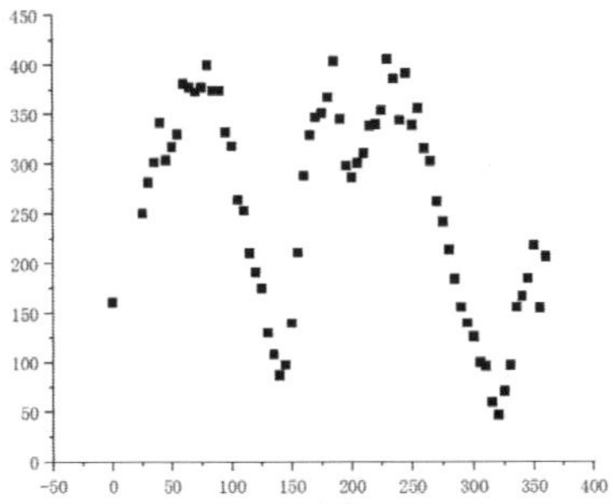

Figure 10. Relationship between frequency and phase of fai-n discharge in pure 2% material.

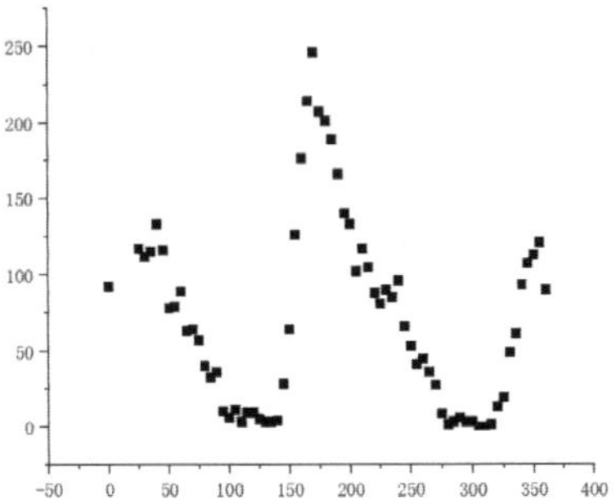

Figure 11. Relationship between discharge times and phase of functionally graded material fai-n.

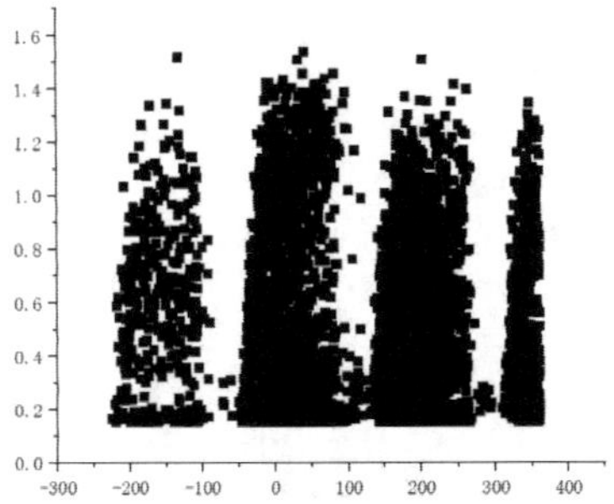

Figure 12. Relationship between discharge and phase in fai-q diagram of pure 2% material.

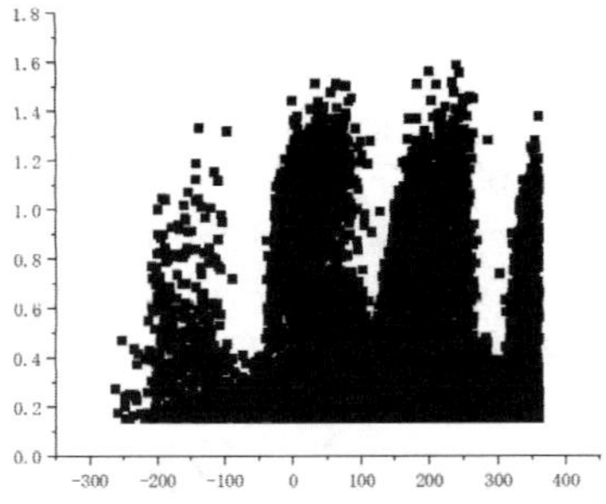

Figure 13. Relationship between discharge and phase of functionally graded material fai-q diagram.

4. Conclusion

In this paper, the properties and correlations of composite foams with different functional gradients are studied. It is found that the leakage current and partial discharge are easy to generate under a high electric field, and the generation of leakage current and partial discharge is closely related to the gradient structure of the material. Specifically, when the gradient structure of functional gradient composite foams is relatively gentle, the leakage current and partial discharge are relatively less. When the gradient structure of the material is steep, the leakage current and partial discharge are relatively greater. In addition, the study also found that the partial discharge of functional gradient composite foams is mainly concentrated at the gradient change, which is easy to causes electric field concentration, resulting in material damage and aging, but it also provides a more excellent and stable practical material in the selection of actual needs. Therefore, the rational design of the gradient structure of functional gradient composite foams and the reduction of unnecessary electric field concentration is of great significance to the safety and durability of the materials. It should be noted that the results of this study are only applicable to functionally graded composite foams, and their applicability to other materials needs further study.

Acknowledgments

This research was funded by the China Southern Power Grid technology project, CGYKJXM20220300.

5. References

[1] Lv JQ, Wang Yichen, Pan XX, et al. Research progress on effect of epoxy resin foaming process on cell morphology [J]. Plastics Industry, 2021, pp. 49 (01):7-12.
[2] Gu Yu, Yang Lin, Zhang Fuzeng, et al. Large size composite post insulator for UHV converter station at high altitude. Dc pollution flashover characteristics [J]. Transactions of China Electrotechnical Society, 2016, pp. 31 (10):93-101.
[3] Fredriksson M, Astbäck J, Pamenius M, etal. A Retrospective Study of 236 Patients with Teeth Restored by Carbon Fiber-reinforced Epoxy Resin Posts [J].The Journal of Prosthetic Dentistry, 1998, pp. 80(2):151-157.
[4] Gu J, Yang X, Lv Z, etal. Functionalized Graphite Nanoplatelets/Epoxy Resin Nanocomposites with High Thermal Conductivity[J]. International Journal of Heat and Mass Transfer, 2016, pp. 92(4):15-22.
[5] Qi Shujing, Yin Liwen, Song Pishuang et al. Production Status and Market Analysis of epoxy resins at home and abroad [J]. Elastomers, 2012, pp. 7 (5):84-89.
[6] Huang C. Preparation and properties of composite foam with ablative material insulation layer [D]. Wuhan: Wuhan University of Technology, 2016.
[7] Wang D Z. Production and Application of epoxy resin (2nd Ed.). Beijing: Chemical Industry Press, 2001.
[8] Chen Xiangbao. Performance resin matrix. Beijing: Chemical Industry Press, 1998.
[9] Chen Xiangbao, ZHANG Baoyan, Xing Liying. Technology development and application status of advanced resin matrix composites [J]. Materials Progress in China, 2009, pp. 11 (6):2-12.
[10] Chen Xiaosong, Gao Weibin, Liu Rixin, et al. Study on preparation of EPDM sponge with expandable microsphere foaming agent [J]. Rubber Industry, 2011, pp. 58(2):101-104
[11] El Husseiny M, Baraka A A, Oraby O, et al. Fabrication of Bimetallic High-Strength Low-Alloy Steel/Si-Bronze Functionally Graded Materials Using Wire Arc Additive Manufacturing[J]. Journal of Manufacturing and Materials Processing, 2023. DOI:10.3390/jmmp7040138.
[12] Nohut S, Schwentenwein M. Vat Photopolymerization Additive Manufacturing of Functionally Graded Materials: A Review[J]. Journal of Manufacturing and Materials Processing, 2022. DOI:10.3390/jmmp6010017.
[13] Fábio M Conde, Coelho P G, José M Guedes. Multi-material and strength-oriented

microstructural topology optimization applied to discrete phase and functionally graded materials[J]. Structural and Multidisciplinary Optimization, 2022, 65(4).
[14] Lee S, Moon J, Kim H, et al. Development of functionally graded austenitic lightweight steel through electrically assisted pressure solid-state joining[J]. Materials Science & Engineering, A. Structural Materials: Properties, Misrostructure and Processing, 2024 (Jan.): 891.

Study on the Correlation Between the Black Sludge and Chemical Oxygen Demand in Wastewater from Cold Rolling Pickling Lines

Xianglai Xu[1,2,3,*], Donghui Li[1], Cui Yan[3], Peng Tian[5 *], Yongyue Liu[4], Xueping Ren[4], Jiangtao Gao[4], Qingwei Jia[2] and Baolei Wang[2]

[1] School of Electrical Automation and Information Engineering, Tianjin University, Tianjin 300072, China.
[2] Wild SC (Ningbo) Intelligent Technology Co., Ltd., Ningbo 315500, Zhejiang, China.
[3] Marine College, Tangshan Normal University, Tangshan 063000, Hebei, China.
[4] School of Materials Science and Engineering, University of Science and Technology, Beijing 100083, China.
[5] Tangshan Donghua Iron and Steel Enterprise Group Co., Ltd., Tangshan 063016, Hebei, China.
Email: *28431663@qq.com

Abstract. Black sludge is commonly found in the acid tanks of cold rolling pickling lines and in the pipelines of acid regeneration units, impacting system efficiency and causing negative effects on equipment and the environment. To address this issue, a method combining black sludge composition analysis and Chemical Oxygen Demand (COD) measurement of rinse water was proposed to study its origin and formation mechanisms. The results showed that the black sludge mainly contains carbon, chlorine, oxygen, and small amounts of iron, with significant organic matter loss between 300-400°C, indicating a high content of alkanes and cycloalkanes. The COD value of rinse water is higher when fresh acid is used, and alkanes and cycloalkanes are the main contributors to the formation of black sludge and the increase in COD values. Strengthening the management of fresh acid can help reduce sludge formation.
Keywords. Cold rolling pickling line; black sludge; Alkanes and cycloalkanes; Chemical Oxygen Demand (COD).

1. Introduction

Cold rolling pickling units are used to remove oxide scale and impurities from the surface of steel, thereby improving the surface quality. However, the pickling process generates black mud containing a variety of complex components. It has been observed that some production lines producing silicon steel generate black mud with a higher silicon content [1]. However, there are also pickling units that do not produce silicon steel at all, yet still generate a large amount of black mud. This black mud not only leads to the waste of pickling solutions but can also cause equipment corrosion and environmental pollution. Acidic black mud often adheres to the inner walls of equipment and tank covers, and can easily fall onto the surface of the strip steel, forming black spots and affecting product quality [2]. More seriously, the black mud may block the mist extraction pipes, leading to hydrogen explosions. When the black mud enters the acid regeneration system, it attaches to the concentrated acid pipes, reducing their inner diameter. To avoid such situations, some large steel plants have even added acid regeneration pipe cleaning systems [3].

Furthermore, COD (Chemical Oxygen Demand) is a key indicator in water quality monitoring, used to assess the degree of organic pollution in water bodies. The higher the COD value, the higher the concentration of organic pollutants in the water, and the more severe the water pollution [4]. Many pickling units have high COD values in their rinsing wastewater, but the constituents affecting the COD value have not been identified [5]. It is suspected that there is a close correlation between the generation of black mud and the changes in the COD value of the rinsing tank wastewater. Therefore, it is of great significance to conduct in-depth research on the composition and formation mechanism of black mud.

2. Experimental Materials and Methods

2.1. Experimental Materials

Black mud samples were collected from two pickling units. Unit 1, a continuous pickling unit producing SPHC steel with 0.1% inhibitor, and Unit 2, a pickling-rolling unit producing mainly low-carbon steel without inhibitors. Samples from the pickling tank covers were named Sample 1 (Unit 1) and Sample 2 (Unit 2), as shown in figure 1. Scheme 1 refers to using concentrated regenerated acid in Unit 1, while Scheme 2 uses fresh acid.

(a) Unit 1 (b) Unit 2

Figure 1. Sampling locations of pickling black sludge.

2.2. Experimental Methods

After collecting the black mud samples, their morphology was analyzed using SEM scanning and energy dispersive spectroscopy. Additionally, thermogravimetric analysis was conducted, and the black mud extracts were analyzed using gas chromatography-mass spectrometry (GC/MS) to identify their components. The equipment used was TGA4000-GCMS 580, under the following conditions: 30 degrees Celsius for 5 minutes, then heated to 250 degrees Celsius at a rate of 5 degrees per minute, held for 5 minutes, with an ion scan range of 30-400 molecular weight. Correspondingly, the potassium dichromate method (standard method) was used to measure the COD values of wastewater at different locations in the rinsing section [6]. The COD value of the fresh acid was measured by first diluting the fresh acid and regenerating acid 100 times, then measuring the COD value, and finally multiplying the measured value by 100 to obtain the actual COD value.

3. Experimental Results

3.1. Morphology and Elemental Analysis

The morphology and elemental composition of Sample 1 and Sample 2, originating from different units, were very similar. Firstly, using scanning electron microscopy, the micro-morphology was observed to be very similar, with the black mud samples exhibiting a rough surface and a certain amount of small pores. During the formation of black mud, gas escapes from the slurry, forming bubbles. When solidified or dried, these bubbles leave behind small pores, as shown in figures 2(a) and 2(b). Subsequently, energy dispersive spectroscopy analysis revealed that the black mud contained a large amount of carbon (C), chlorine (Cl), and oxygen (O), as well as a small amount of iron (Fe). This elemental composition confirms that both black mud samples are the same substance. The high chlorine content is related to the source of the black mud, which is the interior of the acid tank. Oxygen is also one of the main components of black mud, and it may exist in the form of iron oxide or come from functional groups such as alkanes and cycloalkanes.

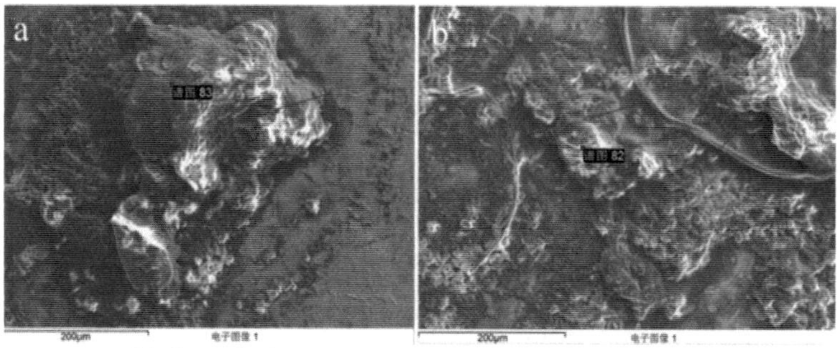

(a) Sample 1 (b) Sample 2

Figure 2. SEM images of pickling Black Sludge.

X-ray fluorescence analysis results were similar to the energy dispersive spectroscopy results, showing that the black mud mainly contained carbon (C), chlorine (Cl), oxygen (O), and a small amount of iron (Fe). The high carbon content in the black mud suggests the presence of organic matter. Therefore, infrared spectroscopy analysis was performed on the black mud.

3.2. Infrared Spectroscopy Analysis

From the infrared spectra of the samples before and after treatment, it can be observed that the infrared curves of the sample before drying (black line) and the sample after drying (red line) are basically consistent in the main characteristic peak positions, but there are also some obvious changes, as shown in figure 3. The broad peak near 3334.5 cm^{-1} is the absorption peak of O-H stretching vibration. This peak is significantly weakened in the dried sample, indicating that the moisture in the sample has been effectively removed [7]. The absorption peaks at 2922.6 cm^{-1} and 2922.7 cm^{-1} correspond to the C-H stretching vibration. These peaks exist both before and after drying, showing the stability of the alkyl structure in the sample. The absorption peak at 1703.7 cm^{-1} is attributed to the C=O stretching vibration.

The absorption peak at 1626.7 cm^{-1} is the characteristic peak of C=C stretching vibration, which disappears or significantly weakens in the dried sample, possibly due to the removal of some volatile components in the sample. The absorption peak near 1270.4 cm^{-1} is the characteristic peak of C-O stretching vibration, which exists both before and after drying, indicating the presence of alcohol or ether structures in the sample [8,9]. The absorption peaks at 1099.6 cm^{-1} and 1096.6 cm^{-1} are attributed to C-O stretching vibration and C-H bending vibration, respectively. These peaks are more pronounced in the dried sample, suggesting that other components or vibration modes in the sample are reduced after drying, making these characteristic peaks more prominent.

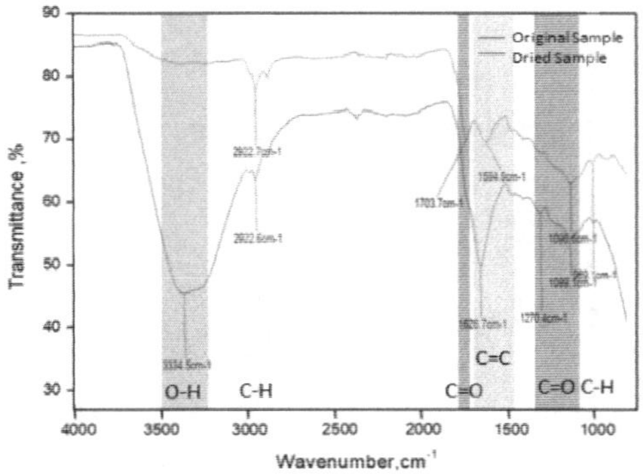

Figure 3. The infrared spectrum of the extract from Black Sludge.

3.3. Distribution of COD Values in Rinse Water

COD is an important indicator for assessing the total amount of organic pollutants in water bodies. It reflects the degree of water pollution by measuring the amount of organic matter in the water sample that requires oxidation by chemical oxidants. The higher the COD value, the higher the concentration of organic pollutants in the water. The phenomenon of black mud and increased COD values also appeared in the pickling-rolling unit without the use of inhibitors, indicating that there is no direct relationship with the addition of inhibitors.

The COD values in the rinsing section of the continuous pickling unit often fluctuate up and down, which is related to the type of acid supplied to the pickling-rolling unit by the acid regeneration unit. The COD values of Scheme 2 are generally lower than those of Scheme 1, during which time the pickling process section uses regenerated acid. The COD value of the demineralized water added to the rinsing tank is 0, ruling out the possibility of organic matter being brought in by external demineralized water. The common feature of Scheme 1 and Scheme 2 is that the COD value gradually decreases from Rinse Tank 1 to Rinse Tank 5, showing a clear directionality. This indicates that organic matter is brought into the rinsing tanks from the pickling section, leading to an increase in the COD values of the waste rinse water and the water in Rinse Tank 1, as shown in table 1.

Table 1. COD values in the rinsing section of the pickling line.

Scheme	Acid Type	Rinse Tank 1, mg/L	Rinse Tank 3, mg/L	Rinse Tank 5, mg/L	Waste Rinse Water, mg/L	Desalinated Water, mg/L
Scheme 1	Raw Acid	320	138	17	359	0
Scheme 2	Regenerated Acid	131	87	5	127	0

4. Analysis and Discussion

4.1. Chemical Composition Analysis

In thermogravimetric analysis (TGA), the sample shows two weight loss steps: one around 100°C, attributed to the evaporation of adsorbed and crystalline water, indicating moisture content [10], and another between 300-400°C, linked to the decomposition of organic matter [11], as shown in figure 4(a).

Through gas chromatography-mass spectrometry (GC-MS) analysis, the retention time results reveal the elution order of each compound in the chromatographic column, allowing for the estimation of the relative concentrations of compounds in the sample. Compounds with shorter retention times (such as undecane) are typically smaller molecules or less polar substances, while compounds with longer retention times (such as dodecane) generally have larger molecular weights or slightly higher polarity. This elution order is consistent with their physicochemical properties, indicating that the analysis process has good separation efficiency, as shown in figure 4(b).

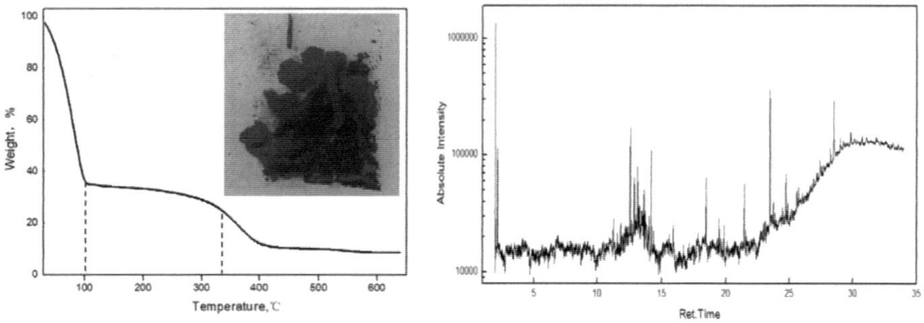

(a) the mass loss (b) GC-MS chromatogram

Figure 4. The mass loss and GC-MS chromatogram of Black Sludge.

Mass spectrometry analysis showed that the main components of the sample are hydrocarbons, including Undecane, trans-Decalin, and Dodecane, identified with retention times of 12.575 to 14.200 minutes. These compounds are primarily alkanes and cycloalkanes. Thermogravimetric and GC-MS analysis revealed significant weight loss around 360°C, indicating the decomposition of alkanes and cycloalkanes. This suggests that the sample is chemically complex, containing not only hydrocarbons but other organic matter as well. [12].

The black mud generated in cold rolling pickling and pickling-rolling units mainly results from the residue and transfer of organic matter from fresh acid during pickling. In addition to chemical dissolution, a physical process called "mechanical stripping" occurs, where hydrogen gas generated during the reaction with iron oxide peels off the scale. In acidic conditions, protonation reactions in organic matter increase the zeta potential, causing sludge aggregation (figure 5(a)). Alkanes and cycloalkanes capture impurities like iron chips and adhere to the black mud surface, making them difficult to leach out. Long-term pickling leads to the accumulation of black mud in the acid tank. Anomalies such as the presence of alkanes and cycloalkanes may be due to contamination during production or transportation (figure 5(b)).

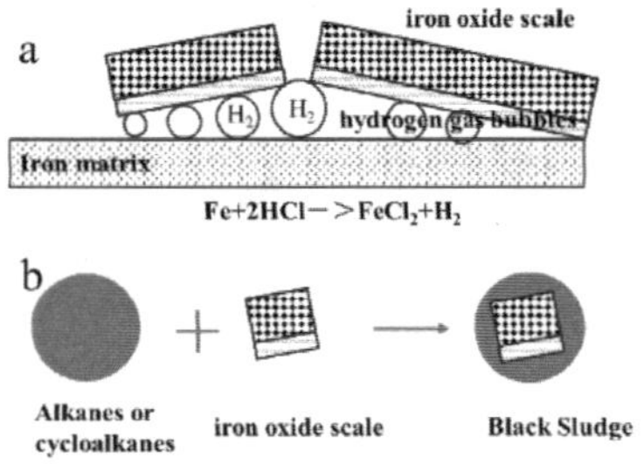

(a) Iron oxide scale peeling process

(b) Organic matter encapsulating iron chips process

Figure 5. Schematic diagram of Black Sludge formation.

4.2. Analysis of Organic Matter Sources

Regenerated acid undergoes high-temperature burning during the acid regeneration process; therefore, it should not contain any organic matter that is not resistant to high temperatures [13]. Based on the appearance of black mud and the fluctuation of COD values, it is speculated that it is related to the composition of fresh acid. The fresh acid supplied to the pickling process section contains alkanes and cycloalkanes, which remain in the acid tank after pickling and are carried into the rinsing tank with the steel strip, resulting in an increase in the COD value of the rinse water. The COD values of the fresh acid from both units were tested, and they were close to 500,000.

The main methods for preparing industrial hydrochloric acid, such as hydrogen chloride gas dissolution, azeotropic distillation, byproduct recovery, electrolysis, and catalytic synthesis, typically produce high-purity acid with minimal organic matter and near-zero COD values [14]. However, in special cases like byproduct recovery or transportation, trace organic impurities may increase the COD value. To reduce black mud generation and control COD values, the pickling process should be optimized with strict control of acid composition to minimize organic matter accumulation during pickling and rinsing, improving production efficiency, product quality, and environmental protection.

5. Conclusions

Through detailed analysis of black mud samples and COD values from continuous pickling units and pickling-rolling units, the following main conclusions are drawn:

1) Black mud mainly contains elements such as carbon (C), chlorine (Cl), oxygen (O), and a small amount of iron (Fe). The combustion reaction of black mud at 300°C indicates that it does contain organic matter.

2) Black mud contains alkanes and cycloalkanes, which originate from the fresh acid used in the pickling process. These organic substances remain in the acid tank after pickling and are carried into the rinsing tank with the steel strip, leading to an increase in the COD value of the rinse water.

3) The abnormal presence of alkanes and cycloalkanes in the fresh acid may be due to equipment contamination during the production process or contamination of the fresh acid during production, transportation, or storage.

Acknowledgment

This paper is funded by the Tangshan Talent Support Project A202301001.

6. Reference

[1] Qs Li 2009 *Research on the Silicon Sludge Cleaning Process of Silicon Steel Automated Pickling Line* (Hunan University) Master's Thesis

[2] Xu X L, Ma G Z and Lv C B 2018 Microanalysis of "black spot" defects in pickling products *Rolling Steel* **35(02)** 81-84.

[3] Xu X L, et al. 2018 Technological improvements in pickling and acid regeneration processes *Rolling Steel* **35(1)** 85-87.

[4] Lambertz S, Franke M, Stelter M and Braeutigam P 2024 Determination of chemical oxygen demand with electrochemical methods: a review *Chemical Engineering Journal Advances* 100615.

[5] Jiao W L and Cai Z D 2011 Research on the deep purification and reuse of steel pickling wastewater *Environmental Science and Management* **36(10)** 92-94.

[6] Paul Worsfold CPMM. *Encyclopedia of Analytical Science* 2019.

[7] Collazos-Escobar G A, Hurtado-Cortés V, Bahamón-Monje A F, et al. 2024 Water sorption isotherms and mid-infrared spectra of dried parchment coffee beans (Coffee arabica L.) processed by wet and semi-dry postharvest methods. A dataset for estimating water sorption and thermodynamic properties *Data in Brief* **57** 110958.

[8] Liu S, Fang L, Wang S and Hu C 2024 Accurate determination of alcohol-based diesels using optimal chemical factors *Spectrochimica Acta Part A: Molecular and Biomolecular Spectroscopy* 125140

[9] Adil M, Giri B R, Elkhazraji A and Farooq A 2023 Mid-IR absorption spectra of C1-C4 alkyl acetates at high temperatures *Journal of Quantitative Spectroscopy and Radiative Transfer* **300** 108522.

[10] Saldo J, Sendra E and Guamis B 2022 Changes in water binding in high-pressure treated cheese, measured by TGA (thermogravimetrical analysis) *Innovative Food Science & Emerging Technologies* **3(3)** 203-207.

[11] Dittmann D, Saal L, Zietzschmann F, et al. 2022 Characterization of activated carbons for water

Journal of Physics: Conference Series **2941** (2025) 012043 doi:10.1088/1742-6596/2941/1/012043

treatment using TGA-FTIR for analysis of oxygen-containing functional groups *Applied Water Science* **12(8)** 1-13.

[12] Wijayanti W and Sasongko M N 2024 The effects of limonene and eugenol additives in n-heptane and low-octane gasoline on the emission characteristics and fuel consumption of single-cylinder gasoline engine *Engineering Science and Technology, an International Journal* **51** 101648.

[13] Regel-Rosocka M 2010 A review on methods of regeneration of spent pickling solutions from steel processing *Journal of Hazardous Materials* **177(1)** 57-69.

[14] Wang W Z, Sheng M Q and Wei L 2024 The Influence of Roasting Process Parameters on Acid Regeneration Systems and Iron Oxide Powder *Baosteel Technology* **(01)** 28-32.

Study on the Effect of a Modified Phosphorus Flame Retardant on the Properties of Epoxy Resin

Ling Xiao*, Yu Li , Jie Xin and Xuan Wang

Department of Basic Course, Naval University of Engineering, Wuhan, China

Email: 359086326@qq.com

Abstract. In this study, we designed and synthesized a phosphorus flame retardant based on DOPO (9,10-Dihydro-9-oxa-10-phosphaphenanthrene 10-oxide) and obtained a modified phosphorus flame retardant (MDOPO). We investigated how the amount of MDOPO affects the flame retardancy and mechanical properties of epoxy resin and characterized the material's microstructure using SEM. The results demonstrated that the flame retardancy of MDOPO/E44 composites increased with increasing MDOPO content. When 11.43 g of MDOPO was added to 100 g of E44 matrix, the oxygen index of MDOPO/E44 composites increased by 24.7%. The oxygen index increased gradually with further increases in MDOPO content. When the phosphorus content in E44 was 0.6%, the oxygen index of MDOPO/E44, molten DOPO/E44, and powdered DOPO/E44 were 27.3%, 26.8%, and 25.8%, respectively, with tensile strengths of 39.43 MPa, 26.65 MPa, and 18.7 MPa, respectively. Scanning electron microscopy revealed no significant phase separation in the cross-section of MDOPO/E44 flame retardant material.

Keywords: Phosphorus flame retardant; Epoxy resin; Flame retardancy; Mechanical properties

1. Introduction

As the Chinese navy's capabilities continue to grow, the focus of naval weapon and equipment design has shifted from imitation and improvement to independent research and development. The application of high-performance polymer materials plays a crucial role in modern naval equipment [1–5]. Epoxy resin, owing to its excellent overall performance, is commonly used as a matrix material in applications such as thermal insulation coatings and stealth materials [6-10].

Epoxy resin-based polymer materials are prone to combustion, have a low limit oxygen index, and can lead to fires and significant loss of life and property when used. This risk is particularly concerning for naval vessels operating in the open ocean, where fires and explosions can result in immeasurable losses. Therefore, enhancing the flame retardancy of materials is of paramount importance in preventing fires, reducing casualties, and minimizing property damage during material development.

In this investigation, a phosphorus-based flame retardant, specifically 9,10-Dihydro-9-oxa-10-phosphaphenanthrene 10-oxide (DOPO), was systematically designed and chemically modified, resulting in the creation of a modified DOPO (MDOPO). Subsequently, the structural characteristics of MDOPO were assessed using IR. Furthermore, the study employed E44, a bisphenol A epoxy resin, as the matrix material, with T31 phenolic aldehyde serving as the epoxy curing agent. Additionally, powdered DOPO, molten DOPO, and MDOPO were incorporated into the matrix through physical

Journal of Physics: Conference Series **2941** (2025) 012044 doi:10.1088/1742-6596/2941/1/012044

blending. This allowed for an examination of how varying amounts of flame retardant influenced the performance of the E44 matrix.

2. Experiment

2.1. Instruments and Reagents

Instruments: YZS-B oxygen index instrument (Ji'nan Ainuo Instrument Co., Ltd.). CMI-4204 universal testing machine (Jiangdu Tianhui Experimental Machinery Co., Ltd.). VHX-S550E 3D stereoscopic microscope (KEYENCE Co., Ltd.). 36–111 digital caliper (STANLEY Co., Ltd.). Nicolet380 FTIR spectrometer (Guangdong Nicolet Scientific Instruments Co., Ltd.)

Reagents: 9,10-Dihydro-9-oxa-10-phosphaphenanthrene 10-oxide (DOPO, Type 852) (Shandong Jiuchong Chemical Co., Ltd.). 1,4-Butanediol diglycidyl ether (Guangdong Weng Jiang Reagent Co., Ltd.). E44 epoxy resin (Baling Petrochemical Corp) (Zhengzhou Guangchen Trade Co., Ltd.). T31 epoxy curing agent of phenolic amines (Zhengzhou Guangchen Trade Co., Ltd.)

2.2. Preparation of Modified Flame Retardant

First, 20.4 g of 1,4-Butanediol diglycidyl ether (BDDE, with a purity of 99%) was added to a 250 mL three-neck flask equipped with a mechanical stirrer and a reflux condenser. The mixture was then heated in a 120°C oil bath. Simultaneously, 21.6 g of DOPO was heated to 120 °C until it melted and transformed into a liquid state. This DOPO was then added to the three-neck flask in batches, stirred, and heated for reflux.

At specific intervals, 0.50 g of the reaction mixture was extracted and transferred to a 250 mL ground conical flask for titration, which measured the epoxy number in the reaction and allowed us to assess the progress of the reaction. After the reaction was complete, the liquid was allowed to cool naturally to room temperature, resulting in a faint yellow, transparent, and viscous liquid. This product is referred to as the modified DOPO (MDOPO) in Figure 1.

The change in epoxy number over time during the reaction is illustrated in Figure 2. It is evident that the epoxy number of the reaction mixture gradually decreased as the reaction progressed. After 720 min of reaction, the epoxy number essentially reached a constant value. The resulting modified product is known as MDOPO.

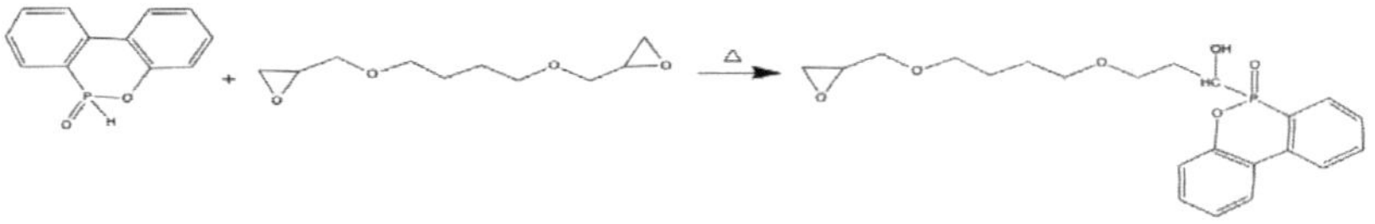

Figure 1. Synthesis reaction of modified DOPO (MDOPO)

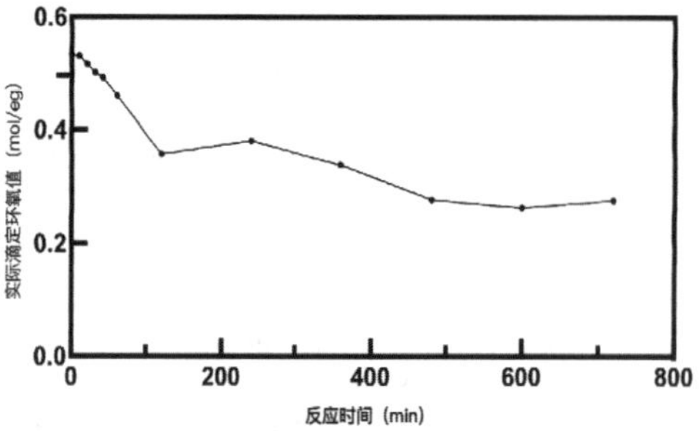

Figure 2. Titration curve of epoxy value in the reaction solution

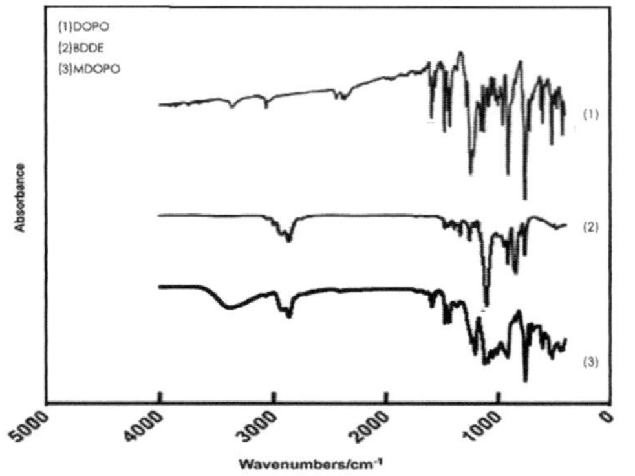

Figure 3. FTIR spectra of DOPO, BDDE and MDOPO

As depicted in Figure 3, the P-H peak located at 2,387 cm–1 vanished subsequent to the interaction with DOPO. A novel, broad -OH stretching vibration peak, situated at 3,410 cm–1, emerged as a result of the ring opening phenomenon stemming from the reaction between the epoxy group and reactive hydrogen. Additionally, the symmetrical and asymmetric -CH2 peaks at 2,930 cm–1 and 2,870 cm–1, the P =O peaks at 1,204 cm–1 and 1,238 cm–1, the C=C (AR) peak at 1,601 cm–1, and the bending vibration peak associated with the epoxy group at 911 cm–1 remained unchanged.

The reactive H in the P-H bond actively engaged in the reaction. Simultaneously, the ring opening of the epoxy group led to the generation of OH. These observations confirm that the resulting product was indeed MDOPO.

3. The Effect of MDOPO on the Properties of Epoxy Resin

3.1. Preparation of MDOPO/E44

Samples were prepared using 100 g of E44 as the matrix, varying the amount of MDOPO, and adjusting the T31 curing agent's amount accordingly. The specific amounts of MDOPO and curing agent are detailed in Table 1.

Table 1. The amount of MDOPO and curing agent added in E44

No.	The amount of MDOPO, in g	The amount of curing agent, in g	Phosphorus content, in %
1#	11.43	29.53	0.6
2#	15.77	30.11	0.8
3#	20.43	30.74	1.0
4#	25.43	31.41	1.2
5#	30.83	32.13	1.4

3.2. The Effect of the Amount of MDOPO on the Flame Retardancy of Epoxy Resin

In accordance with GB/T 2408, the specimens were prepared with dimensions of 120 mm×13 mm×3.2 mm for the vertical combustion test. Each specimen underwent five repeated tests, and the average of these results was computed. The results of the vertical combustion test conducted on MDOPO/E44 are presented in Table 2.

As per GB/T 2406, a limit oxygen index (LOI) test was carried out using standard specimens sized at 120 mm×10 mm×4 mm. Each standard specimen underwent five repeated tests, and the average of these test results was calculated. The obtained LOI test results are provided in Table 2.

Based on the results of the vertical combustion test conducted on MDOPO/E44, it was observed that the epoxy resin specimen, when separately mixed with MDOPO, still exhibited molten drops and liquid seepage on its surface after ignition.

Table 2. Results of the vertical combustion test on MDOPO/E44

No.	T1	T2	The molten drop can light a degreasing cotton	The fire spreads to the clamp	Test phenomenon	Flame retardant rating
1#	-	-	×	✓	There is much dark smoke, as well as molten drops and vacuoles on the surface	No
2#	-	-	×	✓	There is a little dark smoke, as well as molten drops and vacuoles on the surface	No
3#	-	-	×	×	There is a little dark smoke, as well as molten drops and vacuoles on the surface	No
4#	-	-	×	×	There is a little dark smoke, as well as liquid seepage on the surface	No
5#	-	-	×	×	There is a little dark smoke, as well as liquid seepage on the surface	No

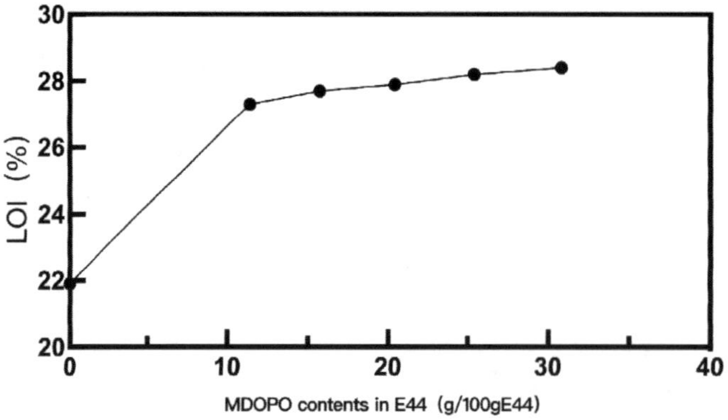

Figure 4. Relationship between LOI of MDOPO/E44 and the amounts of MDOPO

As depicted in Figure 4, the matrix, when blended with MDOPO, exhibited a significantly elevated oxygen index. Specifically, the oxygen index of E44, with the incorporation of 11.43 grams of MDOPO (possessing a phosphorus content of 0.6%), surpassed that of E44 in its flame-retardant-free state. Moreover, it was observed that the oxygen index of E44, as additional quantities of MDOPO were introduced, exhibited a gradual increase.

3.3. The Effect of the Amount of Modified DOPO on the Mechanical Properties of Epoxy Resin

According to GB/T 528–2009, a universal testing machine was employed to assess the mechanical properties of MDOPO/E44. The elongation rate was set at 500 mm/min. Each specimen underwent 5 repeated tests, and the results were averaged. As depicted in Figure 5 and Figure 6, the tensile strength and breaking elongation rate of the MDOPO/E44 composite gradually decreased with increasing MDOPO content. As the MDOPO content increased, the phosphaphenanthrene group content in the system also increased, resulting in an increase in the breaking space of the molecular chain and a gradual decrease in the mechanical properties of the composite. When the amount of MDOPO blended with 100 g of E44 exceeded 11.43 g, the mechanical properties of the composite significantly weakened.

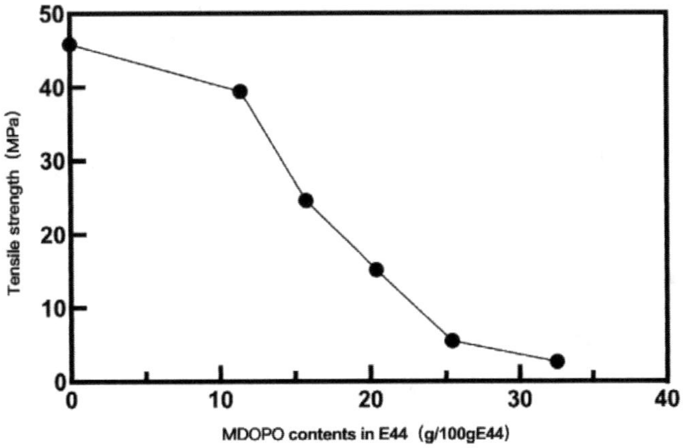

Figure 5. Relationship between Ts of MDOPO/E44 and the amounts of MDOPO

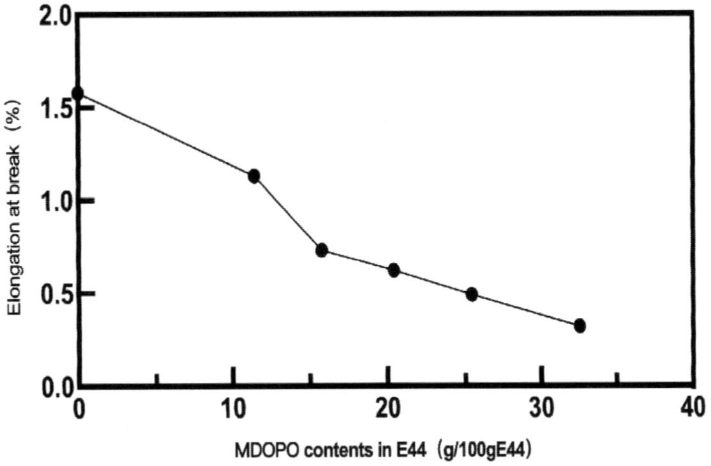

Figure 6. MDOPO/E44 Relationship between Eb and the amounts of MDOPO

3.4. Microstructure of the MDOPO/E44

SEM was utilized for specimen characterization. Consequently, the microstructures of the tensile cross-sections of 11.43 g of molten MDOPO/E44 were obtained, as depicted in Figure 7. The tensile cross-sections resemble well-distributed stereoscopic petals, with no discernible phase interface between the fillers and the matrix.

(a) Specimen (1000X) (b) Specimen (5000X)

Figure 7. SEM picture of MDOPO/E44 flame retardant material

3.5. Comparison Between the Properties of MDOPO/E44 and DOPO/E44

In this study, three types of phosphorus flame retardants—powdered DOPO, molten DOPO, and MDOPO—were employed. They were individually added to the E44 matrix to create three distinct composites, allowing us to assess both flame retardancy and mechanical properties. Figure 8 and Figure 9 depict the variations in oxygen indices and tensile strengths of these three composites as a function of phosphorus content.

As indicated in Figure 8, when the phosphorus contents in the powdered DOPO, molten DOPO, and MDOPO were 0.6%, the oxygen indices of the MDOPO/E44, molten DOPO/E44, and powdered DOPO/E44 composites were 27.3%, 26.8%, and 25.8%, respectively. As demonstrated in Figure 9, when the phosphorus contents in the powdered DOPO, molten DOPO, and MDOPO were 0.6%, the tensile strengths of the MDOPO/E44, molten DOPO/E44, and powdered DOPO/E44 composites were 39.43 MPa, 26.65 MPa, and 18.7 MPa, respectively. Upon comparing composites with the same phosphorus content (Figure 8 and Figure 9), it is evident that the composite using MDOPO as the flame retardant exhibits superior flame retardancy and tensile strength.

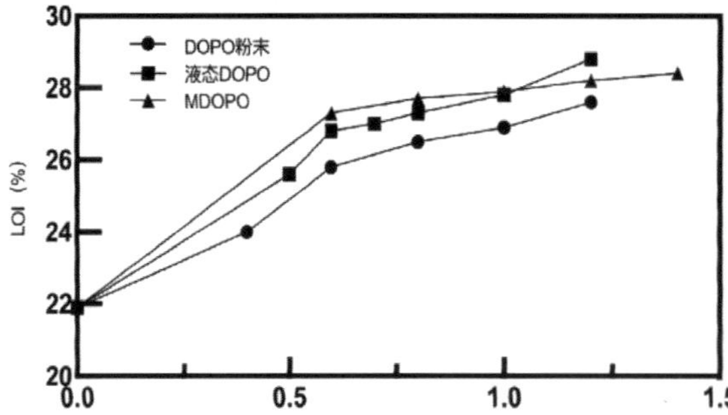

Figure 8. Effect of Phosphorus Flame Retardant Content on LOI of E44

Journal of Physics: Conference Series **2941** (2025) 012044 doi:10.1088/1742-6596/2941/1/012044

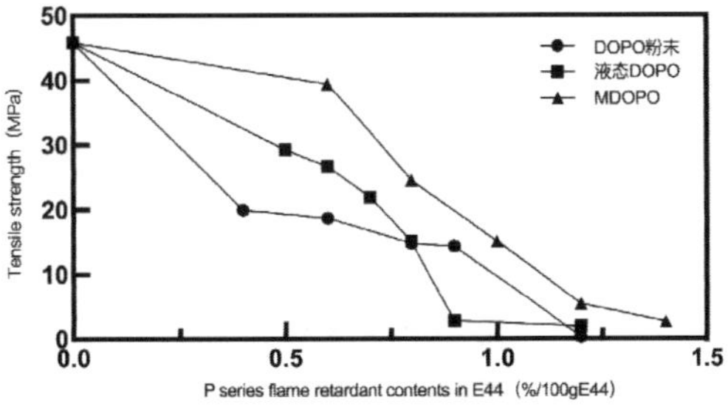

Figure 9. Effect of Phosphorus Flame Retardant Content on Ts of E44

4. Conclusions

(1) The flame retardancy of E44 can be significantly enhanced by increasing the addition of MDOPO. The oxygen index of E44, when combined with 11.43 g of MDOPO, is 24.7% higher than that of the sample without any flame retardant. However, when the MDOPO content added to E44 surpasses 11.43 g, the oxygen index of the composite increases at a slower rate.

(2) MDOPO exhibits a more pronounced capacity for enhancing the properties of E44 in comparison to molten DOPO. When the phosphorus content in the E44 matrix is 0.6%, the oxygen indices of the three composites, namely MDOPO/E44, molten DOPO/E44, and powdered DOPO/E44, are 27.3%, 26.8%, and 25.8%, respectively. Additionally, the tensile strengths of these three composites are 39.43 MPa, 26.65 MPa, and 18.7 MPa, respectively.

(3) SEM analysis of the cross-section microstructures of MDOPO/E44 reveals no noticeable phase separation system.

References

[1] Wen Q Z, Yao S R, Wang Y S. Application and development of new polymer in shipbuilding[J]. Ordnance Materimal Science and Engineering, 2001(06): 58-60+63.

[2] Liu Y L. Development progress of function materials in ships engineering area[J]. Equipment Manufacturing Technplogy, 2014(04): 275-278.

[3] Shi J, Huang Z. Application of composite material in the marine structures[J]. Fiber Reinforced Plastics/Composites, 2012(S1):269-273.

[4] Fang Z F, Liu B, Li G M, Li J. Requirement and development analysis of warship equipment materials system[J]. Materimals China, 2014, 33(07): 385-393.

[5] Brzeska J，Morawska M，Heimowska A， et al. The influence of chemical structure on thermal properties and surface morphology of polyurethane materials[J]. Chemical Papers， 2018， 72(05)： 1249-1256.

[6] dToldy A， Anna P， Csontos I， et al. Intrinsically flame retardant epoxy resin – Fire performance and background – Part I[J]. Polymer Degradation & Stability, 2008, 92(12): 2223-2230.

[7] Erekkath S， Sreejalekshmi K G. Theoretical predictions on Microphase Separation in Polyurethane： Combinatorial Design， Synthesis and Demonstration of Shape Memory Property[J]. Materials Today Communications, 2018, 16(03): 71-80.

[8] Miao Jun Xu， Si Yu Xia， Chuan Liu. Preparation of Poly(phosphoric acid piperazine) and Its Application as an Effective Flame Retardant for Epoxy Resin[J]. Chinese Journal of Polymer Science, 2018, 36(05): 655-664.

[9] Han Z Z, Guo X J, Cui C C. Delopment of High-solid epoxy ballast tank paint[J].Surface technology, 2017， 46(12)： 28-33.

[10] Hoge J， Leach C. Epoxy resin infused boat hulls[J]. Reinforced Plastics， 2016， 60(04)： 221-223.

Finite Element Method-Based Safety Assessment of a Gas Pipeline with Bulge Defect

Zhi Xiang[1, 2], Xiatao Tang[1, 2], Quanbing Liu[1, 2*], Yin Chen[1, 2], Zhongyan Wang[3] and Jintao Li[3]

[1] Zhejiang Academy of Special Equipment Science, Hangzhou 310020, Zhejiang, China.
[2] Key Laboratory of Special Equipment Safety Testing Technology of Zhejiang Province, Hangzhou 310020, Zhejiang, China.
[3] Zhejiang Provincial Natural Gas Development Co., Ltd. Hangzhou 310051, Zhejiang, China.
Email: 80757595@qq.com; txt@zjut.edu.cn; *quanbing21@126.com; chenyin091011@163.com; wangzy46@pipechina.com.cn; 359090255@qq.com

Abstract. A safety assessment was conducted on a natural gas pipeline tee section containing a bulge defect. Geometric data of the bulge were sampled to establish a finite element model of the defective pipe section for stress analysis, which was compared to a defect-free model. The load-bearing capacity of the defective section was predicted using conventional design and analytical design methods, respectively. The results showed that under the design pressure, the maximum stress in the defective section was 71% of the allowable stress, indicating a safe stress level. The bulge did not increase local stresses, suggesting a limited impact on the load-bearing capacity. The defective section could withstand pressures exceeding the design pressure, with the analytical design method yielding a significantly higher predicted value.
Keywords. Natural gas pipeline; bulge defect; finite element analysis; safety assessment.

1. Introduction

With the continuous increase in the proportion of natural gas in the clean energy structure, its transportation safety has become a key link in ensuring the safe operation of natural gas stations. As the main tool for energy transportation, maintaining the structural integrity of pipelines is crucial. Bulging is a common defect in pressure pipelines, which not only affects the appearance but may also reduce the pipeline's load-bearing capacity and safety. Defects can be divided into service defects and manufacturing defects. There are not many research reports on the bulging defect of three-way pipe sections. Liu et al. [1] analyzed the bulging defects on the outer arc side of induction heating bends used in gas pipelines and found that there was mechanical damage on the inner surface of the bulging defects. At the same time, the grain size of the structure in this area was significantly coarsened compared to other areas, and the yield strength exceeded the standard upper limit.

The methods for evaluating pipeline defects include macroscopic inspection, non-destructive testing [2,3], mechanical evaluation, strength verification [4,5], and fatigue life assessment. For large deformations such as bulges, deformation measurement, wall thickness measurement, and strength verification are often used for safety assessment. The conventional design method can calculate the ultimate pressure-bearing capacity of pressure equipment. With the increasing size and complexity of pressure equipment, the analytical design method is being valued for its ability to optimize performance, save materials, and solve complex problems. The finite element method is widely used in pipeline stress calculation due to its strong adaptability, wide application, and high economic benefits, and has become one of the effective methods for pressure pipeline design and safety assessment [6,7].

This study conducts a safety assessment of bulging defects in the three-way section of natural gas pipelines, explores their impact on the safety and functionality of pipeline structures, and calculates the ultimate bearing capacity of internal pressure. The research results can provide theoretical guidance for the safety assessment of bulging defects in natural gas pipelines.

2. Simulation Calculation and Analysis

2.1. Geometric Measurement

During the regular inspection of the pressure pipeline at a natural gas station, it was found that there was a significant bulge at the connection between the branch pipe and the main pipe of a certain three-way pipe section, as shown in figure 1. After macroscopic inspection and magnetic particle testing (ZCM-DA1206A), it was found that the anti-rust paint on the drum and surrounding areas was intact without surface cracks. After the pipeline was put into use, the anti-rust paint was not treated, and it was preliminarily determined that the bulging was a manufacturing defect rather than a result of service.

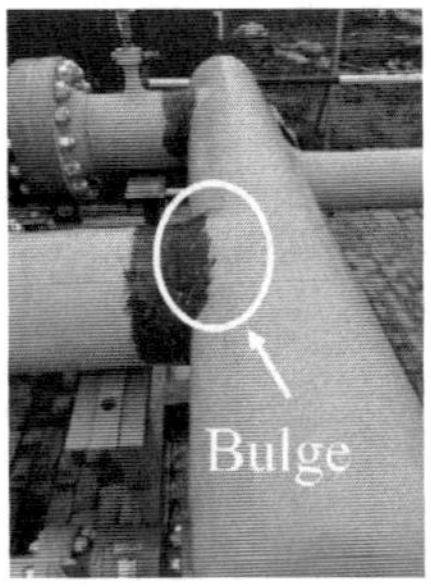

Figure 1. On-site physical image.

According to the quality certification documents, the pipeline was built in 2017. The main pipe material of the three-way section is X60M, the branch pipe material is 16Mn, the design pressure is 4.0 MPa, and the design temperature is 50 °C. The chemical composition and properties of the material are shown in tables 1 and 2. The relevant mechanical properties comply with the requirements of GB/T 20801-2020 Pressure Piping Code - Industrial Piping Part 2: Materials [8] and GB/T 150.2-2011 Pressure Vessels Part 2: Materials [9]. The nominal diameter of the supervisor is Φ 323.9 mm, and the nominal wall thickness is 25.0 mm; The nominal diameter of the branch pipe is Φ 273 mm, and the nominal wall thickness is 15.0 mm. No other defects were found in the pipeline near the bulge.

The actual measurement on site shows that the outer diameter of the main pipe is Φ 327.9 mm and the wall thickness is 26.0 mm; The outer diameter of the branch pipe is Φ 276.9 mm, and the wall thickness is 16.3 mm. The surface expression of the bulge is unknown, and the thickness is uneven throughout. To obtain an accurate drum shape and size, several measurement points were set on the surface of the drum. To describe these measurement points, first establish a coordinate system: the *X*-axis coincides with the highest busbar of the main pipe position, the *Y*-axis is located in the same vertical plane as the branch pipe axis, and the *Z*-axis is vertically upward, as shown in figure 2.

On the XY plane, set a grid point matrix of 9 × 15 with a row spacing of 20 mm and a column spacing of 19 mm. Name the grid points P_{ij}, representing the grid points in row *i* and column *j*, as shown in figure 3. Coordinate measurement and wall thickness measurement are carried out at the projection points along the vertical direction on the surface of the pipeline, and then modeling is performed based on the measurement results.

Figure 2. Schematic diagram of the coordinate system.

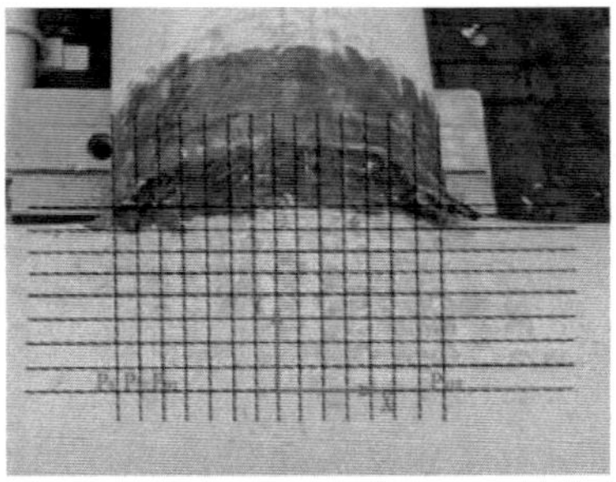

Figure 3. Schematic diagram of the set measurement points.

Table 1. Chemical composition of materials.

Materials	Element content (wt.%)								
	C	Si	Mn	S	P	Cr	Ni	Cu	Fe
X60M	0.61	0.26	0.6	≤0.035	≤0.035	≤0.25	≤0.25	≤0.25	Bal.
16Mn	0.16	0.4	1.45	≤0.030	≤0.030	≤0.30	≤0.30	≤0.25	Bal.

Table 2. Physical properties and performance of materials.

Materials	Density /kg·m^{-3}	Elastic modulus /MPa	Poisson's ratio	Yield strength/MPa	Tensile strength/MPa
X60M	7850	201000	0.3	425	532
16Mn	7850	200000	0.3	348	497

2.2. Modeling

Modeling of pipe sections with bulging defects, and strength verification are carried out using conventional design methods in accordance with GB/T 20801.3-2020 Pressure Piping Code - Industrial Pipelines - Part 3: Design and Calculation, and finite element calculations are performed using the first strength theory (maximum tensile stress theory).

By comparing the measured thickness value with the nominal thickness value, it can be concluded that there is no corrosion thinning in this pipe section. However, in order to ensure safety, it is still assumed that the corrosion amount of this pipe section to the next inspection cycle is 1.0 mm. Based on geometric measurement results, surface fitting is performed according to the coordinates of measurement points on the outer surface to obtain the outer surface model. Then, subtract the corrosion amount from the actual wall thickness measurement to obtain the wall thickness of the model, and then model the inner surface, adjust the transition between surfaces, and generate a conservative solid model, as shown in figure 4.

Grid the geometric model with Solid 186 hexahedron 20 node cells, and set at least 3 layers of cells in the wall thickness direction. After analysis, it is known that when the unit size is set to 5 mm, it can balance calculation accuracy and efficiency. At this time, the number of unit nodes is 711970, the number of units is 620210, and the average grid quality is 0.76. The grid division results are shown in figure 5.

According to GB/T 20801.2-2020, the allowable stress for metal materials is lower than the yield strength at the design temperature. Therefore, the finite element modeling adopts an ideal elastic material model, which only requires the elastic modulus E and Poisson's ratio μ. The drum is located in the main pipe, and the elastic modulus E and Poisson's ratio μ of the main and branch pipe materials are similar. Therefore, the model is simplified and the performance of the main pipe material X60M is uniformly used Apply boundary conditions and loads to the model.

It is necessary to apply symmetrical constraints on the supervisor's end face. Furthermore, frictionless constraints (constraining normal displacement) must be applied on the end face of the branch pipe. Vertical displacement constraints must be set in the area where the bottom of the pipe fittings contacts the bracket. Finally, a gravity acceleration of 9.8 m/s^2 and a design pressure load of 4.0 MPa must be applied on the inner surface, with boundary conditions set as shown in figure 6.

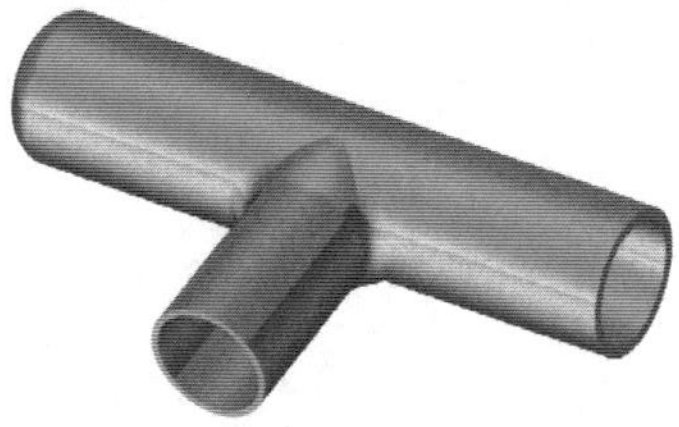

Figure 4. Geometric model of three-way pipe with bulging defects.

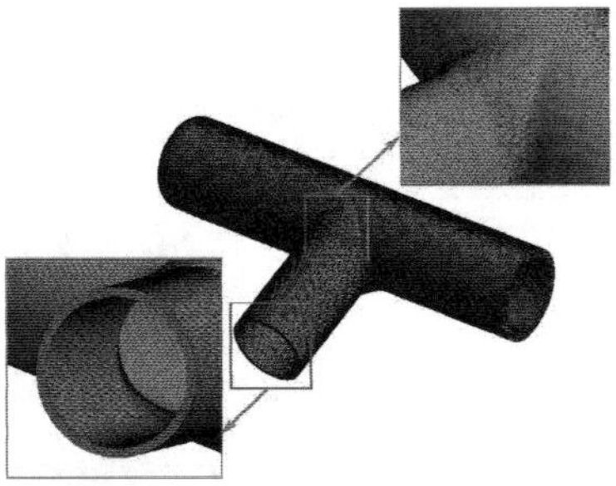

Figure 5. The result of grid division.

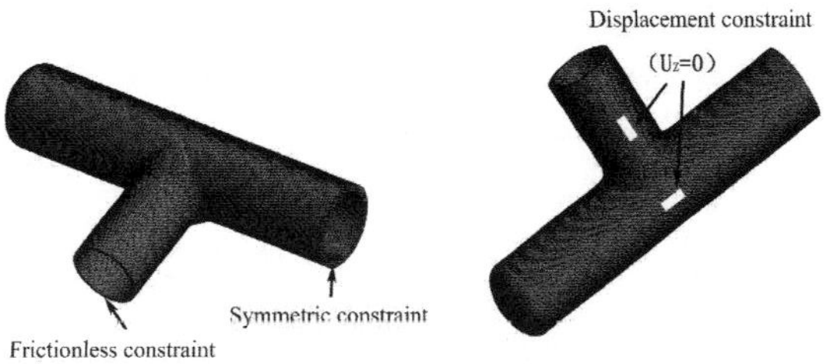

Figure 6. Boundary condition.

2.3. Simulation Results and Analysis

Under the operating conditions described in the previous section, the stress distribution (maximum principal stress) of the pipe section was calculated, as shown in figure 7. The maximum stress value of the bulging part is about 55.3 MPa. The maximum stress occurs at the connection between the main and branch pipes throughout the entire pipe section, with a value of 122.2 MPa, as shown in figure 8.

According to GB/T 20801.2-2020, the allowable stresses for the main pipe material (X60M) and branch pipe material (16Mn) at the design temperature are 181 MPa and 173 MPa, respectively. Therefore, the maximum stress of the pipe section with bulging defects is lower than the allowable stress of the material, and the maximum stress value is only about 71% of the allowable stress of the material. This pipe section can be used.

In order to investigate the influence of bulging defects on stress distribution, a pipe segment model without bulging defects was constructed and finite element analysis was conducted. The stress distribution of the non-bulging pipe section under design pressure is shown in figure 9, and the location of the maximum stress is shown in figure 10.

By comparing the analysis results of whether there is a bulging pipe section, it can be seen that the maximum stress values of the two are almost equal and occur at the same position. The local stress at the bulging part did not increase but instead decreased compared to the same position without the bulging pipe section, which may be due to stress redistribution. From this, it can be seen that the impact of bulging defects on the load-bearing capacity of three-way pipe sections is limited.

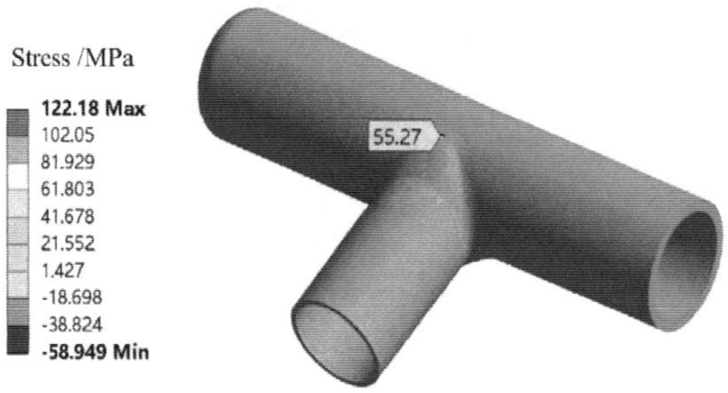

Figure 7. Stress distribution of pipe sections with bulging defects under design pressure.

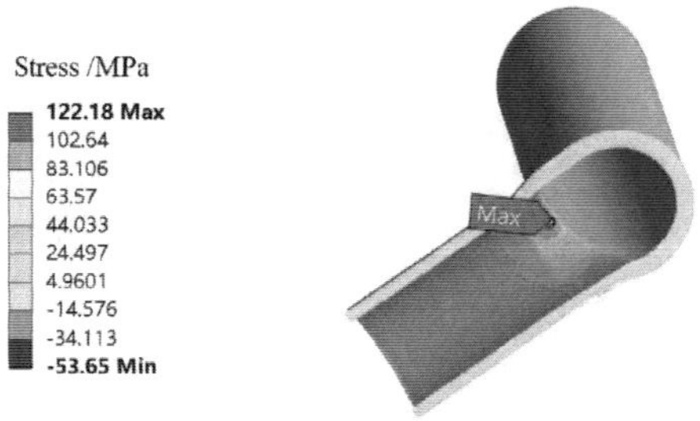

Figure 8. The maximum stress value of the pipe section with bulging defects.

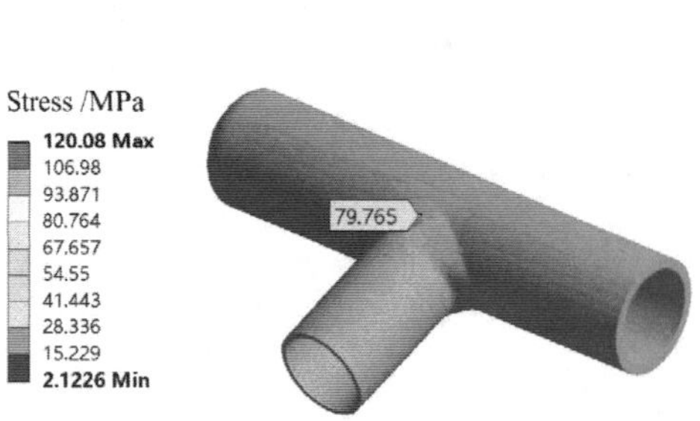

Figure 9. Stress distribution of pipe sections without bulging defects under design pressure.

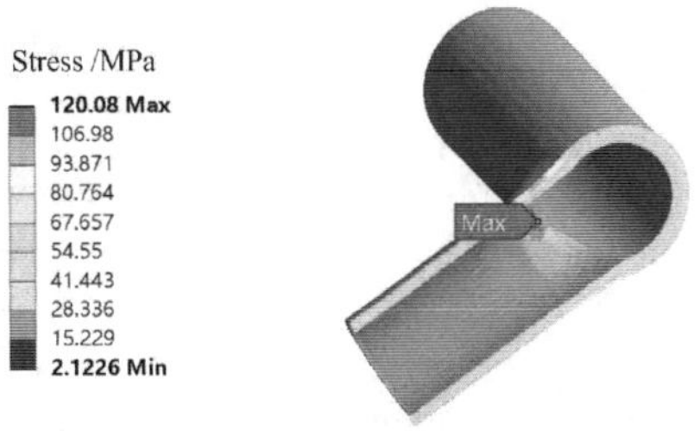

Figure 10. The maximum stress value of the pipe section without bulging defects.

3. Calculation of Ultimate Pressure-Bearing Capacity

According to the analysis in section 2.3, the maximum stress of the pipe section with bulging defects under design pressure is lower than the allowable stress of the material and still has a certain margin. Therefore, this pipe section can withstand greater internal pressure loads. This section studies its ultimate bearing capacity under internal pressure.

3.1. Conventional Design

Due to the use of an ideal elastic material model, the maximum stress is proportional to the load. Under a pressure of 4.0 MPa, the maximum stress is 122.2 MPa. Therefore, when the maximum stress is equal to the allowable stress of 173 MPa, the corresponding pressure load is 5.66 MPa (calculation formula: 173 MPa/122.2 MPa × 4.0 MPa).

In the finite element model of the pipe section with bulging defects established above, the internal pressure load is set to 5.66 MPa, and the stress distribution is shown in figure 11. At this time, the maximum stress of the branch pipe does not exceed the allowable stress of its material. Therefore, according to conventional design methods, the internal pressure ultimate bearing capacity of this pipe section is 5.66 MPa.

Figure 11. Stress distribution of pipe sections with bulging defects under ultimate load.

3.2. Analysis and Design

At present, China has released JB/T 4732-1995 Steel Pressure Vessels - Analysis and Design Standards for the analysis and design of pressure vessels, but there is no relevant standard for pressure pipelines. This study only analyzes the three-way pipe section with bulging defects and does not need to consider the overall flexibility of the pipe system. Therefore, reference can be made to the standard JB/T 4732-1995 for analysis and design. This method has been applied multiple times in the design of irregular pipelines [10,11]. This section uses analytical design methods to calculate the internal pressure ultimate bearing capacity of pipe sections with bulging defects, and the strength verification adopts the third strength theory, namely the maximum shear stress theory.

The stress intensity distribution (Tresca stress) of the bulging pipe section under design pressure is shown in figure 12. Classify the stress at the dangerous section, where the stress intensity is highest.

Set a path along the thickness direction for equivalent stress linearization, as shown in figure 13. The distribution of local thin film stress S_{II} and secondary stress S_{IV} along the wall thickness on the selected path is shown in figure 14. The allowable limits for S_{II} and S_{IV} are 1.5 KS_m and 3 KS_m, respectively, where K is the load combination factor, taken as 1.0; S_m is the design stress intensity, and the calculated design stress intensity for the main pipe material X60M is 204 MPa, while the design stress intensity for the branch pipe material 16Mn is 191 MPa. The stress classification and evaluation are shown in table 3.

According to table 3, under the design conditions, S_{II} and S_{IV} are less than their respective allowable limits. When conducting stress verification according to JB/T 4732-1995, the elastic nominal stress is used as the stress. The nominal stress of elasticity is proportional to the load. Therefore, when the S_{II} of path 1 is equal to the allowable limit of 306 MPa, the corresponding internal pressure ultimate bearing capacity is 19.18 MPa (calculation formula: 306 MPa/63.8 MPa × 4.0 MPa). Therefore, according to the analysis and design method, the internal pressure ultimate bearing capacity of this pipe section is 19.18 MPa.

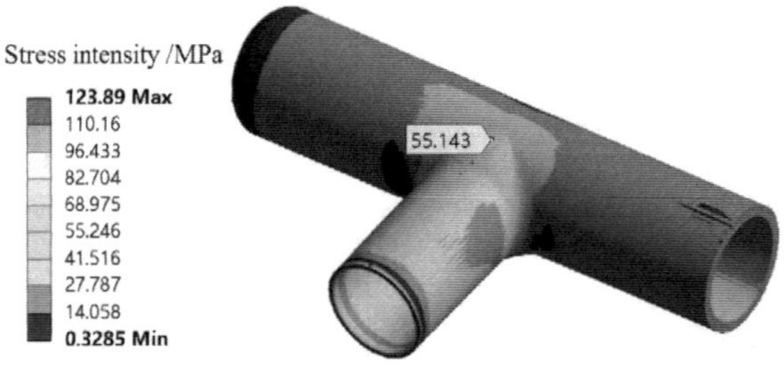

Figure 12. The stress intensity distribution of pipe sections with bulging defects under design pressure.

Figure 13. Path of stress linearization.

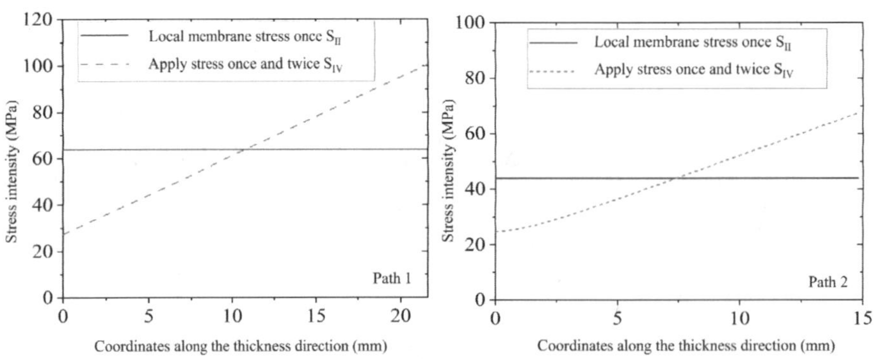

Figure 14. Linearization stress diagram under different paths.

Table 3. Stress classification and evaluation.

Path	Stress classification	Stress intensity /MPa	Permissible limit /MPa	Evaluation results
Path 1	Local membrane stress once S_{II}	63.8	1.5×204=306	Pass
	Apply stress once and twice S_{IV}	103.9	3×204=612	Pass
Path 2	Local membrane stress once S_{II}	43.9	1.5×191=286.5	Pass
	Apply stress once and twice S_{IV}	67.5	3×191=573	Pass

4. Conclusion

This study conducted a detailed safety assessment of the three-way section of natural gas pipelines with bulging defects, and obtained the following conclusions:

(1) Despite the bulging caused by manufacturing defects, the stress level of the three-way section is still within the safe range under design pressure, and the maximum stress value is only about 71% of the allowable stress of the material. This indicates that even with bulging defects, the pipeline system can still maintain its functionality.

(2) Comparing the stress distribution of pipe sections with and without bulging defects, it can be seen that bulging defects do not significantly increase the stress level of the pipe section. This indicates that the bulging defect has a limited impact on the load-bearing capacity of the three-way pipe section.

(3) According to the calculation of ultimate pressure bearing capacity, the three-way pipe section can withstand internal pressure loads higher than the design pressure. Both conventional design and analytical design methods were used for calculation, and the results showed that the ultimate internal pressure-bearing capacity was higher than the design pressure. The predicted ultimate bearing capacity under internal pressure using conventional design methods is 5.66 MPa, which is approximately 1.41 times the design pressure; The predicted ultimate bearing capacity under internal pressure by the analysis and design method is 19.18 MPa, which is equivalent to 4.79 times the design pressure and significantly higher than the predicted value by conventional design methods. This result provides an additional safety margin for the use and management of pipelines.

(4) The methods and findings of this study have important reference value for the safety assessment of pipelines in similar natural gas stations, which can help improve the safety and reliability of natural gas transportation systems. Although finite element analysis provides predictions for the performance of pipe fittings under ideal conditions, in practical applications, the effects of environmental factors, material aging, and other potential changes in service conditions must be considered.

Acknowledgments

The authors thank the Zhejiang Provincial Administration for Market Regulation for its funding of the Science and Technology Innovation Project (ZC2023023).

5. References

[1] Liu Y L, Xu Y, Wang G F, Nie X H, Feng Z J and Li L 2018 Analysis of cause of drum of bend pipe for gas pipeline *Pipeline Technique and Equipment* **(06)** pp 46-51.

[2] Waqar M, Memon A M, Sabih M and Alhems L M 2024 Composite pipelines: analyzing defects and advancements in non-destructive testing techniques *Eng. Fai. Anal.* **157** p 107914.

[3] Carvalho A A, Rebello J M A, Souza M P V, Sagrilo L V S, and Soares S D 2008 Reliability of non-destructive test techniques in the inspection of pipelines used in the oil industry *Int. J. Pres. Ves. Pip.* **85** (11) pp 745-751.

[4] Tong Y H, Gao X, Cao L, Xie Y L, Tang N, Tao L, Yi W, Liu X M, Li Ch, Tao C X, and Liu Y

Journal of Physics: Conference Series **2941** (2025) 012045 doi:10.1088/1742-6596/2941/1/012045

2023 Structural design and strength analysis of the calorimeter pipeline joint for the negative neutral beam injection system of China fusion engineering test reactor *Fusion Eng. Des.* **192** p 113767.

[5] Bouledroua O, Zelmati D, Hafsi Z, and Djukic M B 2024 Hydrogen embrittlement effects on remaining life and fatigue crack growth rate in API 5L X52 steel pipelines under cyclic pressure loading *Eng. Fai. Anal.* **166** p 108917.

[6] Shuai Y, Wang X H, Wang J Q, Yin H G, and Cheng Y F, 2021, Modeling of mechanical behavior of corroded X80 steel pipeline reinforced with type-b repair sleeve *Thin Wall. Struct.* **163,** p 107708.

[7] Terán G, Capula-Colindres S, Velázquez J C, Fernández-Cueto M J, Angeles-Herrera D, and Herrera-Hernández H 2017 Failure pressure estimations for pipes with combined corrosion defects on the external surface: a comparative study *Int. J. Electrochem. Sc.* **12**(11) pp 10152 -10176.

[8] Administration, Pressure piping code—Industrial piping GB/T 20801-2020 (2020).

[9] Administration, Pressure Vessels, GB/T 150-2011 (2011).

[10] Singh R N, Porter I, and Hematian J, 2001, Finite element analysis of three-way roadway junctions in longwall mining *Int. J. Coal Geol.* **45** (2) pp 115-125.

[11] Arumugam T, Karuppanan S, and Ovinis M, 2020, Finite element analyses of corroded pipeline with single defect subjected to internal pressure and axial compressive stress *Mar. Struct.* **72** p 102746.

Preparation of nano-TiO_2 by Biological Template Method and Its Visible Light Catalytic Performance

Jiayi Cao[1], Yihong Yang[1], Wanpeng Li[1], Zhixing Zhan[1], Yuyi Liu[1], Li Li[1] and Fuchang Peng[1,2,3,*]

[1] School of Vanadium & Titanium, Panzhihua University, Panzhihua 617000, Sichuan, China.

[2] State Key Testing Laboratory of Vanadium & Titanium, Panzhihua 617000, Sichuan, China.

[3] Vanadium and Titanium Resource Comprehensive Utilization Key Laboratory of Sichuan Province, Panzhihua 617000, Sichuan, China.

Email: 1186432502@qq.com; 2680237349@qq.com; 3521644697@qq.com; 3312459655@qq.com; 2557411139@qq.com; 1398071920@qq.com; *pzhupfc@163.com

Abstract. Using tetrabutyl titanate as the titanium raw material and kapok flower core as the biological template, KBT-TiO_2 samples with special morphology were prepared by a simple sol-gel process. The KBT-TiO_2 samples were characterized by X-ray diffraction, scanning electron microscopy, UV-Vis and other techniques. Taking the photocatalytic degradation of methylene blue (MB) as a probe reaction, the visible light catalytic performance of different TiO_2 samples was studied. The influence of calcination temperature and sample dosage on the photocatalytic performance of TiO_2 were explored. The results showed that compared with the samples (pure-TiO_2) synthesized by the non-template method, KBT-TiO_2 samples retained and inherited the biological relic structure of the kapok flower core. Due to special morphology, the dispersion of KBT-TiO_2 is improving, and its specific surface area is increased, it is beneficial to improving the visible light catalytic performance of the samples. As the calcination temperature increases, the catalytic performance of the sample first increases and then decreases. When the calcination temperature is 550°C, the calcination time is 3 hours, the initial concentration of MB is 10 $mg \cdot L^{-1}$, and the sample input is 0.2 g, the photocatalytic performance of KBT-TiO_2 samples reaches a maximum of 98.6%. The apparent first-order kinetic constant of the photocatalytic degradation of MB of KBT-TiO_2 is 1.8 times that of pure-TiO_2.

Keywords. Biological template method; kapok flower core; nano TiO_2; photocatalysis.

1. Introduction

TiO_2 nanomaterials possess strong oxidation capacity and photocatalytic performance, good stability, are low cost and are non-toxic and harmless [1,2], thus having broad application potential. However, because the band gap width of TiO_2 is 3.0-3.2 eV, it can only absorb ultraviolet light, which accounts for about 4% of sunlight, resulting in a very low utilization rate of solar energy. In addition, the specific surface area of nano-TiO_2 is reduced due to easy agglomeration, and its photogenerated electron-hole center is easily recombined, which greatly limits the practical application of TiO_2 [3]. To improve the photocatalytic activity of TiO_2, the researchers investigated metal doping [4,5], noble metal recombination [6,7], dye sensitization [8], non-metal doping [9], construction of heterojunctions [10], and generation of appropriate oxygen vacancies [11] for modification. The biological template method is a novel composite material preparation method to obtain nanomaterials with bioresidual structures, which have the advantages of simple operation, low raw material price, environmental protection, and excellent structure [12-14]. Kapok, also known as the Panzhihua flower, has bright and beautiful flower colors. Its flower core is slender and elongated, with natural surface grooves and rich multi-level pore structures. Using this as a biological template to prepare nanomaterials, it can be constructed with a large specific surface area and special morphology, which is beneficial for improving the photocatalytic performance of nanomaterials. Currently, there are few reports in the academic community on the use of kapok flowers as biological templates for the preparation of nano TiO_2.

In this work, Panzhihua stamen was selected as the biological template to prepare nano-TiO_2 samples with special morphology by sol-gel method [15-17]. The performance of photocatalytic degradation MB of different TiO_2 samples under visible light was studied, and the effects of calcination temperature and sample amount on visible light catalytic activity of TiO_2 were investigated.

2. Experimental Method

Tetrabutyl titanate ($Ti(OC_4H_9)_4$) was buyed from Tianjin Kemiou Chemical Reagent Co., Ltd. Acetylacetone ($C_5H_8O_2$) was purchased from Jining Xingchi Chemical Co., Ltd. Methylene blue ($C_{16}H_{18}ClN_3S\cdot 3H_2O$) was purchased from Tianjin Fuchen Chemical Reagent Factory, Anhydrous ethanol was purchased from Chengdu Kelong Chemical Co., Ltd.. Distilled water is self-made in the laboratory. Panzhihua stamens are collected from roadside trees within the school campus.

Firstly, 3 mL acetylacetone was added in a 150 mL beaker with 60 mL deionized water, then 5 mL butyl titanate was slowly dripped in the beaker under stirring and heating conditions at 70°C. pH value of solution was adjusted to 2-3 with 2.5 $mol\cdot L^{-1}$ HNO_3. TiO_2 sol was obtained after stirring for 10-12 h. Secondly, the collected stamens of Panzhihua were sorted to remove debris. Then it was washed with deionized water 3 times, and washed with absolute ethanol 2 times. The stamens was dried at 80°C for 8 hours to obtain the stamens template of Panzhihua. The kapok flower core template was put into TiO_2 sol. TiO_2 sol was aged for 72 h and then put into a blast drying oven at 80°C for 12 h. The dried samples were grind and calcined at different temperatures for 3 h. KBT-TiO_2 samples were obtained after grinding again. The above experimental steps were repeated to prepare pure TiO_2 samples without a kapok flower core template.

D8-Advanle X-ray diffractometer (Bruker, Germany) and JSM-IT800SHL Emission Scanning Electron Microscope (Japan Electronic) were used to analyze the samples. UV-1200 UV-Vis Spectrophotometer (Shanghai Meixi Instrument Co., Ltd.) was used to measure the ultraviolet-visible light spectrum.

50 mL of 10 $mg\cdot L^{-1}$ MB solution was put into a 100 mL beaker, added a certain mass of TiO_2 sample to form a suspension. The beaker was placed on the photocatalytic reactor and stirred in the dark for 30 min. Thus, adsorption and desorption can be achieved to equilibrium. The 30 W fluorescent lamp was turned on to continue the photocatalytic reaction. 5 mL of suspended samples were taken every 30 minutes, and MB concentration was determined. The degradation rate was calculated by following the equation.

$$\eta(\%)=(A_0-A_t)/A_0\times 100\%$$

In the formula, η is the degradation rate, A_0 is the initial absorbance of MB, and A_t is the absorbance of MB at time t.

3. Results and Analysis

3.1. XRD Analysis

The XRD spectra of the calcined KBT-TiO_2 samples at different temperatures are shown in figure 1. As can be seen from figure 1, KBT-TiO_2 is the characteristic diffraction peak of TiO_2(A or R) at different temperatures, and there are no other impurity peaks, which indicates that the pure sample TiO_2 is prepared. All the diffraction peaks of samples calcined at 500℃belong to anatase, with the increase of the calcination temperature, the intensity of the diffraction peak of anatase gradually increases, when the calcination temperature is 600°C, the intensity of the diffraction peak is the highest, and the characteristic peak of rutile appears, indicating that when the calcination temperature is 600°C, the anatase gradually transforms into a rutile crystal form, and the half-peak of the anatase phase gradually narrows, which indicates that the grain size of the sample is increasing, the crystal particles are getting thicker, and the effective specific surface area of the sample is decreasing.

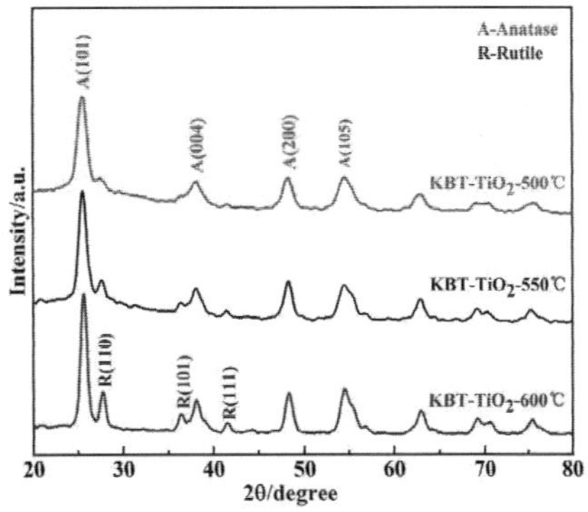

Figure 1. X-ray diffraction patterns of KBT-TiO_2 at different roasting temperatures.

3.2. SEM Analysis

The SEM image of KBF-TiO_2 sample prepared at a calcination temperature of 550 °C is shown in Figure 2. From Figure 2, it can be seen that the morphology of TiO_2 samples prepared by the biological template method better preserves the biological relic morphology of the stamens in Panzhihua. The morphology of the sample is slender rod-shaped, the shape is clear, and the TiO_2 particles are evenly distributed, which is conducive to enhancing the adsorption of TiO_2 samples to polluting molecules and promoting the photocatalytic reaction, which is verified from the subsequent photocatalytic performance experiments.

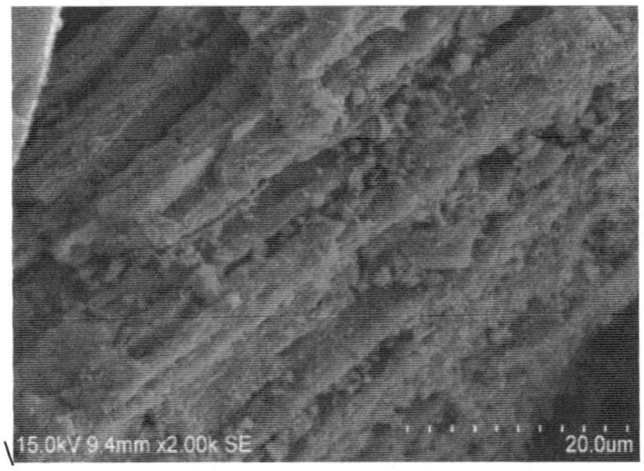

Figure 2. SEM image of KBT-TiO_2 sample.

3.3. Performance Analysis

3.3.1. Photocatalytic Performance of TiO_2 Samples. The experimental results of visible light catalytic degradation of MB in different TiO_2 samples are shown in figure 3. Figure 3(a) shows the UV-Vis spectra of the MB solution with an initial concentration of 10 mg/L in the KBT-TiO_2 sample prepared after calcination at 550°C for 3 h at different times. It can be seen that with the extension of the reaction time, the intensity of the maximum absorption peak (λ_{max}=664 nm) of MB decreases gradually, indicating that the organic molecules in the MB solution are degraded and mineralized into inorganic small molecules. The inset in figure 3(a) is a visual rendering of the color change during the degradation process of MB, and it can be seen that as the reaction time prolongs, the color of the MB solution gradually becomes lighter and nearly becomes colorless after 180 minutes, indicating that KBT-TiO_2 has a significant effect on the degradation of the MB solution.

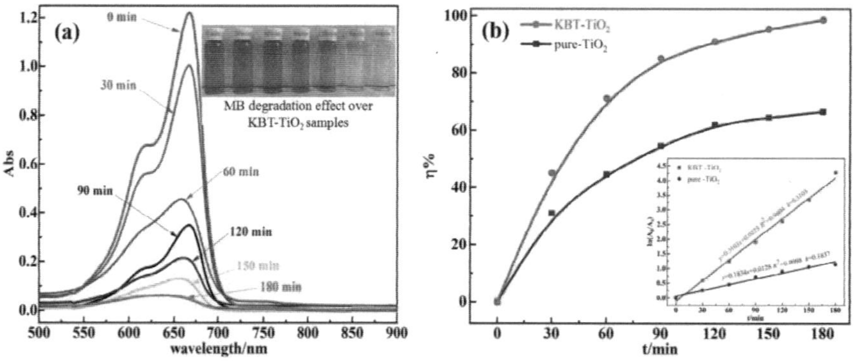

Figure 3. Visible light catalytic performance of TiO_2 samples. (a) UV-Vis spectra of MB at different reaction times, the insert is effect photos, (b) The performance of TiO_2 samples for degrading MB, the insert is MB photodegradation kinetic fitting curves.

Figure 3(b) is the performance diagram of KBT-TiO_2 and pure-TiO_2 samples for degrading the MB solution. From figure 3(b), it can be seen that the photocatalytic performance of both samples shows an upward trend with the increase of time. However, by comparing the two curves, it can be found that the photocatalytic activity of KBT-TiO_2 is higher than that of pure-TiO_2 at the same time. The degradation efficiency of KBT-TiO_2 samples can reach up to 98.6% within 3 h under 30 W fluorescent lamp irradiation, while the degradation efficiency of pure-TiO_2 is only 62.5% under the same conditions, indicating that the photocatalytic performance of KBT-TiO_2 is significantly better than that of pure-TiO_2 samples. The inset in figure 3(b) is the fitting curve of the photocatalytic degradation of MB for the two samples. It can be seen from the inset in figure 3(b) that the process of visible photocatalytic degradation of MB solution by visible light independent of the dark adsorption stage of the two samples is in line with the first-order reaction kinetic characteristics, and the apparent first-order reaction kinetic constant of KBT-TiO_2 is 1.8 times that of pure-TiO_2, indicating that KBT-TiO_2 samples have better photocatalytic performance than pure-TiO_2. This may be due to the fact that the construction of the stamen template of Panzhihua imparts nano-TiO_2 with a special biological relic structure and a large specific surface area, thereby enhancing the adsorption of MB and thus improving the rate and effect of photocatalytic reactions [18,19].

3.3.2. Effect of Roasting Temperature on the Photocatalytic Performance of KBT-TiO_2. The calcination temperature has a significant impact on the process of preparing TiO2 photocatalytic materials using biomass as a template. Therefore, the photocatalytic performance of KBT-TiO_2 samples prepared at different calcination temperatures was studied. The results are shown in figure 4. It can be seen from figure 4, the photocatalytic performance of the sample increases first and then decreases with the increase of calcination temperature, and the photocatalytic performance of the sample is better than that of 500°C and 600°C at the calcination temperature of 550°C. From the previous XRD analysis, it can be seen that the intensity of the diffraction peak of KBT-TiO_2 is relatively low at 500°C, indicating that the crystallization is not complete, while at 600°C, the crystal form of KBT-TiO_2 is transformed, and rutile crystal phase appears, and at 550°C, it has a suitable crystal structure, so the photocatalytic performance is the best.

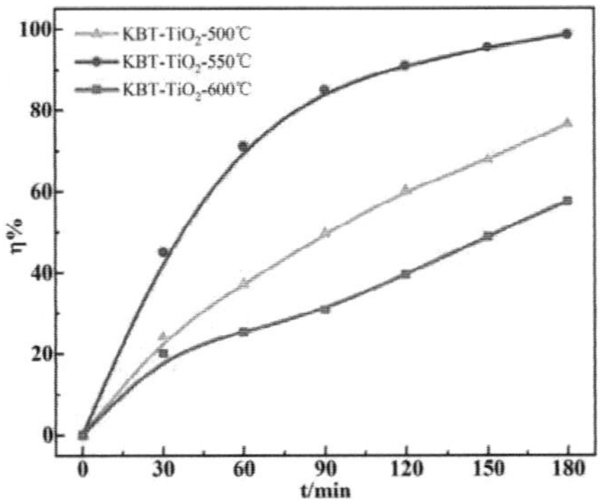

Figure 4. Effect of different calcination temperatures on the performance of KBT-TiO_2 samples.

3.3.3. Effect of Sample Dosage on Photocatalytic Performance of KBT-TiO_2. In order to explore the effect of different sample dosages on the photocatalytic performance, during the experiment of photocatalyst-catalyzed degradation of MB in KBT-TiO_2 samples0.05 g, 0.10 g, 0.15 g, 0.20 g, and 0.25 g were selected as the experimental conditions to study the photocatalytic performance of KBT-TiO_2, as shown in figure 5. It can be seen from figure 5 that the photocatalytic performance increases first and then decreases with the increase of sample input, and when the sample amount is less than 0.2 g, the photocatalytic efficiency increases with the increase of sample input. When the sample amount was more than 0.2 g, the photocatalytic efficiency decreased with the increase of sample input. This is because as the amount of sample is used, the turbidity of the entire solution system increases, and some light transmittance decreases, which weakens the photocatalytic performance [20,21]. Therefore, the optimal sample addition amount is 0.2 g·$[50\ mL]^{-1}$.

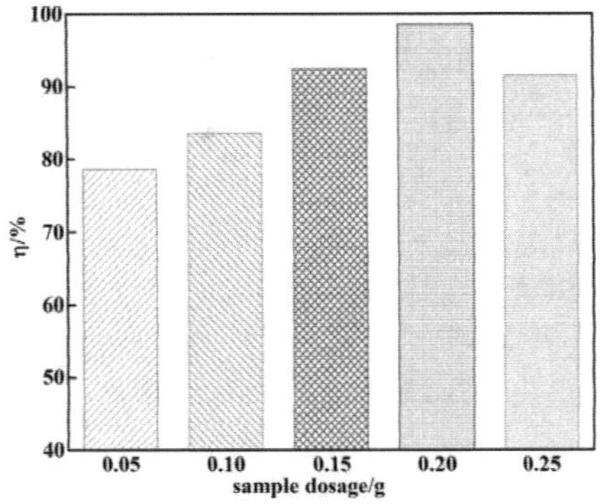

Figure 5. Effect of sample dosage on the performance of the KBT-TiO_2 sample.

4. Conclusion

Using butyl titanate as the titanium source and the stamen of Panzhihua as the biological template, the nano-TiO_2 photocatalytic samples were prepared by a simple sol-gel method, and the KBT-TiO_2 inherited the biological structure of the stamens of Panzhihua, with slender morphology, clear contour and uniform distribution, and the samples showed excellent visible light photocatalytic performance to MB solution. When the calcination temperature was 550°C and the sample dosage was 0.2 g, the KBT-TiO_2 sample had the best photocatalytic performance, and the photocatalytic degradation efficiency of the MB solution reached up to 98.6% within 3 h under fluorescent lamp irradiation, and the first-order kinetic constant of the photocatalytic reduction reaction was 1.8 times that of pure TiO_2.

Acknowledgments
This work was supported by The Opening Project of Vanadium and Titanium Resource Comprehensive

Utilization Key Laboratory of Sichuan Province (2023FTSZ11), and the Sichuan Provincial College Student Innovation and Entrepreneurship Training Project (S202411360024).

5. References

[1] Seifi A, Salari D, Khataee A, Çoşut B, Arslan L C and Niaei A 2022 *Ceram. Int.* **49** 1678
[2] Wang J, Gao X, Mi X, Zhou W and Yang X 2020 *China Pet. Process. Petrochemical Technol.* **22** 118
[3] Fang Y L, Sun M X, Wang Y, Sun S F, Ling X J, Hu M Y and Wang W T 2016 *Mat. Rep.* **30** 24.
[4] Zhu J F, Deng Z G, Chen F, Zhang J L, Chen H J, Masakazu A, Huang J Z and Zhang L Z 2006 *Appl. Catal. B-Environ.* **62** 329.
[5] Liang D D, Liu S M, Wang Z N, Gao Y, Jiang W W, Chao Q, Wang H L, Wang N, He M, Wang L and Xu S C 2019 *Mater. Sci.* **30** 12619.
[6] Song H, Jiang H, Liu X and Meng G 2006 *J. Photoch. Photobio. A* **181** 421.
[7] Sun A H, Zhang G C and Xu Y M 2005 *Mater. Lett.* **59** 4016.
[8] Zhao X F, Zhang Z, Meng X F, Liu L and Jian D Z 2003 *J. Mol. Catal.* **17** 292.
[9] Pstrowska K, Czapor-irzabek H, Borowiak D and Burchacka E 2021 *Catalysts* **11** 1193.
[10] Dong S J, Lee G J, Zhou R and Wu J J 2020 *Sep. Purif. Technol.* **250** 117.
[11] Zhang W, Tian M, Jiao H M, Jiang H Y and Tang J W 2022 *Chinese J. Catal.* **43** 2321.
[12] Zan G T and Wu Q S 2016 *Adv. Mater.* **28** 2099.
[13] Fei J B and Li J B 2015 *Adv. Mater.* **27** 314.
[14] Yang X Y, Chen L H, Li Y, Rooke J C, Sanchez C and Su B L 2017 *Chem. Soc. Rev.***46** 481.
[15] Wu G S, Xie T, Yuan X Y, Li Y, Yang L, Xiao Y H and Zhang L D 2005 *Solid State Commun.* **134** 485.
[16] Ottone C, Bejtka K and Chiondoni A 2014 *New J. Chem.* **38** 2058.
[17] Wang Z and Li H L 2002 *Appl. Phys. A* **74** 201.
[18] Fang W, Deng Y C, Tang L, Zeng G M, Zhou Y Y, Xie X, Wang J J, Wang Y and Wang J J 2017 *J. Colloid Interf. Sci.* **490** 834.
[19] Qin N, Liu Y, Wu W, Shen L, Chen X, Li Z and Wu L 2015 *Langmuir* **31** 1203.
[20] Liu L, Zhang X N, Yang L F, Ren L T, Wang D F and Ye J H 2017 *Nat. Sci. Rev.* **4** 761.
[21] Sudhaik A, Raizada P, Shandilya P, Jeong D Y, Lim J H and Singh P 2018 *J. Ind. Eng. Chem.* **67** 28.

Wrinkle Assessment Method of Space Membrane Structures Based on Strength Factor

Zhibo Yang[1], Haoran Ji[2,*], Yunqing Nie[2], Bing Wang[1] and Da Huo[1]

[1] DFH Satellite Co. Ltd., Beijing 100080, China.
[2] Intelligent Game and Decision Lab, Beijing 100080, China.
Email: yangzhibo212@163.com; *haoranji@139.com; nieyq@nudt.edu.cn; wangbinghit@163.com; huodada1991@163.com

Abstract. The membrane structure is characterized by its small loading volume and high storage ratio, offering vast potential for applications in constructing large-scale spatial structures. However, membranes exhibit low bending stiffness, which makes them prone to wrinkling, thus affecting the surface accuracy of the structure and potentially leading to functional failure of the spatial structure. This paper employs a wrinkling prediction method based on the strength factor (i.e., the pseudo strain energy of wrinkles) as an indicator parameter to characterize the degree of wrinkling deformation in spatial membrane structures. Based on this method, the study predicts wrinkling deformation in a large solar sail structure, focusing on the impact of cable tension, edge reinforcement, and connecting elements on the degree of wrinkling deformation. Simulation results indicate that reducing cable tension can significantly decrease the degree of wrinkling deformation, but deviation of the cable direction from the membrane's centroid can drastically increase the degree of wrinkling deformation. Adding reinforcement bands and connecting elements at the membrane's edges can improve wrinkling deformation to some extent; however, the number of connecting elements on the support rods needs to be chosen rationally to achieve optimal wrinkle elimination. The wrinkling prediction method and analysis conclusions presented in this paper can provide references for design schemes and system improvements of spatial membrane structures.
Keywords. Space membrane structures; wrinkle; tension field theory; strength factor; finite element method.

1. Introduction

Membrane structures have features such as small loading volumes and high storage ratios, offering vast application prospects for constructing large-scale space structures in future space missions, such as solar sails [1] and membrane antennas [2]. However, the thickness of membranes is usually on the micron scale, with a large size-to-thickness ratio (up to 10^8 in lateral size-to-thickness ratio), resulting in low out-of-plane bending stiffness. This makes them highly susceptible to wrinkling deformation under external loads. Wrinkling deformation decreases the surface accuracy of space membrane structures, which can have severe impacts on spacecraft performance. For example, wrinkles on membrane antennas may reduce surface accuracy; in solar sail spacecraft, they could cause uneven heating of the membrane surface, further affecting the static and dynamic characteristics of the spacecraft structure [3]. Therefore, predicting and effectively mitigating wrinkles in membrane structures is crucial for spacecraft design involving large space membrane structures.

Currently, there are mainly two theories for predicting membrane wrinkling deformation: bifurcation theory [4] and tension field theory [5,6]. Bifurcation theory assumes that the membrane has non-zero bending stiffness and can withstand certain compressive stresses. When compressive stress reaches the buckling critical load, out-of-plane configurations induced by in-plane displacements result in wrinkling deformation. Bifurcation theory mainly uses finite element methods (FEM) for calculations and can predict the critical load, wavelength, amplitude, and wrinkle evolution characteristics when a membrane undergoes wrinkling deformation. It can be divided into nonlinear buckling methods [7,8], direct perturbation methods [9], and explicit dynamic methods based on mechanical models [10]. The nonlinear buckling method first conducts linear buckling analysis to obtain the buckling mode of the structure, then uses a linear combination of buckling modes as initial geometric imperfections added to the membrane model for nonlinear buckling analysis [7,11]. The direct perturbation method applies a set of small random loads perpendicular to the membrane plane with opposite directions on the membrane model, causing small random out-of-plane deformations in the membrane as initial geometric imperfections, followed by nonlinear buckling analysis [12]. The explicit dynamic method requires using characteristic modes or perturbation forces to generate small disturbances in the structure and solving the motion equations through explicit dynamic methods [13,14]. Generally, membrane wrinkling prediction methods based on bifurcation theory require an FEM model and analysis, offering high prediction accuracy but also having drawbacks such as complex model processes, poor convergence, and long computational times, making them unsuitable for the preliminary design phase in engineering.

Tension field theory assumes that the membrane's out-of-plane bending stiffness is approximately zero, and the membrane releases energy through out-of-plane deformation under pressure loads. It establishes a constitutive model of wrinkling deformation by modifying the stiffness matrix in the membrane's constitutive model to eliminate compressive stress [6]. There are two types of methods for modifying the elastic stiffness matrix: one determines the shrinkage amount using stress conditions and modifies the deformation gradient tensor, and the other introduces variable Poisson's ratio to modify the elastic stiffness matrix [15,16]. In 2007, Natori and colleagues [17] from the Japan Aerospace Exploration Agency (JAXA) proposed a new method to modify the elastic stiffness matrix using projection matrices based on tension field theory, enabling fast prediction of wrinkling deformation in membrane structures when combined with finite element methods. Kogiso and others [18] built a wrinkle strength factor based on this method to evaluate the degree of wrinkling deformation, using it as an objective function to optimize the design of the membrane structure's surface, achieving an ideal wrinkle elimination effect. Akita and colleagues [5] conducted a parameter sensitivity analysis on membrane structure wrinkling deformation using this method and also verified the feasibility of using the strength factor as an objective function for optimizing membrane structures. Wrinkling prediction methods based on tension field theory can predict the regions, directions, and stress-strain states after wrinkling occurrence, offering the advantages of simple modeling and fast computation, suitable for wrinkle analysis in the preliminary design phase of engineering.

In light of the above considerations, the first part of this paper introduces the wrinkling prediction method for space membrane structures based on the strength factors. It simplifies the membrane structure to obtain a membrane model, performs mechanical analysis based on tension field theory, and derives the constitutive model of wrinkling deformation, subsequently introducing the strength factor (i.e., pseudo strain energy of wrinkles) to characterize the degree of wrinkling deformation. The second part combines the wrinkling prediction theory with finite element methods to establish a numerical method for predicting wrinkles. The third part uses this numerical method to predict wrinkling deformation in a large membrane solar sail structure, analyzing the impact of cable tension, edge reinforcement, and connecting elements on structural wrinkling deformation.

2. Wrinkle Assessment Method

2.1. Space Membrane Structure

The membrane structure studied in this paper is a typical five-point connected square solar sail structure [19], as shown in figure 1(a). Four triangular membranes are tensioned by four mutually perpendicular support rods, each with a length of 106 meters. The membrane material is polyimide, and its material parameters are listed in table 1. Considering the symmetry of the membrane structure, a quarter of the solar sail membrane structure is taken for wrinkling analysis, as shown in figure 1(b). The origin of the coordinate system is at the center of the membrane structure, and the coordinate axes coincide with the two perpendicular edges.

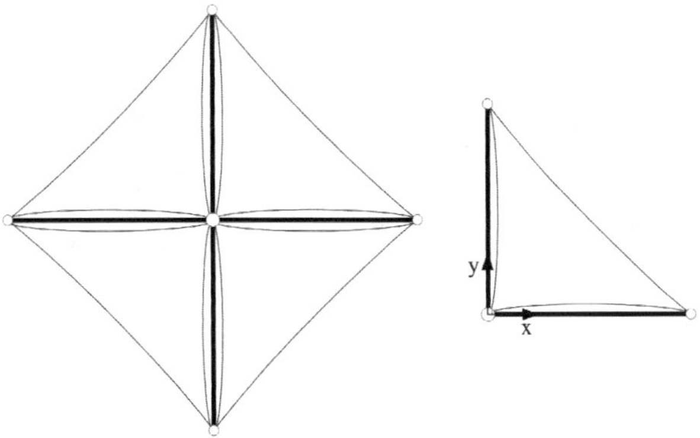

(a) Solar sail model (b) Solar sail 1/4 modal

Figure 1. Five-point connected sail.

Table 1. Parameters of polyurethane.

E/GPa	Poisson ratio	Density/(kg/m^3)	Thickness/μm
2.48	0.34	1572	2.5

The deployed span of large space membrane structures generally ranges from a few meters to several hundred meters, with membrane thickness on the micron scale and relatively low out-of-plane bending stiffness. Based on the characteristics of the membrane structure, the following basic assumptions are made:

(1) The membrane thickness studied in this paper is 2.5 um, with an in-plane size of 106 m. Since the bending stiffness of thin shells is proportional to the cube of the thickness, it is assumed that the out-of-plane bending stiffness is zero [6].

(2) With an elastic modulus of 2.48 GPa and a thickness of only 2.5 um, the membrane's out-of-plane bending stiffness is very low. It is assumed to only withstand in-plane tensile forces and cannot endure in-plane compressive stresses, which are released through wrinkling deformation.

(3) The membrane's thickness dimension is much smaller than its in-plane dimensions, and the load is applied in the plane of the membrane, which can be approximated as a plane stress state [20].

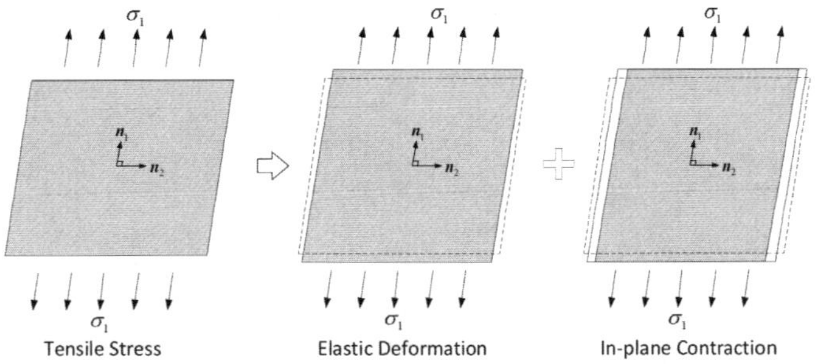

(a) Load Condition (b) Elastic Deformation (c) In-plane Contraction

Figure 2. Deformation process of wrinkling.

Taking an arbitrary infinitesimal element in the membrane, as shown in figure 2(a), a mechanical model of the membrane is established. According to the above assumptions, the stress state of the infinitesimal element, when wrinkling occurs, can be equivalently considered as a plane stress state. It can only withstand uniaxial tensile loads due to its inability to endure compressive stresses. Vectors representing the directions of maximum and minimum principal stresses form the local coordinate system of the infinitesimal element.

2.2. Tension Field Theory

According to assumption (2), since the membrane cannot withstand compressive stress, the stress state of the infinitesimal element can be equated to a uniaxial tensile state. The tensile stress magnitude equals the maximum principal stress, and the tensile direction aligns with the principal stress direction. Under uniaxial tensile loading, the deformation process of the membrane undergoing wrinkling can be equated to two steps [17]: elastic deformation (figure 2(b)) and wrinkle contraction deformation (figure 2(c)).

Under uniaxial tensile loading, the membrane undergoes elastic deformation. The constitutive relationship between the stress tensor and the strain tensor in the plane stress state is given by:

$$\boldsymbol{\sigma} = \begin{bmatrix} \sigma_x \\ \sigma_y \\ \tau_{xy} \end{bmatrix} = \frac{E}{1-v^2} \begin{bmatrix} 1 & v & 0 \\ v & 1 & 0 \\ 0 & 0 & (1-v)/2 \end{bmatrix} \begin{bmatrix} \varepsilon_x \\ \varepsilon_y \\ \varepsilon_{xy} \end{bmatrix} = \boldsymbol{D\varepsilon} \tag{1}$$

where, $\boldsymbol{D}$ is the elastic stiffness matrix, E is the elastic modulus, and v is the Poisson's ratio.

According to elastic mechanics [20], the stress and strain tensors can also be represented by principal stresses and strains as follows:

$$\boldsymbol{\sigma} = \sigma_1 \boldsymbol{n}_1 + \sigma_2 \boldsymbol{n}_2 \tag{2}$$

$$\boldsymbol{\varepsilon} = \varepsilon_1 \boldsymbol{s}_1 + \varepsilon_2 \boldsymbol{s}_2 \tag{3}$$

where σ_1 and σ_2 are the maximum and minimum principal stresses, $\boldsymbol{n}_1$ and $\boldsymbol{n}_2$ are the direction vectors of the principal stresses, ε_1 and ε_2 are the maximum and minimum principal strains, and their direction vectors are $\boldsymbol{s}_1$ and $\boldsymbol{s}_2$. According to elastic mechanics [20], the above vectors can be expressed as

$$\boldsymbol{n}_1 = [\sin^2\theta \quad \cos^2\theta \quad \sin\theta\cos\theta]^T \tag{4}$$

$$\boldsymbol{n}_2 = [\sin^2\theta \quad \cos^2\theta \quad -\sin\theta\cos\theta]^T \tag{5}$$

$$\boldsymbol{s}_1 = [\sin^2\theta \quad \cos^2\theta \quad 2\sin\theta\cos\theta]^T \tag{6}$$

$$\boldsymbol{s}_2 = [\sin^2\theta \quad \cos^2\theta \quad -2\sin\theta\cos\theta]^T \tag{7}$$

where θ is the angle between the original and principal coordinate systems. It can be expressed as

$$\theta = \arctan\left(-\frac{2\tau_{xy}}{\sigma_x - \sigma_y}\right)/2 \tag{8}$$

We define the vector $\boldsymbol{n}_3 = \boldsymbol{n}_1 \times \boldsymbol{n}_2$. Using the inner product, θ and ε has the following relation

$$f(\theta, \boldsymbol{\varepsilon}) = \boldsymbol{n}_3^{\mathrm{T}} \boldsymbol{\sigma} = \boldsymbol{n}_3^{\mathrm{T}} \boldsymbol{D\varepsilon} = 0 \tag{9}$$

When the membrane undergoes elastic deformation under uniaxial tensile loading, the Poisson effect causes contraction deformation. This contraction releases internal compressive stress, leading to the formation of wrinkles. The external load on the membrane model can be equivalently represented by a uniaxial tensile load, with the uniaxial stress tensor σ_{t} denoted as

$$\boldsymbol{\sigma}_{\mathrm{t}} = \sigma_1 \boldsymbol{n}_1 + 0\boldsymbol{n}_2 = \sigma_1 \cdot \boldsymbol{n}_1 \tag{10}$$

The strain energy due to wrinkle deformation is zero. Let the wrinkle strain tensor of the membrane be $\boldsymbol{\varepsilon}_{\mathrm{w}}$, then we have

$$\boldsymbol{\varepsilon}_{\mathrm{w}}^{\mathrm{T}} \boldsymbol{\sigma}_{\mathrm{t}} = 0 \tag{11}$$

From equations (4) and (7), it is known that the base vectors $\boldsymbol{s}_2$ and $\boldsymbol{n}_1$ are orthogonal, so the wrinkle strain can be expressed as

$$\boldsymbol{\varepsilon}_{\mathrm{w}} = \varepsilon_{2\mathrm{w}} \boldsymbol{s}_2 \tag{12}$$

The total strain of the membrane is the sum of the elastic strain and the wrinkle strain, $\boldsymbol{\varepsilon} = \boldsymbol{\varepsilon}_{\mathrm{e}} + \boldsymbol{\varepsilon}_{\mathrm{w}}$. Thus, the elastic strain of the membrane can be expressed as follows

$$\boldsymbol{\varepsilon}_{\mathrm{e}} = \boldsymbol{\varepsilon} - \varepsilon_{2\mathrm{w}} \boldsymbol{s}_2 \tag{13}$$

Since wrinkle deformation only causes compressive stress release and does not alter the tensile stress, from equations (1) and (13) we can obtain

$$\boldsymbol{\sigma}_{\mathrm{t}} = \boldsymbol{D}(\boldsymbol{\varepsilon} - \varepsilon_{2\mathrm{w}} \boldsymbol{s}_2) \tag{14}$$

Substituting equation (14) into equation (11) gives

$$\boldsymbol{\varepsilon}_{\mathrm{w}}^{\mathrm{T}} \boldsymbol{\sigma}_{\mathrm{t}} = \varepsilon_{2\mathrm{w}} \boldsymbol{D}(\boldsymbol{\varepsilon} - \varepsilon_{2\mathrm{w}} \boldsymbol{s}_2) = 0 \tag{15}$$

From the above equation, the wrinkle strain of the membrane can be determined as:

$$\boldsymbol{\varepsilon}_{\mathrm{w}} = \frac{\boldsymbol{s}_2 \boldsymbol{s}_2^{\mathrm{T}} \boldsymbol{D}}{\boldsymbol{s}_2^{\mathrm{T}} \boldsymbol{D} \boldsymbol{s}_2} \boldsymbol{\varepsilon} \tag{16}$$

The elastic strain of the membrane is

$$\boldsymbol{\varepsilon}_{\mathrm{e}} = \boldsymbol{P}\boldsymbol{\varepsilon} \tag{17}$$

where $\boldsymbol{P} = (\boldsymbol{I} - \dfrac{\boldsymbol{s}_2 \boldsymbol{s}_2^{T} \boldsymbol{D}}{\boldsymbol{s}_2^{T} \boldsymbol{D} \boldsymbol{s}_2})$.

From equations (1) and (17), the relationship between stress and total strain after the membrane undergoes wrinkle deformation is obtained as

$$\boldsymbol{\sigma}_{\mathrm{t}} = \boldsymbol{D}\boldsymbol{P}\boldsymbol{\varepsilon} \tag{18}$$

2.3. Wrinkle Strength Factor

The degree of wrinkle deformation of the membrane is related to the wrinkle strain. With reference to elastic strain energy, the pseudo wrinkle strain energy density ϕ_{w} is defined as follows [17]

$$\phi_{\mathrm{w}} = \frac{1}{2} \boldsymbol{\varepsilon}_{\mathrm{w}}^{\mathrm{T}} \boldsymbol{D} \boldsymbol{\varepsilon}_{\mathrm{w}} \tag{19}$$

The pseudo wrinkle strain energy density uses the elastic stiffness matrix to convert the wrinkle strain tensor, which reflects local wrinkle deformation, into a scalar form. This can predict the degree of local wrinkle deformation in the membrane. The greater the pseudo wrinkle strain energy density, the stronger the local wrinkle deformation. The strength factor U_w of the entire membrane structure (i.e., pseudo wrinkle strain energy) is defined as [17]

$$U_w = \int_V \phi_w \mathrm{d}V \tag{20}$$

From the definition of the wrinkle strength factor of the entire membrane structure, we know that performing a volume integral of the pseudo wrinkle strain energy density over the membrane structure yields the strength factor that reflects the degree of wrinkle deformation of the entire structure. The larger the strength factor, the stronger the wrinkle deformation of the overall structure. The strength factor provides a direct assessment index for predicting the degree of wrinkle deformation of the overall structure, with clear physical significance and simple calculation.

3. Numerical Method

3.1. Finite Element Model

According to the stress characteristics of membrane structures, only the tensile force at the corner points is considered. In the finite element model, the support rod model is ignored. A displacement constraint is applied to the central point of the 1/4 solar sail model, and a tensile load with a magnitude of 1 N is applied to the other two corner nodes, with the loading direction at an angle of 22.5° to the right-angle side, as shown in figure 3.

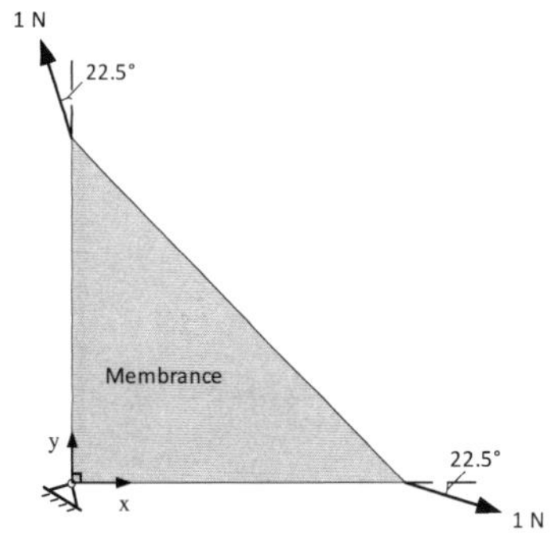

Figure 3. Constraint and load conditions.

3.2. Incremental Constitutive Model

In finite element numerical methods, an incremental constitutive model is generally used, so the Jacobian matrix is derived. The Jacobian matrix $\boldsymbol{J}$ is defined as

$$\boldsymbol{J} = \frac{\Delta\sigma}{\Delta\varepsilon} \tag{21}$$

where $\Delta\sigma$ is the stress tensor increment, $\Delta\varepsilon$ is the strain tensor increment.

According to the different deformation states of the film, the Jacobian matrix is assigned as follows

(1) When $\sigma_2 \geq 0$, The membrane is in a tension state, elastic deformation occurs. The Jacobian matrix $\boldsymbol{J}$ is

$$\boldsymbol{J} = \boldsymbol{D} \tag{22}$$

(2) When $\sigma_2 < 0$ and $\varepsilon_1 \leq 0$, the membrane is in a relaxed state. The Jacobian matrix $\boldsymbol{J}$ is

$$J = O \tag{23}$$

(3) When $\sigma_2 < 0$ and $\varepsilon_1 > 0$, the membrane is in a wrinkled state.

Deriving the stress-strain relationship (18) of the membrane in the wrinkled state with respect to time yields

$$\dot{\boldsymbol{\sigma}}_t = \boldsymbol{D}\boldsymbol{P}\dot{\boldsymbol{\varepsilon}} + \boldsymbol{D}\dot{\boldsymbol{P}}\boldsymbol{\varepsilon} \tag{24}$$

where $\dot{\boldsymbol{P}}$ can be obtained using equation (17)

$$\dot{\boldsymbol{P}} = -\frac{\dot{\boldsymbol{s}}_2\boldsymbol{s}_2^{\mathrm{T}}\boldsymbol{D} + \boldsymbol{s}_2\dot{\boldsymbol{s}}_2^{\mathrm{T}}\boldsymbol{D}}{\boldsymbol{s}_2^T\boldsymbol{D}\boldsymbol{s}_2} = \frac{2(1-\nu^2)}{E}(\boldsymbol{n}_3\boldsymbol{s}_2^{\mathrm{T}}\boldsymbol{D} + \boldsymbol{s}_2\boldsymbol{n}_3^{\mathrm{T}}\boldsymbol{D})\dot{\theta} \tag{25}$$

$\dot{\theta}$ can be calculated using equation (19)

$$\dot{\theta} = -\left(\frac{\partial f}{\partial \varepsilon} / \frac{\partial f}{\partial \theta}\right)\dot{\varepsilon} = \frac{\boldsymbol{n}_3^{\mathrm{T}}\boldsymbol{D}}{(\boldsymbol{s}_1 - \boldsymbol{s}_2)^{\mathrm{T}}\boldsymbol{D}\boldsymbol{\varepsilon}}\dot{\boldsymbol{\varepsilon}} \tag{26}$$

Substituting equations (25) and (26) into equation (24), and setting $\dot{\boldsymbol{\sigma}}_t \approx \frac{\Delta\boldsymbol{\sigma}_t}{\Delta t}$, $\dot{\boldsymbol{\varepsilon}} \approx \frac{\Delta\boldsymbol{\varepsilon}}{\Delta t}$, we obtain the Jacobian matrix in the wrinkled deformation state

$$\boldsymbol{J} = \boldsymbol{D}\boldsymbol{P} + \frac{\nu\varepsilon_1 + \varepsilon_2}{\varepsilon_1 - \varepsilon_2}\boldsymbol{D}\frac{\boldsymbol{n}_3\boldsymbol{n}_3^{\mathrm{T}}\boldsymbol{D}}{\boldsymbol{n}_3^{\mathrm{T}}\boldsymbol{D}\boldsymbol{n}_3} \tag{27}$$

3.3. Numerical Method for Calculating Wrinkle Strength Factor

By performing iterative calculations on the incremental constitutive model, the wrinkle strain field of the entire membrane structure can be obtained. The volume integral of the pseudo wrinkle strain energy density in *k*-th element is

$$U_{\mathrm{w}}^k = \int_{V_{\mathrm{e}}} \phi_{\mathrm{w}} \mathrm{d}V \approx \frac{1}{2}\boldsymbol{\varepsilon}_{\mathrm{w}}^{k\,\mathrm{T}}\boldsymbol{D}\boldsymbol{\varepsilon}_{\mathrm{w}}^k V_{\mathrm{e}} \tag{28}$$

where V_{e} the volume of the element, $\boldsymbol{\varepsilon}_{\mathrm{w}}^k$ is the wrinkle strain tensor of the *k*-th element. The element wrinkle strength factor reflects the degree of wrinkle deformation of the element. Summing up all the element wrinkle strength factors, we get the overall wrinkle strength factor of the membrane structure

$$U_{\mathrm{w}} = \sum_{k=1}^{N_{\mathrm{e}}} U_{\mathrm{w}}^k \tag{29}$$

where N_{e} is the total number of elements, U_{w}^k represents the wrinkle strength factor of the *k*-th element.

4. Numerical Simulation Result

4.1. Convergence Analysis

The convergence of the numerical model for this wrinkle prediction method is analyzed, examining the influence of element density on convergence. The element type used is a three-node plane stress element. The calculation results for different element scales are shown in table 2. for large membrane structures, the Mises stress at the geometric centroid of the membrane is an important indicator for assessing its tensioned state, and the Mises stress at the centroid is used as the convergence criterion. As seen in table 2, when the mesh element size is reduced to 1 m, good convergence and computational accuracy are achieved. Therefore, the element scale used in the subsequent analysis is 1 m.

Table 2. The results of different mesh scales.

Mesh Scale/m	Mises Stress of Centroid/MPa	Convergence Error/%
4	5640.89	-
2	5508.83	-2.39
1	5401.08	-1.99
0.5	5409.63	0.15

4.2. Effect of Cable Stress on Wrinkle

Large membrane structures in space require the imposition of certain prestress through cable elements at the corners to maintain the flatness of their surfaces. Prestress is applied at the corner nodes of the solar sail membrane, and by altering the length and angle of the cables, the magnitude and direction of the prestress can be adjusted.

Keeping the direction of the cable stress constant, i.e., at an angle of 22.5° with the right-angle side of the membrane, as shown in figure 3, and varying the magnitude of the cable stress results in different wrinkle strength factors of the membrane structure under various prestress levels, as shown in figure 4(a). The wrinkle strength factor of the membrane structure increases significantly with the increase in cable stress, indicating that the degree of wrinkle deformation of the structure intensifies significantly. The magnitude of the cable stress is a key parameter affecting the degree of wrinkle deformation of the membrane structure. During design, the prestress should be set within a reasonable range to prevent excessive prestress from causing large wrinkle deformations, as well as to avoid too low prestress leading to structural relaxation.

Maintaining the cable stress magnitude at 1N and altering the cable stress direction, the wrinkle deformation is calculated for angles of 15.0°, 22.5°, and 23.0° with the right-angle side of the membrane, as shown in figure 4(b). When the prestress direction is 22.5°, which points towards the geometric centroid of the triangular solar sail, the wrinkle deformation is minimal; however, when the prestress direction deviates from the sail's centroid, the wrinkle deformation increases rapidly. The direction of prestress has a significant effect on the degree of wrinkle deformation of the membrane structure, with a 7.5° change in direction angle potentially increasing the strength factor by four times. Therefore, the direction of the cables should be reasonably designed and adjusted to minimize wrinkle deformation in the membrane structure.

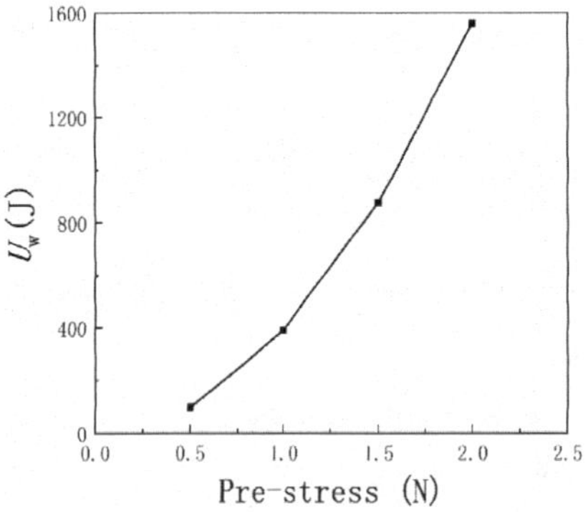

(a) Pre-stress magnitude

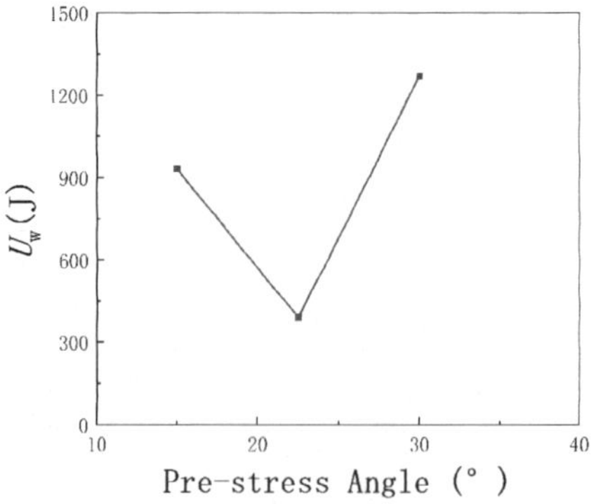

(b) Pre-stress angle

Figure 4. The relation between pre-stress and wrinkled intensity.

4.3. Methods to Eliminate Wrinkle

In engineering, there are two main methods commonly used to eliminate wrinkles in membrane structures. The first method involves attaching reinforcement strips to the membrane to make the stress distribution more uniform and improve structural strength. The second method involves adding connecting elements between the support rods and the membrane structure to improve the internal stress distribution of the membrane.

4.3.1. The Effect of Reinforcing Tape. As shown in figure 5, three reinforcement strips are arranged along the edges of the solar sail membrane, using pressure-sensitive tape as the material for the strips. The effects of the width and thickness of the reinforcement strips on wrinkle deformation were investigated. In the finite element model, the reinforcement strips were modeled using two-node linear beam elements, which were connected with the triangular elements through common nodes.

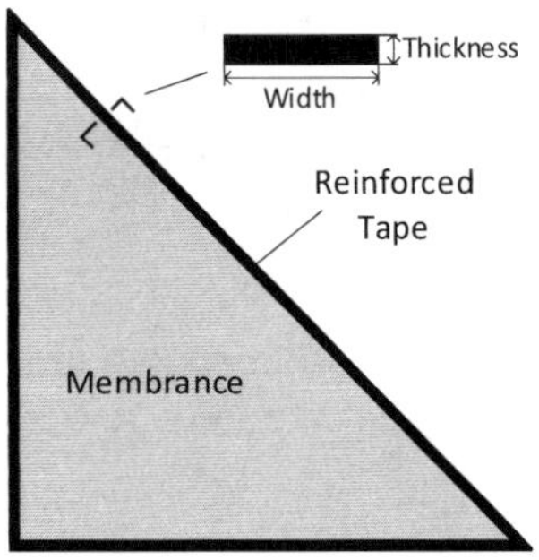

Figure 5. The layout of reinforced type.

First, the effect of reinforcement strip thickness on wrinkle deformation was examined. Keeping the strip width at 0.3 m and varying the strip thickness, figure 6(a) shows the wrinkle strength factors for strips of different thicknesses. As the thickness of the reinforcement strips increased, the wrinkle strength factor of the membrane structure gradually decreased, which is beneficial for wrinkle elimination.

Next, the effect of the width of the reinforcement strip on wrinkle deformation was investigated. Keeping the strip thickness at 40 μm and varying the strip width, figure 6(b) shows the curve of wrinkle strength factors for strips of different widths. As the width of the reinforcement strips increased, the wrinkle strength factor of the membrane structure also decreased. Increasing the width of the reinforcement strips helps reduce wrinkle deformation in the membrane structure.

The mass increment caused by increasing the thickness of the reinforcement strips is much less than that caused by increasing the width and increasing the thickness is more effective for eliminating wrinkles. Therefore, to eliminate wrinkles, it is recommended to increase the thickness while maintaining an appropriate width for the reinforcement strips. Modifying the thickness and width of the reinforcement strips has positive effects on wrinkle elimination in membrane structures. However, the reinforcement strip arrangement scheme studied in this paper has a general effect on wrinkle elimination. The primary function of edge reinforcement strips is to improve the internal stress distribution and enhance the strength of the membrane, and further investigation is needed on the effect of internal reinforcement strips on wrinkle elimination.

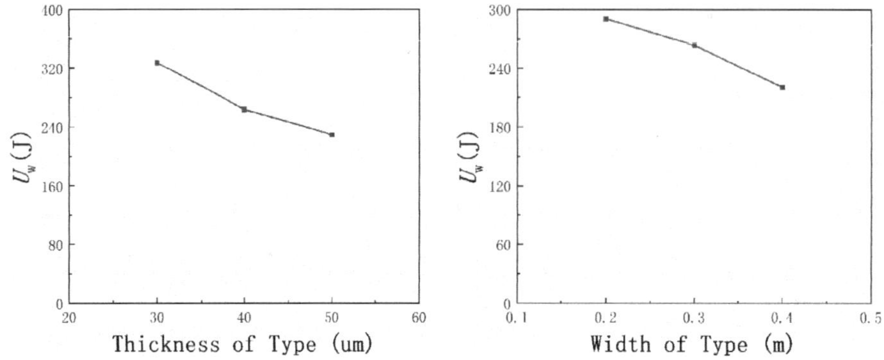

(a) Thickness of type (b) Width of type

Figure 6. The relation of reinforced type and wrinkled intensity.

4.3.2. The Effect of Reinforcing Cable and Connected Element. For the space membrane structure analyzed in this study, using cable elements connected at the membrane nodes alone is insufficient to meet performance requirements. Typically, other connecting elements need to be installed on the support rods to ensure structural reliability. The wrinkle deformation degree of the membrane was analyzed using three different connection schemes, as shown in figure 7. One, two, or three additional connecting elements were arranged on each support rod, with each element applying a tension of 0.1 N perpendicular to the reinforcement strip.

Figure 8 shows the wrinkle strength factors corresponding to different quantities of additional connecting elements. The results indicate that scheme 1, where one additional connecting element is arranged on the support rod, yields the smallest wrinkle strength factor, thus having the best effect on eliminating wrinkles in the membrane structure. As the number of connecting elements increases, the wrinkle strength factor gradually increases, reducing the effectiveness of wrinkle elimination. In the practical design of membrane structures, the overall configuration characteristics of the membrane should be considered comprehensively to determine the appropriate number of connecting elements while balancing structural strength and wrinkle deformation.

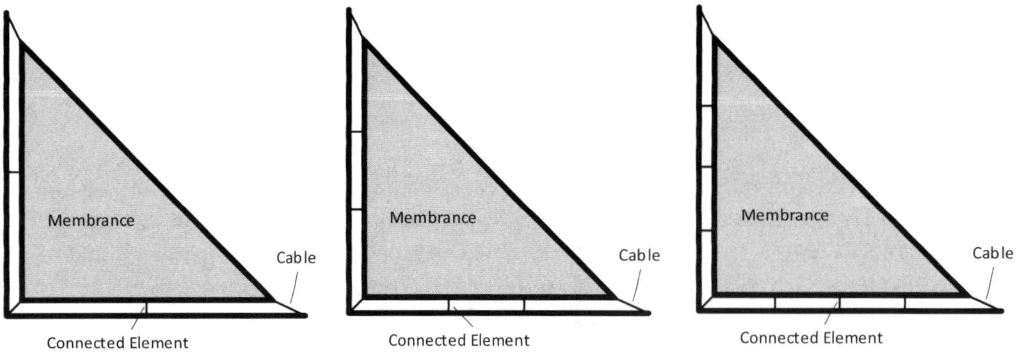

Figure 7. The layout of connected elements.

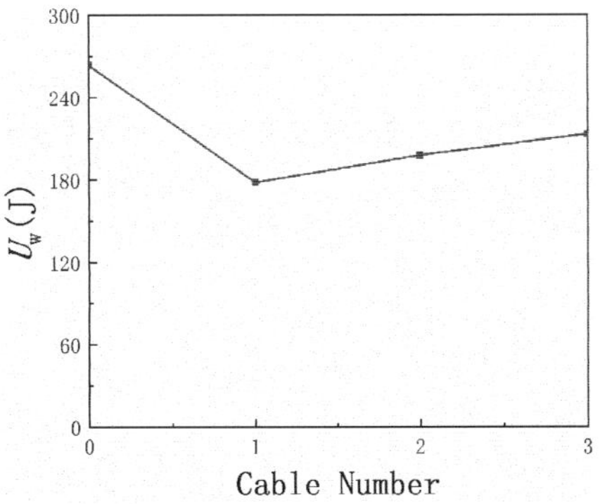

Figure 8. The relation of connected elements and wrinkled intensity.

5. Conclusion

In this paper, a wrinkle prediction method based on strength factors was applied to predict wrinkle deformation in a large membrane solar sail structure, leading to the following conclusions:

(1) Reducing cable tension can significantly decrease the degree of wrinkle deformation, but deviations in cable direction from the centroid of the membrane significantly increase this degree. In engineering design, it is necessary to adjust the magnitude of cable tension to reduce wrinkles while maintaining certain prestress, and aim the cable direction towards the centroid of the membrane as much as possible;

(2) Applying reinforcement strips and connecting elements at the membrane edges can effectively improve wrinkle deformation, with connecting elements having a more significant effect on wrinkle elimination. Optimal wrinkle elimination can be achieved by reasonably selecting the number of connecting elements on the support rods.

6. References

[1] Spencer D A, Johnson L and Long A C 2019 Solar sailing technology challenges *Aerospace Science and Technology* **93** 105276.

[2] Zhang H, Liu H W, Li D Y and Peng F 2018 Study structures of dynamics simulation on large space deployable membrane *Chin J Space Sci* **38(1)** 101-108.

[3] Murphey T, et al. 2002 A method to quantify the thrust degradation effects of structural wrinkles in solar sails.

[4] Wang C G, Wei J Z and Liu Y Y 2020 Some advances in technologies of aerospace flexible deployable structure and their applications *Journal of Astronautics* **41(6)** 761-769.

[5] Akita T and Natori M C 2008 Sensitivity analysis method for membrane wrinkling based on the tension-field theory *AIAA Journal* **46(6)** 1516-1527.

[6] Wong S P 2006 Wrinkled membrance part III: numerical simulations.

[7] Wang C G, Du X W and Wan Z M 2007 Non-linear buckling finite element analysis of membrane wrinkling *Chinese Journal of Computational Mechanics* **24(3)** 269-274.

[8] Yang C, et al. 2019 Static and dynamic evaluations for large square solar sail concept based on scalable prototype validation *Acta Astronautica* **159** 258-266.

[9] Wei J, et al. 2018 Modal analysis and identification of deployable membrane structures *Acta Astronautica* **152** 811-822.

[10] Xiao X, Zhang Q and Guan F L 2020 Finite element analysis of elastic strain energy and principal stress for rectangular shear aerospace wrinkling membrane *Spatial Structure* **26(02)** 3-10.

[11] Tessler A, Sleight D and Wang J 2003 Nonlinear shell modeling of thin membranes with emphasis on structural wrinkling *44th AIAA/ASME/ASCE/AHS/ASC Structures, Structural Dynamics, and Materials Conference* p. 1931.

[12] Li Y L, Lu M Y and Tan H F Initial imperfection sensitivity analysis of membranes buckling based on dynamic disturbing method *Engineering Mechanics* 2010 **27(08)** 94-99.
[13] Zhang J, Yang Q S and Tan F 2010 Analysis of wrinkled membrane structures by thin-shell elements *Engineering Mechanics* **27(08)** 28-34.
[14] Wang J, et al. 2004 *Simulating Nonlinear Deformations of Solar Sail Membranes Using Explicit Time Integration.*
[15] Liu M J 2018 *Wrinkling Deformation Characteristics and Control Method of Tensioned Planar Membrane Structure* (Xidian University).
[16] Li Y L, Tian Z H and Tian H F 2008 Review of methods of wrinkle based on tension field theory *Mechanics in Engineering* **(04)** 8-14.
[17] Akita T, et al. 2007 A simple computer implementation of membrane wrinkle behavior via a projection technique *International Journal for Numerical Methods in Engineering* **71(10)** 1231-1259.
[18] Kogiso N, Toyama K and Akita T 2008 Wrinkle intensity minimization design of space membrane structure *12th AIAA/ISSMO Multidisciplinary Analysis and Optimization Conference* p. 6030.
[19] Sidi M J 1997 *Spacecraft Dynamics and Control: A Practical Engineering Approach* Cambridge University Press.
[20] Sadd M H 2009 *Elasticity: Theory, Applications, and Numerics* Amsterdam: Press.

Experiment and Durability Decision of Dry Cutting 316L Stainless Steel with Coated Circular Saw Knife

Biwen Li[1,a], Jun Chen[1], Yunqiao Dong[1,*], Jiantao Gu[2] and Hongzhe Peng[3]

[1] School of Mechanical Engineering, University of South China, Hengyang 421001, China.
[2] Nuclear Power Institute of China, Chengdu 610213, Sichuan, China.
[3] Hengyang technician College, Hengyang 421001, Hunan, China.
Email: [a]libiwen68@163.com

Abstract. Melting and decontamination of decommissioned waste metals from nuclear power plants requires cutting and disintegration first, dry cutting process can effectively avoid the generation of waste liquid and aerosol in large amounts and a complex treatment process. Low-level radioactive 316 L austenitic stainless steel decommissioned from nuclear power plants has a large amount and poor cutting performance, decontamination enterprises need to balance and improve productivity, cost, and waste knife disposal in the sawing disintegration process. Taking AlCrN coating circular saw knife and TiN thin coating circular saw knife as the research object, the durability test of dry sawing 316L stainless steel pipe with two factors and two levels was carried out. The tool wear mechanism and the applicability of the maximum durability of AlCrN coated circular saw knife are analyzed, the empirical formula of tool durability in the form of a selective subset equation is established, a list algorithm of maximum productivity durability and minimum cost durability based on TableCurve 3D interpolation function is proposed. The results show that under the cutting conditions of low rotational speed and large feed rate, the two kinds of coated knives have the highest productivity durability and the lowest cost durability, at this time, the AlCrN coated tool has slightly better durability and slightly worse economy. The empirical formula of tool durability in the form of a selective subset equation has the advantages of fewer sample points, diverse expressions, and high model accuracy. The list algorithm overcomes the limitations of the traditional algorithm that the feed per tooth needs to be selected first, the relationship between cutting speed and durability must be expressed as a power function, the calculation results need to be rounded, and the equation is likely to have no solution. The calculation results are in line with engineering practice.
Keywords. Dry cutting; coated circular saw knife; tool durability model; TableCurve 3D; list algorithm.

1. Introduction

Smelting, decontamination, and recycling of decommissioned waste metals from nuclear power plants with radioactive specific activity lower than the national standard clean control level is a relatively effective safe volume reduction treatment method at present [1]. The dry cutting process is used to cut and disintegrate the decontaminated waste metal on the surface [2], which can avoid the waste liquid produced by the wet cutting process and the aerosol produced by the gas cutting process. In view of the characteristics of stainless steel, such as radiation resistance, corrosion resistance, good mechanical properties at high temperatures, good sealing, easy cleaning, and low cost, the nuclear industry uses it as the cladding tube of some reactor types, the body material of control rod and the material of steam generation system. The use of austenitic stainless steel as the structural material or lining material of the main pipeline and valve can reduce the corrosion and corrosion products of the circuit [3,4]. Although the strength and hardness of austenitic stainless steel material are not high, it is easy to produce work hardening and low thermal conductivity. Therefore, the cutting force is large and the cutting temperature is high, which is easy to cause the surface scratch of the knife, the adhesive wear of the rake face, and the boundary wear of the flank face [5,6]. In this paper, AlCrN coated circular saw knife and TiN thin-coated circular saw knife are taken as the research objects, through the tool durability test and data processing of dry sawing 316L stainless steel pipe, the tool wear mechanism is revealed, that the empirical formula of the relationship between durability and cutting parameters is established, and the maximum productivity durability and minimum cost durability are calculated. It provides a basis for enterprises to formulate cutting parameters for radioactive contaminated metal decontamination and recycling in the nuclear field.

2. Tool Durability Test of Dry Cutting 316L Stainless Steel Pipe

2.1. Experimental Design

(1) Machine tool: The X62W horizontal milling machine features a main rotational speed map that is identical to that of the XA6132; however, the feed rate map differs [7].

(2) Workpiece material: Low-speed 316L austenitic stainless steel is widely used in the nuclear industry [8], and it is suitable for intermittent cutting. The size of the pipe is 25×3.5.

(3) Tool specification: The substrates are gray-black AlCrN coated circular saw knives and golden TiN thin-coated circular saw knives of high-performance high speed steel W6Mo5Cr4V2Co5 (M36), and uncoated circular saw knives crafted from tungsten-molybdenum general high-speed steel W6Mo5Cr4V2 (M2). All knives conform to the specifications of 150×2×72.

The cutting edges of the three tools are not precisely ground. The test results of the geometric parameters of the cutter teeth, the thickness of the coating, the surface roughness of the cutter side, the surface hardness of the cutter surface, and the adhesion force of the film base (the critical load of coating failure) are shown in table 1.

Table 1. Test results of three kinds of circular saw knives.

	Top edge rake angle (°)	Top blade back angle (°)	Tooth height (mm)	Coating thickness (μm)	Tool side surface roughness (*Ra*) (μm)	Tool surface hardness (HV)	The critical load of coating failure (N)
AlCrN coating knife	2	30	2.91	2.5	0.136	1351.7	13.5
TiN thin coating knife	3	25	2.90	1.5	0.320	1180.5	8.6
M2 uncoated knife	1	14	2.84	—	0.758	793.4	—

(4) Clamping method: Circular saw knife with key installation, do not use beams and hangers; the center distance between the workpiece and the tool is 37.5 mm, the vise is clamped, the extension amount is 50 mm, the thickness of the cut sample is 10 mm, and the actual cutting distance is 29.057 mm.

(5) Selection of cutting parameters: Using the reverse milling method, feed in accordance with the X-axis direction. The range of cutting speed v is 20-60 m /min, and the range of feed per tooth f_z is

0.01-0.02 mm /T [9]. Four groups of dry sawing cutting schemes are worked out as shown in table 2.

Table 2. Dry cutting parameter scheme.

	Rotational speed *n* (r/min)	Feed rate *f* (mm/min)	Cutting speed *v* (m/min)	Feed per tooth f_z (mm/T)
Scheme1	75	75	35.34292	0.01389
Scheme2	75	118	35.34292	0.02185
Scheme3	118	75	55.60619	0.00883
Scheme4	118	118	55.60619	0.01389

(6) Process observation: The wear and damage forms of coating, rake face, flank face, flank face, and knife teeth were observed in real-time, as well as the change of knife blunt radius and slip phenomenon, and the reduction of tooth height was measured by sampling. Observe the evolution of chip color and shape, vibration, and squeal to divide the tool wear stage and determine the blunt standard [8].

2.2. *Test Phenomena and Results*

(1) Tool wear: In the process of dry sawing, the three types of circular saw knives exhibit significant flank wear, resulting in an increase in the blunt radius of the knife and a gradual reduction in tooth height. The coating from the generation of grooves until uniform wear, but there is no lift and local peeling phenomenon, the non-coated knife has a slight amount of adhesive wear and the coated knife is not obvious, furthermore, there were no macroscopic hot cracks, side edge damage, crater wear on the rake face, as well as deformation and chipping of the cutter teeth.

(2) Evolution of chip color: The chips all undergo the color evolution of silver white → slightly yellow (initial wear stage) → light yellow → yellow → yellowish brown → purple → deep purple → purple band blue (normal wear stage) → blue → bluish white (sharp wear stage), reflecting the evolution of cutting temperature. It is verified that the proportion of heat directly transmitted from the cutting area during dry cutting is very low, which also indicates that the effect of diffusion wear on its color is not obvious.

(3) Tool durability statistics: Without considering the machining accuracy and surface quality of the workpiece, the number of workpieces processed at the time of tool failure is used as the basis for calculating the durability time. The statistical data are shown in table 3.

Table 3. Durability statistics of three kinds of tool schemes.

	AlCrN coating knife	TiN thin coating knife	M2 uncoated knife
Scheme1	609 (235.94235 min)	373 (144.50985 min)	241 (93.36963 min)
Scheme2	354 (87.17082 min)	349 (85.93959 min)	171 (42.10794 min)
Scheme3	299 (115.84033 min)	187 (72.44864 min)	37 (14.33476 min)
Scheme4	232 (57.12890 min)	99 (24.37828 min)	33 (8.12609 min)

2.3. *Analysis of Dry Cutting Performance of Coated Knives*

The maximum durability of the two coated knives for dry cutting 316L stainless steel is much higher than that of the M2 uncoated tool. The analysis mainly comes from the following three aspects.

(1) Coating: The surface Vickers hardness of AlCrN coated knife and TiN thin coated knife is 70.36 % and 48.79 % higher than that of the M2 uncoated tool, respectively. The high-temperature friction coefficients of 0.35 and 0.40 are 22.22 % and 11.11 % lower than that of M2 0.45 respectively [10,11], which can effectively reduce cutting force and cutting temperature. After the initial oxidation temperature of the AlCrN coating exceeds 900°C, the Al2O3 inert protective film effectively increases the anti-oxidation diffusion and anti-oxidation ability, thereby reducing the tool friction coefficient

[12,13]. The TiN + Ti2N two-phase region formed after the initial oxidation temperature of the TiN coating exceeds 600°C has better thermal hardness, but when the temperature exceeds 960°C, Ti2N will denitrify to form low hardness TiO2 [14], thus losing the advantage of thermal hardness.

(2) Matrix material: The hardness of M2 at room temperature is 63-66 HRC, while that of M36 is increased to 66-68 HRC. Although the bending strength and impact toughness of M36 decrease slightly, the coated knife matrix can still meet the application requirements even when the knife teeth become pointed. The high temperature hardness of M2 at 600°C is 47-48HRC and M36 is increased to 54HRC. Therefore, the coating knife matrix still has good wear resistance after the coating is worn out [15]. The matrix of the three knives has good anti-adhesive wear resistance, and M36 is better than M2.

(3) Improvement of geometric parameters and surface quality: The ion plating process reduces the surface roughness R_a value of the tool side, and meets the requirements for film-based bonding force. Additionally, increasing the rake angle and the back angle of the top edge enhances the sharpness of the tool [16]. This has changed the traditional view that the coated knife is not suitable for small feed cutting and heavy load cutting because of its sharpness and spalling resistance [8].

3. Establishment of An Empirical Formula for the Relationship between Durability and Cutting Parameters

3.1. The Significance of Establishing the Empirical Formula of the Relationship between Durability and Cutting Parameters

The circular saw knife is a thin-edged tool serving under a high-pressure dynamic load. There is a large contact pressure between the flank face and the workpiece surface. When cutting 316L stainless steel, a plastic material with high work hardening performance, with a small feed per tooth, the severe friction between the tool and the chip and the workpiece produces a very high cutting temperature. The rotational speed n determines the cutting speed v, and v affects the feed per tooth with the relationship $f_z=f\,D\pi\ /\ 1000zv$ and then affects the cutting thickness. The average normal stress σ_{av} on the rake face is affected by the cutting thickness, cutting speed, the real-time changing hardness of the workpiece material, and the rake angle of the tool. The significant change in the cutting temperature caused by the cutting speed and the wear value of the flank affects the shear yield strength τ_s of the workpiece material [17]. Therefore, the average friction coefficient μ on the rake face calculated according to the internal friction law $\mu=\tau_s\ /\sigma_{av}$ is actually a variable. Low friction coefficient can reduce the cutting force and cutting temperature, so as to effectively improve the tool's durability. However, it is very difficult to derive the mathematical relationship between cutting parameters and tool life based on theoretical analysis methods. Therefore, it is necessary to establish an empirical formula for the relationship between them based on experimental data, this approach will facilitate the prediction of tool life and the determination of cutting parameters, considering the perspectives of environmental protection, productivity, and economic efficiency.

3.2. Previous Modeling Method

According to the traditional modeling method, it is necessary to draw the v-T relationship curve between the cutting speed v and the tool life T in a certain range of cutting speed in the double logarithmic coordinate system [18]. Following linear fitting, this relationship is expressed as a linear equation in the form of $vT^m=C_0$. Then the $f_z\,T^{m1}= C_1$ form linear equation expressing the f_z -T relation is obtained by a similar method. The empirical formula $T=C_T/v^{1/m}f_z^{1/m}$ in the form of a nonlinear power function expressing the general relationship between cutting parameters and durability is obtained. This modeling method requires a lot of sample points, and its expression also determines that the accuracy of the model is not high.

3.3. Modeling Method Based on TableCurve 3D

Given the substantial workload associated with dry cutting tests and the limited number of test sample points for the two factors and two levels, it is decided to model based on the three-dimensional curve fitting software TableCurve 3D [19,20]. By thoroughly considering the continuity, smoothness and shape preserving of three-dimensional graphics [21,22], the durability models of three kinds of circular saw knife dry cutting 316L stainless steel with correlation index $R^2 \geq 0.99969$ and residual standard error $\sigma \leq 1.49694$ are obtained. The three-dimensional diagram of the three models is shown in figure 1.

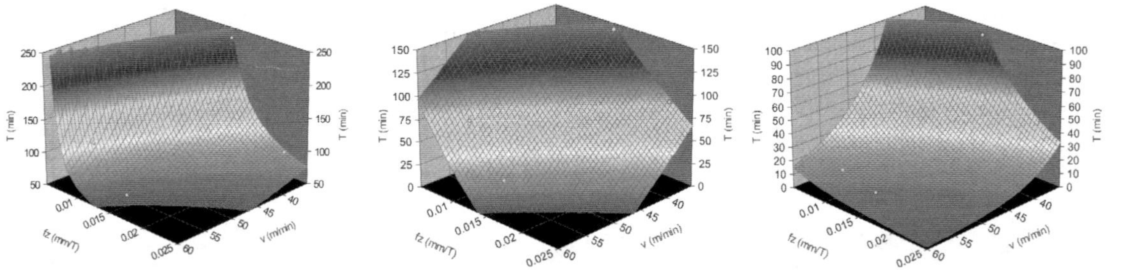

(a) AlCrN coating knife (b) TiN thin coating knife (c) M2 uncoated knife

Figure 1. Three-dimensional diagram of durability model of dry cutting 316L stainless steel by three kinds of circular saw knives.

The expression formulas of the durability model of the AlCrN coated knife, TiN thin coated knife, and M2 uncoated knife are as follows:

$$T=1/(a+bv^2+cf_z^{-0.5})=1/(0.030662288+7.2781503\times10^{-6}v^2-0.0041856061f_z^{-0.5}) \tag{1}$$

$$T=a+bv+c/\ln f_z=859.99323-5.9207548\,v+2162.0759/\ln f_z \tag{2}$$

$$T=e^{a+b\ln v+cf_z\ln f_z}=e^{25.783715-5.4104394\ln v+32.969714\,f_z\ln f_z} \tag{3}$$

Although the expression formulas for the three circular saw knife durability models differ, they all belong to the selective subset equation [23,24]. While ensuring continuity, smoothness, a shape preserving, and high fitting accuracy, unifying their expression formula forms proves to be challenging.

4. The Determination of the Maximum Productivity Durability and Its Cutting Parameters

4.1. Application Occasions and Calculation Formulas

According to the results of the dry cutting test, in the field of nuclear radioactive pollution metal decontamination and recycling-related enterprises if based on the principle of maximum durability or waste minimization perspective, three kinds of circular saw knives should undoubtedly choose scheme 1, but low feed rate is bound to increase the cutting hours and reduce productivity [25]. The maximum productivity durability is measured by the maximum number of products produced per unit of time or the minimum number of working hours per process. To adapt to the rapid expansion of production capacity, urgent production tasks, and rising processing costs, priority will be given to selecting the associated cutting conditions, even if this results in an increase in the cost of individual pieces.

The single process time t_w is:

$$t_w=t_m+t_{ct}\frac{t_m}{T}+t_{ot} \tag{4}$$

T_m—The cutting time of the process;

t_{ct} — The time consumed by changing the knife once;
t_m/T — Number of changing knives;
t_{ot} — Other auxiliary process time except for tool change time.

In this article, a circular saw knife with a diameter of D=150 mm is employed to cut through a workpiece with a diameter of d=25 mm, positioned at a center distance of 37.5 mm. Consequently, the relevant formula can be derived.

$$t_m = \frac{\sqrt{(D/2+d/2)^2-(D/4)^2}-\sqrt{(D/2-d/2)^2-(D/4)^2}}{nzf_z} = \frac{0.19018}{vf_z} \quad (5)$$

According to the current statistics from dry cutting tests, the time for the cutting tool t_{ct} is set at 1.2 min. Additionally, the duration for auxiliary processes t_{ot} is set at 0.5 min, which encompasses activities such as rapid knife return, adjusting the workpiece protrusion, and cleaning chips.

4.2. The Traditional Calculation Method of the Maximum Productivity Durability

The traditional solution method involves several steps: first, a larger f_z value is selected, then, $vT^m=C_0$ is substituted into formula (5), transforming t_m into the form of $t_m=AT^m$. This result is subsequently substituted into formula (4), leading to the expression form of $t_w=AT^m+t_{ct}.AT^{m-1}+t_{ot}$. By setting $dt_w/dT=0$, we can determine the value of T that corresponds to the highest productivity, $T_P=(1-m)t_{ct}/m$, along with the associated cutting speed v_p. However, this process appears overly idealistic and raises several concerns.

(1) Determining the value of f_z is challenging, as it is influenced not only by f but also by v. As illustrated in tables 2 and 3, the impact of these factors on T is notably significant.

(2) In order to obtain a straight-line equation representing the v-T relationship in the form of $vT^m=C_0$, extensive experimental work will require significant manpower and material resources. In the lower cutting speed zone, the built-up edge may protect the cutting edge during a relatively stable period and increase the durability, it may be unstable and produce fragments, thereby reducing durability [26]. These two conditions determine that the v-T relationship curve is not a single straight line.

(3) If the relationship of v-T is not expressed in the form of $vT^m=C_0$, then when $dt_w/dT=0$, the equation may lack a solution.

(4) Even though T_p and v_p can be obtained, the machine tool speed n is a discrete value, which means that the converted cutting speed v is also discrete. If the principle of selecting the nearest value is applied, the value of v_p will vary [27]. Additionally, the pre-selected discrete value of the feed per tooth f_z significantly restricts the range of possible values for v_p, based on the relationship $f_z=f\,D\,\pi/1000zv$.

4.3. List Algorithm Based on TableCurve 3D Interpolation Function

In light of the limitations associated with traditional calculation methods, this article presents a list algorithm aimed at obtaining productivity durability, utilizing the TableCurve 3D interpolation calculation function. Notably, this method eliminates the need to pre-select the value of f_z; instead, it employs f_z in conjunction with T as the variable for t_w.

Taking the AlCrN coated knife as an example, the formula (1) is changed to:

$$v = \sqrt{\frac{T^{-1}-a-cf_z^{-0.5}}{b}} \quad (6)$$

With the transformed formula (1) substituted into formula (5), and then into formula (4), formula (6) is obtained after simplification.:

$$t_w = t_m + t_{ct}\frac{t_m}{T} + t_{ot} = \frac{0.19018\ (T+1.2)}{f_z T\sqrt{\frac{T^{-1} - a - cf_z^{-0.5}}{b}}} + 0.5 \tag{7}$$

Through programming calculations, the t_w values under 4 sets of dry cutting plans can be easily obtained and the values of T, v, and f_z can be matched. Considering that the speed of the X62W horizontal milling machine of 95 r/min and the feed rate of 95 mm/min were not included in the test, the XA6132 horizontal milling machine also has two feed rates of 80 mm/min and 100 mm/min within the test value range. To enhance practicality, all these parameters were integrated and combined into new cutting plans, resulting in a total of 15 groups. The T values of the new schemes can be derived using the interpolation calculation function of TableCurve 3D, while the t_w values are computed through the program. The methods for obtaining the values for the other two cutting tools are analogous. By listing the t_w and T values for the 15 groups of schemes in table 4, it is possible to determine the highest productivity durability along with the corresponding cutting parameters.

Table 4. Durability and single-piece process time of 15 cutting schemes of 3 kinds of circular saw knives.

	n	f	v	f_z	AlCrN coating knife		TiN thin coating knife		M2 uncoated knife	
	(r /min)	(mm /min)	(m/min)	(mm/T)	T (min)	t_w(min)	T (min)	t_w (min)	T (min)	t_w (min)
Scheme1	75	75	35.34292	0.01389	235.907	0.88937	145.175	0.89060	93.364	0.89238
Scheme2	75	118	35.34292	0.02185	87.431	0.74965	85.274	0.74974	42.120	0.75329
Scheme3	118	75	55.60619	0.00883	115.958	0.89134	73.625	0.89364	14.384	0.91964
Scheme4	118	118	55.60619	0.01389	56.651	0.75144	25.201	0.75795	8.041	0.78298
Scheme5	75	95	35.34292	0.01759	122.034	0.80892	115.625	0.80909	63.547	0.81169
Scheme6	95	75	44.76770	0.01096	189.832	0.89006	115.911	0.89162	36.055	0.90051
Scheme7	95	95	44.76770	0.01389	102.731	0.80942	89.373	0.80995	25.986	0.81997
Scheme8	95	118	44.76770	0.01725	74.738	0.75022	62.396	0.75101	18.302	0.76242
Scheme9	118	95	55.60619	0.01118	73.632	0.81090	49.620	0.81331	10.877	0.83967
Scheme10	75	80	35.34292	0.01481	186.576	0.86567	137.478	0.86651	84.614	0.86849
Scheme11	95	80	44.76770	0.01170	152.606	0.86595	108.875	0.86709	33.120	0.87625
Scheme12	118	80	55.60619	0.00942	99.589	0.86745	67.287	0.86955	13.385	0.89562
Scheme13	75	100	35.34292	0.01852	111.148	0.79369	108.713	0.79376	57.936	0.79657
Scheme14	95	100	44.76770	0.01462	94.054	0.79428	83.245	0.79476	24.030	0.80508
Scheme15	118	100	55.60619	0.01177	68.559	0.79567	44.050	0.79850	10.167	0.82488

Table 4 illustrates that the three knives exhibit the smallest single-piece process time t_w in scheme 2, with the corresponding T and v values being designated as T_p and v_p, respectively, while the feed rate f is maximized. This outcome contrasts with scheme 1, which was selected based on the principles of maximum durability and waste minimization. Notably, schemes 4 and 8 also have the highest feed rates f. Although their t_w values are very similar to that of scheme 2, the T values differ significantly, making these schemes less favorable. Among all the schemes, scheme 3 has the largest t_w value, indicating that t_w is influenced not only by f but also significantly by variations in v related to f_z. If a tool is to be selected based on scheme 2, the AlCrN coating knife is recommended due to its slightly superior t_w and T values compared to the TiN thin coating knife.

Figure 2(a) illustrates the graphical relationship represented by formula (6). Figures 2(b) and 2(c) similarly depict the graphical relationships of the t_w expressions for the TiN thin coated knife and the M2 uncoated knife, derived using a comparable method. It is evident from the figures that even when the f_z value is selected first if $dt_w/dT=0$, the equation yields no solution. For various empirical formulas relating durability to cutting parameters, the list algorithm based on the TableCurve 3D interpolation calculation function can effectively determine the maximum productivity durability. This approach not only overcomes the limitations of traditional methods but also demonstrates significant engineering practicality.

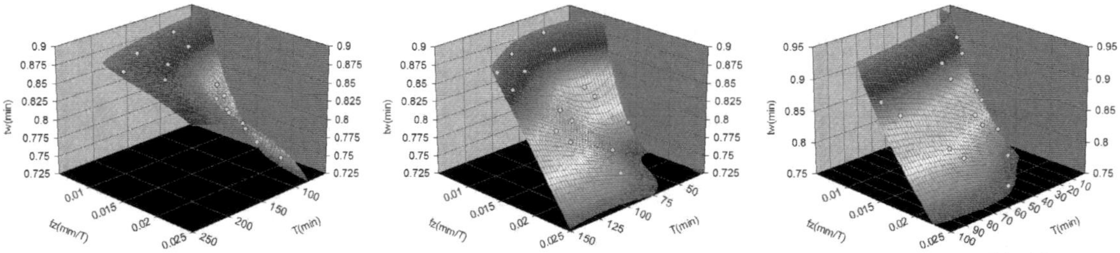

(a) AlCrN coating knife (b) TiN thin coating knife (c) M2 uncoated knife

Figure 2. The relationship between t_w T and f_z of three kinds of circular saw knives.

5. Determination of Minimum Cost Durability and Cutting Parameters

5.1. Application Occasions and Calculation Formulas of Minimum Cost Durability

The maximum productivity and durability and its cutting amount are not values during normal production. From an economic perspective, maximum productivity durability should also be correlated with machine tool costs, knife costs, and labor costs [28]. The minimum cost durability utilized in normal production is determined by the principle of achieving the lowest processing cost for each product or process.

The cost of a single process C is:

$$C = t_m M + t_{ct}\frac{t_m}{T}M + \frac{t_m}{T}C_t + t_{ot}M \tag{8}$$

M — Expenses borne by the process in unit time;

C_t — Cost of disposable coated tools;

Calculation based on enterprise budget, M is set at 2.52 yuan, while the unit cost of AlCrN coated knives, TiN thin coated knives and M2 uncoated knives are 150 yuan, 75 yuan and 55 yuan respectively.

5.2. List Algorithm and Result Analysis

The traditional method for calculating the single-piece process cost C is analogous to the approach outlined in Section 4.2 and shares the same limitations. The final result indicates that under the selected feed rate f_z, the minimum cost durability T_C is obtained, and the cutting speed corresponding to T_C is v_C.

$$T_C = \frac{1-m}{m}\left(t_{ct} + \frac{C_t}{M}\right) \tag{9}$$

In order to avoid the limitations of traditional calculation methods, a list algorithm based on TableCurve 3D interpolation function is also used. Taking the AlCrN coated knife as an example, the formula for calculating its C value is:

$$C = \frac{0.30429\ T + 28.89215}{f_z T \sqrt{\dfrac{T^{-1} - a - c f_z^{-0.5}}{b}}} + 0.48 \tag{10}$$

Table 5 presents the T and C values obtained from three types of knives across 15 cutting schemes. To facilitate analysis and comparison, the cutting parameters in the table are organized into blocks according to the rotation speed group.

Table 5. Durability and single-piece process cost of 15 cutting schemes for three kinds of circular saw knives.

	n	f	v	f_z	AlCrN coating knife		TiN thin coating knife		M2 uncoated knife	
	(r /min)	(mm/ min)	(m/min)	(mm/T)	T(min)	C (yuan)	T(min)	C (yuan)	T (min)	C (yuan)
Scheme1	75	75	35.34292	0.01389	235.907	2.488	145.175	2.444	93.364	2.477
Scheme10	75	80	35.34292	0.01481	186.576	2.474	137.478	2.372	84.614	2.425
Scheme5	75	95	35.34292	0.01759	122.034	2.414	115.625	2.237	63.547	2.310
Scheme13	75	100	35.34292	0.01852	111.148	2.392	108.713	2.201	57.936	2.283
Scheme2	75	118	35.34292	0.02185	87.431	2.312	85.274	2.110	42.120	2.220
Scheme6	95	75	44.76770	0.01096	189.832	2.549	115.911	2.498	36.055	2.861
Scheme11	95	80	44.76770	0.01170	152.606	2.539	108.875	2.435	33.120	2.811
Scheme7	95	95	44.76770	0.01389	102.731	2.486	89.373	2.298	25.986	2.714
Scheme14	95	100	44.76770	0.01462	94.054	2.465	83.245	2.265	24.030	2.694
Scheme8	95	118	44.76770	0.01725	74.738	2.385	62.396	2.189	18.302	2.661
Scheme3	118	75	55.60619	0.00883	115.958	2.747	73.625	2.647	14.384	3.799
Scheme12	118	80	55.60619	0.00942	99.589	2.733	67.287	2.596	13.385	3.749
Scheme9	118	95	55.60619	0.01118	73.632	2.667	49.620	2.512	10.877	3.663
Scheme15	118	100	55.60619	0.01177	68.559	2.641	44.050	2.507	10.167	3.651
Scheme4	118	118	55.60619	0.01389	56.651	2.546	25.201	2.643	8.041	3.657

By analyzing table 5, the following rules are found:

(1) Comparison at different rotational speeds: When the feed rate remains constant, a slower rotational speed results in a smaller C value.

(2) Comparison at the same rotational speed: When the rotational speed remains constant, a larger feed rate results in a smaller C value, with the exception of the TiN thin coated knife and the M2 non-coated knife in the fast-rotational speed block plan 4.

(3) Comprehensive comparison: The smallest C values for the three types of knives are found in scheme 2, which exhibits the minimum rotational speed and the maximum feed rate. The C values decrease in the order of AlCrN coated knives, M2 non-coated knives, and TiN thin-coated knives. The minimum C value corresponds to the T_C value.

Table 5 shows that the economy of the circular saw knife with the maximum durability and the cheapest unit price is not necessarily good. In view of the excellent performance of the TiN thin coating knife in Scheme 2, it is recommended to use, but this is inconsistent with the result selected according to the maximum productivity durability. When comparing the significance of the difference between the t_w value in table 4 and the C value in table 5 for the two coated knives, it becomes evident that selecting the minimum cost durability, along with its cutting parameters, serves as a more scientific and rational basis for determination.

6. Conclusion

Utilizing M36 as the matrix material for the coated circular saw knife enhances high-temperature red hardness and wear resistance, while simultaneously maintaining strength and toughness. The improvement of sharpness and surface roughness makes the coated knife adapt to small feed and heavy dry cutting. Compared with TiN thin coating knife, AlCrN coating knife has better coating high temperature stability, stronger oxidation resistance and a smaller friction coefficient, so it has better dry cutting applicability. The maximum productivity durability and minimum cost durability of the two coated knives correspond to the same cutting parameters; however, the AlCrN coated knives exhibit slightly better durability while demonstrating marginally lower economic efficiency. The empirical formula of tool durability in the form of a selective subset equation has the advantages of requiring fewer sample points and high model accuracy. The list algorithm for determining the maximum productivity durability and the minimum cost durability, utilizing the TableCurve 3D interpolation function, overcomes the limitations of traditional algorithms. This approach not only possesses theoretical significance but also yields results that are consistent with engineering practice. The economic advantage of TiN thin coated circular saw knives is closely linked to the reduction in unit price resulting from the thinner coating. While TiN circular saw knives with standard coating thickness, which are more readily available, offer superior wear resistance, their economic benefits may not be as pronounced. In the case of higher environmental requirements, it is necessary to reduce the workload of decontamination and volume reduction of waste cutters or to avoid midway tool change when the cutting distance is long. It is recommended to use AlCrN coated circular saw knife with higher durability.

7. Reference

[1] Luo S G, Zhang Z T and Zhang H 2010 *Decommissioning of nuclear and radiation facilities* (Beijing: China Environmental Publishing House)

[2] Xu L J, Shen H Y and Yang J 2022 Present situation and Technical research prospect of nuclear facilities decommissioning *J. Huazhong Univ. Sci. And Tech. (Nat. Sci. Ed.)* **50** (65-76.doi:10.13245/j.hust.221010.)

[3] Mathew M D, Lietzan L M, Murty K L and Shah V N 1999 Low temperature aging embrittlement of CF-8 stainless steel *Mater. Sci. Eng. A* **269** 186-196

[4] Lian P S and Luo Q 2024 *Atomic energy industry* (Beijing: China Atomic Energy Publishing House)

[5] Jiang X Y, Li W Y and Tang A M 2009 Experimental Study on Turning Austenitic Stainless Steel with Coated Carbide Tool *Tool Eng.* **43** 58-60

[6] Gerth J, Gustavsson F, Collin M, Andersson G, Nordh L G, Heinrichs J and Wiklund U 2014 Adhesion phenomena in the secondary shear zone in turning of austenitic stainless steel and carbon steel *J. Mater. Process. Technol.* **214** 1467-1481

[7] East China Textile Institute, Harbin Institute of Technology and Tianjin University 1984 *Machine tool design atlas* (Shanghai: Shanghai Science and Technology Press)

[8] Chen R Y 2002 *Theory of metal cutting* (Beijing: Mechanical Industry Press)

[9] Yang W M 2014 Selection and Use of cutting parameters in dry cutting *Equip. Manuf. Technol.* **5** 64-65

[10] Michalak M, Michalczewski R, Osuch-Słomka E, Maldonado-Cortés D and Szczerek M 2016 The effect of temperature on wear mechanism of the AlCrN coated components *Key Eng. Mater.* **4008** 233-238

[11] Zemzemi F, Rech J, Ben S W, Dogui A and Kapsa P 2007 Development of a friction model for the tool-chip-workpiece interfaces during dry machining of AISI4142 steel with TiN coated carbide cutting tools *Int. J. Mach. Machinab. Mater.* **2** 361-377

[12] Cai Z H, Di Y L and Zhang P 2010 Microstructure and high temperature oxidation resistance of CrAlN coating on piston ring surface *China Surf. Eng.* **23** 15-19

[13] Feng Y P 2014 *Thermal stability, high temperature oxidation behavior and corrosion resistance*

of AlN-based hard coatings were studied (Changsha: Central South University) Master's Thesis

[14] Xu J P, Wang G X, Wang J J, Wang C, Yin Z, Wang Y and Chen J 2020 High temperature oxidation characteristics of TiN coating by multi-arc ion plating *Weld. Technol.* **49** 14-16

[15] Peng H Z, Li B W, Zhou N, Gu J T and Fang T 2021 Study on the applicability of AlCrN coated circular saw blade in dry cutting 316L stainless steel *Equip Manuf. Technol.* **12** 30-34+51

[16] Kang H J 2015 *Numerical simulation and experimental study on turning of ZL109 aluminum alloy* (Zhuzhou: Hunan University of Science and Technology) Master's Thesis

[17] Zhang P, Huang C, Zhu H, Wang J, Yao Y and Yao P 2021 The research of tool wear criterion in micro cutting using the elastic recovery ratio of high-strength elastic alloy 3J33B *Int. J Adv. Manuf. Tech.* **114** 1767-1776

[18] Ren J L, Liu Z F and Tang W X 2012 *Dry cutting theory and processing technology* (Beijing: Mechanical Industry Press)

[19] Li B W 2008 Curve fitting in mechanical design analysis *Mod Manuf. Eng.* **05** 72-74+24

[20] Mosnegutu E, Panainte-Lehadus M, Nedeff V, Tomozei C, Barsan N, Chitimus D and Jasinski M 2022 Identify the mathematical models generated by the Table Curve 3 D program **1**

[21] Zhou Y C 2020 *Design of visual process control system for double-sided submerged arc welding of thick-walled components with gap* (Hengyang: University of South China) Master's Thesis

[22] Irimia O, Tomozei C and Lehăduş M P 2016 A mathematical model on the efficiency of the mechanical water filtering process *Int. Multi - disc. Sci. GeoConf. - SGEM.* **2** 695-702

[23] Vo A D, Elraghy A and McAuley K B 2018 Parameter Subset Selection in Differential Equation Models with Dead Time *IFAC POnL* **51** (18) 661-666

[24] Wang D, Zhu X, Pedycz W, Yu Z and Li Z 2021 Development of granular fuzzy relation equations based on a subset of data *IEEE/CAA J. Autom. Sinica* **8** 1416-1427

[25] Zhang Y P and Gao D 2000 A practical algorithm for optimal cutting speed and feed rate. *Coal Min. Mach.* **1** (04) 3-5

[26] Fan S T, Fan Y F, Xu X J and Wang J 2023 Cutting force and tool wear test of titanium alloy Ti750 J. Jiangsu Univ.(Nat. Sci. Ed.) **44** 426-430+436

[27] Bai Q S, Sun J M, Liang Y C 2017 *Mechanical optimization design (6th edition)* (Beijing: Mechanical Industry Press)

[28] Songmene V, Khettabi R and Kouam J 2012 Dry high-speed machining: a cost effective and green process *Int. J. Manuf. Res.* **7** (3) 229-256

Quantitative Analysis on Altitude Affecting Heavy-Duty Diesel Vehicle Real Driving Emissions Based on Engine-in-the-Loop Methodology

Xiaowei Wang[a, *], Lin Zhang[b], Tao Gao[c], Yongzhan Huo[d] and Guangyou Li[e]

CATARC Automotive Research and Inspection Center (Tianjin) Co., Ltd., Tianjin 300300, China.

Email: [a]wangxiaowei@catarc.ac.cn; [b]zhanglin2015@catarc.ac.cn; [c]gaotao@catarc.ac.cn; [d]huoyongzhan@catarc.ac.cn; [e]liguangyou@catarc.ac.cn

Abstract. A dump truck with a maximum designed total mass of 25000 kg was selected to measure the real driving emission characteristics of pollutants and carbon dioxide (CO_2) under different altitudes of 0 meters, 800 meters and 1600 meters, with the same simulated road, driver and vehicle model but different vehicle loading of 0% and 40% by using the engine-in-the-loop (EIL) methodology. The results indicate that the EIL methodology can qualitatively and quantitatively analyze the impact of altitude on real driving emissions. The emissions of carbon monoxide (CO) and total hydrocarbons (THC) show a trend of first decreasing and then increasing with increasing altitude, while the emissions of particulate number (PN) show a trend of first sharply increasing and then slightly decreasing with increasing altitude. The variation pattern of nitrogen oxide (NOx) emissions with altitude is not significant. Under different altitude conditions, the specific emissions of CO_2 and THC decrease with increasing vehicle loading, while the specific emissions of PN and CO increase with increasing vehicle loading. However, the variation pattern of NOx emissions with vehicle loading is not obvious. In addition to being strongly correlated with exhaust temperature, the peak value of NOx emissions is highly correlated with road slope and vehicle acceleration. Cold start increases the specific emissions of CO_2 and pollutants, especially for PN and CO.

Keywords. Heavy duty diesel vehicle; engine-in-the-loop; altitude; vehicle loading; pollutant emissions

1. Introduction

With the sharp increase in the number of automobile owners, the environmental problems brought about by it have also aroused more and more concerns. The exhaust pollution from diesel vehicles, especially heavy-duty diesel vehicles, is even more serious in China. China Mobile Source Environmental Management Annual Report (2023) shows that carbon monoxide (CO), hydrocarbons (HC), nitrogen oxides (NOx), and particulate matter (PM) from trucks will be 1.914 million tons, 405,000 tons, 4.354 million tons, and 46,000 tons in 2022 in China, respectively, which will account for 29.7%, 28.6%, 84.6% and 91.1% of the total emissions from automobiles [1].. Therefore, controlling NOx and PM emissions from diesel vehicles is a top priority for motor vehicle pollution prevention and control. "Pollutant Emission Limits and Measurement Methods for Heavy Duty Diesel Vehicles (China Stage VI)" was issued in 2018 [2]. A notable difference in the China VI standards compared to China V is the addition of real driving emissions (named PEMS). The PEMS test should conducted in a wider range of environmental conditions with an altitude requirement of below 1700 meters for China 6a and below 2400 meters for China 6b, as well as an ambient temperature requirement between -7°C and 38°C.

Due to the low atmospheric pressure and low air density, the amount of air inhaled into the cylinder is reduced when the diesel engine is running at a high altitude, which results in imperfect combustion and deteriorated emissions [3-6]. Ge et al. [7] studied the effects of altitude particle emissions from diesel vehicles and found that the PN emissions of vehicles in the plateau environment were higher at low and high speeds, while the PN emissions gradually decreased with the increase of altitude at medium speeds. Chen et al. [8] found that the NOx emissions of diesel vehicles first increased and then decreased with the increase of altitude, and increased with the increase of road slope. Du et al. [9] studied the impact of load in the plateau environment on the emission for heavy-duty vehicles and found that the NOx emissions of vehicles gradually decreased with the increase of load, but PN emissions of different vehicles exhibit different trends. Wang et al. [10] studied the real driving emission characteristics of heavy-duty vehicles with different technical routes in different plateau areas and found that the adoption of EGR will lead to an increase in particulate matter emissions, and PN emissions are strongly correlated with engine load in PEMS tests. Even in different areas at the same altitude, there is a large variation in PEMS test results.

However, the above studies are based on real driving tests. The real driving test is greatly affected by factors such as traffic conditions, environmental conditions, and driving behavior, and it is difficult to ensure the repeatability and consistency of the test, which is not suitable for the study of the impact of a single variable on emissions. In this context, this paper selects a dump truck with a maximum design total mass of 25,000 kg to study the driving emission characteristics of pollutants and carbon dioxide (CO_2) under different altitudes in the engine laboratory with the same simulated road, driver and vehicle model but different vehicle loading by using the engine-in-the-loop (EIL) methodology, then analyze quantitatively the effects of altitude on emissions.

2. Experimental Set Up

2.1. Methodology and Test Equipment

EIL has been approved as a high-accuracy methodology to investigate the impact of single parameters on real driving emissions [11-17]. EIL methodology uses the engine as the actual physical hardware while simplifying the vehicle, driver, and road as a model, then conducts development work of vehicle performance and emissions on the engine bench. Due to the high precision of environmental control in the laboratory, and high consistency of road and driving behavior, EIL methodology can eliminate interference from driving, traffic conditions, road conditions, and the environment.

The EIL test platform built in this paper is shown in figure 1. The vehicle and driver models were built using the AVL VSM™ real-time system. The real-time system calculated the required speed and torque of the engine through the theory of vehicle dynamics with the input driving road, vehicle model, and driver model, then sent it to the bench control system. The bench control system achieved the demand engine speed and torque by controlling the dynamometer and engine throttle. Simultaneously, the sensor installed on the engine bench collected the real engine speed and torque and then transmitted them back to the real-time system to evaluate the consistency between demand vehicle speed and real vehicle speed to determine the demand values in the next calculation step, thus constituting a closed-loop control.

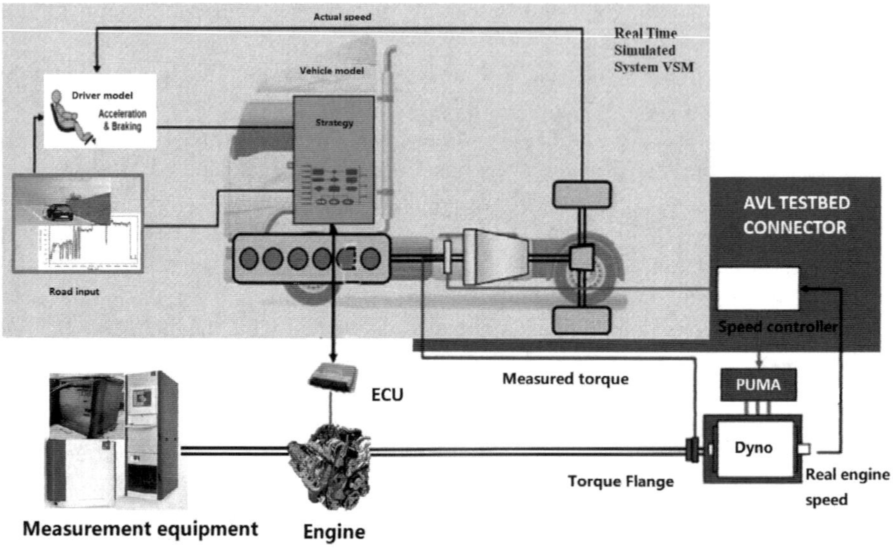

Figure 1. Engine-in-the-loop test platform

The environmental conditions were controlled by a special engine bench chamber. The intake and exhaust altitude simulation system of ETS Corporation is used to simulate the intake pressure at different altitudes. In order to prevent oil leakage of the turbocharger caused by the pressure difference between the crankcase and inlet, the oil-gas separator is connected to the inlet pipe. The pollutant emissions for each test were obtained through the emission analyzer. Details of the test equipment and software are shown in table 1.

Table 1. Test equipment and software.

Equipment name	Equipment Type	Accuracy
AC dynamometer	AVL	Torque: ±0.3%F.S., speed: ±1rpm
Gaseous emission equipment	AVL AMA i60	±2%
Particle number (PN) meter	AVL 489	±10%
Fuel consumption meter	AVL 740	±0.12%
Vehicle model	AVL VSM™	-
Real-time system	AVL testbed connect™ (RT)	-
Coast down evaluation system	AVL coast down manager	-
Altitude simulation	ETS	Static≤±0.5 kPa ， Dynamic≤±2.0 kPa
Temperature chamber	ETS	Dynamic≤±1.5°C

2.2. Vehicle and Engine Parameter
The experimental prototype is a 6L diesel engine that meets the China VI emission standards, with a rated power of 180kW and a maximum torque of 970Nm. The emission control technology route is exhaust gas recirculation (EGR)+diesel oxidation catalyst (DOC)+particulate filter (DPF)+selective catalytic reduction (SCR)+ammonia catalyst (ASC). The corresponding vehicle is an N3 dump truck with a curb weight of 9000 kg and a maximum total weight of 24500 kg. The gearbox includes 10 forward gears. The detailed parameters of the engine and vehicle modeling are shown in table 2.

Table 2. Main parameters of tested engine and vehicle model.

Parameter	Value

Engine capacity	6.234L
Cylinder numbers	6
Compression ratio	17.2
Rated power @ speed	180 kW@2300 rpm
Maximum torque @ speed	970 Nm@1100 -1800rpm
Idle speed	650 rpm
Emission control technology route	EGR+DOC+DPF+SCR+ASC
Emission standards	China VI
Vehicle curb weight	9000 kg
Axle	6×2
Tire specification/numbers	11R22.5/6
Maximum total mass	25000 kg
Transmission system	10-speed manual

2.3. Test Route

The selected road is the corresponding PEMS test route for N3 non-urban vehicles, as shown in figure 2. The PEMS test has a total time of 4950 seconds, a total mileage of 62.84km, and an average speed of 45.8 km/h. The maximum slope is 4.14% and the minimum slope is -4.02%.

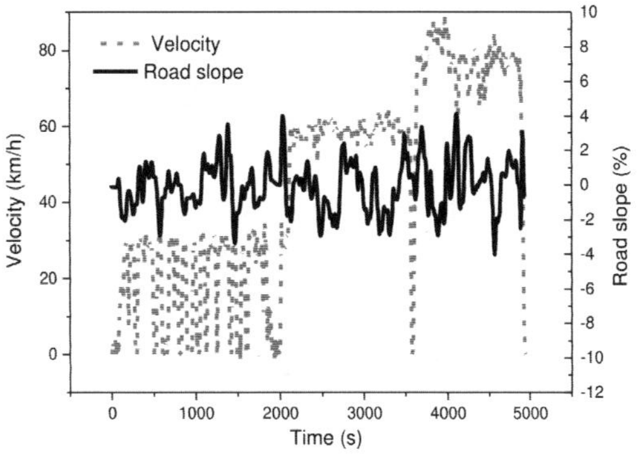

Figure. 2 Speed curve of PEMS trip.

2.4. Test Matrix

The tests were conducted under different altitudes of 0m, 800m, and 1600m and different loading conditions of 0% loading and 40% loading, while the ambient temperature was controlled at 25 °C. The test matrix is shown in table 3. The sliding coefficients under different loads were obtained from actual sliding on the test track road according to the requirements of GB27840-2021. After simulating sliding on the engine bench using the sliding coefficients, they were evaluated and adjusted using AVL's Coastdown Manager software. For each load, the same vehicle model except the set of weight and sliding coefficients and the same driving strategy were used for testing. The calculation methods for pollutant emissions and CO_2 emissions can be found in GB17691-2018 [2]. The calculation method for fuel consumption per 100 kilometers can be found in GB27840-2021.

Table 3. Test matrix.

Parameter	**Value**
Environmental temperature	25°C
Altitude set	0m, 800m, 1600m
Loading set	0%, 40%

3. Results and Analysis

3.1. Effects of Altitude on Real Driving Emissions

Table 4 shows the power, total CO_2 and pollutant emissions, and fuel consumption of the entire PEMS test under 0% loading conditions. As the altitude increases, the total power shows a slight decrease, and the total CO_2 emissions and fuel consumption per 100 kilometers also decrease. The CO_2 specific emissions at an altitude of 1600 meters are about 3.2% lower than those at an altitude of 0 meters. This may be because the demand for torque under 0% loading condition is not high, and the engine has a large margin to combine with the transmission gear to meet the vehicle's power requirements. The total NOx emissions under 0% loading condition gradually decrease with increasing altitude, while the total PN emissions increase by 4.3 times with increasing altitude from 0m to 800m. Continuing to increase the altitude to 1600m resulted in a slight decrease in total PN emissions. The emissions of CO and THC show a trend of first decreasing and then increasing with increasing altitude.

Table 4. PEMS test results under 0% loading.

	0m	800m	1600m
Power/kw	38.73	38.21	37.56
CO_2/g	29812	28762	27989
NOx/g	22.40	14.70	12.46
PN/#	4.29E+11	2.28E+12	2.17E+12
CO/g	0.54	0.51	0.66
THC/g	1.03	0.86	1.49
Fuel consumption/L/100km	18.26	17.62	17.16

Table 5. PEMS test results under 40% loading.

	0m	800m	1600m
Power/kw	49.46	48.95	49.30
CO_2/g	34750	34188	34076
NOx/g	15.43	24.25	21.67
PN/#	9.81E+11	9.06E+12	6.28E+12
CO/g	0.93	0.70	1.22
THC/g	0.76	0.51	1.17
Fuel consumption/L/100km	21.31	20.97	20.91

Table 5 shows the total power, CO_2 and pollutant emissions, and fuel consumption of the entire PEMS test under 40% load condition. As the altitude increases, the total power shows a trend of first decreasing and then increasing, while CO_2 emissions and fuel consumption per 100 kilometers gradually decrease. This indicates that the increase in altitude does not result in a significant decrease in power or a significant increase in fuel consumption. The variation trend of CO, THC, and PN emissions are consistent with those under 0% loading conditions. As the altitude increases, the emissions of CO and THC show a trend of first decreasing and then increasing, while the total PN emissions increase by 8.2 times from 0m to 800m. Continuing to increase the altitude to 1600m resulted in a slight decrease in total PN emissions. However, NOx emissions are not consistent with that 0 % loading condition. As altitude increases, NOx emissions first increase and then decrease.

The fact that the final emission of NOx is a comprehensive result of the raw emission and SCR conversion efficiency. The conversion efficiency of NOx was calculated based on the data from NOx sensors installed downstream and upstream, as shown in table 6. As the altitude increases, the raw NOx emission decreases and the exhaust NOx emission also decreases. However, the trend of SCR conversion efficiency varies for different loading. Under 0% loading condition, the SCR conversion efficiency gradually increases with altitude. Under 40% loading condition, the SCR conversion efficiency gradually decreases with increasing altitude. In addition, the raw NOx emission under a 40% loading condition is much higher than that under a 0% loading condition, so the exhaust NOx emission under the 40% loading condition is also slightly higher than that under a 0% loading condition. This indicates that the final NOx emission is not positively correlated with the loading. Another finding from the result is that the NOx sensor data exhibits a significant difference compared to the results of the bench emission testing equipment, which is generally 1.3-2.3 times higher than the measured value. The reason may be that the engine has been running for a long time, and the testing accuracy of the NOx sensor may have decreased. This also indicates that the durability of NOx sensors needs to be taken seriously attention if the NOx sensor data were used to evaluate the NOx emission in remote monitoring.

Table 6. NOx emission and conversion ratio based on OBD NOx sensor data.

	Raw emission/g	Exhaust emission/g	Conversion ratio%
0% loading under 0m	296.79	29.02	90.22
0% loading under 800m	272.15	22.31	91.80
0% loading under 1600m	245.06	19.42	92.08
40% loading under 0m	445.43	35.03	92.14
40% loading under 800m	373.95	33.98	90.91
40% loading under 1600m	329.02	30.67	90.68

The transient and cumulative emissions of PN and NOx, vehicle speed, road slope, and exhaust temperature at different altitudes under 0% loading conditions are shown in figure 3. As the altitude increases from 0m to 800m, the PN emissions during the high-speed phase rapidly increase. Particulate matter emissions are mainly generated under high temperature and oxygen deficiency conditions. As the altitude increases, it is easy to cause uneven mixture in the cylinder and deterioration of combustion due to the decrease in air density and oxygen content, which is more pronounced under high-speed and high torque conditions. It can be found that the peak of NOx emissions is closely related to slope and vehicle acceleration. When the slope rises and the vehicle accelerates, NOx is prone to significant peaks. The average exhaust temperatures at altitudes of 0m, 800m, and 1600m are 234 °C, 243 °C, and 251 °C, respectively. The higher the altitude, the higher the average exhaust temperature, which brings a higher NOx conversion efficiency.

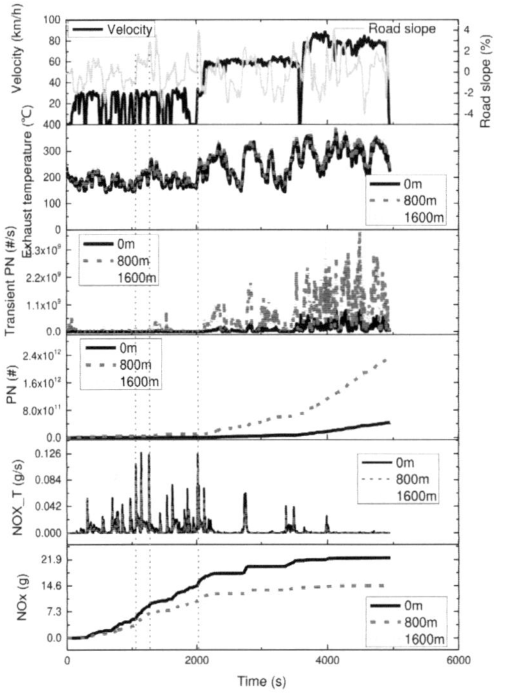

Figure 3. Transient and accumulated NOx and PN emissions, velocity, road slope, and exhaust temperature under different altitudes at 40% loading.

The transient and cumulative emissions of PN and NOx, vehicle speed, road slope, and exhaust temperature at different altitudes under 40% loading conditions are shown in figure 4. Similar to the 0% loading condition, the PN emissions during the high-speed phase rapidly increase as the altitude increases from 0m to 800m. The NOx emissions are lowest at an altitude of 0m and highest at an altitude of 800m. The difference in NOx emissions is mainly reflected in the first 1000s, mainly due to the higher raw emissions at an altitude of 800m. The peak of NOx emissions is closely related to slope and vehicle acceleration. The average exhaust temperatures at altitudes of 0m, 800m, and 1600m are 250 °C, 262 °C, and 273 °C, respectively. The average exhaust temperature also shows a trend of increasing with altitude.

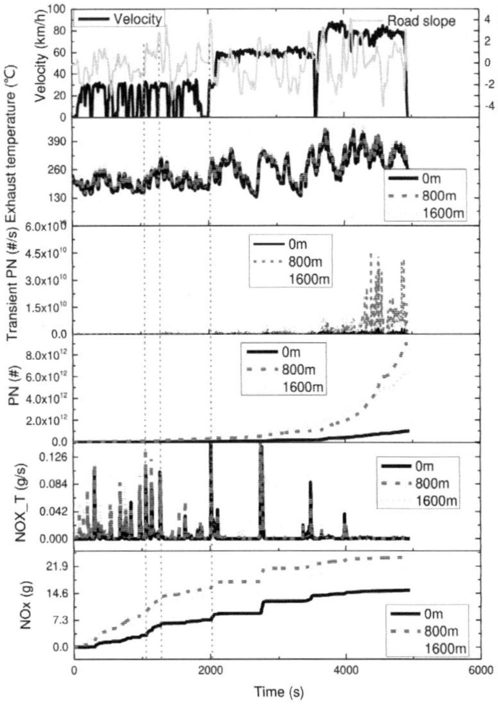

Figure 4. Transient and accumulated NOx and PN emissions and velocity, road slope, and exhaust temperature under different altitudes at 40% loading.

3.2. Effects of Loading on Real Driving Emissions

The specific emissions of CO_2 and pollutants under 0% and 40% loading conditions at altitudes of 0m, 800m, and 1600m are shown in figure 5. It can be seen that there is a very consistent trend except for NOx. The specific emissions of CO_2 and THC decrease with increasing loading, while the specific emissions of PN and CO increase with increasing loading. As the loading increases, the torque of the engine increases, and more high load conditions lead to an increase in incomplete combustion, resulting in an increase in CO and PN emissions. The specific emissions of NOx decrease with increasing loading at 0m but increase with increasing loading at 800m and 1600m. This is mainly because the increase in loading will change the NOx raw emissions and NOx conversion efficiency.

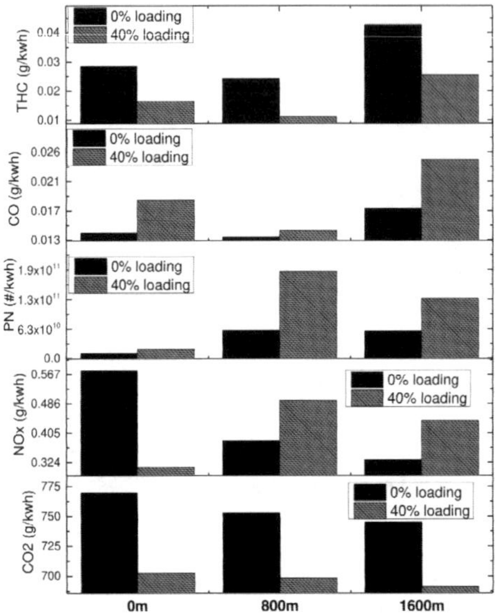

Figure 5. Brake-specific emissions of CO_2 and pollutants under different altitudes and loading.

3.3. Effects of Cold Start on Real Driving Emissions

The specific emissions of CO_2 and pollutants from a cold start and hot start under 0% loading conditions in altitude environments of 0m, 800m, 1600m, and temperature environment of 25 °C are shown in figure 6. It can be seen that the regularity is very consistent, that is, a cold start increases the specific emissions of CO_2 and pollutants. With the increase in altitude, the deterioration of PN and CO emissions caused by cold start intensifies. At 0m, 800m, and 1600m, the PN emissions from cold start are 1.17 times, 6.41 times, and 19.51 times higher than those under hot conditions, respectively. The CO emissions from the cold start are 5.84 times, 7.39 times, and 8.73 times higher than those under hot conditions, respectively.

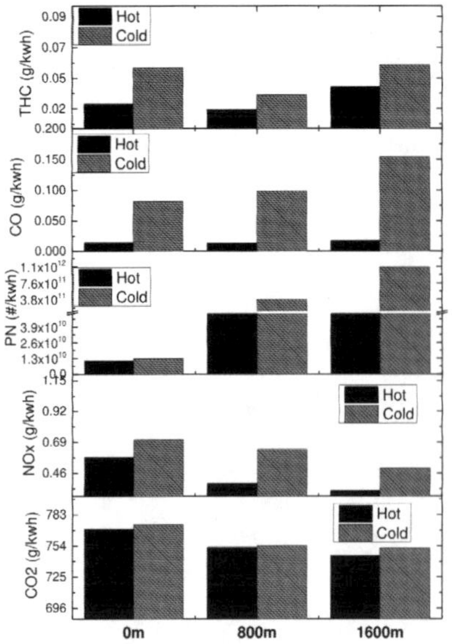

Figure 6. Brake specific emissions of CO_2 and pollutants under cold and hot start.

4. Conclusions

(1) Under low loading conditions, the increase in altitude does not result in a significant decrease in power or a significant increase in fuel consumption for the PEMS test.

(2) The emissions of CO and THC show a trend of first decreasing and then increasing with altitude, while PN emissions show a trend of first sharply increasing and then slightly decreasing with altitude. The variation of NOx emissions with altitude is not obvious.

(3) Under different altitude conditions, the specific emissions of CO_2 and THC decrease with increasing loading, while the specific emissions of PN and CO increase with increasing loading. However, the variation of NOx emissions with loading is not obvious.

(4) The peak NOx emissions are strongly correlated to slope and vehicle acceleration.

(5) Cold start increases the specific emissions of CO_2 and pollutants, especially for PN and CO.

5. Acknowledgments

This research was financially supported by the National Key Research and Development Program of China (2022YFC3703601)

6. References

[1] China Mobile Source Environmental Management Annual Report [R]. Ministry of Ecology and Environment 2022.

[2] GB17691-2018 Limits and measurement methods for emissions from diesel fueled heavy-duty vehicle (Chinese VI) [S]. Ministry of Ecology and Environment 2018.

[3] Di Sarli V, Di Benedetto A, Combined effects of soot load and catalyst activity on the regeneration dynamics of catalytic diesel particulate filters [J]. *AIChE*, 2018 **64** (5): 1714-1722.

[4] Qi Z Y, Gu M, Cao J G, Zhang Z W, You C Z, Zhan Y, Ma Z W and Huang W. The effects of varying altitudes on the rates of emissions from diesel and gasoline vehicles using a portable emission measurement system[J]. *Atmosphere* 2023, **14**, 1739.

[5] Wang X, Ge Y S, Yu L X and Feng X Y, Effects of altitude on the thermal efficiency of a heavy-duty diesel engine[J]. *Energy* 2013, **59**, 543-548.

[6] Pedro B, John A and Andrés A, Effect of altitude and palm oil biodiesel fuelling on the performance and combustion characteristics of a HSDI diesel engine[J]. *Fuel* 2009, **88**(4), 725-731.

[7] Liu J, Ge Y S, Wang X, Hao L J, Tan J W, Peng Z H, Zhang C Z, Gong H M and Huang Y. On-board measurement of particle numbers and their size distribution from a light-duty diesel vehicle: Influences of VSP and altitude [J]. *JES* 2017, **57** (7): 238-248.

[8] Chen J J, Song G F, Li C M, Li J Q and He C. Research on NOx emission of diesel vehicles on plateau based on AWM[J]. *Automobile Technology*, 2018**(7)**: 32-35.

[9] Du J B, Xu D D,Guo Y, Gao D Z and Gao Z M. Study on the influence of load on the emission of China VI heavy-duty vehicles in high altitude environment [J]. *Small Internal Combustion Engine and Vehicle Technique*, 2020,**49**(05):42-46+61.

[10] Zhang D, Wang X W, Gao T, Jing X J, Huo Y Z and Li G. Plateau PEMS emission characteristics analysis of heavy-duty diesel vehicles with different technical routes.*2022 ESAT p 3048*.

[11] Wang X W, Fu T Q, Wang C Q and Lin J. Fuel consumption and emissions at china automotive test cycle for a heavy duty vehicle based on engine-in-the-loop methodology. *2020 ESEAT p 1549*.

[12] Wang X W, Jing X J, Gao T, Gu X J, Zhang Y Y and Wu L L. Study on real driving fine particles emission characteristics for a heavy-duty diesel vehicle based on engine-in-the-loop methodology[J]. *Automotive Engineering*, 2022,**44**(1):58-63. .

[13] Wang X W, Ling J and YAN F. Review on engine-in-the-loop simulation: application and development [J]. *Small Internal Combustion Engine and Vehicle Technique*,

Journal of Physics: Conference Series **2941** (2025) 012049 doi:10.1088/1742-6596/2941/1/012049

2019,**48**(5):79-84.

[14] Wang Z Y, Wang X W, Gao T, Jing X J, Zhang Y Y and Li G. Pollutant emission and CO_2 emission at various vehicle loading for a heavy-duty commercial vehicle by EIL methodology.*2022 ESAT p 3015.*

[15] Wang X W, Jing X J, Gao T, Li G, Gu X J and Zhang Y Y. Low load running characteristics and NOx emission under low load for heavy-duty diesel vehicles [J], *China Environmental Science*, 2022,**42**(11):5063-5071.

[16] Okui N. A study on alternative test method of real driving emissions for heavy-duty vehicle by using engine in the loop simulation. *SAE Technical Paper 2021-01-1227, 2021, doi:10.4271/2021-01-1227.*

[17] Tang Y, Rodgers J, McCallum J, Zhang, Y and Fujii Y. Real-time hydro-mechanical transmission system simulations for model-guided assessment of complex shift sequence. *SAE Technical Paper 2021-01-0715, 2021, doi:10.4271/2021-01-0715.*

Research on Carbon Reduction Potential of the SMA Asphalt Mixture in Life Cycle

Zhihuo Qian[1], Miaomiao Tian[2*], Tao Fang[1], Zheng Huang[1], Haifeng Zhang[1] and Kun Zheng[1]

[1] Anhui Transportation Holding Group CO., LTD, Hefei 230041, Anhui, China.
[2] China Academy of Transportation Sciences, Beijing 100013, China.
Email: 2687059496@qq.com; *67503830@qq.com; 13705692500@126.com; 33598066@qq.com; 78172217@qq.com; 251174163@qq.com

Abstract. To research the carbon emissions of SMA asphalt mixture, SMA-13 and AC-13 asphalt mixtures were selected for comparative analysis, and the carbon emission amounts during the construction period and operation maintenance period were calculated. The research results indicated that the carbon emissions of SMA-13 during the construction period increased by 20.63% compared to AC-13; however, due to the improved performance, SMA-13 could reduce one preventive maintenance and one repair maintenance project compared to AC-13 within the design life. Therefore, the carbon emissions during the operation and maintenance period of SMA-13 could be reduced by 21.26% compared to AC-13, meanwhile with better performance. Based on the life cycle theory, SMA-13 could reduce comprehensive carbon emissions by 11.40% compared to AC-13. Therefore, SMA has better long-term carbon reduction capabilities and is an ideal durable green low-carbon technology for roads. It should be widely promoted and applied in national roads construction.
Keywords. SMA-13 asphalt mixture; AC-13 asphalt mixture; construction period carbon emissions; maintenance period carbon emissions; carbon reduction potential; comparative analysis.

1. Introduction

Global warming has become one of the most pressing environmental issues in recent years, China announced that the carbon dioxide emissions will be peaked before 2030 and carbon neutrality will be achieved before 2060 [1]. At this background, transportation industry should be developed in a low-carbon way. Highway infrastructure construction requires a large amount of materials and energy consumption, which is an important aspect for the green and low-carbon transformation of the transportation industry [2]. Highway durability technology is an important component of building green and low-carbon highways. By extending the service life of highway infrastructure, maintenance cycles could be extended, and maintenance activities and carbon emissions from the operation stage could be reduced. Promoting highway durability technology is an important approach to achieve low-carbon transformation and development of highway construction in China [3].

Stone Mastic Asphalt (SMA) is currently internationally recognized as an asphalt surface mixture with strong deformation resistance and good durability [4]. Some scholars have conducted analysis on the environmental benefits of SMA asphalt pavement. Lan et al. [5] studied the energy consumption and carbon emissions of SMA regenerated pavement and proposed that in-situ thermal regeneration technology has good comprehensive benefits. Gao [6] conducted an environmental impact study on the use of "Poured Asphalt + SMA" and double-layer epoxy asphalt for steel bridge deck pavement layers, different effects on energy consumption, global warming potential, acidification, and other aspects has been compared. However, there are few research on the impact of carbon emissions during the construction and operation periods of the SMA asphalt pavement. Differences in carbon emissions between the SMA and the most commonly used asphalt concrete (AC) are still unclear. Therefore, based on the life cycle theory, carbon emissions of the SMA and AC asphalt mixture have been compared in this paper, and carbon reduction potential of the SMA asphalt pavement has been presented, which would provide more effective support for the further promotion of SMA technology and the construction of green and low-carbon highways.

2. Life Cycle Theory and Carbon Emission Calculation Method

The life cycle assessment (LCA) is currently recognized as the core method for product environmental impact assessment by the International Organization for Standardization (ISO), and is also the main basis for determining the boundary of carbon emissions calculation and establishing calculation models for highway engineering. Based on LCA theory, this paper adopts a simplified life cycle assessment method, the life cycle carbon emissions are divided into construction and operation period, including material production, transportation, construction and maintenance activities. Meanwhile, as the off-site transportation of materials has been included in the carbon emissions of transportation vehicles, in order to avoid double counting, this article only calculates the carbon emissions from on-site transportation [7].

According to the International Intergovernmental Panel on Climate Change (IPCC), the emission factor method was adopted as the calculation method, which reflected the carbon emission level of a certain activity by its consumptions and corresponding emission factors [8]. Emission factors could be obtained from relevant national standards, IPCC and other mainstream databases [9].

3. Carbon Emission Calculation of the Pavement Construction Period

The SMA-13 asphalt mixture was selected as the research sample, in which main raw materials were SBS I-C asphalt, basalt crushed stone, and 0.3% lignin fiber, with an optimal oil stone ratio of 6.3%. Meanwhile, the AC-13 asphalt mixture was selected for comparative research, in which main raw materials were SBS I-C asphalt and crushed limestone, with the optimal oil stone ratio of 4.9%. The design thickness of the upper layer of asphalt pavement of the two different grading structures is both 4 cm. Carbon emissions of SMA-13 and AC-13 are calculated based on 1 m^2 as the analysis unit. Due to the fiber used in the SMA was less than 1%, which would not be included in the calculation [5]. Carbon emission calculation results during construction period of two types of asphalt mixtures are shown in table 1.

Table 1 indicated that carbon emission intensity during the construction period of the SMA-13 was 20.63% higher than the AC-13. As the SMA-13 grading design required more mineral powder, surface area was raised, and also with the addition of lignin fibres, more asphalt was absorbed. Besides, SMA's mixing time was longer than the AC's, therefore the carbon emissions during the material production stage are higher than the AC-13 asphalt mixture.

Table 1. Comparison of carbon emissions during the construction period of the SMA-13 and AC-13.

Item	Carbon emission factor (kg/unit)	Carbon emissions (kg/m^2)	
		SMA-13	AC-13
	Materials		
SBS modified asphalt	320 kg/t [10]	1.86	1.58
Mechanized sand	3.77 kg/m^3 [11]	0.02	0.06
Mineral powder	84.4 kg/t [11]	0.70	0.45
Basalt gravel	6.3 kg/m^3 [12]	0.31	/
Limestone gravel	3.27 kg/m^3 [12]	/	0.14
	facilities		

Item	Carbon emission factor (kg/unit)		Carbon emissions (kg/m^2)	
			SMA-13	AC-13
On site transportation	3.82 kg/kg (diesel oil [12-14])	0.581 t/MWh (electricity [15])	0.13	0.11
Mixing			3.69	3.23
Paving			0.04	0.03
Compaction			0.15	0.12
Total	/		6.90	5.72
Increment	20.63%			

4. Carbon Emission Calculation of the Pavement Operation Period

4.1. Determination of Maintenance Plan

SMA asphalt mixture could extend the curing period and then reduce maintenance construction activities. According to the results of mechanical and fatigue performance tests, the fatigue life of SMA-13 was 1.23 times than the AC-13 [16], which showed that the ability to resist fatigue cracking was improved, and the corresponding years of pavement diseases such as cracks were delayed, thus the SMA-13 maintenance cycle could be delayed by at least 2-3 years compared to the AC-13.

During the operation period, preventive and repair maintenance should be conducted for asphalt pavement. Choose epoxy surface technology as the pre maintenance method for the SMA and AC pavement, and overlay and re-pavement as the repair and maintenance method [17]. According to China's "Design Specification for Asphalt Pavement of Highways (JTG D50-2017)"[18], the design service life of highways is 15 years. According to engineering research, AC type asphalt pavement generally requires preventive maintenance for 3 years after the road is opened to traffic, and repair and maintenance for 5-10 years. Due to factors such as climate and heavy traffic, most highways need to be milled and repaired once every 8-10 years of operation [8]. The preventive maintenance time limit for SMA pavement can be extended to 5 years after traffic opening, the corrective maintenance (overlay) can be extended to once every 7 years, and the milling repair can be extended to once every 10 years. The timing of preventive maintenance and repair maintenance for SMA-13 and AC-13 during the design period is shown in figure 1. SMA could reduce one time preventive maintenance and one time repair maintenance compared to AC pavement.

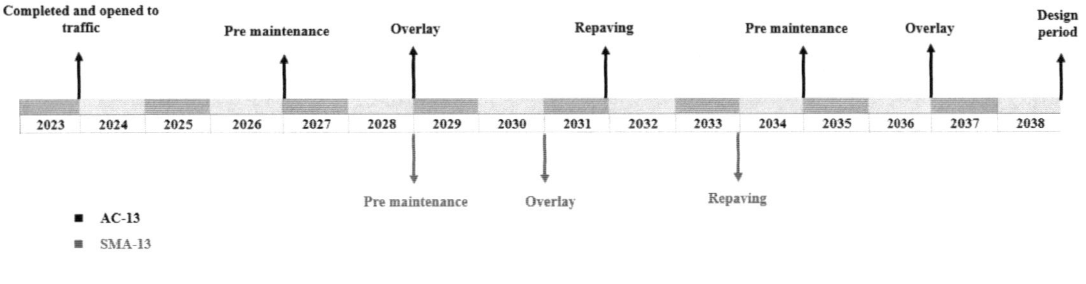

Figure 1. Maintenance activities of SMA-13 and AC-13 during the pavement design period.

4.2. Differences in Carbon Emissions during the Operation Period

Firstly, for preventive maintenance, both SMA-13 and AC-13 asphalt pavements should use single-layer medium particle epoxy surface treatment technology [19]. 1 m^2 was chosen as the analysis unit, and calculation results are shown in table 2.

Table 2. Carbon emission intensity of epoxy surface preventive maintenance.

Item	Quantity/m^2	Carbon emission factor (kg/unit)	Carbon emissions (kg/m^2)
Epoxy modified asphalt	0.55 kg	296 kg/t [20]	0.16
Modified emulsified asphalt	0.3 kg	258 kg/t [20]	0.08
Crushed stone	3.5 kg	2.51 kg/t [12]	0.01
Synchronous gravel sealing truck	0.00034 rating per machine-team	3.82 kg/kg diesel oil [13-15]	0.17
9~16t Tire roller	0.00030 rating per machine-team	3.82 kg/kg diesel oil [13-15]	0.04
Total	/	/	0.46

Secondly, for repaired maintenance, materials and construction facilities used were same as those in construction period, and carbon emissions from milling process were 0.37 kg/m^2 [5].

Therefore, carbon emissions of SMA-13 and AC-13 asphalt pavements during maintenance and operation period could be summarised and the results were shown in table 3. Compared to the AC-13, the SMA-13 could reduce more than 20% carbon emissions, which is sufficient to offset the carbon emissions during the construction phase.

Table 3. Comparison of carbon emissions during the operation and maintenance period of SMA-13 and AC-13 asphalt pavement.

Item		Carbon emissions (kg/m^2)			
		SMA-13		AC-13	
		Once	Frequency	Once	Frequency
Preventive maintenance		0.46	1	0.46	2
Repaired maintenance	Overlay	6.9	1	5.72	2
	Re-pavement	7.27	1	6.22	1
Total		14.63		18.58	
Carbon reduction ratio		21.26%			

5. Comprehensive Analysis of Carbon Emissions

Carbon emissions during construction and operation period of the SMA-13 and AC-13were shown in figure 2. SMA-13 required high-quality raw materials and more strict construction processes, its carbon emissions during material production, transportation, and construction stages are all slightly higher than AC-13. However, due to the durability advantage, road maintenance activities have been reduced, carbon emissions during the preventive, overlay and repaving maintenance stages were all significantly decreased, by nearly 50% of the AC-13.

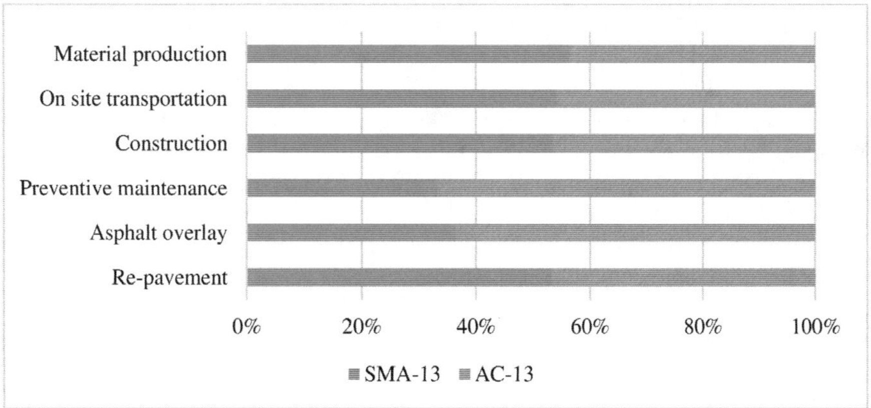

Figure 2. Carbon emissions comparison of the SMA-13 and AC-13 at different stages.

Carbon emissions in LCA period was shown in table 4, the SMA-13 reduced 11.40% carbon emissions compared with the AC-13, which proved that it would provide better energy-saving and emission reduction effects in road engineering.

Table 4. LCA carbon emissions of the SMA-13 and AC-13.

Item	Carbon emissions (kg/m^2)	
	SMA-13	AC-13
Construction	6.9	5.72
Operation	14.63	18.58
Total	21.53	24.30
LCA carbon reduction	11.40%	

6. Conclusion

This paper has conducted a research on carbon emissions during both construction and operation periods of the SMA-13 and AC-13 asphalt pavement. Carbon emission calculation method based on the life cycle theory was elaborated and a case study was presented. During construction period, carbon emissions of the SMA-13 was 20.63% higher than the AC-13, but during the operation period, carbon emissions was 21.26% lower than the AC-13, due to its durability advantage, which would offset the carbon emissions during the construction period, therefore in the LCA period, the SMA-13 could reduce 11.40% carbon emissions in total. Moreover, on this basis, the LCA carbon emissions of the SMA used in different climate and traffic loads should be continued to study in future. Overall, from the perspective of carbon emissions, promotion and application of the SMA should be accelerated.

Acknowledgement

This work is supported by the Anhui Transportation Holding Group CO., LTD. (Grant Nos. JKKJ-2021-17).

7. References

[1] Liu N, Wang Y, et al. 2022 Road life-cycle carbon dioxide emissions and emission reduction technologies: A review *Journal of Traffic and Transportation Engineering (English Edition)* **9(4)** 532-555.

[2] Yan Q, Yi K, Li P, et al. 2021 Analysis of the impact of asphalt pavement structure types on carbon emissions during the construction period *Chinese and Foreign Highways* **41(02)** 41-48.

[3] Zhang J, Su C, Wang C, et al. 2022 Overview of energy conservation and emission reduction

issues and technologies in road infrastructure construction *Journal of Beijing Institute of Technology* **48(03)** 243-260.

[4] Xu X and Lin K 2017 SMA pavement construction management and maintenance control during operation period *Journal of Chengdu Institute of Technology* **20(04)** 42-46.

[5] Lan Y, Gong Y, Hong S, et al. 2020 Analysis of comprehensive maintenance benefits of SMA pavement in situ thermal regeneration throughout its entire life cycle *Modern Transportation Technology* **17(05)** 5-12.

[6] Gao H 2018 *Research on Environmental Impact of Steel Bridge Deck Pavement Based on Life Cycle Analysis* (Nanjing: Southeast University).

[7] Lu Y, Yang K, Liu A, et al. 2023 A review of carbon emission calculation methods for highway engineering infrastructure construction *Transportation Energy Conservation and Environmental Protection* **19(01)** 18-22.

[8] Liu Y 2019 *Research on the Theory and Method of Carbon Dioxide Measurement in the Life Cycle of Highways Based on ALCA (*Xi'an: Chang'an University).

[9] Huang S, Gao S and Liu X 2019 A review of research on lifecycle inventory databases and software in the field of highway engineering at home and abroad *Jilin Transportation Technology* **(3)** 14-16.

[10] Zhang H, Chen H, Xu S, et al. 2021 Energy consumption and carbon emissions evaluation of rubber modified asphalt mixture pavement construction *Highway Engineering* **46(03)** 154-164.

[11] GB/T 51366-2019 Calculation standard for building carbon emissions.

[12] Greenhouse gas emission coefficient library for the entire lifecycle of Chinese products [EB/OL] (2022). http://lca.cityghg.com/.

[13] EU EFD [EB/OL]. https://data.epmap.org/product/efdb.

[14] GB/T 2589-2020, General rules for comprehensive energy consumption calculation.

[15] Ministry of Ecology and Environment Guidelines for Accounting and Reporting of Greenhouse Gas Emissions in Enterprises—Power Generation Facilities [EB/OL]. (2022). https://www.mee.gov.cn/xxgk2018/xxgk/xxgk06/202203/t20220315_971468.html.

[16] Wang Y and Wan C 2023 Design and performance analysis of heavy duty traffic asphalt pavement based on SMA-13 grading *Comprehensive Utilization of Fly Ash* **37(06)** 70-75.

[17] Chen J 2020 Overview of preventive maintenance technologies for highways and strategies for improving their effectiveness *Transportation Energy Conservation and Environmental Protection* **16(02)** 109-112.

[18] JTG D50-2017, Design Specification for Asphalt Pavement of Highways.

[19] T/CHTS 10032-2021, Technical Guidelines for Surface Treatment of Epoxy Composite Modified Asphalt for Highways.

[20] Blomberg T, Barnes J, Bernard F, et al. 2011 Life cycle inventory: Bitumen Brussels, Belgium: The European Bitumen Association.

Research and Engineering Application on Impact and Abrasion Resistance of CW Elastic Epoxy Mortar

Liang Zhang[1,*], Xiaomei Shao[1] and Faquan Li[2]

[1] Yangtze River Science Institute of Water Resources Commission, Wuhan 430010, Hubei, China.

[2] Wuhan Changjiang Science and Technology Development Co., Ltd., Wuhan 430010, Hubei, China.

Email: *409500448@qq.com

Abstract. Using epoxy mortar for coating protection is one of the effective ways to improve the impact and abrasion resistance of hydraulic structures. For the sake of improving the impact and abrasion resistance of ordinary epoxy mortar, high-toughness epoxy resin, low-exothermic curing agent, graded fillers, and other raw materials were used to prepare CW elastic epoxy mortar and the mechanical properties of elastic epoxy mortar were compared with other epoxy mortars. Moreover, the structural characteristics of elastic epoxy mortar were analyzed using SEM, and the engineering application cases of the material were introduced. Verified by the practical results, CW elastic epoxy mortar has an excellent performance in impact and abrasion-resistant protection for large-flow and high-water head structures such as stilling basins and plunge pools and has good application prospects.

Keywords. Hydraulic structure, impact and abrasion resistance, elastic epoxy mortar, engineering application.

1. Introduction

Water release structures and conveyance structures, represented by flood-discharge tunnels, stilling basins, diversion tunnels, and aqueducts, are often located in a service environment with fast water flow, large flow, and high water head. Affected by bed load and cavitation, such structures are prone to impact and abrasion or cavitation damage, resulting in spalling of the concrete surface, and even exposed and corroded steel bars, affecting structural stability and operation safety [1].

To address the problem of impact abrasion and cavitation of hydraulic structures, there are two solutions at present: the first is to adopt a mixed design to enhance the impact and abrasion, such as adding plastic fibre [2], wear-resistant aggregate [3], anti-abrasion admixtures [4], etc. into concrete; the second is to develop an impact and abrasion resistance material system as a protective coating on the surface of concrete. Polymer-based impact and abrasion-resistant materials have been widely studied and applied due to their excellent comprehensive properties and simple construction [5]. Polyurea elastomer and epoxy mortar are common polymer-based impact and abrasion resistance materials. For example, Ma et al. [6] have introduced the construction and application situation of single-component polyurea impact and abrasion-resistant materials on dam overflow surfaces in alpine regions in Xinjiang; Feng et al. [7] have prepared aliphatic polyester and successfully applied it to navigation lock wall of the Three Gorges Dam; Yang and Hao [8] have introduced flexibility into the continuous phase of epoxy resin to form a spherical dispersed "island structure", thus greatly reducing the abrasion rate of epoxy mortar; Zhang et al. [9] have studied the repair and protection application of two kinds of modified epoxy mortar in the Overflow Dam section of Zangmu Dam. Compared with polyurea, on the premise of ensuring the impact and abrasion resistant effect, epoxy mortar has excellent aging resistance, strong environmental adaptability, and good use economy, which is a kind of ideal impact and abrasion resistant protective material. Therefore, epoxy mortar has been more widely used and has been successively showing good performance in Gezhou Dam [10], Xiaolangdi [11], Xinjiang [12], Ertan [13], and other water conservancy and hydropower projects.

However, it has been revealed by domestic and foreign engineering practices that the common epoxy mortar used for impact and abrasion resistant protection still has the problems of low toughness, high viscosity and poor deformation, so it must be modified for large-scale applications. Toughening and strengthening of epoxy resin is an important modification method. Although there are a lot of studies on epoxy toughening [14-16]· toughening treatment generally greatly weakens the adhesive properties and compressive strength of epoxy resins, and modification studies that take into account flexibility and strength are relatively insufficient. Although there are a lot of studies on epoxy toughening [14-16], toughening treatment generally greatly weakens the adhesive properties and compressive strength of epoxy resins, and modification studies that take into account flexibility and strength are relatively insufficient. In addition, existing research has mainly focused on adjusting the aggregation state of the epoxy matrix and changing the interface phase structure of the toughened particles, but there have been few researches on modification design from the perspective of molecular structure. Aiming to further improve the toughness and strength of epoxy resin, in this thesis, the molecular structure of the epoxy resin was adjusted to form an island structure epoxy resin, and then the CW elastic epoxy mortar with high toughness, high strength and low heat release was prepared using modified curing agent and graded filler to study its properties and structure. Finally, this material was applied to projects such as Goupitan Hydropower Station and Shatuo Dam and achieved good defect repair and impact and abrasion-resistant protection effects, which can also provide a reference for similar project treatment.

2. Material Preparation

2.1. Raw Material

Epoxy resin: modified bisphenol Type-A epoxy resin. The main chain structure was retained, the hard segments were cut to reduce cross-linking density, flexible segments containing active groups were introduced to improve flexibility, and the viscosity of the epoxy resin was reduced to 30-70 Pa·s (25°C), so that the active group could be combined with the epoxy group to precipitate spherical particles during the curing reaction, forming an island structure dispersed on the epoxy continuous phase.

Curing agent: modified Armeen; compared with commonly used curing agents, its curing exothermic peak temperature [17] was significantly reduced, as shown in table 1.

Padding: 80-300 mesh emery and nano-quartz sand were selected for granular composition through the orthogonal test.

Table 1. Comparison of curing exothermic peak temperatures between elastic epoxy mortar curing agents and common curing agents).

Type of curing agent	CW elastic epoxy mortar curing agent	NE epoxy mortar curing agent	Diethylenetriamine	Hydroxyalkyl polyamines	Polyamide
Curing exothermic peak /°C	108	105	250	190	125

2.2. Preparation Method
First of all, the island structure epoxy resin, padding of granular composition, and coupling agent were weighed by electronic analytical balance based on the ratio of raw materials determined by orthogonal test, then they were stirred in the GFJ-0.4 high-speed disperser at a slow speed for 5 min to 10 min; later, aids were added and stirred at high speed for 3 min to 6 min; finally, the modified curing agent was added and stirred at high speed for 4 min-8 min to obtain CW elastic epoxy mortar.

2.3. Performance Test
Mechanics Performance Testing: the AG-IC/100kN electronic universal testing machine and quartz dilatometer were used for testing in accordance with "Technical Specifications for Epoxy Resin Mortar" DL/T 5193-2004 and "Determination of Tensile and Shear Strength of Adhesives" GB/T 7124-2008, the test items included: compressive strength, tensile strength, bond strength, tensile shear strength, elongation at break and linear expansion coefficient.

Impact and abrasion resistance testing: the HKS high-speed underwater steel ball grinding testing machine was used for testing in accordance with the "Hydraulic Concrete Test Procedure" SL352 - 2006 "4.20". After the concrete specimen was formed and cured, a 10-mm thick epoxy mortar was applied on the surface of the specimen, and then the impact and abrasion resistance testing was carried out. As required by the testing regulations, the impact and abrasion duration was 72h, but considering that the difference in the impact and abrasion resistance of various materials could not be compared in 72h, the mass loss of the specimen was measured after three impact and abrasion cycles of 72h, 144h, and 216h respectively.

Microtopography measurement: CW elastic epoxy mortar samples before and after the impact and abrasion testing were treated with liquid nitrogen, and an S-570 scanning electron microscope was used to observe the microstructure of the section after metal spraying on the sample section.

3. Results and Analysis

3.1. Mechanical Properties
Table 2 shows a comparison of the mechanical properties of ordinary, modified, and CW elastic epoxy mortars.

Table 2. Comparison of mechanical properties of some types of epoxy mortar.

Material Type	Testing Items					
	Tensile shear strength (28d, MPa)	Elongation at break (%)	Coefficient of linear expansion ($\times10^{6}$/K, 25°C)	Compressive strength (28d, MPa)	Tensile Strength (28d, MPa)	Bonding strength (28d, MPa)
Ordinary epoxy mortar	11.0	1.3	18.9	82.3	11.9	4.3
Modified epoxy* mortar*	14.5	9.4	13.0	105.3	13.8	3.6
CW elastic epoxy mortar	22.0	18.0	9.2	78.5	12.9	4.5

*The mechanical properties of modified epoxy mortar come from the literature [18].

As shown in table 2, compared with ordinary and modified epoxy mortars, the tensile shear strength and elongation at break of CW elastic epoxy mortars were greatly improved. The tensile shear strength increased by 100% and 52% respectively, and the elongation at break increased to 14 times and 2 times respectively; the coefficient of linear expansion decreased significantly and was close to that of concrete (8-1210-6/K); the tensile strength decreased by 4.6% and 25.4% respectively, but it was still much higher than the design strength of 50-70 MPa of common impact and abrasion resistant concrete; the tensile strength and bonding strength of the three epoxy mortars were similar. The test results showed that CW elastic epoxy mortar improved flexibility while ensuring strength.

3.2. Impact and Abrasion Resistance

We can see figure1 for the impact and abrasion resistance test results of CW elastic epoxy mortars, and see table 3 for the impact and abrasion resistance results of three epoxy mortars.

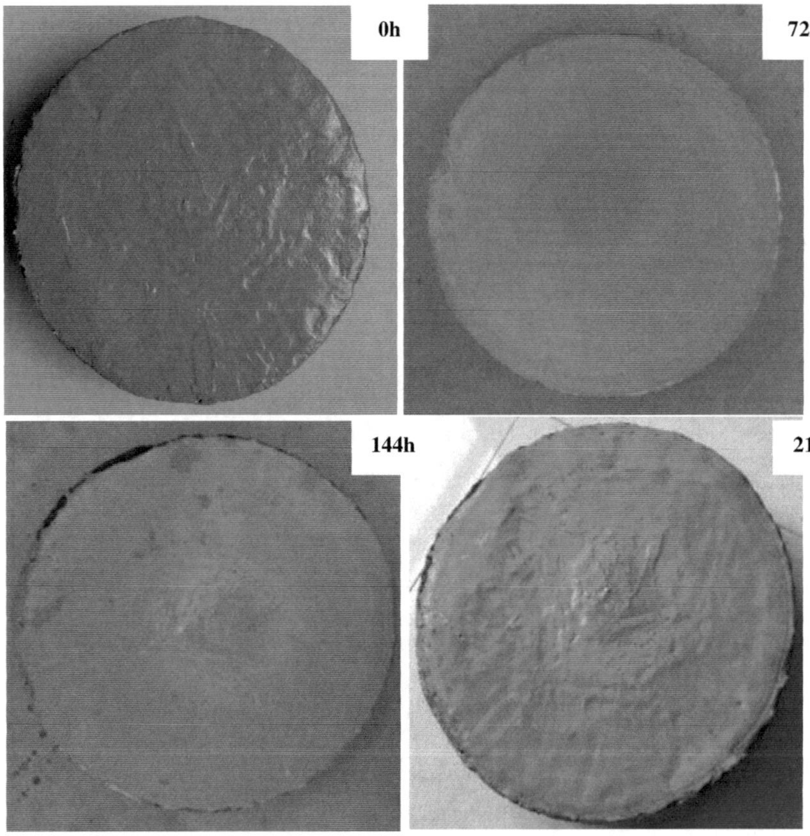

Figure 1. The effect picture of the impact and abrasion resistance test of CW Elastic epoxy mortar.

Table 3. Comparison of impact and abrasion resistance of epoxy mortars.

Material Type	Amount of loss (g)				Strength of impact and abrasion resistance ($h/(kg/m^2)$)			
	0h	72h	144h	216h	0h	72h	144h	216h
Ordinary epoxy mortar	0	73.90	113.00	183.90	0	47.4	46.5	47.6
Modified epoxy* mortar*	0	69.18	/	/	0	75.0	/	/
CW elastic epoxy mortar	0	42.61	66.10	103.40	0	82.5	79.9	82.1

*The mechanical properties of modified epoxy mortar come from the literature [18].

Figure 1 shows that with the increase of the impact and abrasion duration, the color of the impact and abrasion protective coating on the surface of the concrete specimen gradually became darker, that is, the abrasion of the coating gradually increased. After 72h of impact and abrasion, the surface of the coating was smooth, with no cracks, pits, and other defects, and there was only a small amount of damage on the edge; after 144h of impact and abrasion, a small number of pits appeared on the coating surface, and the edge damage increased; after 216h of impact and abrasion, the coating surface became uneven. According to the results in table 3, the results of the impact and abrasion testing were consistent with the test effect picture. After 72h of impact and abrasion, the abrasion of CW elastic epoxy mortar decreased by 40.5% compared with ordinary epoxy mortar, 38.4% compared with modified epoxy mortar, and the strength of the impact and abrasion resistance increased by 72.4%; after 216h of impact and abrasion, the mass loss rate was only 0.08%, and the strength of the impact and abrasion resistance could reach 82.05h /(kg/m2), which was much higher than that of ordinary epoxy mortar.

3.3. Micromorphology

The SEM images of CW elastic epoxy mortar before impact and abrasion testing and after 216h of impact and abrasion testing are shown in figure 2.

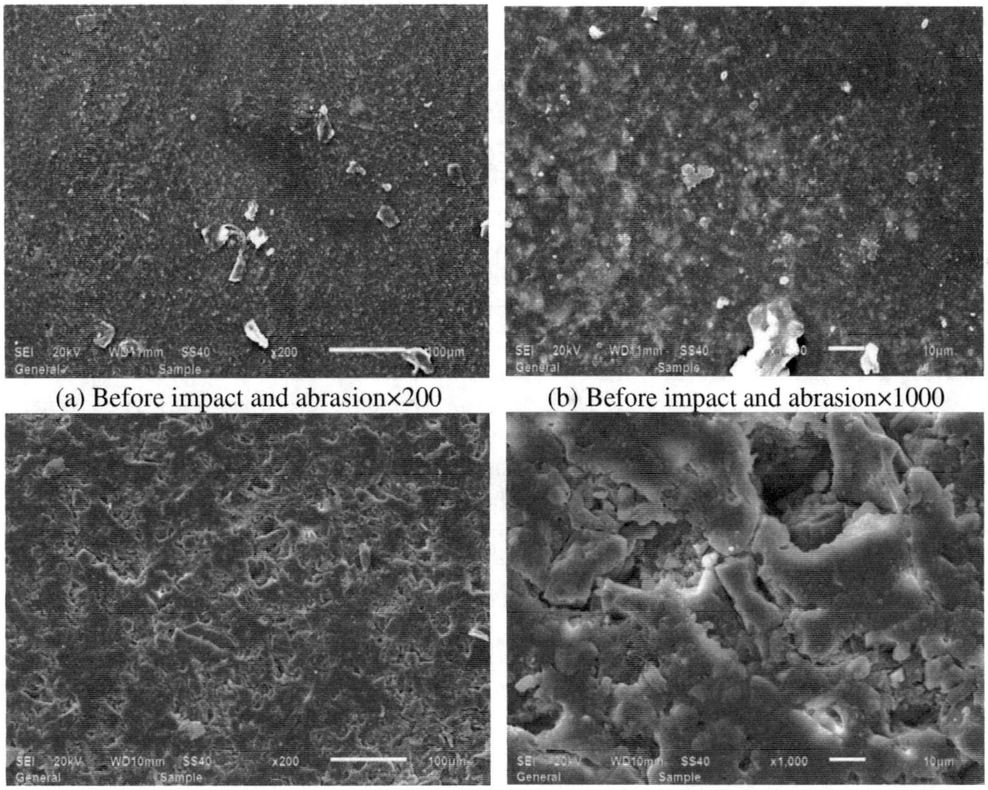

(a) Before impact and abrasion×200 (b) Before impact and abrasion×1000

(c) After impact and abrasion×200 (d) After impact and abrasion×1000

Figure 2. The SEM images before and after impact and abrasion testing of CW elastic epoxy mortar.

It can be seen from figure 2 that the CW elastic epoxy mortar was smooth and flat before impact and abrasion testing, and the structure deformed after impact and abrasion testing, with more wrinkled interfaces and a few pits, but the whole was still tightly combined and presented a relatively complete continuous phase, which indicates that the impact and abrasion resistance of CW elastic epoxy mortar has good impact resistance and can well resist impact and abrasion damage. As shown in figures 2(a) and 2(b), before the impact and abrasion testing, light-colored spherical particles (about 5-10 μm) were uniformly dispersed on the dark epoxy resin continuous phase to form an island structure; moreover, a small amount of plate-like filler crystals were also visible on the continuous phase, and the crystal size was about 10-30 μm. As shown in figures 2(c) and 2(d), after the impact and abrasion testing, the cross-section of the CW elastic epoxy mortar increased, both the epoxy resin phase and the spherical particle phase underwent wrinkle deformation and a dimple structure appeared, the cracks developed in multiple directions, and the filler crystals disappeared. The analysis results of impact and abrasion resistant mechanism are as follows: under the action of external forces such as bed load impact and cutting, the spherical particles and padding with relatively small elasticity modulus induced the epoxy resin matrix to produce folds, cracks, dimples, and other plastic deformation. After the cyclic process of expansion, obstruction, turning, and re-expansion, the more the number of fold interfaces, the larger the surface area, the greater the energy absorbed by deformation, which is macroscopically manifested in the improvement of the toughness and the enhancement of the impact and abrasion resistance of the material system.

4. Engineering Application Cases

4.1. Restoration and Treatment Engineering of Plunge Pool at Goupitan Hydropower Station in Wujiang River, Guizhou Province

The Plunge Pool at Goupitan Hydropower Station in Wujiang, Guizhou adopts the structure of flat bottom closed pumping and drainage, the cross-section is a compound trapezoid, with a net length of about 332m, the standard section bottom width of 70m, and the bottom plate elevation of 420m-412m. The Plunge Pool is a reinforced concrete structure; the 50cm surface thickness below 430m altitude is made of C50 anti-abrasion concrete, and the rest is made of C25 concrete. After the dam was put into operation, the water damage was serious in the easy-to-impact and abrasion parts of the floor and side wall of the plunge pool. The construction contents mainly include repairing the impact and abrasion parts of the floor of the plunge pool, repairing the corner cut between the floor and side wall, and abrasion treatment at the bottom of the side wall, crack treatment, etc.

The floor and side walls were repaired and protected using CW elastic epoxy mortar. This engineering started in February 2015, and lasted more than 50 days, and a total of 4278 m2 of quantities were completed. In view of the large area to be repaired, the repair work was carried out by area division, because division of construction was conducive to avoiding the shrinkage and curing thermal expansion of epoxy mortar or cement causing cracking and hollowing of the epoxy mortar layer itself. The edge length of the divided blocks was controlled within 3m, and the space seam between the construction blocks was reserved for 30 mm to 50 mm, and the space seams were fixed after 1 to 2 days. The epoxy mortar protective coating after filling and sealing with modified epoxy mortar or epoxy cement was smooth and tight, and the coating was firmly bonded to the concrete matrix. The results of the impact and abrasion resistance testing in the laboratory showed that the mass loss rate was only 0.08% after 216h of impact and abrasion, the strength of impact and abrasion after 216h was up to 101h/(kg/m2), and the leakage condition of the plunge pool was significantly improved. After the restoration, there was no leakage in the plunge pool. The plunge pool was ensured to be operated normally before the flood in 2015, thus ensuring the safe flood control work of Goupitan Hydropower Station.

4.2. Restoration and Construction of Flood Discharge and Energy Dissipation Flow Surface of Shatuo Dam in Wujiang River, Guizhou Province

The flood-releasing structure of Shatuo Dam in Wujiang River, Guizhou Province is located on the main channel in the middle of the river bed, using the energy dissipation method of the combination of the broad tail pier and stilling basin. The stilling basin and apron are arranged along the downstream river, with an apron length of 100m, a width of 135m, an excavation elevation of 282 m-284 m, and a top elevation of 287m. After the dam was put into operation, defects such as cracks, craters, and abrasions appeared on the flow surface such as the stilling basin. The construction content mainly includes surface cleaning of flood discharge and energy dissipation surface, defect elimination treatment of the stilling basin flushing pit, abrasion treatment, sealing of various vertical holes, repair of structural joints and water stops in the stilling basin, crack treatment and maintenance, etc.

The cracks on the flow surface, structural joints, and the floor of the stilling basin were repaired and protected using CW elastic epoxy mortar. The engineering started in March 2015 and lasted for 1 month, and the restoration area of 1200 m2 was completed. The epoxy mortar material applied between the floors of the repaired area and the unrepaired was connected smoothly and strictly controlled the elevation and flatness of the repaired area and the connection with the unrepaired area. Based on the excellent performance materials and reasonable processes, the flood discharge overflow surface and stilling basin were put into use in May 2015 after passing acceptance. A quality follow-up visit after 2 years of service showed that there was no damage, abrasion, swelling, or peeling of the epoxy mortar layer in the repaired parts. It is still in good operation and up to date.

5. Conclusion

Firstly, aiming at the requirements of impact and abrasion resistance of the hydraulic structure, CW elastic epoxy mortar has been prepared using high-toughness island structure epoxy resin, low-exothermic modified curing agent, graded wear-resistant padding, etc.

Secondly, CW elastic epoxy mortar has excellent impact and abrasion resistance, in which the 28d tensile shear strength is 22.0MPa, the elongation at break is 18.0%, the linear expansion coefficient is consistent with that of concrete, and the 216h strength of impact and abrasion resistance is 82.1h /(kg/m^2).

Finally, the application effect of CW elastic epoxy mortar in the plunge pool of Goupitan Hydropower Station and the stilling basin of Shatuo Dam shows that the material can effectively slow down the abrasion damage of concrete, and is an ideal material for the repair of hydraulic concrete defects and impact and abrasion resistant protection.

Acknowledgments

Thanks to the National Natural Science Foundation of China “Research on Aging Mechanism between Coating Materials and Concrete Interfaces” (51579019) and Wuhan High-tech Industry Science and Technology Innovation Team Plan Project (2016070204020166).

6. References

[1] Zhang Z Z, Chen L and Wang Z Q, et al. 2016 Development status of impact and abrasion resistance for hydraulic discharge structures *New Chemical Materials* **(10)** 230-232.

[2] Thong M P, Harrison H, Maw K K, Yan Z, Duong T T, Des V and Tung M T 2024 Comparative impact behaviours of ultra high performance concrete columns reinforced with polypropylene vs steel fibres *Defence Technology* **(10)** 138-153

[3] Katelyn K, Lucio S de S and Naser P S Julie V and Lev K 2023 Strength determination for early-opening decision making of high early strength concrete pavement *Journal of Traffic and Transportation Engineering (English Edition)* **(2)** 291-303

[4] Guo J J, Gao M H and Zhang P 2022 Mechanisim and influential variables on the abrasion resistance hydraulic concrete *Nanotechnology Reviews* **(1)**.

[5] Xu X F, Bai Y and Yu Y 2012 Overview of impact and abrasion resistant polymer coating material for hydraulic discharge structures *Yangtze River* **(S1)** 181-183+202.

[6] Ma Y, Ren L and Feng Q, et al. 2017 Application of Single-component Polyurea Material in the Protection of Flow Surface of a Dam in Cold Area *Journal of China Institute of Water Resources and Hydropower Research* **(1)**.

[7] Feng J, Han W and Li Z, et al. 2012 Research on the Development and Engineering Application of New Polyurea Concrete Protection Materials *Journal of Yangtze River Scientific Research Institute* **29(2)** 64-67.

Journal of Physics: Conference Series **2941** (2025) 012051 doi:10.1088/1742-6596/2941/1/012051

[8] Yang W C and Hao J T 2006 Research on Impact and Abrasion Resistance Testing of Epoxy Mortar with "Island Structure" *Journal of China Institute of Water Resources and Hydropower Research* **4(4)**.
[9] Zhang D, Cui T H, Xiao C J, Chen L and Wang Z Q 2015 Application of CW Modified Epoxy Mortar in Impact and Abrasion Resistant Protection in Tibet *Journal of Yangtze River Scientific Research Institute* **32(11)** 102-104.
[10] Sun J J 1989 Research on maintenance materials of Erjiang sluice gate of Gezhou Dam *Yangtze River* **(3)** 12-17.
[11] Zhang T, Xu S Z 2001 Application of new epoxy resin mortar in hydropower projects *Thermosetting Resin* **16(6)** 26-28.
[12] Mai S F, Chen X L and Yao B 2004 Research on aging status of epoxy mortar coatings and development of elastic epoxy materials *Dam & Safety* **(5)**.
[13] Wan X W, Li B X and Min S H, et al. 2007 Research and Application of JME modified epoxy mortar impact and abrasion resistant repair material *Construction Technology* **36(5)** 94-96.
[14] Chen B, Wang X J and Wang X Z 2017 Research Progress on Toughening and Modification of Epoxy Resins *China Adhesives* **(2)** 55-58.
[15] Xu X F and Cai Y B 2011 Research on New Nano-Al2O3/ZrO2 Impact and Abrasion Resistant Coating Materials [J]. *New Building Materials* **38(5)** 63-65.
[16] Fan X, Zhang G C and Li J T, et al. 2017 Research on Microcellular Foaming of CTBN Toughened Epoxy Resin *Engineering Plastics Application* **45(4)** 53-58.
[17] Chen P., Liu S Epoxy Resin [M]. Beijing: Chemical Industry Press 1999.
[18] Kuang Y L, Tan X S and Han W, et al. 2016 Research and construction technology of a new epoxy resin impact and abrasion resistant material *New Chemical Materials* **(10)** 225-226.

Localized Corrosion Behavior of T2 Copper and H62 Brass in NaCl Solution for Semiconductor Applications: An Electrochemical Study Using Wire Beam Electrode

Chengqun Pan, Tingxuan Zhang, Yidan Xing*, Chuan Chen and Shan Zhu

AI School, Shanghai Normal University Tianhua College, Shanghai 201800, China.
Email: panchengqun@163.com; 2655695679@qq.com; *229749033@qq.com; 172311458@qq.com; zs2441@sthu.edu.cn

Abstract. T2 copper and H62 brass are widely used in semiconductor applications, particularly in interconnects and packaging components of integrated circuits. With increasing demands for performance and reliability, the corrosion resistance of these materials, especially in saline environments, becomes critical. Localized corrosion can significantly reduce electrical conductivity, impacting the long-term reliability of semiconductor devices. This study investigates the localized corrosion behavior of T2 copper and H62 brass in a 3.5 wt.% NaCl solution using the wire beam electrode (WBE) technique. Electrochemical measurements, including potential, current density, and localized corrosion intensity (LP), were conducted over varying immersion times. Results show that T2 copper exhibits superior corrosion resistance and lower localized corrosion sensitivity compared to H62 brass, which shows more pronounced corrosion, particularly at the electrode edges. As immersion time increases, H62 brass exhibits higher potential differences and LP values. These findings highlight the impact of localized corrosion on the conductivity of copper alloys and underscore the importance of selecting materials with enhanced corrosion resistance for semiconductor applications.
Keywords. Copper; copper alloys; localized corrosion; WBE.

1. Introduction

Copper and its alloys, such as T2 copper and H62 brass, are essential materials in semiconductor applications, particularly in the interconnects and packaging components of integrated circuits (ICs). As semiconductor technology advances, the performance requirements for these materials have become more demanding, particularly concerning their corrosion resistance in aggressive environments. In semiconductor packaging and interconnect systems, the corrosion of materials—especially localized corrosion—can significantly reduce electrical conductivity, thereby compromising the reliability and performance of the devices [1-3].

Traditional corrosion studies typically focus on homogeneous electrodes (pure metals) or micro-heterogeneous electrodes (uniform alloys), where uniform corrosion is the predominant form of degradation. However, under real-world industrial and environmental conditions, localized corrosion is more prevalent. Localized corrosion tends to concentrate in specific regions of the material's surface, resulting in forms such as pitting, crevice corrosion, droplet corrosion, stress corrosion, fatigue corrosion, and selective corrosion [3,4]. A key characteristic of localized corrosion is the distinct separation between the anode and cathode regions. On a microscopic scale, these anode and cathode sites are confined to small areas, and their positions change randomly over time. Corrosion occurs when the relatively small anode region is connected to the larger cathode region, leading to a much faster corrosion rate compared to uniform corrosion. Therefore, understanding the underlying mechanisms of localized corrosion is critical for improving material performance [5,6].

Copper initially forms a Cu_2O oxide layer when exposed to corrosive environments, which offers a certain degree of protection against further degradation. However, in specific environments, such as marine conditions, this protective oxide layer may be compromised. According to phase film and adsorption theories, aggressive species can penetrate the copper(I) oxide film and engage in electrochemical reactions with the copper substrate, eventually leading to the breakdown of the protective oxide layer. Furthermore, the strong adsorption properties of copper oxide can cause it to interact with metal ions, displacing passivating ions and causing the film to lose its protective function. This results in the destruction of the oxide film and the initiation of localized corrosion on the copper alloy surface. Localized corrosion, characterized by concentrated degradation in specific areas, can disrupt the uniformity of conductivity. This corrosion is often driven by the accumulation of corrosion products or the breakdown of protective oxide films [7,8]. Understanding the localized corrosion behavior of copper and its alloys is essential for enhancing the long-term reliability and performance of electronic components in semiconductor applications.

This study investigates the localized corrosion behavior of T2 copper UNS No. C11000 and H62 brass UNS No. C28000 in a 3.5 wt.% NaCl solution using the wire beam electrode (WBE) technique. By evaluating electrochemical parameters, including potential, current density, and the localized corrosion intensity index (LP), this research aims to provide valuable insights into the corrosion characteristics of copper alloys in semiconductor environments. The findings will guide material selection and contribute to the development of more reliable semiconductor components [9-11].

2. Experimental

2.1. Experimental Setup

Figure 1 illustrates the schematic of the wire beam electrode (WBE) system used in this study. The working electrode was fabricated by preparing 100 identical copper wires, made from T2 copper and H62 brass, with each wire having the same dimensions. These wires were then encapsulated in epoxy resin, resulting in a working area of 0.79 cm^2. Prior to testing, the electrodes were sequentially polished using sandpaper, ranging from 200# to 2000#, followed by ultrasonic cleaning with ethanol. The electrodes were then dried in an oven before use. As shown in the wire beam electrode layout diagram, the stability of each connection was confirmed before commencing the tests. Corrosion data were collected in a 3.5 wt.% NaCl solution. The wire beam electrodes were fixed in a controlled environmental chamber, kept horizontally and vertically, with their surfaces submerged parallel to the solution, which was maintained at 400 mL. To ensure the accuracy of the measurements, the scanning process for each electrode was repeated 3 times. The experiments were conducted at room temperature over a period of 48 hours.

Figure 1. Schematic diagram of the distribution of single copper and copper alloys in WBE.

2.2. Electrochemical Tests

In situ potentiostatic and galvanostatic scans were performed using a 10×10 wire beam electrode setup with the CST520 potentiostat/galvanostat, as shown in figure 2. The testing protocol followed the principles of the instrument's microcontroller-based control system, completing the required potential/current cycling tests within the specified time frame. The data collection interval for each point was set to 1 second. The copper alloy wire beam electrode was used as the working electrode, while a saturated calomel electrode (SCE) served as the reference electrode.

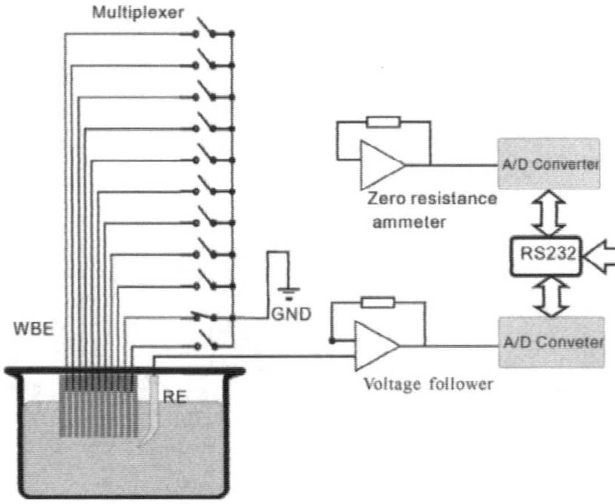

Figure 2. High-impedance conversion and surface potential scanning circuit in CST520.

3. Results and Discussion

3.1. Electrochemical Behavior of Copper and Copper Alloys

At room temperature, pretreated T2 copper alloy wire beam electrodes were immersed in a 3.5 wt.% NaCl solution for 48 hours. Electrochemical analysis, including potential and current measurements, was performed to quantitatively assess the corrosion resistance and localized corrosion sensitivity of the T2 alloy.

Figure 3 shows the variation in surface potential and current with time for the T2 copper alloy wire beam electrode in the 3.5 wt.% NaCl solution. During the initial immersion period, as shown in figure 3 (a1, a2), the potential and current at different electrode positions exhibited a clear correspondence. Specifically, more negative potential values were associated with higher anodic currents, indicating stronger corrosion activity. The electrochemical data demonstrated variability in corrosion characteristics at different locations on the electrode, suggesting the occurrence of localized corrosion. At position (3, 8), the most negative potential of approximately -0.1835 V was observed, corresponding to the maximum anodic current of 1.320×10^{-7}A/cm^2, indicating the greatest corrosion tendency at this site.

As the immersion progressed to the middle stage, as shown in figure 3 (b1, b2, c1, c2), significant changes in the electrochemical information at different positions were observed. The anodic corrosion zones became more concentrated, with new locations such as (5, 7), (6, 7), (9, 10), and (10, 2) displaying distinct anodic currents. The maximum anodic current value reached 2.690×10^{-7} A/cm^2. Other regions were predominantly protected by cathodic currents. The localized corrosion features on the surface of the entire electrode became more pronounced.

After 48 hours of immersion, as shown in figure 3 (d1, d2), the corrosion process reached a stable state. The locations of the anodic corrosion points did not change significantly; however, the electrochemical parameters varied slightly, indicating a stabilization of the corrosion activity. It is hypothesized that this stabilization is due to the differential nature of the oxide protective film on the copper alloy. Areas with a weaker protective film were progressively damaged and became stable anodic corrosion points during the later stages of immersion.

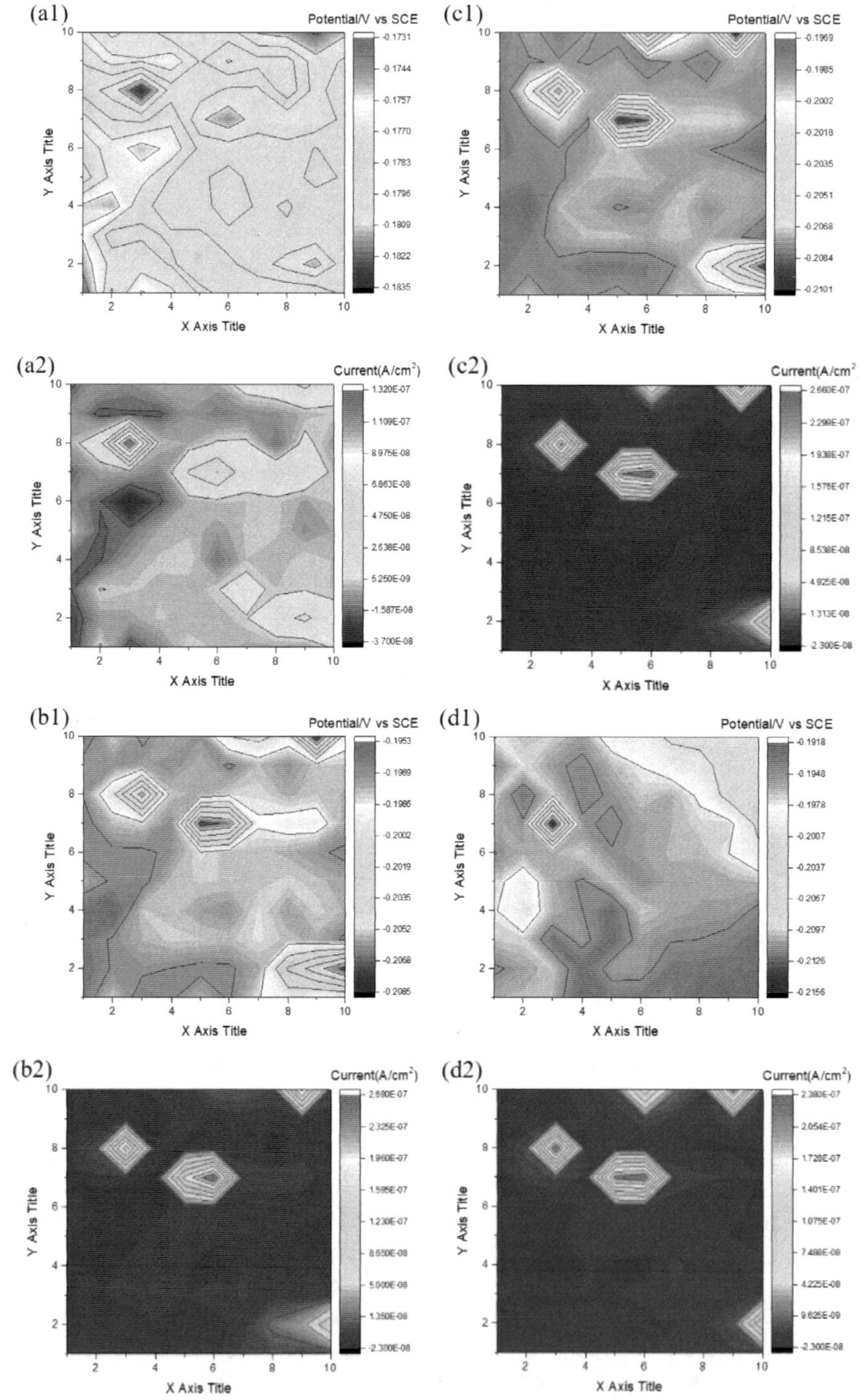

Figure 3. The corrosion potential (a1, b1, c1, d1) and galvanic current (a2, b2, c2, d2) distribution maps in 3.5wt.% NaCl solution for T2; (a1, a2) for initial period; (b1, b2) for 12h; (c1, c2) for 24h; (d1, d2) for 48h.

Figure 4 shows the variation of surface potential and current with time for the H62 brass wire beam electrode in a 3.5 wt.% NaCl solution. During the initial immersion period, as shown in figure 4 (a1, a2), corrosion was concentrated at the edge regions, with the positions (1,10), (2,9), (2,10), (3,10), and (4,10) showing anodic currents, the maximum value of which was 1.395×10^{-6} A/cm². In addition, some other locations exhibited corrosion, but the corresponding anodic currents were smaller, by approximately two orders of magnitude compared to the edge regions, indicating significant localized corrosion. At position (2,10), the potential was the most negative, approximately -0.3820 V, indicating the highest corrosion tendency at this point. As immersion progressed into the middle stage, as depicted in figure 4 (b1, b2, c1, c2), the electrochemical information of the different points showed distinct changes. The previously identified anodic regions shifted; positions (1,4), (1,5), (1,6), and (3,9) became anodic corrosion sites, while later positions such as (1,2), (2,1), (2,3), (2,7), and (3,9) gradually emerged as areas of concentrated anodic current, with the maximum value reaching 4.740×10^{-7} A/cm². Most of the remaining locations exhibited cathodic currents, indicating protection. The localized corrosion features across the entire electrode surface became more pronounced. After 48 hours of immersion, as shown in figure 4 (d1, d2), corrosion continued to develop, with an increase in the number of anodic corrosion sites, such as at positions (5,3), (7,4), (8,4), (8,9), and (8,10). The localized corrosion characteristics became more evident, confirming the progression of the corrosion process.

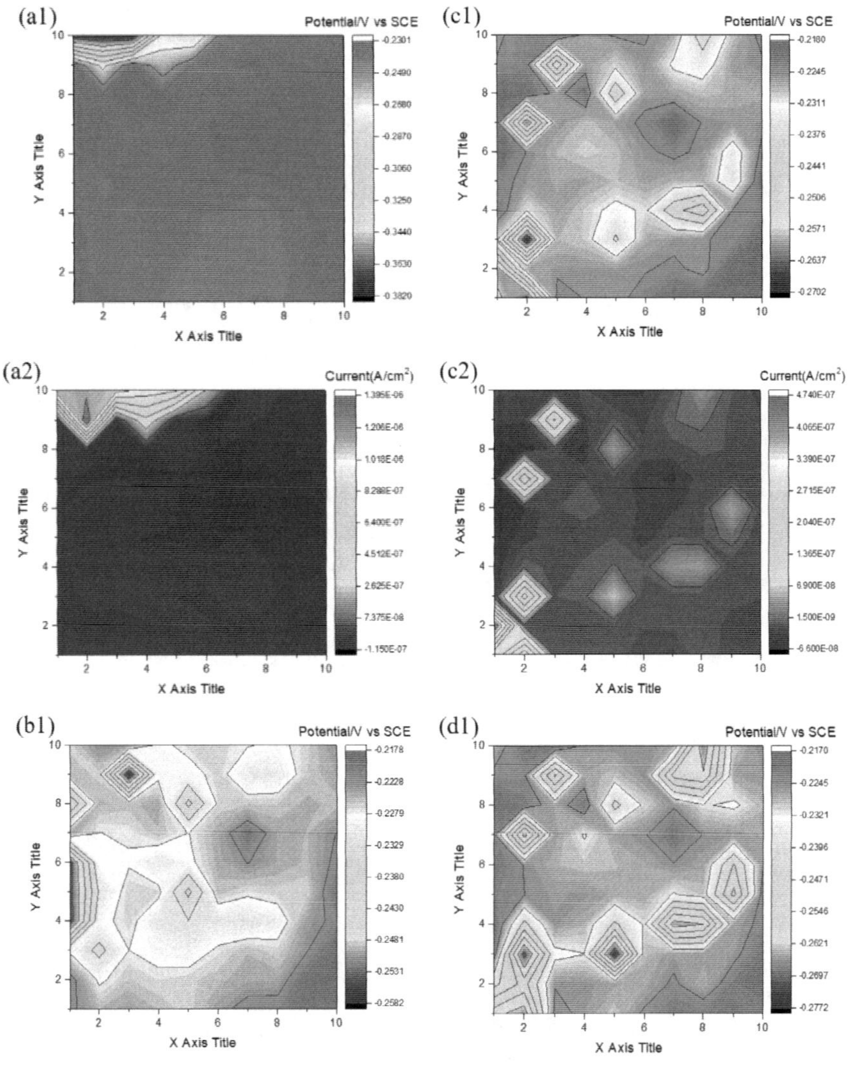

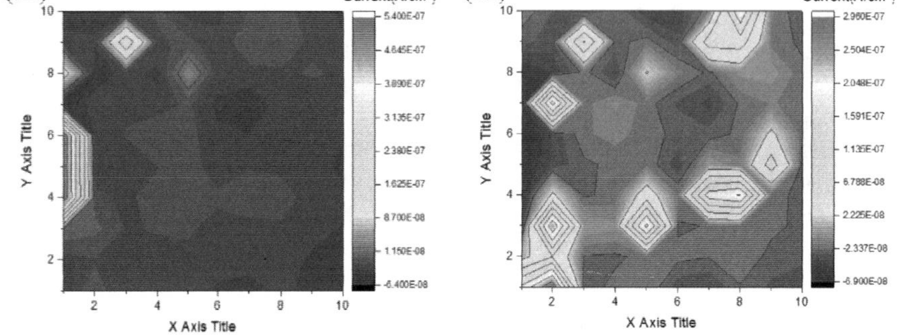

Figure 4. The corrosion potential (a1, b1, c1, d1) and galvanic current (a2, b2, c2, d2) distribution maps in 3.5wt.% NaCl solution for H62; (a1, a2) for initial period; (b1,b2) for 12h; (c1,c2) for 24h; (d1, d2) for 48h.

3.2. Localized Corrosion Sensitivity Analysis

Figure 5 shows the Localized Corrosion Intensity Index (LP) was calculated to quantify corrosion sensitivity. For T2 copper, the LP value remained relatively low and stable throughout the immersion period, indicating minimal localized corrosion. Conversely, the LP for H62 brass initially increased, reaching a maximum of 3.92, and then decreased as immersion continued, reflecting higher localized corrosion intensity, especially at the electrode edges, as shown in figure 6.

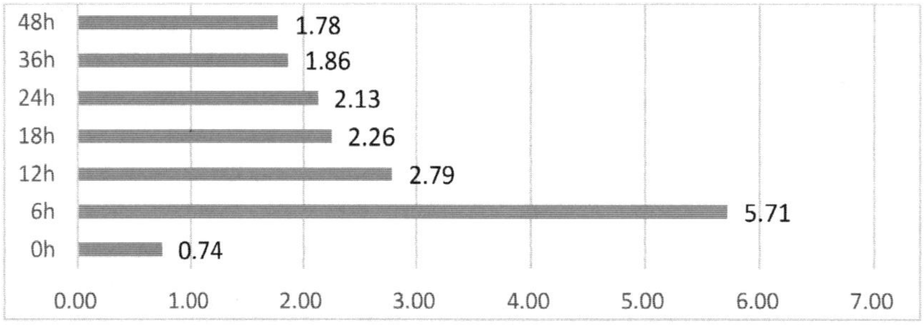

Figure 5. Localization parameter of T2.

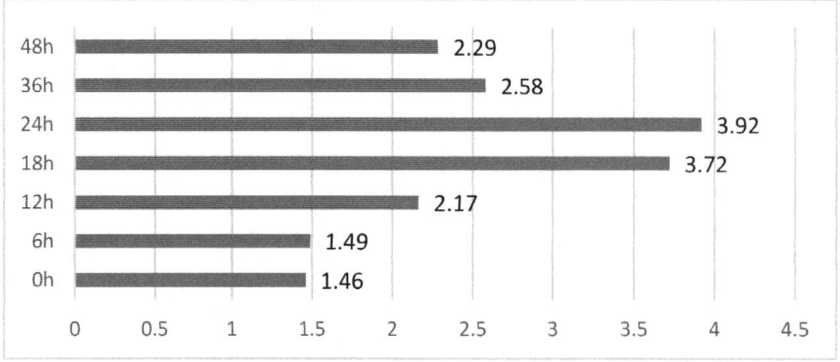

Figure 6. Localization parameter of H62.

3.3. Electrical Conductivity Analysis

Electrical As shown in figure 7, T2 copper, being pure copper, exhibits excellent conductivity, with an initial conductivity (prior to immersion) of approximately 5.8×10^7 S/m. In contrast, the conductivity of

H62 brass is lower than that of T2 copper, around 3.7×10^7 S/m. Over time during immersion, the oxidation layer on the copper surface may thicken, and copper chloride or other corrosion products may begin to form, especially in the case of H62 brass, which contains zinc. Zinc, being more susceptible to corrosion than copper, is prone to react with chloride ions in NaCl solution to form zinc salts.

During the initial phase of immersion, a rapid formation of a rust layer occurs, resulting in a significant decrease in conductivity. As the corrosion products accumulate, the decline in conductivity slows down, and the system eventually approaches a relatively stable state. The initial formation of corrosion products has the most considerable impact on conductivity. However, as corrosion products continue to accumulate and the material adapts to this trend, the reduction in conductivity becomes less pronounced over time. Nevertheless, the corrosion products continue to affect the electrical conductivity of the material throughout the immersion process. This reflects the fact that the corrosion system—comprising the materials, the NaCl solution, and the ongoing electrochemical reactions—has a long-term impact on the conductivity of the materials.

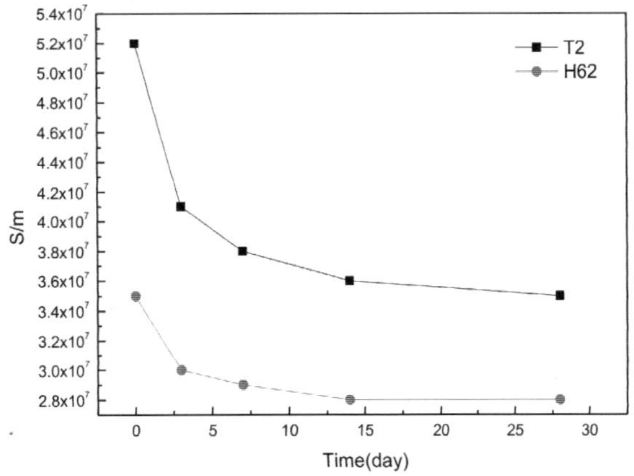

Figure 7. Changes in electrical conductivity of T2 copper and H62 brass during immersion in NaCl solution.

4. Conclusions

The immersion tests of pure T2 copper revealed that localized corrosion was not initially apparent. However, as the corrosion progressed towards stabilization, areas with weaker protective oxide films, due to inherent differences in the quality of the oxide layer, were progressively attacked by chloride ions (Cl^-). These areas eventually became stable anodic corrosion sites. Throughout the process, these sites remained unchanged, with variations observed only in the electrochemical parameters. The potential difference stabilized at approximately 13 mV, indicating the system had reached equilibrium. Over the course of the immersion, the localized corrosion intensity index (LP) gradually decreased, signifying a reduction in localized corrosion sensitivity.

In contrast, H62 brass exhibited a significantly higher potential difference compared to T2 copper, indicating a much greater susceptibility to localized corrosion. The localized corrosion intensity of the brass alloy increased initially and then decreased with extended immersion time, reaching a maximum LP value of 3.92, which corresponded to the highest observed sensitivity to localized corrosion. In the later stages of immersion, the LP for H62 brass remained higher than that of T2 copper, further confirming its increased vulnerability to localized corrosion.

These findings provide valuable insights into the localized corrosion behavior of copper and its alloys under aggressive environments. Understanding the underlying mechanisms of localized corrosion is essential for predicting material failure in real-world applications. This research underscores the importance of selecting appropriate materials and protective measures to mitigate corrosion risks, thus enhancing the durability and reliability of components, particularly in semiconductor packaging and other critical applications.

Acknowledgments
The author sincerely thanks the participants in the WBE experiment.

5. References

[1] Chen S and Zhang D 2018 Effects of metabolic activity of sulphate-reducing bacteria on heterogeneous corrosion behaviors of copper in seawater *Mat. Corros.* **69** 8.
[2] Dewanjee S, Singhal G and Li J 2023 *Proc. 22nd IEEE Intersociety Conf. on Thermal and Thermomechanical Phenomena in Electronic Systems (ITherm)* (Orlando, FL, USA) (Piscataway, NJ: IEEE) P 10.
[3] Siddiqi Z M and Amin M M 2017 Effect of South China Sea Water on Corrosion Behaviour of Copper Alloy and Mild Steel *Asian J. Water Environ. Pollut.* **14** 1.
[4] Ko S J, Lee Y H, Nam K S, Park E H and Kim J G 2023 Effect of surface conditions on the localized corrosion of copper tubes *Mater. Chem. Phys.* **302** 127747.
[5] Tahara A and Kodama T 2000 Potential distribution measurement in galvanic corrosion of Zn/Fe couple by means of Kelvin probe *Corros. Sci.***42(4)** 655-673.
[6] Liu H, Wang Z N, Qiao L J, Su Y J and Yan Y 2023 Role of protein in crevice corrosion of CoCrMo alloy: An investigation using wire beam electrodes *Corros. Sci.* **215** 111028.
[7] Tan Y J, Aung Nn and Liu T 2012 Evaluating localised corrosion intensity using the wire beam electrode *Corros. Sci.* **63(5)** 379-386.
[8] Chen L, Du Y X, Liu Y, Zhu Z H, Liang Y and Zhang L 2024 Investigation on localized corrosion behavior of pipeline steel under AC interference in alkaline earth metal environment using wire beam electrode *Electroanal. Chem.* **973** 118668.
[9] Tan Y J, Bailey S and Kinsella B 2001 Mapping non-uniform corrosion using the wire beam electrode method. III. Water-line corrosion *Corros. Sci.* **43(10)** 0-1937.
[10] Suter T and Böhni H. Microelectrodes for studies of localized corrosion processes 1998 *Electrochim. Acta* **43(19-20)** 2843-49.
[11] Pan C Q, Zhong Q D, Yang J, et al. 2021 Investigating crevice corrosion behavior of 6061 Al alloy using wire beam electrode *JMR& T.* **14** 93-104.

Study on the Morphological Structure Characteristics of the Modified Preparation Fe-N/BC Catalysts

Yang Yang[1], Ting Zhou[1], Tianxing Du[1], Wenyi Dong[1,2,3] and Hongjie Wang[1,2,3,*]

[1] School of Civil and Environmental Engineering, Harbin Institute of Technology Shenzhen, Shenzhen 518055, Guangdong, China.
[2] Shenzhen Key Laboratory of Water Resource Utilization and Environmental Pollution Control, Shenzhen, 518055, Guangdong, China.
[3] State Key Laboratory of Urban Water Resource and Environment, School of Environment, Harbin Institute of Technology, Harbin 150090, Heilongjiang, China.
Email: 284212494@qq.com; 1226133395@qq.com; 1169777374@qq.com; dwy1967@qq.com; whj1533@qq.com; [*]whj1533@qq.com

Abstract. In this study, the citric acid iron and melamine was used to modify the selected biomass, preparing Fe-N/BC iron-nitrogen co-doped biochar catalyst. The BC, N/BC, and Fe-N/BC were characterized by SEM, XRD, FTIR, Raman, XPS, etc., to research the aspects of micromorphology, crystal structure, surface functional group types, and element composition. The results showed that Fe and N elements could be doped into the biochar successfully, making the particle size of Fe-N/BC smaller than that of BC, possessing a bigger surface area with a more abundant pore structure. Additionally, the graphitization degree of Fe-N/BC was enhanced, introducing active sites including pyridinic nitrogen, graphitic nitrogen, and $Fe\text{-}N_x$, enhancing the catalytic activity and electron transfer performance of the catalyst, enabling Fe-N/BC to simultaneously act on catalytic reduction and catalytic oxidation systems.
Keywords. Iron-nitrogen-doped biochar; Fe-N/BC catalyst; characterization; structural features; catalytic activity.

1. Introduction

Carbon-based catalysts can catalyze reducing agents such as sodium borohydride ($NaBH_4$) and hydrazine hydrate to construct catalytic reduction systems for degrading difficult-to-oxidize organic compounds like nitrobenzene [1]; they can also catalyze oxidants such as persulfates (PS) and peroxy disulfate (PMS) to construct catalytic oxidation systems for treating hard-to-degrade organic compounds like aniline [2-4]. Therefore, theoretically, catalysts with both catalytic reduction and catalytic oxidation functions can be developed based on carbon-based materials.

Compared with other carbon-based catalysts, biochar had lower costs and was more environmentally friendly. Biochar contained active sites such as graphite carbon and oxygen-containing functional groups, which could be used to catalyze PMS and construct catalytic oxidation systems. However, unmodified biochar catalysts cannot be applied to catalyze $NaBH_4$ reduction systems, and they could exhibit low catalytic activity towards PMS. Doping biochar with Fe could form Fe active sites in biochar; Fe(0) exhibited high catalytic activity towards $NaBH_4$, while Fe(II) showed high catalytic activity towards PMS [5]. Furthermore, Fe doping could promote the formation of graphite carbon structures in biochar. Further nitrogen doping of Fe-loaded biochar

Journal of Physics: Conference Series **2941** (2025) 012053 doi:10.1088/1742-6596/2941/1/012053

introduced high catalytic activity sites for PMS such as Fe-N_x, pyridinic nitrogen, pyrrolic nitrogen, and graphitic nitrogen. Nitrogen doping could reorganize electron arrangements, strengthen electron transfer processes in the catalytic system, and also enhance the catalytic effect on $NaBH_4$.

For this study, Fe-N/BC iron-nitrogen co-doped biochar catalyst was made through co-doping modification of biochar with citric acid iron and melamine. The catalysts were characterized through SEM, XRD, FTIR, Raman, XPS, etc., to study their microstructure, crystal structure, surface functional groups, element composition, etc.

2. Preparation of Fe-N/BC Catalyst

2 g of biomass powder was weighed in a beaker, 10 mL of hydrochloric acid with a concentration of 1 mol/L was then added, all of above were placed on a magnetic stirrer, and stirred at a speed of 350 r/min for 8 hours, then filtered and washed with pure water 3 times. The filter cake was transferred to a beaker containing 25 mL of citric acid iron solution with a concentration of 8 g/L, which was sonicated at a frequency of 40 kHz for 30 minutes. The melamine was added with a doping ratio of 5 g/g biomass, stirred evenly, and sonicated again at the same frequency for 30 minutes. Then the mixture was transferred into a centrifuge tube (50 mL) and centrifuged at 6000 r/min for 5 minutes. The solid obtained after centrifugation was washed with pure water, and placed in a 60°C hot air oven until completely dried. The solid was removed and grind evenly. Then it was placed in a tube furnace, heated through a nitrogen atmosphere at 5°C/min to 900°C for high-temperature pyrolysis for 1 hour. At last, the biochar was grind and sieved through a 100-mesh sieve to obtain the Fe-N/BC catalyst.

3. Characterization Methods of Catalyst Appearance Structure

(1) SEM analysis was carried out by observing the micro-morphology of the catalyst using a field emission scanning electron microscope.

(2) BET analysis was carried out by determining the N_2 adsorption-desorption isotherms of the catalyst using a fully automatic specific surface area and pore size analyzer (Micromeritics APSP 2460). Prior to testing, the sample is subjected to vacuum degassing at 300 °C for 8 hours, and the testing temperature is 77.3 K. Calculating the sample's specific surface area using the BET method and determining the pore size distribution and pore volume of the sample based on the Barret Joyner Halenda (BJH) curve.

(3) Raman analysis was carried out by determining the structural changes and defects of the catalyst using a Raman spectrometer at 532 nm excitation wavelength and 50-4000 cm-1 detection wave range.

(4) XRD analysis was carried out by analyzing the crystal structure of the catalyst using an X-ray diffractometer, with Cu-K_α X-ray source, a 40 kV accelerating voltage, a 40 mA current, a 10-90° scanning angle range, and a 10 °/min speed.

(5) FTIR analysis was carried out by analyzing the surface functional group structure characteristics of the catalyst using a Fourier transform infrared spectrometer, with a 400-4000 cm^{-1} testing wave range.

(6) XPS analysis was carried out by analyzing the elemental composition, content, and chemical states of the catalyst using an X-ray photoelectron spectrometer. Using Al-K_α X-ray source (h_v=1486.6 eV), a 12 kV working voltage, and a 6 mA filament current. The full spectrum scan parameters: 100 eV pass energy, 1 eV step size; 50 eV pass energy, 0.05 eV step size.

4. Results and Discussion

4.1. SEM

Figure 1 illustrates the apparent images of BC and Fe-N/BC. Figure 1(a) illustrates the surface of raw biochar which was rough with numerous irregular folds, indicating that the onion biochar had a large surface area, which was beneficial for doping iron and nitrogen elements. However, its particle size was larger and was not conducive to dope impurities. Thus, the particle size of the biochar after doping

Journal of Physics: Conference Series **2941** (2025) 012053 doi:10.1088/1742-6596/2941/1/012053

was significantly reduced, as illustrated in figure 1(b). Compared with figures 1(e) and 1(f), it could be observed that after iron and nitrogen modification, the number of micropores on the biochar catalyst surface increased significantly, which might be due to the decomposition and release of gas during the pyrolysis of melamine [6,7], leading to an increase in porosity.

(a) BC(1000 times) (b) BC(10000 times) (c) BC(50000 times)

(d) Fe-N/BC(1000 times) (e) Fe-N/BC(10000 times) (f) Fe-N/BC(50000 times)

Figure 1. SEM images of BC and Fe-N/BC catalyst.

4.2. BET Analysis

The N_2 adsorption-desorption isotherms and the corresponding characteristics of the biochar catalysts (BC and Fe-N/BC) are shown in figure 2 and table 1.

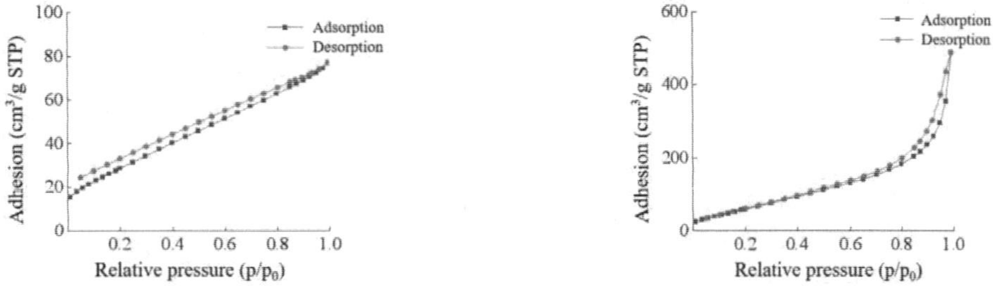

(a) BC (b) Fe-N/BC

Figure 2. BET analysis of BC and Fe-N/BC.

Table 1. Specific surface areas and pore characteristics of BC and Fe-N/BC.

Catalyst	Specific surface area (m^2/g)	Total pore volume (cm^3/g)	Average pore diameter (nm)
BC	114.78	0.13	3.61
Fe-N/BC	276.66	0.76	7.94

As illustrated in figure 2 and table 1, catalysts of BC and Fe-N/BC belonged to type IV adsorption isotherms, exhibiting obvious hysteresis loops under high pressure and indicating the presence of ordered mesoporous structures in both of the two catalysts.

After modification with citric acid iron and melamine, the specific surface area and total pore volume of the catalysts were significantly increased from the initial values of 114.78 and 0.13 cm^3/g to 276.66 and 0.76 cm^3/g, respectively. During the pyrolysis process, melamine generated nitrogen-containing gases such as N_2 and NH_3, with NH_3 reacting with C to produce other gaseous products and forming microporous structures inside the biochar [5,8]. Reductive gases such as NH_3 and H_2 could react with the doped Fe during pyrolysis, converting it to more catalytically favorable Fe(0) for $NaBH_4$. Biomass modified with citric acid iron and melamine led to a more developed pore structure and larger surface area of the biochar catalyst, facilitating the adsorption of drugs and pollutants and exposing more reactive sites.

4.3. XRD Analysis

Figure 3 illustrates the XRD characterization of the catalysts (BC, N-BC, and Fe/N-BC) to analyze their crystal phase structures. Two diffraction peaks at 26.3° and 44.1° were obtained in the XRD patterns of the three catalysts, corresponding to the (002) and (101) crystal planes of graphite, indicating the formation of a high-temperature-treated graphite-like structure [3,6]. Graphitic carbon structures possessed excellent electron transfer capability, facilitating the electron transfer process in the catalytic system. The XRD spectrum also revealed that the iron-nitrogen doping did not form a crystalline structure in the biochar, suggesting a lower degree of crystallization in the catalyst. This might be due to the small loading of Fe resulting from the impregnation method, falling below the detection limit of XRD. Further characterization using FTIR, Raman spectroscopy, and XPS was needed to investigate the improvement effect of iron-nitrogen doping on the catalyst structure and catalytic activity.

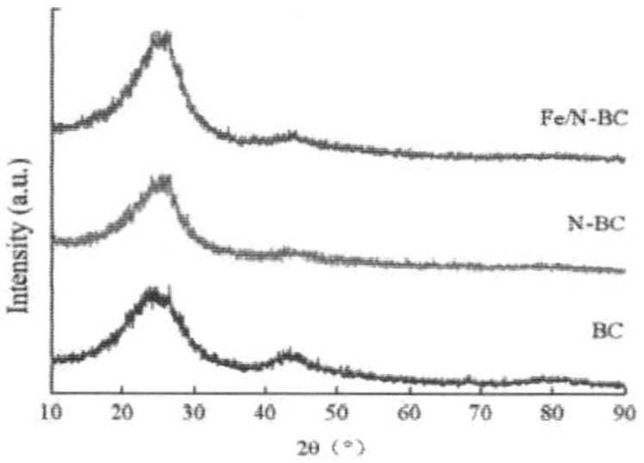

Figure 3. The XRD images of three different biochar catalysts.

4.4. FTIR Analysis

FTIR could reflect the relevant information on catalyst chemical bonds and functional groups, thereby providing further insight into the structure of materials. The Fourier transform infrared spectra of the three catalysts are illustrated in figure 4. The results showed that the number of surface functional groups on biochar significantly increased after N doping. The peaks of Fe-N/BC were similar to those of N/BC, with Fe-N/BC exhibiting stronger absorption peak intensities. At 1340, 1630, 2803, 2930, 3050 cm^{-1}, there were common FTIR characteristic peaks of Fe-N/BC and N/BC, where the absorption peaked at 1340, 1630 cm^{-1} corresponded to C-H bonds and C=C of alkenes; the group at 2803 cm^{-1} corresponded to the stretching vibration of aldehyde and ketone proton; the peak at 2930 cm^{-1} corresponded to C-H bonds of alkanes; the broad and scattered absorption peak at 3050 cm^{-1} corresponded to N-H bonds [9].

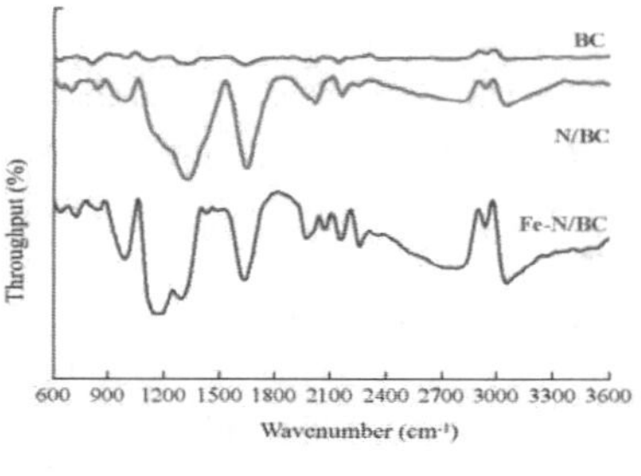

Figure 4. Fourier transform infrared spectra of three different catalysts.

4.5. *Raman Spectroscopy*
The Raman analysis results of the three catalysts are shown in figure 5.

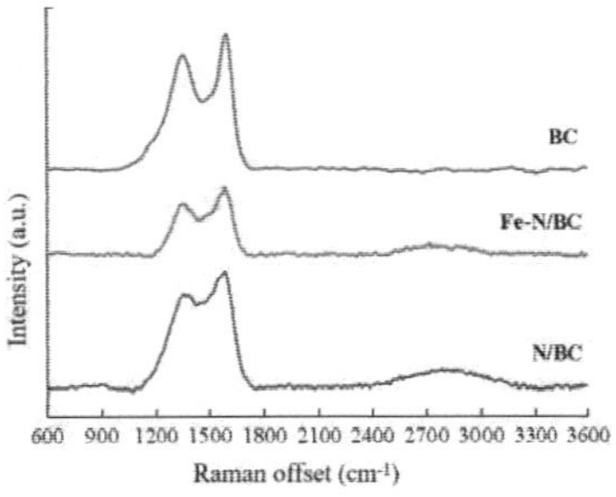

Figure 5. Raman spectraof BC, Fe-N/BC, N/BC.

In the Raman spectra, characteristic peaks were observed at 1340 cm^{-1} (D peak) and 1584 cm^{-1} (G peak) for all samples. The D peak was attributed to the symmetric stretching breathing mode of sp^2 carbon atoms in the aromatic ring, corresponding to structural defects and disordered carbon in carbon materials; while the G peak was caused by the stretching vibration between sp^2 carbon atoms, corresponding to graphitic carbon structures [7,8]. The intensity ratio of I_D/I_G reflected the degree of material defects, with a larger ratio indicating greater disorder and defects, and a lower graphitization degree. The I_D/I_G values were 1.42 for BC, 1.68 for N/BC biochar catalyst, and 1.25 for Fe-N/BC. After triazine doping modification, the I_D/I_G of the material increased, indicating successful nitrogen doping leading to carbon grid deformation. Furthermore, after citric acid iron modification, the I_D/I_G of the material significantly decreased, suggesting that iron citrate doping enhanced the graphitization degree of biochar. The graphitic carbon structure in biochar helped improve the electron transfer ability of the catalyst and provides Fe-N_x active sites. Graphitic carbon itself could also act as catalytic active centers, enhancing the catalytic activity of the catalyst.

4.6. *XPS*
The element composition and content of Fe-N/BC and N/BC surfaces were analyzed using XPS, as shown in figure 6.

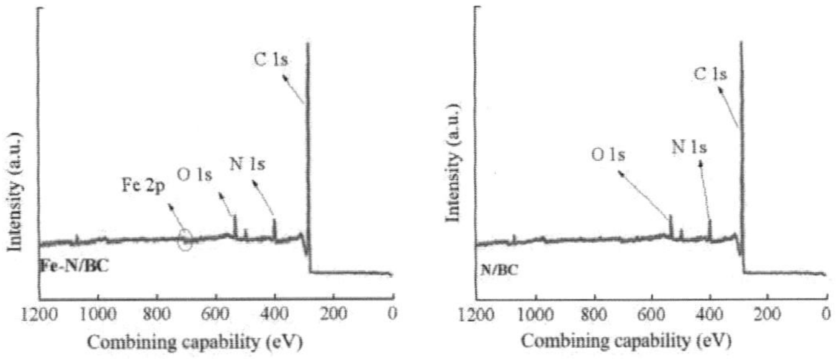

(a) Fe-N/BC (b) N/BC

Figure 6. XPS spectra of Fe-N/BC and N/BC.

As shown in figure 6, characteristic peaks of C 1s, N 1s, and O 1s were observed at the binding energies of 284.8, 400.5, and 532.4 eV, respectively. In addition, a characteristic peak of Fe 2p was also observed in the range of 700-742 eV in Fe-N/BC, indicating the successful doping of Fe and N elements into the biochar. Fine spectral analyses were conducted for Fe and N elements, with the Fe 2p fine spectrum of Fe-N/BC shown in figure 7(a).

In the Fe 2p spectrum of Fe-N/BC, peaks corresponding to Fe^0 $2p_{3/2}$ (relative content 2.7%, 706.4

eV), Fe^{2+} $2p_{3/2}$ (relative content 40.2%, 709.3 eV), Fe^{3+} $2p_{3/2}$ (relative content 24.9%, 711.5 eV), Fe^{2+} $2p_{1/2}$ (relative content 13.6%, 723.6 eV), Fe^{3+} $2p_{1/2}$ (relative content 18.6%, 725.4 eV), and their satellite peaks were observed [7]. No Fe 2p characteristic peaks were obtained in N-BC, indicating that Fe has been successfully loaded on the biochar. Combined with the fine spectrum analysis of N 1s, it could be inferred that Fe existed in the forms of Fe^0, Fe^{2+}, Fe^{3+}, and Fe-N_x.

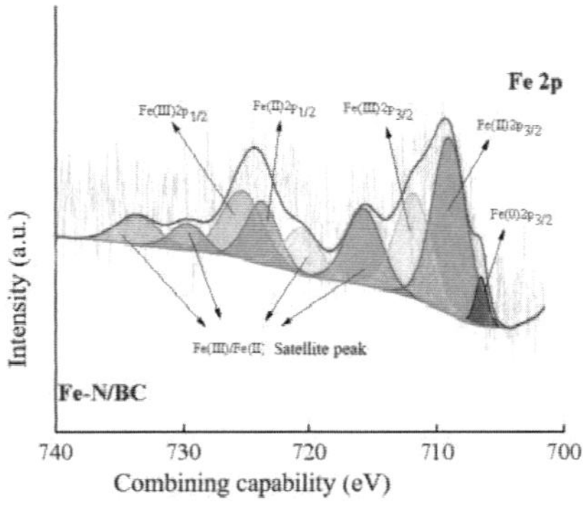

(a) Fe 2p

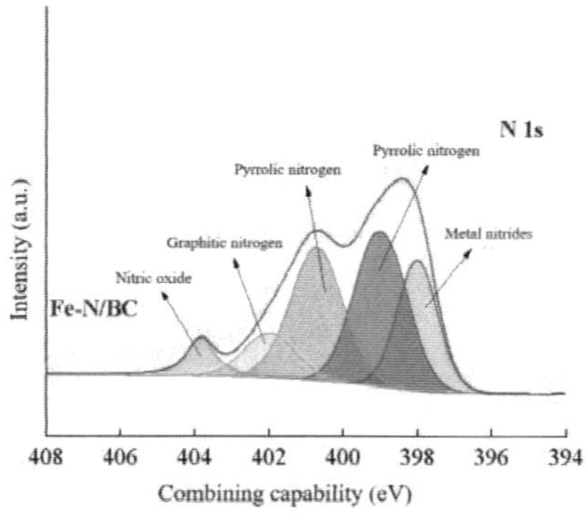

(b) N 1s

Figure 7. The fine spectrum of Fe-N/BC.

The fine spectrum of Fe-N/BC's N 1s is shown in figure 7(b). After peak fitting of the Fe-N/BC spectrum, pyridinic nitrogen (relative content 29.6%, 398.1 eV), pyrrolic nitrogen (relative content 27.8%, 399.1 eV), Fe-N_x (relative content 22.8%, 400.5 eV), graphitic nitrogen (relative content 13.0%, 401.9 eV), and oxidized nitrogen (relative content 6.8%, 403.8 eV) were observed [8,9]. Among them, the pyridinic nitrogen structure was conducive to the catalyst adsorbing oxygen, and its lone pair electrons easily form Fe-N_x structure with Fe; Fe-N_x could adsorb and activate PMS or $NaBH_4$ through radical/non-radical pathways to enhance the catalytic activity of the catalyst; graphitic nitrogen could form large π bonds, lowering the bond energy, enhancing the catalyst activity, and providing more active oxygen sites, thus promoting the generation of surface activated complexes and active oxygen species.

The results of the XPS spectra above indicated that co-doping biomass with iron and nitrogen successfully introduces Fe-N_x, pyridinic nitrogen, graphitic nitrogen, and other catalytically active structures, significantly improving the efficiency of the catalyst.

5. Conclusion

(1) Both Fe and N elements could be doped into the biomass successfully, resulting in Fe-N/BC with smaller particle size compared to BC, possessing a more abundant porous structure and specific surface area.

(2) The graphitization degree of Fe-N/BC was enhanced, introducing active sites such as pyridinic nitrogen, graphitic nitrogen, and Fe-N_x, which enhanced the catalytic activity and electron transfer properties of the catalyst, enabling Fe-N/BC to act simultaneously in catalyzing reduction and oxidation systems.

Acknowledgment

This work was supported by the National Natural Science Foundation of China (Grant No. 42177044).

6. References

[1] Das D, Sarkar R, Mukherjee S, et al. 2021 Sodium borohydride assisted reduction of toxic pollutants by silver coordinated melamine based polymeric material *Materials Today: Proceedings* **44** 444-452.

[2] Olmez-Hanci T, Arslan-Alaton I, Gurmen S, et al. 2018 Oxidative degradation of bisphenol a by

carbocatalytic activation of persulfate and peroxymonosulfate with reduced graphene oxide *J Hazard Mater* **360** 141-149.

[3] Long Y, Huang Y, Wu H, et al. 2019 Peroxymonosulfate activation for pollutants degradation by Fe-N-codoped carbonaceous catalyst: structure-dependent performance and mechanism insight *Chemical Engineering Journal* **369**: 542-552.

[4] Liu J, Kang X, He X, et al. 2019 Temperature-directed synthesis of N-doped Carbon-based nanotubes and nanosheets decorated with Fe (Fe3O4, Fe3C) nanomaterials *Nanoscale,* , **11(18)**: 9155-9162.

[5] Sun M, Wang X, Li Y, et al. 2022 Bifunctional Pd-Ox center at the liquid–solid–gas triphase interface for H2O2 photosynthesis *ACS Catalysis* **12(4)**: 2138-2149.

[6] Peng X, Fan Z, Wang Q, et al. 2024 Recent advances on the role of high-valent metals formed during persulfate activation *Separation and Purification Technology* **345**.

[7] Song M, Nguyen Q B, Kim C, et al. 2023 Sustained Activation of Persulfate by Slow Release of Fe(II) from silica-coated nanosized zero-valent iron for in situ chemical oxidation *Water Res* **246** 120715.

[8] Hu J, Zhou S, Liao S, et al. 2023 Effect of oxidized pyrite activating persulfate on the degradation of aniline aerofloat: mechanism and degradation pathway *Minerals Engineering* **201**.

[9] Chen Y, Long L, Luo Y, et al. 2023 Insights into persulfate activation by FeO for phenol removal: the production and effect of Fe(Iv) *Chemical Engineering Journal* **468**.

A FRP rebar winding method and experimental research on fiber stretching

Shuai Li, Guorong Gan, Daohui Lai, Zhengcui Wang

College of Vocational and Technical Education, Guangxi Science and Technology Teacher's College, Laibin, Guangxi 546199

Corresponding author and Email: Gan Guorong 594712354@qq .com

Abstract: In order to improve the efficiency of basalt fiber tendon preparation, a basalt fiber winding scheme is provided. To solve the singularity of current winding methods and optimize the flexibility of basalt tendons to cope with various types of environment winding. Four groups of fibers were tested in tensile testing using a fiber material tensile testing machine. The results showed that the single tensile force was 47.88 N and the elongation at break was 0.61. The winding device of FRP bars was then designed, and the winding pitch, spiral lift angle, and winding density of FRP bars were adjusted by adjusting the rotating nut scale, and the laws and connections between the three physical quantities were identified.

Keywords: RPF Reinforcement; Basalt Fibers; Tensile Testing

1. Introduction

Basalt fiber has excellent physical and mechanical properties and is a new high performance inorganic material that is green and environmentally friendly. Basalt fiber is made from continuous fiber extracted from natural basalt. Basalt continuous fiber is a kind of both high strength and excellent insulation, corrosion resistance and high temperature resistance, which has a wide application prospect in the construction field[1]. In addition, the production process of basalt fiber determines that it produces little waste and almost no pollution to the environment, and the products produced can be directly degraded without causing any harm to the environment, which is a green material in the true sense[2]. Basalt fiber is one of the four key fibers developed in China, namely, carbon fiber, aramid, UHMWPE and basalt fiber, which have been industrially produced. Basalt continuous fibers are widely used in fiber reinforced composites, friction materials, marine materials, insulation materials, automotive industry, anti-high temperature filter cloth, and protection[3].

2. Introduction of experimental equipment and materials

As shown in Figure 1, the fiber material tensile testing machine is produced by Shenzhen New Sans (SANS) Material Testing Co.Ltd[4]. with a maximum experimental force of 30KN, power supply 220V, power 4kw, accuracy level 1[5]. Basalt fiber tensile length of 20mm, monofilament diameter of 7μm ~ 15μm, relative density of 2.8g/cm3 ~ 3.3g/cm3, the use of temperature of -260°C~600 °C, tensile strength of 300Mpa ~ 4800Mpa[6].

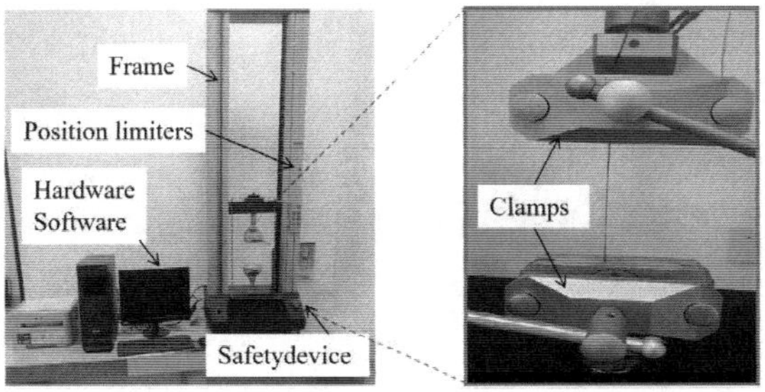

Figure 1. Introduction of fiber material tensile testing machine

3. Basalt fiber mechanical properties testing

As shown in Figure 2, The two collets of the tensile tester are put into the specimen separately, and the basalt fiber monofilament is clamped through the collet and is subjected to uniform force in the collet of the tensile tester until the specimen reaches the required tensile strength. During the test, it should be ensured that both ends of the specimen are at the same horizontal position, otherwise it will lead to uneven tension at both ends of the specimen. Stop stretching after the specimen reaches the desired tensile strength and record the elongation at break of thespecimen on the tester.

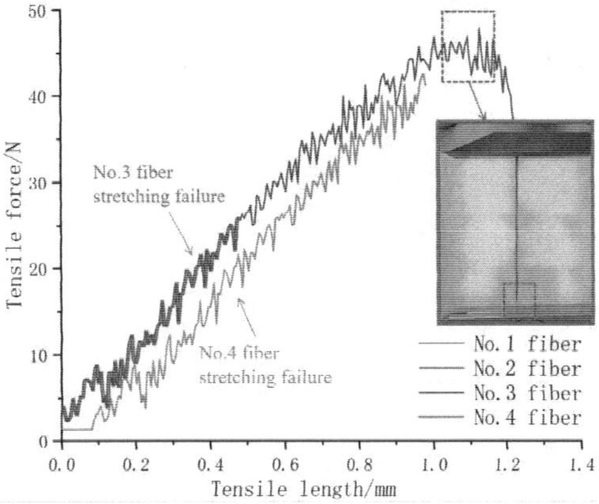

Figure 2. Mechanical properties of fiber materials test results

4. RBF steel fabrication

Basalt fiber has much higher tensile strength than ordinary steel reinforcement, so under the same conditions, the load-bearing capacity of basalt fiber concrete is 30% to 50% [7]higher than ordinary concrete. Basalt fiber has better tensile strength, which can make the volume of reinforcement significantly reduced under the same load-bearing capacity. When cement and basalt fiber and other materials are mixed into the reinforcement at the same time, it can increase the load-bearing capacity of the reinforcement by 10% to 15%[8]. If basalt fiber is wound on the reinforcement, then basalt fiber can improve the tensile strength of the reinforcement (but not its flexural strength This is mainly because basalt fiber is denser compared with ordinary steel reinforcement and is not suitable for use as reinforcement with high tensile strength[9]).Basalt fibers are wound around the steel reinforcement to make it basalt fiber reinforcement, basalt fiber reinforcement not only has good flexibility, but also can absorb more external energy, so basalt fiber reinforcement has a good protective effect on the concrete. After adding basalt fiber reinforcement to concrete, due to its good tensile strength and self-healing ability and other properties[10], it can form a protective layer in concrete to stop the external environment from eroding the concrete structure.

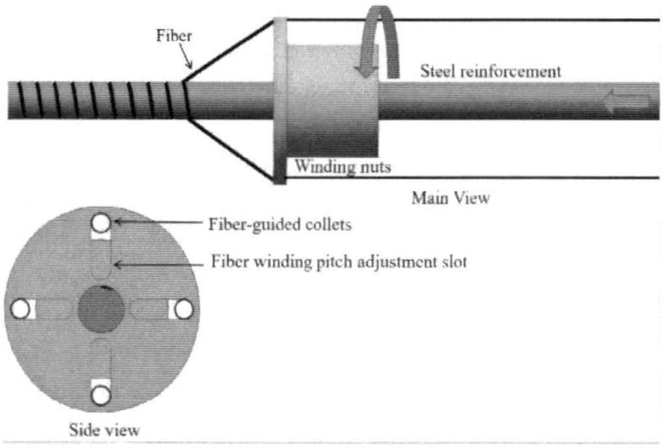

Figure 3. Winding method of RBF reinforcement

Basalt fibers are used to wrap around the reinforcement bars, thereby enhancing the overall performance of the bars. As shown in Figure 3, the basalt fibers are wrapped around the rebar by means of a swivel nut. On the one hand, the rotating nut rotates itself at a uniform speed and wraps the basalt fibers evenly around the rebar through the rotating nut guide joint. On the other hand, the reinforcing bar moves forward through the nut at a uniform speed, and finally completes the winding of the fibers on the reinforcing bar.

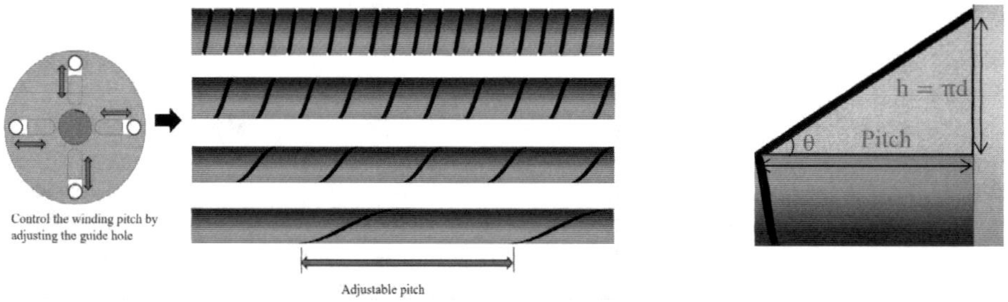

Figure 4. Principle of fiber winding pitch adjustment

As shown in Figure 4, the size of the winding fiber pitch is adjusted by adjusting the fiber guide hole scale. The pitch of the fiber is directly related to the winding effect and the overall strength of the rebar. Therefore, it is necessary to choose the rebar with different winding pitch according to the needs of different environments.

As shown in Figure 5, the winding spiral lift angle and winding pitch have the following relationship. When the spiral lift angle is in the range of 0°~ 5°, the change of pitch reduction is most significant. However, it is not easy to adjust the exact pitch because the difference between spiral lift angle and pitch ratio is too sensitive. When the spiral lift angle is in the range of 5°~ 20°, the spiral lift angle to pitch ratio is appropriate at this time, and it is most conducive to adjusting the winding pitch, so it is the most suitable area. When the spiral lift angle is in the range of 60°~ 90°, it is not conducive to pitch adjustment. As shown in Table 1， Take 6π as an example, other calculation methods are the same.

Table 1. Take 6π as an example, other calculation methods are the same

Spiral lift angle tangent tanθ	Tangent Correspondence Ratio	Corresponding pitch p/mm
Tan1°	6π/P=0.017	1108.24
Tan2°	6π/P=0.035	538.29
Tan3°	6π/P=0.052	362.30
Tan4°	6π/P=0.070	269.14
Tan5°	6π/P=0.110	216.55
Tan6°	6π/P=0.123	171.27
Tan7°	6π/P=0.141	153.17
Tan8°	6π/P=0.158	133.62
⋮	⋮	⋮
Tan89°	6π/P=0.33	57.29

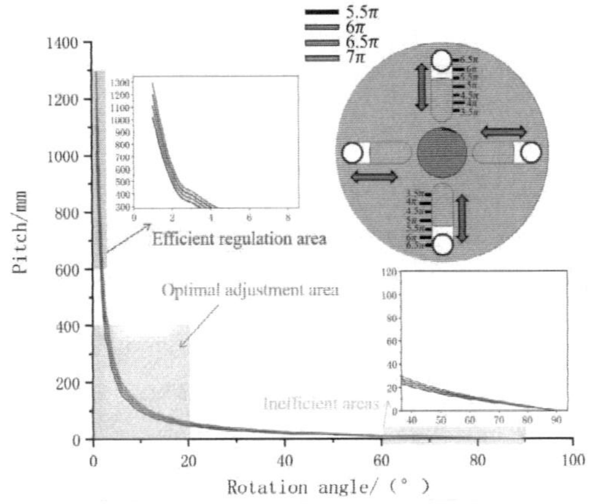

Figure 5. Principle of adjusting the pitch of fiber winding

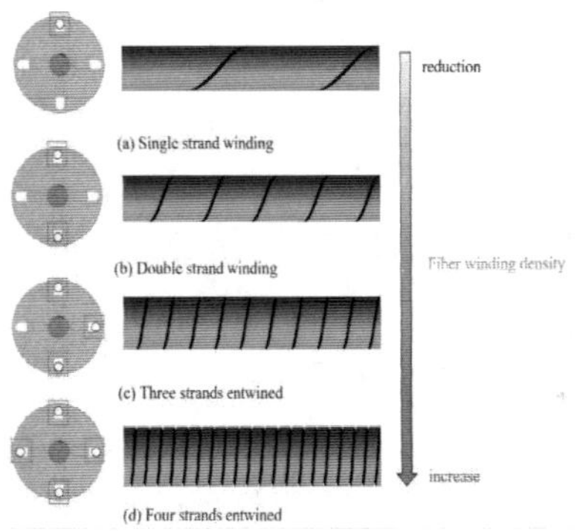

Figure 6. Specific working principle of fiber winding

As shown in Figure 6, depending on the winding requirements, the fibers can be wound by adjusting the winding density through the individual guide joints of the swivel nut. The specific fiber winding method is as follows: when the rebar needs to be wound in a single strand, the fiber is threaded into the single guide joint of the rotary joint, and the rotary joint starts to rotate at a uniform speed to complete the single strand fiber winding. Similarly, when the rebar requires double, triple or quadruple winding, the fibers are inserted into the corresponding guide joint of the rotary joint, and the rotary joint starts to rotate at a uniform speed to complete single strand fiber winding. The more guide joints are involved in the work, the higher the winding density of the rebar.

As shown in Figure 7, the pitch and helix angle of fiber winding are adjusted by changing the scale of the rotating nut, so as to adjust the winding parameters and accomplish the desired winding effect. For example, if the maximum pitch of fiber winding is 100mm, by changing the scale of the rotary nut, the corresponding winding spiral angle and pitch will be changed accordingly. When the spiral angle is close to 90°, the scale of the rotary nut reaches the limit and can no longer be adjusted.

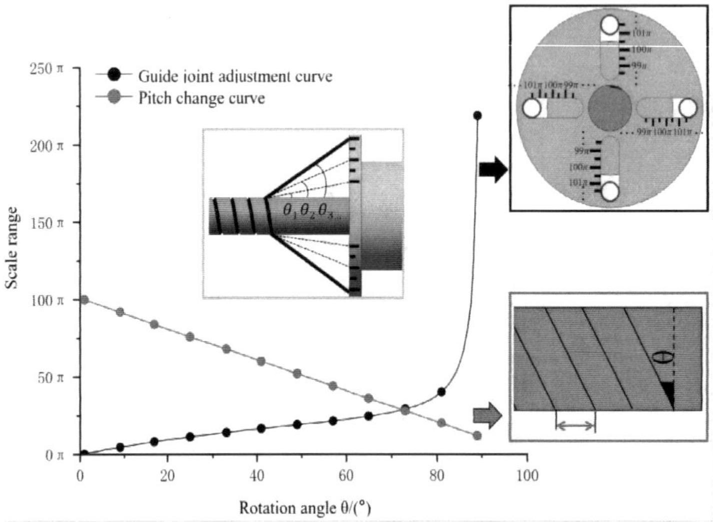

Figure 7. Helical lift angle in relation to the position of the guide joint

5. Conclusion

This paper firstly introduces the value of RPF use and its application areas. Secondly, tensile testing of the wound fibers of RPF is performed to prepare for the subsequent preparation of RPF reinforcement. Finally, a device for the preparation of RPF reinforcement is designed, and the adjustment of the winding parameters and the laws between the parameters are given in order to provide a reference for the subsequent preparation of RPF reinforcement.

Funding:

Guangxi Science and Technology Plan Project Funding Project (GUI KE AD22080060) and Guangxi Normal University of Science and Technology high-level Talent Project (GXKS2021GKY001)

Reference:

[1] Bian X, Zeng L, Deng Y, et al. The Role of Superabsorbent Polymer on Strength and Microstructure Development in Cemented Dredged Clay with High Water Content[J]. *Polymers*, 2018, 10(10): 1069-1085.

[2] Sidiq A, Gravina R, Giustozzi F. Is concrete healing really efficient? A review[J]. *Construction and Building Materials*, 2019, 205: 257-273.

[3] Li Hao. Research on the Development and Application of Concrete Highly Absorbent Resin (SAP) Internal Curing Agent for Northwest Arid Regions[D]. Lanzhou: Lanzhou Jiaotong University, 2020.

[4] Massana J, Reyes E, Bernal J, et al. Influence of nano- and micro-silica additions on the durability of a high-performance self-compacting concrete[J]. *Construction and Building Materials*, 2018, 165(20): 93-103.

[5] Mukharjee B B, Barai S V. Influence of Incorporation of Colloidal Nano-Silica on Behaviour of Concrete[J]. *Iranian Journal of Science and Technology - Transactions of Civil Engineering*, 2020, 44(02): 657-668.

[6] Yu B, Huang JM, Li QM, et al. Quantitative design method for durability of concrete structures under natural carbonation conditions[J]. *Concrete*, 2020, 372(10): 38-42.

[7] Wang X H, Val D V, Zheng L, Jones M R. Influence of loading and cracks on carbonation of RC elements made of different concrete types[J]. *Construction and Building Materials*, 2018, 164(10): 12-28.

[8] Li G, Hu W J, Cui H Y, et al. Long-term effectiveness of carbonation resistance of concrete treated with nano-SiO2 modified polymer coatings[J]. *Construction and Building Materials*, 2019, 201: 623-630.
[9] Wu X N, Zhao B, Wang L, et al. Superhydrophobic PVDF membrane induced by hydrophobic SiO2 nanoparticles and its use for CO 2 absorption[J]. *Separation and Purification Technology*, 2018, 190: 108-116.
[10] Hui Y, Zhu L, Li W, et al. Anchoring CeO2 nanoparticles on monodispersed SiO 2 spheres to construct hydrophobic polymer coating with enhanced UV absorption ability[J]. *Chemical Engineering Journal*, 2017, 321: 268-276.

Journal of Physics: Conference Series **2941** (2025) 012055 doi:10.1088/1742-6596/2941/1/012055

Research on the Vehicle Carbon Emission Inventory in Qinghai Province

Yi Zhang*, Ran Zhou, Shitao Peng, Zhaoyu Qi and Hui Tong

Tianjin Research Institute for Water Transport Engineering, Ministry of Transport, Tianjin300456, China.

Email: *zhangyi@tiwte.ac.cn; zhouran@tiwte.ac.cn; pengshitao@tiwte.ac.cn; qizy@tiwte.ac.cn; tongh@tiwte.ac.cn

Abstract. This study focused on the vehicle carbon emission control needs in Qinghai Province, which was characterized by high altitude and a fragile ecological environment. A bottom-up method was adopted to obtain data on the vehicular activity level in the province through extensive field research. Local comprehensive carbon emission factors for vehicles were obtained through "basic data + actual testing + environmental correction", and then coupled to generate the 2023 provincial vehicle carbon emission inventory, which could provide useful reference and guidance for the carbon emission control of vehicles in Qinghai Province and other high-altitude areas.
Keywords. Vehicle carbon emission; emission inventory; activity level; emission factor.

1. Introduction

Under the severe challenge of global climate change, reducing greenhouse gas emissions has become a widespread concern of the international community [1]. As one of the important sources of greenhouse gas emissions, the impact of motor vehicle exhaust emissions on environmental quality was becoming increasingly significant [2-4]. Especially in Qinghai Province, where the altitude was high and the ecological environment was fragile, with the rapid development of the economy and the acceleration of urbanization, the number of vehicles had also increased rapidly, making carbon emission control particularly important [5]. As an important foundation for conducting carbon emission research and management, the carbon emission inventory of vehicles refers to the process of statistics and accounting for the carbon emissions of a certain region or type of vehicle over a certain period of time. This had important practical significance and practical value for comprehensively understanding the actual situation of vehicular carbon emissions in Qinghai Province, evaluating their impact on the environment, and supporting the development of scientific and reasonable emission reduction measures [6].

The construction of carbon emission inventories for vehicles mostly adopted two methods [7]: one was the "top-down" method, which converted the overall fuel sales or consumption of vehicles in society into CO_2 emissions [8]; Another one was the "bottom-up" approach, which first classified vehicles reasonably, and then combined activity level data such as the number of classified vehicles and driving mileage, as well as fuel consumption data per unit driving mileage, to obtain the total fuel consumption, and finally calculated the CO_2 emissions [9,10]. Compared to the "top-down" method, the "bottom-up" method had higher temporal and spatial resolution, and the results were relatively more accurate, but it required higher data volume [11]. Constructing a carbon emission inventory for vehicles in Qinghai Province using a bottom-up approach was a complex and systematic project that required comprehensive consideration of multiple factors such as the quantity, type, mileage, and emission factors of vehicles [12,13]. Meanwhile, due to the unique geographical and climatic conditions in Qinghai Province, the characteristics and patterns of carbon emissions from vehicles may differ from other regions [14]. Therefore, in the process of building an inventory, it was necessary to fully consider these factors to ensure the accuracy and reliability of the inventory. At present, there is relatively little research on the carbon emission inventory of vehicles in high-altitude areas such as Qinghai Province in literature reports.

Based on the demand for carbon emission control of vehicles in Qinghai Province, this study adopted a "bottom-up" method, obtained data on the activity level of vehicles in the province through a large amount of field research, used "basic data + actual testing + environmental correction" to obtain localized comprehensive carbon emission factors for vehicles, and then coupled them to generate a provincial motor vehicle carbon emission inventory. The aim was to provide useful reference and guidance for the carbon emission control of motor vehicles in Qinghai Province and other high-altitude areas.

2. Methodology

2.1. Methodology for Calculating Vehicle Carbon Emissions

This study focused on the carbon emission control needs of vehicles in Qinghai Province and selected the "bottom-up" method to construct a vehicle carbon emission inventory model with certain local adaptability, which was used to calculate the CO_2 emission inventory of vehicles in the province in 2023.

The model structure is as follows.

$$E = \sum_i EF_i \times AL_i = \sum_i EF_i \times AAM_i \times Q_i \tag{1}$$

where, E: The total carbon emissions of vehicles, g; i: The type of vehicle i; AL_i: The activity level of vehicle type i; EF_i: CO_2 emission factor for vehicle type i, g/km; AAM_i: The average annual mileage of vehicle type i, km; Q_i: The ownership of vehicle type i.

The technical roadmap is shown in figure 1.

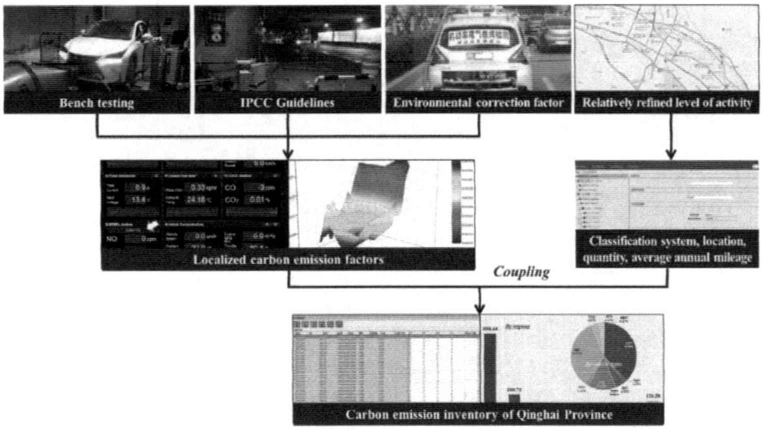

Figure 1. Technical route for constructing a carbon emission inventory for vehicles in Qinghai Province.

2.2. Acquisition of Vehicle Activity Levels

Referring to official guidance documents such as *The Technical Guidelines for the Compilation of Air Pollutant Emission Inventory for Road Vehicles (Trial)* [15] and *The Technical Guidelines for the Compilation of Integrated Emission Inventory of Air Pollutants and Greenhouse Gases (Trial)* [16], the types of vehicles in Qinghai Province were divided into 10 types: mini-duty vehicle (MIDV), light-duty vehicle (LDV), middle-duty vehicle (MDV), heavy-duty vehicle (HDV); mini-duty truck (MIDT), light-duty truck (LDT), middle-duty truck (MDT), heavy-duty truck (HDT); BUS, TAXI. This classification could meet vehicular environmental supervision needs and was suitable for the development of vehicle carbon emission models.

By visiting and investigating departments such as transportation and environmental protection, consulting annual inspection data on vehicle environmental protection, environmental statistics, pollution source census reports, and relevant yearbooks, as well as conducting questionnaire surveys on vehicle drivers in typical parking lots, toll booths, residential areas, and roadside areas in various cities, the activity level data of the above-mentioned types of vehicles, including their location, ownership, and average annual mileage, were obtained.

2.3. *Generation of Vehicle Carbon Emission Factors*

Emission factors refer to the carbon emissions per unit mileage of vehicles, and their main influencing factors include vehicle attributes, fuel quality, road driving characteristics, as well as environmental conditions such as temperature, humidity, and altitude [17-19]. As a typical high-altitude region, the vehicle carbon emissions in Qinghai Province had their own characteristics, so directly copying and referencing common data may result in significant errors. Therefore, based on the recommended CO_2 emission coefficient in *the Technical Guidelines for the Compilation of Integrated Emission Inventory of Air Pollutants and Greenhouse Gases (Trial)* [16], this study comprehensively used a large number of bench tests in laboratory, tunnel tests, and on-board tests in multiple typical areas, to obtain vehicular CO_2 emission data based on real driving conditions, and generated the CO_2 basic emission factor BEF [20,21]. At the same time, coupled with the local environmental correction factor φ, the comprehensive vehicle CO_2 emission factor CEF in Qinghai Province was finally obtained.

The structure of the vehicle CO_2 emission factor model for in Qinghai Province is as follows:

$$CEF_{i,j} = BEF_i \times \varphi_j = BEF_i \times (\varphi_{\text{Temp}} \times \varphi_{RH} \times \varphi_{\text{Height}}) \tag{2}$$

where, $CEF_{i,j}$: CO_2 comprehensive emission factors of type i vehicle in region j; BEF_i: CO_2 basic carbon emission coefficient for type i vehicle; φ_j: The environmental correction factors in region j, including the temperature correction factor φ_{Temp}, a humidity correction factor φ_{RH}, and altitude correction factor φ_{Height}, which were obtained based on measured vehicle CO_2 emissions in different typical regions and literature research.

3. Results and Discussion

3.1. *Vehicle Activity Level*

In 2023, there were a total of 1.50×10^6 vehicles in Qinghai Province (figure 2). According to the types of vehicle, the amount of LDV was the highest, with a number of 1.20×10^6 accounting for 80.10%, followed by LDT with a number of 1.87×10^5 accounting for 12.45%. According to the regions, Xining C had the largest number of vehicles, with 8.23×10^5 vehicles accounting for 54.75%, which was much higher than other regions.

The average annual mileage (AAM) of various types of vehicles was an important parameter for calculating the vehicle carbon emissions. Different types of vehicles had different average annual mileage due to factors such as vehicle type, fuel type, vehicle age, usage, and operating environment. This study was based on investigation and statistical analysis to obtain the average annual driving mileage of different types of vehicles in Qinghai Province, as shown in figure 3. Overall, the average annual mileage of public transportation vehicles was relatively higher than that of private cars.

Journal of Physics: Conference Series **2941** (2025) 012055 doi:10.1088/1742-6596/2941/1/012055

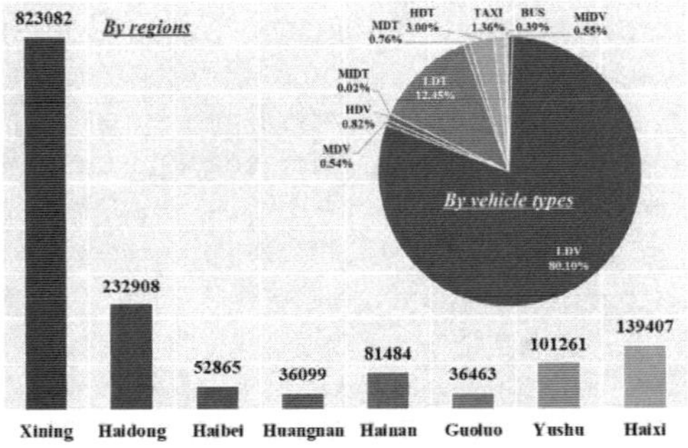

Figure 2. The vehicle ownership in Qinghai Province in 2023.

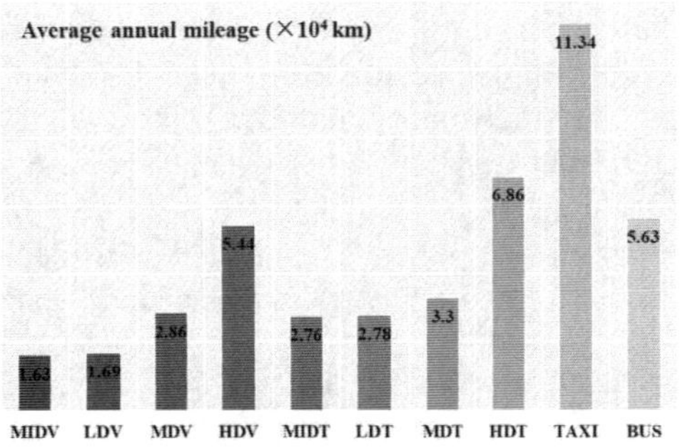

Figure 3. The average annual mileage of different types of vehicles in Qinghai Province.

3.2. Vehicle Carbon Emission Factors

Based on the basic emission factors and environmental correction factors, a localized dataset of CO_2 emission factors for different types of vehicles in Qinghai Province was obtained, as shown in figure 3. From figure 4, it can be observed that the larger the appearance and load of the vehicle, the greater the fuel consumption and the higher the corresponding CO_2 emission level. In addition, the comprehensive emission factors of CO_2 from different types of vehicles were generally higher than their basic emission factors. This was because compared to the eastern plains of China, Qinghai Province had a higher altitude, lower annual temperature, and a dry climate. The resulting increase in fuel consumption and the probability of insufficient fuel combustion would lead to an increase in the CO_2 emission level of single vehicles.

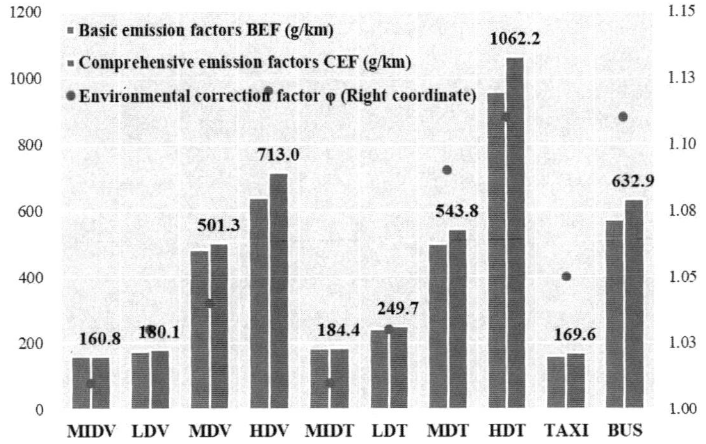

Figure 4. CO_2 emission factors of vehicles in Qinghai Province.

3.3. Vehicle Carbon Emission Inventory

By using the "bottom-up" method and coupling the vehicle activity level with carbon emission factors, the total carbon emissions of vehicles in Qinghai Province in 2023 could be calculated to be 9.67×10^6 t. The carbon emissions in Qinghai Province by vehicle types and regions are shown in figure 5. According to the types of vehicles, LDV had the highest carbon emissions (3.67×10^6 t), with an emission sharing rate of 37.90%, followed by HDT (3.28×10^6 t) and LDT (1.30×10^6 t) with emission sharing rates of 33.93% and 13.43%, respectively; The proportions of carbon emissions from other types of vehicles were relatively small. According to the regions, Xining City had the highest vehicle carbon emissions with a total of 4.90×10^6 t, which exceeded half of the total vehicle emissions in the province and was much higher than other cities. This basically conformed to the economic development level and population change patterns of each region.

Journal of Physics: Conference Series **2941** (2025) 012055 doi:10.1088/1742-6596/2941/1/012055

The changes in the carbon emission sharing rate of different types of vehicles compared to the proportion of ownership in Qinghai Province are shown in figure 6. It could be observed that compared to the proportion of their own ownership, except for the decrease in carbon emission sharing of MIDV and LDV, all other types of vehicles had shown an increase, with HDT showing the most significant increase. This indicated that in the carbon emission control work of vehicles, in addition to focusing on the types of vehicles with large ownership, vehicles with high carbon emission coefficients or activity levels, such as HDT, should also be given special attention.

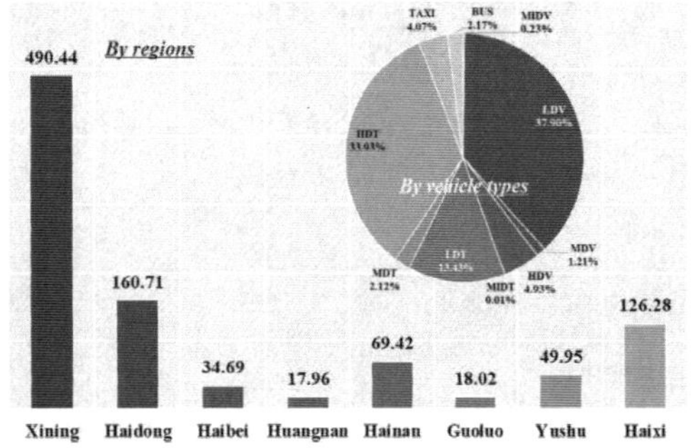

Figure 5. The carbon emissions of vehicles in Qinghai Province in 2023.

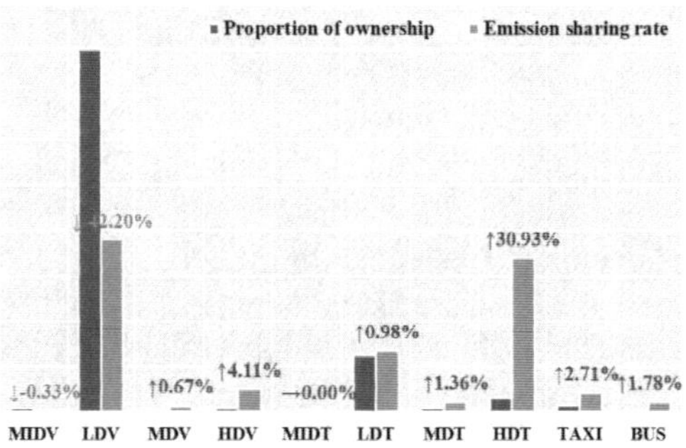

Figure 6. The changes in the carbon emission sharing rate of different types of vehicles compared to the proportion of ownership in Qinghai Province.

4. Conclusions

This study used a "bottom-up" method to generate a provincial vehicle carbon emission inventory by coupling the vehicle activity level in the province based on a large amount of field research data, and localized vehicle comprehensive carbon emission factors obtained through "basic data + actual testing + environmental correction". The calculation results showed that in 2023, there were a total of 1.50×10^6 vehicles in Qinghai Province, with a cumulative carbon emissions of 9.67×10^6 t. According to the types of vehicle, LDV had the largest carbon emission (3.67×10^6 t), with an emission sharing rate of 37.90%, followed by HDT (3.28×10^6 t) and LDT (1.30×10^6 t), with emission sharing rates of 33.93% and 13.43%, respectively. From a regional perspective, Xining City had the highest vehicle carbon emissions, accounting for more than half of the total emissions in the province and far higher than other cities. This basically conformed to the economic development level and population change patterns of each region.

It has also been observed that factors such as altitude, and temperature had a significant impact on vehicle carbon emissions. In the future, it is suggested to further strengthen the actual testing of vehicle carbon emissions in high-altitude areas such as Qinghai Province, construct localized vehicle carbon emission factors based on real road driving conditions, and create a more refined activity level database. This would further enhance the precise control of vehicle carbon emissions by improving the accuracy of emission inventory.

5. References

[1] Lei Z, Song Y and Ma Z 2024 Simulation study on the diffusion of atmospheric pollutants from ships in inland waterways *Journal of Waterway and Harbor* **45(1)** 131-139.
[2] Yin F and Mei S 2023 The national conference on transportation work in 2023 was held *Journal of Waterway and Harbor* **44(2)** 312.
[3] Yin F and Mei S 2022 Discussion on sustainable development at the 2022 world transportation conference *Journal of Waterway and Harbor* **43(6)** 720
[4] Yin F and Mei S 2019 The outline for building a strong transportation country *Journal of Waterway and Harbor* **40(6)** 628
[5] Ma Q and Liu D 2020 A brief discussion on the prevention and control of mobile source

pollution *Qinghai Environment* **30(4)** 165-170.

[6] Guo X, Liu L and Chen Y 2022 Methodological study on voluntary greenhouse gas reduction for unmanned container truck projects in ports *Journal of Waterway and Harbor* **43(3)** 396-400.

[7] Liu H, Zhang H and Tang H 2023 Research on ship black smoke control and supervision *Journal of Waterway and Harbor* **44(3)** 494-498+512

[8] Yin F and Mei S 2022 Green transportation standard system release system planning for green and low carbon standardization development of transportation *Journal of Waterway and Harbor* **43(4)** 560.

[9] Hu J, Zhang H and Peng S 2022 Feasibility analysis of satellite remote sensing evaluation of the implementation effect of ship air pollutant emission control zones *Journal of Waterway and Harbor* **43(5)** 660-667.

[10] Lin Y, Liu C and Zhang H 2020 Comparison and reference of green port evaluation systems at home and abroad *Journal of Waterway and Harbor* **41(5)** 613-618.

[11] Zhong W, Lei Z and Fu Z 2023 Development status and prospects of green hydrogen energy engineering technology in China's offshore areas *Journal of Waterway and Harbor* **44(1)** 124-130.

[12] Cui P, Liu G and Chen Z 2022 Research on risk grading assessment of marine environment pollution from ships in ports *Journal of Waterway and Harbor* **43(5)** 668-676.

[13] Gu M 2021 The impact of dust accumulation at dry bulk cargo terminals on solar heating systems *Journal of Waterway and Harbor* **42(3)** 370-375.

[14] Zhai J, Song C and Xiao J 2019 Preliminary exploration of the application of motor vehicle PEMS detection in high-altitude environments *Qinghai Technology* **26(6)** 15-21

[15] Ministry of Environmental Protection 2015 The technical guidelines for the compilation of air pollutant emission inventory for road vehicles (trial) p 22.

[16] Ministry of Ecology and Environment 2024 The technical guidelines for the compilation of integrated emission inventory of air pollutants and greenhouse gases (trial) p 35.

[17] Zhang Y, Wu L and Zou C 2017 Development and application of urban high temporal-spatial resolution vehicle emission inventory model and decision support system *Environmental Modeling & Assessment* **22(5)** 1-14.

[18] Zhang Y, Zhou R and Peng S 2022 Development of vehicle emission model based on real-road driving conditions *Atmosphere* **13** 595.

[19] Han Z Si Y and Wang X 2024 Battery mode selection and carbon emission decisions of competitive electric vehicle manufacturers *Mathematics* **12** 162472.

[20] Liu J Luo Y and Zhu Y 2024 Research on carbon emission characteristics and mitigation pathways in the vehicle fuel cycle: a case study of Guangdong province *Atmosphere* **15(1)** 235-248

[21] Lou P Zhou Z and Zeng Y 2024 Vehicle routing problem with time windows and carbon emissions: a case study in logistics distribution *Environmental Science and Pollution Research* **31(29)** 41600-41620

Intelligent Dust Control Technology Path for the Entire Area of Dry Bulk Cargo Terminal

Yi Zhang*, **Ran Zhou, Shitao Peng, Ningning Hong and Yin Ye**

Tianjin Research Institute for Water Transport Engineering, Ministry of Transport, Tianjin 300456, China.
Email: *zhangyi@tiwte.ac.cn; zhouran@tiwte.ac.cn; pengst@tiwte.ac.cn; hongnn@tiwte.ac.cn; yey@tiwte.ac.cn

Abstract. In the current macro context of tightening national and local atmospheric environment control policies, accelerating the construction of world-class green and smart ports, and increasing social attention, the requirements for dust pollution control at bulk cargo terminals are also increasing. Overall, the understanding of dust generation mechanisms from wind erosion in bulk cargo yards and during loading and unloading operations in the domestic port industry and academic community was still not comprehensive enough; the technical means of comprehensive dust intelligent monitoring in the yard area were still lacking; and the application of intelligent and refined control for the entire process still needed further improvement and promotion. Based on the requirements for the construction of smart and green ports in the new era, and an in-depth analysis of the key links, technical means, and weaknesses of dust pollution in bulk cargo terminals, this paper proposed intelligent control technology countermeasures and suggestions for dust pollution from three aspects: accurate characterization of dust under different operating conditions, intelligent three-dimensional monitoring of point surface integration, and intelligent decision-making system for the entire field and operation process.
Keywords. Bulk cargo terminal; dust pollution; accurate characterization; intelligent monitoring; accurate control.

1. Introduction

Ports were the hub of comprehensive transportation and strategic resources for economic and social development. The ports and terminals of coal, ore, and other dry bulk cargo were crucial links and important support for energy safety and supply, which were related to the stability of the national energy supply chain and the construction of the domestic and international "dual circulation" development pattern [1]. However, due to the limitations of material characteristics and port transportation processes, coal, ore, and other materials could generate certain dust pollution during loading, unloading, storage, and transportation, which had become the main environmental problem faced by dry bulk port terminals, seriously affecting the port area, surrounding environment, and people's health [2]. Every year, millions of tons of materials such as coal powder and mineral powder are lost due to wind erosion, storage, and loading and unloading, resulting in economic losses of billions in China. In the current macro context of tightening national and local atmospheric environment control policies, accelerating the construction of world-class green and smart ports, and increasing social attention, the environmental supervision requirements for dry bulk port terminals were also increasing. Therefore, how to form a dust pollution prevention and management system for the entire site and operation process, and effectively curbing dust particle pollution from the root had become an urgent task for dry bulk port terminals to achieve high-quality and sustainable development [3]. In response to the demand for dust pollution control in dry bulk cargo port terminals under the new situation, this paper deeply analyzed the current key links, technical means, and weaknesses of dust pollution in dry bulk cargo port terminals, and systematically proposed innovative paths for targeted intelligent control technology.

2. Key Emission Links of Dust Pollution in Dry Bulk Cargo Terminal

The dry bulk cargo was inevitably affected by mechanical disturbances or external meteorological factors during port operations, resulting in dust pollution, which made dust particles the main pollutant in dry bulk cargo terminals [4]. Generally speaking, the dust removal process in dry bulk cargo terminals mainly included train dumper operations, belt conveyor line, belt conveyor transfer, stacking and reclaiming operations (including truck loading and unloading), ship-loading operations, ship-unloading operations, and yard stacking according to the workflow, shown as figure 1. Overall, it could also be divided into two categories: dynamic dust removal and static dust removal [5].

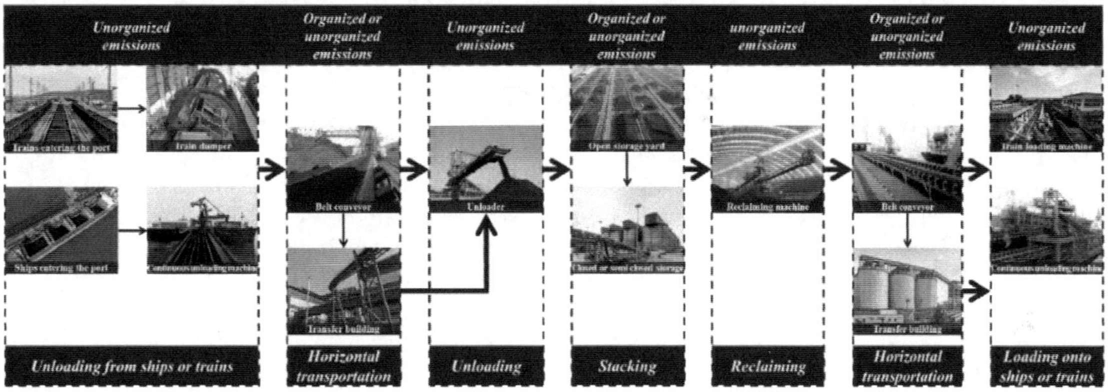

Figure 1. Typical operational processes and dust production links of specialized dry bulk cargo terminals.

Train dumper operations: The train dumper refers to large mechanical equipment used to dump railway open trains, which could flip or tilt trains to unload materials [6]. It was widely used in modern large dry bulk terminals. The dumping process of the train dumper generated a large and concentrated amount of dust, which was the main link for dust generation in loading and unloading operations. During this process, about 70% of the dust (Particle size < 100 μm) would be disturbed. The train dumper was generally set up in a semi-enclosed workshop and could complete the unloading of the entire carriage within a few minutes. After adopting environmental protection processes such as watering or spraying, dry dedusting, and setting a lower funnel, the impact of dust from the dumping operation on the outside of the workshop was not obvious, and the main dust control focus was the dust from the lower tunnel of the workshop building.

Belt conveyor line: Most of the existing specialized port belt conveyors were located in closed corridors, which were only partially open during maintenance [7]. Therefore, the amount of dust generated during the normal operation of the belt conveyor could be ignored. Therefore, this process was not a key aspect of dust control in loading and unloading operations.

Belt conveyor transfer: When coal, ore, and other dry bulk cargo were transported from one conveyor belt to another, there was usually a drop of several meters, resulting in the lifting of dust particles and overflow from the inlet and outlet, guide plates, and other places [8]. This process was generally completed in enclosed or semi-enclosed transfer towers, which were equipped with dry or wet dust removal equipment, so the external impact was very small. The main focus was on dust control inside the transfer tower.

Stacking and reclaiming operations (including truck loading and unloading): Due to the height difference of the stacking operation and the mechanical disturbance of the reclaiming operation, dust would be generated during the stacking and reclaiming process in the yard [9]. Although the amount of dust generated in this process could be reduced by environmental protection measures such as controlling the fall or spraying water, such operations were generally carried out in the open air so that the dust pollution generated would directly affect the external atmospheric environment. Therefore, the dust in this stage should be one of the key dust production links in loading and unloading operations.

Ship-loading operations: Similar to the stacking operation in the yard, there was also dynamic dust in the dock ship-loading operation due to the height difference. In the early stages of ship-loading operations, the loading machine penetrated deep into the interior of the ship's hold with less dust-generating [10]. However, when the cabin was loaded to a certain extent, there would be a certain height difference between the chute of the loading machine and the top of the material pile due to the departure of the cabin. In this process, the dust production condition was similar to the material stacking condition in the yard. Therefore, dust production during ship-loading operations should also be one of the key links in dust prevention and control.

Ship-unloading operations: For professional dry bulk terminals, ship unloaders were generally used to carry out loading and unloading operations on arriving ships [11]. The loading machine was connected to the belt conveyor at the forefront of the terminal and achieved uninterrupted unloading operations. This working method generally had a small amount of dust and was easy to protect. For general dry bulk terminals, the loading and unloading of ships were mostly noncontinuous operations, that was, the bulk cargo was first transferred from the ship's hold to the front yard of the port through bridge-type grab unloaders, and then loaded onto trucks by loaders and transported to the rear yard for storage. During this process, there was a certain operational drop after the grab bucket falls, which could cause significant dust pollution.

Yard stacking: At present, China's dry bulk cargo terminals mainly adopted open-air storage, with only a small number using fully enclosed or semi-enclosed storage methods such as silos and strip silos [12]. The material piles in open-air storage yards would continuously release a large amount of dust under wind or other external disturbances. Static dust from open-air storage yards was one of the main sources of dust in dry bulk cargo terminals.

In summary, it could be observed that the key dust production processes in the loading and unloading operations of dry bulk cargo terminals mainly included train dumper operations, stacking and reclaiming operations (including truck loading and unloading), ship loading and unloading and operations, and yard stacking.

3. Current Status and Shortcomings of Dust Pollution Control Technology for Dry Bulk Cargo Terminal

The dust pollution control of dry bulk cargo terminals was a complex decision-making process, and the characterization, monitoring, and control of pollution required a series of technical means as support. Overall, research and application in the field of port dust pollution in European and American countries started earlier and gradually formed a system. With the increasing emphasis on port environmental protection, Chinese researchers have also made some effective breakthroughs. However, overall, for the port industry and academic community in China, the understanding of dust generation mechanisms from wind erosion in bulk cargo yards and during loading and unloading operations was still not comprehensive enough, the technical means of comprehensive dust intelligence monitoring in the yard area were still lacking, and the application of intelligent and refined control for the entire process still needed further improvement and promotion.

3.1. Unclear Mechanism-Prominent Shortcomings in Accurately Characterizing and Tracing Dust from Dry Bulk Cargo Terminals

After more than half a century of experimental research, researchers have achieved fruitful results in calculating the static dust emission of particulate matter in storage yards and the dynamic dust emission during loading and unloading operations [13]. Based on on-site measurements and wind tunnel experiments, researchers generally believed that material moisture content, wind speed, and material type were important factors affecting dust generation from wind erosion in the yard and loading and unloading operations. The following consensus had been reached: under the premise of constant moisture content, when the wind speed was greater than the starting wind speed, the amount of dust generated increased exponentially with the wind speed; under the premise of constant wind speed, the amount of dust generated decreased negatively exponentially with the increase of moisture content; the degree of oxidation and coalification of different types of bulk cargo could affect the amount of dust generated to a certain extent. Based on the discovered dust generation patterns of particulate matter, researchers further constructed dust generation calculation models and static and dynamic total dust generation accounting methods. Among them, the AP-42 model released by the US EPA and *the Environmental Impact Assessment Specification for Port Construction Projects* (JTS 105-1-2011) (hereinafter referred to as the "Specification") and *the Technical Guidelines for Compiling Dust Particle Emission Inventory (Trial)* (hereinafter referred to as the "Guidelines") had the most extensive applications [14]. The dust emission calculation formula recommended by the US EPA model only provided a limited range of values for several materials, and key parameters such as ground roughness and friction wind speed were related to material characteristics, therefore, using the EPA calculation formula to estimate the actual dust emissions from ports may have significant limitations. The dust calculation method in the "Specification" was relatively simple, but the influence of local wind farms had not been considered yet, and the applicability of factor localization was relatively poor [15]. The formula recommended in the "Guidelines" was relatively close to the source intensity calculation formula proposed by the US EPA in terms of basic theory and manifestation [16]. This calculation model emphasized more on the source intensity of a dust event and explained and deduced the phenomenon of particulate matter starting and drifting under the action of wind at the micro level [17]. However, due to the complexity of actual environmental and meteorological conditions, it was difficult to provide a suitable range of values for many parameters involved in the formula, such as ground roughness, wind speed conversion at different heights, and particle size coefficient. At the same time, the above parameters were extremely sensitive to the estimation of dust emission, and there were still certain application problems. At the same time, looking at existing research, there were many studies on the dust generation mechanism and laws of static storage of dry bulk particulate matter, but there was relatively little research on dynamic loading and unloading operations, and there was a lack of relevant research on the dust generation laws and estimation models for new specialized equipment loading and unloading processes such as train dumper.

Therefore, it was imperative to deeply explore the influencing factors of dust generation from wind erosion in bulk cargo yards and loading and unloading operations, clarify the dust generation mechanisms under different operating procedures, and develop a locally applicable dust generation calculation model. This could provide theoretical data support for the fine traceability of port dust emissions and also provide a basic guarantee for achieving intelligent control of port dust.

3.2. Insufficient technology - the main weakness affecting the dynamic linkage of dust pollution monitoring and control in dry bulk cargo terminals

Accurately monitoring the dust pollution situation at the dry bulk cargo terminal site was an important foundation for carrying out fine pollution prevention and control and formulating relevant policies. At present, the mainstream dust concentration monitoring methods mainly include the oscillating balance method β Ray method, and the light scattering method. Due to differences in testing mechanisms, the applicability and limitations of existing dust monitoring equipment varied under different port operating conditions. For example, the oscillating balance method was often limited by the working conditions and experimental conditions at the port during use; β ray method had low accuracy in testing high-concentration particles; the accuracy of the light scattering method for testing fine particles was relatively low. To meet the dual needs of environmental protection department supervision and port enterprise self-inspection, many researchers have conducted case studies on online monitoring systems based on port layout, bulk cargo volume, dust generation patterns, etc., mainly including the selection of monitoring points and equipment, development of monitoring plans, and construction of monitoring platforms. However, there was relatively little further research on dust monitoring technology and equipment. The shortcomings of dust monitoring technology seriously hindered the comprehensive intelligent monitoring of port dust, which was a key technical problem that urgently needs to be solved at this stage.

On the basis of achieving distributed grid monitoring of dust, the development of dynamic perception technology for the port environment was an important prerequisite for achieving intelligent and precise control of port dust. At the same time, it was also a strategic direction for accelerating the construction of port intelligence in the face of cutting-edge information technology and major industry development needs.

3.3. Deficiency Application-A Key Bottleneck Hindering Precise Dust Suppression and Intelligent Control of Dry Bulk Cargo Terminal Operations

The existing port dust removal technology and equipment had developed rapidly and could achieve good dust suppression effects in multiple scenarios and different operating conditions to a certain extent. However, at present, the dust removal and suppression work in most ports is still stuck in the manual or semi-automatic stage, and is mostly passive, extensive, and non-standard, resulting in insufficient intelligent application, low work efficiency, high energy consumption, poor applicability, and timeliness. These modes lacked intelligent integration and application of automation, the Internet of things, and information technology, as well as thinking about the mechanism of dust formation, its source, and collaborative control of the entire operation. In addition, there were currently many dust suppression measures used for a single operation link, but there was still insufficient research on the comprehensive dust prevention and suppression system that achieved full coverage of port point sources and surface sources. There was also a lack of a complete and good system between the dust removal and suppression processes, technologies, equipment facilities, management, and systems, resulting in insignificant governance effects. With the continuous development of 5G, big data, Internet of Things Technology, and cloud computing platforms, unmanned monitoring of port dust in the future has become possible.

Therefore, based on intelligent, automated, and information-based methods, constructing a comprehensive collaborative dust control system for the entire operation process of dry bulk cargo terminals was of great significance for supporting the continuous improvement of air quality and the construction of smart and green ports, targeting key links such as unloading, transfer, stacking, and retrieval.

4. Suggestions on Intelligent Control Technology Path for Dust Pollution in Dry Bulk Cargo Terminal

Based on the shortcomings of dust control technology in dry bulk cargo terminals and in combination with the requirements of smart green port construction, an innovative system for intelligent dust control technology in dry bulk cargo terminals was proposed, which included precise characterization, intelligent monitoring, and precise control for all aspects of the site and the entire operation process. The recommended technical path is shown in figure 2.

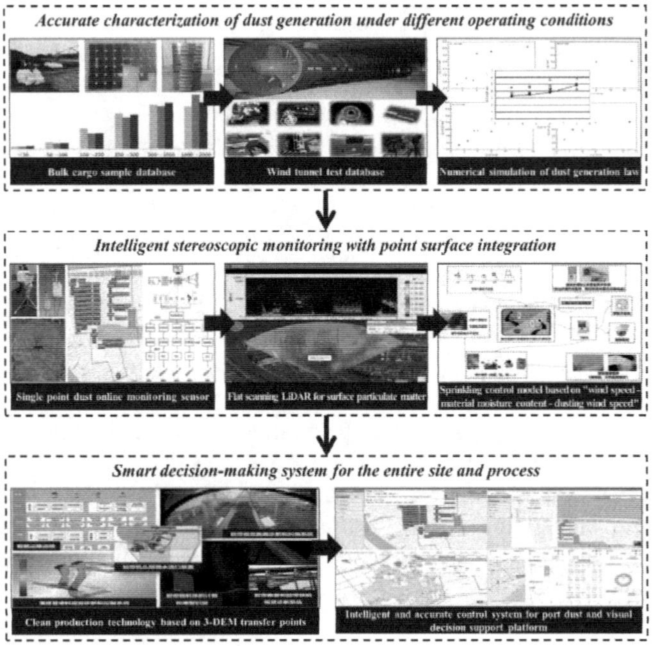

Figure 2. Intelligent control technology path for dust pollution in dry bulk cargo terminals for the entire site area.

4.1. Strengthen the Precise Characterization of Dust Generation under Different Operating Conditions in Dry Bulk Cargo Terminals

In response to the main bottlenecks in current research and refined control of dust pollution in dry bulk cargo terminals, such as the incomplete sample database of bulk cargo types, unclear dust generation rules in different scenarios, weak local adaptability of calculation models, insufficient backtracking ability, and fragmented research, it was recommended to continuously improve the sample database of bulk cargo types in dry bulk terminals, and use the coupling method of "large database + wind tunnel test + numerical model" to systematically analyze the inherent relationship among particle size composition, wind speed, moisture content and dust generation of bulk cargo materials, and reveal the dust generation mechanism of wind erosion (static) and loading and unloading operation (dynamic) in the terminal yard, and further develop localized dust generation calculation models and unorganized emission traceability models for intelligent monitoring, providing algorithm basis and means support for intelligent dust monitoring of dry bulk terminals and environmental impact assessment of related construction projects.

4.2. Improve the Intelligent Three-Dimensional Monitoring of Dust Points and Surfaces in Dry Bulk Cargo Terminals

In response to the shortcomings in improving the dust reduction efficiency and economic benefits of dust suppression water resources in dry bulk cargo terminals, it was recommended to use a combination of "point monitoring + area monitoring" and a new non-contact monitoring technology of "single point dust online monitoring instrument + particle level scanning LiDAR" to continuously improve the intelligent three-dimensional dust monitoring system covering the entire area of dry bulk cargo terminals. Furthermore, based on the fitting algorithm and model of "wind speed - material moisture content - dust removal wind speed", a high-precision spray water dust reduction control strategy for stacking was suggested to be developed to achieve dual precision control of dust removal points and spray water volume, thereby supporting comprehensive dynamic perception and real-time intelligent feedback of port dust pollution. In addition, for new specialized unloading systems such as train dumper, non-contact monitoring technology for the moisture content of bulk cargo in the workshop based on the "moisture - infrared signal" fitting model was suggested to be developed, and a precise watering control strategy for the bottom layer of the workshop guided by environmental and economic benefits improvement could be formulated.

4.3. Build an Intelligent Decision-Making System for Dust Pollution in Dry Bulk Cargo Terminals for the Entire Site and Process

In response to the dust control bottleneck in the process of picking up, transferring, and stacking dry bulk cargo terminals, it was recommended to apply the 3-DEM transfer points technology to focus on breaking through problems such as impact, scattering, wear, blockage, deviation, and tape tearing. Furthermore, a series of clean production technologies such as diagnosis and correction of tape deviation, identification and control of material stacking, visual diagnosis of tape tearing, and diagnosis and control of material blockage were used to fundamentally solve the problem of dust pollution during bulk cargo transportation operations. On this basis, it was suggested to integrate the monitoring data of the whole process dust concentration, bulk cargo moisture content, operating system operation, etc. in the whole site, and develop a decision support platform by using 5G mobile internet, internet of things and dynamic database linking means, realizing the comprehensive perception of port dust, automatic pollution warning, intelligent evaluation and fine control, and supporting the real-time dynamic monitoring of port dust and the intelligent and accurate control decision-making of the whole process.

5. Conclusion

In the current macro context of tightening national and local atmospheric environment control policies, accelerating the construction of green and smart ports, and increasing social attention, the requirements for dust pollution control in dry bulk cargo terminals were also increasing. The control of dust pollution in dry bulk cargo terminals was a complex decision-making process. It was recommended to further construct an intelligent control technology system from three aspects: precise characterization, intelligent monitoring, and precise control, to achieve coordinated control of dust pollution in all aspects of the site and the entire operation process.

6. References

[1] Yin F and Mei S 2023 Global Sustainable Transportation Summit held in Beijing *Journal of Waterway and Harbor* **44(05)** 850.
[2] Xuan H, Liu Z and Zhu L 2020 Analysis of the management issues of port pollutant discharge permits *Journal of Waterway and Harbor* **41(05)** 625-629.
[3] Li M, Liu H and Liu Q 2023 Theoretical and technological progress and prospects of the Yangtze River ecological waterway *Journal of Waterway and Harbor* **44(06)** 917-923.
[4] Yin F and Mei S 2022 Systematic planning for green, low-carbon, and standardized development of transportation *Journal of Waterway and Harbor* **43(04)** 560.
[5] Lin Y, Liu C and Zhang H 2020 Comparison and reference of green port evaluation systems at home and abroad *Journal of Waterway and Harbor* **41(05)** 613-618.
[6] Hao L and Xu H 2023 Research on the evaluation system of green and low-carbon schemes for waterway regulation engineering design *Journal of Waterway and Harbor* **44(05)** 772-779.
[7] Liu S, Gong H and Wang G 2023 Development and application of digital twin system for dry bulk cargo terminals *Journal of Waterway and Harbor* **44(04)** 667-673.
[8] Gu M The impact of dust accumulation in dry bulk cargo terminals on solar heating systems s *Journal of Waterway and Harbor* **42(03)** 370-375.
[9] Zhong W, Lei Z and Fu Z 2023 Current status and prospects of green hydrogen energy engineering technology development in China's offshore areas *Journal of Waterway and Harbor* **44(01)** 124-130.
[10] Sun L and Liu K 2020 Design of intelligent water spraying and dust removal control system for ore terminal yard *Journal of Waterway and Harbor* **41(03)** 360-366.

Journal of Physics: Conference Series **2941** (2025) 012056 doi:10.1088/1742-6596/2941/1/012056

[11] Liu H, Zhang H and Tang H 2023 Research on Ship Black Smoke Control and Supervision *Journal of Waterway and Harbor* **44(03)** 494-498+512.
[12] Hu J, Zhang H and Peng S 2022 Feasibility analysis of satellite remote sensing evaluation of the implementation effect of ship air pollutant emission control zones *Journal of Waterway and Harbor* **43(05)** 660-667.
[13] Cui P, Liu G and Chen Z 2022 Research on risk grading assessment of marine environment pollution from ships in ports *Journal of Waterway and Harbor* **43(05)** 668-676
[14] Liu X, Yang T and Li X 2024 Research on intelligent control technology of spray parameters for coal cutter dust source tracking based on real time monitoring of dust concentration *Coal Mining Machinery* **45(03)** 172-175.
[15] Tarannum N, Rathore N and Natwadiya A 2024 Evaluation of the effects of dust pollution on specific plant species near and around the marble mining site in Rajasthan, India *Environmental Science and Pollution Research* **31(23)** 33515-33529
[16] Wang Z, Zhou W and Jiskani I M 2022 Annual dust pollution characteristics and its prevention and control for environmental protection in surface mines *Science of The Total Environment* **825** 153949
[17] Yin J Lei J and Fan K 2023 Integrating image processing and deep learning for efective analysis and classifcation of dust pollution in mining processes *International Journal of Coal Science & Technology* **10(4)** 17-33

Design and Application of Efficient Ecological Combination Process in Advanced Treatment of Industrial Tailwater

Yang Liu, Weimian Wang, Sushuang Luo, Fangling Li, Wu Lin, Xiaojiang Wang* and Qi Han#

Shenzhen Academy of Environmental Sciences, Shenzhen 518001, Guangdong, China.
Email: yangtze_as@163.com; 5592548@qq.com; 110368773@qq.com; 469071317@qq.com; *110368773@qq.com; #469071317@qq.com

Abstract. In order to solve the advanced treatment problem of industrial wastewater and simultaneously satisfy the needs of ecological and park mode construction, the process of "efficient three-dimensional ecology + constructed wetland" was designed and applied. The technological process, ecological design, operational effect, and technical and economic analysis of the project were systematically described. After stable operation for 12 months, the average removal rates of COD_{cr}, BOD_5, SS, NH_3-N, TN, and TP were 86.13%, 81.19%, 90.86%, 97.97%, 84.03%, and 80.85%, respectively. The effluent of the system could meet the Class III standard requirements of "Environmental Quality Standards for Surface Water" (GB3838-2002), achieving high-quality effluent and resource utilization.
Keywords. Three-dimensional ecology; constructed wetland; advanced treatment; industrial tailwater.

1. Introduction

The designing treatment scale of the wastewater was 0.5 million m^3/d. The designing influent water quality was based on the effluent water quality of the industrial tailwater (as shown in table 1). The effluent water quality should meet the Class III standard of "Environmental Quality Standards for Surface Water" (GB3838-2002) so as to be reused for landscape water replenishment, green irrigation, and road flushing.

Table 1. The designing influent and effluent water quality (mg/L)

Water quality	COD_{cr}	BOD_5	SS	NH_4^+-N	TN	TP
Influent	90	20	60	10	15	0.5
Effluent	≤20	≤4	≤6	≤1	10	≤0.2

2. Process Flow and Designing Parameters

2.1. Process Flow

The main process of advanced wastewater treatment was the combined ecological process of "efficient three-dimensional ecology + constructed wetland", and the process flow diagram is shown in figure 1.

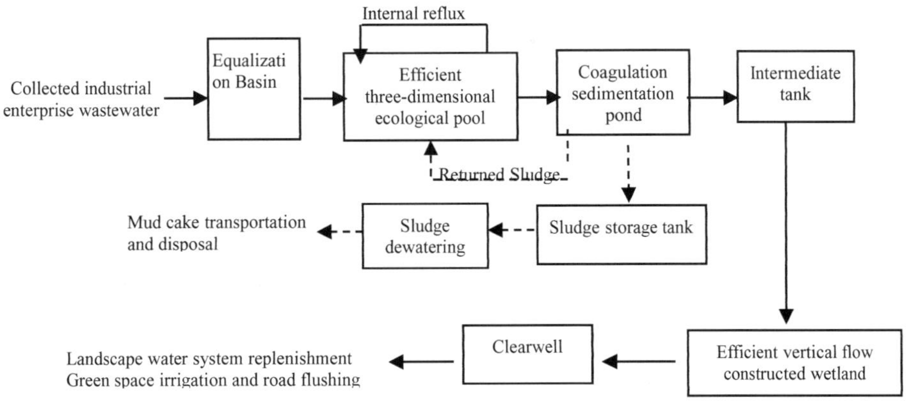

Figure 1. Process flow diagram.

After being collected through the pipeline network, the tailwater of the industrial zone first entered the regulating tank of the treatment system for temporary storage. It was homogenized and evenly distributed in the regulating tank, and then pumped up to the three-dimensional ecological tank. In the three-dimensional ecological tank, anaerobic, anaerobic, and aerobic biochemical reactions were carried out in sections to remove organic matter, ammonia nitrogen, and phosphorus from the wastewater. After the treatment of the three-dimensional ecological pool, the effluent entered the coagulation and sedimentation tank for reaction and sedimentation. The effluent was lifted by a lifting pump to a high-efficiency vertical flow constructed wetland for further treatment. In the constructed wetland, the sewage was further purified through microbial degradation, filtration, physicochemical reactions, plant absorption, and other functions. The effluent was used as landscape water system replenishment and green space irrigation water.

2.2. Plane Layout

The engineering plan was divided into two areas: the three-dimensional ecological treatment area and the constructed wetland. The three-dimensional ecological treatment area included a regulating tank, a three-dimensional ecological tank, a coagulation and sedimentation tank, an intermediate water tank, and an equipment room. The treatment structures and equipment rooms were all underground, with the control room and exhibition hall being located on the ground floor. The areas covered by the three-dimensional ecological treatment area and the artificial wetland were 1300 and 3700 m^2, respectively. The overall layout design of the project is shown in figure 2.

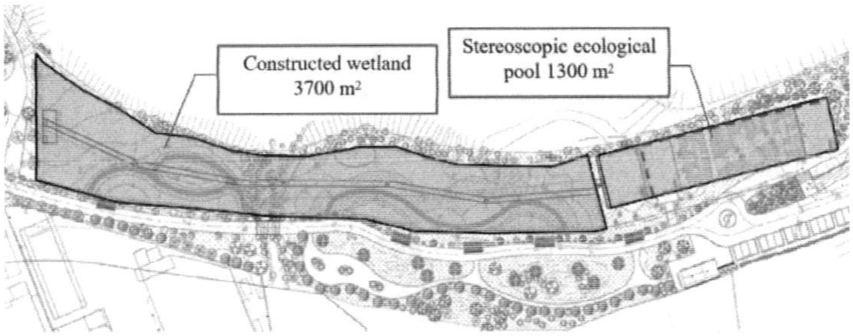

Figure 2. The overall plane layout of the project.

2.3. Design of the Ecological Process

(1) Regulating tank

The tailwater discharged from the industrial area was collected through the pipe network and entered the regulating tank (16.3×8.3×5.2 m). The regulating tank regulated the water quantity and water quality, reducing the impact of the incoming water. The retention time of the regulating tank was designed of 3 hours. Two submersible circulation mixers were installed in the tank, along with online monitoring facilities for COD_{cr}, ammonia nitrogen, total phosphorus, and fluoride.

(2) Efficient three-dimensional ecological pool

The efficient three-dimensional ecological pool could enhance the removal capacity for difficult-to-degrade wastewater with a low B/C ratio in the industrial area. It adopted a horizontal anaerobic, two-stage anoxic, and multi-stage aerobic process. Vertically, it used an upper layer of aquatic plants, a surface deodorization module, and a lower layer of simulated root fill structure, as shown in figure 3.

Anaero Anoxic Anoxic Aerobi Aerobi Aerobi Aerob

Figure 3. The multistage treatment process of a three-dimensional ecological pool.

The retention time of the three-dimensional ecological pond was 13.6 hours, and the ratio of anaerobic, anoxic, and aerobic zone retention time was 1:2:4. It was designed as 2 groups, with each group size of 39.5×8.0×5.2 m. By dividing, segmenting, and zoning horizontally, a series of anaerobic and aerobic reactors were formed to optimize the biological phase in the treatment system, achieving targeted removal of pollutants and concentrating advantageous microbial strains. The anoxic zone was designed in two stages, and the microbial activity of the microorganisms in the plant roots was enhanced to transform difficult-to-degrade organic matter. The coupling fillers made it easier for the wastewater to adsorb and react with the enriched microorganisms, accelerating the hydrolysis and acidification of difficult-to-degrade organic matter, and greatly improving the biodegradability of organic matter. The organic carbon produced by the plant roots provided a carbon source for denitrification, enhancing the removal efficiency of nitrate [1,2]. The aerobic zone was designed with multiple stages, and the microorganisms at different concentration levels fully utilized the metabolism of high, medium, and low-concentration organic pollutants, thereby achieving low-concentration effluent of organic matter. Two low-speed flow pushers were installed in the anaerobic zone; coarse bubble aerators were used in the anoxic zone; and fine bubble aerators were used in the aerobic zone for a total air supply volume of 792 Nm^3/h. Dissolved oxygen online analyzers and nitrate nitrogen online analyzers were installed in the anoxic zone and aerobic zone, respectively, with a nitration liquid reflux ratio controlled at 50-100% [3,4].

The vertical design of the three-dimensional ecological pond used a combination of plant-coupled simulated root fillers. The upper layer was wetland plants, with their roots extending into the biochemical pond water to a depth of 1-2 m. The surface was adsorption fillers made of ceramic particles, which also served as planting carriers for plants. The lower layer was a three-dimensional network of woven polypropylene simulated root fillers, with a filler load of 0.062 $kgBOD_5/m^3$ of filler and a height of 2.5 m. By fixing the plant roots and simulated root fillers that grew in the system, a coupled bio-filler was formed, which utilized its large specific surface area to adsorb and enrich microorganisms. The transfer and release of oxygen in the root zone promoted the degradation and transformation of organic matter and ammonia nitrogen in wastewater and improved the removal efficiency of organic matter and ammonia nitrogen [5]. The plant roots absorbed and utilized nutrients from the sewage, and the root secretions absorbed and converted toxic and harmful substances in industrial wastewater through chelation, adsorption, ion exchange, and other pathways, thereby improving the ecological and safety of the effluent [6,7]. The system used a ceramic particle deodorization module to adsorb, absorb, and degrade the odor gases overflowing from the wastewater, synergistically purifying the odor.

(3) Coagulation sedimentation tank and intermediate water tank

The coagulation sedimentation tank consisted of a coagulation zone and a flocculation zone, with each section having a retention time of about 26 minutes. The sedimentation section used an inclined tube sedimentation tank, with a sedimentation load of 1.08 m^2/m^3h. There were 2 groups of coagulation sedimentation tanks, with each group size of 12.0×8.0×5.0 m. The surface of the coagulation sedimentation tank and the intermediate water tank was covered with soil to a depth of 0.3 m, and then grass was planted to harmonize with the surrounding ecology.

(4) Efficient vertical flow constructed wetland

The high-efficiency vertical flow constructed wetland had a hydraulic load of 0.74 m^3/m^2d and a total area of approximately 3700 m^2. Aquatic plants were planted in combination with various shrubs and herbaceous plants. The depth of the wetland media layer was 2 m, and various fillers were used, including graded gravel fillers (particle size ranging from 2 to 40 mm), natural zeolite (particle size ranging from 4 to 8 mm), and composite functional fillers (particle size ranging from 4 to 8 mm). Natural zeolite fillers were added to the upper layer of the media, which had a high specific adsorption capacity for NH_4^+ and enhanced the nitrification-denitrification reaction. The root cells of the plants absorbed and utilized nitrogen in the form of ammonium salts and nitrates, promoting plant growth and improving the overall efficiency of the constructed wetland system in removing ammonia nitrogen and total nitrogen [8]. An independently developed patented filler, enhanced phosphorus removal composite functional filler, was added to the upper layer of the wetland media. It was made of oyster shells, aluminum sludge from water treatment plants, and fly ash, and had a Ca-Al-Fe active component. The filler had a phosphorus adsorption capacity greater than 3.0 mg/g, with multiple active adsorption sites. Ca^{2+} could exchange ions and precipitate with phosphate ions, and Fe^{3+} and Al^{3+} could form complexes with phosphorus through ligand exchange. The aluminum salts formed hydroxy compounds in the wastewater, which could adsorb and sweep the generated colloidal particles, thereby enhancing the removal of phosphorus from the wastewater [9]. The combined fillers enhanced the physical and chemical reactions, adsorption, and trapping of pollutants in the wetland, thereby greatly improving the removal efficiency of nitrogen and phosphorus pollutants in the wastewater.

(5) Clearwater tank

The clear water tank functioned as a water storage and recycling system, with a retention time of 1 h and a size of 20×4.0×2.5 m. The clear water tank was located below the wetland floor, and the water was pumped up for reuse in landscape lake replenishment, park green irrigation, and road flushing.

(6) Sludge treatment system

The sludge treatment system consisted of a sludge storage tank and sludge dewatering. The sludge concentration time was 48 h, and the sludge dewatering was carried out using a chamber filter press with a filtration area of 20 m^2. The moisture content of the dewatered sludge was about 80%.

3. Operation Effect of the Project

3.1. Operating Mode Description

(1) In the three-dimensional ecological pond, the anaerobic zone and three-stage aerobic zone maintain DO levels at around 0.5-1.0, 1.5, and 2 mg/L, respectively. The nitration liquid's internal reflux ratio

was controlled at 50%-100%. Plant pruning and harvesting were conducted in batches based on growth potential, typically 1-2 times per year.

(2) During the initial adjustment phase of the vertical flow artificial wetland, the wetland operated at a high water level to meet the rapid growth needs of plants. Once normal operation was established, the wetland switched to a low water level. The inflow to the vertical flow constructed wetland was divided into 4 groups, with inflow controlled independently by electric valves, utilizing a pulse intermittent inflow operation.

(3) When the influent quality showed TP < 0.5 mg/L and fluoride < 1 mg/L, coagulation sedimentation tanks were used solely for sedimentation without the addition of chemicals. Phosphorus removal primarily relied on biological phosphorus removal in the three-dimensional ecological pond and enhanced phosphorus removal in the artificial wetland system.

(4) In cases where the influent concentration of TP was ≥ 0.5 mg/L, PAC and PAM were added to the coagulation sedimentation tank for chemical phosphorus removal, reducing the subsequent phosphorus load on the artificial wetland. When the influent fluoride concentration was ≥ 1 mg/L, dedicated defluorination agents and PAM were added sequentially to lower the fluoride concentration to below 1 mg/L through chemical reactions, coagulation, precipitation, and adsorption [10].

3.2. Water Quality Testing Results

The project has been continuously sampled and monitored since it was put into operation for a year. The average water quality data of the main unit of the system are shown in table 2.

Table 2. The water quality results from influent and effluent (mg/L).

Project	**COD_{cr}**	**BOD_5**	**SS**	**NH_3-N**	**TN**	**TP**	**Fluoride**
Inlet	79.50	14.41	42.53	8.61	12.34	0.47	0.79
Effluent of the ecological pool	14.62	3.72	40.15	0.89	4.61	0.45	0.78
Effluent of sedimentation tank	14.51	3.66	9.31	0.87	4.38	0.41	0.72
Effluent of vertical flow constructed Wetland effluent (system effluent)	11.02	2.71	2.97	0.17	2.13	0.09	0.686
Total removal rate	86.13%	81.19%	90.86%	97.97%	84.03%	80.85%	13.16%
Class III standard for surface water	20	4	/	1	/	0.2	1.0

The system's effluent quality had stabilized to meet Class III standards of the "Environmental Quality Standards for Surface Water" (GB3838-2002), with average removal rates of COD_{cr}, BOD_5, SS, NH_3-N, TN, and TP reaching 86.13%, 81.19%, 90.86%, 97.97%, 84.03%, and 80.85% respectively. The influent B/C ratio averaged at 0.18. The three-dimensional ecological system achieved over 80% COD_{cr} removal and close to 90% NH_3-N removal, while the vertical flow constructed wetland could remove an average of 80% of TP from wastewater.

3.3. Technical and Economic Indicators

The total investment in the project was approximately 32 million yuan, with a unit water construction investment of 0.64 yuan/m^3 and an operational cost of 0.342 yuan/m^3, making the project cost-effective. The project covered an area of about 0.5 hectares, utilizing land intensively and efficiently to create a diverse ecological park (as shown in figure 4). It also featured environmentally friendly visiting and exhibition areas, serving as a demonstration of environmental protection education [6-8].

(a) Efficient three-dimensional ecological pool

(b) Vertical flow constructed wetland

Figure 4. Pictures of ecological wastewater treatment combination process.

4. Conclusion

(1) The project adopted a combination of "efficient three-dimensional ecology + constructed wetland" ecological process for deep treatment of industrial tailwater, ensuring effluent quality consistently met the Class III standard of the "Environmental Quality Standards for Surface Water" (GB3838-2002), achieving high-quality effluent and resource utilization.

(2) The three-dimensional ecological system horizontally employed anaerobic-two-stage anoxic-multi-stage aerobic zonation, and vertically utilized aquatic plants-deodorization module-simulated root filling for construction. Even under poor biodegradability of industrial effluent, it could achieve up to 90% NH_3-N removal and over 80% COD_{cr} removal, efficiently eliminating organic matter and nitrogen from wastewater.

(3) The design of the vertical flow constructed wetland optimally combined denitrification and phosphorus removal materials. By incorporating zeolite and a developed Ca-Al-Fe composite functional material, the system enhanced the removal efficiency of nitrogen and phosphorus pollutants in wastewater, with removal rates for NH_3-N and TP reaching 80%.

5. References

[1] Zhong C Q, Xu D C, Zhang J, et al. 2019 Pilot scale research on the advanced treatment of integrated circuit industrial tail water *Industrial Water Treatment* **39(01)** 29-32.

[2] Chen W M, Tang Y Q, Mori K and Wu X L 2012 Distribution of culturable endophytic bacteria in aquatic plants and their potential for bioremediation in polluted waters *Aquat Biol* **15(2)** 99-110.

[3] Jan V, Wei T, Zhao Y Q, et al. 2021 Counting the roles of plants in constructed wetlands for wastewater treatment *China Water & Wastewater* **37(02)** 25-30.

[4] Zhou X S, Lou S, Radnaeva L D, et al. 2022 Advances in heavy metal accumulation characteristics of plants in soil *Asian Journal of Ecotoxicology* **17(3)** 400-410.

[5] Zhao Y H, Hu Z, Hao Z Y, et al. 2024 Revealing the size effect mechanisms of micro(nano)plastics on nitrogen removal performance of constructed wetland, *Journal of Hazardous Materials* **480** 136409.

[6] Truong H B, Nguyen T H, Tran Q B, et al 2024 Algae-constructed wetland integrated system for wastewater treatment: A review *Bioresource Technology* **406** 131003.

[7] Kamilya T, Majumder A, Saidulu D, et al. 2023 Optimization of a continuous hybrid moving bed biofilm reactor and constructed wetland system for the treatment of paracetamol-spiked domestic wastewater *Chemical Engineering Journal* **477** 147139.

[8] Fu J M, Zhao Y Q, Yang Y, et al. 2023 A glance of configuration-operational strategies and intensification of constructed wetland towards land-effective occupation *Journal of Water*

Process Engineering **56** 104473.
[9] Zhang J., Qin Y., Zhang Y., et al. 2023 Comparative assessment of the spatiotemporal dynamics and driving forces of natural and constructed wetlands in arid and semiarid areas of northern China *Land* **12** 1980.
[10] Wang Y., Hao A., Quan Y., et al. 2023 Analysis of the degradation characteristics of chlorpyrifos in an electrochemically constructed wetland coupled system under different pesticide exposure conditions and microbial community analysis *Sustainability* **15** 15958.

Analysis and Suggestion on the Abnormal Concentrations in Influent of a Wastewater Treatment Plant (WWTP) in the South

Xiaojiang Wang, Wu Lin, Weimian Wang, Sushuang Luo, Fangling Li, Yang Liu* and Qi Han*

Shenzhen Academy of Environmental Sciences, Shenzhen 518001, Guangdong, China.
Email: 110368773@qq.com; 5592548@qq.com; 729536934@qq.com; yangtze_as@163.com; 469071317@qq.com; *yangtze_as@163.com; 469071317@qq.com

Abstract. To address the issue of abnormal influent concentration at a wastewater treatment plant in the southern region, this study established 4 key monitoring points along the influent pipeline network. Over a specific period, 12 water quality monitoring activities were carried out to assess the variations in water quality throughout the sewer network, identify the factors contributing to abnormal influent concentration, and propose mitigation measures for the plant. The results revealed that the primary contributors to excessive concentration at the plant were COD and total nitrogen, with the most significant points of excess being #2 and #4. These areas, located in close proximity to the main excessive points, were densely populated by industrial establishments, particularly in sectors such as electronic circuit manufacturing, metal surface treatment, and PCB production, which may present risks of inadequate treatment or unauthorized discharge. Recommendations included optimizing operational parameters at the wastewater treatment plant, enhancing the source-to-sewer rainwater diversion network by the water department, and implementing improved early warning, forecasting, and source tracking measures by the environmental protection department to ensure overall enhancement of the water quality in the city.
Keywords. Wastewater treatment plant; abnormal concentration; influent; effluent.

1. Introduction

In order to solve the problem of black and smelly rivers in the city, Shenzhen has made great efforts to construct and improve the sewage interception network during the 13th Five-Year Plan period. The city has implemented rainwater and sewage diversion, as well as source control actions, diverting all sewage to sewage treatment plants for processing. At the same time, most of the sewage treatment plants have been upgraded and some new ones have been built. These measures have not only improved the sewage treatment capacity but also improved the water quality of the effluent. The high-quality effluent is injected into the rivers, further eliminating the black and smelly phenomenon and protecting the water quality of the rivers.

However, the water environment quality in Shenzhen is still unstable, and there are still some problems in the management work. Among them, there are frequent occurrences of abnormal inflow concentration in sewage treatment plants. Although rainwater and sewage diversion, as well as source control actions, have improved the concentration of water quality in most sewage treatment plants, the routine indicators of inflow water quality in some plants are consistently higher than the design values. The concentration of heavy metals in the inflow water quality of some sewage treatment plants often exceeds the standard, affecting the subsequent treatment effect of biochemical reactions. There are also some sewage treatment plants where the inflow water quality concentration consistently falls below the design values, which affects the normal operation and treatment efficiency of the plants. These phenomena impact the effectiveness of the treatment process within the plants, and some plants even struggle to maintain stable compliance with pollutant standards. This poses significant challenges to the environmental benefits and regulatory efforts of municipal sewage treatment plants.

This study aims to examine the current status of abnormal inflow concentration at a specific sewage treatment plant. It involves tracking and monitoring the water quality of the sewage network, analyzing the reasons behind abnormal inflow concentration in typical sewage treatment plants, and proposing appropriate solutions and recommendations. The goal is to offer technical assistance in maintaining the stable operation of sewage treatment plants, enhancing pollution removal efficiency, and achieving standard discharge levels.

2. Basic Information of the Sewage Treatment Plant

The design scale of the sewage treatment plant is 350,000 tons/day, and the design effluent standard is Grade IV surface water standard. The main treatment processes are multi-stage enhanced AAO process and magnetic coagulation, with an average MLSS of 4000 mg/L and an air-water ratio (average) of 6:1. The specific process flow is shown in figure 1.

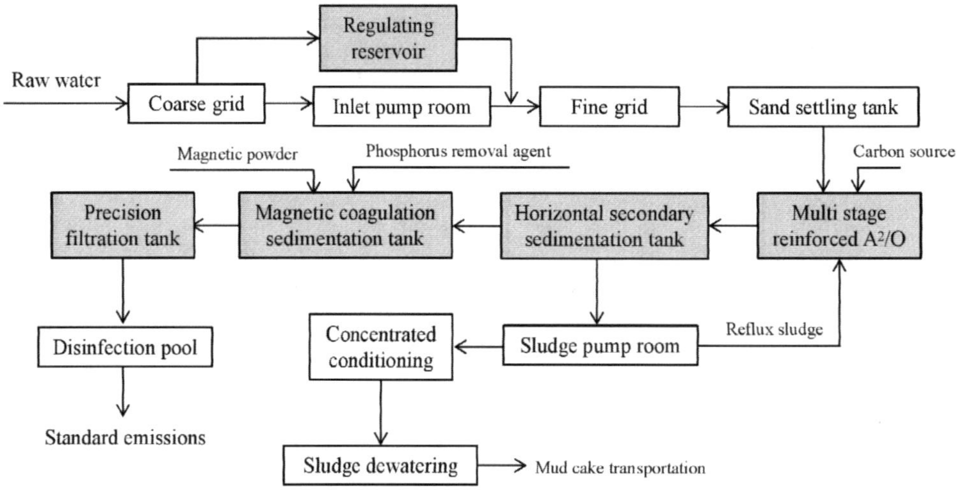

Figure 1. Process flow diagram of the sewage treatment plant.

The main indicators of the inflow and outflow water of the sewage treatment plant are shown in table 1.

Table 1. The designing of influent and effluent water quality (mg/L).

Water quality	COD_{cr} (mg/L)	BOD_5 (mg/L)	SS (mg/L)	TN (mg/L)	TP (mg/L)	NH_4^+-N (mg/L)	Fecal coliform count (CFU/L)	pH	Anionic surfactant
Influent	280	150	200	45	6	35	/	/	/
Effluent	30	6	10	15	0.3	1.5	1000	6-9	0.3

3. Arrangement of Monitoring Points

Based on the construction of the sewage network supporting the sewage treatment plant, the flow direction of the incoming water network, the location of the convergence points, and the flow direction of sewage and wastewater inside and outside the plant have been determined. Combined with the distribution of pollution sources in the area, 4 monitoring points (1#, 2#, 3#, and 4#) have been set up at the pollution source convergence points in industrial areas, residential areas, commercial/educational areas, and pumping stations.

4. Water Quality Monitoring Data Analysis

The main water quality monitoring indicators were selected based on the changes in the inflow water quality of the sewage treatment plant and the characteristics of industrial pollutants in the area. These indicators include pH, total chlorine, BOD_5, COD_{cr}, SS, ammonia nitrogen, total nitrogen, total phosphorus, fluoride, total nickel, total chromium, total lead, and total copper. A total of 6 monitoring sessions were conducted over a period of 2 months. The monitoring results of each water quality indicator are shown in figure 2.

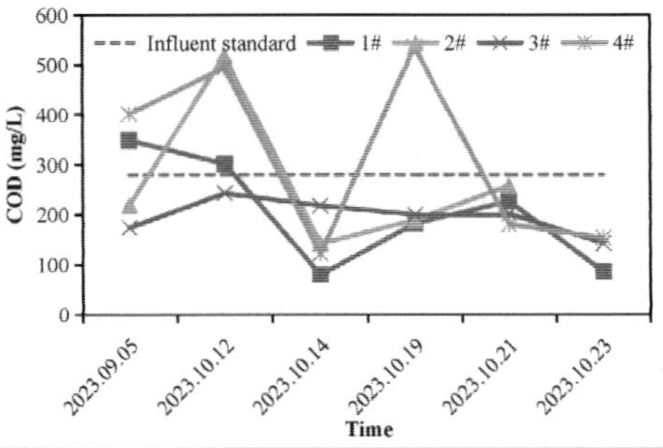

(a) COD

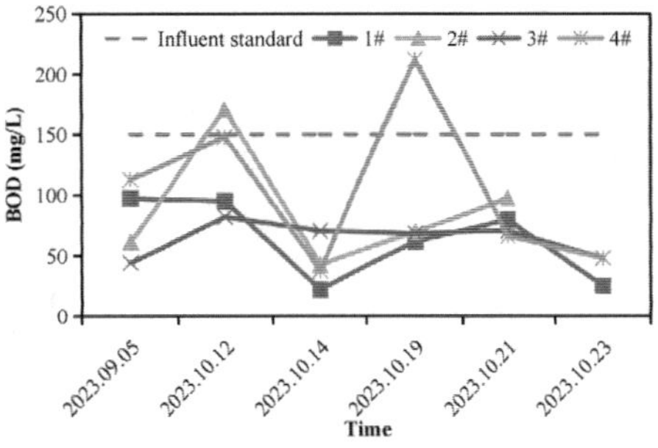

(b) BOD

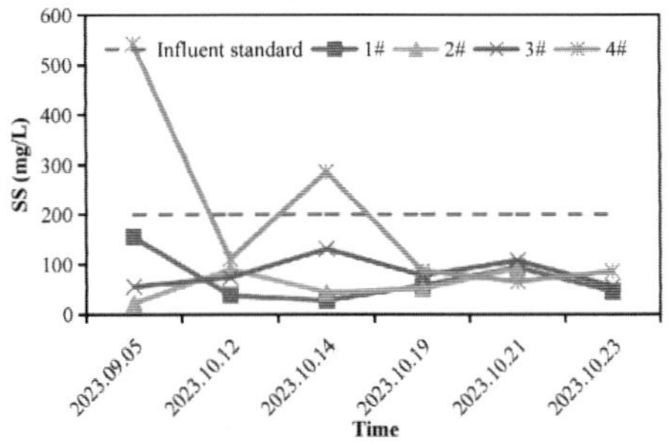

(c) SS

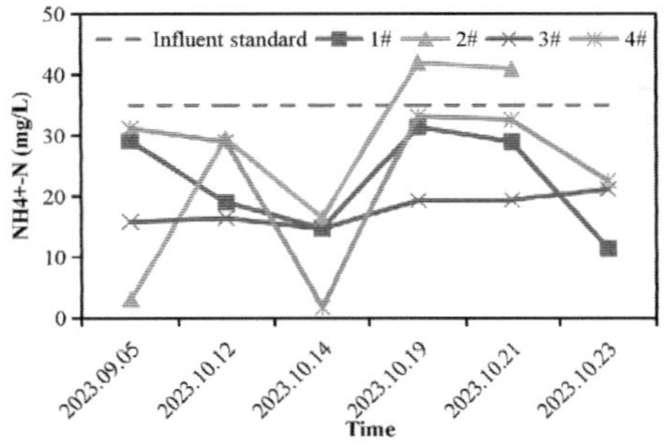

(d) NH_4^+-N

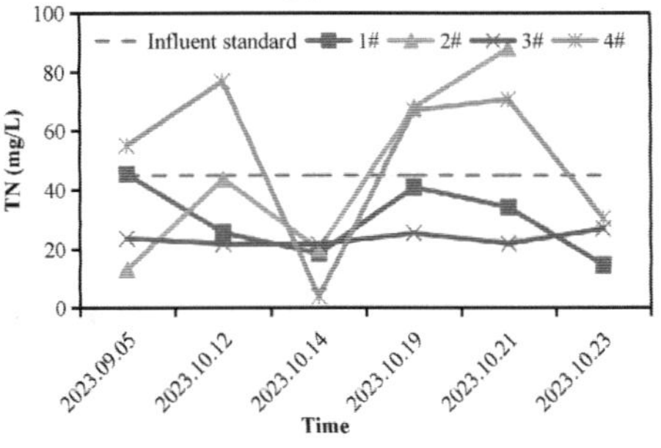

(e) TN

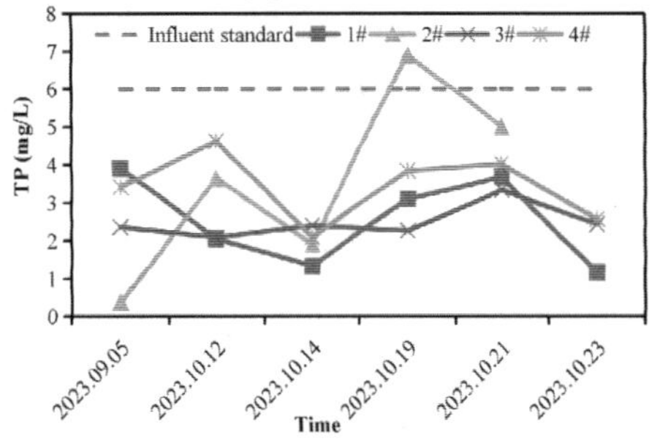

(f) TP

Journal of Physics: Conference Series **2941** (2025) 012058 doi:10.1088/1742-6596/2941/1/012058

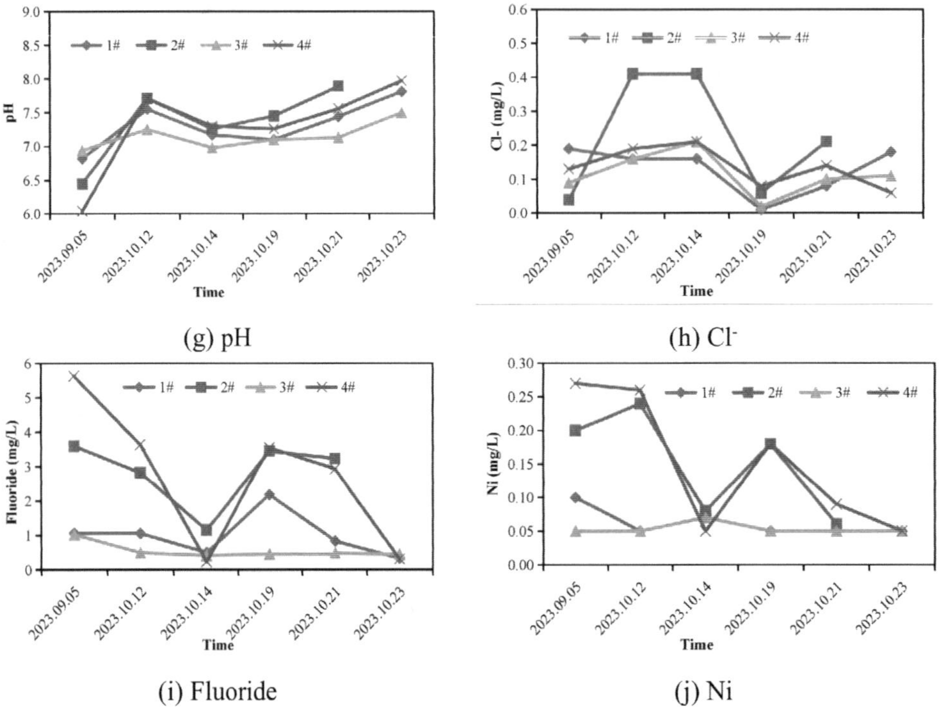

Figure 2. Analysis of water quality monitoring in the sewage treatment plant.

According to figure 2, the range of each indicator in the tracing water quality monitoring of the sewage treatment plant is as follows: pH ranges from 6.04 to 7.97, total chlorine ranges from 0.02 to 0.41 mg/L, BOD_5 ranges from 22.3 to 212 mg/L, COD ranges from 80 to 518 mg/L, SS ranges from 23 to 542 mg/L, ammonia nitrogen ranges from 1.71 to 42.0 mg/L, total nitrogen ranges from 13.3 to 88.1 mg/L, total phosphorus ranges from 0.37 to 6.90 mg/L, fluoride ranges from 0.330 to 5.63 mg/L, total nickel ranges from 0.06 to 0.27 mg/L, total chromium ranges from 0.04 to 0.14 mg/L, total lead is basically undetectable, and total copper ranges from 0.06 to 0.24 mg/L.

Compared with the inflow standards of the sewage treatment plant, the main factors that exceeded the standard in this tracing water quality test are COD and total nitrogen. Specifically, COD exceeded the standard 6 times, with a 20% probability of exceedance. The highest concentration recorded was 518 mg/L, exceeding the standard by 1.85 times. The main exceedance points were 1# and 3#. The total nitrogen exceeded the standard 7 times, with an exceedance probability of 23.3%, the highest concentration was 88.1 mg/L, and the exceedance multiple was 1.96 times. The main exceedance points were 2# and 4#.

5. Analysis of the Causes of Abnormal Inflow Concentration

The main exceedance points of the sewage treatment plant are 2# and 4#. According to the distribution of key pollution sources near these points (as shown in figure 3), it can be observed that there are many industrial enterprises near these two points, including over a dozen sewage enterprises primarily involved in electronic circuit manufacturing, metal surface treatment, and circuit board industries (marked in red and blue). Therefore, it is determined that the abnormal inflow concentration of the plant is caused by industrial source pollution. If the wastewater treatment of industrial enterprises does not meet the standards or if there are violations of illegal discharge, it will directly impact the water quality at these two points, potentially resulting in abnormal inflow concentration in the water plant.

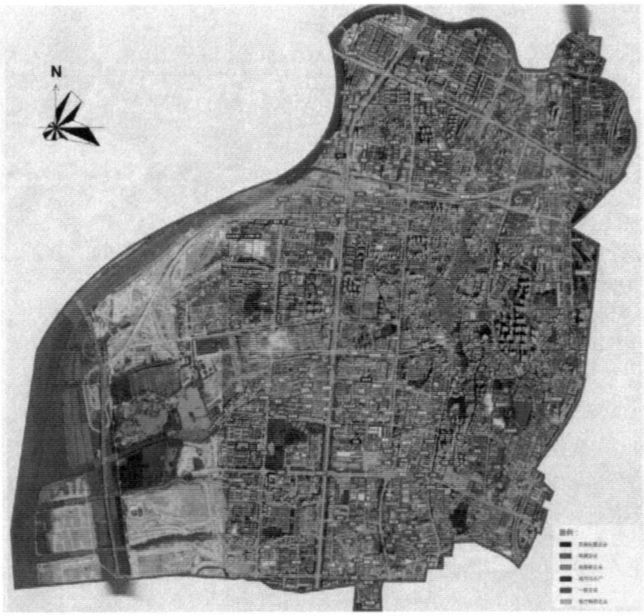

Figure 3. Distribution map of key pollution source enterprises.

6. Countermeasures and Suggestions for Abnormal Inflow Concentration

The sewage treatment plant should adjust its emergency plan and develop standardized emergency operating procedures based on the concentration and frequency of past exceedances of inflow concentration in the area covered by the plant. By adjusting the operating parameters, the impact of toxic and harmful substances on the system can be minimized, and the stability and compliance of the effluent can be maximized.

(1) For the sewage treatment plant, measures to be taken for possible inflow concentration exceedance include adjusting operating parameters, reporting to the water and environmental authorities in a timely manner, and installing online warning systems at the inflow end of the water purification plant.

(2) For the water authorities, measures to be taken for possible inflow concentration exceedance include continuous construction of rainwater and sewage diversion drainage pipe networks, source control reconstruction, along-river diversion system operation survey and overflow pollution control special actions, and implementation of more comprehensive sewage scheduling projects.

(3) For the ecological environment authorities, measures to be taken for possible inflow concentration exceedance include the use of high-tech tracing methods, such as the establishment of a water fingerprint database and monitoring and warning tracing systems, promotion of third-party evaluation of pollution treatment facility efficiency, implementation of expert full-process verification mechanisms, and improvement of industrial source management mechanisms and systems.

7. Conclusion

(1) After conducting a 2-month tracing of water quality sampling and monitoring at the sewage treatment plant, the analysis results show that the main factors contributing to exceedances are COD and total nitrogen. The main points of exceedance are found at 2# and 4#.

(2) The analysis of the causes of abnormal inflow concentration shows that there are dense industrial enterprises near the main exceedance points of the sewage treatment plant, mainly engaged in electronic circuit manufacturing, metal surface treatment, and circuit board industries, which may pose risks of non-compliant treatment or illegal discharge.

(3) For sewage treatment plants experiencing abnormal inflow concentrations, it is recommended to implement measures that involve the operation and management of the plant, as well as

collaboration with water authorities and ecological environment authorities. These actions aim to comprehensively enhance the treatment efficiency of the sewage treatment plant and improve the overall quality of the city's water environment.

Acknowledgment
This research was funded by the Shenzhen Science and Technology Program (No. JCYJ20220531091403007).

8. References

[1] Wang Y, Yuan Y, Xue H, et al. 2024 Analysis on operation and water quality characteristics of centralized wastewater treatment plants of industrial parks in yellow river basin *China Water* **16** 806.
[2] Zhang W., Zhao J., Quan P., et al. 2023 Prediction of influent wastewater quality based on wavelet transform and residual LSTM *Applied Soft Computing* **148** 110858.
[3] Gaona À., Solís B., Guerrero J., et al. 2023 Nitrite pathway in A_2/O WWTPs: Modelling organic matter reduction, operational cost and N_2O emissions *Journal of Cleaner Production* **414** 137453.
[4] Andreides M., Dolejš P., Bartáček J. 2022 The prediction of WWTP influent characteristics: Good practices and challenges *Journal of Water Process Engineering* **49** 103009.
[5] Lu Q., Mao J., Xia H., et al. 2022 Effect of wastewater treatment plant discharge on the bacterial community in a receiving river *Ecotoxicology and Environmental Safety* **239** 113641.
[6] Wang Q, Wang M,Yang Q X, et al. 2025 The role of bacteriophages in facilitating the horizontal transfer of antibiotic resistance genes in municipal wastewater treatment plants *Water Research* **268(B)** 122776.
[7] Mao Z G, Li X Q, Zhang X, et al. 2024 Optimization of effluent quality and energy consumption of aeration process in wastewater treatment plants using artificial intelligence *Journal of Water Process Engineering* **63** 105384.
[8] Wen Y, Wang J Q, Wang F, et al. 2024 Recent advances in membranes modified with plant polyphenols in wastewater treatment: A review *Separation and Purification Technology* **334** 125861.
[9] Sim J, Lee J, Rho H, et al. 2023 A review of semiconductor wastewater treatment processes: Current status, challenges, and future trends *Journal of Cleaner Production* **429** 139570.
[10] Shi Y P, Liu T, Quan X, et al. 2023 Enhanced nitrogen removal in the upgrading of municipal wastewater treatment plants by using zero-valent iron-modified biofilm carriers and clinoptilolite-modified biofilm carriers *Chemical Engineering Journal* **477** 146948.

Ecological Stoichiometry of Carbon, Nitrogen and Phosphorus in Coffee Leaves and Soil from Different Origins in Yunnan Province

Yu Lv[1], Qianrui Huang[1,3,*], Shirong Meng[1], Xiaohua Zhou[2,3], Mingxing Hou[2,3],Youxian Bao[2,3], and Duoju Zhu[1]

[1] Insititute of Resource and Environmental, Yunnan Land and Resources Vocational College, Kunming 2652501, Yunnan, China.
[2] Central Laboratory of Yunnan Geological Survey and Development Bureau, Kunming 650200, Yunnan, China.
[3] Key Laboratory of Sanjiang Mineralization Resource Exploration and Utilization, Ministry of Natural Resources, Kunming 650200, Yunnan, China.
Email: 724385973@qq.com; *hqr0423@foxmail.com; 523419358@qq.com; 48149771@qq.com; houmingxing087@163.com; 980985683@qq.com; 1512510573@qq.com

Abstract. This study examines the ecological stoichiometry of carbon(C), nitrogen(N) and phosphorus(P) in coffee from Baoshan, Pu'er, and Xishuangbanna in Yunnan. The results show that there are significant differences in soil carbon (C) content and C:N ratios among different regions, and the coffee leaves from Pu'er have the highest carbon content.The threshold analysis indicates that the growth of coffee in Pu'er is limited by both nitrogen and phosphorus, while the growth of coffee in Baoshan and Xishuangbanna is limited by phosphorus.The correlation analysis shows strong positive correlations between soil carbon and nitrogen, as well as between leaf carbon and soil nitrogen.Soil management should focus on N and P fertilizers for Pu'er, and P fertilizer for Baoshan and Xishuangbanna to enhance coffee growth and yield.
Keywords. Ecological stoichiometry; coffee; leaves; soil.

1. Introduction

Coffee is a perennial economic crop belonging to the Rubiaceae family (C. arabica L.), characterized by its rapid growth, high yield, and high value, with a wide range of sales channels. It is known as one of the world's three major beverage crops along with tea and cocoa [1], and possesses extensive industrial value and social benefits. Yunnan is the earliest province in mainland China to introduce coffee cultivation and is currently the main coffee-producing area in China, constituting over 98% of the national total.It boasts abundant high-quality natural resources. By 2021, the coffee planting area in Yunnan Province had reached 92,860 hectares, with a production and output value of 108,700 tons and 2.643 billion yuan, respectively [2], making it an important industry in the tropical regions of Yunnan. Some scholars have found that the history of coffee in Yunnan can be traced back to 1884. Due to Yunnan's excellent geographical and climatic resources, the protection and utilization of coffee resources in this area have become a research field of great concern to many scholars.

Studies have shown that the stoichiometry of nutrients in the soil significantly affects the nutrient stoichiometry of plant leaves, primarily through the soil's physical and chemical properties and the availability of nutrients [3]. Currently, research in the field of coffee has mostly focused on the commercial quality and cup quality of coffee beans, as well as the impact of altitude on the chemical quality of coffee. [4] The research by Chen Xiao and others found that the levels of nitrogen (N) and phosphorus (P) in coffee leaves are directly proportional to the age of the tree,while it is inversely proportional to the carbon (C) content, changing as the tree age increases [5,6]. These findings have important practical significance for guiding the cultivation and fertilization of coffee. This study focuses on three major coffee-producing areas in Yunnan Province, and by measuring and analyzing the carbon(C), nitrogen(N) and phosphorus(P) content in the coffee leaves and in the soil, aims to provide scientific data for the nutrient balance and ecological stoichiometric features of Yunnan coffee soil, thus providing a reference for managing the fertility of coffee soil.

2. Experimental Materials and Research Methods

2.1. The Collection of Samples

In January and February 2024, We collected samples from three main coffee-producing areas in Yunnan Province: Lujiangba New Village Base in Baoshan City (altitude 900-1100 m), Simao District in Pu'er City (altitude 1100-1200 m), and Jinghong City in Xishuangbanna Prefecture (altitude 500-600 m). Each main producing area had three standard plots, and each plot had three 10m×10m quadrats, which served as three biological replicates, and the "S" shaped sampling method was adopted. (table 1 for meteorological elements of the sample production areas)

Table 1. Climate conditions where the coffee samples were collected.

Sampling position	Climatic conditions						
	Annual solar radiation amount / (kcal /cm^2)	The average annual temperature is /°C	Low-temperature extreme value /°C	Yearly precipitation /mm	Annual sunshine hours /h	Relative humidity /%	Average wind speed /(m.s^{-1})
Lujiang Dam, Baoshan City	133.4	21.5	0.2	751.4	2333.7	70	1.3
Simao District, Pu'er City	129.8	17.8	-2.5	1514.1	2131.3	82	1.1
Jinghong City, Xishuangbanna Prefecture	128	24.3	9.3	627.5	1966.2	83	1

Simultaneously with the procurement of soil specimens, botanical samples were also gathered. Specifically, from the identical coffee tree from which the soil specimens were extracted, a collection of 30 non-diseased, fully-ripened leaves was conducted, spanning both the periphery and the interior of the tree's canopy across the cardinal points—east, south, west, and north. This collection yielded a total of 27 leaf specimens. These specimens were carefully enclosed in plastic bags for preservation and transportation back to the lab. [7]

2.2. Determination Items and Methods

The soil and coffee leaf specimens that were gathered were subjected to laboratory analysis to ascertain the levels of carbon, nitrogen, and phosphorus. The soil organic carbon (SOC) and the carbon content within the plant material were quantified employing a high-temperature digestion technique with potassium dichromate, utilizing an HH-S model thermostatic oil bath for the procedure; soil TN was determined using the automatic nitrogen determinator method; plant N was determined using the H_2SO_4-H_2O_2 digestion and Nessler's colorimetric method; soil TP was determined using the $HClO_4$-H_2SO_4 fusion-colorimetric method, and P in plants was determined using the H_2SO_4-H_2O_2

Journal of Physics: Conference Series **2941** (2025) 012059 doi:10.1088/1742-6596/2941/1/012059

digestion technique coupled with the molybdenum-antimony anti-colorimetric method for sample analysis,with definite methodological approaches referring to "Methods of Soil Agricultural Chemical Analysis" [8].

2.3. Statistical Data Analysis and Interpretation

For the comprehensive data, SPSS 21.0, a highly regarded and industry-standard tool in the field of data processing, was deployed. Specifically, with regard to the stoichiometric relationships, the ratios of C:N, C:P, and N:P in both the soil samples and the leaves were meticulously calculated and presented in the form of mass ratios, thereby providing a clear and quantifiable representation of the elemental compositions and their interconnections within the studied ecological context [9].

3. Results and Analysis

3.1. Contents of Soil Carbon (C), Nitrogen (N), and Phosphorus (P) along with Their Stoichiometric Features across Diverse Geographical Origins

As shown in table 2, the quantified levels of soil carbon (C), nitrogen (N), and phosphorus (P) were precisely determined to be 7.20-11.55, 0.94-1.42, and 0.56-0.74 g/kg, respectively. The carbon (C) stock in the soil of Baoshan coffee was significantly lower than that of Xishuangbanna coffee ($P<0.05$), while upon meticulous statistical analysis, no statistically significant difference was discerned in carbon (C) content between Baoshan coffee soil and Pu'er coffee soil, or between Pu'er coffee soil and Xishuangbanna coffee soil ($P>0.05$). Upon meticulous statistical analysis, no statistically significant difference was discerned in N and P content among the soils of Baoshan coffee, Pu'er coffee, and Xishuangbanna coffee ($P>0.05$). [10]

The stoichiometric ratios of carbon (C) to nitrogen (N), carbon (C) to phosphorus (P), and nitrogen (N) to phosphorus (P) within the soil matrix were meticulously analyzed and monitored were 5.96-11.61, 13.74-20.17, and 1.61-2.81, respectively. The C:N ratio in the soil of Baoshan coffee was significantly lower than that in the soils of Xishuangbanna and Pu'er coffee ($P<0.05$). Upon meticulous statistical analysis, no statistically significant difference was discerned in C:P among the soils of Baoshan coffee, Pu'er coffee, and Xishuangbanna coffee ($P>0.05$). The N:P ratio in the soil of Baoshan coffee was significantly higher than that in the soil of Pu'er coffee ($P<0.05$).

Table 2. Carbon (C), nitrogen (N) and phosphorus (P) content and ecological stoichiometric ratios of coffee soil in different regions.

Place of production	C (g/kg)	N (g/kg)	P (g/kg)	C:N	C:P	N:P
Lujiang Dam, Baoshan City	7.20±3.01Ab	1.27±0.41Aa	0.56±0.25Aa	5.96±2.78Ab	13.74±4.88Aa	2.81±1.62Aa
Simao District, Pu'er City	9.43±6.07Aab	0.94±0.62Aa	0.58±0.29Aa	11.56±5.20Aa	16.71±8.35Aa	1.61±0.87Ab
Jinghong City, Xishuangbanna Prefecture	11.55±5.76Aa	1.42±0.90Aa	0.74±0.42Aa	11.61±9.54Aa	20.17±16.82Aa	2.24±1.64Aab

3.2. Leaf Carbon (C), Nitrogen (N), and Phosphorus (P)Content and Stoichiometric Characteristics of Coffee from Different Origins.

Table 3 illustrates the concentration ranges of carbon (C), nitrogen (N), and phosphorus (P) in coffee, which are 79.08 to 119.41, 7.42 to 8.99, and 0.45 to 0.52 g/kg, respectively. The carbon content in the leaves from Pu'er coffee regions markedly exceeds that found in leaves from Baoshan and Xishuangbanna, with highly significant variations observed across the three regions (p-value < 0.01). The nitrogen content in Baoshan coffee leaves is notably higher compared to those from Pu'er and Xishuangbanna, indicating a significant difference (p-value < 0.05). However, no significant variation is detected in phosphorus content among leaves from the three coffee-growing regions (p-value > 0.05).

The stoichiometric ratios of C:N, C:P, and N:P in the leaves are within the ranges of 8.88 to 16.02, 152.88 to 257.26, and 15.98 to 17.57 /kg, respectively. The C:N ratio in Pu'er coffee leaves is considerably higher than in leaves from Baoshan and Xishuangbanna, with extremely significant differences among the three regions (p-value < 0.01). No significant differences are found in the C:P and N:P ratios across leaves from Baoshan, Pu'er, and Xishuangbanna (p-value > 0.05).

Table 3. carbon(C), nitrogen(N) and phosphorus(P) content and ecological stoichiometric ratios of coffee leaves from different origins.

Place of production	C (g/kg)	N (g/kg)	P (g/kg)	C:N	C:P	N:P
Lujiang Dam, Baoshan City	79.08±5.26Cc	8.99±0.86 Aa	0.52±0.051 Aa	8.88±1.39Cc	152.88±9.07Aa	17.57±3.55 Aa
Simao District, Pu'er City	119.41±4.20 Aa	7.48±0.45 Ab	0.48±0.08Aa	16.02±1.22 Aa	257.26±67.63 Aa	15.98±3.45 Aa
Jinghong City, Xishuangbanna Prefecture	89.44±1.66Bb	7.42±1.42 Ab	0.45±0.16Aa	12.36±2.00 Bb	221.29±83.97 Aa	17.54±4.26 Aa

Note: Distinct lowercase letters within the same column signify statistically significant variations in coffee attributes across various regions at the 0.05 significance level, while uppercase letters within the same column denote significant disparities in coffee characteristics among different regions.(P <0.01).

3.3. Coffee Soil and Leaf C, N, P Content and Stoichiometric Characteristics

As shown in table 4, the correlation analysis results indicate that coffee soil C is extremely positively correlated with N, N with P, P with C:P, N:P, C:N with C:P, and C:P with N:P (P < 0.01). As shown in table 5, the correlation analysis results indicate that coffee leaf C is extremely positively correlated with C:N and C:N with C:P (P < 0.01). Among these, the correlation coefficient between C and C:N is the highest, reaching 0.879, indicating a relatively high - level correlative relationship between them.

As depicted in Table 6, the correlation analysis between the coffee leaf and soil parameters reveals that there is a highly significant positive association between leaf carbon (C) and soil nitrogen (N), as well as between leaf nitrogen (N) and soil carbon to phosphorus (C:P) ratio (p-value < 0.01). The correlation dynamics among the remaining nutrient elements and their stoichiometric ratios in both soil and leaves are influenced by the inter-element correlation relationships.

Table 4. Pearson correlation analysis between C, N, P content and stoichiometric ratios of coffee soil.

Soil	SOC	TN	TP	C:N	C:P	N:P
SOC	1					
TN	0.471**	1				
TP	0.351*	0.417**	1			
C:N	-0.177	-0.391**	-0.078	1		
C:P	0.059	-0.037	0.435**	0.653**	1	
N:P	-0.071	0.196	0.692**	0.056	0.637**	1

Table 5. Pearson correlation analysis between C, N, and P contents and stoichiometric ratios of coffee leaves.

Blade	C	N	P	C:N	C:P	N:P
C	1					
N	-0.362	1				
P	-0.022	0.589*	1			
C:N	0.879**	-0.754**	-0.36	1		
C:P	0.512	-0.67**	-0.837**	0.729**	1	
N:P	-0.218	-0.106	-0.845**	-0.042	0.637*	1

Table 6. The correlation of C, N, P and measurement ratios between coffee leaves and soil.

Soil	Blade					
	C	N	P	C:N	C:P	N:P
SOC	-0.196	0.533*	0.314	-0.402	-0.435	-0.116
TN	0.672**	-0.064	-0.094	0.521*	0.458	0.102
TP	-0.092	-0.455	-0.347	0.177	0.242	0.160
C:N	-0.641**	0.335	0.259	-0.646**	-0.587*	-0.176
C:P	-0.292	0.642**	0.345	-0.539*	-0.467	-0.06
N:P	0.603*	0.176	-0.035	0.342	0.365	0.165

Note(Table4-6: The symbol "*" denotes a correlation that is statistically significant at the 0.05 level, whereas "**" signifies a correlation that is highly significant at the 0.01 level, as determined by the Pearson correlation method.

4. Discussion and Conclusion

4.1. Soil Nutrient Elements and Their Correlation in Yunnan Coffee-Growing Regions

Soil serves as the foundation for plant growth, and the contents of C, N, and P elements directly influence soil fertility and coffee growth. In the three coffee - growing regions of Yunnan Province, the average soil C content is 9.78g/kg, which is lower than the national average. The average N content is 1.21g/kg, higher than the national average, and the average P content is 0.66g/kg, also higher than the national average. There is a significant positive correlation among C, N, and P elements in the soil, indicating that these nutrient elements interact with each other in the soil cycle [11,12].

In the coffee cultivation areas of Yunnan Province, the soil's C:N ratio is found to be below the national mean, which is indicative of a slower decomposition rate of organic matter. The C:P ratio is significantly lower than the national average, pointing towards an enhanced phosphorus availability in the soil [13]. The N:P ratio also falls below the national average, suggesting that the soil has a more robust capacity to provide nutrients to plants. Notably, the elevated C:N ratio observed in the coffee soils of Pu'er and Xishuangbanna regions suggests a higher accumulation of organic carbon, whereas the increased N:P ratio in the coffee soils of Baoshan indicates a reduced availability of phosphorus [14].

4.2. Nutrient Utilization and Growth Limitations in Yunnan Coffee Cultivation

This study found no significant correlation between carbon and nitrogen content in coffee leaves, indicating that coffee employs a nutrient utilization strategy distinct from other plants. The average carbon content in leaves from three coffee-growing regions in Yunnan Province is 101.35g/kg, lower than the global average of 462g/kg; the nitrogen content averages 7.76g/kg, below the global average of 12.20mg/g [15]; and the phosphorus content averages 0.48g/kg, below the global average of

1.8g/kg. These lower concentrations may be related to insufficient soil nutrients [16]. The highest carbon content in leaves from Pu'er coffee suggests slower growth and photosynthetic efficiency [17]. The C:N and C:P ratios in leaves are signs of a plant's carbon fixation capacity and nutrient utilization efficiency quotient Pu'er coffee has a significantly higher C:N ratio compared to other regions, while Baoshan coffee has the lowest. The lower nitrogen content in Pu'er coffee leaves may be a strategy to enhance nutrient use efficiency [18]. The N:P ratio is key to assessing the growth-limiting factors for plants [19]; the N:P ratio of Pu'er coffee indicates dual limitations by nitrogen and phosphorus, whereas Baoshan and Xishuangbanna coffee are primarily limited by phosphorus. Therefore, the Pu'er region should increase the application of nitrogen and phosphorus fertilizers, while the Baoshan and Xishuangbanna regions should focus on phosphorus fertilizer application to promote coffee growth.

4.3. Soil-Leaf Nutrient Correlations and Fertilization Impacts on Coffee Plant Dynamics

This cycle is related to leaves and soil. Here, soil organic carbon is positively correlated with leaf nitrogen, which is likely due to the enhanced soil fertility promoting plant growth and nitrogen accumulation in leaves. However, in contrast to most studies which show no correlation between leaf and soil C:N:P ratios, this study found a significant negative correlation between leaf and soil C:N ratios. Further research is required on how varying fertilization, particularly of nitrogen and phosphorus, affects the dynamics of coffee plants and soil.Acknowledgments

This paper was funded by Scientific and technological innovation project of Study on ecogeochemical evaluation system in Plateau specialty coffee producing area of Yunnan Province and Science and technology innovation team fund project of Yunnan Land and Resources Vocational College. (Ecological Environment Geological Science and Technology Innovation Team2021KJT004).

5. References

[1] Li R 2017 Ecological coffee cultivation management techniques in Pu'er County *Southern Agriculture* **11(29)** 11-12.

[2] Huang J X, Li W R, Xia B, et al. 2022 Preliminary exploration of high-quality development strategies for Yunnan coffee industry *Tropical Agricultural Science and Technology* **45(03)** 21-29.

[3] Tian L M, Peng Y, Ke D, Li H, Chen L, Zhang C and Sen L 2020 The effect of Yang Yan Qing E Wan on senescent phenotypes and the expression of β-catenin and p16INK4a in human skin fibroblasts *J Tissue Viability.* **29(4)** 354-358.

[4] Zhang Y, Cui D, Yang Y, Liu H, Yang H and Zhao Y 2021 Ecological stoichiometric characteristics and influencing factors of carbon, nitrogen, and phosphorus in the leaves of Sophora alopecuroides L. in the Yili River Valley, Xinjiang *PeerJ* **9** e11701.

[5] Niu Z, Zhang Y, Han J, Zhao Y, Zhu X and He P 2024 The Soil Ecological stoichiometry characteristics of the highest latitude areas in the main tea-producing regions of China *Agronomy* **14(7)** 1359.

[6] Chen X, Cui X, Luo Y, et al. 2023 Ecological stoichiometric study of C, N, P in coffee leaves and soil of different tree ages *Journal of Tropical Crops* **44(09)** 1801-1809.

[7] Guo P, Shu Z, Wang X and Chi Y 2023 Using ecological stoichiometric characteristics to inform grassland management in the karst desertification area *Agronomy* **13(7)** 78-84

[8] Bao S 2000 *Soil Agricultural and Chemical Analysis* (Beijing: China Agricultural Publishing House) **(30)** 266-270.

[9] Guo X, Li K, Zhang W, Zou K, Hu D and Niu D 2013 Analysis of the bottleneck in the development of Camellia oleifera industry in Jiangxi and strategic thinking *Journal of Economic Forestry Research* **31(2)** 1-6.

[10] Liu Y, Luo W, Mu G, Wu X, et al. 2021 C:N:P stoichiometric characteristics of the soil-vegetation system of three rare tree species growing on Mount Fanjing in Southwest China *Global Ecology and Conservation* **(11)** 44-47

[11] Zhang F, Liu Y, Shi C, et al. 2021 Soil carbon, nitrogen, and phosphorus content and their

ecological stoichiometric characteristics in locust forests of different recovery years *Journal of Ecological Environment* **30(3)** 485-491.

[12] Peng X, Fang X, Yu L, et al. 2016 Chemical stoichiometric characteristics of soil carbon, nitrogen, and phosphorus in four forest types in the subtropical region *Journal of Central South University of Forestry and Technology* **36(11)** 65-72.

[13] Bo S, Feng Z and Li P 2018 Vicent Calatayud.Elevated ozone affects C, N and P ecological stoichiometry and nutrient resorption of two poplar clones *Environmental Pollution* **234** 136-144.

[14] Li Z, Zhou B, Ma T, Ke H, Xu G, Zhang Y, Yu and Cheng Y 2017 Impact of ecological management in the hilly area of the Loess Plateau on soil carbon, nitrogen, phosphorus and their stoichiometric characteristics *Journal of Soil and Water Conservation* **31(6)** 312-318.

[15] Liu N, Yu L, Zhao Q, Wu Y and Yan L 2020 Ecological stoichiometric characteristics of carbon, nitrogen, and phosphorus in the continuum of secondary forest leaves, litter, and soil in the karst plateau desertification area *Chinese Journal of Applied and Environmental Biology* **26(3)** 681-688.

[16] Pham T T, Nguyen N H, Yen P N D, Lam T D and Le N T T 2020 Proposed techniques to supplement the loss in nutrient cycling for replanted coffee plantations in Vietnam *Agronomy* **10** 905.

[17] Tian H Q, Chen G S, Zhang C, Melillo J M and Hall C A S 2010 Pattern and variation of C:N:P ratios in China's soils: A synthesis of observational data *Biogeochemistry* **98(1-3)** 139-151.

[18] Andrade E M, Valbrun W, De Almeida A M M, Rosa G and Da Silva A G R 2020 Land-use effect on soil carbon and nitrogen stock in a seasonally dry tropical forest *Agronomy* **(10)** 147-158

[19] Franzluebbers A J, Sawchik J and Taboada M A 2014 Agronomic and environmental impacts of pasture-crop rotations in temperate North and South America Agric. *Ecosyst. Environ.* **(190)** 18-26.

Evaluation of River Ecosystem Service Function Value Based on Emergy Analysis

Zifen Han[1], Yunyun Zhang[2,*], Chaojiang Shu[3] and Hongyi Zhu[2]

[1] State Grid Gansu Electric Power Company, Lanzhou, Gansu, China.
[2] Gannan Power Supply Company of State Grid, Gansu Cooperation, Lanzhou, Gansu, China.
[3] Electric Power Research Institute, Gansu Electric Power Company, State Grid Lanzhou, Gansu, China.
Email: 7638403@qq.com; *251940059@qq.com; shucj@gs.sgcc.com.cn; 376260555@qq.com

Abstract. Evaluating the value of river ecosystem services can provide a theoretical basis and data support for the ecological protection and development of related watersheds in the survey region. Taking Gansu Province as an example, combined with the characteristics of river ecosystem services in Gansu Province, its river ecosystem service functions are classified into 5 categories at the first level and 15 categories at the second level, and an energy value assessment model is established to assess the value of river ecosystem services in Gansu Province in the years of abundance, flatness, depletion and typical hydrological years. The results show that: the energy value of river ecosystem services in Gansu Province is 902.84×10^{21} sej, 846.91×10^{21} sej, and 797.35×10^{21} sej in abundant, flat, and dry water years respectively, and the annual energy value currency is 197.13×10^{10} rmb, 184.91×10^{10} rmb and 174.09×10^{10} rmb respectively, and the ordering is High flow year > Mean water year > Low flow year. Mean water year > Low flow year, indicating that river flow has a greater impact on the value of river ecosystem services; Fresh water supply, ecological security, ecological regulation service function energy value, and monetary value in each hydrological year accounted for a large proportion, indicating that the ecological service function of rivers in Gansu Province is mainly manifested in the above three aspects, and material production and cultural recreation accounted for a relatively small proportion.
Keywords. Service function; value assessment; energy value analysis; river ecosystems; Gansu Province.

1. Introduction

The connotation of ecosystem service function includes biodiversity maintenance, nutrient cycling, material supply water conservation, and many other aspects [1]. At present, most of the studies on ecosystem service value are on wetland ecosystems and ecological and economic systems, and there are fewer studies on the value of river ecosystem services. Objective understanding and quantitative evaluation of the value of river ecosystem services is the basis for rational utilization and development of river basins. In view of the different forms of expression and quantitative outline of the value of water resources, the calculation of the total value of river ecosystem services encounters the difficulty of analyzing and comparing the values of various types of ecosystem services in an inconvenient and uniform manner. The energy value analysis method is to take solar energy as the benchmark, through the energy value conversion rate of other forms of various types of energy into a unified standard of

Journal of Physics: Conference Series **2941** (2025) 012060 doi:10.1088/1742-6596/2941/1/012060

solar energy value, so as to compare the contribution of different substances in the river ecosystem [2]. This method has the advantages of simple calculation and convenient data collection, which can well solve the problem that different grades and types of substances cannot be analyzed and compared at the same time, and it can do long-timescale projection [3].

Therefore, this article applies the energy value analysis method to establish an energy value model to calculate the value of river ecosystem services in Gansu Province, which is convenient to visualize the contribution of river ecosystem services in the development of the national economy and provides theoretical support and data support for the formulation of river protection and development policies in Gansu Province.

2. Overview of the Research Area

2.1. Basin Overview

Gansu Province (32°11'-42°57'N, 92°13'-108°46'E) is located at the intersection of the Loess Plateau, the Tibetan Plateau, and the Inner Mongolian Plateau, with a land area of 4.258×10^5km^2, and has 12 prefectural-level municipalities and 2 autonomous prefectures under its jurisdiction. The province has longitudinal rivers, spanning from east to west, covering three major basins of the Yellow River, Yangtze River, and Inland River, involving 11 water systems such as Taohe River, Yellow River, Shule River Jialing River, etc. The average multi-year surface runoff in the province is 1.75×10^{10}m^3, and the theoretical reserves of hydroelectric resources are 1.724×10^7 KW, which is ranked 10th in China. By the end of 2022, Gansu Province, with a permanent population of 24,924,200, will be an important economic development region in China and an important ecological barrier in western China. It is home to the Qilian Mountains, Lanzhou Basin, Hexi Corridor, and other national key ecological reserves, which play a vital role in regional ecological security. However, Gansu Province is located in a fragile area of the ecosystem in northwest China, with arid and semi-arid areas accounting for 75% of the total area. Dry climate, scarce precipitation, unbalanced utilization and distribution of water resources, serious water pollution, and other problems have affected the ecological balance of rivers in Gansu Province, resulting in the deterioration of river ecosystems in Gansu Province, and restricting the sustainable economic and social development of Gansu Province. Objective understanding and reasonable evaluation of the value of river ecosystem services in Gansu Province is the basis for the high-quality development of the social economy in the Gansu basin.

2.2. Typical Year Selection

In this paper, Pearson's three-type frequency analysis was applied to classify the river runoff in Gansu Province for 28 years (1994-2022) into three typical years, namely, High flow year, Mean water year, and Low flow year. Based on the results, this paper takes 2018 as a High flow year, 2011 as a Mean water year, and 2016 as a Low flow year to calculate the value of river ecosystem services in Gansu Province.

3. Research Method

3.1. Service Function Partitioning

River ecosystems have a variety of ecosystem service functions. Referring to the existing research results [4], the river ecosystem service functions in Gansu Province are categorized into five major categories: freshwater supply, material production, ecological safeguard, ecological regulation, and cultural and recreational, and further subdivided into 15 categories: socio-economic water supply, hydroelectric power generation, water resource storage, nutrient cycling, provision of ecological environment, sediment transport and cultural and recreational, and so on.

3.2. Energy Value Analysis Methods

The energy value is a new concept and measurement method proposed by H T Odum [5], an American ecologist, which can compare the actual value and contribution of different levels of energy in ecosystems by converting various energies of natural ecosystems into a unified standard energy value based on solar energy. The conversion relationship between energy, matter, etc. and energy value is as follows [6]:

$$Em = \tau B \tag{1}$$

Style: *Em* is the energy value (sej); τ is the emergy transformity (sej/J or sej/g) [7]; *B* is the amount of energy or matter, J or g.

Conversion of energy values into money through the energy/currency ratio, i.e.

$$E_{DR} = EM_u / GDP \tag{2}$$

Style: E_{DR} is the energy value/money ratio (sej/rmb); EMu is the total energy value used by the regional economic system throughout the year (sej); and GDP is the gross national product of the corresponding region for the year (RMB) [8].

Energy-value analysis overcomes the shortcomings of energy analysis, which is unable to measure the contribution of nature to human economic and social development in monetary terms and tends to ignore the characteristics of energy quality [9].

3.3. Emergy Evaluation Model

Referring to the existing research results [10-12] and combining them with the characteristics of river ecosystems in Gansu Province, the river ecosystem service function energy value evaluation model is established as follows.

3.3.1. Fresh Water Supply Function Value Evaluation Model. The freshwater supply function includes four categories of water supply for economic life, water supply for hydropower generation, water supply for shipping, and water supply for ecological environment, etc., and the energy value assessment model is established as:

$$\begin{cases} E_{SE} = \sum_{i}^{n} W_i \cdot \rho \cdot G \cdot T_{SEi} \\ E_{HP} = Q_{HP} \cdot T_{HP} \\ E_{SH} = D \cdot M_g \cdot T_{\text{g}} \\ E_{EE} = Q_{EEi} \cdot T_{SEi} \end{cases} \tag{3}$$

Style: E_{SE} is Energy value of water supply for economic life (sej); W_i is the water supply for economic life in category i (m^3); ρ is the density of rainwater 1.03×10^6 (g/m^3); *G* is freshwater Gibbs free energy 4.94 (J/g); T_{Sei} is the emergy transformity of water supply for category i (sej/j);E_{HP} is Value of energy supplied for hydroelectric power generation (sej); Q_{HP} is Generation of electricity actually supplied (kWh), T_{HP} is emergy transformity of an electric quantity (sej/J)； E_{SH} *i*s Value of water supply for shipping (sej); M_g is transported by shipping mode i (t/km or people/km); *D* is Kilometers of waterways (km); T_g is emergy transformity for shipping mode i (sej/people/km or sej/t/km); E_{EE} is Energetic value of the function of supplying water to the ecosystem (sej); Q_{EE} is Water supply for ecosystems (m^3); T_{EE} is emergy transformity for water supply to ecosystems (sej/m^3)

3.3.2. Functional Value Evaluation Model of Material Production.

$$E_{AF} = 4186\mu \cdot K \cdot Q \cdot T_{AF} \tag{4}$$

Style: E_{AF} is the energy value of fish production (sej), Q is the total fish production (g), K is the dry weight proportionality coefficient, μ is the standard calorific value of the animal, 1kcal=4186J, and T_{AF} is the emergy transformity of aquatic product (sej/g);

3.3.3. Ecological Security Function Value Evaluation Model. Ecological protection functions include water resources storage, nutrient cycling, biodiversity maintenance and habitat provision. The emergy evaluation model is established as follows:

$$\begin{cases} E_{SW} = W_{SW} \cdot T_{SW} \cdot \rho \cdot G \\ E_{AN} = \sum_{i}^{n} P_{ANi} \cdot T_{ANi} \\ E_{BM} = N_{BM} \cdot T_{BM} \\ E_{LE} = SV_{L} \cdot F \cdot T_{LE} \end{cases} \tag{5}$$

Style: E_{SW} is the energy value of water resources storage (sej), W_{SW} is the amount of water resources stored in the study area (m^3), T_{SW} is the emergy transformity for freshwater (sej/j), ρ, G is the same as above; E_{AN} is the energy value of nutrient uptake cycle (sej), P_{ANi} is the amount of uptake of the ith nutrient element (g), and T_{NCi} is emergy transformity of the ith type of element (sej/g); E_{BM} is the energy value of the function of maintaining biodiversity (sej), N_{BM} is the number of biological species (species), T_{BM} is the emergy transformity of the species (sej/species); E_{LE} is the energy value of the function of providing habitats (sej), SV_L is the value of the benefits obtained from the provision of habitats per unit of area [13] (3892.98 rmb/hm^2).

3.3.4. Ecological Regulation Function Evaluation Model. Ecological regulation functions include flood regulation and storage, air purification, water purification and sediment transport. The emergy evaluation model is established as follows:

$$\begin{cases} E_{AF} = V_{AF} \cdot \rho \cdot G \cdot T_{AF} \\ E_{AP} = W_{C} \cdot T_{C} + W_{O} \cdot T_{O} \\ +W_{S} \cdot T_{S} \\ E_{WP} = E_{W} \cdot T_{K} \\ E_{ST} = Q_{TS} \cdot T_{TS} \end{cases} \tag{6}$$

Style: E_{AF} is the energy value of flood storage (sej), T_{AF} is the volume of reservoir flood storage (m^3), ρ, G is the same as above; E_{AP} is the energy value of air purification function (sej), W_C is the amount of fixed carbon dioxide (g), W_O is the amount of released oxygen (g), W_C is the amount of fixed sulfur dioxide (g), T_C is the amount of CO_2 (sej/g), T_O is the emergy transformity of O_2 (sej/g), and T_S is the emergy transformity of SO_2 (sej/g); E_{WP} is the energy value of the water purification function (sej), E_W is the same as above, and T_k is the emergy transformity of the water purification (sej/g); E_{ST} is the energy value of sediment transport (sej), and Q_{TS} is the annual average of the river channel sediment transport (t), T_{TS} is the emergy transformity of sediment (sej/t).

3.3.5. Cultural Leisure Function Value Evaluation Model. Cultural leisure functions include scientific research, education and leisure and entertainment. The emergy evaluation model [14] is as follows:

$$\begin{cases} E_{CR} = P_{CR} \cdot T_{CR} \\ E_{LR} = IT_{LR} \end{cases} \tag{7}$$

Style: E_{CR} is the energy value of research and education (sej), P_{CR} is the average annual number of academic papers published in the watershed (pages), and T_{CR} is the paper energy conversion ratio (sej/page); E_{LR} is the energy value of recreation and entertainment (sej), I is the annual income from tourism in the study area (rmb), and T_{LR} is the local energy-to-currency ratio (sej/rmb);

4. Emergy Evaluation and Result Analysis

4.1. Data Source and Processing

The relevant data in this paper are from the Gansu Water Resources Bulletin, Gansu Development Yearbook, China Energy Statistical Yearbook, and China Statistical Yearbook. The habitat area of Gansu Province [15] is 10.90×10^4 km^2, and the nutrient absorption and cycling function in the ecological protection function is insufficient. Due to the lack of specific aquatic plant quantity in rivers in Gansu province, the paper takes the nutrient recycling quality of grassland in Gansu province as a substitute, and the grass yield in Gansu Province in 2014 was 55t. The economic benefits of the scenic spots in the study area were replaced by the average domestic tourism income of Gansu Province from 2011 to 2022. The emergy conversion rate selected the research results of LAN Sheng-fang [6]; The currency emergy ratio of Gansu Province, calculated by Han [16], was 4.58×10^{11} sej/ RMB.

4.2. Result Analysis

The emergy value of various river ecosystem service functions in Gansu Province and the emergy evaluation results of river ecosystem service functions in Gansu Province are shown in tables 1 and 2:

Table 1. Value of service functions of river ecosystem in Gansu Province.

Year	High flow year		Mean water year		Low flow year	
Sort	energy value ($\times10^{21}$sej)	Portion (%)	energy value ($\times10^{21}$sej)	Portion (%)	energy value ($\times10^{21}$sej)	Portion (%)
Fresh water supply	303	33.6	326	38.47	313	39.3
Material production	1.83	0.2	2.31	0.27	1.99	0.25
Ecological guarantee	295	32.72	221	26.08	184	23.03
Ecological regulation	239	26.48	234	27.68	235	29.5
Cultural leisure	63	6.99	63.5	7.5	63.2	7.93
Total	902.84	100	846.9	100	797.35	100

Table 2. Emergy evaluation results of river ecosystem service function in Gansu Province.

Year	Value of ecosystem services ($\times10^{21}$ sej)	Ecosystem services can be valued in currency ($\times10^{10}$ rmb)
High flow year	902.84	197.13
Mean water year	846.90	184.91
Low flow year	797.35	174.09

The results show that:

- The variation law of river ecosystem service value is consistent with the law of runoff in abundant, flat, and dry hydrological years, and runoff is the main influencing factor of river ecosystem service value. The research results can provide references for determining the appropriate ecological flow of rivers in different hydrological years.
- The sum of the proportion of solar energy value and monetary energy value of freshwater supply function, ecological security function, and ecological regulation function in each hydrological year is more than 90%, indicating that the classification of river ecosystem

functions is reasonable and the value of river ecosystem services is huge.

- The freshwater supply function accounted for the largest proportion in each hydrology year, indicating that water resource supply, hydropower generation, and inland shipping contributed the largest social and economic value to the river ecosystem. The ecological regulation and ecological protection functions take second place, indicating that water purification, sediment transport, and other functions play a vital role in river ecosystems. The sum of material production and cultural leisure function is the smallest, less than 10%, indicating that the development potential is large;

Through the reasonable division of river ecosystem service functions and emergy analysis and emergy model, the great value of river ecosystem service functions and the proportion of various service functions in Gansu Province are obtained, which provides data support and theoretical support for formulating relevant river ecological protection and development policies in Gansu Province, and provides a reference for other regions.

5. Conclusion and Prospect

At present, the evaluation of ecosystem service value is mostly qualitative. In this article, emergy analysis method is used to build emergy evaluation model to evaluate the river ecosystem service value in Gansu province, and quantitative results are obtained, which shows the significance of emergy analysis for ecological protection in a more intuitive way and can provide a certain reference for the calculation of river ecosystem service value in other regions. Evaluation of the energy value of river ecosystem services can be added to the planning stage of future water conservancy project construction, so as to quantify the impact of water conservancy project construction on river ecosystems. Due to the lack of observational data on the quantity of specific aquatic plants in rivers in Gansu Province, the quality of nutrients absorbed by grassland in Gansu Province was used instead, which would slightly affect the accuracy of the calculation results. In the follow-up study, if a long series of data of continuous observation can be collected, the accuracy of the model calculation can be further improved, and a more accurate reference can be provided for regional ecological protection and economic development.

Acknowledgement

This work was supported by the the State grid Gansu Power Company 2023 science and technology project (522713230003).

6. References

[1] Ouyang Z, Wang R and Zhao J 1999 Evaluation of ecosystem service function and its eco-economic value *Chinese Journal of Applied Ecology* **(5)** 635-640.

[2] Martin, J. F. 2002. Emergy valuation of diversions of river water to marshes in the Mississippi River Delta *Ecological Engineering* **18(3)** 265-286.

[3] Li L, Wang X, Luo L, et al. 2018 Review on methods of ecosystem service value assessment *Chinese Journal of Ecology* **37(4)** 1233-1245. (in Chinese).

[4] Zhang D, Shen C and Yu G 2019 Study on evaluation model of river ecosystem service value based on in-channel discharge *Journal of Economics of Water Resources* **37(5)** 16-20.

[5] Odum H T. 1989 Ecological engineering and self-organization. *Ecological Engineering: An Introduction to Ecotechnology* **101** 79-101.

[6] Lan S, Jing Y, Lu H et al. 2002 *Emergy Analysis of Eco-Economic System* (Beijing, Chemical Industry Press).

[7] Brown M T and Ulgiati S 2004 Energy quality, emergy, and transformity: HT Odum's contributions to quantifying and understanding systems *Ecological Modelling* **178(1-2)** 201-213.

[8] Wu Z, Lu C, Hu C, et al. 2013 Evaluation method and application of eco-economic value of regional water resources *Basin Water Cycle and Water Security: Proceedings of the 11th*

China Water Forum 112-118.

[9] Li W 2018 *Emergy-Based Evaluation and Management of Water Resources in Reservoirs* (Wuhan: Wuhan University) Doctor's Thesis.

[10] Yang T, Zhang D, Shen C, et al. 2023 Assessment of ecosystem service function based on emergy analysis: A case study of Dongjiang River Basin *Chinese Journal of Water Ecology* **44(1)** 9-15.

[11] Li L, Hui S, Song H, et al. 2013 Evaluation of ecosystem service function and energy value of Shuangtai estuary wetland in Panjin *China Environmental Science* **33(8)** 1454-1458.

[12] Zou L and Sun Q 2016 Ecological service function evaluation of Datong Lake wetland based on emergy analysis *Fujian Forestry Science and Technology* **43(4)** 104-110.

[13] Fu J, Zhang D, Zhang H, et al. 2023, Optimal scheduling of ecological service value of Manwan Reservoir based on emergy analysis *Hydropower Energy Science* **41(7)** 94-97.

[14] Fu J F 2023 *Emergy Analysis and Optimal Operation of Ecosystem Service Value in Reservoir Group* (Kunming: Kunming University of Science and Technology) *Master's Thesis*.

[15] Zhao X, Wang J, Su J, et al. 2020 Evaluation of habitat quality and degradation degree in Gansu Province based on InVEST model and Moran index *Transactions of the Chinese Society of Agricultural Engineering* **36(18)** 301-308.

[16] Han Z, Zhu H, Yuan C, et al. 2023 Emergy analysis and evaluation of eco-economic system of water resources in Gansu *Value Engineering* **42(34)** 35-37.

Experimental Study of Leakage Simulation in Gas Extraction Boreholes

Ruxiang Qin[1,] *, Xiaoxin Qin[2] and Mingyang Yu[3]

[1] School of Safety Science and Engineering, Anhui University of Technology, Huainan 232001, Anhui, China.
[2] School of Safety Science and Engineering, Anhui University of Technology, Huainan 232001, Anhui, China.
[3] College of Artificial Intelligence, Nankai University, Tianjin 300350, China.
Email: *rxqin@aust.edu.cn; 2457045117@qq.com; hlhymy5201314@163.com

Abstract. The main issues with gas extraction via drilling are the reduction in gas concentration, decreased extraction efficiency, and prolonged time to meet extraction standards caused by drill hole leakage. To determine the location of leaks in gas extraction drill holes, this paper establishes a simulation system, substituting CO_2 for gas, to study the characteristics of gas concentration and pressure distribution within the drill holes under various leakage scenarios, thereby inferring the leakage points. Results indicate that during drill hole leakage, the CO_2 concentration sharply decreases from the borehole entrance to the leakage point, while the concentration from the leakage point to the bottom of the hole remains high. The difference between the internal pressure of the hole and the entrance pressure is denoted by ΔH. A stable ΔH value indicates effective sealing of the drill hole; a decrease in ΔH at any point within the hole suggests a leakage, with a reduction of up to 3 Pa observed in the experiments. Based on similar experiments, we propose a gas extraction drill hole leakage criteria and apply it in the detection of interlayer drill hole leakage at the Pingdingshan 15-17-33190 working face. The leakage is located in the sealed section. After improving the sealing parameters, the gas concentration at the borehole mouth increases from 10% to 45%.
Keywords. Gas extraction; drill hole leakage; gas concentration; ΔH value; gas extraction drill hole leakage criteria.

1. Introduction

Gas extraction is a crucial technological method for addressing coal mine gas disasters, utilizing coalbed methane resources, and reducing environmental pollution. During the extraction process, the entry of underground air into the drill hole under negative pressure causes a reduction in gas concentration, thereby increasing the cost of gas extraction, reducing its efficiency, and prolonging the time to meet extraction standards [1-5]. In practical production, the effectiveness of gas extraction from a drill hole is typically determined by the gas concentration at the mouth of the hole. When this concentration falls below 10%, the hole is deemed ineffective, and extraction is stopped, often necessitating the drilling of new holes to extract gas from the area [6-9]. There are two scenarios that can lead to a reduction in gas concentration at the mouth of the drill hole: one is leakage in the drill hole, which increases the amount of underground air entering; the other is no significant change in the leakage, but prolonged extraction reduces the methane content in the coal seam within the extraction area. Both drill hole leakage and the reduction in gas content change over time, presenting uncertainties. Therefore, determining whether leakage has occurred in an extraction drill hole remains a challenging issue in gas extraction from drill holes [10-12].

To determine whether gas extraction drill holes are leaking, numerous scholars have conducted extensive research, including studies on drill hole sealing theory, field tests for drill hole leakage, and similar simulation experiments. Hu [13] believes that drill hole leakage can be categorized into gaps between the extraction pipe and the hole wall, leakage of the sealing material itself, leakage through the surrounding rock fissures of the drill hole, and leakage due to interconnection with nearby drill holes. Zhang [14] analyzed the mechanisms of gas leakage and derived a dynamic leakage model. Hong [15] established equations for the area of the leakage ring and the amount of leakage, concluding that the drill hole diameter, seal strength, and other parameters determine the amount of gas leakage. Wang and others [16] studied the air leakage process during coalbed methane extraction under various negative pressures. Zhang [17] proposed a method based on pressure gradients to detect and locate leaks in gas extraction pipelines. Ba [18], Fan et al. [19], Hao [20], and Wang et al. [21] developed devices to detect leakage in drill holes by testing the gas pressure and methane concentration within drill holes in experimental mines to determine the sealing quality based on how gas concentration varies with drill hole depth. Zheng [22] derived a fully coupled mathematical model to study the performance of drill hole gas leakage and provided optimal sealing measures. Gao [23] established a fully coupled gas-air mixed flow model to study the effects of extraction time, different negative pressures, and gas leakage on gas concentration. Zhang [24] determined the optimal sealing depth for drill holes by establishing a mathematical model of gas leakage. Zhou [25] used similarity theory to conduct experiments on gas leakage diffusion, using carbon dioxide as a substitute for methane to study the effect of leakage quantity on gas leakage diffusion. Zou [26] created a Multiphysics device for coal and gas outbursts, simulating coal and gas outbursts under various initial geological stresses, gas pressures, and temperatures.

Field tests for determining drill hole leakage are time-consuming and susceptible to environmental interference and equipment precision limitations, resulting in lower accuracy. However, laboratory simulations under controlled conditions can provide stable experimental conditions, making it easier to identify the impact of leaks on gas parameters within the hole, thus inferring the characteristics of the drill hole leakage. To investigate the distribution of gas concentration and pressure characteristics under various conditions such as leak position and quantity in drill holes, this paper utilizes a simulated experimental apparatus for gas extraction drill holes to study the effects of leaks on the gas parameters within the hole, thereby deducing the position of the leaks based on their characteristics.

2. Principle of Simulation Test

During the pre-drainage of gas from coal seam drill holes, the concentration of extracted gas is primarily controlled by two factors: the outflow of gas from the drill hole and the leakage from the sealed section of the hole. At the initial stage of extraction, both the concentration and the purity of the gas are at high levels, with concentrations at the mouth of the hole reaching up to 90%. As the extraction time progresses, both the concentration and the purity of the gas exhibit varying degrees of decline. As the extraction time progresses, both the concentration and the purity of the gas exhibit varying degrees of decline. Based on the principles of fluid dynamics in pipes, the drill hole during extraction can be likened to a conduit, where at any given moment at a specific location in the drill hole, the expression for the gas concentration within the extraction conduit is as follows:

$$C = \frac{Qgas}{Qgas+Qair} \times 100\% \tag{1}$$

where C is the gas extraction concentration at a certain drill hole location at a specific time, expressed in %; Q_{gas} is the volume of gas passing through a certain drill hole location at a specific time, measured in m^3/min; Q_{air} is the volume of air passing through the same location at the same time, also in m^3/min.

Consequently, in a short period, by adjusting the relevant parameters of the device to simulate the gas emission rate and maintain its stability, it is feasible to use the differences in gas concentration to determine the leakage status between various test points within the pipe, and thus analyze the causes and amounts of leakage. This method allows for the analysis of drill hole leakage in a controlled environment, providing more reliable data support and a theoretical basis for gas extraction.

3. Gas Leakage Simulation Test of Extraction Borehole

3.1. Design and Manufacture of Test Equipment

The device consists of five parts: the gas supply system, the gas extraction system, the flow regulation system, the piping system, and the monitoring system.

(1) Gas supply system: Since methane is a combustible gas and a controlled substance in laboratory experiments, this test uses carbon dioxide as a substitute for methane. The gas supply system consists of carbon dioxide cylinders, air cylinders, and pressure-reducing valves, with the purpose of introducing carbon dioxide and air into the piping system.

(2) Gas extraction system: Composed of a vacuum pump, vacuum ball valve, flow meter, and pressure gauge, its purpose is to provide negative pressure for the entire system and extract the gas from within the pipes.

(3) Flow regulation system: Consists of a flow meter. It adjusts the flow meter' s values to achieve the desired simulated gas flow rates.

(4) Piping system: Modeled after the interlayer drill holes in the floor extraction gallery of Pingmei Shenma Mine' s $_{15\text{-}17}$-33190, this coal seam drilling employs polyurethane sealing. It uses eight 4m-long extraction pipes, with a sealing section length of 10m and a depth of 12m, a bag-type sealing belt, and an extraction negative pressure of 14kPa. The device is designed based on the working face parameters and is divided into three sections from left to right as shown in figure 1: the mouth section, approximately 0.2m long with an external diameter of about 0.032m; the sealing section, about 1m long with an external diameter of about 0.16m; and the extraction section, about 2m long with an external diameter of about 0.24m. On the extraction pipe, five leak points A_1, A_2, A_3, A_4, and A_5 are sequentially arranged according to their distance from the mouth, along with five CO_2 intake points C_1, C_2, C_3, C_4, and C_5 on the same cross-section.

(5) Monitoring system: Based on the construction parameters of the drill hole, six different characteristic points are set for monitoring, each equipped with a CO_2 sensor. The first characteristic point, marked as B_6 concentration testing point, is positioned in the middle of the sealing section. Five concentration testing points are established in the extraction section, and one concentration testing point is placed at equal intervals along the piping, marked as B_1, B_2, B_3, B_4, and B_5.

As shown in figure 2, the carbon dioxide from the cylinder is regulated by a pressure-reducing valve to deliver the CO_2 required for the experimental design. The connecting pipeline enters the system device through five branch ports: C_1, C_2, C_3, C_4, and C_5. Similarly, the air cylinder is connected in the same manner and enters the system device through five branch ports: A_1, A_2, A_3, A_4, and A_5. After the two gases are mixed, they are finally extracted by a negative-pressure vacuum pump.

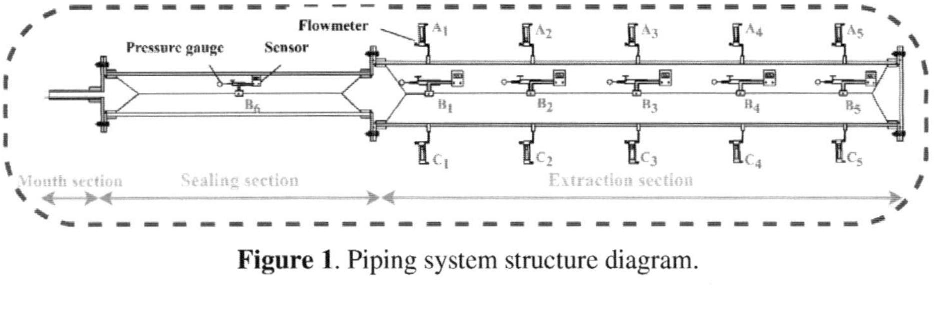

Figure 1. Piping system structure diagram.

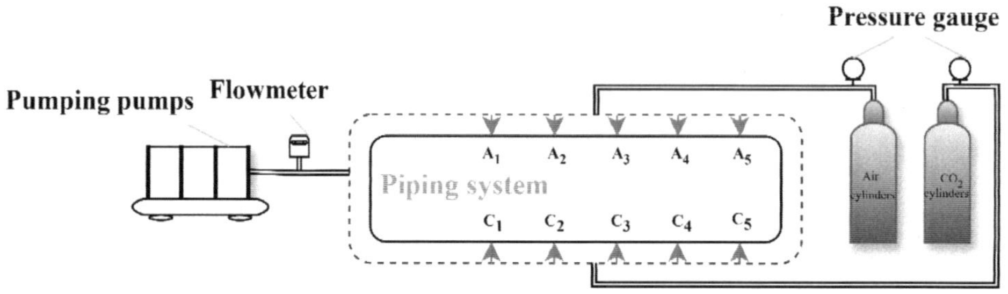

Figure 2. Overall schematic diagram of the experimental system.

3.2. Test Working Condition

Due to the poor contact between the sealing material and the drill hole wall, air infiltrates the drill hole, a phenomenon defined as internal leakage. When leakage occurs inside the hole, external air flows in from the point where the sealing grout section meets the extraction piping or the drill hole. Under negative pressure, the external air flows out through the first section of screen holes in the extraction piping. To simulate the internal leakage scenario, the experiment introduces air at point A_1.

After the drill hole is formed, if the sealing material cannot effectively close the fractures in the broken zone, air will enter the extraction drill hole through these fractures under the effect of negative pressure, resulting in gas leakage. This phenomenon is defined as external leakage. After the drill hole is formed, if the sealing material cannot effectively close the fractures in the broken zone, air will enter the extraction drill hole through these fractures under the effect of negative pressure, resulting in gas leakage. This phenomenon is defined as external leakage. To simulate the external leakage scenario, the experiment introduces air at points A_2 and A_3.

During the interlayer drilling process, due to the weight of the drill rod, the complex underground geological environment, or limitations of the construction equipment, the drilling trajectory may deviate. Different drill holes may intersect, leading to internal interconnection, a scenario defined as deep interconnection leakage. If one drill hole experiences leakage, it can cause a concurrent decline in the extraction efficiency of associated drill holes. If one drill hole experiences leakage, it can cause a concurrent decline in the extraction efficiency of associated drill holes.

The simulated drilling leakage experiment is set up with three conditions, with each condition being tested three times. The detailed experimental parameters are as follows:

Condition 1: The air intake at port A_1 is set to 10 L/min, and each of ports C_1-C_5 is set to 0.2 L/min. The total air intake is 10 L/min, and the total CO_2 intake is 1 L/min. This setup studies the impact of internal leakage on the gas extraction concentration within the hole.

Condition 2: The air intakes at ports A_2 and A_3 are each set to 5 L/min, and each of ports C_1-C_5 is set to 0.2 L/min. The total air intake is 10 L/min, and the total CO_2 intake is 1 L/min. This setup studies the impact of external leakage on the gas extraction concentration within the hole.

Condition 3: The air intakes at ports A_4 and A_5 are each set to 5 L/min, and each of ports C_1-C_5 is set to 0.2 L/min. The total air intake is 10 L/min, and the total CO_2 intake is 1 L/min. This setup studies the impact of interconnection leakage on the gas extraction concentration within the hole.

4. Test Results and Analysis

4.1. Analysis of Gas Concentration Distribution in Hole

The gas concentration distribution in the hole under three test conditions of gas leakage is shown in figure 4.

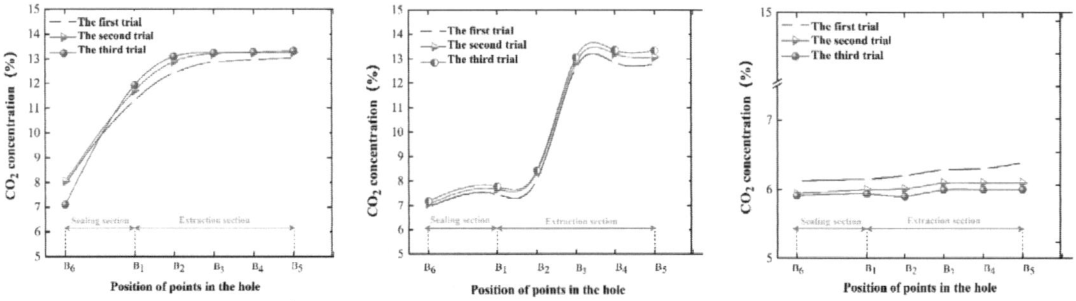

(a) Condition 1 (b) Condition 2 (c) Condition 3

Figure 3. Distribution of gas concentration inside the tube under different leakage conditions.

As shown in figure 3(a), in Condition 1, the CO_2 concentration at the bottom of the hole remains relatively high, with the concentration at the mouth being only 50% of that at the bottom. The CO_2 concentration inside the hole decreases gradually from the bottom to the mouth, with varying rates of decay at different depths. The greatest variation in CO_2 concentration occurs between B_6 and B_2. Between B_2 and B_5, the fluctuation in CO_2 concentration is minimal, and the concentration remains at a higher level. In the simulated deep parts of the drill hole between B_2 and B_5, the surrounding rock of the drill hole wall is in the stress recovery section of the high-stress zone and the original rock stress zone. The development of composite fractures in the coal is low, making it difficult for outside air to leak into the extraction drill hole, resulting in fewer leakage channels. The leakage point within the hole is determined to be near the B_1 range, indicating poor stability of the sealing section of the extraction drill hole, with possible damage to the extraction pipe or issues with the pipeline connection.

As shown in figure 3(b), a significant decrease in CO_2 concentration is observed at test point B_2, which can be attributed to two factors: Firstly, point A_1 may be an area susceptible to external air influence, allowing external air to easily penetrate into the strata where point A_2 is located. As shown in figure 3(b), a significant decrease in CO_2 concentration is observed at test point B_2, which can be attributed to two factors: Firstly, point A_1 may be an area susceptible to external air influence, allowing external air to easily penetrate into the strata where point A_2 is located. The CO_2 concentration at test point B_2 is significantly lower than at test points B_4 and B_5. The small fluctuation range of CO_2 concentration between B_3 and B_5 suggests instability and deformation of the drill hole sealing section, possibly due to interconnected fractures in the coal layer or insufficient sealing depth.

As shown in figure 3(c), the overall CO_2 concentration inside the hole is significantly lower compared to Conditions 1 and 2, with smaller fluctuations in concentration, suggesting the presence of leakage channels in the deep coal seam, or that the original coal seam gas concentration is low.

The CO_2 concentration inside the hole gradually decreases from the bottom of the drill hole to the mouth, showing a pattern of lower concentration at the mouth and higher concentration at the bottom. When there is a leak at a certain position inside the hole, under the influence of extraction negative pressure, there will be a sudden change in the CO_2 concentration distribution. The concentration decreases sharply from the mouth to the leak position, while from the leak position to the bottom of the hole, the CO_2 concentration remains high.

4.2. Analysis of Gas Pressure Difference Distribution Law in Hole

The distribution diagram of gas pressure difference in the hole in the case of air leakage under three test conditions is shown in figure 4.

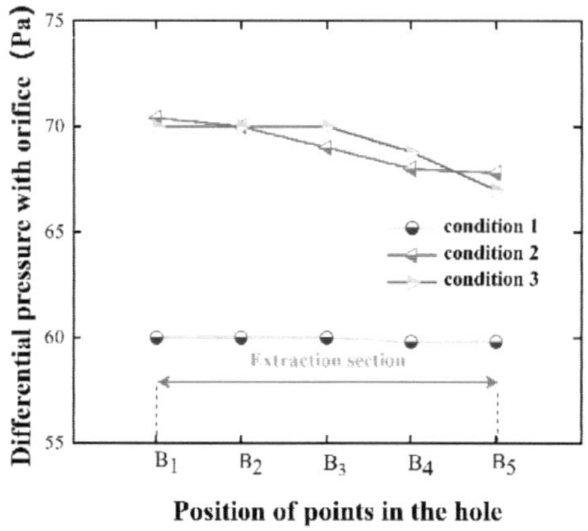

Figure 4. Distribution of pressure difference inside the hole under different air leaks.

As shown in figure 4: ΔH represents the difference in pressure between the inside of the hole and the mouth. A stable ΔH value indicates good drill hole sealing, while a decrease in ΔH at any point inside the hole indicates a leak. In Condition 1, Δ H shows no significant fluctuations within the extraction section, and the trend of decrease in negative pressure along the length of the hole is minimal, indicating good drill hole sealing in this condition. In other conditions, the drill hole sealing is poor, with a greater downward trend of Δ H along the length of the hole in the extraction section. In Condition 3, the highest decrease reaches 3pa. This is due to the impact on gas flow within the hole, which alters the gas pressure distribution. Additionally, leakage in the drill hole is accompanied by increased wall roughness and reduced cross-sectional area, increasing both the frictional and local resistance, further affecting the extraction performance.

4.3. Extraction Borehole Leakage Judgment Criteria

During the interlayer drilling process, gas from the coal seams diffuses into the surrounding rock, and the rock layers contain gas within their natural fracture networks. Based on the established model and in conjunction with practical mine operations, the arrangement of measurement points is shown in figure 5. Point B_6 is located in the middle of the sealing section. Beyond the sealing depth, a measurement point is placed every 4 meters, with five points arranged in total.

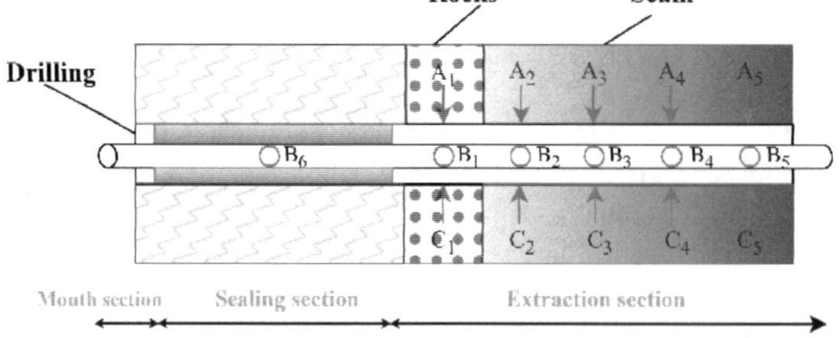

Figure 5. Layout of on-site measurement points.

Based on the measurement points arranged as shown in figure 5, the gas concentration and negative extraction pressure at different depths are measured to analyze and determine the sealing quality and leakage location of the extraction drill hole.

Based on the concentrations between B_6, B_1, and B_2, the stability of the sealing section of the extraction drill hole can be assessed. If the concentration at B_6 is less than at B_1 and B_2, and the concentrations at B_1 and B2 are similar, it indicates poor quality of the extraction pipe, with air from the tunnel flowing into the end of the extraction pipe along the sealing section and entering the extraction system.

Based on the concentrations between B_1, B_2, and B_3, if the concentration at B_1 is less than at B_2 and B_3, and the concentrations at B_2 and B_3 are similar, it indicates instability and deformation damage in the sealing section of the drill hole.

Based on the concentrations between B_3, B_4, and B_5, it is possible to determine if there is leakage in the deep coal layer inside the drill hole. If the concentration at B_4 is less than at B_5, and the concentrations at B_3 and B_4 are similar, it indicates that leakage has occurred in the deep parts of the coal seam.

5. Field Application and Effect Analysis

The test location was selected at the Pingmei Shenma Mine' s $_{15\text{-}17}$-33190 working face, where there are 18 drill holes per group, arranged in two rows in a staggered distribution. Based on the on-site experimental conditions, Drill Hole No. 8 was selected for the experiment with 6 measurement points arranged: B_6 at 7m from the extraction pipe mouth, B_1 at 14m from the mouth exceeding the sealed section, B_2 at 18m from the pipe mouth, another B_2 at 22m, B_4 at 26m, and another B_4 at 30m inside the hole, situated in the deep part of the drill hole. During the detection period, a sealing quality detector was used to measure the gas volume fraction and changes in negative extraction pressure at different measurement points to determine the extraction performance of the drill hole. The test results are shown in figure 6.

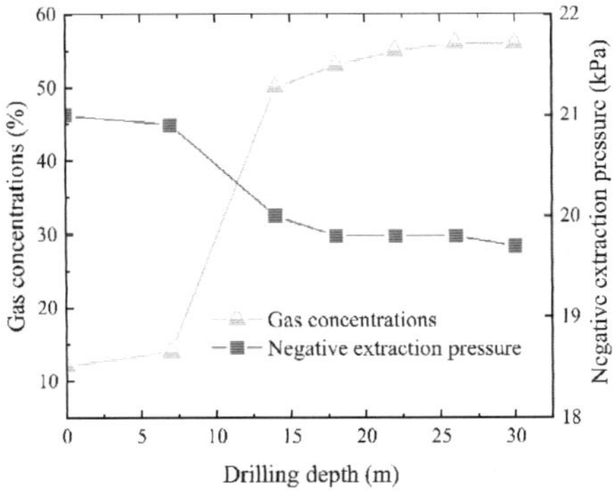

Figure 6. Distribution of drill hole detection parameters along hole length.

According to figure 6, there is a sudden change in gas concentration within the 7m-14m range, indicating a leakage in this drill hole' s sealing section, suggesting poor stability or poor quality of the sealing. Improvements were made to the sealing process and parameters, extending the length of the sealing section to 12m and using a thickened bag-type seal. Secondary grouting was performed on the drill hole, and as shown in figure 7, the gas concentration at the mouth of the hole increased to 45%. This indicates that the low concentration of gas extraction from pre-drilled holes on this working face is significantly affected by insufficient sealing length and depth. Increasing the sealing length or depth can enhance the extraction performance.

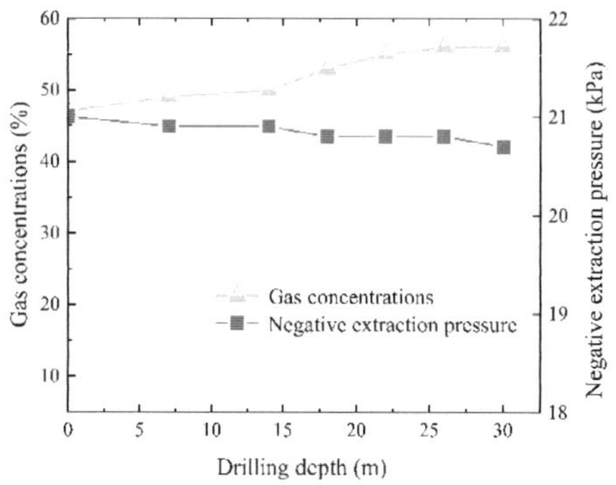

Figure 7. Improved distribution of drill hole detection parameters along hole length.

6. Conclusion

1) This paper employs a laboratory similar physical simulation method, substituting CO_2 for methane, to study the characteristics of gas concentration and pressure distribution under different drilling leakage modes, thereby determining the sealing quality and leakage location of extraction drill holes. When leakage occurs, a significant sudden change in the distribution of CO_2 concentration is observed inside the hole, with a sharp decrease in CO_2 concentration from the mouth to the leakage point.

2) The ΔH value serves as an index parameter to determine the sealing quality of gas extraction drill holes. A stable ΔH value indicates good drill hole sealing, while a decrease in ΔH at any point inside the hole indicates a leak. In the experiment, the decrease in ΔH reaches 3pa. For well-sealed drill holes, the trend of decrease in negative pressure along the length of the hole is smaller. For poorly sealed drill holes, the trend is larger.

3) The criteria for determining the location of leaks are proposed and applied to the interlayer drilling leakage detection at the Pingdingshan $_{15\text{-}17}$-33190 working face. The leakage type on this working face is in the sealing section. Improving the sealing parameters significantly enhances the extraction performance.

7. References

[1] Wang F, Zhang C and Liang N., 2017 Gas permeability evolution mechanism and comprehensive gas drainage technology for thin coal seam mining. *Energies* **10** p. 1382.

[2] Shan R, Wei Y, Wang C, Li Z, Zhao Y, Tong X and Xiao S., 2023 Research on the evolution of the deviatoric stress and control of a gas extraction roadway under mining influence *Environ. Earth Sci* **82** p. 425.

[3] Xue Y., Liu S., Chai J., et al, "Effect of water-cooling shock on fracture initiation and morphology of high-temperature granite: application of hydraulic fracturing to enhanced geothermal systems," *Appl. Energ.*, vol. **337**, p. 120858, May 2023.

[4] Liu T., Lin B., Fu X., et al, "Modeling air leakage around gas extraction boreholes in mining-disturbed coal seams. *Process. Saf. Environ*, vol. **141**, pp. 202-214, September 2020.

[5] Cai J., Wu J., Yuan S., et al, "Numerical analysis of multi-factors effects on the leakage and gas diffusion of gas drainage pipeline in underground coal mines," *Process. Saf. Environ*, vol. **151**, pp. 166-181, July 2021.

[6] Tang Y., Yang S., and Wang X., "A new parameter of cleanability to determine the clean potential of coal resources," *Acta Geol. Sin-Engl*, vol. **91**, pp. 2338-2339, 2017.

[7] Tang Y., Wang X., H. H. Schobert, et al, "Cleaning potential of selected coals in Shanxi Province, China: evaluation of geological factors affecting the cleaning potential," *Energ. Fuel*, vol. **34**, pp. 1396-1407, January 2020.

[8] Yi M., Wang L., Liu Q., et al, "Characteristics of Seepage and Diffusion in Gas Drainage and Its

Application for Enhancing the gas utilization rate," *Transport Porous Med*, vol. **137**, pp. 417-431, March 2021.
[9] Wei Y., Chen K., Kang J., et al, "Policy and management of carbon peaking and carbon neutrality: a literature review," *Engineering*, vol. **14**, pp. 52-63, July 2022.
[10] Wang K., Lin S., and Zhao W., "Reviews on simulation studies of coalbed gas recovery and transport processes," *Energ. Fuel*, vol. **37**, pp. 14622-14644, September 2023.
[11] Yi M., Wang L., Hao C., Liu Q., and Wang Z., "Method for designing the optimal sealing depth in methane drainage boreholes to realize efficient drainage," *Int. J. Coal. Sci. Techn*, vol. **8**. pp. 1400-1410, July 2021.
[12] Kai D., Shang L., Qian S., et al, "Development and performance testing of the new sealing material for gas drainage drilling in coal mine," *Powder Technol,* vol. **363**, pp. 152-160, March 2020.
[13] Hu S., and Liu H., "Leakage mechanism of coal seam gas drainage borehole and its application research progress," *Saf. Coal. Mines*, vol. **47**, pp. 170-173, 2016.
[14] Jun Z., Bo L., and Yu S., "Dynamic leakage mechanism of gas drainage borehole and engineering application," *Int. J. Min. Sci. Techno*, vol. **28**, pp. 505-512, May 2018.
[15] Hong Z., Xiong Z., and Wang C., "Gas drainage sealing mechanism and optimization of key leakage parameters: analysis, experiment, and field verification," *Arab. J. Geosci*, vol. **15**, p. 1431, August 2022.
[16] Xiang Z., Bo L., and Yu S, "Dynamic leakage mechanism of gas drainage borehole and engineering application," *Int. J. Mining. Sci. Technol*, vol. **28**, pp. 505-512, May 2018.
[17] Wang Z., S. Hurter, You Z., et al, "Influences of negative pressure on air-leakage of coalseam gas extraction: laboratory and CFD-DEM simulations," *J. Petrol. Sci. Eng*, vol. **196**, pp. 107731, January 2021.
[18] Zhang H., Shen M., Huo Z., et al, "Research on Gas Drainage Pipeline Leakage Detection and Localization Based on the Pressure Gradient Method," *Processes*, vol. **12**, pp. 1590, June 2024.
[19] Ba Q, "Research on gas leakage mechanism and detection technology of coal mine gas drainage," IOP Conference Series: Earth and Environmental Science. London: *IOP Publishing*, 2020.
[20] Fan X. and Ma Q., "Application test of hole sealing quality testing device in Wangpo Coal Mine," *Coal Mine Saf*, vol. **50**, pp. 133-135, August 2019.
[21] Hao Q., Chen H., Xu G., et al, "Development and Application of Unsealed Borehole Leakage Detection Device Based on Flow Method," *Appl. Sci*, vol. **14**, pp. 6684, June 2024.
[22] Wang N., Zhang T., Fan J., et al, "Testing technology and application of sealing hole quality in coal mine gas extraction," *Coal Technology*, vol. **38**, pp. 87-89, January 2019.
[23] Zheng C., Chen Z., M. Kizil, S. Aminossadati, et al, "Characterisation of mechanics and flow fields around in-seam methane gas drainage borehole for preventing ventilation air leakage: a case study," *Int. J. Coal. Geol*, vol. **162**, pp. 123-138, May 2016.
[24] Wang H., Wang E., Li Z., et al, "Study on sealing effect of pre-drainage gas borehole in coal seam based on air-gas mixed flow coupling model," *Process Saf. Environ*, vol. **136**, pp. 15-27, April 2020.
[25] Zhang X., Gao J., Jia G., et al, "Study on the influence mechanism of air leakage on gas extraction in extraction boreholes," *Energ. Explor. Exploit*, vol. **40**, pp. 1344-1359, January 2022.
[26] Zhou Z., Zhao W., Zhang Y., et al, "Study on the Diffusion Law of Heavy Gas Leakage in Complex Scenarios Based on Scaled-Down Experiments," *Acs Omega*, vol. **9**, pp. 31533-31545, July 2024.

Research on Haze Pollution Network of Beijing-Tianjin-Hebei Region

Yubo Ma[1,*] and Siyu Cai[2]

[1] Research Institute of Electronic Science and Technology, University of Electronic Science and Technology of China, Chengdu, Sichuan, China.
[2] College of Mathematics, Sichuan University, Chengdu, Sichuan, China.
Email: caisiyu_cd@163.com; *yuboma_stat@163.com

Abstract. To investigate the dynamic haze pollution correlation among cities in the Beijing-Tianjin-Hebei (BTH) region, a dynamic directed weighted network of BTH haze pollution is established. In this network, nodes represent nine cities in the BTH region, and the weight of the edge is the haze pollution correlation between two cities, which is characterized by the proposed asymmetric upper-tail correlation coefficient formula based on the copula function. The evolution process of the dynamic network is analyzed through the network average weight, clustering coefficient, and network efficiency. The analysis of real haze pollution data shows that the overall correlation extent of the BTH haze pollution network is higher in the early and late stages of the entire time series. The general edge linking the state of the BTH haze pollution network reflects that the haze pollution from other cities has a positive impact on Tianjin and Cangzhou. In most periods, haze pollution from Baoding has affected most cities except Xingtai. Finally, this article gives some suggestions about the joint prevention and control of haze pollution based on the evolution of the haze pollution network.
Keywords. BTH region; copula function; dynamic asymmetric tail dependence; haze pollution network.

1. Introduction

In recent years, haze weather has occurred frequently and air quality has deteriorated. In response to the increasingly severe problem of haze pollution, the country has successively introduced relevant environmental protection policies and taken various measures to prevent and control air pollution. As the source of life that humans rely on for survival, air quality directly affects the quality of life. Therefore, the issue of haze pollution has become the primary problem that urgently needs to be solved. As a major focus of this blue sky defense battle, deepening comprehensive management of haze pollution and strengthening joint prevention and control of regional haze pollution have strategic significance.

With the deepening of people's attention to issues related to air pollution and haze pollution, some scholars have conducted research on the regional characteristics of haze pollution. Ma and Zhang [1] concluded that there is a significant spatial positive correlation between haze pollution in China by calculating the global Moran's I index. Pan et al. [2] used the Granger causality test and generalized impulse response function to confirm the spatial spillover effect of haze pollution between adjacent cities in the BTH region, while Beijing has a leading lagging relationship with other surrounding cities. Besides, Aliki et al. [3] investigate both local and regional aerosol pollution emissions of Greater Cairo.

Haze pollution is a continuous dynamic evolutionary process, and there is a certain correlation between haze pollution concentrations at adjacent time points. On the other hand, due to the meteorological, terrain, and other objective factors, the degree of mutual influence of haze pollution between cities is not consistent, so the study of the dynamic asymmetric correlation of haze pollution between cities has become one of the primary goals of this paper.

The study of dynamic asymmetric correlation is often used in the field of finance, and the main research methods are the AG-DCC-MVGARCH model [4] and the copula function [5]. Early studies on Copula methods mainly focus on static, single-parameter copula functions. However, financial data is often time series data, which is dynamically changed from moment to moment by changes in the external environment or changes in the internal market. Therefore, the time-varying copula function is more widely used. Mike et al. [6] used the copula-GARCH model with rattan structure to study the change characteristics of the yield of five Hong Kong blue chip stocks over time. Evkaya et al. [7] aim to investigate a regular Vine copula approach to estimate the complex inter-dependence patterns of the Istanbul Stock Exchange index.

In this article, a dynamic asymmetric upper-tail correlation coefficient between two cities in the BTH region is derived first. On this basis, the dynamic asymmetric correlation coefficient matrix of nine cities in the BTH region is obtained, and a dynamic directional weighted network is constructed. Finally, the evolution state of the haze pollution network in the BTH region is analyzed. The remainder of this article is organized as follows. Section 2 briefly reviews some related works and introduces the dynamic asymmetric tail correlation model and network. In Section 3, a comprehensive analysis is conducted on real haze pollution data. Finally, Section 4 gives the conclusion.

2. Methodologies

2.1. Copula Function

The copula function can represent the multi-dimensional joint distribution by the function combination of the one-dimensional edge distribution according to Sklar's theorem, which can be written as [8]:

$$H(x,y) = C\big(F(x), G(y)\big) \tag{1}$$

where $H(\cdot,\cdot)$ is joint probability density function (PDF), $F(\cdot)$ and $G(\cdot)$ are two marginal PDFs, and $C(\cdot,\cdot)$ is the copula function.

According to the two-dimensional Sklar theorem, the upper tail conditional probability can be expressed as:

$$P(X > x | Y > y) = \frac{1-F(x)-G(y)+C\big(F(x),G(y)\big)}{1-G(y)} \tag{2}$$

Copula contains many distribution families, where the elliptic distribution family and the Archimedean distribution family are two classical families [9]. The family of elliptic distributions includes normal distributions and t-distributions, while the three main copula functions in the Archimedean family are Gumbel copula, Clayton copula, and Frank copula [10]. Since the concentration distribution of PM2.5 is characterized by a thick tail, the Gumbel copula function, which is more sensitive to the upper tail, is adopted in this article.

2.2. Dynamic Asymmetric Correlation Coefficient

Since the current PM2.5 concentration of a city is also related to the previous stage, this article constructs an upper-tail correlation coefficient to depict the asymmetric dynamic tail correlation of haze pollution between two cities. The asymmetric upper-tail correlation coefficient between cities A and B at the t stage can be respectively expressed as:

$$\lambda_u^{A\to B} = P(B_t | B_{t-1}, A_t) - P(B_t | B_{t-1}) \tag{3}$$

$$\lambda_u^{B\to A} = P(A_t | A_{t-1}, B_t) - P(A_t | A_{t-1}) \tag{4}$$

where $\lambda_u^{A\to B}$ and $\lambda_u^{B\to A}$ are the asymmetric correlation coefficients from A to B and from B to A, respectively.

2.3. Three Network Characteristics

This article adopts the Pearson correlation coefficient to reflect the correlated weight w_{ij} of haze pollution between two cities i and j. To depict the features of an urban haze pollution network, three kinds of network characteristics are selected, including average weight, clustering coefficient, and network efficiency.

2.3.1. Average Weight. The average weight of the network can effectively present the strength of the average connectivity among nodes, which is expressed as:

$$\overline{w} = \frac{1}{n^2}\sum_{i=1}^{n}\sum_{j=1}^{n} w_{ij} \tag{5}$$

2.3.2. Clustering Coefficient. The clustering coefficient reflects the tightness of the internal connections of the network [11]. The higher clustering coefficient corresponds to the denser network structure, and vice versa. The weighted cluster coefficient of node i in the directed weighted network is expressed as:

$$C(i) = \frac{1}{k_i(k_i-1)}\sum_{j=1}^{k_i}\sum_{k=1}^{k_i}(\widetilde{w}_{ij}\widetilde{w}_{jk}\widetilde{w}_{ik})^{\frac{1}{3}} \tag{6}$$

where k_i denotes the degree of node i, $\widetilde{w}_{ij} = w_{ij} + w_{ji}$, $\widetilde{w}_{jk} = w_{jk} + w_{kj}$, $\widetilde{w}_{ik} = w_{ik} + w_{ki}$.

Then, the clustering coefficient of the network is denoted as:

$$C_N = \frac{1}{N}\sum_{i=1}^{N} C(i) \tag{7}$$

2.3.3. Network Efficiency. The network efficiency can be used to reflect the overall connectivity of the network, and the better the connectivity of the network, the higher the network efficiency [12]. Network efficiency is defined as:

$$E = \frac{1}{N(N-1)}\sum_{i=1}^{N}\sum_{j=1,j\neq i}^{N}\frac{1}{d_{ij}^{w}} \tag{8}$$

where $d_{ij}^{w} = \frac{w_{ij}\,w_{ji}}{w_{ij}+w_{ji}}$.

3. Data Analysis

This article selects the daily average PM2.5 concentration data of 9 cities in the BTH region from October 28, 2013, to October 9, 2018, for a total of 1800 days as samples. The unit is $\mu g/m^3$. All PM2.5 concentration data can be acquired from the national air quality real-time release platform (https://air.cnemc.cn/).

3.1. Analysis of BTH Haze Pollution Dynamic Network

According to the asymmetric upper-tail correlation coefficient, the correlation coefficient matrix of each stage is calculated. Taking the 1st, 718th, and 1435th stages as examples, the asymmetric correlation coefficient matrices and dynamic haze pollution network structures of nine cities are shown in figure 1. At the first and 1435th stages, the overall matrix color is dark red, indicating a strong correlation among most cities in the BTH region. While at the 718th stage, the color is light yellow, indicating a weak correlation among cities.

From the perspective of network structure, the evolution process of the haze pollution network in the BTH region shows that the number of edges in the 718th stage was significantly less than that in the 1st and 1435th stages. Besides, cities with high node degrees in the BTH network are Tianjin, Baoding, and Tangshan.

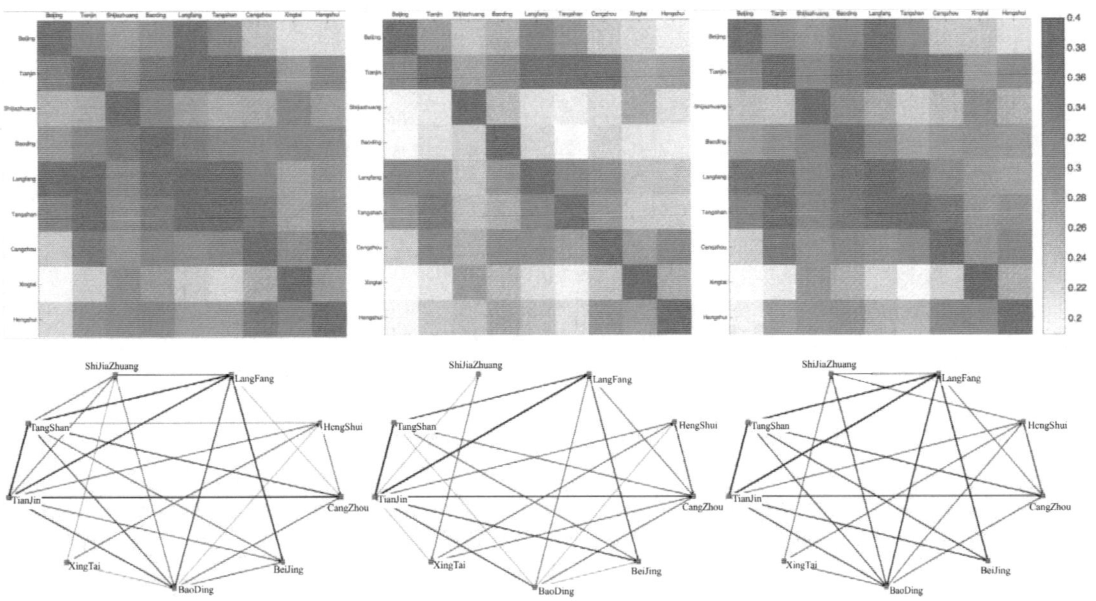

(a) 1st stage (b) 718th stage (c) 1435th stage

Figure 1. The asymmetric correlation coefficient matrices and dynamic haze pollution network structure of 9 cities.

3.1.1. Characteristic Analysis of BTH Haze Pollution Network. The BTH haze pollution network characteristics under three stages are obtained (table 1).

Table 1. Haze pollution network characteristics of BTH cities under three stages.

Stage	Average Weight	Clustering Coefficient	Network Efficiency
1st	0.3424	0.2068	0.1458
718th	0.3241	0.1757	0.1405
1435th	0.3688	0.1914	0.1541

As shown in table 1, the average weight of the 1435th stage is the highest, while that of the 718th is the lowest. It indicates that the average degree of interaction among cities is the highest at the end of the time series and is slightly weaker in the middle period. In the third column, the cluster coefficient of the first stage is slightly larger than that of the 1435th stage, and the cluster coefficient of the 718th stage is the smallest. This means that the overall connections within the network are sparse in the middle period, while the connections within the network are dense in the early and late periods. Meanwhile, it can be found that the network efficiency of the 718th stage is the lowest among the three states. This indicates that the correlation between internal nodes is stronger in the latest stage, and the correlation between internal nodes is weakest in the middle stage.

To observe the evolution processes of three characteristics of the BTH haze pollution network under all stages, the average edge weight, cluster coefficient, and network efficiency evolution diagrams of the BTH haze pollution network are generated (as shown in figure 2). The blue solid line, red dotted line, and black solid line represent the variation curves of the average weight, clustering coefficient, and network efficiency, respectively.

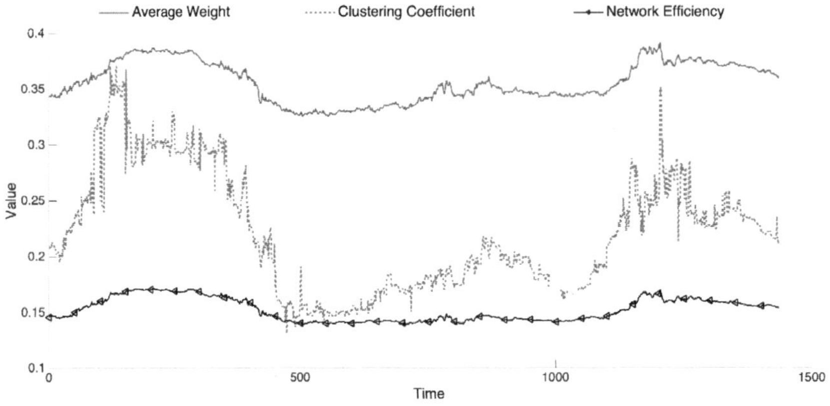

Figure 2. The evolution of average weight, clustering coefficient, and network efficiency in haze pollution network

It can be seen from figure 2 that the three characteristics possess the same trend generally, which indicates that the overall correlation of the BTH haze pollution network is high in the early and late stages, and low in the middle stages. Besides, the clustering coefficient fluctuates sharply compared with the average weight and network efficiency, which means the topological structure of the BTH network varies significantly with the change of time, which can be understood as the state of the entire network is not very stable. The reason for this phenomenon is probably related to objective factors such as the geographical environment and real-time meteorological conditions in the BTH region.

4. Conclusion

This article focuses on the time-varying and asymmetric characteristics of the haze pollution interaction among cities. Combining copula function and time series, dynamic asymmetric tail dependences between cities are derived, and a dynamic weighted BTH haze pollution network is constructed. Finally, the evolution process of the BTH network is analyzed. According to the variations of three network characteristics, it is found that the overall correlation among cities in the BTH region is relatively high at the early and late stages. While in the middle stage, it stabilizes at a relatively low correlation. Besides, haze pollution emission has a positive impact on Tianjin and Cangzhou, while haze pollution in Baoding affects most cities except Xingtai.

Due to the boundaryless and overflow characteristics of the transmission of haze pollutants, no city in the region can be self-sufficient in the prevention and control of haze pollution. Therefore, it is imperative to promote the integration of haze pollution prevention and control within the region. The focus of joint prevention and control in the entire region should be placed on important nodes and directional edges in the urban haze pollution network of the BTH region. Adopt stricter prevention and control policies for key nodes such as Baoding, Langfang, and Tangshan, and take more active measures to slow down or block the transmission of pollutants on important transmission routes such as Beijing-Langfang.

5. References

[1] Ma L and Zhang X 2014 The spatial effect of China's haze pollution and the impact from economic change and energy structure *China Industrial Economics* **(4)** 19-31. (in Chinese)

[2] Pan H, Wang X and Zhang S 2015 Duration and spatial spillover effects of haze pollution—evidence from Beijing-Tianjin-Hebei region *China Soft Science* **(12)** 134-143. (in Chinese)

[3] Aliki C, Spyros B, Efstratios B, et al. 2024 Submicron aerosol pollution in Greater Cairo (Egypt): a new type of urban haze *Environment International* **186** Art. No.108610.

[4] Huang W, Zhuang X, Yao S, et al. 2016 A financial network perspective of financial institutions' systemic risk contributions *Physica A: Statistical Mechanics and Its*

Applications **456(2)** 183-196.

[5] Solea E and Li B 2022 Copula Gaussian graphical models for functional data *Journal of the American Statistical Association* **117(538)** 781-793.

[6] Mike K and Cherry Y 2014 Vine-copula GARCH model with dynamic conditional dependence *Computational Statistics & Data Analysis* **(76)** 655-671.

[7] Evkaya O, Guer I, Kuelekci B, et al. 2024 Vine copula approach to understand the financial dependence of the Istanbul Stock Exchange index *Computational Economics*.

[8] Samadian M, Asadi E, Ghorbani M, et al. 2024 Joint analysis of drought affected by climate change in Zarinehrood watershed, Iran, using copula functions *Acta Geophysica* **72(5)** 1-15.

[9] Su H, Wang D, Xu S, et al. 2024 Exploring frequency-domain nonstationary of seismic ground motions using evolutionary power spectrum and Archimedean copula functions *Soil Dynamics and Earthquake Engineering* **177** Art. No.108364.

[10] Yang X, Chen Z and Qin M 2023 Joint probability analysis of streamflow and sediment load based on hybrid copula *Environmental Science and Pollution Research* **30(16)** 46489-46502.

[11] Chen B, Hou G and Li A 2024 Temporal local clustering coefficient uncover the hidden pattern in temporal networks *Physical Review E* **109(6)** Art. No.064302.

[12] Vragovic I, Louis E and Diaz G 2005 Efficiency of informational transfer in regular and complex networks *Physical Review E* **71(3)** 036122-036131.

Electrochemical-Activated Carbon Combined Process for Treating Municipal Landfill Leachate Membrane Concentrate: A Pilot Study

Liping Yang[1], Zhaoqi Liang[1] and Qi Han[2,*]

[1] Shenzhen Expressway Environment Co., Ltd., Shenzhen 518001, Guangdong, China.
[2] Harbin Institute of Technology (Shenzhen), School of Civil and Environmental Engineering, Shenzhen 518055, Guangdong, China.
Email: oy5086@163.com; liangzq@sz-expressway.com; *469071317@qq.com

Abstract. This study investigated the pilot treatment efficiency of the electrochemical-activated carbon combined process for the membrane concentrate of municipal landfill leachate in a southern city in China. The number of electrode plates, current density, and reaction time were optimized, and the stability of pollutant removal during continuous operation was examined. Based on this, the treatment effects and costs of the process were analyzed. The results show that, at the pilot scale, the optimal parameters for the electrochemical oxidation process are three pairs of electrode plates, a current density of 20 mA/cm^2, and a reaction time of 30 minutes, resulting in an effluent COD of 573 mg/L and an energy consumption of 40.9 kW·h/kgCOD. During continuous operation treating the membrane concentrate, the effluent COD stabilized around 570 mg/L, conductivity around 19.4 mS/cm, and pH around 8.6. The pollutant removal effect was close to that from small-scale tests, indicating stable continuous operation. After treatment with the electrochemical oxidation-activated carbon combined process, the effluent meets MVR inflow standards, with a treatment cost of 19.95 yuan/t of water.

Keywords. Municipal landfill leachate; membrane concentrate; electrochemical-activated carbon combined process; treatment efficiency.

1. Introduction

With the rapid development of "zero waste city" construction, municipal landfill leachate has become a key challenge in waste management. At present, the typical treatment method in China is a combination of "pre-treatment-biological treatment-depth treatment" [1]. Among these, depth treatment often utilizes membrane methods primarily based on nanofiltration (NF) or reverse osmosis (RO) [2], generating a certain amount of membrane concentrate (approximately 25%-45% of the treated water volume) [3]. Due to the high pollutant concentration and salinity in the membrane concentrate, biological treatment is difficult [4], leading to the common use of physicochemical methods. Mechanical vapor recompression (MVR) is a widely used energy-efficient technology in the evaporation and crystallization systems for membrane concentrates, aiming for wastewater reduction, resource recovery, and even zero discharge [5]. However, the "Technical Standards for Municipal Solid Waste Leachate Treatment" (CJJ/T150-2023) stipulates design inflow indicators for MVR used for resource recovery from membrane concentrates (hereinafter referred to as "MVR inflow standards"). Since membrane concentrates contain a significant amount of organic matter, ammonia nitrogen, and hardness ions, direct entry into the MVR evaporator can lead to scaling, impure salt crystallization, and substandard condensate [6]. Therefore, effective treatment of the membrane concentrate is required before entering the MVR system to meet the MVR inflow requirements.

Currently, to address the gap between the water quality characteristics of membrane concentrates (such as high organic content, high color, high hardness, etc.) and the relevant standards for MVR inflow, advanced oxidation, ion exchange resins, and other processes are commonly employed [7,8].

For example, for organic matter removal, methods such as ozone oxidation and electrochemical oxidation are typically used; however, standalone advanced oxidation methods tend to have high costs and need to be integrated with other technologies [9,10].

This study primarily conducts pilot experiments on the electrochemical-activated carbon combined pre-treatment process, further optimizing the number of electrode plates, current density, and reaction time, while examining the stability of continuous operation. Ultimately, an overall process effectiveness evaluation is carried out based on pollutant removal efficiency, sludge production, reagent dosage, and energy consumption, providing a reference for subsequent MVR treatment of municipal landfill leachate membrane concentrates.

2. Experimental Section

The pilot experimental device in this study mainly consists of a power supply, electrolytic cell, and electrode plates. The electrolytic cell is made of acrylic panels and consists of four reaction tanks with dimensions of 25 cm×10 cm×30 cm, connected in series, with a maximum volume of 30 L. The electrode size is 15 cm×10 cm×0.2 cm, with an inter-electrode distance of 2 cm. The electrochemical oxidation pilot experimental device is shown in figure 1.

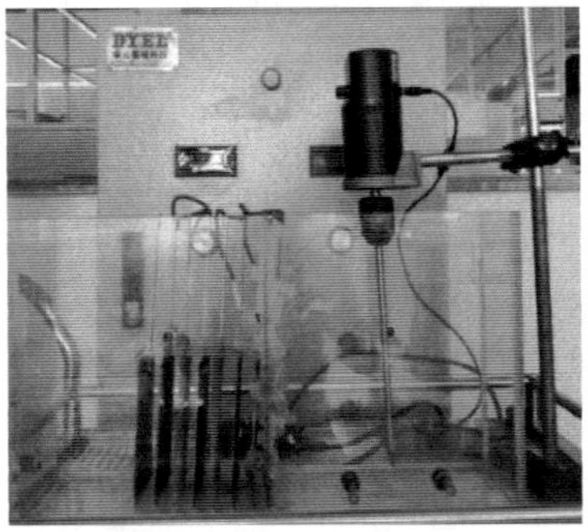

Figure 1. Experimental setup diagram.

3. Results and Discussion

3.1. Optimization of Electrochemical Oxidation Process Conditions

3.1.1. Number of Electrode Plates. Increasing the number of electrode plates can reduce the system impedance; however, as the number of plates increases, the current density decreases, and after a certain number, the impact of the number of plates on system impedance diminishes. Therefore, optimization of the number of plates is necessary. The results are shown in figures 2(a) and 2(b).

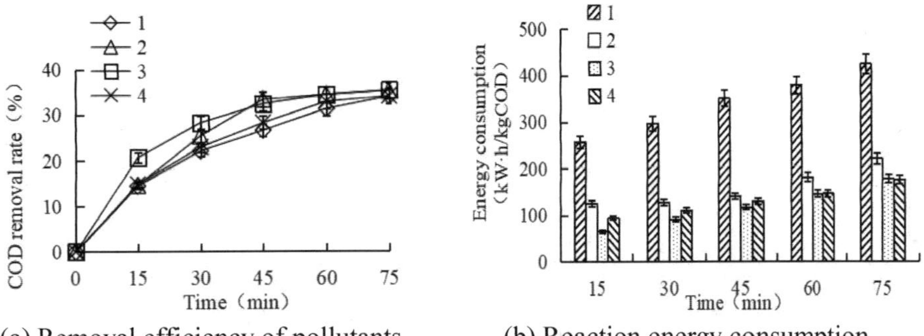

(a) Removal efficiency of pollutants (b) Reaction energy consumption

Figure 2. Optimization of extreme board numbers.

From figure 2(a), it can be seen that within the first 60 minutes, the COD removal rates when using 2, 3, and 4 pairs of plates are all higher than when using only 1 pair of plates. This may be because, when the input current remains constant, a smaller electrode area results in higher current density. Once the current density exceeds a certain level, secondary reactions increase, converting some electrical energy into heat. Additionally, the substantial generation of active substances on the electrode surface within a unit of time and the localized high concentration also somewhat affect the mass transfer efficiency of the active substances [5,6]. The COD removal rate stabilizes after 60 minutes.

From figure 2(b), it can be observed that the overall energy consumption of the system decreases as the number of electrode plates increases. This is because an increased number of plates leads to a reduction in the overall resistance of the reaction system. By calculation, the average resistance of the reaction system for 1, 2, 3, and 4 pairs of plates in the first 60 minutes is 3.37 Ω, 1.75 Ω, 1.41 Ω, and 1.34 Ω, respectively. The smaller the system resistance, the fewer secondary reactions occur, significantly reducing energy consumption [8]. Considering that the pollutants are mostly removed after 60 minutes and that increasing the number of plates beyond 3 pairs does not significantly reduce system resistance and energy consumption while decreasing pollutant removal efficiency due to reduced current density, the optimal choice is to use 3 pairs of plates, with a COD removal rate of 34.4% and energy consumption of 43.2 kW·h/kgCOD at 60 minutes.

3.1.2. Current Density and Reaction Time. Due to variations in system impedance and electrode area, it is necessary to determine the optimal current density and reaction time. The results are shown in figure 3.

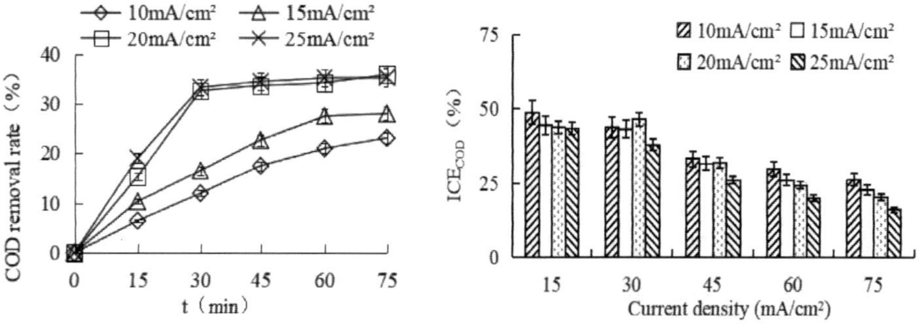

(a) Removal efficiency of pollutants (b) Current efficiency

Figure 3. Optimization of current density and reaction time.

From figure 3(a), it is evident that a higher current density leads to a faster approach to equilibrium in COD removal. When the current density is 20 mA/cm², further increasing the current density does not significantly enhance the pollutant removal rate, indicating that 20 mA/cm² is a suitable current density. At this current density, the COD removal rate stabilizes after 30 minutes.

From figure 3(b), it can be seen that during the first 30 minutes, the current efficiency is higher at current densities of 10-20 mA/cm². However, further increases in current density or time lead to a decrease in current efficiency, as the degradable organic matter is nearly removed, making it difficult to further reduce COD. When the current density is 20 mA/cm², both the pollutant removal rate and current efficiency are high, and the reaction time is relatively short. Therefore, the selected parameters are a current density of 20 mA/cm² and a reaction time of 30 minutes, resulting in a COD removal rate of 32.7% and energy consumption of 40.9 kW·h/kgCOD.

3.2. Investigation of Continuous Operation Characteristics

To explore the characteristics of the electrochemical oxidation process during continuous operation, this section focuses on the electrochemical oxidation of nanofiltration membrane concentrate in a sequential batch treatment process, evaluating the stability of continuous operation based on pollutant removal effects, with results shown in figure 4.

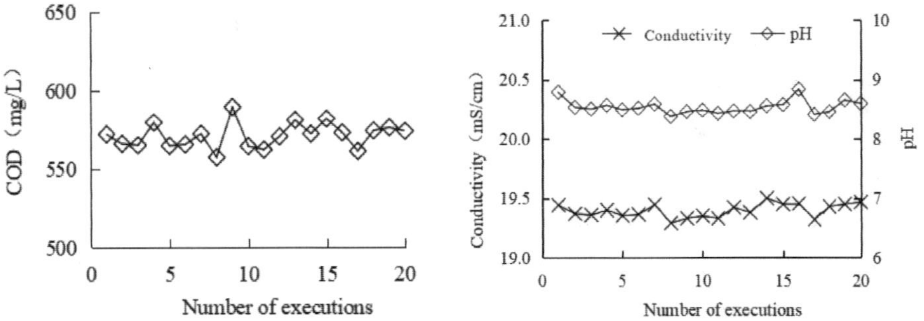

(a) Removal efficiency of pollutants (b) Changes in conductivity and pH

Figure 4. Continuous operation conditions in the experiment.

From figure 4(a), it can be seen that the effluent COD stabilizes at around 570 mg/L, with various indicators close to the effluent from the pilot test, indicating that the pilot-scale electrochemical oxidation unit maintains stable pollutant removal efficiency under continuous operation conditions. This is because the effluent stability of the electrochemical oxidation process is mainly related to the performance of the anode material. Due to the dimensionally stable anode (DSA), no dissolution occurs during the electrochemical oxidation process, and gas is continuously generated during the reaction, allowing bubbles to timely remove fouling substances attached to the electrode surface as they rise, resulting in stable effluent quality [9].

From figure 4(b), the effluent conductivity remains around 19.4 mS/cm, indicating that electrochemical oxidation has little impact on the removal of salt substances, thus not affecting subsequent MVR salt crystallization. The effluent pH is maintained around 8.6, similar to the pilot results.

3.3. *Overall Process Performance Evaluation*

3.3.1. Analysis of Pollutant Removal Efficiency. The water quality changes and standards during the overall process of treatment are shown in table 1. It can be seen from the table that the COD and color of the NF concentrate are mainly reduced through electrochemical oxidation and activated carbon adsorption stages; the NH_3-N concentration in the membrane concentrate is low and is synchronously removed during electrochemical oxidation; the overall process does not require pH adjustment, and the process effluent pH is 8.2. After treatment with electrochemical oxidation-activated carbon adsorption, the process effluent meets the MVR inlet standard.

Table 1. Water quality changes and standards during the overall process of treatment.

Indicator	COD (mg/L)	NH_3-N (mg/L)	Color (PCU)	Total Hardness (mg/L)	SS (mg/L)	pH
NF Concentrate	6500	7	5600	305	0	7.5
After Electrochemical Oxidation	575	0.2	18	88	0	8.5
After Activated Carbon Adsorption	481	0	15	80	0	8.2
Standard	500	20	50	100	10	8-10

3.3.2. Cost Calculation. When calculating the cost of chemicals, market prices are used for conversion. Saturated activated carbon can be recycled through heating, biological, and other regeneration methods, showing significant economic efficiency [3]. The regeneration rate of spent activated carbon can reach 80%, with low losses, and the regeneration cost is about 60% of the cost of replacing new carbon. The cost calculation assumes new carbon at 5 yuan/kg, with the regeneration cost of activated carbon being about 60% of the cost of replacing new carbon, at 3 yuan/kg. The total cost is shown in table 2. When treating NF concentrate water with electrochemical oxidation-activated carbon adsorption combined process, the treatment cost is 19.95 yuan/t of water, and the effluent meets the MVR inlet standard.

Table 2. Cost calculation table.

Process	Item	Usage (kg/t water)	Market Price (yuan/kg)	Cost (yuan/t water)
Electrochemical Oxidation	Electricity	11.36	0.7	7.95
Activated Carbon Adsorption	Activated Carbon	4	3	12
Total Cost				19.95

4. Conclusion

This study conducted pilot-scale electrochemical oxidation experiments, optimized the number of electrodes, current density, and reaction time, and examined the stability of pollutant removal efficiency during continuous operation of the device. At last, the treatment effects and costs of each process stage were analyzed. The main conclusions are as follows:

(1) The optimal parameters for the pilot-scale electrochemical oxidation process are 3 pairs of electrodes, a current density of 20 mA/cm^2, and a reaction time of 30 minutes. At this point, the effluent COD is 573 mg/L, and the reaction energy consumption is 40.9 kW·h/kgCOD. The pollutant removal efficiency is close to that of the pilot test, with a slight reduction in reaction energy consumption.

(2) When continuously treating NF concentrate water, the effluent COD stabilizes at around 570 mg/L, conductivity at around 19.4 mS/cm, and pH at around 8.6. The pollutant removal efficiency is close to that of the pilot test, showing stable continuous operation.

(3) After treating NF concentrate water with an electrochemical-activated carbon process, the effluent meets the MVR inlet standard, with a treatment cost of 19.95 yuan/t of water.

5. References

[1] Zhang X 2019 Comparison and analysis of Anaerobic+MBR+NF+RO and two-stage DTRO technology for landfill leachate treatment. *Technology of Water Treatment* **45(9)** 126-129+132.

[2] Keyikoglu R, Karatas O, Rezania H, et al. 2021 A review on treatment of membrane concentrates generated from landfill leachate treatment processes *Separation and Purification Technology* **259** 118182.

[3] Ahmed Z, Yusoff M S, Kamal N, et al. 2021 Optimization of the humic acid separation and coagulation with natural starch by RSM for the removal of COD and colour from stabilized leachate *Waste Management & Research* **39(11)** 1396-1405.

[4] Long Y, Xu J, Shen D, et al. 2017 Effective removal of contaminants in landfill leachate membrane concentrates by coagulation *Chemosphere* **167** 512-519.

[5] Jiang C, Cui S, Han Q, et al. 2019 Study on application of activated carbon in water treatment *IOP Conference Series Earth and Environmental Science* **237(2)** 022049.

[6] Dong Y, Li Y, Kong F, et al. 2020 Source, structural characteristics and ecological indication of dissolved organic matter extracted from sediments in the primary tributaries of the Dagu river *Ecological Indicators* **109(Feb.)** 105776.

[7] Yan C, Tian Y, Cheng Z, et al. 2021 Simultaneous desalination and removal of recalcitrant organics from reverse osmosis leachate concentrate by electrochemical oxidation *ACS Omega* **6(24)** 16049-16057.
[8] Chen W, He C, Zhuo X, et al. 2020 Comprehensive evaluation of dissolved organic matter molecular transformation in municipal solid waste incineration leachate *Chemical Engineering Journal* **400** 126003.
[9] Ren T F, Lu K C, Yin M X, et al. 2024 Single-atom cobalt enables efficient catalytic ozonation for advanced wastewater treatment: Mechanisms and application *Chemical Engineering Journal* **500** 156805.
[10] Svante R, Sanne J. S, Lutz A, 2024 Suspect and non-target screening of per- and polyfluoroalkyl substances (PFAS) and other halogenated substances in electrochemically oxidized landfill leachate and groundwater *Journal of Hazardous Materials* **480** 136316.

Study on the Coupling and Coordination Relationship between Ecological Services and Community Development in Nature Reserves

Dongping Xu[1], Yao Tong[1], Qiuling Lu[2,*] and Liang Tian[1]

[1] School of Civil Engineering and Architecture, Xi'an University of Technology, Xi'an, Shaanxi, China.
[2] Northwest Surveying, Planning Institute of National Forestry and Grassland Administration, Key Laboratory National Forestry and Grassland Administration on Ecological Hydrology and Disaster Prevention in Arid Regions, Xi'an, Shaanxi, China.
Email: 379821411@qq.com; 1124827992@qq.com; *253087478@qq.com; 18374457238@163.com

Abstract. Starting from the perspective of the harmonious coexistence of man and land, we construct a comprehensive evaluation index system of coupling coordination, study the coupling coordination relationship of "ecological service-community development" in nature reserves, and analyze the driving factors affecting the coupling coordination relationship. The study shows that: (1) the coupling degree indicates that the "ecological service-community development" of the nature reserves in the case shows a high level of coupling, and the main factors affecting the coupling state are mainly the sub-indicators of the ecological service system. (2) The degree of coordination shows significant differences in the degree of coordination of the nature reserves. The driving factors of protected areas in different coordinated development states are different, the driving factors of highly coordinated protected areas are mainly the proportion of tourism and the degree of spatial aggregation of the community; the driving factors of moderately coordinated nature reserves are mainly the proportion of tourism and the abundance of green tourism resources; and the driving factors of lowly coordinated nature reserves are mainly the degree of aggregation of the community and the average elevation. (3) The results of the synergistic effect show that only Yinghu Wetland Provincial Nature Reserve is the leading type of "community development"; the leading type of "ecological service" mainly includes Micangshan National Nature Reserve and Nangongshan National Geopark; The "synchronized development" type is mainly Hanzhong Crested Ibis National Nature Reserves, Shiquan Yanxiang Cave Geopark and Liping National Forest Park.
Keywords. Nature reserves; community development; ecological services; coupling degree; grey correlation analysis.

1. Introduction

Nature reserves aim to achieve long-term conservation of natural resources, ecosystem services and cultural values, and are of primary importance in maintaining national ecological security [1]. Nature reserves not only carry the important mission of maintaining ecological balance and protecting biodiversity but also bear the burden of ecosystems as the main resource carrier of life production, agricultural cultivation, tourism and other human activities, which inevitably produce conflicts of interest between the main parties. How to harmonize the relationship between nature reserves and their internal or neighboring communities and balance the interests of all parties is still the focus of attention and a practical problem that needs to be solved urgently.

Ecological services, as an important part of the many complex functions of nature reserves, have always been a research hotspot in the academic community, and Costanza [2] systematically proposed for the first time that nature reserves have 17 ecosystem services, including soil and water conservation, biological control, water supply, genetic resources and other ecosystem services. "The implementation of the Millennium Ecosystem Assessment (MA) project signaled that ecosystems and human well-being have become the core issue and development direction of ecological research. In the new period of China, the focus of research shows a trend of shifting from macro-scale to micro-scale, and many scholars began to focus their research on rivers, mountain ranges, and nature reserves, etc. [3-5]. A large number of studies have shown that ecological service functions help to maintain the health and stability of natural ecosystems and improve ecosystem service functions; moreover, in serving society, they can provide high-quality ecological products for the people and public services such as scientific research and recreation for the whole society [6,7].

Nature reserves have ecological preservation as their primary purpose, at the expense of community interests [8]. As the academic debate over whether to protect or utilize the resources of nature reserves continues to intensify, community development has begun to receive attention. Some scholars affirm the benefits that nature reserves bring to the ecology but also point out that only the neigh-boring communities bear the cost of protecting the ecology, which seriously hinders the backward development groups around the reserves from sharing the dividends of the economic development of the whole society [9,10]. How to harmonize nature reserves and community development has become a difficult problem for scholars and managers.

The report of the 20th Party Congress proposes to comprehensively promote rural revitalization, which puts forward the dual requirements of ecological protection and rural revitalization for nature reserves and their surrounding communities, and many scholars have begun to conduct multi-dimensional research on the relationship between nature reserves and community development. Based on different perspectives, many scholars' studies on the relationship between nature reserves and local development show that conservation and development should be viewed in a unified and dialectical way [11]. Ecological services, as the foundation of sustainable community development, have rarely been used as a perspective to explore the synergistic relationship with rural communities in the process of development.

Based on this, this paper takes the Qinba mountainous area as the research object, explores the coupled and coordinated development status of "ecological service-community development" in different types of nature reserves, and scientifically identifies the main factors constraining the coupled and coordinated development of the system, to provide scientific references for the ecological environmental protection of the nature reserves, improvement of people's livelihoods, and enhancement of the community's awareness.

2. Overview of the Study Area and Analysis of the Coupling Coordination Mechanism

2.1. Overview of the Study Area

In order to better study the "ecological service-community development" coupling and development status of nature reserves, this paper chooses the Qinba mountainous area, which has a large number and rich variety of nature reserves, as a research case study, and selects a variety of types of nature reserves as the research object, mainly including nature reserves, Forest parks, wetland parks, geological parks, etc. Specifically, there are six nature reserves: Liping National Forest Park, Nangong Mountain National Geological Park, Crest-ed Ibis National Nature Reserve, Micang Mountain National Forest Park, Shiquan Yanxiang Cave Geological Park, Yinghu Wetland Provincial Nature Reserve and so on, and the details of the nature reserves are shown in figure 1.

Journal of Physics: Conference Series **2941** (2025) 012064 doi:10.1088/1742-6596/2941/1/012064

Figure 1. Schematic location of typical nature reserves in the Qinba Mountain area.

2.2. Coupled Coordination Mechanisms

Coupling coordination refers to the interaction between two or more elements and movement systems through material transfer and energy transfer and is now mostly used in geography environmental science and other research. In the study of human-land relations, it is used to express the complex interactive relationship between people and the environment, and the interdependence, influence, feedback and unification between different systems [12]. Correctly recognizing the coupling and coordination mechanism of "ecological services and community development" and coordinating "ecological services and community development" is the key to realizing the harmonious coexistence of human beings and nature.

The relationship between ecological services and community development is characterized by the unlimited demand for resources and the limited capacity to supply ecological services. From the perspective of community development, whether it is community construction, agricultural production and other means of living, land use, tourism development or other production activities, all need to obtain resources from the ecological environment of the nature reserve and return some of the resources to the ecological environment system as garbage. From the perspective of ecological services, after the level of community development has increased, there are sufficient financial resources and technology to provide power to support the management of the environment, the restoration of ecosystems, the alleviation of environmental pressures and the protection of living organisms.

3. Research Methodology and Data Sources

3.1. Research Methods

3.1.1. Coupling Coordination Model. The degree of mutual influence between systems or between elements within a system is called the degree of coupling. In this paper, the human-land relationship system is mainly "ecological services - community development" in nature reserves, and its coupling coordination degree model is calculated as follows:

$$C = \left\{ \frac{f(x) \times f(y)}{\frac{f(x)+f(y)}{2}} \right\}^{\frac{1}{2}} \tag{1}$$

$$f(x) = \sum_{i=1}^{n} a_i x_i, i = 1, 2, \ldots, n \tag{2}$$

$$f(x) = \sum_{i=1}^{n} b_i y_i, j = 1, 2, \ldots, n \tag{3}$$

Explanation: *C* represents the system coupling degree; ai and *bj* represent the weight of each indicator of nature reserves and rural community subsystems, respectively; *f(x)* and *f(y)* represent the development index of nature reserves and rural community subsystems, respectively, which is weighted by the indicators. The coupling degree judgment, specifically shown in table 1, divides the coupling degree into four levels, the larger *C* means the stronger the relevance of the two systems, the higher the level of system coupling, and vice versa, the lower.

Table 1. Coupling Degree C judgment table.

Coupling Degree	(0,0.3]	(0.3,0.5]	(0.5,0.8]	(0.8,1.0]
Typology	Low-level coupling	Antagonistic coupling	Break-in phase	High-level coupling

In order to judge the overall level of development of the subsystems, it is also necessary to introduce the degree of coordination and the coefficient of synergistic development to reflect more comprehensively and accurately the overall level of development between the "ecosystem" and the "community development" as well as the degree of leadership of the system, and the formulas for calculating the degree of coordination and the coefficient of synergistic development are shown in equation (4)-(6).

$$T = \alpha f(x) + \beta f(y) \tag{4}$$

$$D = \sqrt{C \times T} \tag{5}$$

$$U = f(x)/f(y) \tag{6}$$

Explanation: *C* is the degree of coupling, *D* is the degree of coordination, *T* is the comprehensive coordination index of the two systems, and *U* is the coefficient of synergistic development. α and β are the weight coefficients, $\alpha > 0, \beta > 0$, and $\alpha + \beta = 1$. Concerning related research, the level of coordination is divided into five types, and synergistic development is divided into three types, and combined with the degree of coordination - synergy unified classification, such as the "high degree of coordination - community development leading type" is recorded as, "highly coordinated - community development leading type" is labeled as "V-Uc" type, and the rest are also the same (tables 2 and 3).

Table 2. Coordination Degree D judgment table.

Terms of judgment	(0, 0.2]	(0.2, 0.4]	(0.4, 0.6]	(0.6, 0.8]	(0.8, 1]
Coordinated state	Dysregulated state	Near-disorder coordination	Low coordination	Moderate coordination	High coordination
Classification	I	II	III	IV	V

Table 3. Synergy development factor U judgment table.

Terms of Judgment	(0,0.8]	(0.8,1.2]	(1.2, ∞]
Synergy State	Community Development	Community Development	Ecological Service
Classification	Uc	Uce	Ue

3.2. Data Sources and Processing

Vector data for this study were obtained from the China Basic Geographic Information Database (http://www.ngcc.cn); water resource abundance, biological abundance index, primary productivity, population kilometer grid data, and per capita income were obtained from the Center for Resource and Environmental Science and Data of the Chinese Academy of Sciences (http://www.dsac.cn/DataProduct/Index); nighttime lighting data from two satellite data sources, DMSP/OLS and NPP/VIIR, from the National Oceanic and Atmospheric Administration's National Geophysical Data Center (htps://ngdc.noaa.gov/eog/dmsp/downloadV4composites.html); DEM, slope data from the geospatial database according to the cloud platform (http://www.gscloud.ocn); integrated spatial data of nature reserves from the Northwest Survey and Planning Design Institute of the State Forestry and Grassland Administration; and social data such as village area, per capita basic farmland area, and the construction of public service facilities from the database of the master plan of key districts, counties, and townships in Hanzhong City and the field research. The study obtained first-hand research data on village settlements through the study of image maps, combined with information on the current resident population of Ankang Hanzhong City and field research on some village settlements. In order to eliminate the differences between the indicators, dimensionless processing is carried out so that the data can achieve uniformity.

$$x_i^* = \frac{x_i - x_{min}}{x_{max} - x_{min}} \tag{7}$$

$$x_i^* = \frac{x_{max} - x_i}{x_{max} - x_{min}} \tag{8}$$

Equation (7) applies to positive indicators and equation (8) applies to negative indicators.x_i and x_i^* denote the values before and after data normalization, and x_{min},x_{max} denote the minimum and maximum values in the sample data, respectively. After normalization, all the quantitative data are changed into dimensionless data in the interval [0, 1].

3.3. Construction of the Indicator System

Based on the principles of systematicity, scientificity, comprehensiveness, hierarchy and data accessibility, this paper combines the existing research results to construct a total of 19 tertiary evaluation indicators of the coupled rating index system of ecological services and community development of nature reserves, as shown in table 4. Considering the different degrees of different indicators on the role of the coupling, we utilize the hierarchical analysis of hierarchical analysis (AHP) method for the empowerment of the indicators, and we invite eight experts to make importance judgments on the indicators sequentially and construct a judgment matrix to test the consistency of the judgment matrix. A total of eight experts were invited to make important judgments on the indicators in turn and constructed a judgment matrix to test the consistency of the judgment matrix, which resulted in the random consistency ratio CR=CI/RI=O.0107<0.05, indicating that the indicators of each layer passed the consistency test, and the final calculation was made to obtain the weight value of each indicator layer.

Table 4. Indicator system and weighting.

Dimension	Categorization	Indicator	Indicator interpretation /formulation	Unit	Indicator weights	Direction
Ecological Service	Background conditions	Average Elevation (A1)	Average elevation of each protected area	m	0.0625	-
		Average Slope (A2)	Average slope of each reserve	°	0.0625	-
		Water	Regional average	/	0.1250	+

Journal of Physics: Conference Series **2941** (2025) 012064 doi:10.1088/1742-6596/2941/1/012064

Dimension	Categorization	Indicator	Indicator interpretation /formulation	Unit	Indicator weights	Direction
		Resource Abundance (A3)	water index			
		Biological Richness (A4)	Regional species counts and comprehensive assessment of nationally protected flora and fauna	/	0.1250	+
		Abundance of Green Tourism Resources (A5)	Abundance of tourism resources	/	0.1250	+
		Forest Cover (A6)	Forest cover/total protected area	/	0.1362	+
	Natural resource conditions	The First Productive Force (A7)	The amount of energy fixed or organic matter produced by green plants through photosynthesis per unit area and per unit time	/	0.1481	+
		Green Space Per Capita (A8)	Area of green space/total population of the protected area	/	0.1114	+
		Vegetation Cover (A9)	Area covered by vegetation/total protected area	/	0.1044	+
		Population Density (B1)	Number of people in protected communities	person/km^2	0.099	+
Community development	Social development	Farmland Production Potential (B2)	The ability to maximize the yield and quality of agricultural products that can be achieved under specific natural conditions and human management	/	0.1359	+
		Village Area (B3)	Village area is the total area of geographic space occupied by a village	m^2	0.084	+

Journal of Physics: Conference Series **2941** (2025) 012064 doi:10.1088/1742-6596/2941/1/012064

Dimension	Categorization	Indicator	Indicator interpretation /formulation	Unit	Indicator weights	Direction
		Building Land Per capita (B4)	Total area of built-up land/total population of the community	m^2	0.0906	+
		Degree of Spatial Aggregation of Communities (B5)	Refers to the spatial concentration of communities within a community	/	0.0906	+
		Basic Farmland Per Capita (B6)	Basic farmland area/total population	m^2	0.0834	+
		Commercial Forest Area Per Capita (B7)	Commercial forest area/total population	m^2	0.0568	+
		Share of Tourism (B8)	Proportion of agricultural income to total income	/	0.1016	+
	Economic development	Income Space Data (B9)	On the basis of national sub-county income statistics (taking into account the closely related income interaction pattern of human activities)	$yuan/km^2$	0.1183	+
		Nighttime Lighting Data (B10)	Human activity characterization (based on United States satellite data sources)	dm	0.1399	+

4. Empirical Results and Analysis

4.1. Analysis of the Harmonization of "Ecological Services-community Development" Coupling

Based on the evaluation index system constructed in this paper, the standardized data were substituted into the coupling coordination model to obtain the evaluation results of the coupling coordination between "ecological services" and "community development" in six nature reserves, as shown in table 5. The results are shown in table 5, where $f(x_i)$ denotes the comprehensive level of ecological services and $f(y_i)$ denotes the comprehensive level of community development, C_i denotes the degree of coupling between the systems of "ecological services-community development", which is used to measure the correlation between the two. D_i denotes the degree of coordination between the "ecological services-community development" systems, the higher the value, the higher the level of coordination between the two. U_i is the ratio between the "ecological services-community development" system, which is used to measure the extent to which "ecological services" and "community development" are leading or lagging behind each other. U_i denotes the ratio between the "ecological services" and "community development" systems.

Table 5. Nature reserves the "ecological services-community development" coupling harmonization index.

Nature Reserves	Indexes						Typology			
	$f(x_i)$	$f(y_i)$	T_i	C_i	D_i	U_i	Coupling Degree	Coordination Degree	Synergistic Effect	Type Label
Liping National Forest Park	0.232	0.251	0.367	0.992	0.603	0.926	High	Moderate	Synchronized Development	IV-Uce
Nangong Mountain National Geopark	0.246	0.100	0.223	0.971	0.465	2.468	High	Low	"Eco-services" Lead	III-Ue
Crested Ibis National Nature Reserve	0.461	0.442	0.673	0.997	0.819	1.042	High	High	Synchronized Development	V-Uce
Micangshan National Nature Reserve in Xi'an	0.336	0.090	0.258	0.844	0.467	3.717	High	Low	"Eco-services" Lead	III-Ue
Shiquan Yanxiangdong Geopark	0.155	0.145	0.223	0.971	0.665	1.068	High	Moderate	Synchronized Development	IV-Uce
Yinghu Wetland Provincial Nature Reserve	0.215	0.328	0.435	0.995	0.659	0.656	High	Moderate	"Community Development" Lead	IV-Uc

4.1.1. Coupling Evaluation Results. The results of the coupling degree evaluation show that the coupling degree of ecological services and community development in the six study cases all show a high level of coupling development (above 0.8), which indicates that the degree of interaction between "ecological services" and "community development" is more significant in the nature reserves in the Qinba Mountain area. This high level of coupling indicates that the interaction between "ecological services" and "community development" in the nature reserves in the Qinba Mountain area at this stage is relatively significant. At the same time, due to the limitations of the coupling degree index, the coupling degree evaluation cannot effectively reflect the coordinated development status between "ecological services" and "community development" of nature reserves, and needs to be further combined with the analysis of the degree of coordination to determine the degree of interaction between "ecological services" and "community development". It needs to be further analyzed with the degree of coordination to judge the level of coordinated development between "ecological services" and "community development".

4.1.2. Coordination *Evaluation Results.* Based on the evaluation of the coupling degree, we further evaluated the coordination degree of "ecological service" and "community development" of nature reserves in the Qinba Mountain area. The results show that only Hanzhong Crested Ibis National Nature

Reserve is in a high degree of coordinated development (0.819), Yinghu Wetland Provincial Nature Reserve (0.659), Shiquan Yanxiang Cave Geopark (0.665) and Lai National Forest Park (0.604) are in a medium degree of coordinated development, while Micangshan National Nature Reserve (0.467) and Nangongshan National Geopark (0.465) are in a medium degree of coordinated development, while Micangshan National Nature Reserve and Nangongshan National Geopark (0.465) are in the medium degree of coordinated development. The three nature reserves, such as Micangshan National Nature Reserve (0.467) and Nangongshan National Geopark (0.465), are in a state of medium coordinated development, while the two nature reserves are in a state of low coordinated development. This shows that although the "ecological services" and "community development" of nature reserves are coupled at a high level, the "ecological services" and "community development" of each reserve are in a low degree of coordinated development. However, there are obvious differences in the degree of coordination between "ecological services" and "community development" in each protected area, indicating that the development pace of each subsystem of the protected areas is not convergent at this stage. It is necessary to comprehensively examine the synergistic effect of "ecological services" and "community development" and analyze the degree of leadership of each subsystem in the development of nature reserves under different degrees of coordination.

4.1.3. Synergy Evaluation Results. Based on the evaluation of the degree of coordination, the systematic effects of "ecological services" and "community development" in the development of nature reserves were identified. The evaluation results show that the nature reserves in the study case exhibit three different synergistic development states: "ecological service" leading, system development and "community development" leading. Among them, three nature reserves are in the state of synchronized development, two are in the state of "ecological service" leading development, and one is in the state of "community development" leading development. The nature reserves in a synchronized development state show that the coordination between "ecological services" and "community development" is good, and the two can promote the coordinated development of the system as a whole; the nature reserves with unilateral development in the lead have certain aspects of "ahead of the curve" and "ahead of the curve" in some aspects. The nature reserves with unilateral leading development have certain aspects that are "ahead" or "lagging", and future development may jeopardize the coordinated development of the nature reserves, so it is necessary to strengthen the shortcomings of the development or conditionally adjust the factors of ahead development, to promote the sustainable and coordinated development of the nature reserves.

4.2. Analysis of Major Influencing Factors

In this paper, we use the gray correlation co-evaluation model to explore further the main factors affecting the degree of coupling and coordination of "ecological services-community development", and the results are shown in figure 2.

4.2.1. Main factors influencing the coupling degree. The main factors affecting the "ecological service-community development" coupling degree C of nature reserves in the Qinba Mountains were, in descending order of influence, primary productivity (A7), proportion of tourism (B8), average elevation (A1), production potential of farmland (B2), richness of green resources (A5) and average slope (A2). Average slope (A2). Among them, the number of sub indicators of the "ecological services" sub-system accounts for 66.7%, and the contribution rate of the indicators accounts for 68.26%, which indicates that the "ecological services" sub-system plays a major role in influencing the development status of the coupled nature reserves in the case of Qinba mountainous areas, and the "community development" sub-system plays a major role in influencing the development status of the coupled nature reserves in the case of Qinba mountainous areas. The "community development" subsystem plays a secondary role. When formulating relevant policies and development plans, it is necessary to pay attention to the development status and development trend of the "ecological service" subsystem, and to comprehensively balance the key factors of the "community development" subsystem, so as to better promote the development of the "ecological service-community development" subsystem of nature reserves. Ecological service-community development" system coupling development.

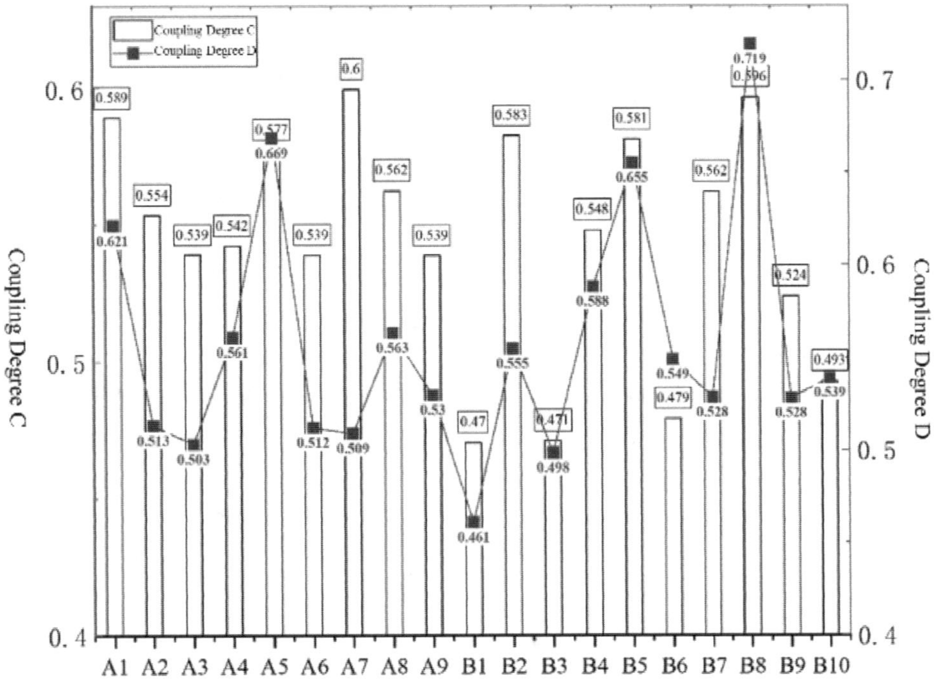

Figure 2. Numerical analysis of the main factors affecting the degree of coupling and coordination.

4.3. Main Factors Affecting the Coordinated Development of Nature Reserves

Using the gray correlation model to further analyze the main factors affecting the coordinated development of each nature reserve, the main factors affecting the coordinated development of each nature reserve under the three development states of "high coordination", "medium coordination" and "low coordination" were obtained respectively. The main factors affecting the coordinated development of each nature reserve under the three development states of "high coordination", "medium coordination" and "low coordination" were obtained respectively, as shown in table 6.

Table 6. Main factors affecting the coordinated development of the various nature reserves.

Impact Factor		Highly Coordinated Nature Reserves	Moderately Coordinated Nature Reserves			Lowly Coordinated Nature Reserves	
		Hanzhong Crested Ibis National Nature Reserve	Yinghu Wetland Provincial Nature Reserve	Liping National Forest Park	Shiquan Yanxiangdong Geopark	Nangong Mountain National Geopark	Micangshan National Nature Reserve
Ecological Service Factor	A1	0.428	0.383#	0.508	0.415	0.788*	0.780*
	A2	0.333#	0.484	0.509	0.671	0.434#	0.649
	A3	0.860*	0.545	0.404#	0.779*	0.546	0.484#
	A4	0.694	0.434	0.693*	0.392#	0.468	0.684
	A5	0.694	0.817*	0.737*	0.828*	0.468	0.467#

	A6	0.516	0.451	0.508	0.381#	0.603	0.613
	A7	0.756	0.629	0.404#	0.550	0.668	0.646
	A8	0.790*	0.383#	0.444#	0.550	0.666	0.545
	A9	0.651	0.469	0.508	0.381#	0.567	0.601
	B1	0.355#	0.551	0.508	0.398#	0.468	0.485
	B2	0.694	0.576	0.425	0.602	0.565	0.467#
	B3	0.694	0.480	0.409#	0.424	0.517	0.467#
	B4	0.561	0.545	0.717*	0.588	0.468	0.650
Community Development Factor	B5	0.881*	0.383#	0.424#	0.590	0.822*	0.830*
	B6	0.694	0.383#	0.717*	0.440	0.497	0.562
	B7	0.653	0.383#	0.421#	0.550	0.655	0.507
	B8	0.938*	0.775*	0.862*	0.550	0.622	0.567
	B9	0.449	0.545	0.606	0.600	0.468#	0.500

Note: * and # denote the driving factor and weak factor, respectively, affecting the degree of coordination of "ecological service-community development".

4.3.1. Highly Coordinated Development Impact Factor. Based on the gray correlation model, the influence factors of the highly coordinated protected area-Yangxian Crested Ibis nature reserves were analyzed. The main driving factors of the highly coordinated nature reserves of the Crested Ibis in Yangxian County are the proportion of tourism (B8), the degree of spatial aggregation of the community (B5), the abundance of water resources (A3) and the per capita area of green space (A8). Due to the abundant green space vegetation and water sources in the Crested Ibis Reserve, it provides a suitable habitat for rare species such as the crested ibis. The community actively promotes Crested Ibis viewing and science education activities and develops the ecotourism industry while balancing conservation and development, transforming natural resources into economic advantages, promoting community economic growth, providing more employment opportunities, and realizing ecotourism sustainability and environmental friendliness. In addition, the local community actively explores the optimal balance of the degree of community agglomeration, adopts a regional zoning management approach, and balances the relationship between population size and resource utilization according to the specific conditions and carrying capacity of each region, thus minimizing the pressure and negative impacts on the ecological environment, improving the quality of life and well-being of the community's inhabitants, and laying the foundation for the revitalization of the countryside in the nature reserves.

4.3.2. Moderately Coordinated Development Impact Factor. Overall, the driving factors for the moderately coordinated nature reserves Yinghu Wetland Provincial Nature Reserve, Liping National Forest Park and Shiquan Yanxiang Cave Geological Park are mainly the proportion of tourism (B8) and the abundance of green resources (A5), which implies that the development of eco-tourism plays an important role in the sustainable development of the Yinghu Wetland Provincial Nature Reserve and the neighboring communities. In addition, the influence factors of different nature reserves are also different: (1) The main influence factors of the coordination of Yinghu Wetland Provincial Nature Reserve are the abundance of green tourism resources (A5) and the proportion of tourism (B8). As an area with beautiful natural landscapes and abundant wildlife resources, Yinghu Wetland Provincial Nature Reserve has the potential to develop ecotourism. Communities can develop various types of ecotourism projects such as wetland sightseeing, adventure activities, ecological education, etc., and carry out ecological civilization publicity and education activities to enhance public awareness of ecological environmental protection while promoting tourism to increase income. In the process of tourism development, attention should be paid to strengthening the protection of tourism resources, maintaining the original appearance of Yinghu Wetland, and avoiding over-development and commercialization, to realize a virtuous cycle of protection and development, thus promoting the coordinated development of the protected area and the community, and realizing a win-win situation in terms of economic benefits and ecological benefits. (2) The coordination degree of Liping National Forest Park is influenced by multiple factors, including the proportion of tourism (B8), the abundance of green tourism resources (A5), per capita construction land (B4) and per capita basic farmland (B6). These factors together affect the coordinated development between Liping National Forest Park and the community. Therefore, in order to realize the coordinated development of the protected area and the community, in addition to giving full play to local characteristics and focusing on the development of eco-tourism, it is also necessary to take into account the construction of basic farmland in the community to ensure the basic livelihood of community residents. Reasonable planning of land, to avoid resource depletion and environmental problems caused by over-development of the community due to land use, so as to enhance the ecological value of the protected area. (3) The coordination degree of Shiquan Yanxiang Cave Geopark is mainly affected by factors such as the abundance of water resources (A3) and the abundance of green tourism resources (A5). Shiquan Yanxiang Cave Geopark is famous for its rich karst geomorphic landscapes such as stalactites, stalagmites, and underground rivers, which are rich in recreational value. In order to achieve long-term conservation and sustainable development of the reserve, it is necessary to make full use of tourism resources and water resources, develop eco-tourism as well as improve the hydrological cycle, and contribute to the economic development of the local communities and the improvement of the lives of the residents.

4.3.3. Lowly Coordinated Development Impact Factor. Overall, the driving factors of the low coordinated conservation sites, Micangshan National Nature Reserve and Nangongshan National Geopark, are mainly the degree of community aggregation (B5) and average elevation (A1). Due to the high elevation of these two nature reserves, rural communities face a series of challenges in the development process, including restricted land use, difficulties in road construction, and inconvenient transportation, which limit the development of agricultural production and residents' lives. In addition, geographic conditions have led to the dispersal of rural communities and a low level of community aggregation. This leads to a lack of communication and cooperation between communities, difficulty in forming scale effects, and low output efficiency, which in turn affects community development. At the same time, due to the sparseness of the community, it is difficult to form a complete ecological reserve and ecological service pattern, which is prone to irrational land use and waste of resources, and easy to reduce the provision of ecological services such as the first productivity, vegetation cover, and so on, which makes the development of both sides of the system is not high. For such communities, the option of ecological relocation can promote ecological integrity while fostering sustainable community development and improving the current situation.

5. Conclusions and Recommendations

This paper constructs the evaluation index system of "ecological service-community development" of nature reserves, selects and introduces the coupling and coordination model, and conducts empirical research on the degree of coupling (C), degree of coordination (D), and synergy (U) of six different types of nature reserves in Qinba Mountain area, and obtains the following conclusions: First, the results of coupling degree indicate that the coupling degree of ecological service and community development of the study case sites all show a high level of coupling development. The coupling degree results show that the coupling degrees of ecological services and community development in the study case sites all show a high level of coupling development. Second, the results of the coordination degree show that there are obvious differences in the coordination degree between "ecological services" and "community development" in each protected area. Only Hanzhong Crested Ibis National Nature Reserve is in a high degree of coordinated development, Yinghu Wetland Provincial Nature Reserve, Shiquan Yanxiang Cave Geopark, and Li National Forest Park are in a medium degree of coordinated development, and the two nature reserves, Micang Mountain National Nature Reserve and Nangong Mountain National Geopark, are in a low degree of coordinated development. Thirdly, the results of the synergistic effect evaluation show that the three nature reserves of Hanzhong Crested Ibis National Nature Reserve,

Shiquan Yanxiang Cave Geopark, and Liping National Forest Park are of "synchronized development"; the two nature reserves of Mount Michaoshan National Nature Reserve and Nangong Mountain National Geopark are of "ecological service" leading type; only two of them are of "ecological service" leading type. Two nature reserves, namely, Micangshan National Nature Reserve and Nangongshan National Geopark, are in the "ecological service" leading category, while only Yinghu Wetland Provincial Nature Reserve is in the "community development" leading category. Fourth, the driving factors of nature reserves with high coordinated development are mainly the proportion of tourism and the degree of spatial agglomeration of communities; the driving factors of nature reserves with medium coordinated development are mainly the proportion of tourism and the abundance of green tourism resources; and the driving factors of nature reserves with low coordinated development are mainly the degree of aggregation and the average elevation of communities. The main drivers for the low coordinated development of nature reserves are community aggregation and average elevation.

Based on the results of the above analysis, in order to realize the revitalization of the countryside and build a coordinated and innovative regional development, this paper suggests:

Firstly, the development of ecotourism. Eco-tourism, as an important tool for rural revitalization, can effectively promote the rational use of resources and socio-economic development.

Secondly, balancing the degree of community aggregation. High levels of community aggregation will bring enormous carrying pressure to the surrounding ecological environment, leading to increased ecosystem damage and loss of natural resources. On the contrary, a low degree of community aggregation will lead to a lack of communication and cooperation between communities, which will prevent the formation of scale effects. Therefore, a reasonable layout of community planning is needed to ensure the rational allocation of resource space.

Thirdly, ecological migration should be implemented by local conditions. Eco-migrant relocation can reconstruct the relationship between community industrial development and ecological protection, and at the same time, it can reduce the impact of human activities on the environment, help to enhance the ecological value of nature reserves, and provide communities with a more superior development environment.

Acknowledgments

Work was supported by National Soft Science Project: Research on Integrated Development of Nature Reserves and Communities (2023131020); Joint Fund Project: Research on Integrated Development of Nature Reserves and Communities under the Perspective of Territorial Space (2021JLM-44); Innovative Entrepreneurship Training Program for College Students (202410700073).

6. References

[1] Zhuang Y B 2018 IUCN protected area management classification research and reference *China Garden* **34(7)** 6.

[2] Costanza R and Arge Groot R 1997 The value of the world's ecosystem services and natural *Capital Nature* **387(15)** 253-260.

[3] Zheng Y, Yu F L and Sang X F 2023 Evaluation method of ecological service value of water system in the Yangtze River Basin in 2023 *Journal of China Institute of Water Resources and Hydropower Research* **21(03)** 222-235.

[4] Yang H J, Gou X H and Ma W Z 2023 Progress of research on valuation of forest ecosystem service functions in China in the last 10 years *Journal of Ecology* 1-13.

[5] Zhang Y, Zhao Z B and Li X Y 2023 Impacts of spatial changes of "three living things" on the value of social and ecological services in peri-urban protected areas--The case of Qionghai National Wetland Park *Journal of Ecology* **2023(16)** 1-13.

[6] Ren D F, Cao A and Du W J 2023 Changes in the value of ecosystem services and food supply and demand in Liaoning Province in 2023 *Soil Bulletin* **54(03)** 527-535.

[7] Seppelt R, Dormann C F and Eppink F V 2011 A Quantitative Review of Ecosystem Service Studies: Approaches, Shortcomings and the Road Ahead *Journal of Applied Ecology* **48** 630-

Journal of Physics: Conference Series **2941** (2025) 012064 doi:10.1088/1742-6596/2941/1/012064

636.
[8] Nakakaawa C, Moll R and Vedeld P 2015 Collaborative resource management and rural livelihoods around protected areas: A case study of Mount Elgon National Park Uganda *Forest Policy & Economics* **57** 1-11.
[9] Li S, Li B Y and Liu W W 2021 Exploring community development paths in national parks based on the interests of community residents in 2021 *Forestry Economic Issues* **41(03)** 320-327.
[10] Runte A 2010 National Parks-The American Experience *Geographical Review* **85(4)**.
[11] Li J X, Wu C Z and Tao C 2022 Research on spatial coupling analysis method of human-land relationship in nature reserves--Taking Tianmu Mountain Nature Reserve as an example *Chinese Garden* **38(05)** 74-79.
[12] Liu H M and Fang C L 2019 Basic concepts and framework of the "coupled magic formula" of urbanization and ecology *Acta Geoscientica Sinica* **74(8)** 1489-1507.

Liquid Chromatograph-High Resolution Mass Spectrometry Analysis of the Fluorinated Proteins in Antarctic (*Euphausia Superba*) Krill Muscle

Zhongquan Jiang and Xiaosheng Shen*

East China Sea Fisheries Research Institute, Chinese Academy of Fishery Sciences, Shanghai 200090, China.

E-mail: jiangzq@ecsf.ac.cn; *foodsmc98@126.com

Abstract. This study aims to analyze the fluorinated proteins in Antarctic krill (*Euphausia superba*) muscle using liquid chromatography-high resolution mass spectrometry (LC-HRMS) to explore the binding form, composition, binding mechanism, and molecular weight range of fluorine with proteins, providing a scientific foundation for improving the defluorination efficiency of Antarctic krill. Polypeptides from Antarctic krill muscle were enzymatically extracted using trypsin, fractionated by ultrafiltration and gel filtration chromatography (GFC), and purified by C18 reversed-phase high-performance liquid chromatography (RP-HPLC). Fluorine content and molecular weights of the polypeptide fractions were determined by ion chromatography (IC) and HPLC-Q/Orbitrap MS. Among the three separated polypeptide fractions AKP-I, AKP-II, and AKP-III, AKP-II had the highest fluorine content of 1219.52 μg/g, accounting for 63% of the total fluorine, followed by AKP-I with 495.04 μg/g, accounting for 26%, and AKP-III with the lowest fluorine content of 209.44 μg/g, accounting for 11%. The molecular weight of fluorophosphorus peptides in AKP-II was measured to be approximately 2000 Da. The results indicate that fluorine in Antarctic krill muscle is accumulated through ion exchange adsorption between fluorapatite (FHAP) and the amino acid side chains, maintaining a stable fluorine content. This study provides new insights into the binding mechanism of fluorine in Antarctic krill muscle and offers a scientific basis for the development of targeted defluorination methods.

Keywords. Antarctic krill protein; fluorine; ultrafiltration; Sephadex G-25 Sephadex column chromatography; high performance liquid chromatography; ion chromatography.

1. Introduction

Antarctic krill *(Euphausia superba)* is the largest undeveloped high-quality protein resource in the world, and has been considered "the future food protein library of the world" and a carrier for the development of emerging industries in the 21st century [1,2]. However, the high-fluorine abnormality in its muscle significantly limits its development and utilization [3,4]. Studies have shown that the fluorine content in the whole Antarctic krill can reach up to 2400 μg/g, and the fluorine content in muscle can be as high as 570 μg/g [5]. The US Food and Drug Administration sets the daily fluoride uptake limit of 100 μg/g, taking account of sodium fluoride. The fluorine content of Antarctic krill obviously exceeds the safety limit as a food. Therefore, a variety of fluorine analysis methods have been developed and efforts have been made to understand the defluorination of Antarctic krill in recent years [6-8]. However, most of these studies have been focused on the quantitative analysis. The forms of fluorine are simply classified into water-soluble fluorine, exchangeable fluorine, oxidized fluorine, organic fluorine and residual fluorine, based on the treatment methods. Antarctic krill is usually defluorinated by biosorption and chemical reactions, which usually exhibit low defluorination rates, or cause structural damages to protein, and thus affect the flavor and edible value of Antarctic krill. Due to the lack of information on the binding mechanism between fluorine and proteins in Antarctic krill, no defluorination methods can be developed for specific binding structures.

The formation mechanism of fluorine in the Antarctic krill shrimp remains unclear. Pan et al. (2000) discovered a cyclical variation of fluorine in Antarctic krill muscle during the shell formation by the Antarctic krill live culture experiment [9]. They found that, during the shell development, the fluorine content rapidly dropped to the lowest level, and gradually increased to a stable level after the shell completely formed. However, such periodic change mechanism of fluorine in muscle is not included in the work. Based on the rapid dissolution of fluoride in the exoskeleton, Sands et al. (1998) concluded that fluoride might only bind with water-soluble protein in krill [10]. Zhang et al. (1993a) studied the migration of fluoride in Antarctic krill and found that fluorine might bind to the structural protein on the krill surface [11]. The binding mechanism, binding mode, or binding structure of fluorine with protein in Antarctic krill that are critical for the target defluorinations remains unclear. This study is based on the combination of various forms of fluorine in the muscles of Antarctic krill. In the present work, based on the related protein separation work [12-15], polypeptides were enzymatically extracted from Antarctic krill, fractioned with ultrafiltration and GFC based on molecular weight and purified by C18 reversed phase HPLC (RP-HPLC). The fluorine contents and molecular weights of the polypeptide fractions were determined by IC and HPLC-Q/Orbitrap MS to explore the binding form, composition, binding mechanism and molecular weight range of fluorine with protein in Antarctic krill, aiming to provide the scientific foundation for further studying the specific binding form and binding structure of fluorine and amino acid and improving the defluorination efficiency of Antarctic krill.

2. Materials and Methods

2.1. Materials

Antarctic krill were collected from the Antarctic area in December 2017, immediately shelled, transported to the laboratory at -20°C and stored at -80°C before use. Analytic grade trypsin (≥ 50,000 U/g), hydrochloric acid and sodium hydroxide were purchased from Sinopharm Chemical Reagent Co., Ltd. (Beijing, China). Ultrapure water was prepared with an Ultrapure water meter (Millipore, USA) and used in all experiments.

2.2. Experimental Methods

2.2.1. Preparation of Antarctic Krill Polypeptides (AKP). To prepare polypeptides, 100 g fresh Antarctic krill and 1 g trypsin were added into 500 mL ultrapure water (Millipore, USA). The pH of the mixture was adjusted to 8 with 0.1 mol/L NaOH and shaken in a 45°C water bath for 8 h. The enzymatic hydrolysate was collected, heated in boiling water for 15 min to terminate the enzymatic hydrolysis reaction, cooled to room temperature (RT, 25°C) and centrifuged at 5,000 rpm (Hitachi CF, Japan) at 5°C for 15 min. The supernatant was collected and ultra-filtrated with a 30kD ultrafiltration membrane (8400, Merck, USA) based on the molecular weights of the enzymatic polypeptides [16] to remove large proteins and other macromolecular substances due to incomplete enzymatic hydrolysis. The peristaltic pump speed was adjusted to achieve the inlet pressure of 15-20 psi on the ultracentrifuge membrane and the outlet pressure of 5 psi with over 15 psi pressure difference. The filtrate was collected, frozen at -80 °C and then freeze-dried to afford an enzymatically hydrolyzed polypeptide powder of Antarctic krill. The yield of the polypeptide was calculated with the equation: polypeptide yield (%) = weight of enzymatically hydrolyzed peptide powder (g)/weight of fresh shrimp powder (g) × 100%.

2.2.2. Characterization of Antarctic Krill. The protein content of Antarctic krill was determined by the Kjeldahl method (GB 5009.5-2016). The water content, ash, and fat content were measured by the drying method (GB 5009.3-2016), the magnesium acetate method (GB 5009.4-2016) and the Soxhlet extraction method (GB 5009.6-2016), respectively.

2.2.3. Separation and Purification of Krill Peptide.

(1) GFC fractionation

A Sephadex G-25 dextran gel column (Φ2×60 cm, Shanghai Huamei Experimental Instrument Factory, China) was packed and equilibrated with ultrapure water for 2 h. The enzymatic polypeptide solution (0.1 g/mL, 8-10 mL) was loaded into the column and eluted at a constant rate of 2 mL/min. The eluate was collected manually every 5 mL (5 mL/tube).

(2) HPLC separation and purification

The eluates collected above were then respectively analyzed by RP-HPLC (Agilent 1260, America). The HPLC conditions were as follows: detection wavelength, 220 nm; column temperature, 30°C; mobile phase flow rate, 0.1 mL/min; injection volume, 80 uL; mobile phase A, acetonitrile; mobile phase B, ultrapure water; and gradient elution: 0-2 min 5% A (isocratic), 2-12 min 5%-60% A (linear gradient), 12-15 min 60%-80% A (linear gradient), 15-17 min, 80% A (isocratic), 17-17.5 min 80%~5% A (linear gradient), 17.5-20 min, 5% A (isocratic). The eluates with the same chromatograms were combined, concentrated to -5 mL with the rotary evaporator (RE-52A, Shanghai Yarong Biochemical Instrument Factory, China), frozen and freeze-dried for 24 h to obtain the powders of polypeptide fractions of different molecular weight ranges.

2.2.4. Determination of Fluorine Content. The fluorine content was measured with the IC system (850 Professional IC, Metrohm, Swiss) based on the GB/T 5009.18-2003 and the method reported in the literature [17]. Briefly, 0.1g of sample was added to 2 mL hydrochloric acid aqueous solution (11:1, v:v), sealed and acid hydrolyzed for 1 h. The hydrolysate was diluted with ultrapure water to 20 mL, passed through a 0.22 μm needle filter and loaded into the IC-Ag^+ column and IC-Na^+ column. The columns were eluted with 3.2 mmol/L Na_2CO_3 - 1.0 mmol/L $NaHCO_3$ at 0.700 mL/min. The column temperature was 25°C and the injection volume was 20 μL.

2.2.5. Determination of Molecular Weights of Polypeptide Fractions. The molecular weight of each polypeptide fraction was measured with a UHPLC-Q/Orbitrap MS system (Thermo, USA). The mass spectrometry was operated in full MS scanning mode and the spectra were collected in the range of 400-6000 m/z at the resolution of 3000. The sample injection rate was 0.3 mL/min and the capillary voltage was set to 4.0 kV. High purity nitrogen was used as the collision gas. Acetonitrile and ultrapure water (1:1, v:v) at the flow rate of 0.3 mL/min was used as the mobile phase

2.2.6. Data Analysis. The *experimental* data was performed using SPSS v19.0 software. The quantitative experiments were repeated three times and the mean value was reported.

2.3. Flowchart of Experimental Design

Figure 1 shows the flowchart of the whole experiment design.

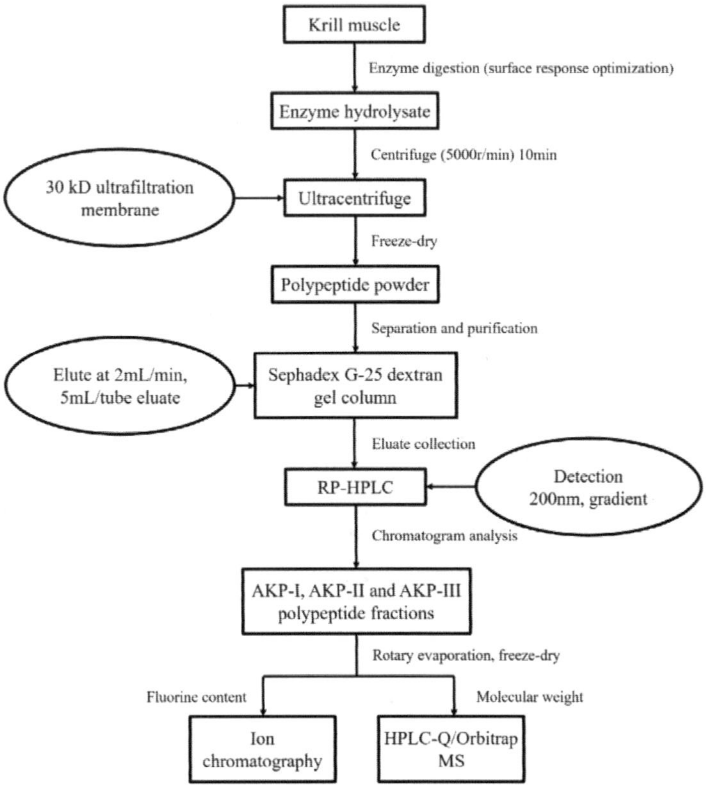

Figure 1. Flowchart of experimental design.

3. Results and Discussion

3.1. Composition of Antarctic Krill Muscle

The composition of the Antarctic krill muscle used in the present work is as follows (table 1). The crude protein content (dry weight), fat content (wet weight), ash (wet weight), and water content (wet weight) of the Antarctic krill muscle are consistent with the results reported by Sun et al. (2008) are 64.40%, 1.30%, 2.76% and 74.69, respectively [18]. Yuan et al. (2012) also reported a crude protein content of 62.89 ± 0.43% in Antarctic krill [19]. These results suggest that the Antarctic krill used in the present work was of high quality.

Table 1. Composition of Antarctic krill muscle.

Component	Protein (dry weight)	Protein (wet weight)	Moisture (wet weight)	Fat (wet weight)	Ash (wet weight)
Content (%)	61.10	21.74	76.49	1.35	2.67
Fluorine (μg/g)	401.71	137.51	/	/	/

3.2. Yield of Polypeptide

About 20 g polypeptide was obtained from the enzymatical hydrolysis, centrifugation, ultrafiltration and lyophilization of 100 g fresh krill muscle. The yield was calculated to be 20%.

3.3. HPLC Analysis of Polypeptides Extracted from Krill Muscle

The enzymatic hydrolysate of Antarctic krill muscle was separated with an HPLC system and detected at 220 nm (figure 2). The hydrolysate was separated into six fractions with well-separated peaks, indicating that the chromatographic conditions were optimal and could be used to further separate and purify the polypeptides extracted from the krill muscle.

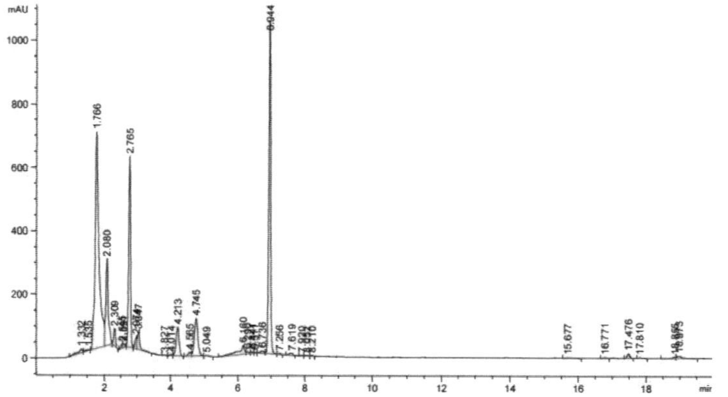

Figure 2. HPLC chromatogram of polypeptide extracted from Antarctic krill muscle.

The eluates obtained from the Sephadex G-25 Sephadex GFC were respectively analyzed by the HPLC system under the optimal conditions mentioned above. It was found the chromatograms of the eluates in the first 25 tubes were similar (figure 3). All of them contain two well-separated peaks and many overlapped peaks that were not completely separated possibly because of their similar molecular weights. However, these incompletely separated peaks of each sample exhibited similar retention times. Therefore, the eluates in the first 25 tubes were combined as a fraction AKP-I. Similarly, the eluates in the 26-48 tubes exhibited similar chromatograms (figure 4) with three distinct peaks and several overlapped peaks at similar retention times and thus were combined as fraction AKP-II. The eluates in the 49-90 tubes only exhibited one strong sharp peak with a good shape (figure 5), suggesting that the eluates were relatively purer and the molecular weight was smaller than those of AKP-I and AKP-II. The eluates were then combined as fraction AKP-III. In all, the GFC separated the enzyme hydrolysate into three fractions, AKP-I, AKP-II and AKP-III, based on the molecular weight.

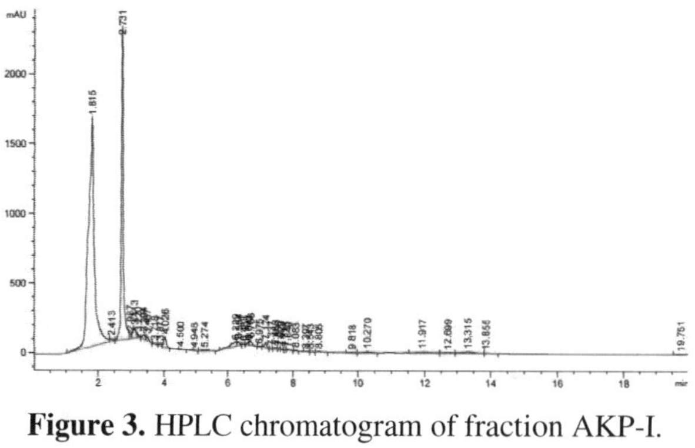

Figure 3. HPLC chromatogram of fraction AKP-I.

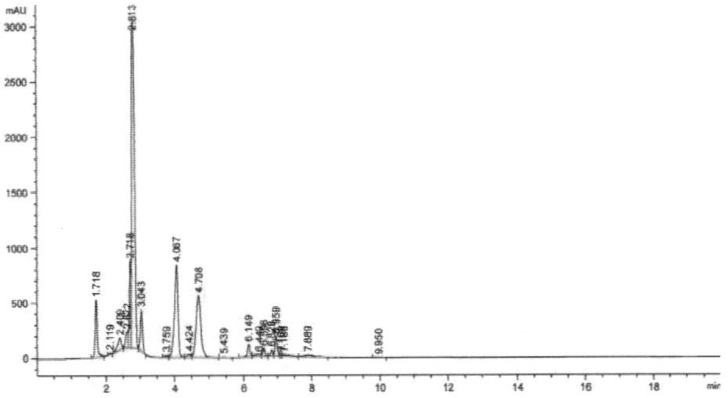

Figure 4. HPLC chromatogram of fraction AKP-II.

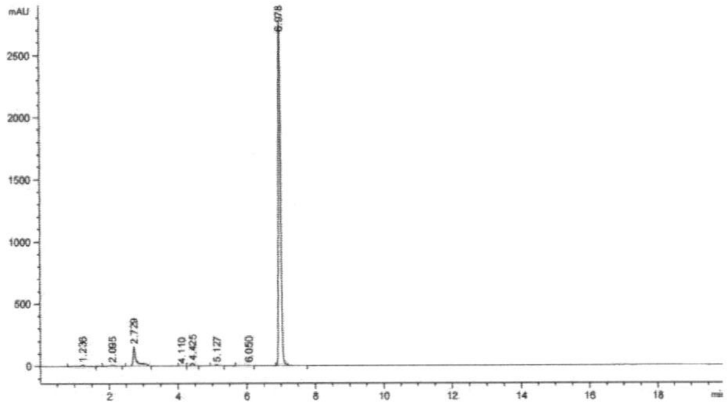

Figure 5. HPLC chromatogram of fraction AKP-III.

3.4. Distribution of Fluorine in the Polypeptides

The fractions AKP-I, AKP-II and AKP-III were then further analyzed by IC for fluorine contents. Based on their peak areas, it can be seen that the fluorine contents of the three fractions are in the order of AKP-II>AKP-I>AKP-III (figures 6-8). Further calculated based on peak area and concentration revealed the fluorine content of 495.04 μg/g for AKP-I, 1219.52 μg/g for AKP-II and 209.44 μg/g for AKP-III, respectively accounting for 63%, 26% and 11% of total fluorine. The average fluorine content was calculated to be 645.25 μg/g. Based on the protein content of Antarctic krill muscle (dry weight), the fluorine content of Antarctic krill muscle (dry weight) was calculated to be 396.10 μg/g, similar to the result (401.71 μg/g) obtained with krill muscle (dry weight) directly, slightly higher than that (325 μg/g) reported by Guo et al. (2018) [20], but lower than that (570.00 μg/g) reported by Soevik et al. (1979) [5]. Based on the polypeptide yield of 20%, the average fluorine content of fresh krill muscle was calculated to 129.05 μg/g, similar to the result (137.51 μg/g) obtained with the krill protein (wet weight) directly, higher than that (66.00~105.00 μg/g) measured by Kim et al. (1990b) and Zhu et al. (2015) [21, 22], but lower than that (226.00 μg/g) reported by Zhang et al. (1993a) [11]. The differences between our results and the results reported in the literature are possibly due to the different storage conditions of Antarctic krill. Studies have shown that the storage condition can significantly affect the migration, forms and content of fluorine in Antarctic krill. With the prolongation of storage time, the fluorine in the shell is more likely to convert into ion-exchangeable fluorine and/or water-soluble fluorine that is easily absorbed by the muscle [22,23]. Therefore, the different sources, harvest times and storage conditions can affect the experimental results. However, the variation is in a reasonable range, and thus our results are reliable.

Journal of Physics: Conference Series **2941** (2025) 012065 doi:10.1088/1742-6596/2941/1/012065

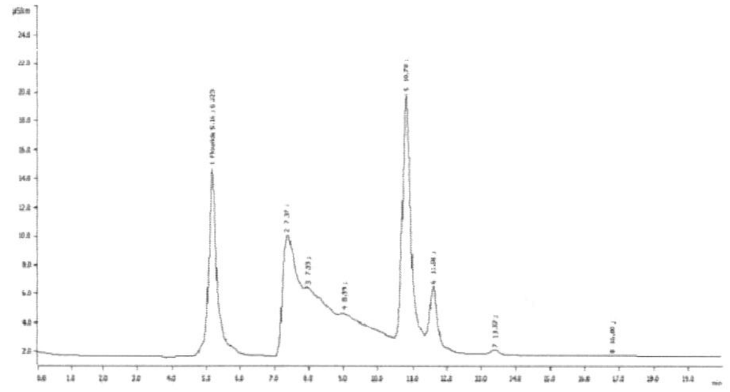

Figure 6. Ion chromatogram of fraction AKP-I.

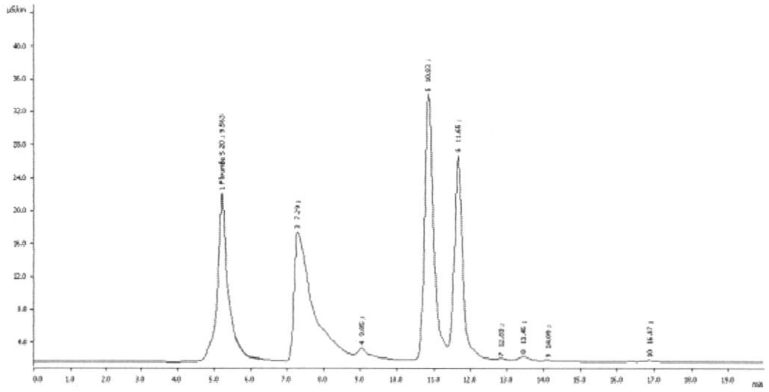

Figure 7. Ion chromatogram of fraction AKP-II.

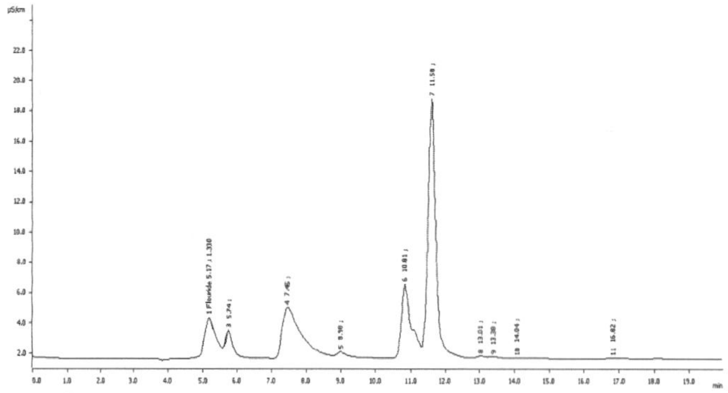

Figure 8. Ion chromatogram of fraction AKP-III.

As demonstrated above, the fraction AKP-II contained the highest amount of fluorine, reaching up to 63% of the total fluorine in the polypeptides extracted from krill muscle. This was also evidenced by other phenomena. For example, the AKP- II solution gradually changed from a colorless transparent liquid (figure 9(a)) to a milky white suspension (figure 9(b)) during the rotary evaporation, while the AKP-I and AKP-III solutions remained colorless and transparent. Based on these results and observations, we deduce that the physical and chemical properties of the fluoropeptides changed significantly during the heating under low pressures. The phosphorylation modification of proteins is of great significance for its stability and biological activity [24,25]. Calcium hydroxyphosphate (HAP) is a protein adsorbent. Under certain circumstances, its surface hydroxyl group instantaneously becomes vacant and forms a binding site with the adjacent Ca^{2+} that can bind some OH-containing amino acids. Eventually, a monolayer of protein can be achieved on its surface [26]. The OH^- of HAP

Journal of Physics: Conference Series **2941** (2025) 012065 doi:10.1088/1742-6596/2941/1/012065

can be easily replaced by F^- due to its strong ion exchange capacity, forming fluorophosphate (FHAP) which is very stable and possesses a stronger protein adsorption capacity [27,28]. Such process has been well-recognized as the accumulation mechanism of fluoride, e.g. $Ca_5(PO4)_3OH+F^-=Ca_5(PO4)_3F+OH^-$ [29]. During the rotary evaporation, the concentration of the amino acids that could bind to FHAP in AKP-II became higher. The temperature and pressure changes could also change the protein conformation or the structures of amino acids, which thereby changed the structure and pH of FHAP and caused isoelectric precipitation. These explain the white precipitates of AKP-II formed during rotary evaporation. No precipitate formed in the solutions of AKP-III and TKP-I during the rotary evaporation, possibly due to their low fluorine contents, as well as the low contents of amino acids that could bind FHAP. Our results indicate that the fluorine in Antarctic krill binds its proteins, mainly the AKP-II peptides, in specific forms. Therefore, the molecular weight range of fraction AKP-II was further investigated.

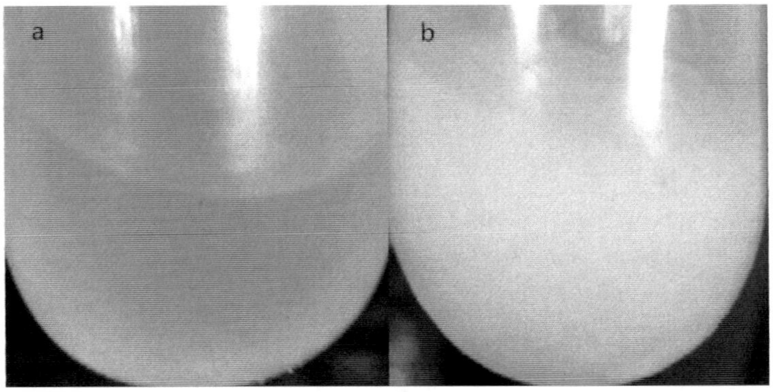

Figure 9. State diagram before and after AKP-III rotary evaporation.

The fraction AKP-I, AKP-II and AKP-III was analyzed with the on-line HPLC-Q/Orbitrap MS system for the molecular weights of its fluorophosphorus peptide components. The molecular weight of AKP-II was measured to be-2000 Da (figure 10), and those of AKP-I and AKP-III were measured to be-6000 Da (figure 11) and-1000 Da (figure 12), respectively, consistent with the results reported by Gao et al. (2016) [30].

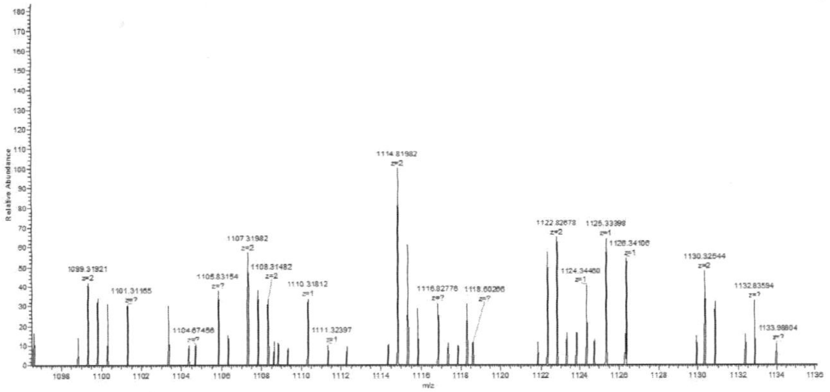

Figure 10. Mass spectrum of fraction AKP-II.

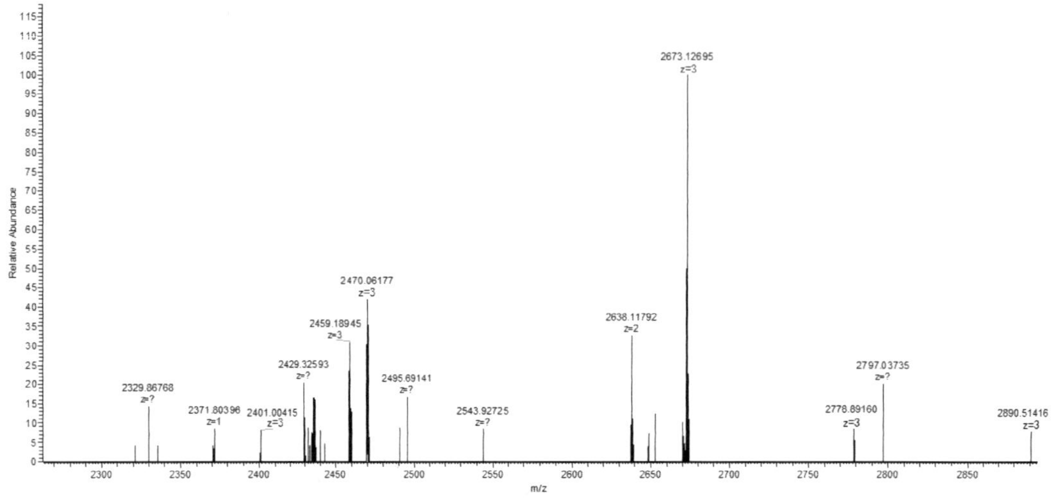

Figure 11. Mass spectrum of fraction AKP-I.

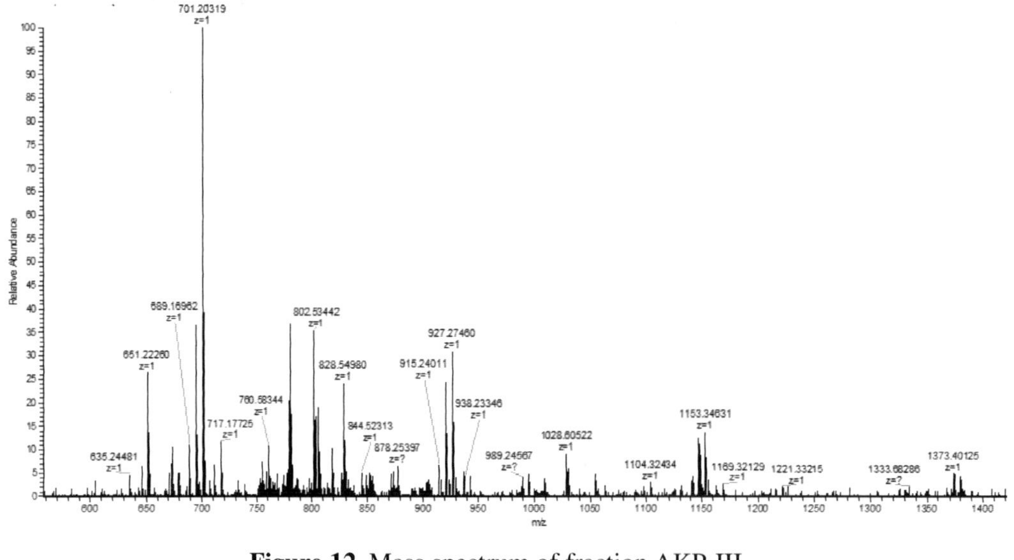

Figure 12. Mass spectrum of fraction AKP-III.

4. Discussion

In this study, the separation of phosphopeptides from Antarctic krill and the determination of the fluorine content of each component of the phosphopeptides resulted in the molecular weight of the three components of the peptide in descending order (AKP-I > AKP-II > AKP-III), and the content of fluorine per unit mass of the three components of the peptide was AKP-II > AKP-I > AKP-III in descending order. It is assumed that fluorine in the Antarctic krill protein exists in a free state. When Sephadex g-25 glucan gel column chromatography is collected, the ionic fluorine will gradually elution down with the eluent, and with the continuous elution of the eluent, the fluorine content in the collected chromatography will gradually decrease until the final value is 0. However, in our experimental results, AKP -I was the 1- 25 tube liquid chromatography that was collected, AKP - II was the 26- 50 tube liquid chromatography that was collected, AKP - III was the 51- 80 tube liquid chromatography that was collected, the pipe number of chromatography liquid of AKP -I and AKP - II was the same (i.e., the same volume) if the hypothesis is true, the AKP -I peptide fluoride component should be of the highest. However, this is not consistent with our experimental results. In the experiment, AKP-II has the highest fluorine content, which is up to 63% of the total fluorine content. Obviously, the hypothesis is not valid. The existing experimental data combined with experimental phenomena indicate that fluorine in the Antarctic krill protein may be bound to the protein in the form of minerals.

5. Conclusions

In the present work, polypeptides were extracted from Antarctic krill muscle by trypsin digestion and fractioned and purified by ultrafiltration, GFC and HPLC to determine their fluorine contents and molecular weights. The trypsin hydrolysate of Antarctic krill muscle was fractioned by Sephadex G-25 Sephadex column chromatography and HPLC into three polypeptide fractions, AKP-I, AKP-II and AKP-III. The IC analysis revealed that the fluorine content of AKP-II was the highest with a value of 1219.52 $\mu g \cdot g^{-1}$, accounting for 63.0% of total fluorine, followed by AKP-I with a value of 495.04 $\mu g \cdot g^{-1}$ that accounted for 26.0% of total fluorine. The fluorine content of fraction AKP-III was the lowest with a value of 209.44 $\mu g \cdot g^{-1}$, accounting for 11.0% of total fluorine. The molecular weights of the fluoropeptides in AKP-II were measured with an online purification UHPLC-Q/Orbitrap MS system to be -2000 Da. Based on these results, it is deduced that fluorine is accumulated in the Antarctic krill via ion exchange adsorption between FHAP and the amino acid side chain to maintain a stable fluorine content.

6. Data Availability Statement

The datasets generated during and /or analyzed during the current study are available from the corresponding author upon reasonable request.

Acknowledgments

This work was supported by the National Natural Science Foundation of China (No. 32172312).

7. References

[1] La H S, Lee H and Kang D 2015 Ex situ echo sounder target strengths of ice krill Euphausia crystallorophias *Chin. J. Oceanol. Limn* **33** 802-808
[2] Song W, Zhu Y F, Wang L M, Jiang K J, Zhang F Y, Ma C Y and Ma L B 2018 Identification and profiling of microRNAs of *Euphausia superba* using Illumina deep sequencing *J. Oceanol. Limn* **36** 2278-2287
[3] Gigliotti J C, Jaczynski J and Tou J C 2008 Determination of the nutritional value, protein quality and safety of krill protein concentrate isolated using an isoelectric solubilization/ precipitation technique *Food Chem* **111** 209-214
[4] Nicol S, Foster J and Kawaguchi S 2012 The fishery for Antarctic krill - recent developments *Fish Fish* **13** 30-40
[5] Soevik T and Braekkan O R 1979 Fluoride in Antarctic krill *(Euphausia superba)* and Antarctic krill (*Meganyctiphanes norvegica*) *J. Fish. Res. Board Canada* **36** 1414-1416
[6] Guo F, Shi W Z and Wang Z H 2016 Study on fluoride removal of Antarctic krill hydrolysate *Sci. Tech. Food Industry* **9** 245-249
[7] Fang B, Wang Z H, Shi W Z, Wang L L and Chen X Y 2018 Study on the removal of various morphological fluorine in Antarctic krill meat by calcium salt *Mod. Food Sci. Tech* **34** 64-68
[8] Huang J W, Cui C, Zheng X J and Chen Z G 2018 Study on fluoride removal of Antarctic krill hydrolysate *Chin. Condiment* **43** 1-8
[9] Pan J M, Zhang H S and Liu X Y 2000 The fluorine enrichment mechanics of Antarctic krill I. The fluorine chemical formation *Acta Oceanol. Sin* **22** 58-64
[10] Sands M, Nicol S and McMinn A 1998 Fluoride in Antarctic marine crustaceans *Mar. Biol* **132** 591-598
[11] Zhang H S, Pan J M, Cheng X and Zhu B 1993a Biogeochemistry research of fluoride in

Antarctic ocean I. The study of fluoride anomaly in krill *Antarctic Res* **4** 55-61 (in Chinese)

[12] Ren J Y, Lai T, Jiang Y, Lu Y, Liu P and Liao W 2015 Enzymatic preparation, separation, purification and antioxidant activity evaluation of whey protein peptides *J.South Chin.Univ. Tech.(Natural Science Edition)* **43** 1-7 (in Chinese)

[13] Ji W, Zhang C H and Ji H W 2017 Purification, identification and molecular mechanism of two dipeptidyl peptidase IV (DPP-IV) inhibitory peptides from Antarctic krill *(Euphausia superba)* protein hydrolysate *J. Chromatogr* **1** 56-61

[14] Zhang Y, He S D and Simpson B K 2017 A cold active transglutaminase from Antarctic krill *(Euphausia superba)*: Purification, characterization and application in the modification of cold-set gelatin gel *Food Chem* **232** 155-162

[15] Cho J Y and Kim J K 2018 Isolation and identification of a novel algicidal peptide from mackerel muscle hydrolysate *J. Chromatogr* **56** 39-46

[16] Liu Y J, Zhang H Y, Liu S H, Wang Y, Han F, Kong C, Wang Y, Zhang X and Shen X S 2018 Optimization of enzyme digestion of proteins in Antarctic krill via response surface and the peptide analysis *Mod. Food Sci. Tech* (in Chinese)

[17] Shen X S, Li Y L, Zhang H Y, Huang X, Feng C and Cai Y L 2013 Distribution and relationship of fluoride and mineral elements in Antarctic krill *Mod. Food Sci. Tech* **23** 2279-2282 (in Chinese)

[18] Sun L, Zhou D and Sheng X 2008 Nutrition and safety evaluation of Antarctic krill *Mar. Fish. Res* **29** 57-64

[19] Yuan Y, Li X, Yang X, Chi H and Huang H 2012 Analysis and comparison on the nutritional components of different Antarctic krill meals *Mar. Fish* **34** 457-463

[20] Guo F, Wang Z H, Shi W Z and Fang B 2018 Distribution characteristics and speciation of fluorine in different parts of Antarctic krill *Food Sci* **39** 237-242 (in Chinese)

[21] Kim, K H, Kim D M, Kim Y H and Yoon H H 1990b Fluoride migration of frozen Antarctic krill according to thawing methods *Korean J. Food Sci. Tech* **22** 168-171

[22] Zhu L L, Zhao Y L, Zhou D Q, Sun Y and Guo T 2015 Migration variation of fluoride during the storage of *Euphausia superba J. Chin. Institute Food Sci. Tech* **15** 81-86 (in Chinese)

[23] Cao M X 2016 *Migration and conversion of fluorine in antarctic krill and its biotoxicity evaluation* (Shanghai: Shanghai Ocean University) Master's Thesis

[24] Yoshita H, Takao T and Satomi Y 2009 Roles of clock phosphorylation in suppression of E-box-dependent transcription *Mol. Cell. Biol* **29** 3675-3686

[25] Shi S Q, Bichell T J, Ihrie R A and Johnson C H 2015 Ube3a imprinting impairs circadian robustness in Angelman syndrome models *Current Biol* **25** 537-545

[26] Ye Q, Hu R, Lin ZY and Lin C 2006 In situ ATR-FTIR study on the interaction of HA with bovine serum albumin *Chem. J. Chin.Univ* **27** 1552-1554

[27] Liu Q, Chen Z, Gu H and Chen Z 2012 Preparation and characterization of fluorinated porcine hydroxyapatite *Dent. Mater. J* **31** 742-750

[28] Dorozhkin S V 2015 Calcium orthophosphate-containing biocomposites and hybrid biomaterials for biomedical applications *J Funct Biomater* **6** 708-832

[29] Zhang H S, Pan J M, Cheng X and Xia W 1993b Biogeochemistry research of fluoride in Antarctic ocean II. The variation characteristics and accumulation cause of fluoride in the cuticle of Antarctic krill *Chin. J. Polar Res* **4** 36-41 (in Chinese)

[30] Gao Y, Wang Y C, Chang Y G, Xue Y, Li Z J and Xue C H 2016 ACE inhibition and antioxidant activities of three types of Antarctic krill *(Euphausia superba)* peptides with low fluorine content *Sci. Tech. Food Industry* **92** 82-87

Research on Land Use and Pollution Source Distribution in the Sanjiang Plain, China

Shaohua Wang[1, 2], Lingfen Guo[3,*] and Yuanxi Huo[3]

[1] Natural Resources Survey Institute of Heilongjiang Province, Harbin 150036, Heilongjiang, China.
[2] Northeast Geological S&T Innovation Center of China Geological Survey, Shenyang 110000, Liaoning, China.
[3] College of Geographic Science, Harbin Normal University, Harbin 150025, Heilongjiang, China.
Email: wshua_8729@126.com; *guolf@hrbnu.edu.cn; huoyuanxi0521@163.com

Abstract. The Sanjiang Plain is a gathering area of the Heilongjiang River, Songhua River and Wusuli River basins. There are numerous rivers, lakes and wetlands within the Sanjiang Plain, and the urban population is dense. The Sanjiang Plain is one of the important agricultural production bases in China. It is of great significance to conduct research on the ecological environment in this region. This article uses remote sensing technology to monitor land use changes, surface water pollution status, pollution source types and distribution characteristics in the Sanjiang Plain using ETM+ data from 1999, and GF-1 and GF-2 data from 2023. It has shown that the main types of land use changes in the Sanjiang Plain region are arable land, swamp land and forest land in the past 20 years. Industrial, agricultural, and domestic pollution sources are relatively concentrated, and surface water pollution is more prominent. The proportion of Class IV and Class V water quality river sections is relatively high, accounting for 32.9% and 62.7% of the evaluated river length, respectively. The discharged wastewater is mainly industrial wastewater, followed by domestic wastewater.
Keyword. The land use status; pollution source; remote sensing; the Sanjiang Plain.

1. Introduction

The Sanjiang Plain is one of the regions with the most drastic changes in land use in China [1], and it is also the region with the strongest human land relationship. It is the most serious area of deforestation and land reclamation in China in the 1990s [2]. The Sanjiang Plain is located in the northeastern of Heilongjiang Province and is a low-lying muddy plain formed by the alluvial and floodplain of the Heilongjiang River, Songhua River, and Wusuli River [3]. The conversion between land use types in the Sanjiang Plain is significant [4], the trend of changes in paddy fields, dry land, and wetland areas is closely related to the three land use transformation processes [5], among them, the areas of dry fields, paddy fields, and construction land are increasing year by year, while the areas of forests, grasslands, and wetlands are decreasing year by year [6]. Agricultural reclamation activities and changes in arable land [7] are the main characteristics and dominant factors of land use change in the region [8], Groundwater is also the main source of water for agricultural irrigation [9]. The vigorous development of irrigated agriculture in response to overexploitation of groundwater has resulted in low utilization of surface water, leading to a series of ecological and environmental problems such as wetland degradation, loss of biodiversity, deterioration of the water environment, and decline in soil quality in the Sanjiang

Plain [10].

The Sanjiang Plain is the largest inland wetland concentrated distribution area in China [11]. Since the founding of the People's Republic of China, the Sanjiang Plain has undergone four large-scale reclamation activities and has gradually become an important production base of commodity grain in China [12], It plays an important role in ensuring China's food security and regional ecological security. In recent years, with the rapid development of modern agriculture and industry, the population has gradually increased, and the resources obtained from nature have exceeded the regenerative capacity of nature. At the same time, the amount of waste discharged into the environment has exceeded the environmental capacity, leading to global resource depletion and serious environmental pollution and destruction, with water pollution being a particularly serious one. The extensive use of fertilizers and pesticides in the agricultural production process, as well as the discharge of solid waste such as coal gangue from coal mining enterprises. The sewage generated by urban heavy industry production leads to a large amount of heavy metals entering the ecosystem through surface runoff and atmospheric deposition, making heavy metal pollution one of the most severe environmental problems [13]. Weakened agricultural ecological environment and prominent agricultural environmental pollution [14]. The content of heavy metals in surface water bodies is much higher than the average content of rivers in the world, with manganese having the highest content, followed by lead and copper, and mercury having the lowest content [15]. Based on remote sensing (RS) and geographic information systems (GIS), many scholars have used multi period remote sensing images to analyze the dynamic changes in land use in the Sanjiang Plain, providing scientific references for ecological protection and restoration in the region [16].

Conducting land use, surface water pollution status, pollution source types and their distribution in the Sanjiang Plain is of great significance for studying the ecological environment of the Sanjiang Plain. This paper is based on remote sensing technology to quickly and effectively identify the current situation and changes in land use in the Sanjiang Plain over the past 20 years, as well as quantitative information on different water quality and river length of surface water pollution, and the types and distribution of surface water pollution sources. It provides preliminary basic data and strong support for ecological environment investigation in the region.

2. Background of the Research Area

The Sanjiang Plain starts from the Heilongjiang River in the north, reaches Xingkai Lake in the south, borders the Xiaoxing'an Mountains in the west, and ends at the Wusuli River in the east. Its administrative areas include Hegang, Jiamusi, Shuangyashan, Qitaihe, Jixi, and Yilan County in Harbin City, with a total area of 102700 km². There are numerous rivers, lakes, and wetlands within the territory, with a dense urban population, making it one of the important agricultural production bases in China. Among them, the Heilongjiang River and Wusuli River are the northern and eastern boundaries of the Sanjiang Plain, respectively. The Songhua River flows through Yilan County and Jiamusi and joins the Heilongjiang River as the boundary between Jiamusi and Hegang [17-20]. Figure 1 shows the geographical location of the Sanjiang Plain.

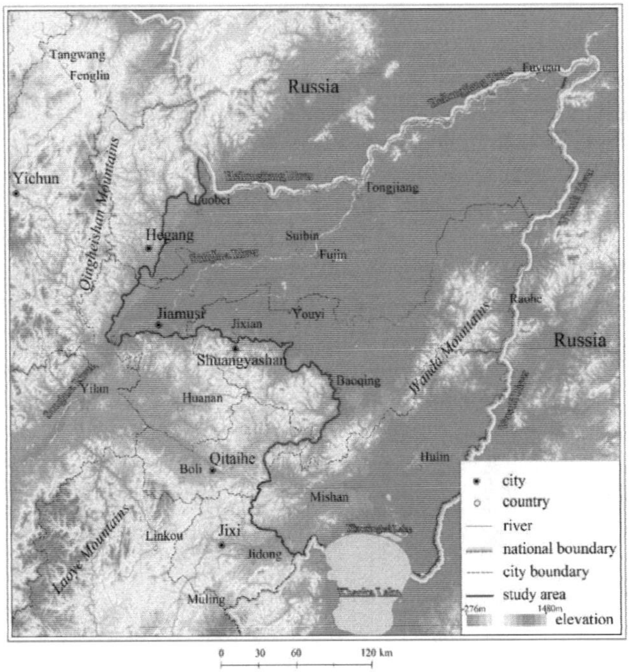

Figure 1. Geographical location map of Sanjiang Plain.

3. Research Methods and Data Processing

The data sources used in this paper include: (1) Remote sensing image data: Landsat-7 ETM+ data in 1999, domestic high-resolution GF-1 and GF-2 satellite data in 2023; (2)1: 50000 digital topographic maps; (3) DEM data generated by interpolation of contour lines extracted from a 1:50000 topographic map. The spatial resolution of the processed ETM+ data in 1999 was 15m, while the spatial resolutions of the domestically produced GF-1 and GF-2 satellite data in 2023 were 2m and 1m, respectively.

The remote sensing data is processed by using Erdas Imagine9.2 image processing software to achieve band combination, geometric correction, orthorectified correction, image fusion, mosaic registration, image cropping, and other tasks of the image. The geometric correction reference standard is derived from a 1:50000 topographic map. According to the 1:50000 remote sensing image plan production standard, the accuracy of correction error is controlled within 0.5 pixels. The remote sensing interpretation work is completed based on the ArcGIS 10.1 software platform. The technical roadmap is shown in figure 2.

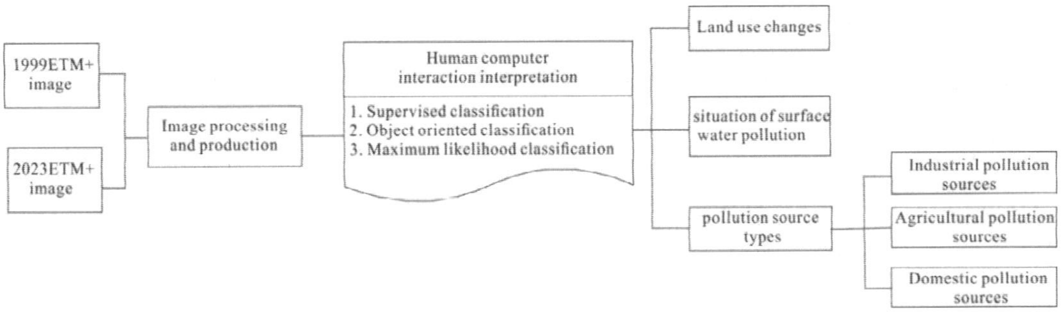

Figure 2. Technical roadmap.

Journal of Physics: Conference Series **2941** (2025) 012066 doi:10.1088/1742-6596/2941/1/012066

4. Result Analysis

4.1. Land Use Changes

The types of land use include cultivated land, forest land, grassland, water area, residential and industrial land, and other unused land, among which the main type of unused land is swamp wetland. In 2023, the total land cover area of Sanjiang Plain was 56986km^2, of which arable land accounted for 61.1%, forest land accounted for 10.5%, grassland area accounted for 2.9%, water area accounted for 8.3%, residential and industrial land accounted for 1.9%, and other unused land accounted for 15.3%.

According to the monitoring results of land use in the Sanjiang Plain region from 1999 to 2023, the types of land use areas with significant changes are mainly paddy fields, swamps, high-density forests, medium density forests, low-density forests and high-density grasslands. Among them, the paddy field area increased by 7453.2 km^2, the high-density forest area increased by 620.1 km^2, the swamp area decreased by 4176.8 km^2, and the forest area decreased by 4389.7 km^2 (figure 3).

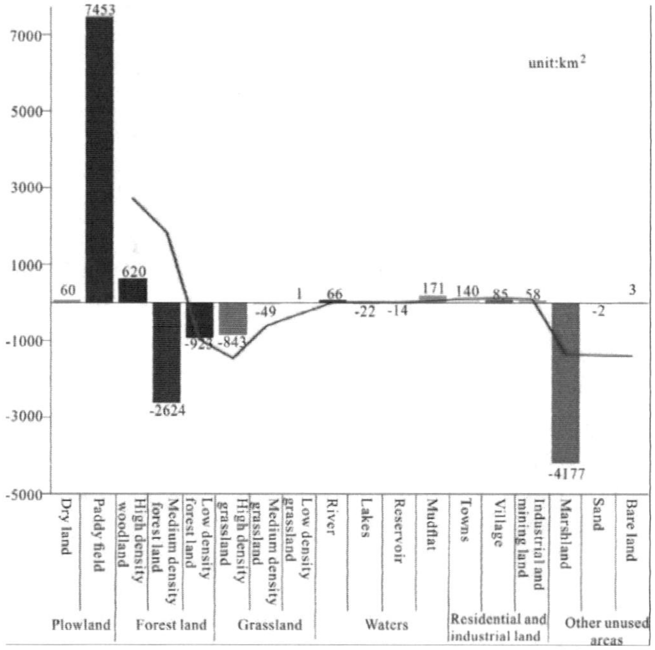

Figure 3. Statistical chart of changes in land use types and occupation area.

The total area of urban residential land increased by 282.6 km^2, with a growth rate of 25.84%. The total area of cultivated land has increased by 7513.5 km^2. Among them, the growth rate of dry land area is 0.23%, and the growth rate of paddy field area is 90.17%. The increase in paddy fields is mainly concentrated in areas such as Fuyuan, Tongjiang, and Fujin. The increased land area of the mine is 57.96 km^2, with a 68% increase for coal mines and a 32% increase for non-metallic mines. The mining activities in the Sanjiang Plain area are very active, and the western edge is mainly dominated by frequent coal mining activities, distributed in the four major coal cities of Jixi, Shuangyashan, Qitaihe, and Hegang in Heilongjiang Province; Nonmetallic mines are densely distributed in the Sanjiang Plain area.

4.2. Situation of Surface Water Pollution

In 2023, based on the water quality monitoring results of the Sanjiang Plain area, a comprehensive evaluation of the water quality in the area was conducted using the Surface Water Environmental Quality Standards (GB3838-2002). There are no Class I or Class II water quality river sections throughout the year; The Class III water quality river section is 353.26km, accounting for 2.0% of the evaluated river length; The Class IV water quality river section is 584.9km, accounting for 32.9% of the evaluated river length; Class V water quality reaches 454.3km, accounting for 2.4% of the evaluated river length; The river section with poor Class V water quality is 804.18km, accounting for 62.7% of the evaluated river length.

The pollutant discharge from the main discharge outlet is 264.772 million m^3. The discharged wastewater is mainly industrial wastewater, with an annual discharge of 194.733 million m3, accounting for 73.5% of the total wastewater discharge; the discharge of domestic sewage is 101.006 million m^3, accounting for 26.5% of the total amount of wastewater discharge. The main pollutants are permanganate index, ammonia nitrogen and chemical oxygen demand (table 1).

Table 1. Statistical table of water quality status of main rivers in the Sanjiang Plain area.

Discharge of wastewater / million m^3	Discharge of Industrial wastewater / million m^3	Discharge of domestic wastewater / million m^3	River Reaches/km				
			ClassIII	ClassIV	ClassV	Inferior Class V	Amount
264.772	194.733	101.006	353.26	584.9	454.3	804.18	2196.64

4.3. Pollution Source Types

The types of surface water pollution sources in the Sanjiang Plain area mainly include three categories: industrial pollution sources, agricultural pollution sources and domestic pollution sources. Through remote sensing interpretation work, a total of 393 pollution source investigation points were discovered, including 281 industrial pollution sources, 64 agricultural pollution sources (including fertilizer agricultural pollution and breeding farm pollution) and 48 domestic pollution sources (solid waste sites and sewage treatment plants) (figure 4).

4.3.1. Industrial Pollution Sources. Industrial pollution sources mainly include waste and wastewater generated during mining, stacking, and processing by coal mining enterprises, as well as oil spills and sewage discharge from gas stations.

4.3.2. Agricultural Pollution Sources. The widespread distribution of farmland in the area, the positioning of the grain base in the Sanjiang Plain and the vigorous development of agriculture have made agricultural pollution in the Sanjiang Plain the main source of pollution. The accumulation of pesticides and fertilizers in large-scale farmland throughout the year will inevitably cause pollution to groundwater, especially in the western diving area of the Sanjiang Plain with poor pollution prevention performance. Due to the coarse particles in the aeration zone, groundwater is highly susceptible to pollution.

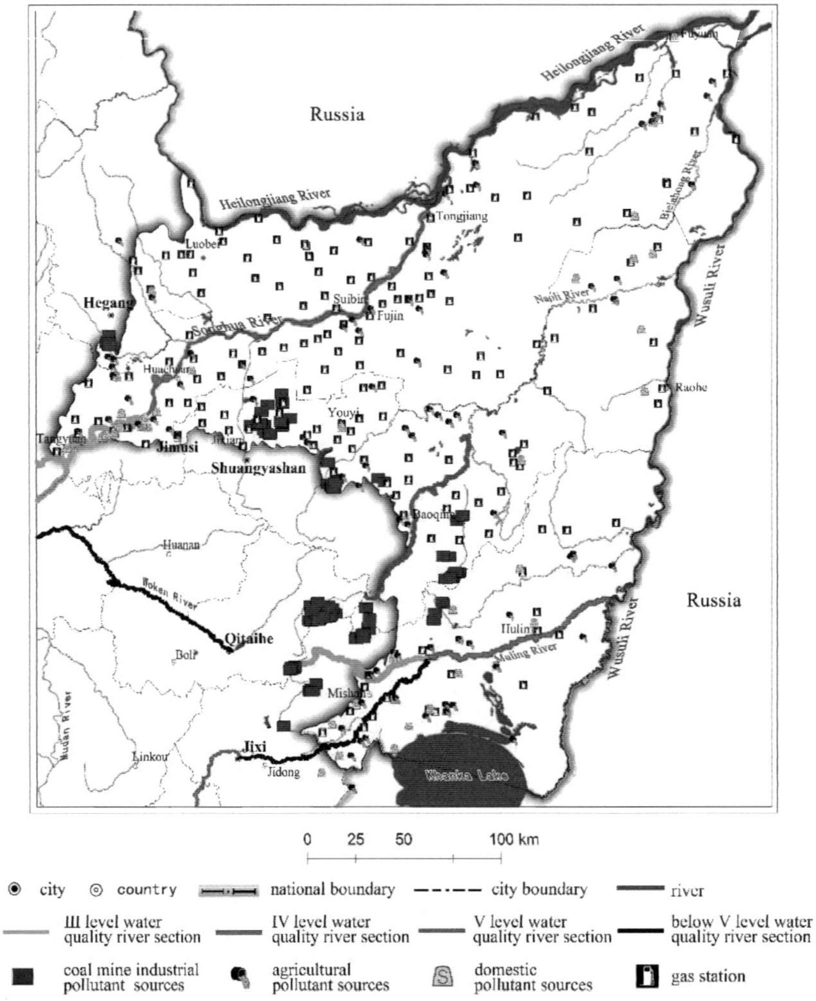

Figure 4. Distribution map of pollution sources in the Sanjiang Plain.

4.3.3. Domestic Pollution Sources. The types of domestic pollution sources mainly include: (1) urban domestic sewage and industrial wastewater discharged from manufacturing industries; (2) Construction decoration waste; (3) Other human household waste. The main sources of domestic pollution in this area are solid waste disposal sites and sewage treatment plants. The solid waste disposal sites are located at the township level, and there are basically no anti-seepage measures in the waste disposal sites. Domestic, industrial and medical waste are mixed and even directly piled up in the surface water, causing serious impacts on surface water and groundwater.

The contribution of three types of pollution sources to surface water pollution in the Sanjiang Plain is as follows: agricultural pollution>industrial pollution>domestic pollution. Specifically, the widespread distribution of farmland in the Sanjiang Plain and the long-term accumulation of pesticide and fertilizer use make agricultural pollution the main source of surface water pollution, with the largest impact range and degree; Next are industrial pollution sources, the distribution of pollution sources in the Sanjiang Plain. The four major coal cities in Heilongjiang Province are all located in the Songhua River Basin, Muling River, and the upper reaches of the Naoli River. The non-compliance of industrial sewage discharge has a huge impact on the surface water in the region; Domestic pollution sources have the characteristics of small individual areas, large quantity, wide distribution range, open-air storage, and no human treatment. They enter the surface water and groundwater systems through surface runoff and atmospheric deposition, and the degree of pollution cannot be ignored.

5. Conclusion

The types of land use that underwent significant changes in the Sanjiang Plain region from 1999 to 2023 were mainly paddy fields, swamps and forests. The main influencing factors are the expansion of urban areas, the transformation of farming methods such as expanding farmland and transforming swamps into paddy fields, and the growth of mining land. The Sanjiang Plain is a densely populated agricultural planting area, with a large number of urban populations and industries. Industrial, agricultural and domestic pollution sources are relatively concentrated, and surface water pollution is more prominent. The proportion of Class IV and inferior Class V water quality river sections in the region is relatively high, accounting for 32.9% and 62.7% of the evaluated river length, respectively. The discharged wastewater is mainly industrial wastewater, followed by domestic wastewater. The main pollutants are permanganate index, ammonia nitrogen and chemical oxygen demand.

Acknowledgments

This work was financially supported by the funding project of Northeast Geological S&T Innovation Center of China Geological Survey (QCJJ2023-37) and Harbin Normal University Doctoral Initiation Fund Project (XKB202013).

6. References

[1] Wang Z M, Guo Z X, Song K S, Liu D W, Zhang B, Zhang S Q, Li F, Jin C, Yang T and Huang N 2009 Effects of land use/cover change on net primary productivity of Sanjiang Plain, during 2000–2005 *Journal of Natural Resources* **24(1)** 136-146.

[2] Gao Z Q 2000 *Comprehensive Analysis of Spatiotemporal Changes and Causes of Land Use and Land Cover Change in China* (Beijing: Institute of Geographical Sciences and Natural Resources Research) Master' s Thesis p1

[3] Cui H Q, Wu T W, Liu J T, Liu W P, Li Z H, Cheng X X and Liu W M 2023 Effects of succession processes of marsh wetland and farmland on groundwater in the Sanjiang Plain *Hydrogeology & Engineering Geology* **50(06)** 51-58.

[4] Lu Z J 2011 Space-time Characteristic analysis of the land use status in the Sanjiang Plain *Heilongjiang Agricultural Sciences* **06** 131-134.

[5] Yu Y X 2023 *Study of Groundwater Recharge in the Sanjiang Plain under the Influence of Land Use Change* (Chuang Chun: Jilin University) Master' s Thesis p27-34

[6] Sun B M, Chen J, Wu X, Chen D, Zhang Y Y and Guo L Z 2014 Ecosystem services value based on land-use changes in the small Sanjiang Plain *Journal of Irrigation and Drainage* **33(4)** 261-265.

[7] Song K S, Liu D W, Wang Z M, Zhang B, Jin C, Li F and Liu H J 2008 Land use change in Sanjiang Plain and its driving forces analysis since 1954 *Acta Geographica Sinica* **63(1)** 93-104.

[8] Wang Z M, Liu D W, Song K S, Zhang B, Zhang S Q, Li F, Jin C and Yang T 2008 Impact of Soil Type on Land Use/Cover Change in Sanjiang Plain *Resources Science* **30(5)** 694-701.

[9] Yang X, Dai C L, Jing J Y 2024 Analysis of influencing factors and estimation methods for groundwater recharge in Sanjiang Plain *Yunnan Water Power* **40(5)**:39-43.

[10] Yang C X, Zheng H, Ouyang Z Y 2020 Changes, effect and driving forces of land use in Sanjiang Plain *Environmental Protection Science* **46(5)** 99-104.

[11] Wang X H 2015 *Research on Joint Simulation and Regulation of Groundwater and Surface Water in Sanjiang Plain* (Beijing: University of Chinese Academy of Sciences) Doctor' s Thesis p4

[12] Fang M 2019 *Formation Process of Inorganic Nitrogen Pollution in Shallow Groundwater of Songhua and Naoli River Basin, Sanjiang Plain* (Chuang Chun: Jilin University) Master' s Thesis p1

[13] Zhang F X 2021 *Characteristic and Ecological Environment Effects of Heavy Metals Pollution*

in Typical Wetland Ecosystem of Sanjiang Plain, Northeast China (Harbin: Northeast Agricultural University) Master' s Thesis p1

[14] Sun Z G, Liu J S, Qin S G, Li X H, Li B and Yang J S 2006 Study on the eco-environmental problems caused by agricultural exploitation of wetland in Sanjiang Plain and regional sustainable development *Journal of Arid Land Resources and Environment* **20(4)** 55-60.

[15] Zhang Z S 2011 *Study on the Distribution Patterns and Biogeochemical Processes of Heavy Metal Elements in Soil of Different Land Use Types in the Sanjiang Plain* (Beijing: University of Chinese Academy of Sciences) Doctor' s Thesis p3

[16] Zhang B, Liu W B and Zhang J H 2021 Landscape ecological risk of the Sanjiang Plain based on land use change *Territory & Natural Resources Study* **(2)** 53-55.

[17] Gao R N, SHI J Z, Fan H and Jia Y H 2024 Study on crop classification in Sanjiang Plain based on random forest *Engineering Journal of Wuhan University* **57(04)** 519-527.

[18] Li S S, Yang S N, Liu X F, Liu Y X and Shi M M 2015 NDVI-Based analysis on the influence of climate change and human activities on vegetation restoration in the Shaanxi-Gansu-Ningxia region,Central China *Remote Sensing* **7(9)**11163-11182.

[19] Zoungrana B J, Conrad C, Thiel M, Amekudzi, Leonard K and ED D 2018 MODIS NDVI trends and fractional land cover change for improved assessments of vegetation degradation in Burkina Faso,West Africa *Journal of Arid Environments* **(153)**66-75.

[20] Yang J and Huang X 2021 The 30 m annual land cover dataset and its dynamics in China from 1990 to 2019 *Earth System Science Data* **13(8)**3907-3925.

Changes and Influencing Factors of Bacterial Communities in Surface Seasonal Frozen Soil at Different Freeze-thaw Stages

Yao Yu, Shirui Hou, Fengyue Yin, Yang Cao and Jianfei Guan*

School of History and Culture, Mudanjiang Normal University, Mudanjiang, Heilongjiang, China.
Email: 3076637889@qq.com; 1240114177@qq.com; 980912100@qq.com; bsycaoyang@126.com; *bangeshiji@126.com

Abstract. Microorganisms are of great significance to prediction of the response of frozen soil to the climate, as well as the mechanism behind. In order to clarify the impacts of freeze-thaw cycle of soil in terms of different stages on the microorganisms in frozen soil, we sampled the topsoil from typical seasonally frozen ground in Heilongjiang, China, and investigated the samples by using the Illumina MiSeq high-throughput sequencing to clarify the structural composition of the bacterial community. The results showed that different freeze-thaw stages had different impacts on the three bacterial groups (abundance taxa, abundance, and rare taxa, rare taxa). At the phylum level, the abundance of Acidobacteria was significantly higher in the FTP stage than in the TP stage ($P<0.05$), while the Actinobacteria exhibited negligible variations in different freeze-thaw stages. At the genus level, significant differences were mainly detected in the FP stage. Total phosphorus and total nitrogen in soil were key soil properties affecting bacterial community structure, followed by organic carbon in soil, while pH had negligible impacts on their community structures.
Keywords. Seasonal frozen soil; bacterial communities; rare taxa; soil properties.

1. Introduction

Global warming is accelerating the degradation of permafrost. The total area covered by seasonal frozen soil is larger than that covered by permafrost, and seasonal frozen soil contains abundant microbial resources and nutrients [1]. Therefore, the study of seasonal frozen soil in the context of global warming is as important as the permafrost. Microorganisms in frozen soil are active participants and sensitive responders of material circulation and energy conversion in permafrost ecosystems [2,3], it is an important biological index to predict the potential feedback mechanism between frozen soil and climate. [4,5]. Microorganisms in the environment have a typically skewed distribution, with several dominant species and various minor species [6]. The existing research mainly focuses on abundance groups, often ignoring the role of rare groups. It has been demonstrated that the rare microbial groups are of great significance to stabilizing community structure and function [7], driving geochemical cycles [8], indicating environmental changes [9], and maintaining ecosystem health [10].

2. Material and Method

2.1. Soil Sample Collection

In this study, the typical seasonal frozen soil (46°34'N, 123°16'E) surface soil (0 ~ 20 cm) in Heilongjiang, China was employed as the subject. Soil samples were collected in August 2020 (complete thawing period, TP), November 2020 (freeze-thaw cycle period, FTP), and January 2021 (complete freezing period, FP). A total of 36 soil samples were collected and placed in sealed bags. Part A was air-dried indoors for the determination of soil properties, and Part B was transported by dry ice to complete the high-throughput sequencing analysis of fungal communities.

2.2. Soil Property Determination and High-throughput Sequencing

The moisture content, the pH value, the organic carbon, the total nitrogen, and total phosphorus of the soil were determined using a vacuum oven (NY 525-2012/5.6), the potentiometric method (NY/T 1377-2007), the $K_2Cr_2O_7$ oxidation-spectrophotometric method (HJ 615-2011), the semi-micro Kjeldahl method (NY/T 53-1987), and the NaOH fusion-MoSb anti-colorimetric method (NY/T 88-1988), respectively.

High throughput sequencing analysis was conducted as reported elsewhere [11].

2.3. Data Analysis

In order to distinguish the roles of different abundance groups in the bacterial communities, we classified the OTU data. Part one, abundant taxa, relative abundance of OTU > 0.1% in three periods. In part two, abundant and rare taxa, the relative abundance of OTU fluctuated around 0.1 % in three periods. Part three, rare taxa, relative abundance of OTU < 0.1% in three periods [12,13].

Software such as Microsoft Excel, Origin2021, IBM SPSS Statistics 27, and Wekemo Bioincloud (https://www.bioincloud.tech) were used to process and analyze the data. One-way ANOVA was employed to determine the differences in the bacterial abundances among multiple soil samples ($P<0.05$) [14]. LDA Effect Size analysis was utilized to achieve comparisons among bacterial communities in different freeze-thaw stages and identified species with significant differences in groups. RDA was employed to elucidate the correlation between bacterial communities and soil properties [15].

3. Result and Analysis

3.1. Analysis of Bacterial Community Composition in Different Freeze-thaw Stages at Phylum Level

At the phylum level, the abundance taxa, the abundance-rare taxa, and the rare taxa belong to 7, 8, and 19 bacterial phyla, respectively. Actinobacteria was the most abundant bacterial phylum in the three groups, but it did not vary with the stage. Among the three groups, Acidobacteria showed the highest abundance in the FTP stage and its abundance varied significantly in the TP stage ($P<0.05$). The abundance of Chloroflexi was consistent in the abundance and rare groups despite the freeze-thaw stage, and its abundance was significantly higher in the FTP stage than in the TP stage ($P<0.05$). Firmicutes, Nitrospirae, and Proteobacteria showed consistent trends of abundance in both abundance and rare taxa, and their abundances were significantly higher in the FP stage significantly than in the TP and FTP stages ($P<0.05$). Gemmatimonades, on the other hand, showed consistent trends of abundance in both abundance-rare and rare taxa, and their abundances were significantly higher in the FTP stage than in the other two stages ($P<0.05$). Bacteroidetes and Verrucomicrobia only existed in rare groups, and their abundances did not vary significantly among different freeze-thaw stages. Overall, the freeze-thaw stage had influences on bacterial abundance groups, abundance-rare groups, and rare groups, and the influences varied with the stage, as shown in figure 1.

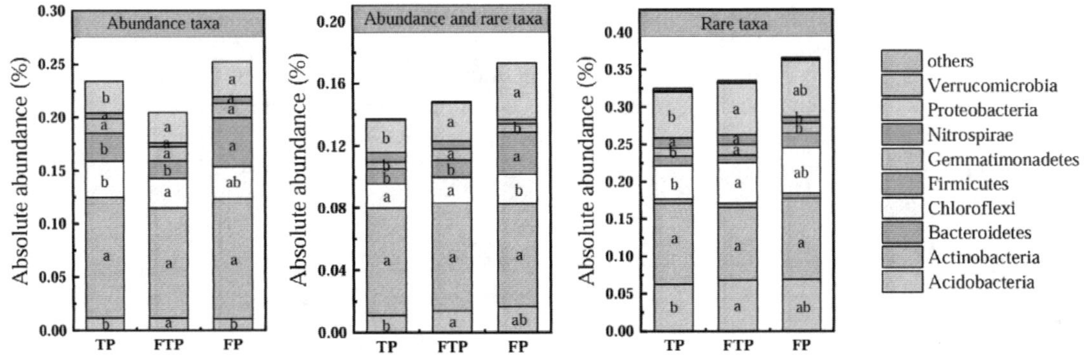

Figure 1. Compositions of bacterial communities in different stages of the freeze-thaw cycle at the phylum level.

3.2. Analysis of Bacterial Community Differences in Different Freeze-thaw Stages at Genus Level

Compositions of bacterial communities in different stages of the freeze-thaw cycle at genus level are shown in figure 2. Among the abundance taxa, *Mycoplasma* in TP, *Bacillus* and *Skermanella* in FTP, and *Kaistobacter* in FP were bacterial genera with significant differences in three stages. Among the abundance-rare groups, most bacterial genera with significant differences were originated from the FP stage. Among the rare groups, few bacterial genera showing significant differences in each freeze-thaw stage were present, *Terribacillus* and *Skermanella* in the FP stage were the only cases. Overall, bacteria in different genera showed different responses to the freeze-thaw stage, and the differences were significant in the FP stage.

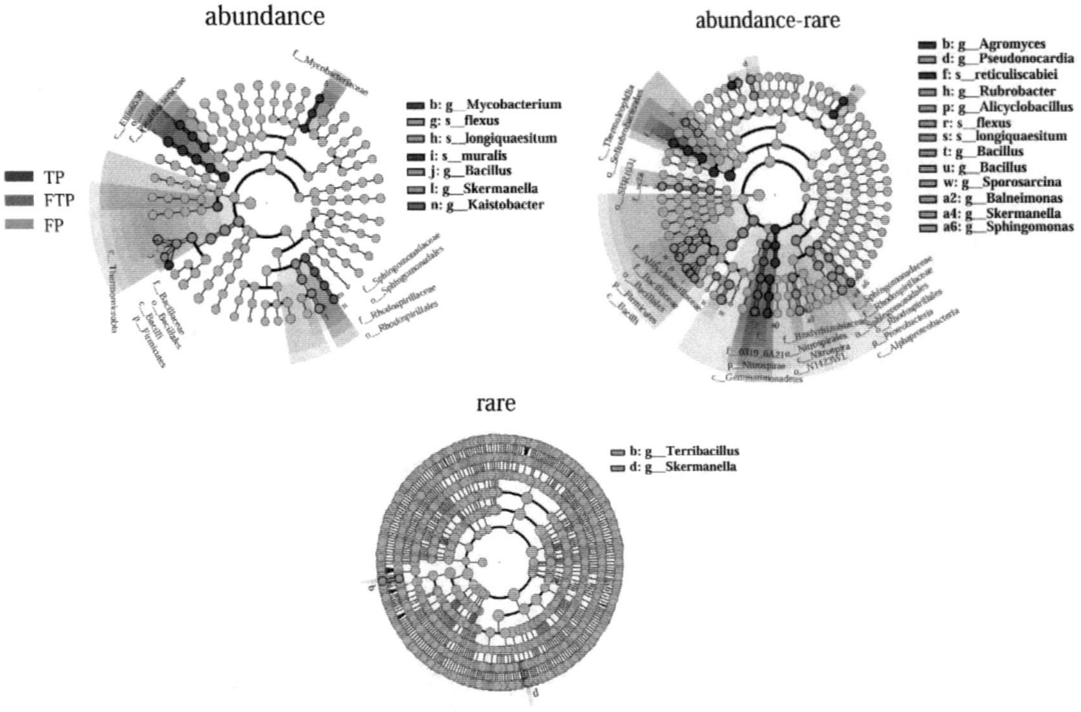

Figure 2. Compositions of bacterial communities in different stages of the freeze-thaw cycle at the genus level.

3.3. Relationships Between Bacterial Community and Soil Properties

According to the results of RDA analysis (figure 3), TP and TN in soil were the main soil parameters affecting the structure of bacterial communities in abundance taxonomic groups, or abundance-rare taxonomic groups, and rare taxonomic groups. Soil organic carbon (SOC) had significant effects on the structure of the three taxonomic groups, while pH had negligible effects on the community structure of each group.

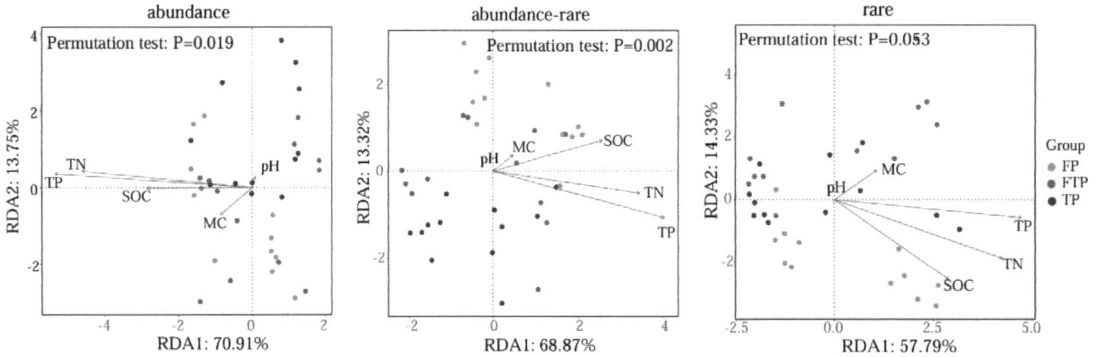

Figure 3. RDA analysis.

Among the rich groups, Gemmatimonadetes, Actinobacteria, and Nitrospirae had significant correlations with TN and TP. Firmicutes had significant correlations with TN, TP and SOC. Acidobacteria, Chloroflexi and Proteobacteria had significant or extremely significant correlations with MC. Among the abundant and rare groups, Actinobacteria, Nitrospirae, Firmicutes, Proteobacteria and Verrucomicrobia were significantly correlated with TN, TP and SOC in an extremely manner. Acidobacteria were significantly correlated with MC. Among the rare groups, Firmicutes, Chloroflexi, Nitrospirae and Verrucomicrobia were significantly correlationed with TN, TP and SOC in an extremely manner. Proteobacteria and Actinobacteria had significant or extremely significant correlations with TN and TP. Soil MC had significant effects on rare types, Acidobacteria, Chloroflexi, Proteobacteria, Bacteroidetes, Gemmatimonadetes and WS3 had significant or extremely significant correlations with MC.

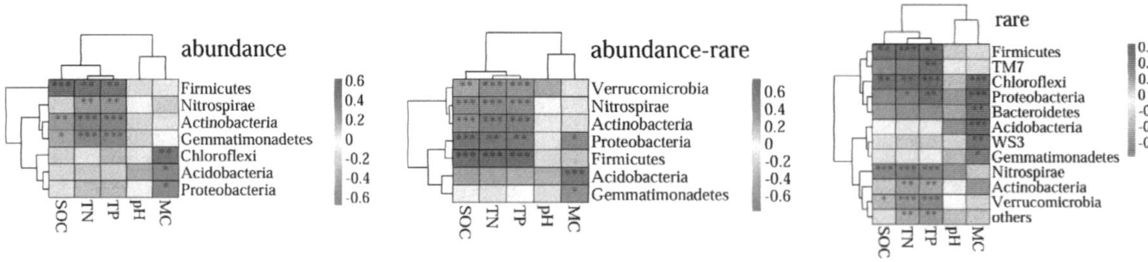

Figure 4. Correlation heatmap.

4. Conclusion

Surface soil samples collected in a typical seasonal frozen soil area in Heilongjiang, China were analyzed to clarify the bacterial community structure (abundance group, abundance-rare group, rare group) in frozen soil at different stages (TP, FTP, FP) of the freeze-thaw cycle. As indicated, bacteria in different genera had different responses to the stage, and the differences were extremely significant in the FP stage. TP, TN, and SOC were the main soil properties affecting the structure of bacterial communities, while pH had negligible influences on the community structure. Actinobacteria and Nitrospirae had significant correlations with TN and TP, MC had significant effects on the rare group.

Acknowledgments
This project was supported by the Research Project on Higher Education Teaching Reform in Heilongjiang Province (General research projects for 2024—Research on the reform strategy of undergraduate thesis topic selection based on OBE Concept) and the Basic Research Project of Heilongjiang Provincial Department of Education (1454QN026, 1451TD020, 1452TD001).

5. References

[1] Chen M, Yang Z, Abulaizi M, Hu Y, Tian Y X, Hu Y P, Yu G L, Zhu X P, Yu P J and Jia H T 2023 Soil bacterial communities in alpine wetlands in arid Central Asia remain stable during the seasonal freeze-thaw period *Ecol. Indic.* **156** 111164.

[2] Han C L, Gu Y J, Kong M, Hu L W, Jia Y, Li F M, Sun G J and Siddique K H M 2018 Responses of soil microorganisms, carbon and nitrogen to freeze-thaw cycles in diverse land-use types *Appl. Soil Ecol.* **124** 211-217.

[3] Edwards K A, McCulloch J, Kershaw G P and Jefferies R L 2006 Soil microbial and nutrient dynamics in a wet Arctic sedge meadow in late winter and early spring *Soil Biol. Biochem.* **38** 2843-2851.

[4] Carla P M, Beat F and Aline F 2020 Functional and structural responses of arctic and alpine soil prokaryotic and fungal communities under freeze-thaw cycles of different frequencies *Front. Microbiol.* **11** 982.

[5] Etienne Y and George A K 2008 Responses of Antarctic soil microbial communities and associated functions to temperature and freeze-thaw cycle frequency *Environ. Microbiol.* **10** 2223-2235.

[6] Nemergut D R, Costello E K, Hamady M, Lozupone C, Jiang L, Schmidt S K, Fierer N, Townsend A R, Cleveland C C, Stanish L and Knight R 2011 Global patterns in the biogeography of bacterial taxa *Environ. Microbiol.* **13** 135-144.

[7] Alexandre J, Alexandre B, Antonis C, Laure G, Angélique G, Viola K, Kirsten K, Rillig Matthias C, Rivett Damian W, Salles Joana F, van der Heijden Marcel G A, Youssef Noha H, Zhang X W, Wei Z and Hol W H 2017 Gera Where less may be more: how the rare biosphere pulls ecosystems strings *ISME J* **11** 853-862.

[8] Laurent P, Aymé S, Catherine H, David B, Florian B, Christophe J M, Amadou S and Pierre-Alain M 2013 Loss in microbial diversity affects nitrogen cycling in soil *ISME J* **7** 1609-1619.

[9] Chambers L G, Guevara R, Boyer J N, Troxler T G and Davis S E 2016 Effects of salinity and inundation on microbial community structure and function in a mangrove peat soil *Wetlands* **36** 361-371.

[10] Yachi S and Loreau M 1999 Biodiversity and ecosystem productivity in a fluctuating environment: the insurance hypothesis *PNAS* **96** 1463-1468.

[11] Guan J F, Li X L, Cao Y and Liu Y 2020 Fungal community structure in top layer of frozen soil in Heilongjiang Province *Chin. J. Ecol.* **39** 2904-2911.

[12] Lynch M D J and Neufeld J D 2015 Ecology and exploration of the rare biosphere *Nat. Rev. Microbiol.* **13** 217-229.

[13] Zhang Z M, He P, Hao X X, Liu J J, Ge T and Li L J 2022 Rare microbial populations as sensitive indicators of bacterial community dissimilarity under different agricultural management practices *Agric. Soil Sci.* **69** 1013-1026.

[14] Sang C, Xia Z, Sun L, Sun H, Jiang P, Wang C and Bai E 2021 Responses of soil microbial communities to freeze-thaw cycles in a Chinese temperate forest *Ecol Process* **10** 66.

[15] Chen M, Yang Z L, Abulaizi M, Hu Y, Tian Y X, Hu Y P, Yu G L, Zhu X P, Yu P J and Jia H T 2023 Soil bacterial communities in alpine wetlands in arid Central Asia remain stable during the seasonal freeze-thaw period *Ecol. Indic.* **156** 111164.

Experimental Study on Coal Ash to Improve Southern Paddy Field

Pinyi Liu[1], Ling Bai[2,*] and Yangping Wen[2]

[1] College of Landscape and Art, Jiangxi Agricultural University, Nanchang 330045, Jiangxi, China.
[2] College of Chemistry and Materials, Jiangxi Agricultural University, Nanchang 330045, Jiangxi, China.
Email: m15879138561@163.com; *lingbb716@163.com

Abstract. In order to investigate the effect of a small dosage of pulverized coal ash on the improvement of rice paddies in southern China, the rice paddies in the suburban county of Nanchang City were taken as the research object, and the control group was set up. The effects of two kinds of pulverized coal ash on rice growth yield and straw length under different particle sizes and dosages were observed. The results showed that pulverized coal ash could change the pH value of cultivated soil in a short time. When the same amount of granular ash was added (e.g. 1.94kg/m2), the yield increase rate of rice was small granular ash (47.5%) > medium granular ash (35%) > large granular ash (15%). The yield increase of brown ash residue was higher than that of black ash residue. The effect of ash and slag on straw length is similar to that of rice yield. The increase in rice yield from 0.49kg/m^2 to 0.86kg/m^2 was greater than that from 0.86kg/m^2 to 1.94kg/m^2. The results of SEM, IR, and XRD analysis showed that the soil surface porosity was improved by ash.
Keywords. Ash residue; rice; pH; small doses.

1. Introduction

Acid-alkalinity is a comprehensive reflection of a series of soil chemical properties, especially the state of the salt base, which directly affects soil fertility and crop growth, including the leaching, migration, and enrichment of soil elements, the activity of soil microorganisms, the synthesis and decomposition of soil organic matter, and the transformation and release of soil nutrients such as nitrogen and phosphorus. Jiangxi is a major agricultural province with a large red soil area, acidic soil, low organic matter content, heavy texture, poor water retention, and poor physical and chemical properties, which lead to the difficulty of rice root absorption of nutrients, the difficulty of seedling germination during sowing, and the difficulty of crusting, cracking, and tearing of roots during watering, thus affect the yield and quality of rice [1].

Coal ash is the industrial solid ash discharged from the production process of coal-fired power plants [2]. The main components of coal ash in China's thermal power plants are SiO_2 and Al_2O_3, the total content of which is more than 70%, and the content of other oxides such as Fe_2O_3, CaO, MgO, K_2O, Na_2O, SO_3, and other oxides is not high [3]. Coal ash is more alkaline, and its particles have a porous structure, large specific surface area, and strong adsorption of nutrients and water retention. It is widely used in environmental pollution control, such as the remediation of heavy metal contaminated soil such as Pb, phosphorus [4], PCBs, and ammonia nitrogen wastewater treatment as adsorbents [5], and also for improving the physical and chemical properties of more acidic soil [6].

Journal of Physics: Conference Series **2941** (2025) 012068 doi:10.1088/1742-6596/2941/1/012068

Xing et al. [7] applied a mixture of "20% coal ash+80% filter mud" to improve the red soil in the dry land, which significantly weakened the soil acidity and had a good barley yield increase effect [8]. Coal ash was mixed with other substances to improve soil in the form of compound fertilizer, and had a significant impact on the yield per plant [9]. From the perspective of different particle sizes of coal ash, the effects of small dosages on the growth and yield of rice were little discussed. In this study, the paddy field in the suburban county of Nanchang City was taken as the research object to study the improvement effect of two kinds of coal gangue ash with different particle sizes on the paddy field.

2. Materials and Methods

2.1. Source and Particle Size of Coal Ash

In this study, coal gangue ash and slag came from Anyuan Coal Mine and Tianhe Coal Mine in Jiangxi Province, which were discharged from the furnace after the crushing and burning of coal gangue. The content of SiO_2 in ash residue is relatively high, which is 50%-60%; the content of Al_2O_3 is 20%-30%; and the content of carbon element is 4%-5%, which is higher than the content in soil; and the content of N, P, K and other elements is basically equivalent to the content in soil. It is divided into black and brown ash residue, and four types of ash residue are screened out through the screen with three pore sizes of 1.5*3.0mm, 5.0*7.0mm, 8.0*12.0mm: Small particles <3.0mm, medium particles (3.0-7.0mm), large particles (7.0-12.0mm), coarse particles >12.0mm, coarse particles of ash residue is less likely to be used in the field, this experiment uses small particles, medium particles, large particles of ash residue for the following tests.

2.2. Experiment Design

Farmland in a township of Nanchang City was selected and sorted into each experimental field with an area of 8m*3m, and the original rice field 20M away from the experimental field was used as a control group. As shown in table 1, the control group A was an originally cultivated field, and ash residue was added to each experimental field C1 to G2, thereinto, the experimental fields C1, C2 and C3 were added by small, medium and large particles of black coal ash, where the dosage was 1.94kg/m^2; the experimental fields D1, D2 and E1, E2 were added by small and medium particles of black ash, where the dosage was 0.86kg/m^2 and 0.49kg/m^2, respectively; the experimental fields F1, F2, G1 and G2 were added small and medium particles of brown ash, and where the dosage was 0.86kg/m^2 and 0.49kg/m^2, respectively. The cultivation methods and rice varieties of the control group were the same as those of the experimental group (planting yellow flower sticky rice, which belongs to the temperature-sensitive conventional rice variety).

Table 1. Changes in pH value of cultivated land, the yield of rice and the length of straw under different test conditions.

Samples	Sample characteristics	Particle size	Quantity of delivery /kg/m^2	pH1 (3th days)	pH (fifth day)	pH3 (30th days)	Yield of rice /g/8	Rice growth rate /%	Straw length /cm	Straw growth rate /%
Control group soil A	Original soil	/	/	6.17	6.16	6.10	200	/	82.0±5.6	/
Coal ash B	Black	/	/	7.65	7.66	7.55	/	/	/	/
Coal ash C1	Black	Small	1.94	6.64	6.92	6.47	295	47.5	101.3±5.8	23.5
Coal ash C2	Black	Medium	1.94	6.54	6.68	6.13	270	35.0	99.7±3.0	17.5
Coal ash	Black	Large	1.94	6.29	6.34	6.10	230	15.0	93.9±1.6	14.5

Samples	Sample characteristics	Particle size	Quantity of delivery /kg/m^2	pH1 (3th days)	pH (fifth day)	pH3 (30th days)	Yield of rice /g/8	Rice growth rate /%	Straw length /cm	Straw growth rate /%
C3										
Coal ash D1	Black	Small	0.86	6.65	6.69	6.34	285	42.5	99.1±3.3	20.9
Coal ash D2	Black	Medium	0.86	6.45	6.52	6.17	267	33.5	92.1±1.4	12.3
Coal ash E1	Black	Small	0.49	6.56	6.43	6.28	270	35.0	96.4±2.3	17.6
Coal ash E2	Black	Medium	0.49	6.25	6.33	6.10	255	27.5	90.7±2.5	10.7
Coal ash F1	Brown	Small	0.86	6.65	6.53	6.37	295	47.5	102.7±3.0	25.2
Coal ash F2	Brown	Medium	0.86	6.22	6.24	6.32	264	32.0	100.9±1.5	18.1
Coal ash G1	Brown	Small	0.49	6.59	6.52	6.43	279	39.5	102.2±2.1	20.6
Coal ash G2	Brown	Medium	0.49	6.19	6.24	6.12	261	30.5	93.6±6.5	14.2

2.3. pH
In the control group, rice seedlings were cultivated in the cultivated fields and the experimental fields, and the soil from the surface to the depth of 5cm was taken from the fields on the 3rd, 15th and 30th days, respectively. The soil in the shade was dried and crushed, and 400 grams of soil was stirred with neutral water (pH = 7.0), and pH of soil and coal ash B (undifferentiated particle size) was measured with a pH meter (PH-100 meter).

2.4. Scanning Electron Microscopy and Infrared Spectrum Analysis
On the 30th day of the experiment, the soil at 5cm of the experimental field C1 was analyzed by scanning electron microscopy (SEM) (TESCAN MIRA LMS, Czech Republic), infrared spectrometer (Thermo Scientific Nicolet iS20, USA) and XRD.

2.5. Rice Yield Increase and Straw Growth Rate
Rice yield increase rate (%) = (experimental group yield - control group yield)/control group yield *100%

Straw growth rate (%) = (straw length of experimental group - straw length of control group)/straw length of control group *100% Rice yield was based on different experimental groups. In order to eliminate the differences among different experimental groups, 8 plots of rice were randomly selected in the same experimental group as the harvest calculation unit (g), and the wet weight was measured by fine pallet balance in the field. Straw length refers to the stem and leaf parts remaining after rice ripening and threshing above the surface of the cultivated field.

3. Results and discussion

3.1. Change of pH under Different Experimental Conditions
As table 1 shows, the pH order of the control group was: $pH_{ash\ residue}$ (7.65) > $pH_{original\ soil}$ (6.17), ash residue was weakly alkaline, and original cultivated soil was neutral. In the experimental fields, the pH value of adding small particles of ash residue was higher. When the addition amount was the same, the pH value of adding small particles of ash was also large; When the addition amount increased, the pH

value also increased; In addition, pH2 (day 15) > pH1 (day 3) > pH3 (day 30). These results showed that the addition of coal ash could change the pH value of soil, but with the passing of time, the pH value gradually recovered to the pH of the original soil, indicating that the soil had a large buffer capacity. At the same time, small coal ash particles changed the pH of paddy soil more obviously under the same addition amount. This was inconsistent with the previous report that coal ash significantly changes pH. This study showed that the change and duration of pH of paddy soil are closely related to the amount of coal ash added.

3.2. Change of Rice Yield and Straw Length

In the experimental group with coal ash (C1 to G2) (table 1), coal ash can both increase rice yield and straw length. The average yield increase rate of rice yield was 35.0%, and the average growth rate of straw length was 18.9%. Among them, the minimum yield increase rate of large-particle ash residue was 15%, and the growth rate of straw length was 14.5%. The results showed that alkaline small and medium particles of ash residue could improve soil more obviously, especially C1 (black small particles, dosage 1.94 kg/m2) and F1 (brown small particles, dosage 0.86 kg/m2) increased significantly (47.5%), reflecting the different experimental results caused by the difference of dosage and coal type. In the case of the same dosage, brown small particles of ash residue can bring more rice yield.

3.3. Relationship Between Coal Ash Dosage and Yield Growth

As figure 1 shows, rice yield and straw length both increased with the increase of black or brown ash content (0.49kg/m^2→0.86kg/m^2→1.94kg/m^2). Among them, the growth rate of rice and straw after the small ash content was added was greater than that after the medium ash content was added. Moreover, under the same dosage (0.49kg/m^2 or 0.86kg/m^2), the yield increase of small particle ash residue was much higher than that of medium particle ash residue, and the growth range of rice and straw after brown ash residue was greater than that after black ash residue was added, indicating that the yield increase effect of small particle brown ash residue was better. In addition, when the black ash residue increased from 0.49kg/m^2 to 0.86kg/m^2, the increase in rice yield was greater than that when the black ash residue increased from 0.86kg/m^2 to 1,94kg/m^2, indicating that the yield increase of rice tended to be stable with the increase of dosage. In short, in a small dose of ash (<2.0kg/m^2), small and medium-sized particles of alkaline ash can improve the soil. Adding ash residue to the existing cultivated land can increase the yield of rice. However, the increase in rice yield has a great relationship with the type, particle size, and dosage of coal ash. Small particles of brown coal ash had a good effect on the increase of straw length, with an average increase of 21.76% (figure 1). Under the condition of the same dosage, the small particles of brown ash had a better effect on the growth of straw length. Similar to rice yield, brown cinder had a stronger effect on straw length than black cinder under the premise of a certain dosage.

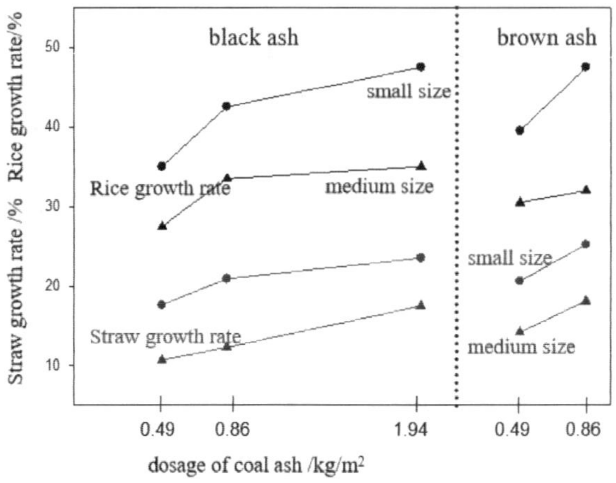

Figure 1. Effect of different sizes and dosages of coal ash (kg.m^{-2}).

3.4. Scanning Electron Microscopy Analysis
The morphology of the soil in Group G of this experiment was studied by SEM. In figure 2, the picture showed the soil's irregular particle shape, which was rather chaotic. It can be seen that coal ash crystals are embedded in the soil, showing a relatively smooth laminar structure on the surface. The soil structure with coal ash was relatively loose, the surface roughness was increased, and the surface area was larger, indicating that coal ash changed the surface structure of the soil, increased the porosity of the soil, and thus affected the growth of crops.

Figure 2. SEM photos of cultivated soil after adding coal ash.

3.5. Infrared Spectroscopy Analysis
An infrared spectrometer was used to test the soil samples of this group G. In the curve (figure 3), the peak around 1034 cm^{-1} was the tensile vibration of Si-O, and the peak here was the most basic characteristic peak forming the soil skeleton. The peak near 3434 cm^{-1} represented the stretching vibration peak of -OH. The bending vibration peak of Al-O is at 1631 cm^{-1}. When the absorption intensity of a peak was higher and wider [10], the relative molecular content of the characteristic vibration mode was larger and the molecular spacing was smaller, indicating that the dust ash can change the relative molecular content and molecular spacing of soil.

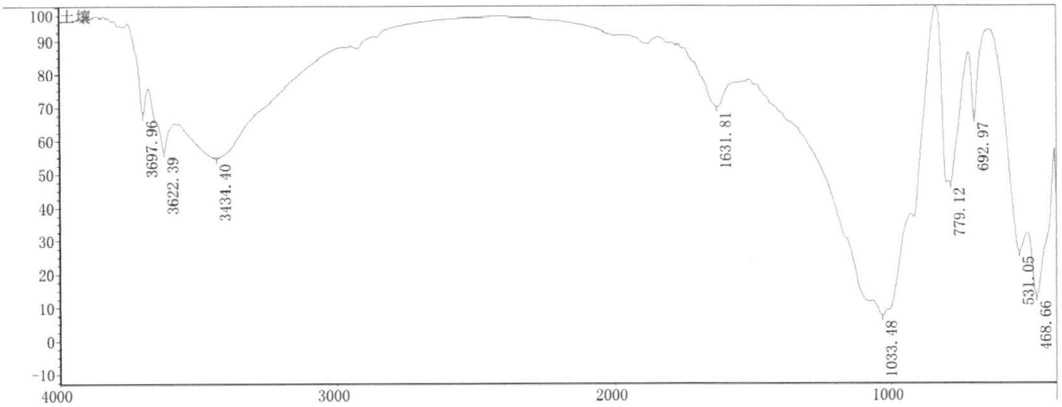

Figure 3. Infrared spectrum analysis of cultivated soil after adding fly ash.

3.6. XRD Analysis

According to the characteristics of the peak position, peak pattern and peak shape in the XRD diffraction pattern, the crystal type, crystal structure, lattice constant, crystal orientation and other information can be determined in the sample, and the relative content of each crystal phase in the sample can be calculated. After testing, the highest peak of cultivated soil after adding coal ash was consistent with the characteristics of SiO_2 (figure 4), which reflected the main composition characteristics of soil improved by coal ash.

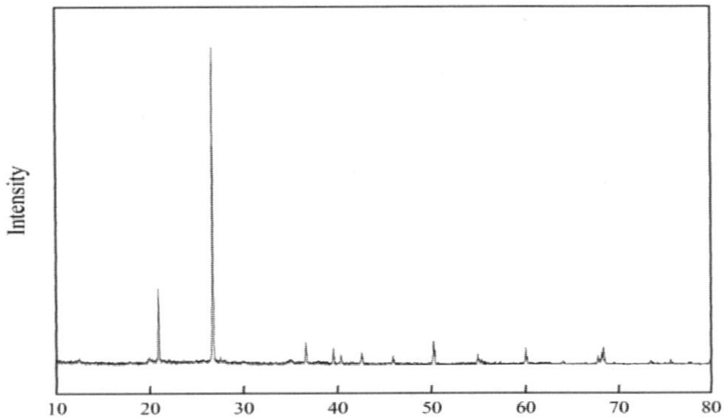

Figure 4. XRD analysis of cultivated soil with fly ash added.

4. Discussion

Due to the different coal types, pretreatment, and combustion processes, the particle size, chemical composition and mineral phase composition of coal ash have certain differences. During the combustion process, almost all trace metal elements are enriched with the decrease of coal ash size. Small particle coal ash has more active components such as calcium sulfate and potassium aluminum silicate than large particle coal ash, and particle size has a significant impact on composition distribution and reactivity [4]. From the point of view of trace elements and active components, the coal ash of small and medium particles in the soil provides more nutrients for the growth of crops. Small and medium particles of coal ash were the modifier of soil aggregate structure. The alkalinity of coal ash reduced the acidity of soil to the range of weakly acidic to weakly alkaline [8,11], improved the characteristics of soil's local interface, reduced soil reducing substances, and increased reoxidation potential [12]. Coal ash promoted the formation of soil microaggregates, increased the soil microaggregates of 0.25-0.05mm grain size, and improved the permeability of clay soil [3]. The increase in porosity had the effect of porosity, permeability, and moisture retention, which was conducive to the activity of soil microorganisms.

The plant rhizosphere is a region where soil and plants exchange substances. Due to plant root exudations and oxygen secretion, soil biochemical properties in the rhizosphere are significantly different from those in the non-rhizosphere [13]. Organic matter content is the main factor affecting the composition of the soil bacterial community in reclaimed soil, and pH is the main factor affecting the composition of the soil bacterial community in reclaimed soil filled with coal ash [14]. There was a significant positive correlation between bacterial diversity index and soil pH [15]. The plant rhizosphere microbiome has many ecological functions, such as promoting nutrient absorption, disease resistance, and resistance, growth, and development. The addition of ash and residue optimized the physical and chemical properties of soil, thus changing the community structure of rhizosphere microorganisms in soil, which was conducive to the growth and extension of seedling roots through absorbing nutrients, and laid a foundation for the subsequent expansion of straw and the increase of rice yield.

5. Conclusions

(1) The addition of small doses of medium and small particles of coal ash (such as less than 2 kg.m^{-2}) in a certain period (about 1 month) changed the pH of cultivated soil, changed the interface characteristics of soil fine structure, improved the living environment of crops, was conducive to the growth of rice seedling roots, and laid the foundation for the subsequent increase of rice yield.

(2) The addition of coal ash in small doses did not significantly change the main characteristics of the original soil, the input cost was low, and the relevant technology was easy to promote and use. Adding ash residue to cultivated land can increase the yield of rice. The increase in rice yield had a great relationship with the type, particle size, and amount of coal ash.

(3) Ash residue of different coal quality had a great difference in the increase of rice yield and straw length, which may be due to the differences in physical and chemical characteristics of coal ash, including the content of trace elements and particle size, which need to be further studied.

Acknowledgments

This research was supported by the National Natural Science Foundation of China, No. 42167004.

6. Reference

[1] Deng P 2002 Discussion over economical benefits for comprehensive utilization of fly ash in agricultural production *Coal Ash* **4(1)** 37-37.

[2] Wang Z, Coyte R M, Cowan E A, et al. 2021 Evaluation and integration of geochemical indicators for detecting trace levels of coal fly ash in soils *Environmental Science & Technology* **55(15)** 10387-10397.

[3] Li Q, Dong Y, Zhu J, et al. 2023 Effect of particle size on mineral phases reactivity of fly ash *Clean Coal Technology* **29 and (S2)** 42-45.

[4] Lei C, Chen T, Zhang Q Y, et al. 2020 Remediation of lead polluted soil by active silicate material prepared from coal fly ash *Ecotoxicology & Environmental Safety* **206** 111409.

[5] Paul V M, Willy V 2003 Removal of PCBs from wastewater using fly ash *Chemosphere* **3(2)** 655-665.

[6] Liu X Z, Yang Z Y 2022 Combined treatment of high ammonia nitrogen wastewater by blow down and fly ash adsorption *Journal of Nanchang University (Engineering &Technology)* **44(1)** 65-72.

[7] Zhang J X, Lu J, Yu J Y 1999 Effects of fly ash and iron tailings on biological characteristics of red soil in orange garden *Bulletin of Science and Technology* **15(4)** 292-295.

[8] Xing S H, Zhou B Q 2003 Effects of the mixture with different proportion of fly ash and filtered mud on red soil properties and barley yield *Journal of Fujian Agriculture and Forestry University (Natural Science Edition)* **32(2)** 240-245.

[9] Ma S C, Lv P, Li C X, et al. 2011 Effect of soil amendments on growth of soybean in coal

gangue contaminated soil *Journal of Ecology and Rural Environment* **27(5)** 101-103.

[10] Hou H, et al. 2016 Remediation effect of soil amendments on gleyed paddy field in Poyang Lake region *Chinese Journal of Soil Science* **47(6)** 1448-1454

[11] Newton A G, Kwon K D, Cheong D K 2016 Edge structure of montmorillonite from atomistic simulations *Minerals* **6(2)** 25.

[12] Ukwattage N L, Lakmalie U V, et al. 2021 Soil and plant growth response and trace elements accumulation in sweet corn and snow pea grown under fresh and carbonated coal fly ash amendment *Agronomy Journal* **113(4)** 3147-3158.

[13] Feng J Y 2020 *Assembly of Root-Associated Microbiomes of Rice (Oryza sativa L.) in Response to Lindane Pollution* (Hangzhou: Zhejiang University) Doctor's Thesis.

[14] Zhang S, Cai H, Zhang Y, et al. 2024 Soil bacterial community structure in coal mining area and its response to different reclamation patterns *Coal Science and Technology* **52(2)** 338-349.

[15] He L, Li Y Q, Li B C, et al. 2017 Effects of different vegetation types and reclamation years on soil bacterial community structure in reclaimed mine areas *Environmental Science* **38(2)** 752-759.

Property and Leaching Toxicity Changes of Fly Ash under Thermal Treatment and the Underlying Mechanism

Jiayi Zheng[1,2,*], Shuijiang Li[1], Jian Su[1], Wuyou Gao[2] and Weizhen Liu[2]

[1] Guangzhou Environmental Protection Investment Group Co., Ltd., Guangzhou 510016, Guangdong, China
[2] School of Environment and Energy, Guangdong Provincial Key Laboratory of Solid Wastes Pollution Control and Recycling, South China University of Technology, Guangzhou 510006, Guangdong, China
Email: *mrzhengjy@foxmail.com; 136816350@qq.com; sujian@gzepi.com.cn; gaowuyou2022@126.com; 181430371@qq.com

Abstract. Thermal treatment can achieve the solidification of heavy metals by changing the physical properties and mineral phases of municipal solid waste incineration (MSWI) fly ash. Taking the fly ash from an MSWI power plant in southern China as an example, this work used X-ray fluorescence, inductively coupled plasma-optical emission spectrometer, toxicity characteristic leaching procedure, thermogravimetric analysis, X-ray diffraction, and energy calculation to study the property and leaching toxicity changes of fly ash under thermal treatment. The results showed that the leaching concentrations of Pb and Cd in fly ash exceed the standard. After thermal treatment at 200-1100°C, the leaching concentrations of Pb, Cu, and Zn first increased and then decreased, while the leaching concentrations of Cr and Cr (VI) constantly increased. Calcium compounds in fly ash gradually decomposed and formed new calcium salts. However, due to the low silicon content, fly ash cannot form an obvious glass body. NaCl and KCl converted heavy metal oxides into chlorides, increasing their solubility while also causing their evaporation at high temperatures. Cr(III) reacted with alkaline substances and was oxidized, resulting in an increasing leaching concentration. This work enhances the understanding of the underlying mechanism of MSWI fly ash thermal treatment and guides the innocuous and beneficial disposal of fly ash.
Keywords. Fly ash; thermal treatment; heavy metal; mineral phases; energy calculation.

1. Introduction

In the past decade, China's incineration capacity for MSW has rapidly increased. In 2022, the incineration treatment volume of MSW in China reached 195 million tons per year. However, incineration typically generates 2-5% fly ash, equivalent to 3.9-9.75 million tons annually. The heavy metals in fly ash, which potentially released into the environment, are particularly concerning [1]. Therefore, controlling the environmental damage caused by heavy metals in fly ash is crucial.

Methods for treating heavy metals include solidification, leaching, and thermal treatment. Solidification involves simple operation and low cost but increases the volume of fly ash, requiring more landfill space [1]. Leaching transfers heavy metals into leachate, but treating the low-concentration, high-salinity leachate poses challenges. Thermal treatment modifies the physical structure or chemical composition, stabilizing heavy metals in the residue and potentially enabling resource recovery. Therefore, thermal processing represents a promising development strategy.

Thermal treatment methods mainly include three types: (i) Low-temperature pyrolysis occurs at 400-600°C, primarily for dioxin decomposition, with insignificant effects on heavy metal immobilization [2]. (ii) Sintering occurs at 700-1200°C, inducing re-crystallization of mineral phases and densifying particle structures to prevent heavy metals from leaching. Products from sintering can also be used as ceramic aggregates [2,3]. (iii) Melting, also known as vitrification, occurs at 900-1400°C, transforming phases into a glassy matrix that immobilizes heavy metals within the glass structure to avoid leaching. Melting products can serve as construction aggregates, but additives such as Si and Al are usually required [3-5].

Thermal treatment of fly ash covers a wide temperature range and involves changes in mineral phases and properties of heavy metals [6-8]. These changes can intensify or mitigate the leaching of heavy metals, thereby affecting the leaching toxicity. Therefore, studying the mechanisms is crucial for understanding the variations in heavy metals leaching concentrations. In this study, fly ash was heated at different temperatures to assess the TCLP toxicity of heavy metals, and changes of mineral phases, and elucidate the mechanisms through techniques such as thermogravimetric analysis and energy calculations. This work aims to provide insights and guidance for the harmless and resourceful treatment of fly ash through thermal treatment.

2. Material and Method

2.1. Raw Fly Ash

The fly ash studied in this work was taken from an MSWI power plant in southern China. The procedure for treating flue gases is "selective non-catalytic reduction (SNCR) + semi-dry desulfurization + dry desulfurization + activated carbon injection + bag dust removal". Fly ash was dried overnight at 105°C for use throughout the experiment. The particles of fly ash are small and grayish-white, with a moisture content of 0.38%.

2.2. Thermal Treatment

In the experiment, 8.000 g of fly ash was placed into a tube furnace and then heated to 200-1100°C (10°C/min), then kept for 1 hour. No vacuum was applied before thermal treatment, and no gas was introduced into the tube furnace during heating. After thermal treatment, when the temperature decreases to about 100°C, the treated fly ash is transferred to a desiccator for cooling. Then the treated fly ash was ground for experiments.

2.3. Toxicity Leaching Procedure

The Chinese National Standard (HJ/T 299–2007), was followed in performing the TCLP for thermally treated fly ash. An acid solution (pH=2.64) was prepared by mixing acetic acid and deionized water. The solid samples and the acid solution were mixed (1 g: 20 mL) in a conical flask and shaken for 18 h. The leachate was extracted by syringe and filtered through a 0.22-μm syringe filter.

According to the Chinese National Standards (GB 18485-2014 and GB 16889-2008), the heavy metals needed to be tested in the leachate, and their toxicity threshold is listed in table 1.

Table 1. The tested heavy metals and their toxicity threshold.

Heavy metal	Be	Hg	Se	Cd	Pb	As
Limit (mg/L)	0.02	0.05	0.1	0.15	0.25	0.3
Heavy metal	Ni	Cr(VI)	Cr	Ba	Cu	Zn
Limit (mg/L)	0.5	1.5	4.5	25	40	100

2.4. Analysis of the Components

X-ray fluorescence (XRF) test was conducted on PANalytical Zetium equipment for semi-quantitatively analyzing the content of fly ash. The content of metal elements was displayed in the form of oxides.

Using an Agilent 730, an inductively coupled plasma-optical emission spectroscopy (ICP-OES) test was used to determine the leachate's heavy metal content. As for the fly ash, aqua regia was used for digestion, and then the pH was adjusted to 2–4 before the ICP-OES test.

The mineral phase composition of solid samples was analyzed by X-ray diffraction (XRD), Bruker D8 Advance with Cu- Kα-radiation (λ=1.5418 Å) at 40 kV and 40 mA. The test step is 0.02°; the scanning time is 1 s/°.

Thermogravimetric analysis (TGA) was conducted to evaluate the thermal kinetics of fly ash. American TA (SDT Q600) was used. TGA test was conducted under temperatures ranging from 35°C to 1100°C with a heating rate of 10°C/min. During the analysis, air was introduced with a flow rate of 50 mL/min to remove decomposition products.

2.5. Reaction Energy Calculation

The Gibbs free energy of the trivalent chromium oxidation process under 50–1400°C with a 50°C interval was calculated using the HSC Chemistry 9.0 program.

3. Results and Discussion

3.1. Property of the Raw Fly Ash

As shown in table 2, the elements with the highest concentration in fly ash were Na, Cl, and Ca (> 20%). S and K elements also had significant concentrations (8%-9%). Na, Cl, and K originated from the high salt content in MSW, with plastics contributing to the chlorine content. Ca and S derive from SO_2 in the incineration flue gas, as well as from the introduction of $Ca(OH)_2$ and CaO during "semi-dry desulfurization" and "dry desulfurization" processes. These components primarily constitute the major mineral phases in fly ash. Additionally, the raw fly ash contained minor elements such as Si, Mg, Al, Fe, and others.

Table 2. The element content of fly ash in the form of oxides

Component	Na_2O	Cl	CaO	SO_3	K_2O	SiO_2	MgO
Content /%	27.51	26.78	22.33	8.97	8.05	2.80	1.00
Component	Al_2O_3	Fe_2O_3	P_2O_5	TiO_2	ZnO	Other[1]	
Content /%	0.80	0.56	0.42	0.29	0.26	0.24	

Note: Other components include Pb, Br, Cu, Co, Sr, Ta, Mn, and Rb.

As shown in figure 1, the mineral phases in the raw fly ash primarily consisted of calcium and chlorine compounds, including $CaSO_4$, $CaCO_3$, KCl, NaCl, CaClOH, and $Ca(OH)_2$, consistent with XRF analysis results. Based on this information, it was hypothesized that during thermal treatment, the decomposition of calcium compounds and the volatilization of chlorine compounds may lead to a decrease in fly ash mass. Simultaneously, these compounds are likely to react with heavy metal elements during thermal treatment, resulting in changes in the leaching toxicity of fly ash.

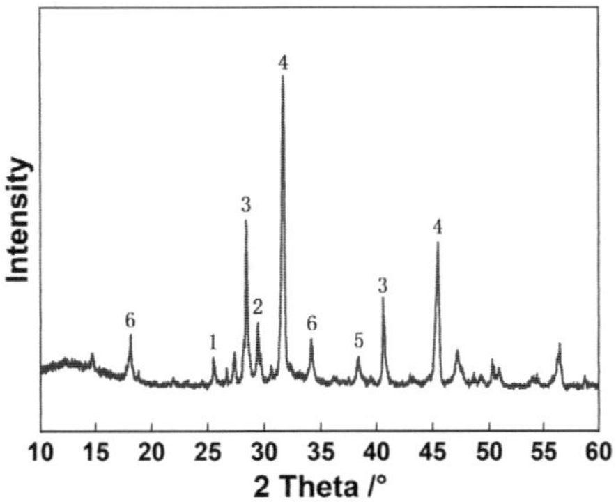

Figure 1. The XRD pattern of the raw fly ash (1- $CaSO_4$; 2- $CaCO_3$; 3- KCl; 4- NaCl; 5- CaClOH; 6- $Ca(OH)_2$).

According to Chinese national standards, the heavy metals of primary concern in fly ash include Be, Hg, Se, Cd, Pb, As, Ni, Cr(VI), Cr, Ba, Cu, and Zn. According to table 3, the highest content of these metals in the raw fly ash was Zn (2425.2 mg/kg), followed by Pb, Cd, Ba, and Cu (200-350 mg/kg). These heavy metals originate from batteries, inks, coatings, electroplated products, and other sources found in MSW. They are significant elements that may pose environmental risks. Other heavy metal concentrations were lower, but they were examined in later leaching experiments.

Table 3. The content of heavy metals in the raw fly ash.

Element	Zn	Pb	Cd	Ba	Cu	Cr
Content (mg/kg)	2425.2	352.9	322.1	235	207.4	48.2
Element	As	Ni	Hg	Se	Be	
Content (mg/kg)	21.5	21.3	13.8	5.6	0.2	

3.2. Leaching Toxicity Changes after Thermal Treatment

After thermal treatment at temperatures ranging from 200 to 1100°C, fly ash underwent TCLP leaching tests. The pH of the leachate and concentrations of heavy metals are shown in figure 2. Prior to reaching 800°C, the pH of the leachate ranged from 5.5 to 6. The initial pH of the TCLP leaching reagent was 2.64, neutralized by the alkalinity. Between 800 and 1100°C, the pH of the leachate increased. This pH increase typically inhibits the leaching.

Pb, Zn, and Cu concentrations in the fly ash leachate first increased and subsequently reduced as the temperature of the thermal treatment increased. When temperatures reached 900°C, the initial Pb-leaching concentration from the raw fly ash exceeded regulatory limitations. Only then did the fly ash become compliant. The leaching concentration of Zn exceeded limits after treatment at 600-700°C but met the standard at other temperatures. Despite its high initial content in fly ash (2425.2 mg/kg) leading to elevated leaching concentration, the exceedance of Zn was mitigated by its higher toxicity threshold. With a lower initial content (207.4 mg/kg) and a higher toxicity threshold, Cu consistently remained within permissible leaching concentrations.

As the thermal treatment temperature increased, the concentrations of Cr and Cr(VI) in the leachate constantly rose. After reaching a thermal treatment temperature of 1000°C, the leaching concentration of Cr(VI) exceeded regulatory limits. Chromium typically exists as Cr(III) or Cr(VI). Comparing figures 2(c) and 2(d), the concentrations of total Cr were close to those of Cr(VI), indicating that the increase in leaching concentration of total Cr was primarily due to the elevated Cr(VI) levels.

The concentrations of Ni and Cd showed no clear pattern of change. However, Cd consistently exceeds regulatory limits. Other heavy metals do not exceed their toxicity threshold.

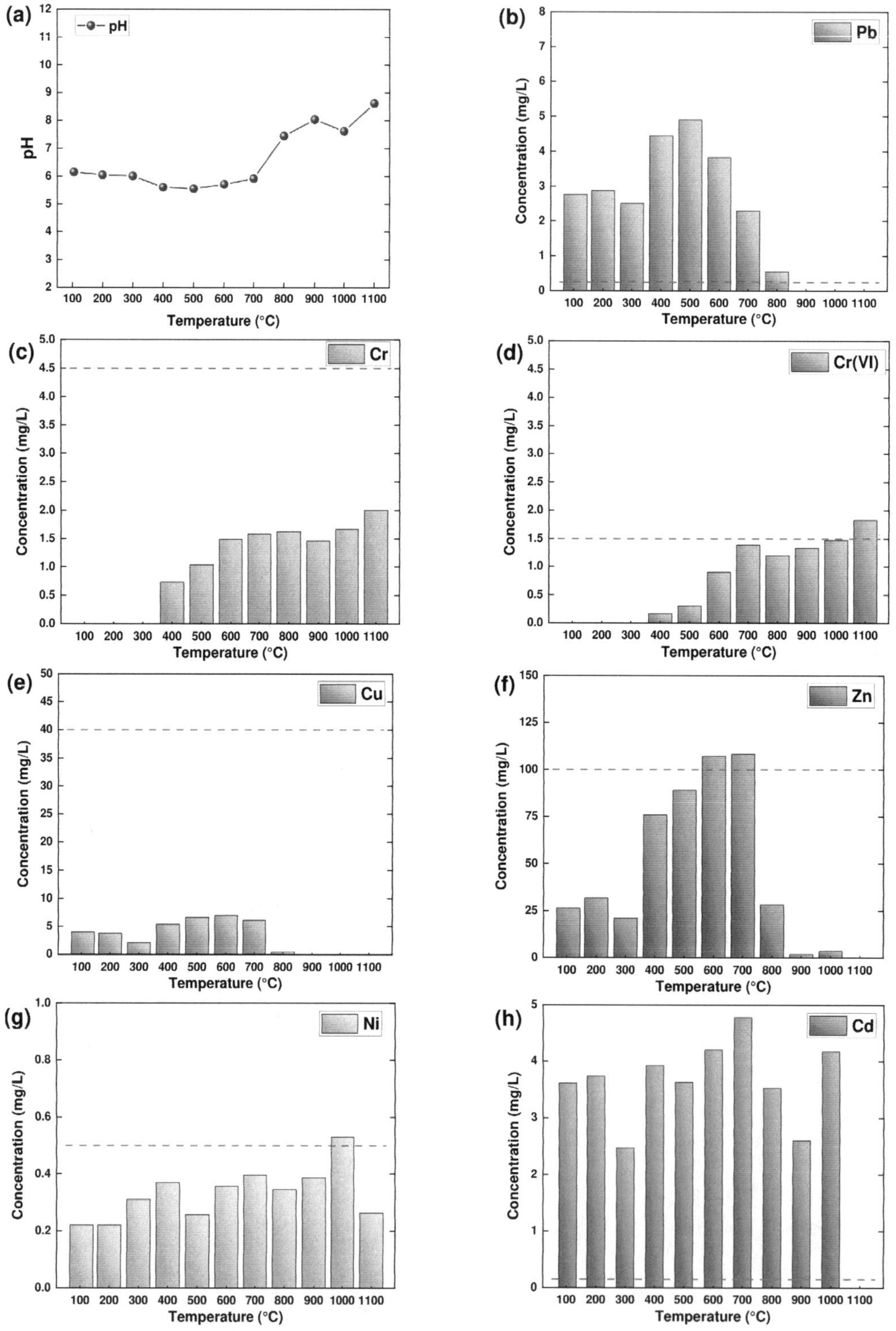

Figure 2. The pH value and heavy metal concentration of the TCLP leachate after thermal treatment from 200-1100°C, with the toxicity threshold labeled by a red dash line.

3.3. Mineral Phase Changes after Thermal Treatment

The overall changes in fly ash under different temperatures were first analyzed using thermogravimetric analysis (TGA). As shown in figure 3, at 105°C, there was a 1% mass loss attributed to moisture evaporation. Between 105°C and 500°C, there was minimal additional mass loss.

After reaching 500°C, the mass begins to decrease rapidly. From the derivative thermogravimetry (DTG) curve, the first peak mass loss rate occurred around 700°C, with the rate decreasing to a minimum at 800°C. From 800°C to 1100°C, the mass loss rate of fly ash peaked around 1000°C. Subsequently, around 1100°C, the weight remained essentially unchanged, with a loss of 49% in mass.

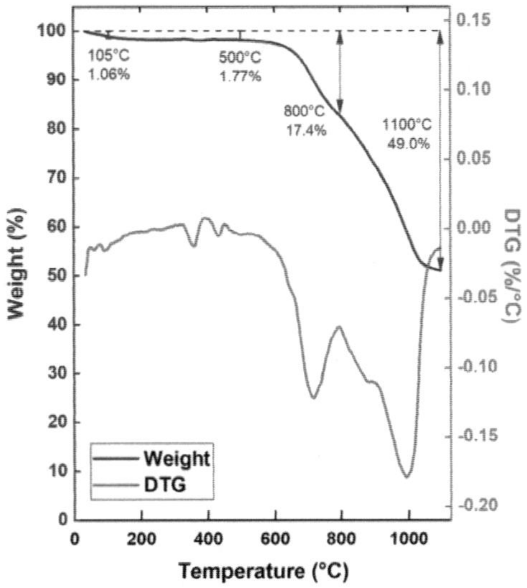

Figure 3. The weight curve and DTG curve of fly ash from 50 to 1100°C (10°C/min).

The thermal treatment temperatures are categorized into three ranges based on the analysis above: low-temperature range (200–400°C), medium-temperature range (500–800°C), and high-temperature range (900–1100°C). Further analysis is conducted to investigate the phase changes of fly ash within these three temperature intervals (figure 4).

In the low-temperature range, most of the calcium-containing and chlorine-containing phases in the raw fly ash remained stable. Upon heating to 400°C, only $Ca(OH)_2$ disappeared, which accounted for the decrease in fly ash mass during this stage.

In the medium-temperature range, CaClOH disappeared after 700°C [6], and calcium salts recrystallized to form $Ca_5(PO_4)_3Cl$. However, $CaSO_4$, $CaCO_3$, KCl, and NaCl remained present, indicating incomplete decomposition or volatilization. At higher temperatures, the heavy metals in the fly ash might react with these salts of calcium and chlorine.

In the high-temperature range, CaO was detected after 900°C. The decomposition of calcium salts into CaO is responsible for the increase in leachate pH. $Ca_5(PO_4)_3Cl$ disappears above 1000°C, while silicate compounds including Ca_2SiO_4 and $Ca_3Al_2(SiO_4)_2Cl_4$ were formed. The formation of silicate compounds can encapsulate heavy metals within the Si-O network, reducing their leaching concentrations. However, XRD patterns indicated that even heating to 1100°C did not result in significant amorphous glass formation. Due to the high calcium content in fly ash, typically silicon is added to promote glass formation at high temperatures [9]. If Al and B are added, the temperature required for fly ash to transform into glass can be further reduced.

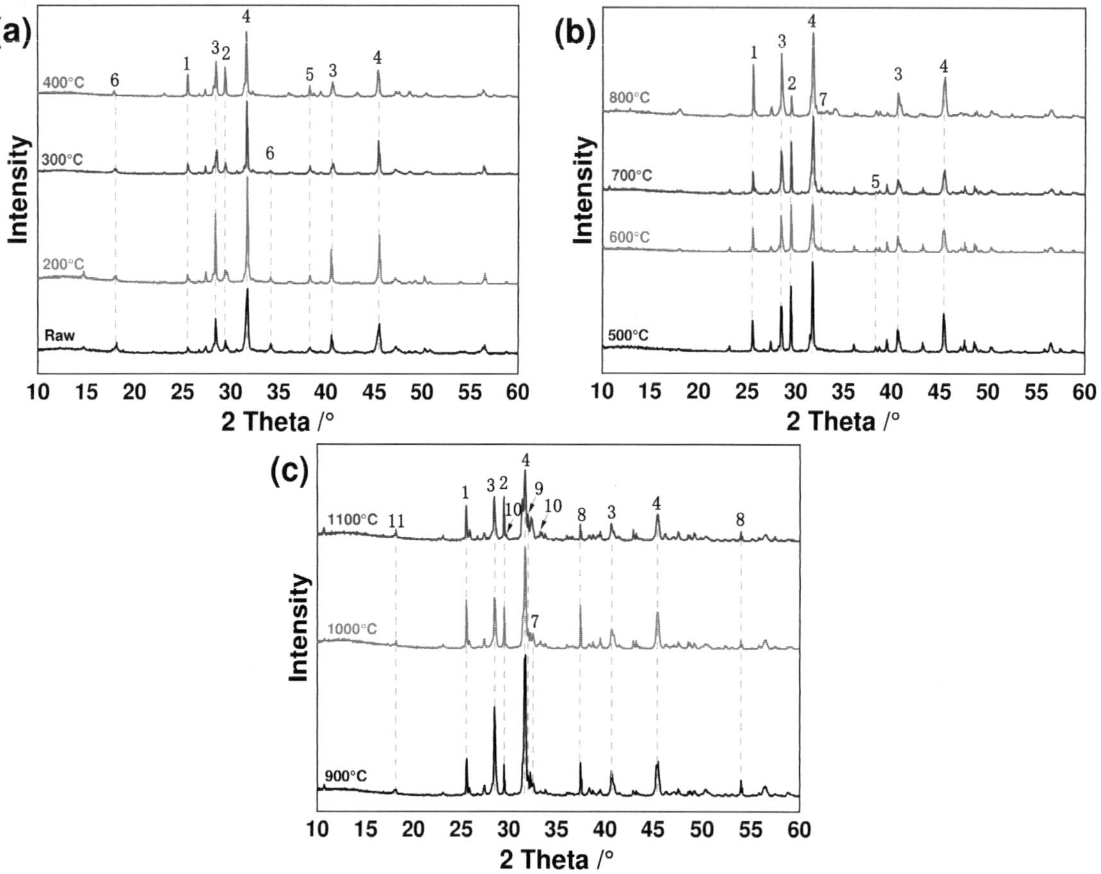

Figure 4. The XRD patterns of fly ashes after thermal treatment under (a) low-temperature range, (b) middle-temperature range; and (c) high-temperature range. (The identified mineral phases: 1- CaSO4; 2- CaCO3; 3- KCl; 4- NaCl; 5- CaClOH; 6- Ca(OH)2; 7- Ca5(PO4)3Cl; 8- CaO; 9- Ca2SiO4; 10- Ca3Al2(SiO4)2Cl4).

3.4. Heavy Metal Changes After Thermal Treatment

After thermal treatment ranging from low to high temperatures, changes in the leaching concentrations of heavy metals were observed (figure 2). Fly ash mostly contains heavy metals in the form of oxides and chlorides. [10]. During the thermal treatment process, heavy metal oxides continuously react with NaCl and KCl to form heavy metal chlorides [6, 7]. Due to the higher solubility of metal chlorides, the leaching concentrations of heavy metals in fly ash increased. Therefore, between 200°C and 800°C, the leaching concentrations of Pb, Cr, Cu, and Zn increased as temperature rose.

Conversely, heavy metal chlorides often have lower boiling temperatures than oxides (figure 5). [11,12]. Therefore, after the transformation of heavy metal oxides into chlorides, they are more likely to volatilize during thermal treatment. This results in a decrease in the metal content in fly ash, thereby reducing its leaching toxicity. As shown in figure 5, except for Cr, the chlorides of Cu, Zn, Pb, Cd, and Ni have boiling points below 1000°C, indicating a tendency to volatilize at the thermal treatment temperatures used in this study.

For instance, $PbCl_2$ and PbO have boiling points of about 900°C (figure 5). The content of Pb in fly ash decreased from 352.9 to 230.1 mg/kg after treatment under 900°C, which was caused by the evaporation of Pb. This explained the decrease in Pb's leaching concentration after treatment above 900°C (figure 2(b)). $CdCl_2$ has a boiling point of 960°C (figure 5). The content of Cd decreased from 322.1 to 64.5 mg/kg after treatment under 1100°C. This explained the decrease in Cd's leaching concentration after treatment above 1100°C (figure 2(h)). Given that fly ash did not form significant glass phases after high-temperature treatment, the volatilization of chlorides is an important reason for the reduction in leaching toxicity of fly ash after high-temperature treatment.

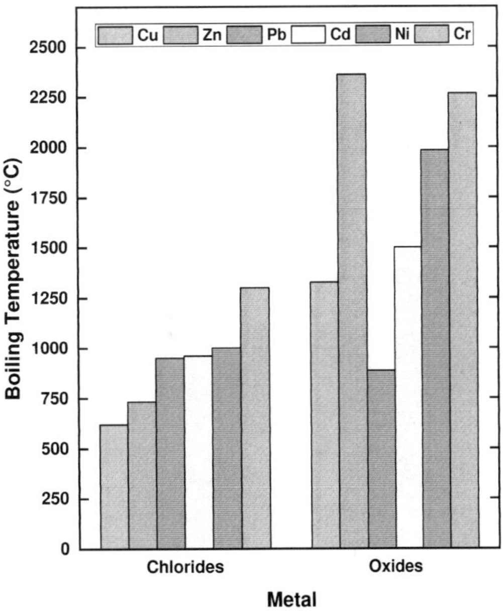

Figure 5. The boiling temperature of different heavy metal chlorides and heavy metal oxides [11,12].

As mentioned above, $CrCl_3$ and Cr_2O_3 have higher boiling points. Cr(VI) is more readily leached compared to Cr(III). Therefore, the Gibbs free energy of the oxidation reaction of Cr(III) is analyzed to assess the tendency of its oxidation. Alkaline substances can promote the oxidation of Cr(III) by O_2. Alkaline substances involved in reactions in fly ash include NaOH and $Ca(OH)_2$. Cr(III) is considered in the forms of $CrCl_3$ and Cr_2O_3. The reaction equations are as follows:

$$4CrCl_3 + 20NaOH + 3O_2 = 4Na_2CrO_4 + 10H_2O + 12NaCl \quad (1)$$

$$4CrCl_3 + 10Ca(OH)_2 + 3O_2 = 4CaCrO_4 + 10H_2O + 6CaCl_2 \quad (2)$$

$$2Cr_2O_3 + 8NaOH + 3O_2 = 4Na_2CrO_4 + 4H_2O \quad (3)$$

$$2Cr_2O_3 + 4Ca(OH)_2 + 3O_2 = 4CaCrO_4 + 4H_2O \quad (4)$$

To normalize, the stoichiometric coefficient of Cr was unified as 1 in the energy calculations. The results are shown in figure 6. For $CrCl_3$, regardless of whether the reactant is NaOH or $Ca(OH)_2$, the Gibbs energy of oxidation reactions remained negative between 50-1400°C. For Cr_2O_3, when NaOH is the reactant, the Gibbs energy of oxidation reactions is also negative. However, when $Ca(OH)_2$ is the reactant, the Gibbs energy of oxidation reactions turns positive when the temperature exceeds 950°C. Nevertheless, as mentioned earlier, Cr_2O_3 continuously transforms into $CrCl_3$ during the thermal treatment process. In summary, during thermal treatment, there is a consistent trend for Cr(III) in fly ash to oxidize to Cr(VI).

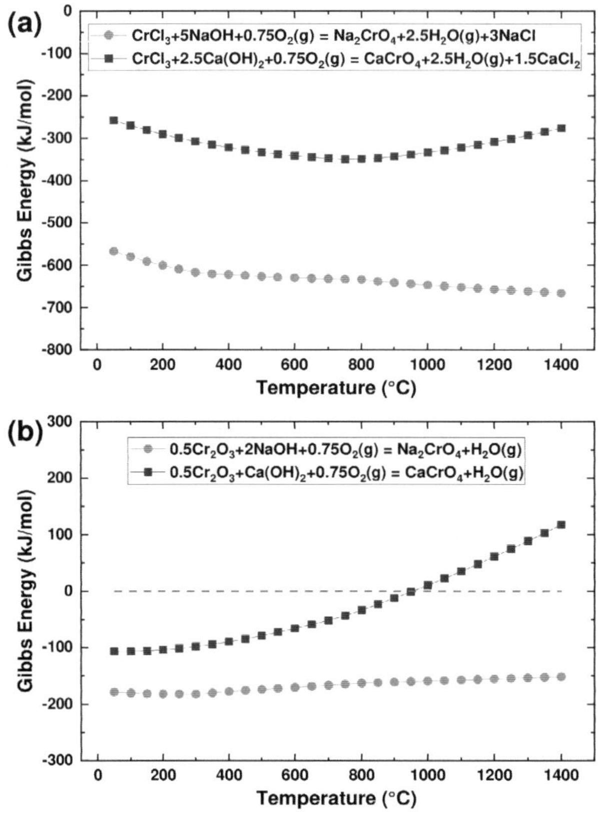

Figure 6. The Gibbs free energy of the oxidation reactions of Cr(III).

4. Conclusion

In this work, the leaching toxicity changes and the underlying mechanism of MSWI fly ash after thermal treatment were studied. The main conclusions are as follows:

(1) The major elements in the fly ash were Na, Cl, Ca, S, and K, with predominant mineral phases being chlorides and calcium salts. The main heavy metals included Zn, Pb, Cd, Ba, and Cu. The leaching concentrations of Pb and Cd in the raw fly ash exceeded their toxicity thresholds. Following thermal treatment, the leaching concentrations of Pb, Cu, and Zn initially increased and then decreased, while those of Cr and Cr(VI) continued to rise. There was no discernible trend of change in the Ni and Cd leaching concentrations. In the low-temperature range (105-400°C), calcium compounds decomposed while chloride compounds remained stable. In the middle-temperature range (400-800°C), calcium salts further decomposed, yielding CaO, along with $Ca_5(PO_4)_3Cl$ formation. In the high-temperature range (800-1100°C), silicate compounds were formed, including Ca_2SiO_4 and $Ca_3Al_2(SiO_4)_2Cl_4$, but no obvious amorphous glass phase was detected.

(2) Under thermal treatment, heavy metal oxides transformed into chlorides. This transformation not only increases the solubility of heavy metals but also facilitates their volatilization at high temperatures. As a special heavy metal, Cr was more easily leached when it was oxidized in the presence of alkaline materials like $Ca(OH)_2$.

In summary, to extract heavy metals from fly ash, it is advisable to lower the thermal treatment temperature to convert the metals into chlorides for leaching. To immobilize these metals, chloride removal through water washing should precede thermal treatment to prevent their volatilization as chlorides at high temperatures, which would complicate flue gas treatment. Furthermore, preserving an oxygen-free or oxygen-deficient environment can prevent Cr(III) from oxidizing to Cr(VI), which will lessen the toxicity of fly ash. This work provides insights into the safe and creative disposal of MSWI fly ash and improves our understanding of its thermal treatment mechanism.

Acknowledgment
This study was supported by Guangzhou Science and Technology Plan Project (No. 202206010054), Guangzhou Environmental Protection Investment Group Co., Ltd. Postdoctoral Innovation Practice Base, Academician Expert Workstation of Guangzhou Environmental Protection Investment Group Co., Ltd. (Innovation Center No. 033, 2021), the National Natural Science Foundation of China (No. 22336006), the National Key Research and Development Program of China (No. 2019YFA0210400), and Guangzhou Science and Technology Project (No. 2024A04J6253).

5. References

[1] Zhao X, Yang J, Ning N and Yang Z 2022 Chemical stabilization of heavy metals in municipal solid waste incineration fly ash: a review *Environmental Science and Pollution Research* **29(27)** pp. 40384-40402.

[2] Xue Y and Liu X 2021 Detoxification, solidification and recycling of municipal solid waste incineration fly ash: A review *Chemical Engineering Journal* **420** pp 130349.

[3] Wang X, Ji G, Zhu K, Li C, Zhang Y and Li A 2021 Integrated thermal behavior and compounds transition mechanism of municipal solid waste incineration fly ash during thermal treatment process *Chemosphere* **264(1)** pp. 128406.

[4] Huang J and Jin Y 2024 Fate of Cl and chlorination mechanism during municipal solid waste incineration fly ash reutilization using thermal treatment: a review *Environmental Science and Pollution Research* **31(3)** pp. 3320-3342.

[5] Lei H, Chen Y, Jia R, Chen P, Zhang Q and Xu K 2023 Study on Phase Transformation of Waste Incineration Fly Ash and Glass Additive Melting Solids (Chinese Edition) *Applied Chemical Industry* **52(07)** pp. 2084-2087+2092.

[6] Yang G, Ren Q, Zhou L, Xu J and Lyu Q 2023 Effect of Atmosphere on HCl Releasement during MSWI Fly Ash Thermal Treatment *Journal of Thermal Science* **32(6)** pp. 2243-2255.

[7] Xie K, Hu H, Xu S, Chen T, Huang Y, Yang Y, Yang F and Yao H 2020 Fate of heavy metals during molten salts thermal treatment of municipal solid waste incineration fly ashes *Waste Manag* **103334** 341.

[8] Wu S, Xu Y, Sun J, Cao Z, Zhou J, Pan Y and Qian G 2015 Inhibiting evaporation of heavy metal by controlling its chemical speciation in MSWI fly ash *Fuel* **158764** 769.

[9] Yuan Z, Cai G, Gao L, Wu M, Kong L, Bai J, Bai Z, Li H and Li W 2023 The physical encapsulation and chemical fixation of Zn during thermal treatment process of municipal solid waste incineration (MSWI) fly ash *Waste Management* **166203** 210.

[10] Li Q, Meng A, Jia J and Zhang Y 2010 Investigation of heavy metal partitioning influenced by flue gas moisture and chlorine content during waste incineration *Journal of Environmental Sciences* **22(5)** pp. 760-768.

[11] Belevi H and Moench H 2000 Factors Determining the Element Behavior in Municipal Solid Waste Incinerators. 1. Field Studies, *Environmental Science & Technology* **34(12)** pp. 2501-2506.

[12] Greenberg R R, Zoller W H and Gordon G E 1978 Composition and size distributions of particles released in refuse incineration *Environmental Science & Technology* **12(5)** pp. 566-573.

Supramolecular Solvent Based Vortex-Assisted Dispersive Liquid-Liquid Microextraction of Phthalate Esters in Beverages

Wei Zhao, Jia Hu*, Lipeng Liu, Yongqi Li and Jingli Huang

Environmental and Biological Engineering College, Wuhan Technology and Business University, Wuhan 430065, Hubei, China.
Email: 2623364970@qq.com; *hj060712@126.com; 2815141369@qq.com; 785295804@qq.com; 3110267068@qq.com

Abstract. A novel method, vortex-assisted dispersive liquid-liquid microextraction technique based on supramolecular solvent (SUPRAS) was developed for the determination of five phthalate esters (PAEs) in beverage samples. The factors influencing SUPRAS type, undecanol and hexafluoroisopropanol (HFIP) volume, pH, vortex time, and standing time were studied and optimized. Under the optimal experimental conditions, PAEs had a wide concentration range (5-1000 ng mL^{-1}) with a good linear relationship ($R^2 \geq 0.9981$), desirable precision (RSDs less than 5.4%), the mean extraction recoveries ranged from 78.1 to 91.2%, and limits of quantification from 0.8 to 1.7 ng mL^{-1}. The proposed methodology was green, sensitive, and economical and was successfully used to detect PAEs in beverage samples.
Keywords. Dispersive liquid-liquid microextraction; supramolecular solvent; phthalate esters; high-performance liquid chromatography.

1. Introduction

Phthalate esters (PAEs) are typical polymer additives to are widely used in food contact materials, carriers of pesticides, building materials and medical instruments [1,2]. PAEs may be transported into the environment as point or non-point sources at all stages of manufacture, transportation and use, for PAEs are physically bound to the polymer chains [3]. PAEs have reproductive and neurotoxicity. Long-term exposure or ingestion of PAEs may cause human reproductive system abnormalities, and even induce cancer [4]. Some PAEs like dimethyl phthalate (DMP), diethyl phthalate (DEP) and Di-n-octylo-phthalate (DNOP) were regulated as priority control pollutants for their potential threats to humans. Therefore, the rapid and accurate analytical methods of PAEs in the complex matrix are significant for the contamination control of PAEs. The content of PAEs in the environmental matrix is usually small, so it needs to be separated and enriched before analysis. At present, many common pretreatment methods such as solid phase extraction (SPE), solid phase microextraction (SPME) and liquid-liquid extraction (LLE) are used for the purification and enrichment of PAEs [5-7]. However, the methods suffer from some drawbacks as follows: high consumption of organic solvents, complicated operation or long extraction time. While vortex-assisted dispersive liquid-liquid microextraction (VA-DLLME) with the advantages of fast extraction speed, high enrichment ratio and reduced cost, it has been applied for the pretreatment of many kinds of samples in recent years [8]. Supramolecular solvent (SUPRAS) with a nanostructure that is immiscible in water and can form micelle aggregate by self-assembly and coacervation of amphiphilic compounds. Thus, SUPRAS shows excellent solvation properties for the extraction of organic compounds with different polarity and is often used as an extraction solvent in VA-DLLME. Hexafluoroisopropanol (HFIP) is a strong hydrophobic perfluorinated alcohol with high density and strong hydrogen-bond donor [9,10]. Less than 10% of HFIP can cause the amphiphilic molecules to agglomerate and form SUPRAS in aqueous media and make the SUPRAS phase facilitate removal for analysis.

Based on the above considerations, different types of SUPRASs comprising HFIP-alkanoic acid /alkanol were chosen to evaluate extraction recovery for PAEs in samples. Extraction volume, pH, vortex time and standing time were investigated carefully. Finally, the novel approach was successfully applied to enrich PAEs in beverages to evaluate the application of this approach.

2. Materials and Methods

2.1. Chemicals and Reagents

hexanoic acid (99%), heptanoic acid (98%), octanoic acid (99%), hexanol (99%), heptanol (99%), undecanol (98%), HFIP (99.5%), alkyldiethyl phthalate (99.5%, DEP), diallyl phthalate (98%, DAP), benzyl butyl phthalate (98%, BBP), dipentyl phthalate (98%, DPP), dicyclohexyl phthalate (99%, DCHP), were purchased from Shanghai Macklin Biochemical Technology Co., Ltd. (Shanghai, China). Chromatographic grade methanol and acetonitrile were purchased from Welch Materials, Inc. (Shanghai, China).

2.2. Sample Preparation

Beverage samples (green tea, black tea and lemon tea) were obtained from the local supermarket. After centrifugation at 4000rpm and a duration of 5 min, the obtained supernatants were passed through utilizing a 0.45 μm microporous membrane for subsequent experiments. The individual stock solutions of five phthalate esters were dissolved by acetonitrile as solvent. The mixed working solutions were varying concentrations by diluting the stock solution with water.

2.3. SUPRAS-Based VA-DLLME Procedure

First, an aliquot of 5 mL aqueous solution containing PAEs was injected into a 10 mL glass conical tube. Subsequently, undecanol (50 μL) and HFIP (300 μL) were added precisely. After vortexing 30s, the mixture was stood for 6 min to ensure the extraction efficiently. Then centrifugation for 1 min at 4000 rpm, the high-density SUPRAS phase was separated from the aqueous solution at the bottom of the test tube. Finally, the deposited SUPRAS phase (50 μL) was withdrawn and measured using HPLC.

2.4. HPLC-UV Analysis

Chromatography was carried out in a Shimadzu Technologies LC-20A series HPLC system (Kyoto, Japan) outfitted by a C18 column (Shim-pack GIST, 250 mm × 4.6 mm × 5 μm) and an SPD-20A detector for chromatographic separation and detection. The linear elution mode of acetonitrile-water (0-10 min, 75% acetonitrile increased to 100% and maintained constant for 15 min) was employed using 1.0 mL min^{-1} of flow velocity and maintaining 30°C of column temperature. The UV wavelength of substances was operated at 225 nm and utilized 10 μL of solution volume.

3. Results and Discussion

3.1. Optimization of Experimental Conditions

Several influencing factors including the composition of SUPRAS, the volume of alkanol and HFIP, the solution's pH, vortex and standing time were investigated in turn. The extraction solution (100 ng mL-1) was taken to explore the optimal extraction conditions.

3.1.1. Effect of SUPRAS. The interaction between the targeted analytes and SUPRAS has strongly affected the efficiency of the microextraction procedure. Here, various compositions of SUPRASs were investigated. The extraction process involved hexanoic acid, heptanoic acid, octanoic acid, hexanol, heptanol and undecanol (50 μL) as hydrogen bond acceptor, along with HFIP (300 μL) as hydrogen bond donor. As can be seen in figure 1, the HFIP-undecanol-based SUPRAS displayed the best performance. So undecanol was selected for subsequent experiments.

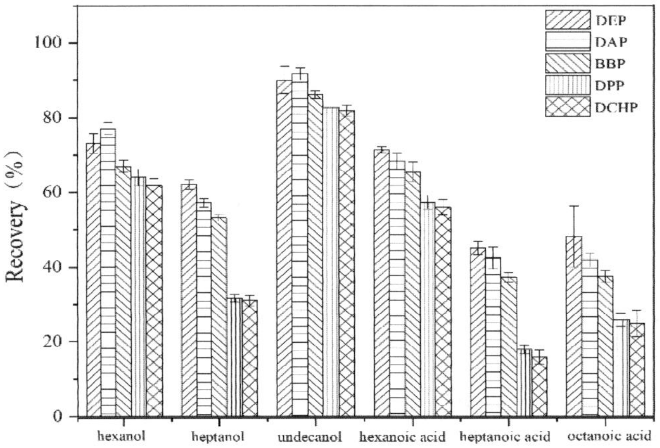

Figure 1. The effect of compositions of SUPRAS on the extraction recovery of PAEs.

3.1.2. Effect of Undecanol Volume. In order to examine the influence of the change in undecanol amount on extraction performance, various amounts of undecanol (50-150 μL) and HFIP (300 μL) were tested. As shown in figure 2, the extraction recovery of PAEs was gradually reduced with the increase of undecanol volume. Because the concentration of amphiphilic molecules has a great effect on the self-assembly process of SUPRAS. A low volume of undecanol can not form the SUPRAS with HFIP, while a large volume is easy to generate the dilution effect. Thereby, 50 μL undecanol was sufficient for extracting PAEs in 5 mL solutions.

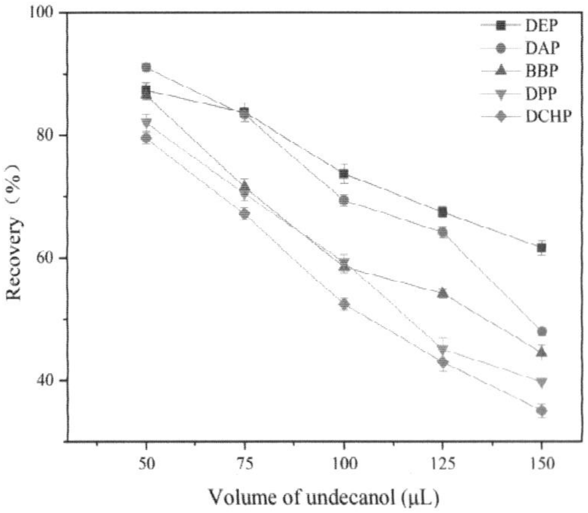

Figure 2. The effect of the volume of undecanol on the extraction recovery of PAEs.

3.1.3. Effect of HFIP Volume. HFIP plays an imperative factor in the isolation process of PAEs. To obtain the best volume of HFIP, we examined different volumes of HFIP (200-400 μL) containing 50 μL undecanol. The results are shown in figure 3. The extraction recovery of PAEs increased significantly with the altering of the volume of HFIP from 200 μL to 300 μL. On the contrary, the volume of HFIP was greater than 300 μL, and the extraction recovery decreased. Because when HFIP volume was 300 μL, HFIP could cause more undecanol to form SUPRAS and increase the extraction efficiency of PAEs. However, a large volume of HFIP would lead to dissolve the corresponding SUPRAS and failure of PAEs to completely precipitation. So 300 μL HFIP was ultimately chosen.

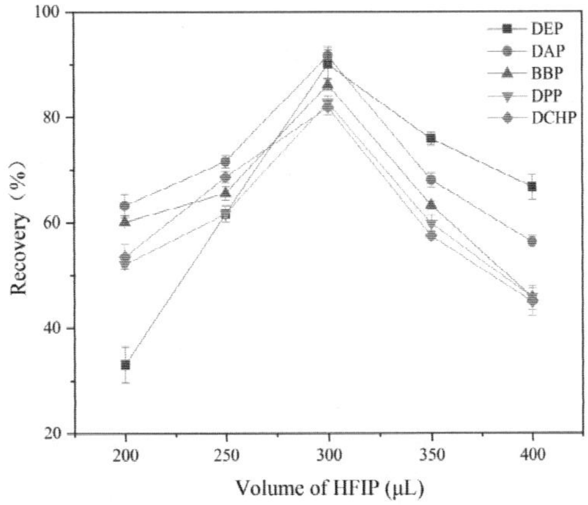

Figure 3. The effect of the volume of HFIP on the extraction recovery of PAEs.

3.1.4. Effect of pH. Variations in the pH of the sample influence the existence of target components in a molecular state. By adjusting the pH of the solution, the extraction efficiency may be changed. The range of pH (2-11) was adjusted by HCl and NaOH solution. The extraction recovery of PAEs practically still unaffected during the pH altering. As a result, no adjustment of pH was used when dealing with samples.

3.1.5. Effect of Vortex. Vortex can accelerate the rate of SUPRAS formation and help mass transfer. Hence, different vortex time (10-60 s) was examined. As seen from figure 4, the extraction recovery of PAEs increases as the vortex time increases, then decreases slightly. Thus, 30 s was selected to reduce the operating time of samples.

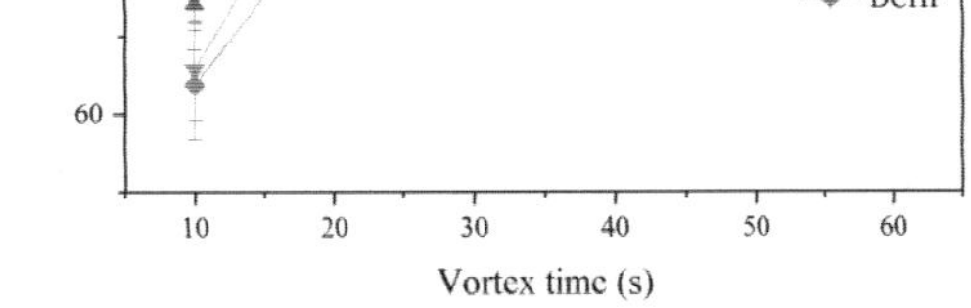

Figure 4. The effect of vortex time on the extraction recovery of PAEs.

3.1.6. Effect of Standing. Standing time can give enough access to the SUPRAS phase and water phase. Therefore, different standing time (0-10 min) was examined with constant experimental conditions. The results are shown in figure 5, the extraction recovery of PAEs was the greatest in the 6 min. According to this result, 6 min was set as the suitable standing time.

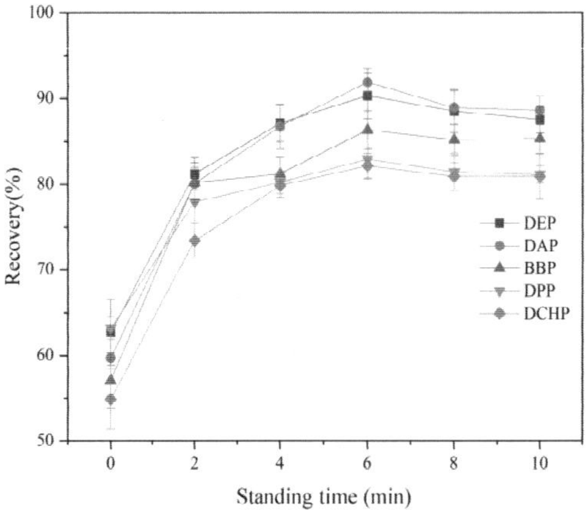

Figure 5. The effect of standing time on the extraction recovery of PAEs.

3.2. Method Validation

To verify the validation of the proposed VA-DLLME procedure under the optimized circumstances, various concentrations of mixed standard solutions were prepared to construct a linear range, the limits of detection (LODs), and the precision of this method. The results of the calibration parameters are summarized in table 1. As seen, the method has excellent linearity for PAEs in the range of 5-1000 ng mL^{-1}, with favorable coefficient of correlation(R^2) ranging from 0.9981 to 0.9993. The LODs were calculated by the standard IUPAC method as 3σranged from 0.8 to 1.7 ng mL-1. As shown in figure 6, the chromatograms of the PAEs have not observed obvious interference.

Table 1. The analytical performances of the developed method.

Analytes	Linear equation	Coefficient of determination (R^2)	Linear range (ng mL^{-1})	RSD (%)	LOD (ng mL^{-1})
DEP	y = 391.45x + 2957.2	0.9981	5-1000	3.9	1.2
DAP	y = 422.82x + 668.89	0.9993	5-1000	2.3	0.8
BBP	y=1930.3x + 5715.7	0.9991	5-1000	2.5	1.3
DPP	y = 1206.8x - 2202.2	0.9987	5-1000	4.7	1.7
DCHP	y = 1490.1x + 5670	0.9986	5-1000	5.4	1.5

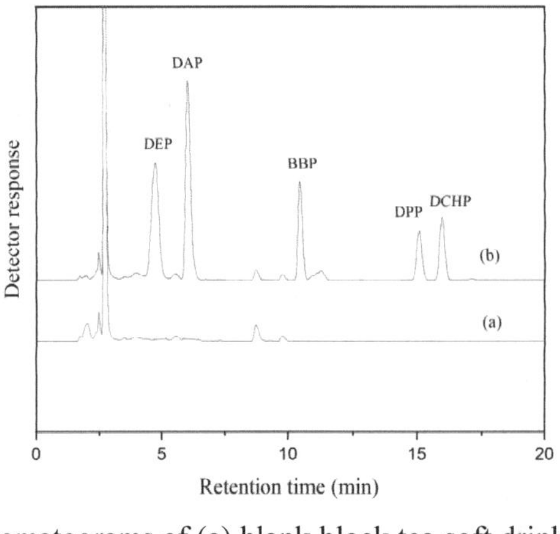

Figure. 6. Typical HPLC chromatograms of (a) blank black tea soft drink and (b) spiked black tea soft drink (100 ng mL^{-1}).

3.3. Analysis of Real Samples

To assess the applicability of the method, the analysis of the green tea soft drink, the black tea soft drink, and Lemon tea soft drink were performed under the optimal experimental conditions. No phthalate esters were detected in the analyzed real samples. Recovery testing was carried out with 50 ng mL^{-1}ls of PAEs. The results are given in table 2. Preliminary results show the ERs of PAEs ranged from 78.1% to 91.2%, and RSDs of PAEs ranged between 2.8 % and 5.3%., indicating that the discussed experimental method is suitable and replicable for detecting phthalate esters in beverage samples.

Table 2. The analytical results for samples.

Analytes	Sample	Recovery (%)	RSD (%)
DEP	Green tea Black tea Lemon tea	87.1	4.7
DAP	Green tea Black tea Lemon tea k	91.2	3.2
BBP	Green tea Black tea Lemon tea	83.5	2.8
DPP	Green tea Black tea Lemon tea	78.1	5.3
DCHP	Green tea Black tea Lemon tea	80.2	4.1

4. Conclusion

This study presents a new, straightforward, and cost-effective approach for the quantitative analysis of PAEs in beverages using SUPRAS coupled with HPLC-UV. SUPRASs based on combinations of HFIP and undecanol were evaluated as extraction solvents of PAEs in beverages. Only microliter volumes of undecanol (50 μL) and HFIP (300 μL) were adopted in the whole operation procedure. Validation studies indicated the characteristics of the approach that fit the quantitation of PAEs. LODs of the method were found in the range of 0.8 to 1.7 ng mL^{-1} and acquired relative recovery values in the range of 78.1%-91.2% with RSDs≤5.3% for PAEs. SUPRAS-VA-DLLME had many advantages in the aspects of greenness and simplicity. It has tremendous potential for monitoring target compounds for sample extraction.

Acknowledgments

S202113242015 is supported by the Innovation Training Program for college students in Hubei Province. A2023016 is supported by Wuhan Technology and Business University Research Program Fund.

5. References

[1] Konieckia D, Wang R, Moody R P and Zhu J 2011 Phthalates in cosmetic and personal care products: concentrations and possible dermal exposure *Environ. Res.* **111** 329-336.

[2] Zota A R, Calafat A M and Woodruff T J 2014 Temporal trends in phthalate exposures: findings from the national health and nutrition examination survey *Environ. Health Persp* **122** 235-241.

[3] Serrano S E, Braun J, Trasande L, et al. 2014 Phthalates and diet: a review of the food monitoring and epidemiology data *Environ. Health Glob.* **13(1)** 43.

Journal of Physics: Conference Series **2941** (2025) 012070 doi:10.1088/1742-6596/2941/1/012070

[4] Hamid N, Junaid M, Manzoor, et al. 2020 Prioritizing phthalate esters (PAEs) using experimental in vitro/vivo toxicity assays and computational in silico approaches *J. Hazard. Mater.* **398** 122851.
[5] Xue L, Zhang D, Wang T X, et al. 2014 Dispersive liquid-liquid microextraction followed by high performance liquid chromatography for determination of phthalic esters in environmental water samples *Anal. Methods* **6** 1121-1127.
[6] Guo Z, Wei D, Wang M, et al. 2010 Determination of six phthalic acid esters in orange juice packaged by PVC bottle using SPE and HPLC-UV: application to the migration study *J. Chromatogr. Sci.* **48** 760-765.
[7] Wang X, Feng J, Tian Y, et al. 2019 Melamine-formaldehyde aerogel functionalized with polydopamine as in-tube solid-phase microextraction coating for the determination of phthalate esters *Talanta* **199** 317-323.
[8] Wen J Q, Chen Q N, Ji H X, et al. 2023 Simultaneous determination of five compounds in Codonopsis Radix using a hydrophobic deep eutectic solvent based on vortex-assisted dispersive liquid-liquid microextraction coupled with HPLC *Microchem. J.* **193** 109197.
[9] Li X, Chen J, Wang H, et al. 2020 Hexafluoroisopropanol-alkanol based high-density supramolecular solvents: Fabrication, characterization and application potential as restricted access extractants *Anal Chim Acta* **1124** 20-31.
[10] Hu W X, Yu Y, Weng Z H, et al. 2024 Supramolecular solvent for the vortex-assisted dispersive liquid-liquid microextraction of aristolochic acids from Aristolochia debilis prior to HPLC-DAD *Microchem. J.* **206** 111608.

Synthesis of $La_2(WO_4)_3$/g-C_3N_4 and Photocatalytic Degradation of Tetracycline

Yuqi Xie, Sini Zhang, Wenjing Liang, Jinqiao Long* and Liansheng Cui#

Guangxi Key Laboratory of Urban Water Environment, College of Chemistry and Environmental Engineering, Baise University, Baise 533000, Guangxi, China.
Email: xiekey@sina.com; 2040816146@qq.com; 3334899204@qq.com; *416874679@qq.com; #409269722@qq.com

Abstract. The overuse of tetracycline antibiotics has brought serious threats to human health and ecological environment. Therefore, it is imperative to find efficient methods to degrade tetracycline. In this paper, $La_2(WO_4)_3$/g-C_3N_4 was synthesized to investigate its photocatalytic degradation effect for tetracythe. The melamine was used as raw material and g-C_3N_4(CN) blue phosphor was synthesized by pyrolysis technology. $La_2(WO_4)_3$ (LWO) was synthesized by high temperature solid phase method, and a composite photocatalyst $La_2(WO_4)_3$-7%CN(LWO-CN) was prepared by precipitation adsorption reaction. The composition, structure and band gap of the phosphor were tested by X-ray diffraction, SEM, TEM and ultraviolet diffuse reflection. Using tetracycline as the degradation target, the catalytic degradation activity of (LWO-CN) under ultraviolet irradiation was studied. The results show that the catalytic activity of LWO-CN is superior to that of LWO. When the amount of composite catalyst is 10 mg, the ultraviolet irradiation time is 90 min, and the concentration of tetracycline hydrochloride is 5 mg/L, the degradation efficiency of tetracycline by LWO-CN is 89.49%, which has a good application prospect.
Keywords. Composite photocatalyst; tetracycline hydrochloride; g-C_3N_4.

1. Introduction

Tetracycline hydrochloride (TCs) belongs to the category of new pollutants, which is mainly used in human, agriculture, animal husbandry and other fields for the prevention and treatment of diseases of antibiotics. Neither humans nor animals fully absorb antibiotics, and most of them are excreted in feces and urine. It causes water pollution and poses serious harm to human health and ecosystem: [1]. The treatment of water polluted by new pollutants is an important subject of current environmental problems. At present, the treatment methods for new pollutants include adsorption: [2], membrane separation: [3], advanced chemical oxidation: [4] and biodegradation: [5]. The photocatalytic technology in advanced chemical oxidation method has the advantages of high efficiency, green, safety and energy saving, and is widely used to degrade new pollutants such as antibiotics: [6]. Among the developed photocatalysts, ZnO and g-C_3N_4: [7,8] have been widely studied for their good biocompatibility, environmental stability and high photocatalytic efficiency. However, this kind of photocatalyst also has the defect that the electric pair-hole pair is easy to recombine, which seriously affects the practical production application. In order to overcome the above defects, researchers concluded that the construction of heterogeneous structures can avoid the phenomenon of electric pair-hole pair recombination. The $FeWO_4$@$ZnWO_4$/ZnO heterojunction composite photocatalyst prepared by Wang et al.: [9] effectively enhanced the absorption capacity of visible light. The "S-type" heterojunction composite photocatalyst formed by $ZnO/BiVO_4$: [10] effectively improves the charge separation efficiency at the catalyst interface, thus enhancing the photocatalytic performance. The heterojunction composite photocatalyst materials prepared by g-C_3N_4, TiO_2, ZnO, $KLa_{0.5}Eu_{0.5}F_4$, Ag_3PO_4, Bi_2WO_6, etc.: [11-15] can reduce the photogenerated electron-hole pair recombination rate, and have a good catalytic effect.

Therefore, in this study, LWO-CN composite photocatalyst was prepared and applied to the degradation of tetracycline hydrochloride in antibiotic wastewater. The structure and composition of the composite catalyst were analyzed, its ultraviolet degradation performance of tetracycline hydrochloride in wastewater was evaluated, and the mechanism of ultraviolet degradation of LWO-CN was discussed.

2. Materials and Methods

2.1. Reagents and Instruments

Lanthanum nitrate hexahydrate (99%, AR), sodium tungstate (AR), anhydrous ethanol (AR), tetracycline hydrochloride (95%), melamine: $C_3H_6N_6$ mass fraction is 99%.

Electric blast drying box, ultrasonic cleaner, magnetic heating agitator, F-7000 fluorescence spectrophotometer (Hitachi High-tech Company), D/max 2500V X-ray powder diffractometer (Nippon Rigaku Corporation), UV-2700 Ultraviolet Visible Spectrophotometer (Shimadzu Instrument (Suzhou) Co., LTD.), JEM-3500 scanning electron microscope (Japan JEOL Corporation), Tecnai G2 F20 type Field Emission transmission electron microscope (United States FEI Corporation).

2.2. Preparation of Materials

2.2.1. Preparation of CN. 10 g melamine was taken and transferred to the Muffle furnace, and gradually heat up at an efficiency of 10 °C per second, and kept the constant temperature for 4h when the temperature gradually rises to 550 °C. Waited for the Muffle furnace to cool to room temperature, removed, after careful grinding to obtain g-C_3N_4 yellow powder.

2.2.2 Preparation of LWO. According to the stoichiometric ratio, 9.8955 g $Na_2WO_4{\cdot}2H_2O$ and 8.6602 g $La_2(NO_3)_3{\cdot}6H_2O$ were added into the mortar and ground for 30 min. Then it was put into the Muffle furnace and heated at 10 °C/min to 500 °C, and kept the constant temperature for 4h. Then the sample was heated to 800 °C and kept the constant temperature for 6 hours, then cooled to room temperature before taken out, and ground. Then the sample powder was wash with the appropriate amount of deionized water and anhydrous ethanol for three times each, and dried at 120°C for 2 h. LWO can be obtained.

2.2.3 Preparation of LWO-CN Composite Photocatalyst. An appropriate amount of g-C_3N_4 was added to the centrifuge tube, and 15 mL of anhydrous ethanol was added to the tube drop by drop. The centrifuge tube was heated in a water bath at 60°C, and steadily dispersed by ultrasound for two hours, and then transferred to a beaker. $La_2(WO_4)_3$ was added to the beaker in proportion. After magnetic stirring at room temperature for 24 h, it was left for 24 h, pumped and filtered under reduced pressure, dried in a blast drying oven at 120°C for 2h, cooled and ground to obtain white precipitated LWO-CN composite photocatalyst.

2.3. Photocatalytic Degradation Experiment

20 mg catalyst was added into 100 mL 10 mg·L^{-1} tetracycline solution, and placed the solid-liquid mixture in a dark room, stir away from light at a constant speed for 30 minutes, so that the adsorption and detachment of tetracycline hydrochloride solution can reach equilibrium under the action of a composite photocatalyst. And then it was place under UV lamp for photocatalytic degradation experiment. Under balanced ultraviolet irradiation, 5 ml tetracycline solution was taken out and

centrifuged at 5000 $r\cdot min^{-1}$ for 3 min after a fixed interval of illumination time (0, 15, 30, 45, 60, 75, 90 min). The absorbance of the solution after catalytic degradation was measured by an ultraviolet spectrophotometer at the wavelength of 357 nm as the remaining concentration of tetracycline hydrochloride in the supernatant. With the absorbance of the solution after the dark reaction as the starting concentration, the photocatalytic degradation efficiency can be calculated according to the following formula

The removal rate of R% is as follows

$$Remove(\%) = \frac{C_0 - C_t}{C_0} \times 100\% \tag{1}$$

2.4. Characterization of Samples

The samples were analyzed by fluorescence spectrophotometer in the wavelength range of 300-750 nm. The crystal structure of the product was characterized by X-ray diffraction (XRD). The morphology was characterized by scanning electron microscope and transmission electron microscope, and the diffuse reflection of the composite catalyst was measured by UV-2700 UV-vis spectrophotometer.

3. Experimental Results and Discussion

3.1. XRD pattern Analysis

Figure 1 shows the XRD patterns of pure CN, LWO and LWO-CN. It can be seen from figure 1 that 2θ is the diffraction peak generated by the accumulation of the $g\text{-}C_3N_4$ aromatic conjugated system. The diffraction peaks of LWO and LWO-CN are very consistent with the coincidence degree in the diffraction standard of *PDF#82-2068*, indicating that the catalyst prepared by high temperature solid phase method is a pure phase, and after adding $g\text{-}C_3N_4$ to the LWO matrix, the diffraction peak of the composite catalyst does not change, and there is no $g\text{-}C_3N_4$ diffraction peak. It may be caused by the low content of $g\text{-}C_3N_4$ in LWO.

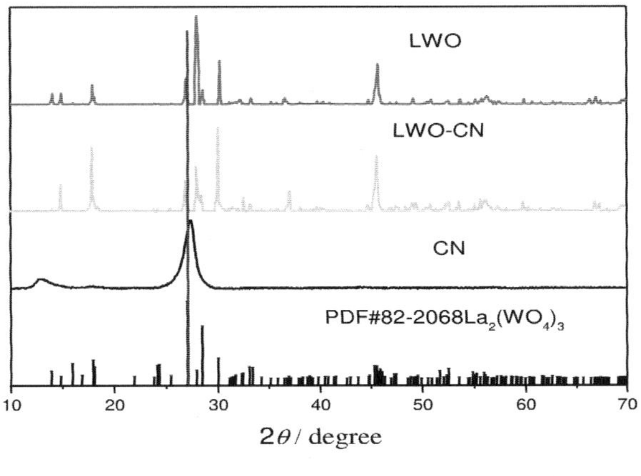

Figure 1. XRD patterns of sample CN, LWO and LWO-CN.

3.2. SEM and TEM Analysis of Catalyst

As can be seen from figure 2(a), LWO is a massive crystal with a smooth surface. Figure 2(b) shows that CN synthesized by polycondensation is loose, such as a sponge. Figure 2(c) shows the SEM diagram of LWO-CN. The surface of the massive crystal of LWO is covered with a layer of CN. Figure 2(d) shows the TEM image of LWO, which is a tetragonal crystal with a smooth surface. Figure 2(e) shows the TEM image of CN. The soft cotton-like structure of CN is clearly observed in the TEM image. Figure 2(f) shows the TEM image of LWO-CN, where the surface of the tetragonal LWO-CN crystal is coated with a layer of CN. According to SEM and TEM images, CN and LWO have been combined to form a new photocatalytic material.

Figure 2. The SEM(a,b,c) and TEM(d,e,f) images of LWO, CN and LWO-CN.

3.3. UV-Vis Spectra and Band Gap Spectra of Catalysts

According to figure 3(a), it can be seen that the obvious semiconductor absorption occurs near 400 nm in the ultraviolet region of LWO, CN, and LWO-CN; the sideband absorption of CN is about 430 nm; and the LWO-CN of LWO-CN is between LWO and CN, with a redshift. As shown in figure 3(b), CN has a higher absorption intensity in the 200-450 nm region, but the absorption intensity of LWO in this region is lower than that of CN. The absorption strength of the composite photocatalyst LWO-CN in this region is enhanced, and it can be used as a photocatalyst under ultraviolet irradiation. According to figure 3(b), it can be obtained that the band gap of CN, LWO, LWO-CN is 2.43, 3.24 and 2.91 eV, and the band gap of the composite photocatalyst LWO-CN is smaller, indicating that the combination of CN and LWO is conducive to promoting the separation of photogenerated carriers.

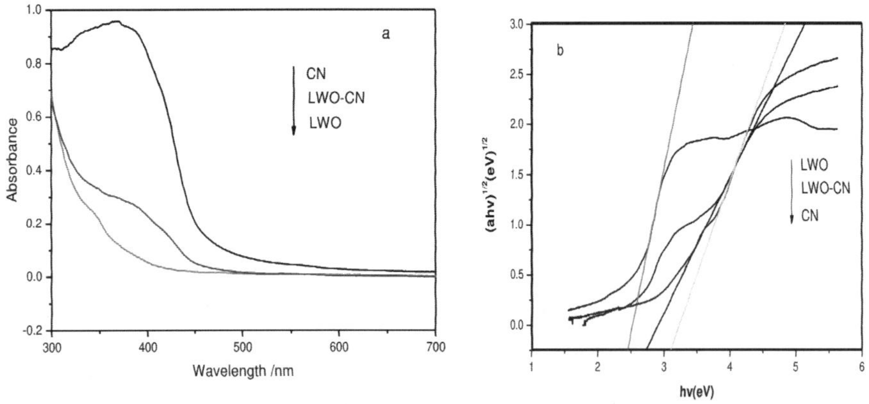

Figure 3. UV-vis spectra and Tauc-Plot of CN, LWO, LWO-CN.

3.4. Photoluminescence Spectra of Catalysts

To determine the efficiency of photogenerated carrier separation. We conducted fluorescence tests on CN, LWO and LWO-CN, as shown in figure 4. Under excitation of 393 nm, LWO does not produce a peak at 450 nm. With the addition of CN, the emission intensity of LWO-CN increased, and the emission intensity of pure CN was the highest at 450 nm. This may be related to the recombination of photogenerated electrons and holes in CN. The photoluminescence intensity of LWO-CN composite is significantly lower than that of CN, and the recombination of photogenerated electrons and holes in LWO-CN composite is greatly inhibited so that a more effective photogenerated carrier separation path can be formed.

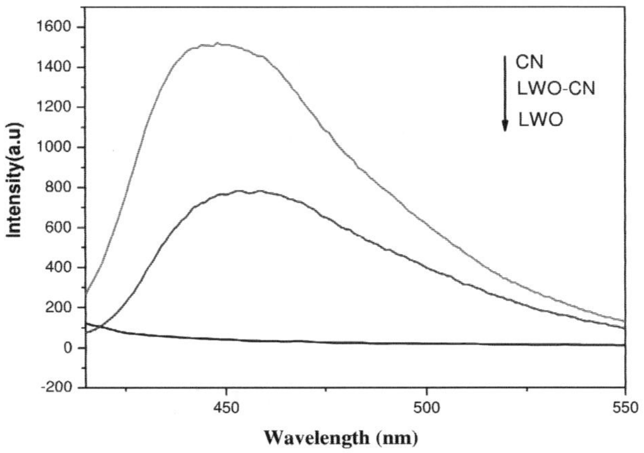

Figure 4. Photoluminescence spectra of CN, LWO, LWO-CN.

3.5. Evaluation of Photocatalytic Activity

Under the same conditions, the degradation effect of different catalysts on tetracycline hydrochloride aqueous solution is shown in figure 5(a). In the dark room, stirring away from light for 30 minutes, the tetracycline hydrochloride solution did not change significantly; under ultraviolet irradiation, the concentration of tetracycline hydrochloride solution

In order to further understand the effect of photocatalytic degradation, degradation data of different catalysts were used for first-order kinetic fitting, calculated according to the equation $\ln(C_t/C_0)=Kt$, where K is the apparent rate constant, C_t is the concentration of tetracycline hydrochloride after a certain period of light exposure, and C_0 is the concentration of tetracycline hydrochloride before light exposure. Through first-order kinetic fitting, the apparent rate constants of LWO and LWO-CN were 0.02563 and 0.02577, respectively, indicating that the degradation rate of LWO-CN was better than that of LWO as shown in figure 5(b).

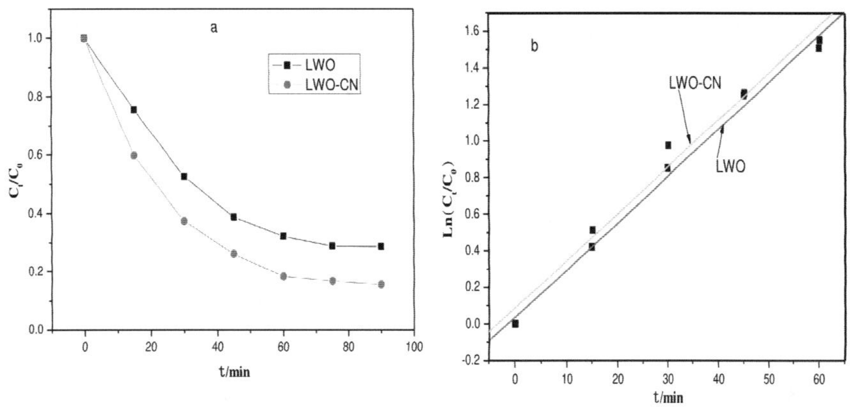

Figure 5. The degradation rate of TC by LWO, LWO-CN.

3.6. Photocatalytic Degradation Mechanism

As shown in figure 6, under ultraviolet irradiation, LWO and CN electrons are excited to make transitions from their respective valence band (VB) to the conduction band (CB), leaving holes in the corresponding valence band (VB) after the transitions. The conduction band potential of LWO is significantly higher than the valence band potential of CN. The electrons in the conduction band of LWO are attracted by the potential difference and can be transferred to the holes in the valence band of CN, which leads to the accumulation of a large number of electrons in the conduction band of CN. To a great extent, the electron-hole recombination rate in the composite photocatalyst can be significantly and effectively alleviated. The effective lifetime of photogenerated electron-hole pairs is also greatly extended. This causes the electrons of the CN conduction band to combine with the oxygen in the water, and the superoxide radical ($\cdot O^{2-}$) is produced accordingly. At this time, compared with $E_0(\cdot OH/H_2O)$=2.48V, it can be found that the potential on the LWO valence band (CB) is higher, so the hole reaction of ·OH on the LWO valence band is generated. Finally, under the common effect of h^+, $\cdot O^{2-}$ and ·OH groups, tetracycline can be catalyzed and degraded to a large extent.

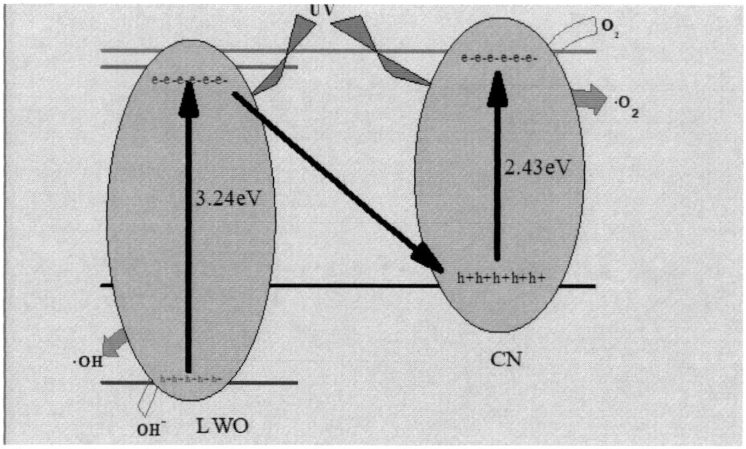

Figure 6. The photocatalytic mechanism of LWO-CN.

4. Conclusion

The synthesized composite photocatalyst $La_2(WO_4)_3$-7%g-C_3N_4 can reduce the recombination rate of electron-hole pairs in the composite photocatalyst, prolong the effective life of the photogenerated electron-hole pairs, and show good catalytic degradation activity. Under UV irradiation, the degradation efficiency of tetracycline by LWO-CN reached 89.49% when the amount of composite photocatalyst was 10 mg, the UV irradiation time was 90 min and the concentration of tetracycline hydrochloride was 5 mg/L.

Acknowledgment

This work was supported by Baise City science and technology plan project (20232045 and 20232037), the Guangxi Natural Science Foundation Project (2020GXNSFAA159036) and 2023 District-level College Students Innovation and Entrepreneurship Plan Project (202310609146).

References

[1] Peng X H, Cao J R, Xie B L, et al. 2020 Evaluation of degradation behavior over tetracycline hydrochloride by microbial electrochemical technology: performance, kinetics, and microbial communities *Ecotoxicology and Environmental Safety* **188(Jan)** 109869.1-109869.7

[2] Yang C, Wang H, Wang Q, et al. 2022 Adsorption of sulfonamides by amino modified magnetic mesoporous silicon *China Water Supply and Drainage* **38(23)** 123-128.

[3] Manoj S and Jonas B 2021 Destruction of emerging organophosphate contaminants in wastewater using the heterogeneous iron-based photo-Fenton-like process *Journal of*

Hazardous Materials Letters **2** 100012.

[4] Liu L M, Gao Z M, Deng Z X, et al. 2022 Activation of persulfate and its mechanism and application in oxidative degradation of antibiotics in water *Environmental Chemistry* **41(5)** 1702-1717.

[5] Duan T, Zeng X Y and Tan S T 2022 Removal of antibiotic resistance genes from waste-water treated by MBR *Environmental Engineering* **40(4)** 8-13.

[6] Zhang F, Liu X N, Li B, et al. 2022 Niobate/titanate nanosheets catalyzed the degradation of ciproloxacin in water under simulated sunlight *Chinese Journal of Environmental Engineering* **16(10)** 3232-3242.

[7] Mirzaeifard Z, Shariatinia Z, Jourshabani M, et al. 2020 ZnO photocatalyst trevisited: Effective Photo-catalytic degradation of emerging contaminants using S-doped ZnO nanoparticles under visible light radiation *Industrial& Engineering Chemistry Research* **59(36)** 15894-15911.

[8] Majdoub M, Anfar Z and Amedlous A 2020 Emerging chemical functionalization of g-C_3N_4: Covalent/noncovalent modifications and applications *ACS Nano* **14(10)** 12390-12469.

[9] Wang F, Li W, Gu S, et al. 2016 Fabrication of $FeWO_4$@$ZnWO_4$/ ZnO heterojunction photocatalyst: Synergistic effect of $ZnWO_4$/ZnO and $FeWO_4$@$ZnWO_4$/ZnO heterojunction structure on the enhancement of visible-light photocatalytic activity *ACSSustainableChemistry & Engineering* **4(12)** 6288-6298.

[10] Wannakan K, Khansamrit K, Senasu T, et al. 2023 Ultrasound-assisted synthesis of a ZnO/$BiVO_4$ S-scheme heterojunction photocatalyst for degradation of the reactive red 141 dye and oxytetracycline antibiotic *ACSOmega* **8** 4835-4852.

[11] Hu Y W, Gao H and Wang X F 2018 Preparation and photocatalytic performance of g-C_3N_4/TiO_2 nanotube arrays *Surface Technology* **47(12)** 113-118.

[12] Gao Y X, Lin X X, Zhao Q, et al. 2019 Preparation and Photocatalytic performance of C_3N_4-ZnO composite photocatalyst *Contemporary Chemical Industry* **48(2)** 261-264.

[13] Zhou Y, Zhou W G, Pan X H, et al. 2019 Study on the synthesis and photocatalytic performance of $KLa_{0.5}Eu_{0.5}F4$@g-C_3N_4 *Chemical Technology and Development* **53(4)** 6-9,13.

[14] Zhu P F, Wang P P, Huang Q, et al. 2020 Preparation and properties of g-C_3N_4-Ag_3PO_4 /Fe_3O_4 magnetic photocatalyst *Applications* **49(9)** 2202-2210.

[15] Liu Y F, Li M L, Luo Y, et al. 2020 Preparation of 3D-C_3N_4/Bi_2WO_6 visible light catalyst and its degradation of tetracycline *Environmental Science and Technology* **43(7)** 84-92.

Research Progress in Microbial Degradation of Plastics

Peibiao Liu[1], Juan Liu[1], Xida Yu[1], Siyu Lv[1], Zhaogang Liu[2,3] and Yudong Wang[1,*]

[1] Faculty of Public Security and Emergency Management, Kunming University of Science and Technology, Kunming 650093, Yunnan, China.
[2] State Key Laboratory of Clean Utilization of Complex Nonferrous Metal Resources, Kunming University of Science and Technology, Kunming 650093, Yunnan, China.
[3] School of Materials and Metallurgy, Inner Mongolia University of Science and Technology, Baotou 014010, Inner Mongolia, China.
Email: 175395866@qq.com;157020626@qq.com;1398485463@qq.com; 3251248324@qq.com; liuzg65@163.com;*wyd@kust.edu.cn

Abstract. With the development of society, plastic is widely used in people's industrial production and daily life, playing an important role in all fields of the world economy. Plastic is notoriously resistant to degradation in natural environments, resulting in severe 'white pollution' from widespread accumulation. Particles smaller than 5 mm are recognized for their ecotoxicological impact, posing a significant risk to both human health and the ecosystem. Urgent action is required to address this pressing issue. Compared with the traditional field incineration, concentrated landfill, and recycling methods, microorganisms are becoming a kind of emerging resources for process recycling and plastic degradation, the article summarizes the microbial degradation research report in recent years, explores the current situation of the microbial degradation of plastics, first introduces the structure of the six big plastics, and then focus on the plastic biodegradation mechanism, summarizes the microorganisms of biodegradable plastics in different environment, provide a theoretical basis for further study of plastic microbial degradation, in order to effectively solve the problem of plastic pollution in the environment.
Keywords. Plastics; biodegradation; microorganisms; enzymes; degradation mechanism; safety and environment.

1. Introduction

Plastics are high-molecular-weight organic polymers derived from different hydrocarbons and petroleum derivatives [1], The most widely used category is polyethylene (Polyethylene, PE), polypropylene (Polypropylene, PP), polyvinyl chloride (Polyvinyl chloride, PVC), Polystyrene (Polystyrene, PS), polyethylene terephthalate (Polyethylene terephthalate, PET), polyurethane (Polyurethane, PUR). The polymer-type demand distribution is shown in figure 1.

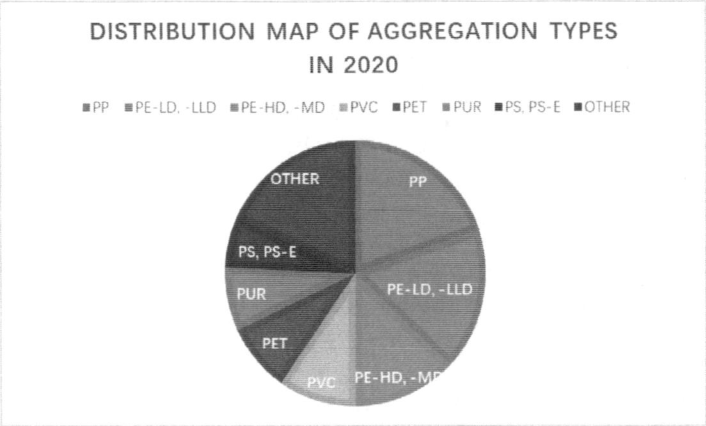

Figure 1. 2020 Polymer type distribution.

Plastic degradation produces a plastic particle that is less than 5 mm in diameter and has ecotoxicological effects [2]. Microplastics may be ingested in the body and exist in human organs, which are carcinogenic and mutagenic, affecting human health [3].

At present, the treatment mode of plastic waste is mainly on-site incineration, centralized landfill and recycling treatment [4]. It is stated that, Plastic can be completely eliminated by incineration, but, Inadequately burned plastic material still exists in the form of solid residue at the bottom of the incinerator, 360 to 102,000 microplastics per ton after incineration (Microplastics, MPs) particles [5], This is a potential source of MPs released into the environment, And plastic in the incineration process will produce a large number of toxic and harmful gases, Such as dioxins, furans, and mercury, into the atmosphere, Cause serious health and environmental hazards [6]. If plastic waste is buried, it not only requires a lot of space, but also produces persistent organic pollutants (Persistent organic pollutants, POP) due to the decomposition of by-products, which has a negative impact on the microbial content in the soil and leads to poor soil [7]. Due to waste contamination and inadequate plastic separation, the plastic recycling process is complex and expensive [8] and cannot be fully recovered. In recent years, the study found that there are microbial biodegradable plastics in nature, this way mainly through the degradation of biodegradation, biological crushing, biological assimilation and mineralization steps, by identifying the effective microbial community to a single species, and then extract their enzymes responsible for plastic degradation [9], can enhance biodegradation without damaging the environment [10].

There have been several studies on the application of microbial degradation of plastic polymers, most of which have emphasized the application of various microorganisms to traditional plastic polymers and microplastics [11]. The main mechanism of biodegradable plastics is the synthesis of modified intermediates by enzyme depolymerization, increasing the accessibility of their cell assimilation [12]. At present, the different organisms that can convert plastic polymers into simple molecules such as carbon dioxide and water are mostly insects with larval form including mealworms, superworms and wax worms [12], which pay less attention to microorganisms. Microbes have the inherent ability to adapt to almost all environments, and various microorganisms in different ecosystems also show the potential for biodegradation [13]. Therefore, this review focuses on the latest discoveries of various microbial degradation of plastics in different environments, and integrates their degradation mechanism, to help the degradation of plastics by microorganisms, so as to eliminate plastic pollution in the ecosystem, reduce the ecotoxicological effects brought by microplastics, and provide a safe living environment.

2. Microbial Degradation Mechanism

The nature of plastic degradation is the fracture of the polymer compounds, through the introduction of weak chemical bonds in the polymer molecular chain or has reaction that is conducive to polymer degradation, molecular chain cracking way is divided into four kinds: depolymerization, random fracture, weak bond separation, side chain or small molecules out. Microbial hydrolysis is when bacteria, fungi algae, and other natural microorganisms secrete depolymerization enzymes, under the action of enzymes, the catalytic polymer chain breaks into relatively small molecular mass fragments, then the debris, depolymerization catalysis in the body, fragments further depolymerization, eventually decomposed into small molecular compounds such as water, carbon dioxide, and methane.

2.1. The Microbial Degradation Mechanism of PE

The biodegradation of polyethylene (PE) is a process by which microorganisms or enzymes break down the polymer's molecular chains at the molecular level. The metabolic pathway of PE degradation involves several steps, including oxidation, dehydrogenation, and the cleavage of C-C bonds, ultimately leading to the entry of the degradation products into the tricarboxylic acid (TCA) cycle. During the microbial degradation of PE, the alkane chains (-CH_2- and -CH_3) and the carbonyl (-CO) regions of the polymer are modified, resulting in the formation of various compounds, including esters, aldehydes, carboxylic acids, aromatic compounds, and ethers [14]. Notable examples of these metabolites are 2-dodecanol and 1,8-nonanediol, which are further converted into aldehyde (-CHO) and hydroxyl (-OH) groups during the biodegradation process [15].

PE is a type of n-alkanes that should also undergo carboxylation through one of the oxidation pathways. In this process, a carboxyl group is introduced into the hydrocarbon backbone of PE to form low molecular weight products such as carboxylic acid [16]. The smaller the PE molecule is, the faster it is consumed by microorganisms. Then the depolymerizing molecules turn into oxidative metabolites (CO_2, CH_4, H_2) [17]

2.2. The Microbial Degradation Mechanism of PVC

At present, there are not many studies on the biodegradation mechanism of PVC, and most of them only have a degradation effect on PVC. When microorganisms degrade PVC polymers, Fourier transform IR analysis (FT-IR analysis) showed a slight stretch of the hydroxyl (-OH) region of the carboxylic acid, indicating the appearance of the-OH group of the carboxylic acid and its derivatives [14]. When PVC is degraded, the CH_2-Structure, CH group and C-Cl group are reduced, and the carbon backbone breaks on the main chain to form O-C and O=C functional groups, generating different products with pyrolysis properties. PVC Under the action of enzymes, oxidation and decomposition into small molecules is used by microorganisms.

2.3. Microbial Degradation Mechanism of PS

When microorganisms degrade PS, they will form biofilms on the polymer surface, secrete degrading enzymes, and decompose the complex polymers into short chains of oligomers, dimers, and monomers. Two major pathways for microbial styrene degradation, One is that styrene is usually metabolized by the oxidation of the vinyl side chains, Vinyl side chains oxidation to initially form styrene oxides, Subsequent isomerization to produce phenylacetaldehyde and oxidation to produce phenylacetic acid, Finally generate phenyl acetyl-CoA, Enter the TCA cycle; Styrene and phenylacetaldehyde can also produce 2-phenyl ethanol; Second, in the direct attack of the aromatic ring, styrene is ethylene glycol, 3-ethylene catechol, Thereafter by ring lysis to generate a muconaldehyde, Ultimately metabolized to pyruvate and acetaldehyde [18, 19]. The metabolic pathways of styrene are shown in figure 2.

Journal of Physics: Conference Series **2941** (2025) 012072 doi:10.1088/1742-6596/2941/1/012072

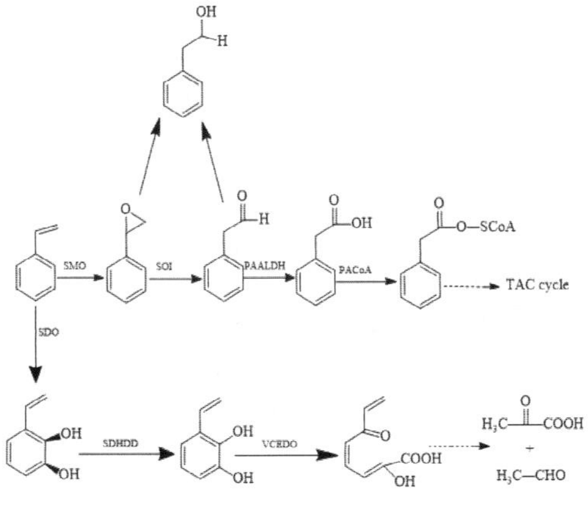

Figure 2. Metabolic pathway of polystyrene.

2.4. The Microbial Degradation Mechanism of PET

The enzymes with PET degradation effect are mainly hydrolases such as lipase, keratinases, and esterases, These enzymes hydrolyze PET to terephthalate (2-hydroxyethyl) diester (Bis by acting on the ester bond (2-hydroxyethyl) terephthalic acid, BHET) and terephthalic acid (2-hydroxyethyl) monoester (Mono (2-hydroxyethyl) terephthalic acid, MHET), After terephthalic acid (2-hydroxyethyl) monoesterase (MHET hydrolases, MHETase), MHET hydrolysis to terephthalic acid (Tereph-thalic acid, TPA) and ethylene glycol (Ethylene glycol, EG) and other water-soluble small molecules, These water-soluble small molecules will eventually be taken up into the microorganism, And then are brought to the cells for further metabolism, The enzyme eventually hydrolyzes it into water, carbon dioxide and other small molecules. The metabolic pathways are shown in figure 3.

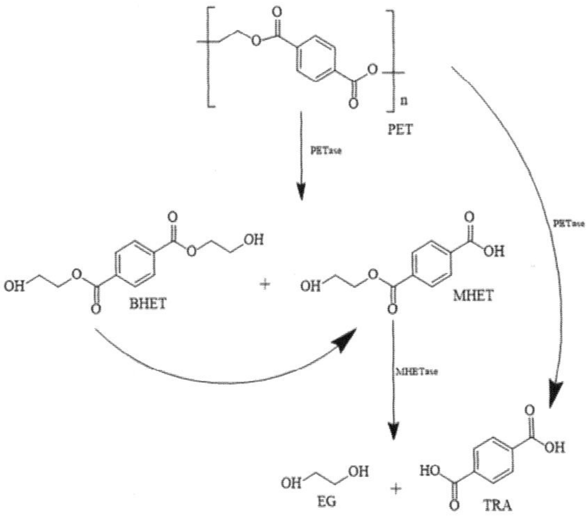

Figure 3. The metabolic pathway of polyethylene terephthalate.

2.5. Microbial Degradation Mechanism of the PUR

The reported degrading enzymes with PUR degradation effect mainly include: esterase, lipase, urease, amidase and protease, esterase can hydrolyze polyurethane ester chain into diethylene glycol and adipic acid, specific degradation mechanism and PUR species, strain characteristics and environmental conditions . PUR molecules contain carbamate bonds, ester, amide, ether, aromatic, urea, enyl, dienyl, the most hydrolyzed is the ester group, followed by urea, ether, the most difficult to degrade is diene group. The first step of polyurethane biodegradation is adhesion to the polyurethane surface by membrane-bound enzymes, followed by hydrolysis of the urethane bond to release PUR fragments or monomers [20]. Studies have shown that the chemical ether bonds of polyether-type PUR are insensitive to the activity of microbial enzymes, while polyester ester bonds are more susceptible to cleavage by hydrolases [21]. FT-IR analysis revealed incomplete hydrocarbon urea hydroxyl formation and increased C-H bond during PUR degradation (Barratt et al, 2003), which represents the new chemical structure of the microbial treated polymers [22].

At present, there are still some deficiencies in the study of microbial degradation of plastics, and the microbial degradation mechanism of PP and PVC is not clear. When PP polymer was vaccinated, FT-IR analysis showed that C=O, O-H hydroxyl, C-H alkyl disappeared, but-CH appeared$_2$methylene, -CH_3 [23] The methyl group and the appearance of new C-H alkyl functional groups (Auta et al, 2018); but the detailed degradation process is not clear; the complete degradation process of PUR is not known and the biodegradation of polyether PUR is not obvious, so the exploration of enzymes for plastics needs to be further strengthened.

3. The Plastic Degradation by Microorganisms in Different Environments

Biodegradation of plastics includes aerobic biodegradation, which usually occurs in nature, carbon dioxide and water, carbon dioxide, water, and methane, and partial aerobic and anaerobic biodegradation in compost and soil. Under the research of many scholars, it is found that there are different types of plastic degradable microorganisms in different environments such as coastal areas, waste treatment sites, high temperature composting fields, soil, and insect gut, as shown in table 1.

Table 1. Plastic degradation by microorganisms varies depending on the environment.

Separation environment	Microorganism	Type	Plastics/ MP	Degradation efficiency (%)	Time	Ref.
Marine	Aspergillus tubingensis VRKPT1; Aspergillus sp; Vibrio sp PD6; Bacillus AIIW2; Aspergillus flavus VRKPT2	Fungi /bacteria	HDPE PE PET PVC	6%-35%	30-90d	[14, 24-27]
Landfill	Arthrobacter sp.GMB5; Pseudomonas sp.GMB7; Pseudomonas; Brevibacillus; Rhodococcus; A.tubingensis HKAS; Brevibacillus sp.btDSCE03 and Brevibacillus brevis btDSCE04; Pseudomonas lini JNU01; Ideonella sakaiensis 201-F6;	Fungi /bacteria	HDPE PE PU PS PET	1.45%-40.5%	21-140d	[15, 22, 28-34]
Compost	Geobacillus stearothermophilus FAFU011; Bacillus vallismortis; Thermus Bacillus	Fungi /bacteria	H DPE LDPE PS	4.2%-75%	56-120d	[35, 36]
Soil	Brevibaccillus borstelensis 707; Rhodococcus ruber C208; Trichoderma harzianum; Bacillus sp.27;	Bacteria	LDPE PP PUR PVC PET	4.0%-40% 11%	30-90d	[23, 37-44]

	Rhodococcus sp.36; B.cereus; Aspergillus glaucus; Pseudomonas		PS			
Insect gut	Acinetobacter sp.AnTc-1; Enterobacteriaceae; Sphingobacteriaceae; Aeromonadaceae; Spiroplasma; Enterobacter asburiae YT1; Klebsiella sp.EMBL-1;	Bacteria	PS PE PVC	12.14%-30.7%	28-60d	[45-48]

3.1. Ocean
Microbes in the Marine environment are widely distributed and abundant, and there are great differences in nutrients, temperature and dissolved oxygen in the ocean [20]. However, due to people's littering, there are more plastic wastes in the Marine environment, which also results in the microorganisms that can degrade plastic in the Marine environment. Many researchers have also found a large number of bacteria with the ability to degrade plastics, which provides the possibility of pollution-free treatment of plastics.

For example, Devi, Kannan, Nivas, Kannan, Chandru and Antony [24] and others found two strains capable of grading HDPE in the Marine environment of plastic waste, Aspergillus us VRKPT1 and Aspergillus flavus (Aspergillus flavus) VRKPT2, which can use HDPE as carbon source and energy, degrade 6.02% and 8.51% respectively within 30d, and HDPE has higher degradation efficiency under the action of mineral oil [25]. Thereafter, Paco and others reported from coastal Taospores, the percentage weight increase was positively correlated with the percentage mass loss of plastic particles. Within 14d, the biomass change was 82.0%, and the PE polymer material quality decreased by 56.7%, i. e., the removal rate was more than 43%. Sarkheldw [26] and other two strains with degradable PET characteristics were isolated from seawater, bacterial vibrio. And the fungal Aspergillus (Aspergillus sp.) PD6,35% and 22% of the PET polymer degradation within 42d at 35°C, respectively. Kumari and others through 3 months of experimental research, found that the Marine bacterium Bacillus AIIW 2 can adhere to polyvinyl chloride (PVC), polyethylene (PE), and then make the polymer material destabilization, degradation, chemical group conversion and plastic mineralization into CO_2 [14].

3.2. Insect Gut
As a rich microbial resource pool, intestinal microorganisms are gradually becoming an important source of strains for degradable plastics. Many scholars have found a large number of microorganisms with the ability to degrade plastics in the intestinal tract of yellow mealworm and wax moth, and have good degradation effect [46]. For example, Xu Jianquan fed the yellow mealworms with PVC plastic. After 60d, analysis showed that the rate of PVC membrane weight reached 41.6%, and the dominant population was plasma (Spiroplasma); accordingly, Wang, Xin, Shi and Zhang [45] isolated from Acinetobacter (Acinetobacter sp.) After 60 days of AnTc-1 in PS, PS showed a 12.14% weight reduction and 13% / 25% (Mw / Mn); in addition, Ren et al. (Ren et al, 2019) isolated Enterobacteria from the gut of the wax moth (Galleria mellonella) (Enterobacter sp.) After 14d of D1 MK934326, D1 culture, obvious depressions, cracks and eroded traces were detected on the surface of the PE membrane, but the presence of base functional groups and ether groups was also detected on the PE membrane. The presence of these two functional groups indicates that the strain can promote the cleavage or formation of new bonds for PE oxidation.

In conclusion, microorganisms in different environments play different degrees of role in grading plastics, bringing different degrees of damage to plastic polymers, which indicates that it is feasible to adopt microorganisms to alleviate the problem of plastic pollution in the environment.

4. Summary and Outlook

Biodegradation of plastics is a promising method. microbial degradation can reduce the molecular weight of the polymer surface of plastics, after which the polymer can be decomposed and converted into available carbon source, promoting bacterial growth and reproduction [49]. Microbes play a leading role in the biodegradation process, and their secreted enzymes are the key to the degradation of polymer. The degradation of polymers is also affected by environmental factors such as temperature, humidity and light, because microorganisms need the optimal conditions to carry out their metabolic processes [50].

To date, researchers have screened some bacteria and fungi that can degrade plastics efficiently from different environments, from which specific functional enzymes are purified. Comprehensive plastic degradation mechanism and degradation effect in different environment, microbial degradation of plastic research is constantly progress, but the known degradation of plastic polymer microorganisms, and the degradation process is quite time-consuming, microbial degradation enzyme secretion is relatively less, the degradation of enzyme synthesis has not yet realized, enzyme degradation plastic mechanism is still not completely clear, most of the research is focused on enzymatic PET hydrolysis. Therefore, further research is needed for the screening of plastic degradable microorganisms and the enzyme degradation mechanism and the performance of the enzyme degradation itself.

In the future studies, The need to clarify the PE, PVC, PP, The specific biodegradation mechanism of the PUR, In order to reduce the environmental pollution caused by plastic, Improve the safety of the living environment; Mass screening of microorganisms with highly efficient degradation capabilities, Developing technologies that can degrade plastics efficiently without affecting the environment, Reduce the "existence of white pollution"; With the help of high-throughput genomics technology and molecular biology methods, key functional genes are identified, and protein engineering technology is utilized to modify degradation enzymes, establish efficient synthetic degradation enzymes, improve the degradation performance of specific functional enzymes, and better solve the problem of plastic pollution in the environment.

Due to the continuous rise of plastic pollution, it puts great pressure on environmental safety belt. While studying how to quickly and effectively degrade plastics, we should reduce the use of plastics and increase the recycling rate of plastics, for example, recycling plastics into daily necessities to improve the utilization rate of materials; biodegradable plastics, the degradation products are environmentally friendly, which can replace the role of traditional plastics in daily life. In the future, the research of microbial degradation should be strengthened to find a way for the rapid degradation of plastics.

Acknowledgments

This work was funded by the 2022 Yunnan University Students Innovation and Entrepreneurship Training Program (S202210674124X), the Open Fund of State Key Laboratory of Clean Utilization of Complex Nonferrous Metal Resources, and the First National Comprehensive Risk Assessment of Natural Disasters in Yunnan Province with Expert Technical Support Project (2021001).

5. References

[1] Kale S K, Deshmukh A G, Dudhare M S and Patil V B 2015 Microbial degradation of plastic: a review *Journal of Biochemical Technology* **6(2)** 952-61.
[2] Wong J K H, Lee K K, Tang K H D and Yap P S 2020. Microplastics in the freshwater and terrestrial environments: Prevalence, fates, impacts and sustainable solutions *Science of the Total Environment* **719** 137512.
[3] Gasperi J, Wright S L, Dris R, Collard F, Mandin C, Guerrouache M, Langlois V, Kelly F J and Tassin B. 2018. Microplastics in air: are we breathing it in? *Current Opinion in Environmental Science and Health* **1** 1-5.
[4] Ignatyev I A, Thielemans W and Vander Beke B 2014 Recycling of polymers: a review *ChemSusChem* **7(6)** 1579-93.
[5] Yang Z, Lü F, Zhang H, Wang W, Shao L, Ye J and He P 2021 Is incineration the terminator of plastics

and microplastics? *Journal of Hazardous Materials* **401** 123429.

[6] Naveen B, Sivapullaiah P and Sitharam T 2013 Disposal options for solid waste of Bangalore city based on its characteristics *International Journal of Environment and Waste Management* **12(1)** 77-88.

[7] Chamas A, Moon H, Zheng J, Qiu Y, Tabassum T, Jang J H, Abu-Omar M, Scott S L and Suh S 2020 Degradation rates of plastics in the environment *ACS Sustainable Chemistry and Engineering* **8(9)** 3494-511.

[8] Gradus R H, Nillesen P H, Dijkgraaf E and Van Koppen R J 2017 A cost-effectiveness analysis for incineration or recycling of Dutch household plastic waste *Ecological Economics* **135** 22-8.

[9] Kawaguchi H, Ogino C and Kondo A. 2017. Microbial conversion of biomass into bio-based polymers *Bioresource Technology* **245** 1664-73.

[10] Qi X, Ren Y and Wang X 2017 New advances in the biodegradation of Poly (lactic) acid *International Biodeterioration & Biodegradation* **117** 215-23.

[11] Morohoshi T, Oi T, Aiso H, Suzuki T, Okura T and Sato S 2018 Biofilm formation and degradation of commercially available biodegradable plastic films by bacterial consortiums in freshwater environments *Microbes and Environments* **33(3)** 332-5.

[12] Zhang J, Gao D, Li Q, Zhao Y, Li L, Lin H, Bi Q and Zhao Y 2020 Biodegradation of polyethylene microplastic particles by the fungus Aspergillus flavus from the guts of wax moth Galleria mellonella *Science of the Total Environment* **704** 135931.

[13] Brooks A N, Turkarslan S, Beer K D, Yin Lo F and Baliga N S 2011. Adaptation of cells to new environments *Wiley Interdisciplinary Reviews: Systems Biology and Medicine* **3(5)** 544-61.

[14] Kumari A, Chaudhary D R and Jha B. 2019. Destabilization of polyethylene and polyvinylchloride structure by marine bacterial strain *Environmental Science and Pollution Research* **26** 1507-16.

[15] Park S Y and Kim C G. 2019. Biodegradation of micro-polyethylene particles by bacterial colonization of a mixed microbial consortium isolated from a landfill site *Chemosphere* **222** 527-33.

[16] Bardají D K R, Moretto J A S, Furlan J P R and Stehling E G 2020 A mini-review: current advances in polyethylene biodegradation *World Journal of Microbiology and Biotechnology* **36** 1-10.

[17] Singh B and Sharma N 2008 Mechanistic implications of plastic degradation *Polymer Degradation and Stability* **93(3)** 561-84.

[18] Mooney A, Ward P G and O'Connor K E 2006 Microbial degradation of styrene: biochemistry, molecular genetics, and perspectives for biotechnological applications *Applied Microbiology and Biotechnology* **72** 1-10.

[19] O'Connor K, Duetz W, Wind B and Dobson A 1996 The effect of nutrient limitation on styrene metabolism in Pseudomonas putida CA-3 *Applied and Environmental Microbiology* **62(10)** 3594-9.

[20] Loredo-Treviño A, Gutiérrez-Sánchez G, Rodríguez-Herrera R and Aguilar C N 2012 Microbial enzymes involved in polyurethane biodegradation: a review *Journal of Polymers and the Environment* **20** 258-65.

[21] Cregut M, Bedas M, Durand M-J and Thouand G 2013 New insights into polyurethane biodegradation and realistic prospects for the development of a sustainable waste recycling process *Biotechnology advances* **31(8)** 1634-47.

[22] Barratt S, Ennos A, Greenhalgh M, Robson G and Handley P 2003 Fungi are the predominant micro-organisms responsible for degradation of soil-buried polyester polyurethane over a range of soil water holding capacities *Journal of Applied Microbiology* **95(1)** 78-85.

[23] Auta H S, Emenike C U, Jayanthi B and Fauziah S H 2018 Growth kinetics and biodeterioration of polypropylene microplastics by Bacillus sp. and Rhodococcus sp. isolated from mangrove sediment *Marine pollution bulletin* **127** 15-21.

[24] Devi R S, Kannan V R, Nivas D, Kannan K, Chandru S and Antony A R 2015 Biodegradation of HDPE by Aspergillus spp. from marine ecosystem of Gulf of Mannar, India *Marine Pollution Bulletin* **96(1-2)** 32-40.

[25] Paço A, Duarte K, da Costa J P, Santos P S, Pereira R, Pereira M, Freitas A C, Duarte A C and Rocha-Santos T A 2017 Biodegradation of polyethylene microplastics by the marine fungus Zalerion maritimum *Science of the Total Environment* **586** 10-5.

[26] Sarkhel R, Sengupta S, Das P and Bhowal A 2020 Comparative biodegradation study of polymer from plastic bottle waste using novel isolated bacteria and fungi from marine source *Journal of Polymer Research* **27** 1-8.
[27] Harshvardhan K and Jha B 2013 Biodegradation of low-density polyethylene by marine bacteria from pelagic waters, Arabian Sea, India *Marine Pollution Bulletin* **77(1-2)** 100-6.
[28] Balasubramanian V, Natarajan K, Hemambika B, Ramesh N, Sumathi C, Kottaimuthu R and Rajesh Kannan V 2010 High-density polyethylene (HDPE)-degrading potential bacteria from marine ecosystem of Gulf of Mannar, India *Letters in Applied Microbiology* **51(2)** 205-11.
[29] Nanda S and Sahu S S. 2010. Biodegradability of polyethylene by Brevibacillus, Pseudomonas, and Rhodococcus spp *New York Science Journal* **3(7)** 95-8.
[30] Bardají D K R, Furlan J P R and Stehling E G 2019 Isolation of a polyethylene degrading Paenibacillus sp. from a landfill in Brazil *Archives of Microbiology* **201** 699-704.
[31] Bhatia M, Girdhar A, Tiwari A and Nayarisseri A 2014 Implications of a novel Pseudomonas species on low density polyethylene biodegradation: an in vitro to in silico approach *SpringerPlus* **3** 1-10.
[32] Khan S, Nadir S, Shah Z U, Shah A A, Karunarathna S C, Xu J, Khan A, Munir S and Hasan F J E p. 2017. Biodegradation of polyester polyurethane by Aspergillus tubingensis *Environmental Pollution* **225** 469-80.
[33] Munir E, Harefa R, Priyani N and Suryanto D 2018 Year Plastic degrading fungi Trichoderma viride and Aspergillus nomius isolated from local landfill soil in Medan; proceedings of the *IOP Conference Series: Earth and Environmental Science*, F, [C]. IOP Publishing, Published.
[34] Kim H W, Jo J H, Kim Y B, Le T K, Cho C W, Yun C H, Chi W S and Yeom S J 2021 Biodegradation of polystyrene by bacteria from the soil in common environments *Journal of Hazardous Materials*, **416**: 126239.
[35] Skariyachan S, Setlur A S, Naik S Y, Naik A A, Usharani M and Vasist K S 2017 Enhanced biodegradation of low and high-density polyethylene by novel bacterial consortia formulated from plastic-contaminated cow dung under thermophilic conditions *Environmental Science and Pollution Research* **24** 8443-57.
[36] Chen Z, Zhao W, Xing R, Xie S, Yang X, Cui P, Lü J, Liao H, Yu Z and Wang S 2020 Enhanced in situ biodegradation of microplastics in sewage sludge using hyperthermophilic composting technology *Journal of Hazardous Materials* **384** 121271.
[37] Hadad D, Geresh S and Sivan A 2005 Biodegradation of polyethylene by the thermophilic bacterium Brevibacillus borstelensis *Journal of Applied Microbiology* **98(5)** 1093-100.
[38] Hadar Y and Sivan A 2004 Colonization, biofilm formation and biodegradation of polyethylene by a strain of Rhodococcus ruber *Applied Microbiology and Biotechnology* **65** 97-104.
[39] Sowmya H, Ramalingappa, Krishnappa M and Thippeswamy B 2014 Degradation of polyethylene by Trichoderma harzianum—SEM, FTIR, and NMR analyses *Environmental Monitoring and Assessment* **186** 6577-86.
[40] Han Y-N, Wei M, Han F, Fang C, Wang D, Zhong Y J, Guo C L, Shi X Y, Xie Z K and Li F M 2020 Greater biofilm formation and increased biodegradation of polyethylene film by a microbial consortium of Arthrobacter sp. and Streptomyces sp *Microorganisms* **8(12)** 1979.
[41] Mukherjee K, Tribedi P, Chowdhury A, Ray T, Joardar A, Giri S and Sil A K 2011 Isolation of a Pseudomonas aeruginosa strain from soil that can degrade polyurethane diol *Biodegradation* **22** 377-88.
[42] Ali M I, Ahmed S, Robson G, Javed I, Ali N, Atiq N and Hameed A 2014 Isolation and molecular characterization of polyvinyl chloride (PVC) plastic degrading fungal isolates *Journal of Basic Microbiology* **54(1)** 18-27.
[43] Auta H, Abioye O, Aransiola S, Bala J, Chukwuemeka V, Hassan A, Aziz A and Fauziah S H 2022 Enhanced microbial degradation of PET and PS microplastics under natural conditions in mangrove environment *Journal of Environmental Management* **304** 114273.
[44] Kathiresan K Year Polythene and Plastics-degrading microbes from the mangrove soil, F, 2004 Published.
[45] Wang Z, Xin X, Shi X and Zhang Y 2020 A polystyrene-degrading Acinetobacter bacterium isolated from the larvae of Tribolium castaneum *Science of the Total Environment* **726** 138564.

[46] Ren L, Men L, Zhang Z, Guan F, Tian J, Wang B, Wang J, Zhang Y and Zhang W 2019 Biodegradation of polyethylene by Enterobacter sp. D1 from the guts of wax moth Galleria mellonella *International Journal of Environmental Research and Public Health* **16(11)** 1941.

[47] Kim H R, Lee H M, Yu H C, Jeon E, Lee S, Li J and Kim D H 2020 Biodegradation of polystyrene by Pseudomonas sp. isolated from the gut of superworms (larvae of Zophobas atratus) *Environmental Science and Technology* **54(11)** 6987-96.

[48] Zhang Z, Peng H, Yang D, Zhang G, Zhang J and Ju F 2022 Polyvinyl chloride degradation by a bacterium isolated from the gut of insect larvae *Nature Communications* **13(1)** 5360.

[49] Gajendiran A, Krishnamoorthy S and Abraham J 2016 Microbial degradation of low-density polyethylene (LDPE) by Aspergillus clavatus strain JASK1 isolated from landfill soil *3 Biotech* **6** 1-6.

[50] Sharma S, Basu S, Shetti N P, Nadagouda M N and Aminabhavi T M 2021 Microplastics in the environment: occurrence, perils, and eradication *Chemical Engineering Journal* **408** 127317.

Analysis of the Impact of Vibro-Compaction on Pore Water Pressure in Coral Sand Strata

Xiong Xu[1] and Can Xiao[2,*]

[1] The Second Engineering Company of CCCC Fourth Harbor Engineering Co., Ltd., Guangzhou, Guangdong, China.
[2] CCCC Fourth Harbor Engineering Institute Co., Ltd., Guangzhou, Guangdong, China.
Email: 260399122@qq.com; *xiaocan2020@163.com

Abstract. This study investigates the effect of pore water pressure during the vibro-compaction stone column construction in coral reef sand strata. Field tests were conducted at a port project in a Southeast Asian country to monitor the pore water pressure during construction. The study focuses on the impact of vibro-compaction on pore water pressure in coral sand strata at different depths, radial distances, and under various vertical drainage conditions. Results show that the greatest pore water pressure changes occur during the drilling stage, with the area closest to the vibro-compaction head experiencing the highest pressure. The peak excess pore water pressure significantly affects existing stone columns. The liquefaction characteristics of coral reef layers are closely related to the surrounding drainage conditions.
Keywords. Vibro-compaction stone columns; pore water pressure; foundation treatment; coral reef sand strata.

1. Introduction

Vibro-compacted stone columns increase the density and compression modulus of foundation soils under vibro-compaction, partially replacing the native soil to enhance the overall strength and bearing capacity of the foundation [1,2]. This method optimizes drainage paths, accelerates consolidation drainage, and reduces settlement stabilization time. Consequently, it is widely used in port engineering projects for weak foundation soils like soft clay and sand [3]. Due to its simple process, convenient material sourcing, and high economic efficiency, it has gained significant attention in engineering circles [4]. Coral reef rock mainly comprises calcium carbonate and is predominantly distributed in the tropical Pacific between the Tropics of Cancer and Capricorn [5,6]. Coral reef rock is a unique type of geotechnical medium, sharing similar properties with sandy soils, such as high porosity, good drainage, and low cohesion, making it highly applicable in various fields. However, limited theoretical frameworks exist for calculations involving coral reef-type geotechnics, and the application of vibro-compacted stone columns in coral sand geological conditions remains rare [7]. The variation law of pore water pressure plays an important role in evaluating the impact range of horizontal treatment of foundation soil [8,9]. In order to promote the study of the variation law of pore water pressure during the construction process of vibrating crushed stone piles under the geological conditions of coral reef sand, this paper conducted on-site tests of pore water pressure during the construction process of crushed stone pile foundation treatment in the coral reef sand formation of a port project in a Southeast Asian country [10]. The focus was on analyzing the growth and dissipation law of pore water pressure at different depths and radial distances of the formation under the action of vibrating equipment and analyzing the influence of vertical drainage channels on the growth and dissipation of pore pressure. This engineering practice has achieved the monitoring of changes in pore water pressure indicators in coral reef sand formations treated with vibratory gravel piles, and based on the analysis of monitoring results, it can effectively improve the efficiency of vibratory gravel pile construction.

2. Project Overview

The test site is located in a port project in a Southeast Asian country. The port features a 630m long container high-pile dock, 3 million cubic meters of dredging, 270,000 square meters of land reclamation, 185,000 square meters of foundation treatment, a container yard, residential buildings, and a backup power station, among other facilities. In the foundation reinforcement zone, stone column foundation treatment methods were employed using both vibro-compaction and vibro-displacement techniques.

The physical and mechanical properties of the soil layers in the test area are detailed in table 1.

Table 1. Physical and mechanical indicators of each soil layer in the experimental area.

Soil layer number	Soil layer category	Thickness /m	Moisture content /%	Density /(kN/m^3)	Pore ratio	Plasticity index	Standard penetration number
1	Miscellaneous fill	1.2-4.0	-	-	-	-	-
2	Silt	1.3-15.2	20.3	19.35	0.761	7.4	0-7
3	Silty clay	3.8-20.5	23.8	23.42	0.769	11.3	1-8
4	Sandy soil	1.0m-22.5m	24.5	19.15	0.668	6.6	5-30
5	Coral reef sand	2.1m-43.5m	20.7	20.78	0.532	0.131	25-50

3. Vibro-Compacted Stone Column Construction and Pore Water Pressure Monitoring Plan

3.1. Vibro-Compaction Stone Column Construction Process

Vibro-compaction stone column construction involves using a crane to operate a vibroflotation device, which is powered by an underwater motor and uses an eccentric block to generate high-frequency vibrations. The high-pressure water flow helps the vibroflotation device penetrate the preset depth of the soil layer. Gravel or other materials are then filled into the hole and compacted to form a stone column, thereby reinforcing the soft foundation. The equipment includes a crane, water pump, and vibroflotation device, with the motor driving the eccentric block's horizontal frequency and amplitude to strengthen the soil structure.

The construction process is as follows:

- Mark the pile location.
- Position the vibroflotation device.
- Start the air compressor.
- Activate the vibroflotation device and begin drilling.
- Continue drilling until the designated depth is reached.
- Raise the vibroflotation device to discharge material.
- Perform back-insertion and vibration compaction.
- Maintain the achieved depth and vibration for a period.
- Continue raising the vibroflotation device and discharging material (repeat until the stone column is completed).

3.2. Pore Water Pressure Monitoring

3.2.1. Pore Water Pressure Measurement Method. Based on the permeability properties of coral reef sand strata and the objectives of monitoring, this study used a vibrating wire piezometer for pore water pressure measurement. The sensor consists of a steel wire attached to a diaphragm. When the sensor interacts with the surrounding liquid, pressure variations lead to diaphragm movement, which changes the steel wire's tension. As the tension is directly related to the square of the resonance frequency, the pore water pressure is calculated by monitoring the frequency.

3.2.2. Piezometer Layout. The piezometers were arranged in an equilateral triangle pattern around the stone columns. For varying distances from the centerline of the stone column and different depths, piezometers were placed at depths of 8.0m and 13.0m, and radial distances of 0.0m, 0.8m, 1.2m, and 1.385m. The layout of the piezometers is shown in figure 1.

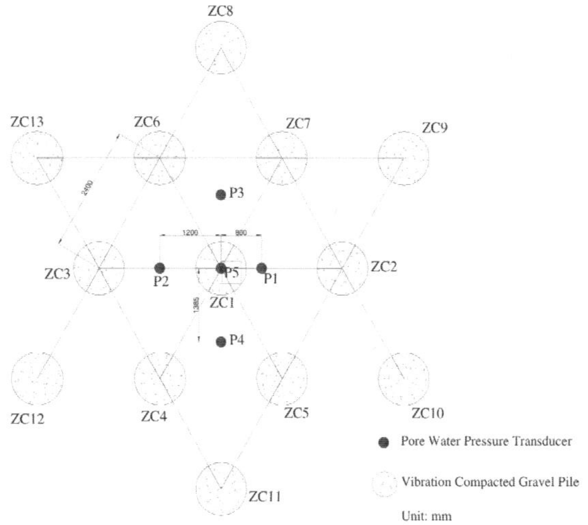

(a) Plan view layout

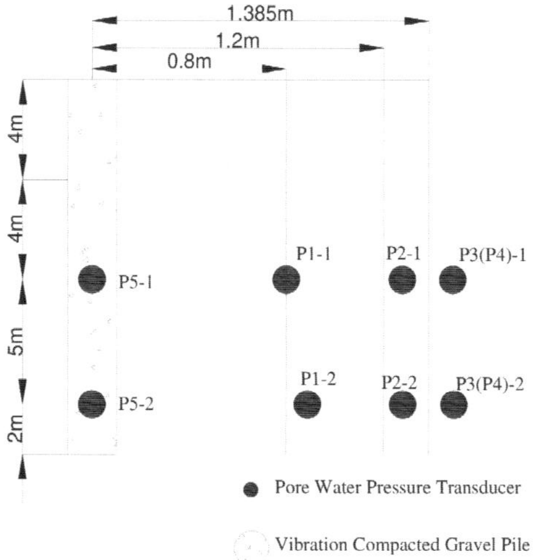

(b) Sectional view layout

Figure 1. Schematic of pore water pressure sensor layout.

The layout of pore water pressure sensors around Column ZC1 is shown in table 2.

Table 2. Layout of ZC1 pile pore water pressure gauge.

Monitoring deployment points	P5-1	P1-1	P2-1	P3-1	P5-2	P1-2	P2-2	P3-2
Radial distance /m	0.0	0.8	1.2	1.385	0.0	0.8	1.2	1.385
Buried depth /m	8	8	8	8	13	13	13	13

3.2.3. Method for Installing Pore Water Pressure Transducers. The accuracy of pore water pressure test results is heavily influenced by the installation technique of the transducer. Various installation methods disrupt the stress conditions of undisturbed soil differently, thereby impacting the transducer's measurements. This test focuses on the characteristics of coral reef sand and adopts drilling installation technology.

- The device casing is made of fine steel wire mesh with moderate rigidity, which can effectively adapt to the deformation of the soil layer and maintain sufficient stability.
- The mesh aperture of the steel wire mesh ranges from 0.01m to 0.015m, which is related to the diameter of the bentonite balls used on-site, ensuring that the bentonite balls do not fall out of the mesh.
- The bottom of the pore pressure device is filled with bentonite balls, with diameters between 0.01mm and 0.015m. A nylon sandbag with a hole size of 0.425mm is placed in the center, followed by a filling with medium-coarse sand.
- The pore water pressure transducer is covered with medium-coarse sand and fixed with steel wire mesh with a mesh diameter of 10mm to 15mm. A schematic of the installation of the pore

water pressure transducer is shown in figure 2.

- The installation process is as follows: (a) Drill to the design elevation and install the first pore pressure device. (b) After installing the pore pressure device at the design position, pull the instrument cable to ensure it adheres to the inner wall of the borehole to avoid interference with subsequent installations. (c) Place an appropriate number of sealing devices according to the design plan to isolate the upper and lower pore pressures. (d) After placing the necessary sealing devices, install the second pore pressure device.

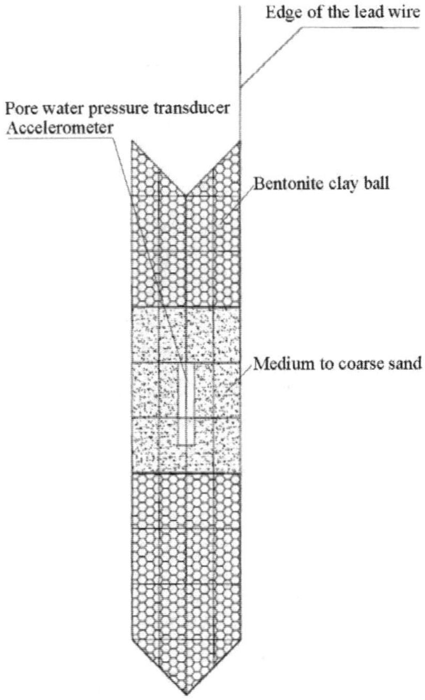

Figure 2. Structure diagram of the pore water pressure transducer installation device.

After the installation of the pore water pressure gauge is completed, based on the equal pore water pressure values at the same depth, the data is corrected and error analysis is carried out to ensure the accuracy and reliability of the measurement results.

4. Results and Analysis

4.1. Analysis of Pore Water Pressure Time Series During Pile Installation

During the construction of vibro-replacement gravel piles, the excess pore water pressure exhibits a spatial correlation with the location of the vibration source. By plotting the pore water pressure time series, the cumulative pore water pressure variation at different locations from the vibration source during the vibro-replacement process can be clearly observed. Taking the gravel pile ZC1 as an example, the pore water pressure variations at different positions during the vibration construction are analyzed. The cumulative pore water pressure time series curve is shown in figure 3.

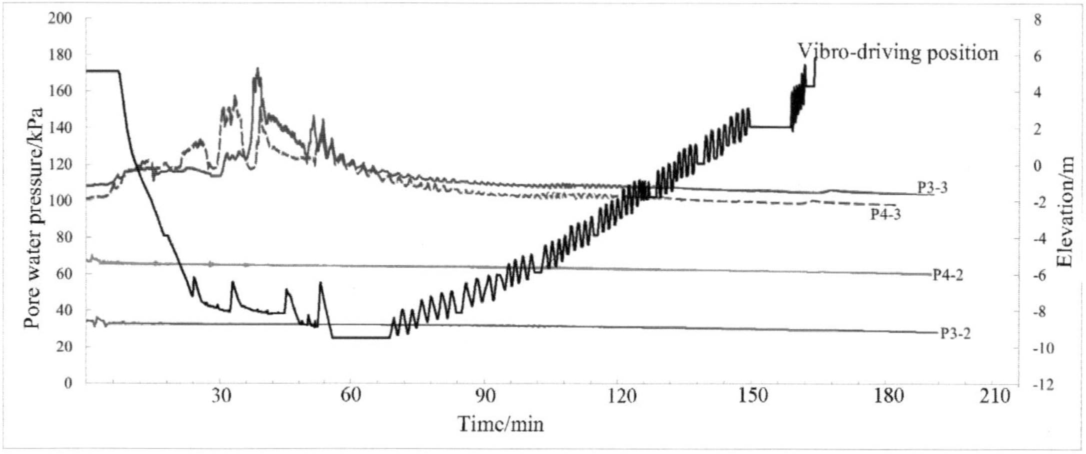

Figure 3. Cumulative pore water pressure time series curve.

Through a comparison of the pore water pressure time series curves at different locations during the vibro-replacement process, the following observations were made:

The change in excess pore water pressure during the entire construction process of the vibro-replacement gravel pile can be divided into two stages: the hole-formation stage and the reverse insertion stage. The excess pore pressure generated during the hole-formation stage is significantly higher than during the reverse insertion stage. This is mainly because, during the hole-formation stage, the tip of the gravel pile has to overcome the natural strength of the sand, compress, and displace the soil, thus requiring more energy. In contrast, during the reverse insertion stage, the vibrating medium consists of disturbed sand or gravel, which has a lower structural strength and shear resistance, requiring less energy. Therefore, the continuous vibration during the hole-formation stage causes a more significant change in excess static pore water pressure, which better reinforces the soil and improves its strength.

The variation in excess pore pressure at different depths of the soil during the hole-formation stage is related to the proximity and duration of the vibration head. The closer the location is to the vibration head and the longer the vibration time, the larger the change in excess static pore water pressure.

During the reverse insertion stage of the vibro-replacement gravel pile, there is some change in the excess static pore water pressure, but overall, the variation is smaller. Since the reverse insertion stage mainly involves the gravel pile being inserted into the borehole, the force applied during the vibration process primarily acts on the gravel pile within the borehole, resulting in less significant reinforcement of the surrounding soil. Although the reverse insertion still has some reinforcing effect on the foundation, the main effect is on the gravel pile itself.

During the hole-formation stage, when the vibration head descends to a shallower depth, negative pore pressure changes may appear at deeper positions (e.g., P3-3). This could be due to several factors:

- The vibration head has not yet affected the deeper soil layers.
- As the hole is formed, a cavity similar to a sump is created, causing groundwater from deeper layers to flow into the hole, leading to a negative pore pressure at the measurement point.

4.2. Variation of Maximum Excess Pore Pressure with Distance

The variation of excess pore pressure with distance is region-dependent. Ideally, the response to vibro-replacement in a sandy soil layer has a certain correlation.

Due to difficulties in optimizing the construction area and zoning, the vibro-replacement process and existing gravel piles have a significant impact on subsequent vibro-replacement activities. Based on on-site test results, the variation of the maximum excess pore pressure with horizontal distance at different depths is analyzed. The variation of maximum excess pore pressure with horizontal distance is shown in figure 4.

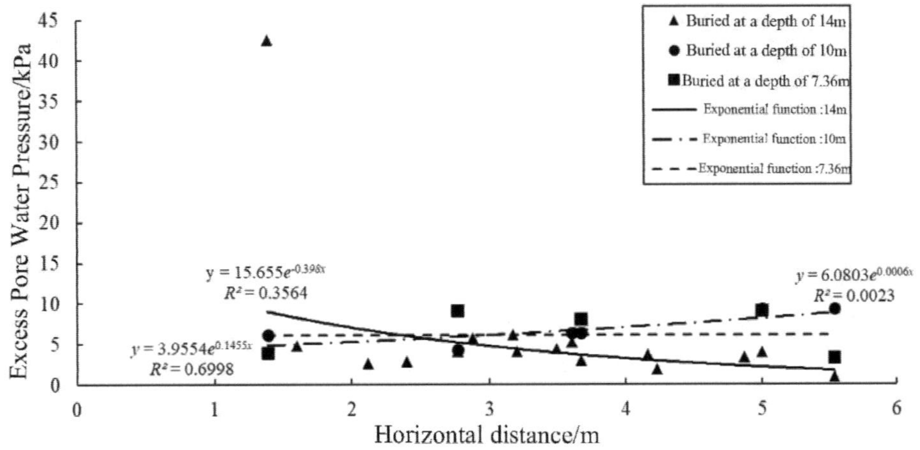

Figure 4. Variation of maximum excess pore pressure with horizontal distance.

From the analysis of the variation of maximum excess pore pressure with horizontal distance, it can be seen that during the vibro-replacement process, the distribution of maximum excess pressure with horizontal distance is relatively small. The reasons for this include:

- The impact range of the vibro-replacement is small; the maximum excess pore pressure outside the 1.5m radius is less than 10kPa, indicating that the vibro-replacement effect is not significant.
- Based on the cumulative pore water pressure time series curve, the groundwater level has a significant effect on the excess static pore water pressure, even more so than the vibro-replacement itself.
- The drainage channels formed by the vibro-replacement gravel piles have a considerable impact on subsequent vibro-replacement tests.
- The soil surrounding the pore pressure sensors is influenced by vibro-replacement activities. Post-construction, the soil's relative compaction and effective stress rise, while the total pore water content diminishes, leading to a reduction in the excess pore pressure induced by vibro-replacement vibrations.
- The relative compaction of the soil at depths of up to 10m in the vibro-replacement test area is high, and the total pore water between the soil particles is relatively low, making it less susceptible to vibro-replacement vibration.

4.3. Time Series Analysis of Excess Pore Pressure Ratio During Pile Installation

Changes in pore water pressure during vibro-replacement correlate strongly with the thickness of the overlying soil. The ratio of excess static pore pressure to pre-consolidation pressure, termed the excess pore pressure ratio, serves as a key parameter. Pre-consolidation pressure, the peak vertical consolidation pressure historically experienced by natural soil, is a vital marker of stress history. Hence, the excess pore pressure ratio acts as a crucial metric for assessing soil stability and deformation properties. The time series of excess pore pressure ratios under different experimental conditions shows the following characteristics:

- When simulating seismic effects with vibro-replacement, the vibro-replacement gravel piles demonstrate good resistance to liquefaction. When there are no completed vibro-replacement gravel piles surrounding the sensor, the peak excess pore pressure ratio is relatively high, with a maximum value of 0.45. However, after a gravel pile is completed around the sensor, the peak excess pore pressure ratio rapidly decreases, and the peak values for all experimental groups are less than 0.1, indicating a significant improvement in soil resistance to liquefaction under vibration.

- At the same depth for each experimental group, the excess pore pressure ratios exhibit strong consistency, with similar values and trends.
- During the extraction process, the vibro-replacement head creates suction in the surrounding soil, accelerating the dissipation of excess pore pressure. The excess pore pressure generated during the penetration stage is significantly higher than during the vibration stage. This suggests that the vibro-replacement head has a better effect on strengthening the soil between piles during the penetration stage than during the vibration stage. Therefore, in practical engineering, increasing the number of penetration cycles can improve the reinforcement effect on the soil between piles.

4.4. Analysis of the Influence of Drainage Channels

The liquefaction characteristics of coral reef strata are related to the surrounding drainage conditions. When constructing new gravel piles, the surrounding completed gravel piles can form drainage channels, which have a certain impact on the changes in pore water pressure generated during the construction of new gravel piles. This article analyzes the variation law of maximum excess pore water pressure under different drainage conditions, and the experiment compares the number of newly constructed gravel piles within different horizontal distance ranges. Research has found that through the analysis of the relationship between the peak overpressure and the drainage channel, it can be concluded that the existing gravel piles in the surrounding area have a significant impact on the construction of vibratory gravel piles, that is, the existing gravel piles play a role in the drainage channel; When there are no drainage channels around, the peak overpressure is relatively high; In the presence of a drainage channel, the peak overpressure generated by the vibration crushed stone pile rapidly decreases, and the peak overpressure drops sharply compared to the situation without a drainage channel. In addition, increasing the number of drainage channels has little effect on drainage efficiency, that is, the peak attenuation of excess pore pressure is relatively small.

5. Conclusion

Based on the above monitoring results and analysis, the following key conclusions can be drawn:

In vibro-replacement construction, the excess static pore water pressure produced during the hole-formation phase surpasses that of the reverse insertion phase. During the hole-formation stage, as the vibrating head moves closer and vibration duration extends, excess static pore pressure in the soil intensifies.

Due to the effects of vibro-replacement, changes in groundwater levels, and the presence of drainage channels from the gravel piles, the correlation between excess static pore water pressure and horizontal distance in the coral reef sand strata is minimal during the vibro-replacement process.

The peak excess pore pressure ratio is significantly influenced by the presence of existing gravel piles.

Vibro-replacement gravel pile construction has a significant impact on existing gravel piles. When no drainage channels exist, the peak excess pore pressure is higher, while the peak excess pore pressure rapidly decreases during subsequent vibro-replacement construction when drainage channels are present.

6. References

[1] Gopinadhanpillai H and George B E 2023 Bearing capacity, elastic modulus, and modulus of subgrade reaction of soft soil improved by floating stone compaction pile group *J. International Journal of Geotechnical Engineering* **17(5)** 570-580.
[2] Bziaz M, Bahi L, Ouadif L, et al. 2023 Determination of vibro-compaction parameters by making a test board in a port site in Morocco *J. Civil and Environmental Engineering* **19(1)** 328-338.
[3] Lin K Q 1984 Discussion on the bearing capacity and settlement calculation of gravel pile foundation *J. Industrial Architecture* **(07)** 3-8.
[4] Chen Z R, Liu N F and Zhang J Q 2010 Application of vibrant crushed stone pile in Shandong

Shidao central fishing port project *J. Water Transport Engineering* **(09)** 124-127.

[5] Yuan Z, Yu G F, Wang Y H, et al. 2016 Research progress on engineering geological characteristics of coral reef rock and soil *J. Tropical Geology* **36(01)** 87-94.

[6] Sun Z X, Zhan W H and Zhu J J 2004 Coral reef rock mass structure and engineering geological zone in the Nansha Islands, South China Sea *J. Journal of Tropical Oceanography* **(03)** 11-20.

[7] Lu P W, Xu X, Huang B, et al. 2019 Research on the construction of vibro replacement stone pile test under coral reef sand geological conditions *J. Construction Technology* **48(23)** 117-121.

[8] Liang Y, Beier N and Sego D C 2022 New method for internal pore-water pressure measurements *J. Geotechnical Testing Journal* **(2)** 45.

[9] Ueda K, Tamaizumi S and Uzuoka R 2022 Effect of fines content on shear deformation properties of sand subjected to pore water inflow after undrained cyclic shearing *J. Journal of Japan Society of Civil Engineers, Ser. A1 (Structural Engineering & Earthquake Engineering (SE/EE))* **78(4)**.

[10] Abdrabbo F and Gaaver K E S 2011 Pore water pressure arising during pile drilling in sand *J. Journal of Civil Engineering and Architecture* **5(4)**.

Arctic Glacial Lake Area Changes and Its Influencing Factors Under Climate Change: A Case Study of Southwest Greenland Ice Sheet

Hui Wang[1,2], Xiang Chen[1,*], Jue Chen[1] and Min Yi[2]

[1] Hunan Institute of Water Resources and Hydropower Research, Changsha 410007, Hunan, China.
[2] Hunan Ecological Environment Monitoring Center, Changsha 410019, Hunan, China.
Email: 1328091148@qq.com; *xiang8234@126.com; 958072394@qq.com; ymtmzgis@163.com

Abstract. Glacial lakes are an important component of ice sheet response to climate change. To understand the effect of climate change on glacial lakes, we studied the change characteristics of the glacial lake area, the spatial distribution of glacial lakes, the inter-annual variation of temperature and precipitation and their relationship with the glacial lake pattern in the Greenland ice sheet. The results showed that the area of glacial lakes in the southwest of the Greenland ice sheet increased from 2000 to 2019. The proportion of glacial lake area was relatively small in the low altitude area near the coast, while the proportion of glacial lake area was gradually increasing in the high altitude area of the ice sheet. During this period, the temperature in the study area showed an overall upward trend, but the regional precipitation fluctuated greatly. From 2000-2019, glacial lakes in the southwest of the Greenland ice sheet had a strong response to temperature, but a weak response to regional precipitation. In this study, the interannual variation of the glacial lake area and the influencing factors are more accurately studied by the remote sensing method, which can provide the scientific basis for understanding global climate change and its potential impact on glacial lakes.
Keywords. Glacial lakes; multi-source remote sensing data; Greenland ice sheet; climate change; temperature and precipitation.

1. Introduction

With the deepening of global warming, natural phenomena such as ice sheet melting and glacier retreat will accelerate [1]. It is expected that the global average temperature will rise by 1.5°C to 2100 compared to pre-industrial [2]. Because of climate change, the glaciers are undergoing widespread retreat [3]. Meltwater runoff will cause sea levels to rise by tens to hundreds of millimeters in the next century [4]. The Greenland ice sheet is the second-largest ice sheet in the world. Glacial lakes are an important component of ice sheet response to climate change. Since the 1980s, there has been a progressive increase in the number and area of glacial lakes in the Greenland ice sheet [5].

In the past few decades, the Greenland Ice Sheet has thinning at low elevations [6] and the glacial lakes range on the Greenland Ice Sheet has also increased [7]. The increased number and size of lakes can accelerate ice flow through reductions in longitudinal buttressing force and basal drag near the ice margin due to buoyancy effects [8]. Zwally et al found that the surface melting of the Greenland ice sheet was synchronized with the transient acceleration of the basement, ice cap uplift, and horizontal displacement [9]. Ignéczi et al predicted that glacial lake water reserves in the Greenland ice sheet increased from 9.8 km3 in 2079 to 12.6 km3 in 2099 by the digital elevation model and the regional climate model, it represented an increase of 113% and 174% compared to 1980 and 2009, respectively [10]. Covering around 10% of the Greenland ice sheet and 5% of the peripheral glaciers and the ice caps [5], glacial lakes exert an important influence on ice dynamics and have become an important indicator of polar region response to climate warming.

The southwest Greenland ice sheet has a long melting period and a larger melting intensity, which makes this region the largest and fastest loss area in the whole Greenland ice sheet. This area has a large settlement of the ice sheet and has become a base for ice sheet field survey activities. The low latitude in this region has better imaging conditions than other regions during the summer, which creates conditions for the accumulation and acquisition of remote sensing data. It provides a unique opportunity to perform a multi-decadal, regional-scale analysis of glacial lake changes for the disparate ice environments of the Greenland ice sheet in 20 years. This study aims to quantify the changes in glacial lake areas during 2000-2019 in southwest Greenland and to investigate the relationship between the glacial lake area change and the climate change (air temperature and precipitation) in Greenland.

2. Materials and Methods

2.1. Study Area
The inland area of the Greenland ice sheet is high in elevation and belongs to the meltwater accumulation area. The elevation of the edge of the ice sheet is relatively low, which is a melting area of ice and snow. The Equilibrium Line of the ice sheet is the boundary between these two types of regions [11].

The Greenland ice sheet has accelerated its material loss in recent decades and has become the single largest ice body contributing to global sea level rise [12]. For the study of glacial lakes of the Greenland ice sheet, many scholars have selected the study area in the melting area of the southwest Greenland ice sheet (figure 1). The Jakobshavn Isbrae, an overflowed glacier in the north, is the most representative of the region, with an ice tongue about 15km long. The southern part of the ice sheet is the Kanger Lussuaq region. From 1997 to 2008, the annual average temperature detected at the Kanger Lussuaq meteorological monitoring station was 10.9°C, the annual average wind speed was 5.3 m/s, the annual average precipitation was 246mm, and the annual average humidity was 64% [13].

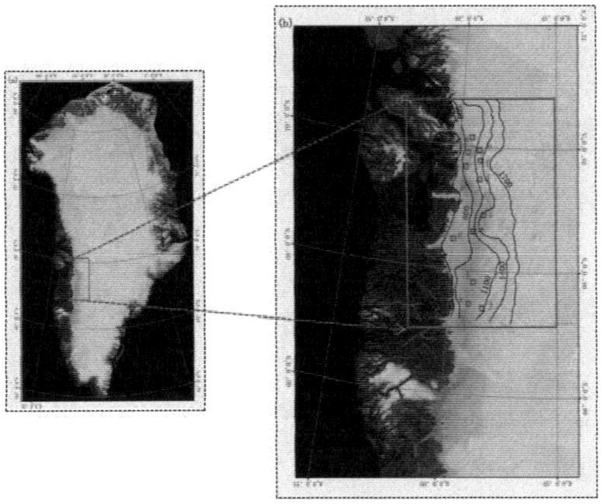

Figure 1. The location of the study area in Greenland. The red square in the thumbnail shows the orientation of the study area in Greenland. (a) The map of Greenland; (b) The range and elevation of the study area.

2.2. Data
Image data for the southwest of the Greenland ice sheet were taken from July to August 2000-2019, and Landsat series data were used for other years except for ASTER in 2005 and 2011. ASTER is an

advanced optical sensor on the Terra satellite, including a total of 14 spectral channels from visible light to thermal infrared, which can provide practical and scientific satellite remote sensing data for the field of Earth's environment and resources research. Landsat enhanced graphic (ETM+) sensor on the Landsat-7 satellite has acquired global images in a 16-day repetition cycle since July 1999. The Landsat-7 ETM+ image is composed of 8 spectral segments, and the spatial resolution of 1-7 band is 30m. Panchromatic band 8 has a resolution of 15m. An operational Land Imager (OLI) image from Landsat-8 was also used to reflect the state of the Greenland ice sheet. The elevation data of the North Pole was obtained from the Tibetan Plateau Data Center with a spatial resolution of 0.8km.

The meteorological indexes selected for the study area in southwest Greenland are temperature and precipitation. Because the recording time of meteorological stations in this region is short, the CRU TS4.04 (The gridded Climatic Research Unit (CRU) Time-series (TS) data version 4.04) is selected in this paper. The dataset is based on observations from the UK National Met Office and others using angular distance weighted interpolation to obtain month-to-month climate change data for the period 2000-2019. The data format is raster with a spatial resolution of about 50km [14].

2.3. Methods

2.3.1. Glacial Lake Area Extraction. The study area is covered by a large amount of ice and snow with a single background. According to the general situation of the research area and previous studies, the index method was selected as the extraction method of glacial lake information in this research area. In the near-infrared band (NIR), the water body has very low reflectance, while in the blue-green band, the water body has a very high reflectance. The normalized water body extraction index (NDWI) was used to extract the water body from the Greenland ice sheet in this paper, the formula is as follows [15]:

$$NDWI = \frac{Band1 - Band2}{Band1 + Band2} \tag{1}$$

Where Band1 is the green band (Green) and Band2 is the near-infrared band (NIR). After the fusion of the panchromatic band and multispectral band, the spatial resolution of these two bands is increased from the original 30m to 15m. The gray-level image obtained by the index method is segmented by density, and the threshold value is selected. The selection of threshold value is determined by referring to the relevant literature and the extraction effect [16]. Both the Greenland ice sheet meteorological data and the digital elevation model from 2000 to 2019 are interpolated data.

2.3.2. Accuracy Verification. Due to the lack of high-resolution satellite remote sensing images in the research area, this study uses the visual comparison method to interpret the marker library with the Antarctic research area to compare the results of water bodies extracted from remote sensing (figure 2). Visual interpretation is used for data correction and accuracy verification for the high accuracy [17]. The results showed that the accuracy range of the frozen lake area extraction is 0.793-0.956, and the overall accuracy is 0.944, the formula is as follows:

$$\text{precision} = \frac{\text{Overlapping area}}{\text{Area from visual interpretation}} \tag{2}$$

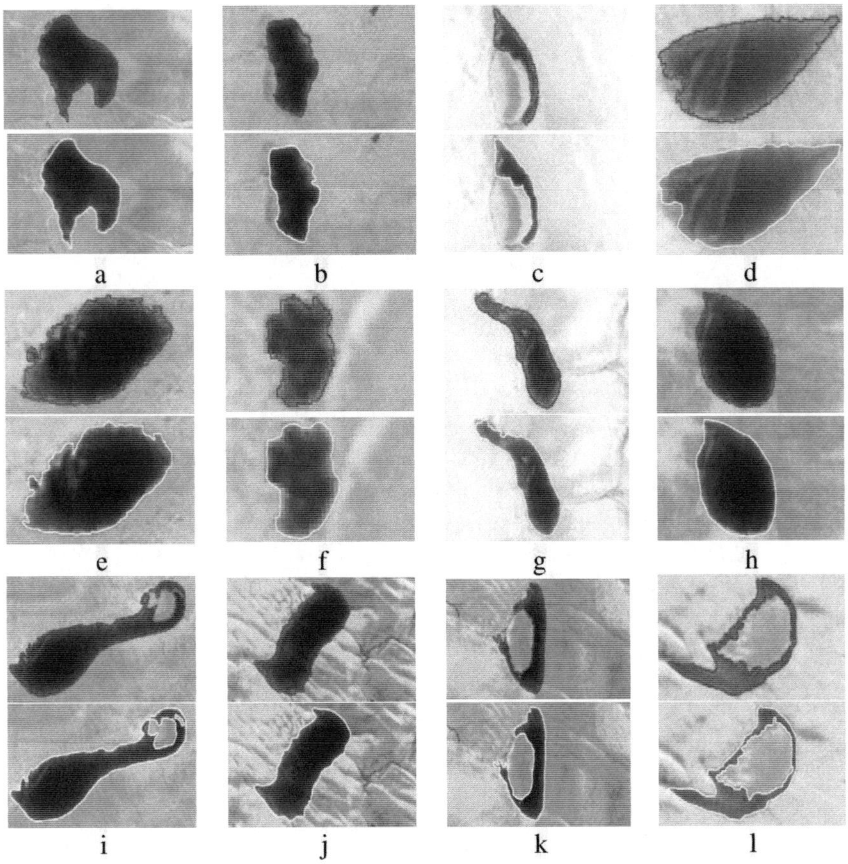

Figure 2. NDWI results were compared with visual interpretation results (red is the NDWI index, and yellow is the visual interpretation).

2.3.3. Correlation Analysis. Pearson's correlation coefficient is used to calculate the correlation between glacial lake change and temperature and precipitation, the formula is as follows:

$$\rho_{XY} = \frac{Cov(X,Y)}{\sqrt{D(X)}\sqrt{D(Y)}} = \frac{E((X-E(X))(Y-E(Y)))}{\sqrt{D(X)}\sqrt{D(Y)}} \quad (3)$$

Where *X* and *Y* represent the input variable, E is the mathematical expectation, D is the variance, and the covariance of the two variables.

3. Results

3.1. Characteristics of the Glacial Lake Area Change

The area of glacial lakes in the southwest Greenland ice sheet showed an overall increasing trend from 2000 to 2019 (figure 2). From 2000 to 2002, The glacial lake area experienced a brief decline, and in 2002, the smallest glacial lake area reached the smallest 72.21 km^2 in the past 20 years. From 2003 to 2008, the glacial lake area showed an increasing trend. From 2009 to 2019, the area of glacial lakes fluctuated on the line, the cycle was roughly 2-3 years. In 2014, there was the largest area of the glacial lake in nearly 20 years, which was 385.33 km^2 (figure 3). Interannual variation of glacial lake area results showed that the quadratic term coefficient of the trend line is negative. From the trend line, the rate of glacial lake increase in this region is slowing down although it is not significant.

Journal of Physics: Conference Series **2941** (2025) 012074 doi:10.1088/1742-6596/2941/1/012074

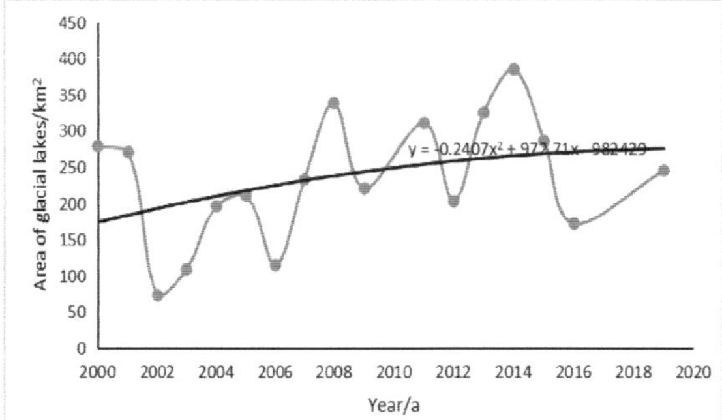

Figure 3. Annual variation of glacial lakes area in southwest Greenland ice sheet from 2000 to 2019.

3.2. The Elevation Distribution Change of Glacial Lakes

The overall topographic characteristics of the Greenland ice sheet are high in the middle, and the elevation becomes lower toward the sea (figure 1). According to the distribution of the glacial lake area in the southwest of the ice sheet and its regional topography, this area is divided into six parts (figure 4). The results showed that the glacial lake area in the study area was smaller and the proportion was less than 10% at the elevation of 500m from 2000 to 2019. From 500-1400m, however, the glacial lake area is gradually larger. At an altitude of 800-1100 meters, the glacial lake had the largest area of 19.24km^2 in 2000, 227.82km^2 in 2011, and 104.4km^2 in 2019, respectively, and the proportion was reached 68.24% in 2000, 73.29% in 2011, and 42.59% in 2019, respectively. However, the area and proportion of the glacial lake decreased in the elevation above 1400m. From 2000 to 2011, the glacial lake area changed more drastically and expanded faster at the high altitude of 800-1400m, while the glacial lake changed less at the altitude below 800m.

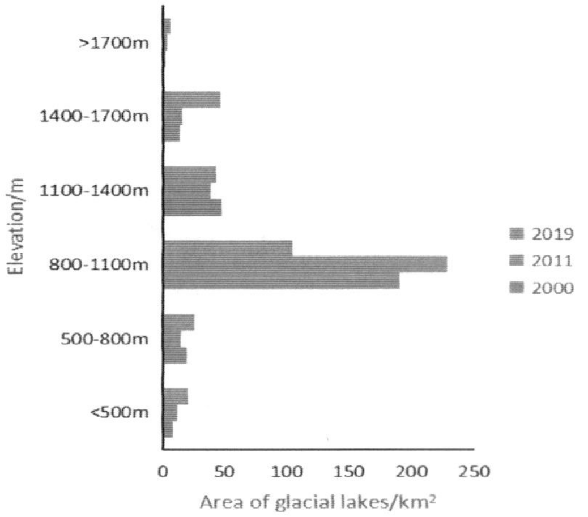

Figure 4. Altitude distribution of glacial lakes in southwest Greenland ice sheet.

3.3. Inter-annual Variation Characteristics of the Temperature

Temperature changes showed that the temperature is gradually rising in the southwest of the Greenland ice sheet. The trend line showed that although the temperature in this region is rising, the rising rate will slow down (figure 5). From July to August 2000-2019, the annual average temperature of the Greenland ice sheet was 3.95°C. From 2000 to 2009, the temperature in this region fluctuated with the cycle of 2-3 years, but the upward trend was very obvious. However, the temperature had not exceeded 3.95°C in 10 years. In 2002, the lowest temperature in 20 years was 2.30°C. Since 2010, the temperature in the study area has exceeded 4°C and reached 4.53°C. From 2010 to 2017, the temperature remained above

3.95°C, with a relatively clear upward trend during this period. By 2018, the temperature had dropped to 3.41°C. In 2019, the temperature rose again, reaching the highest temperature of 5.93°C in nearly 20 years. In figure 5, the quadratic term coefficient of the trend line is negative and the opening is downward, which indicates that the temperature in this region showed an upward trend from 2000 to 2019. Although the slope of the trend is greater than zero, it decreases every year, indicating that although the temperature in the region is rising, the rate of rise will slow down.

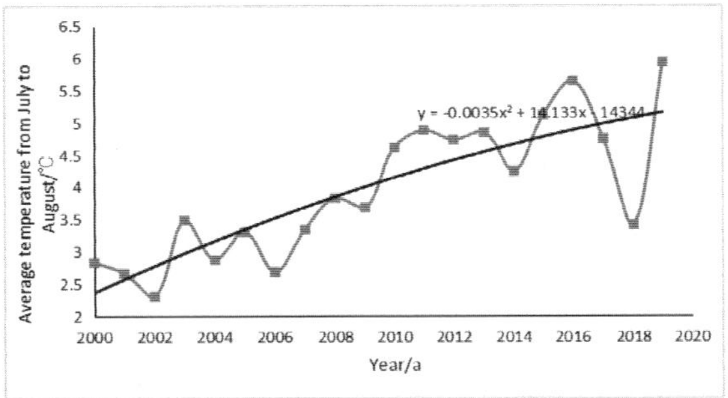

Figure 5. Average temperature from July to August of 2000-2019.

3.4. Inter-Annual Variation Characteristics of the Precipitation

The average precipitation shows that the precipitation in the southwest of the Greenland ice sheet has remained relatively stable from 2000 to 2019 (figure 6). But in 2001, there was a relatively rapid decline in precipitation in this region. Since 2002, the precipitation in the area has been rising for three consecutive years until 2004. The regional precipitation decreased again and dropped to 46.75mm in 2009, then it fluctuated between the values for the next three years until 2013. From 2014 to 2016, the regional precipitation drastically decreased from 90mm to 20mm. Until 2017, the regional precipitation swung around 70mm.

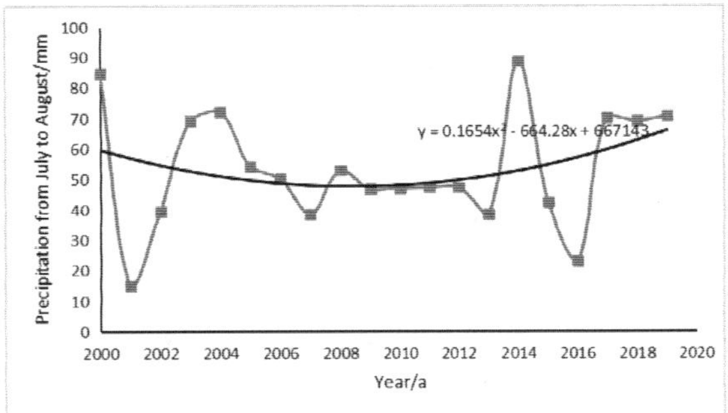

Figure 6. Average precipitation from July to August of 2000-2019.

3.5. Correlation Analysis between the Glacial Lake Area and Air Temperature and Precipitation

The correlation analysis of the lake area and temperature in the Greenland ice sheet shows that the change in lake area from 2000 to 2019 was not completely consistent (r=0.4832) with the temperature change in this region (figure 7a). From 2000 to 2009, the correlation coefficient between glacial lake area and temperature was higher (r=0.6473) (figure 7b). Since 2009, the correlation coefficient between the glacial lake area and temperature decreased rapidly (r=0.0052) (figure 7c).

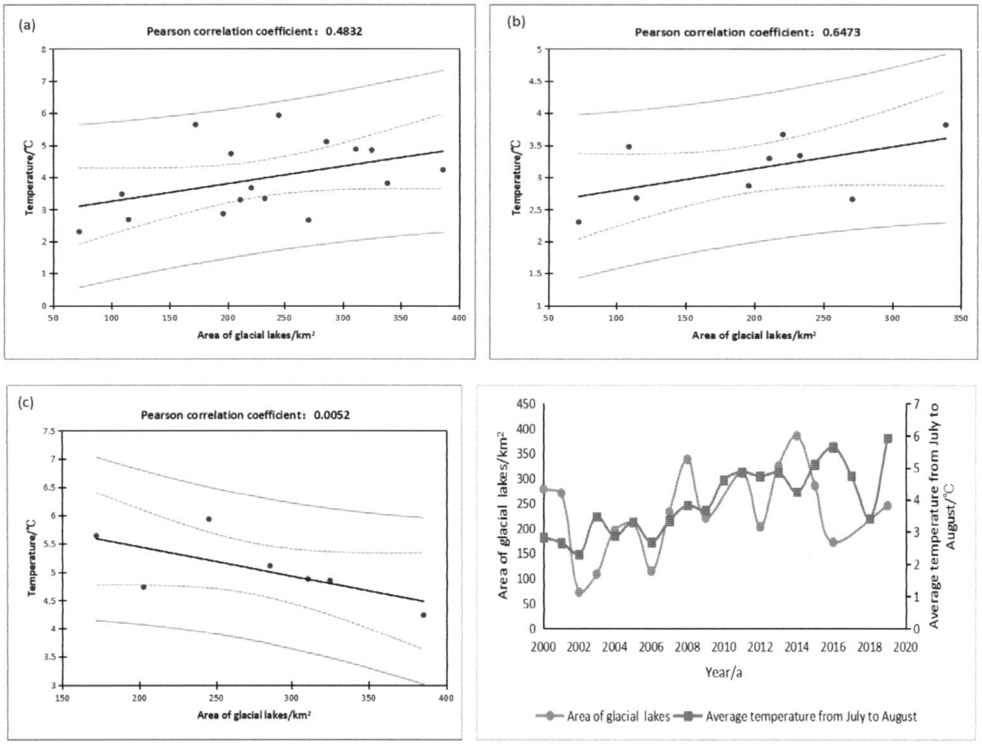

Figure 7. Correlation analysis of the area change of glacial lakes in the Southwest Greenland ice sheet with the average temperature in the region from July to August, a) 2000-2019; b) 2000-2009; c) 2009-2019.

The Greenland glacial lake area and precipitation were less consistent in the years 2000 to 2019 (r=0.4019) (figure 8a). From 2000 to 2009, the relationship between the Greenland glacial lake area and precipitation was poor (r=0.0631) (figure 8b). But from 2009 to 2019, the consistency of the glacial lake area and precipitation trend in this region was relatively good (r=0.6241) (figure 8c). The results showed that temperature is as important as precipitation for the glacial lakes in the region.

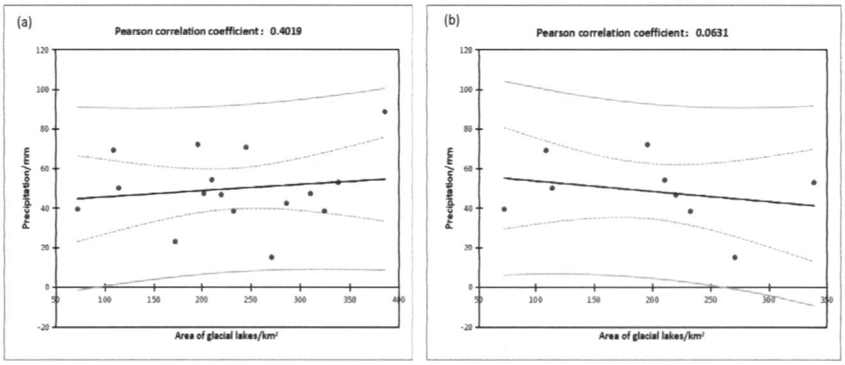

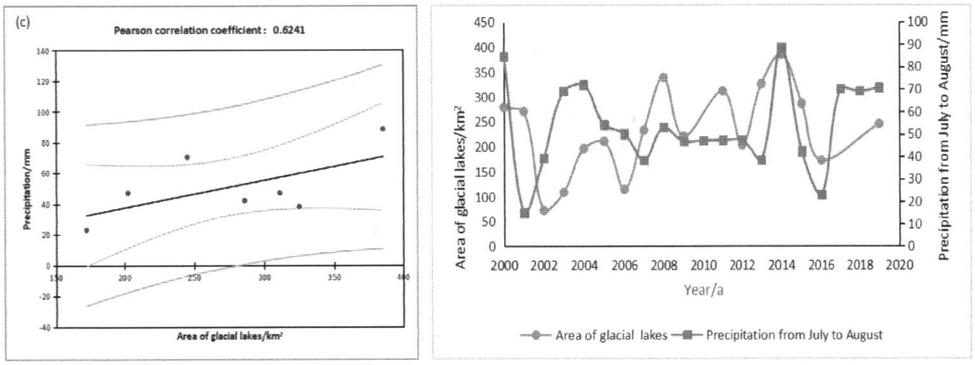

Figure 8. Correlation and comparison analysis of the area change of glacial lakes in the Southwest Greenland ice sheet with average precipitation in the region from July to August. a) 2000-2019; b) 2000-2009; c) 2009-2019.

4. Discussion

The increase in the area of glacial lakes is a direct manifestation of climate change. The analysis of 20 years of meteorological data in the study area shows that from 2000 to 2019, the average temperature from October to November increased by 3°C, with an average annual increase of 0.15°C. Also, the subsurface ocean temperatures along the coast of Greenland increased in the mid-1990s [18]. The average precipitation from October to November increased by 17.44mm, with an average annual increase of 0.88mm. The statistical results of the average daily and monthly temperature in the glacier accumulation area show that the temperature is below zero. However, in the melting season of glaciers from July to October, the expansion and formation of the glacial lake mainly occur when the daily temperature on the surface of the glacier can reach above 0°C, and the relative humidity reaches nearly 100% [19] (Li et al., 2011).

Correlation analysis of regional glacial lake and climate factors shows that during 2010-2019, the correlation between climatic factors and glacial lake area is also relatively close, and the correlation between glacial lake area and temperature is also stronger than that between glacial lake area and precipitation during the same period, which is also consistent with the trend of climate change. Although the glacial lake area had a poor correlation with these two climatic factors before 2010, the glacial lake became more sensitive to the changes in temperature and precipitation with the rise of temperature and precipitation. This is because the glacial lake is not only affected by temperature and precipitation but also by some other climate changes, such as sea surface temperature and wind, which can also be seen as the result of climate change accumulation. Fahrner et al. found that the glaciers in the Greenland ice sheet are shown sensitive to annual sea surface temperature and air temperature in summer [20]. Kang et al. found a glacial lake at the altitude of 5750-5800m in the Panu Glacier [21]. The appearance of the ice river in 2007 caused the acceleration of glacier melting, thus promoting the formation and expansion of the glacial lake. When a glacier melts strongly, it will form a large number of melt pits of different sizes and irregular shapes on its surface, which are not only the manifestation of the intense melting of the glacier but also contribute to the generation of glacial lakes [22]. Doemagaard (2023) studied over 1100 lake water level changes in Greenland from 2003-2023 and found 326 lake water levels exhibit active fluctuations because of the meltwater runoff.

The spatial distribution of the glacial lake shows that the size of the glacial lake increases first and then decreases with the elevation increasing. The spatial patterns distribution of the glacial lakes is similar in the alpine areas of the Kunlun Mountains, Nyainqentanglha Mountains, and Tianshan Mountains [23-25]. This is probably related to the mass balance of glaciers at different altitudes. Wei et al. found that the surface mass loss of the Greenland ice sheet was concentrated below 1500m above sea level, while the surface mass accumulation in the inland area of the ice sheet above 1500m was small, and the mass balance of Greenland ice sheet also showed a similar distribution pattern [26]. In

Greenland, there are more than 3300 glacial lakes and the number is increasing, while changes in their size are less clear [27]. Due to the influence of natural conditions such as cloud cover, the images obtained in some research areas are still missing. With the continuous development of remote sensing technology, the image data and meteorological data from polar regions will become more comprehensive, providing a stronger support for our future research.

5. Conclusion

This research reveals the spatio-temporal variation of glacial lake areas in different periods and elevations, as well as the response of the glacial lake area to climate change. The conclusions are as follows:

(1) The area of the glacial lake in the southwest of the Greenland ice sheet has an overall increasing trend from 2000 to 2019, but the fluctuation trend is obvious in different years.

(2) In the southwest of the Greenland ice sheet, the glacial lake area is relatively small in the low-altitude area near the coast, while the proportion of the glacial lake area in the high-altitude area near the inland of the ice sheet gradually increases.

(3) From 2000 to 2019, the temperature in the southwest of the Greenland ice sheet showed an overall upward trend, with an average annual growth rate of 0.31°C /a. However, the precipitation fluctuated greatly and remained relatively stable on the whole.

(4) From 2000 to 2019, the glacial lake in the southwest of the Greenland ice sheet has a strong response to temperature, but a weak response to regional precipitation.

Acknowledgments

The authors thank Dr. Jun Ding and Shanlong Lu for their mentorship in conceptualizing the study. The authors express their appreciation for Lu Shanlong for his help in collecting climate data. Furthermore, they extend their gratitude to Ding Jun for his suggestions on the logical structure of the paper. In addition, I would also like to express my gratitude for the support received from the Hunan Water Conservancy Science and Technology Projects (XSKJ2023059-21, XSKJ2022068-14); the Open Fund Project of Hunan Dam Safety and Disease Control Engineering Technology Research Center (Hndam2023kf01); and the Excellent Personnel Training Support Project of Hunan Institute of Water Resources and Hydropower Science (06).

6. References

[1] Shen Y, Wang G 2013 Key findings and assessment results of IPCC WGI Fifth Assessment Report *Journal of Glaciology and Geocryology* **35(5)** 1068-1076.

[2] IPCC. AR6 Synthesis Report: ClimateChange 2023 [R] 2023.

[3] Glasser, et al. 2016 Recent spatial and temporal variations in debris cover on Patagonian glaciers *Geomorphology* **273** 202-216.

[4] Hofer, et al. 2020 Greater Greenland ice sheet contribution to global sea level rise in CMIP6 *Nature Communications* **11(1)** 1-11.

[5] Carrivick J L and Quincey D J 2014 Progressive increase in number and volume of ice-marginal lakes on the western margin of the Greenland ice sheet *Global and Planetary Change* **116** 156-163.

[6] Sundal, et al. 2009 Evolution of supra-glacial lakes across the Greenland ice sheet *Remote Sensing of Environment* **113(10)** 2164-2171.

[7] Abdalati W and Steffen K2001 Greenland ice sheet melt extent: 1979-1999 *Journal of Geophysical Research* **106(24)** 33983-33988.

[8] Carrivick J L, How P, Lea J M, Sutherland J L, Grimes M, Tweed F S, et al. 2022 Ice-marginal proglacial lakes across Greenland: Present status and a possible future *Geophysical Research Letters* **49** e2022GL099276.

[9] Zwally, et al. 2002 Surface melt-induced acceleration of greenland ice-sheet flow *Science* **297(5579)** 218-222.

[10] Ignéczi, et al. 2018 Greenland ice sheet surface topography and drainage structure controlled by the transfer of basal variability *Frontiers in Earth Science* **6** 101.
[11] van de Wal, et al. 2008 Large and rapid melt-induced velocity changes in the ablation zone of the Greenland ice sheet *Science* **321(5885)** 111-113.
[12] Smith, et al. 2017 Direct measurements of meltwater runoff on the Greenland ice sheet surface *Proceedings of the National Academy of Sciences of the United States of America* **114(50)** E10622-E10631.
[13] Yang, et al. 2014 *Understanding Greenland Ice Sheet Supraglacial Hydrology Using a Remotely Senced Approach* (Nanjing: Nanjing University).
[14] Ian H, et al. 2020 Version 4 of the CRU TS monthly high-resolution gridded multivariate climate dataset *Scientific Data* **7** 109.
[15] McFeeters, et al. 1996 The use of normalized difference water index (NDWI) in the delineation of open water features *International Journal of Remote Sensing* **17** 1425-1432.
[16] Stokes, et al. 2019 Widespread distribution of glacial lakes around the margin of the East Antarctic Ice Sheet *Scientific Reports* **9(1)** 1-14.
[17] Ma, et al. Dataset of maximum extent and type of glacial lake in the Asia (1980s-2019) *Journal of Global Change Data & Discovery* **202 6(2)** 200-208.
[18] Holland, et al. 2008 Acceleration of Jakobshavn Isbræ triggered by warm subsurface ocean waters *Nature Geoscience* **1(10)** 659-664.
[19] Li, et al. 2011 Seasonal transition characteristics of the westerly jet: Study based on field observations at an altitude of 6900m on the Mt Xixiabangma Dasuopu glacier *Chinese Bulletin* **56(20)** 1642-1650.
[20] Fahrner, et al. 2021 Linear response of the Greenland ice sheet's tidewater glacier terminus positions to climate *Journal of Glaciology* **67(262)** 193-203.
[21] Kang, et al. 2007 Glacier retreating dramatically on the Mt Nyainqentanglha during the last 40 years *Journal of Glaciology and Geocryology* **29(6)** 869-873.
[22] Yao, et al. 2007 Recent rapid retreat of the Naimona'nyi glacier in Southwestern Tibetan Plateau *Journal of Glaciology and Geocryology* **29(4)** 503-508.
[23] Tao, et al. 2021 Spatial changes of the glacial lakes in the western Nyainqentanglha Range *Arid Zone Research* **38(03)** 618-628.
[24] Meng, et al. 2023 Analysis of temporal and spatial characteristics and driving forces of Kunlun glacial lakes *Arid Zone Research* **40(07)** 1094-1106.
[25] Wang N, et al. 2023 Spatiotemporal characteristics and driving forces of glacial lakes in Tianshan Mountains *Arid Land Geography* DOI: 10.12118/j.issn.1000-6060. 673.
[26] Wei J N, Luo K, Chen Y R, et al. 2024 Estimating Greenland ice sheet surface mass balance using remote sensing and ice flux divergence *Remote Sensing for Natural Resources* 1-8.
[27] Shugar, et al. 2020 Rapid worldwide growth of glacial lakes since 1990 *Nature Climate Change* **10** 939-945.

Numerical Simulation Study of PM2.5 Diffusion and Distribution in Building Cluster Microenvironment

Jingjing Zhang

Built Environment and Energy Application Engineering, Shenyang City University, Shenyang, Liaoning, China.
Email: *18809821294@163.com

Abstract. In order to improve the quality of the micro-environment of the architectural complex and ensure that residents enjoy a healthy and comfortable living experience, this paper uses Fluent simulation software to simulate the wind environment under the three common layouts of the architectural complex and the diffusion status of PM2.5. On this basis, green trees are added around the pollution sources to compare and analyze the adsorption and blocking effect of green trees on particulate matter. The results showed that after the addition of green trees, the wind environment of the community was optimized, and the concentration of PM2.5 also decreased significantly. These results provide a reference for improving the micro-environment air quality of the building complex.
Keywords. Fluent Simulation software; complex layout; green trees; PM2.5 concentration; microenvironment of buildings.

1. Introduction

In the era of rapid development of urbanization, urban environmental problems have become increasingly prominent, among which air pollution has become one of the focuses of attention. As a key pollutant in air pollution, PM2.5 has had a profound impact on urban ecosystems and residential life due to its unique physicochemical properties. PM2.5 is small in particle size, which can penetrate the human respiratory system and even enter the blood circulation. A variety of toxic and harmful substances can cause a series of health problems, including respiratory diseases, cardiovascular diseases, and immune system disorders [1]. At the same time, it stays in the atmosphere for a long time and is not easy to dissipate leading to the cumulative effect in the urban environment, seriously affecting the quality of the urban atmospheric environment. As an important part of the city, the diffusion and distribution law of PM2.5 in the microenvironment is of great significance for improving the quality of the urban atmospheric environment.

2. Layout Mode and Advantages and Disadvantages of the Architectural Complex

The layout of the architectural complex includes determinant, enclosed, point, and oblique [2]. In this paper, juxtaposition, mis column, and enclosure were selected as shown in figure 1.

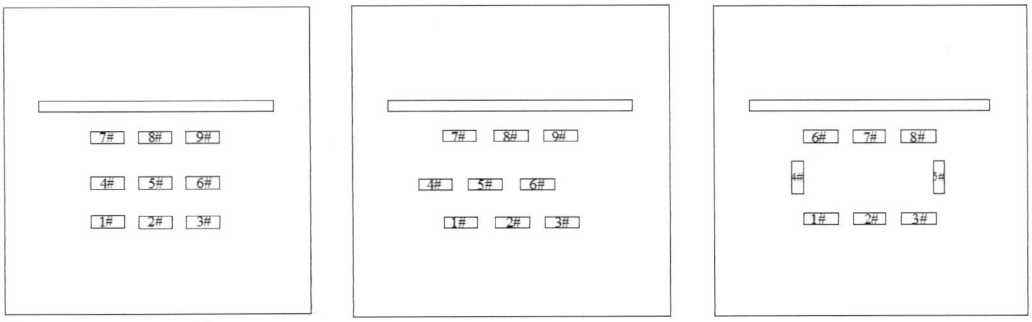

(a) The parallel type (b) The staggered type (c) The enclosed type

Figure 1. The layout of the buildings.

Table 1. The advantages and disadvantages of building layout.

Building layout form	Advantage	Shortcoming
The parallel type	Easy to construct Strong composition Structural economy Easy to obtain good sunlight and ventilation	The space is monotonous and rigid Easy to form through the traffic
The staggered type	Easy to integrate with terrain changes Rich space	Not conducive to energy conservation and structural economy
The enclosed type	Forming a courtyard Cold-proof and warm-keeping Save land use	Sunshine ventilation Difficulty adapting to terrain Complex construction and uneconomical structure

According to the advantages and disadvantages of the layout mode of the architectural complex in table 1, the ventilation effect of the determinant is better. In addition, relevant studies show that under the three combinations of architectural groups studied by the author, the ways conducive to the diffusion and dilution of particulate matter are the staggered type, followed by parallel and enclosed forms (parallel and staggered type belong to determinant) [3].

3. Methods for reducing particulate matter in greening trees

The capture of particulate matter by green trees is the result of the interaction between the surface of tree organs such as leaves, branches and particulate matter.

1) Dust reduction effect, utilizing trees to cover the ground and reduce the source of particulate matter from the source.

2) Dust settling effect. Promote particle settling by reducing wind speed. The dense canopy of trees has a strong effect on reducing wind speed. When the airflow containing atmospheric particles passes through the canopy, some of the atmospheric particles carried in the air are blocked by the branches and leaves and quickly fall to the ground, while the other part remains on the surface of the branches and leaves. But this dust retention method is easily influenced by the external environment, as the wind will blow up atmospheric particles that remain on the leaves, causing them to resuspend [4].

3) Dust retention effect. Utilize the adsorption effect of tree leaves to directly capture particulate matter. The structure of the leaf surface varies among different trees. Some trees have rough and hairy leaf surfaces, while others can secrete viscous oils or juices. When atmospheric particles pass through trees, they attach to the leaves and branches, so trees can adsorb floating particles. Its dust retention effect is relatively stable, and particulate matter is not easily blown up by the wind.

4) Dust blocking effect. Trees have the ability to alter the wind field by planting trees appropriately

to prevent particulate matter from entering local areas. This method is a key focus of community planning, which can reduce particulate matter from its diffusion pathways. The dust retention effect is relatively stable.

5) Dust absorption effect. Particles are stored and absorbed by leaf pore tissues. During photosynthesis and respiration, green tree leaves adsorb particulate matter through stomata, bark pores, and other mechanisms. This method has the most stable dust retention effect.

4. The Establishment of Model

When building the model, CAD is used to model the three layouts of the building community (parallel, staggered, and enclosed) and the plan of the green trees near the side of the street, and then the plan. The format of sat is imported into Gambit to stretch the plane model using the stretch command. Build computational models for the next simulation calculation.

5. The Set of Parameters

In the Fluent simulation, the selection *k-ε* model, the standard *k-ε* model established on the basis of a semi-empirical formula, was proposed by Launder and Spalding in 1972 and widely used due to its wide range, economy, and reasonable accuracy [5]. The corresponding transport equation in the standard *k-ε* model is:

$$\frac{\partial(\rho k)}{\partial t}+\frac{\partial(\rho k u_i)}{\partial x_i}=\frac{\partial}{\partial x_j}[(\mu+\frac{\mu_t}{\sigma_k})\frac{\partial k}{\partial x_j}]+G_k+G_b-\rho\varepsilon-Y_M+S_k \tag{1}$$

$$\frac{\partial(\rho\varepsilon)}{\partial t}+\frac{\partial(\rho\varepsilon u_i)}{\partial x_i}=\frac{\partial}{\partial x_j}[(\mu+\frac{\mu_t}{\sigma_\varepsilon})\frac{\partial\varepsilon}{\partial x_j}]+C_1\frac{\varepsilon}{k}(G_k+C_3G_b)-C_2\rho\frac{\varepsilon^2}{k}+S_\varepsilon \tag{2}$$

In equations (1) and (2), G_k is the generation term of turbulent kinetic energy due to the average velocity gradient; G_b is the generation term of turbulent kinetic energy due to buoyancy; Y_M represents the contribution of pulsation expansion in compressive turbulence; C_1, C_2, C_3 is the empirical constant; σ_ε and σ_k is the number of Prandtl, corresponding to the turbulent kinetic energy k and the dissipation rate respectively ε; S_k and S_ε is the user-defined source term.

In the parameter setting, the inlet wind speed is set to 3 *m/s*, the wind direction is the north wind, the pollution source is the street automobile exhaust, and PM2.5 is stably discharged outward at the mass concentration of 0.08 mg/m^3 and the speed of 0.5 *m/s*. In addition, the existence of green trees has a certain influence on the surrounding airflow field and thus affects the diffusion of PM2.5, the green trees are simplified in size: 6m high, 2m long, 2m wide, and the porosity of the crown is 0.6, the porosity of the trunk is 0.98; considering the presence of porous viscosity, the viscosity coefficient of the trunk is 1.02, the viscosity coefficient of the crown is 1.67, and the resistance coefficient of the crown is 0.2, and the resistance coefficient of the trunk is 1 [6].

6. The Analysis of the Simulation Results

Planting green trees can improve the building wind environment, and then improve the air quality of the community. The simulation obtains the outdoor wind speed distribution map of the complex and the PM2.5 concentration field of the complex. The criteria for determining the wind environment are shown in table 2 Concept of outdoor comfort. The Ministry of Environmental Protection of China issued the Environmental Air Quality Standard (GB3095-2012), which stipulates that the annual average mass concentration limit of PM2.5 is greater than or equal to 35 $\mu g/m^3$, and the 24-hour average concentration limit is greater than or equal to 75 $\mu g/m^3$ [7]. Figures 2-4 show the simulation results.

Table 2. The advantages and disadvantages of building layout outdoor comfort concept (comfortable wind speed < 5 *m/s*).

Wind scale	Description	Wind speed (*m/s*)	The effect of the wind
2	breeze	1.6-3.3	You can feel the wind on your face
3	Gentle wind	3.4-5.4	Hair will be blown out of place
4	Moderate wind	5.5-7.9	Lifting dust, paper being blown up
5	Fresh wind	8.0-10.7	The body can feel the wind pressure
6	a fierce wind	10.8-13.8	It's difficult to hold an umbrella
7	Approaching strong winds	13.9-17.1	I feel inconvenient walking
8	gale	17.2-20.7	Generally, it can prevent progress
9	Strong winds	20.8-24.4	People will be blown down

(a) The wind field of parallel type

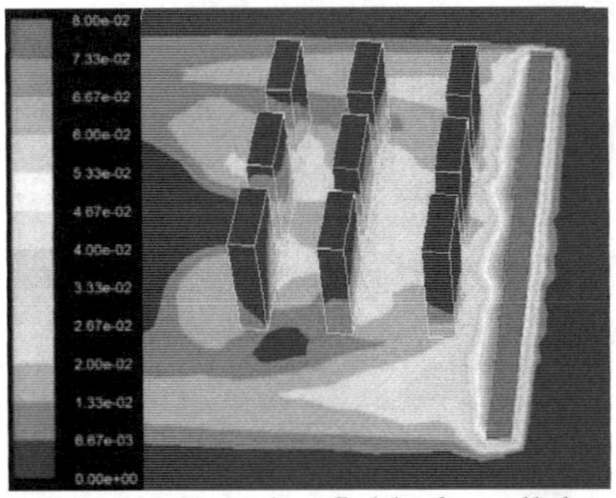
(b) The concentration field of parallel type

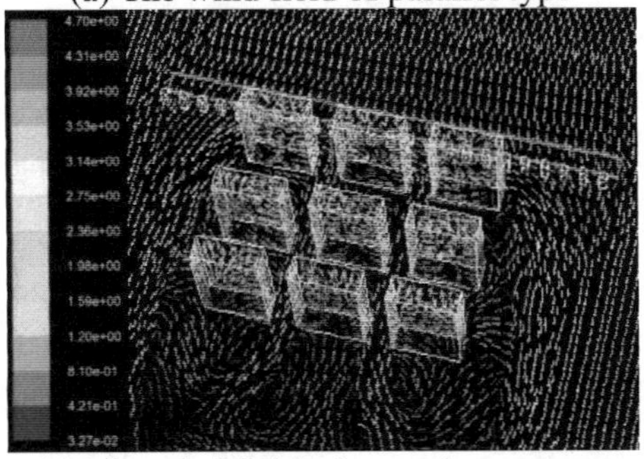
(c) The wind field after setting up trees

(d) The concentration field after setting up trees

Figure 2. The comparison of parallel-type buildings.

From figure 2, it can be seen that in the parallel building complex, there is a certain amount of wind entering the two main roads between buildings 1-3#. Before setting up green trees, the wind speed is basically maintained at around 4.5 *m/s* but after setting up green trees, the wind speed is about 4 *m/s*. In the transverse passage between the first two rows of buildings, due to the obstruction of the front row buildings, the leeward crosswind flow of the building area separates and reattaches, forming a more obvious vorte [8]. It can be determined that the above are all wind shadow areas, and the wind speed is generally lower than 1.0 *m/s*. The horizontal passage between the two rows of buildings at the back forms a relatively large wind comfort zone. After setting up green trees, the mass concentration of PM2.5 next to the street decreased from about 0.0667 mg/m^3 to 0.0533 mg/m^3.

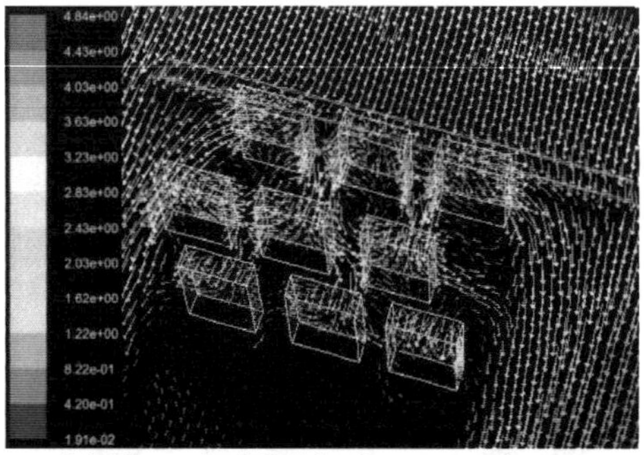

(a) The wind field of staggered type

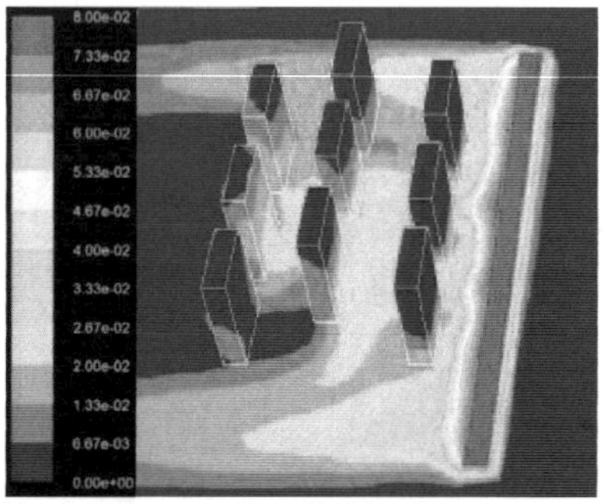

(b) The concentration field of staggered type

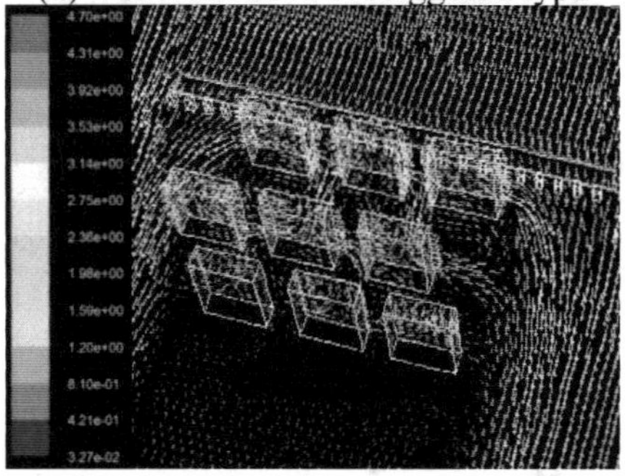

(c) The wind field after setting up trees

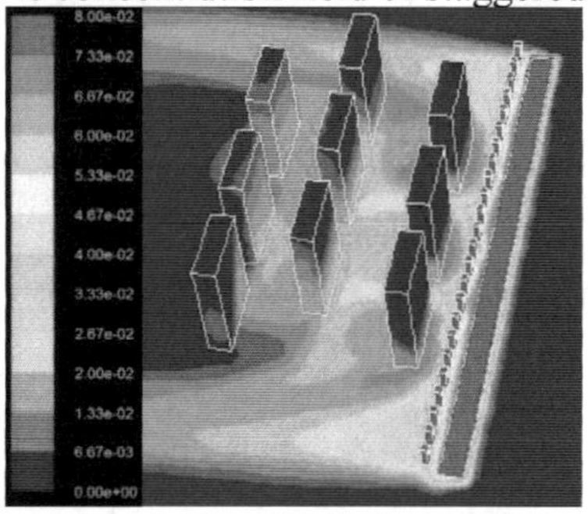

(d) The concentration field after setting up trees

Figure 3. The comparison of staggered type buildings.

As can be seen from figure 3, in the staggered buildings, the wind check area of the front building reaches the maximum because the windward wall reaches the maximum. At the same time as the front row building is blocked, its leeward surface will also be affected, that is, submerged in the shadow area [9]. There is active wind in the channel between buildings 1 and 3 #, running through the whole channel, and the wind speed is maintained at about 4.8 *m/s*. When entering the channel in the middle of the back row of the building, it will turn into the channel between 4 and 6 # of the building respectively, but the wind speed has decreased, basically below 2.8 *m/s*. In the transverse passage formed by the next two rows of buildings, merged into a certain range of wind comfort zone. At the same time, the rear of the building is in the wind shadow area. The degree of flow between each row of buildings is relatively complex, and there is still a relatively obvious vortex phenomenon behind the building in the rear row. After setting up the green trees, the maximum wind speed in the area decreased from 4.84 *m/s* to 4.70 *m/s*, and the mass concentration of PM2.5 on the roadside was also significantly reduced from 0.0533 mg/m^3 to about 0.0333 mg/m^3.

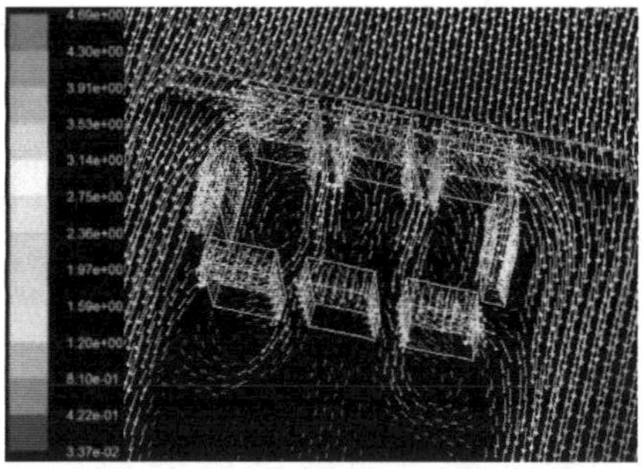

(a) The wind field of enclosed type

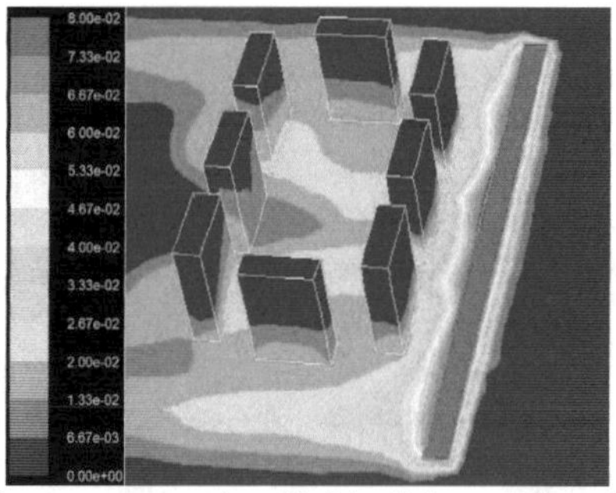

(b) The concentration field of the enclosed type

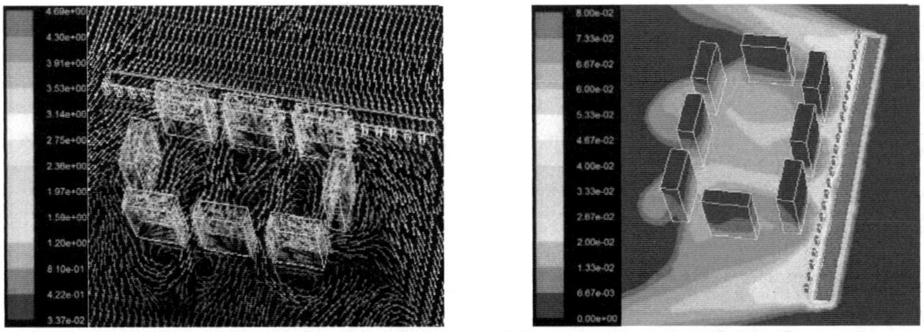

(c) The wind field after setting up trees (d) The concentration field after setting up trees

Figure 4. The comparison of enclosed type buildings.

As can be seen from figure 4, in the enclosed buildings, due to the vertical entry of the wind, the detention area increases, and the formation of the detention area affects the wind environment at the front side of the front row building [10]. In the case of more wind channels, the channel entrance wind speed does not change sharply. It is basically maintained at about 3.5 m/s , and about 3 m/s after setting the green trees. The wind speed decreases to the wind speed that meets the outdoor comfort level, and the wind speed along the downwind direction gradually weakens. At the back of the rear building, two obvious vortices are formed, and the shadow area of the wind accounts for 60~70% of the total internal area. Therefore, the layout and group in the community should be treated carefully. After setting up the green trees, the high-quality concentration area of PM2.5 was significantly reduced due to the adsorption of the trees.

To sum up, in parallel, staggered, and enclosed buildings, in the process of photosynthesis and respiration, the leaves of green trees and pores, green trees can play a role in absorbing and removing particulate matter in the air, and accelerate the settlement of particulate matter.

7. Conclusion

Through the comparative analysis of the advantages and disadvantages of the architectural complex layout, we know that the ventilation effect of the determinant is better. According to the Fluent simulation building group layout parallel, staggered, and enclosure, and the wind field and PM2.5 concentration field after setting up green trees, in the three typical layout methods of building groups, green trees can not only improve the wind environment but also effectively adsorb and remove particles in the air. It is of great significance to choose the appropriate layout mode and greening scheme of the micro-environment of the architectural complex.

8. References

[1] Hu A X, Li R, Chen G and Chen S 2024 Impact of respiratory dust on health: a comparison based on the toxicity of PM2.5, silica, and nanosilica *International Journal of Molecular Sciences* **25 (14)** 7654-7654.

[2] Chen Y F 2021 *CFD Numerical Simulation Study of Wind Environment in Typical High-Rise Residential Communities in Guangzhou* (Guangzhou: Guangzhou University) Master's Thesis p3

[3] Lu X G, Lei Y and Chen C P 2022 Study on the wind environment and pollution diffusion of building group based on LES staggered joint arrangement *Journal of Aerospace Engineering* **35(1)** 78-84

[4] Robert P 2024 Shields against pollution: phytoremediation and impact of particulate matter on trees at Wigry National Park, Poland *International Journal of Phytoremediation* **26(1)** 11-14.

[5] Xie H and Zhao S 2014 PM2.5 Control standard and comparison at home and abroad *Building Science* **20(30)** 37-43.

[6] Kantieva E, Snegireva S and Platonov A 2021 Formation of density and porosity of pine wood

in a tree trunk *IOP Conference Series: Earth and Environmental Science* **17(1)** 875
[7] Xu A and Xie Z 2017 Experimental study of wind effects on two neighboring high-rise buildings *Advances in Structural Engineering* **21(3)** 500-513.
[8] Zhang Y Z 2018 *Numerical Research on Floating Upflow by Large Vortex Simulation and Low Reynolds Number k-ε Model* (Hangzhou: Hangzhou Xidian University) Master's Thesis p20
[9] Huang Y M 2022 Research on the layout pattern of residential buildings based on wind environment simulation evaluation *Settlement* **12(05)** 126-130.
[10] Wang K 2021 *Simulation Study on the Effect of Arbor Layout on PM2.5 Concentration in Shenyang Enclosed Old Residential Area* (Shenyang: Shenyang Jianzhu University). Master's Thesis p37.

Research and Application of Harmless Disposal and Landfill Technology of Coal Gangue Waste in Shendong Mining Area

Ruimin He[1], Pengfei Cao[1], Yuhang Chen[2], Kai Zhang[2,*] and Wen Li[3]

[1] National Energy Shendong Coal Group Co., Ltd, Ordos, Inner Mongolia, China.
[2] School of Chemical and Environmental Engineering, China University of Mining and Technology-Beijing, Beijing, China.
[3] Sinopec Henan Oil Construction Engineering Co., Ltd., Zhengzhou, Henan, China.
Email: ruimin@he.ceic.com; 20022963@ceic.com; 2637038391@qq.com; *zhangkai@cumtb.edu.cn; liwen.osec@sinopec.com

Abstract. Shendong mining area is one of the important coal production bases in China. As a by-product in the process of coal mining, a large accumulation of coal gangue not only occupies land but also causes serious pollution to the environment. In order to solve the problem of coal gangue waste in Shendong mining area as the starting point, the research studies the harmless disposal and landfill technology, aiming to realize the resource utilization and environmentally friendly treatment of coal gangue. First, the physical and chemical properties of gangue, including its composition, structural characteristics and impact on the environment. On this basis, the necessity and feasibility of harmless disposal of coal gangue are discussed. Secondly, for the problem of coal gangue landfill, study the site selection principle, design requirements and construction technology of landfill, as well as the monitoring and management measures after landfill. Finally, through example analysis, the application effect of harmless disposal and landfill technology in Shendong mining area is shown, and the future research direction and policy suggestions are discussed. The results show that the negative impact of coal gangue on the environment can be effectively reduced by the comprehensive development of harmless disposal technology and scientific landfill method.
Keywords. Harmless disposal, landfill technology, coal gangue waste, coal mining, resource utilization

1. Introduction

As the world's largest coal producer and consumer, China's coal-fired power generation accounts for 50.2% of the global total. In the process of coal mining and processing, a large amount of coal gangue is generated, which is a kind of solid waste accompanying mine mining, and its emission accounts for about 10% to 15% of the coal production, becoming one of the major industrial solid wastes in China [1]. Currently, there are 14 billion-tonne coal bases in China, and the coal production from these bases accounts for more than 90% of the total national production [2]. By 2022, the total amount of accumulated coal gangue in China has reach 7 billion tonnes [3], and the improperly handled stockpiling and discharge problems pose a serious threat to the mining environment and have become a key strategic challenge hindering the ecological restoration of mines.

China's coal gangue is mainly concentrated in coal-producing provinces such as Shanxi, Shaanxi, Inner Mongolia, Xinjiang and Guizhou, the gangue production in these areas accounted for 78.74% of the national total [4]. With the progress of science and technology, the gangue can be processed to

Journal of Physics: Conference Series **2941** (2025) 012076 doi:10.1088/1742-6596/2941/1/012076

"turn waste into treasure", and its comprehensive utilization rate is improved and widely used. However, this also puts forward higher requirements for the disposal technology, which needs to be constantly innovated and improved.

For this reason, many scholars at home and abroad have conducted in-depth studies. These studies mainly focus on the energy use of coal gangue, road construction, agricultural improvement and the production of building materials [5]. Wang [6] and other scholars proposed a sanitary landfill method, emphasizing effective seepage control treatment along with scientific site selection; Liu [7] and other scholars designed a set of detailed landfill operation procedures, including sub-cells, layered paving, layered crushing, and day-by-day mulching, and combined with the trench-filling process, they established a feedback mechanism for the effect and a monitoring program. The effectiveness of the process was proved through field tests; Sun [8] effectively avoided the formation of internal low-lying and external high terrain by carrying out a staged and regional terraced landfill strategy for the landfill, realized automatic drainage, and further enhanced the environmental benefits of the landfill.

The harmless disposal and resource utilization of coal gangue has become an important issue for ecological environmental protection and sustainable development in mining areas. This paper takes the harmless landfill disposal technology of coal gangue in Shendong mining area as the research object, aiming to provide a theoretical basis and practical reference for the harmless disposal of coal gangue.

2. Overview of the Study Area

Located in the southwestern part of the Inner Mongolia Autonomous Region, Ordos City is situated in the core of Ordos Plateau, bordering Shanxi, Shaanxi and Ningxia provinces and regions, and is a key energy production base in China. In recent years, with the high-intensity development of coal resources in the city, the massive accumulation of coal gangue has become an urgent environmental problem.

Shendong Group has 13 production mines, which produce up to 19 million tonnes of coal gangue annually, but its utilization rate is only 4.91%. Taking the trial price of fine solid waste disposal in Ningxia (RMB 18 per tonne) as a reference, the annual cost of coal gangue disposal in Shendong mines is as high as RMB 340 million. Supplementary Lianta Coal Mine, which belongs to Shendong Coal Company, is located in Ulanmulun Town, Yijinholo Banner, Ordos City, and is a mine that was reconstructed and expanded in accordance with the mode of "20 million tonnes of coal in one well and two faces with 500 people", and its annual production capacity reaches 20 million tonnes, which is the world's largest single-shaft production capacity of the shaft coal mines, as shown in Figure 1. The gangue field of the mine covers an area of 35.12 hectares, with a closure elevation of 1,286.50 meters, a closure roof area of 33.74 hectares, and a designed gangue discharge of 11 million tonnes [9].

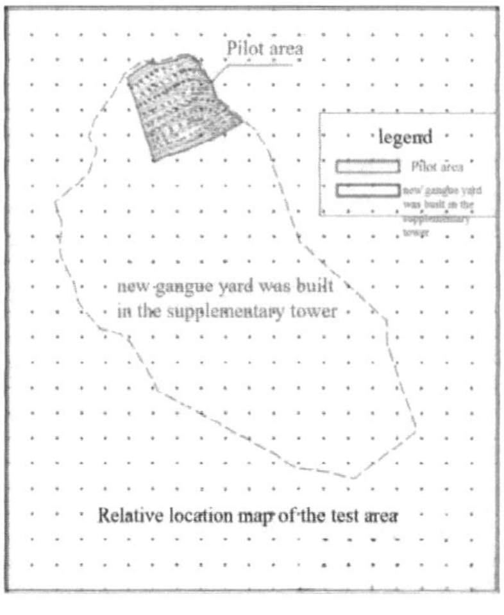

Figure 1. Relative position diagram of the experimental zones.

Journal of Physics: Conference Series **2941** (2025) 012076 doi:10.1088/1742-6596/2941/1/012076

3. Gangue Landfill Disposal Methods and Technical Routes

3.1. Research Method

This study focuses on typical coal gangue and prepares slurry materials for filling by integrating resources such as wind-deposited sandy soil, colluvial loess, fly ash and desulphurization gypsum in the study area. At the same time, the properties of different raw materials were compared to optimize the preparation process of the slurry materials, and the selection and preparation strategy of the slurry materials was formulated with harmless gangue treatment as the core. The developed slurry material has adjustable viscosity and yield stress, and its performance regulation mechanism is clear, it can effectively prevent heavy metal pollution and spontaneous combustion problems when adding additives to fill the gangue, showing the optimal effect [10]. Making the slurry material is not only affordable and easy to construct, but also through the improvement of land remediation, to achieve the resourcefulness of solid waste and environmental protection use [11,12].

3.2. Technical Route

The key technical research and development indexes include: the unconfined compressive strength of the developed slurry material bionic soil is not more than 5 MPa, the pH after the consolidation is 6-9, the water content is controlled below 20%, the soil porosity is not less than 20%, and it has good water and fertilizer retention properties. In addition, the best slurry material construction process matching it is proposed, as shown in Figure 2.

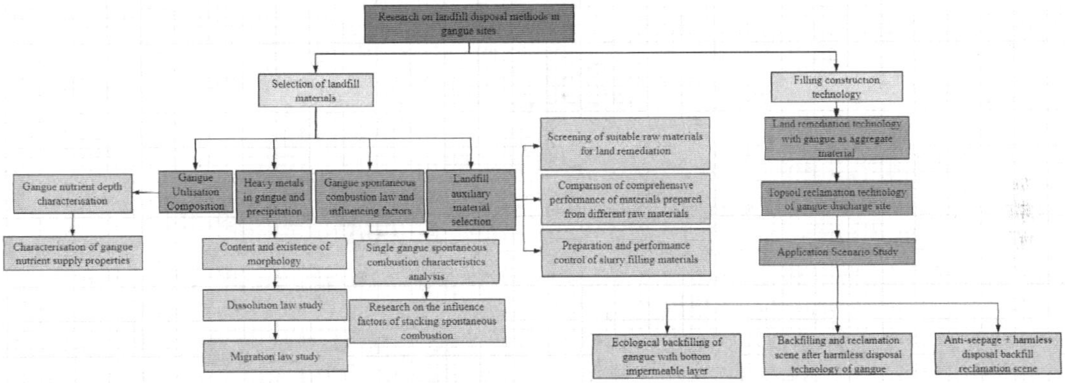

Figure 2. Research technology roadmap of landfill disposal mode of gangue discharge site.

4. Gangue Landfill Construction Technology Research

4.1. Research on the Selection of Landfill Materials

Adopting the local pulverized soil, fly ash and cement in the Shendong area as raw materials, after mixing and proportioning, the preliminary preparation of bionic soil is realized. To comprehensively assess the performance of the bionic soil, a series of tests were carried out, including geotechnical performance tests covering the determination of unconfined compressive strength, pH value and electrical conductivity (EC), as well as analyses using techniques such as X-ray diffraction (XRD), Fourier Transform Infrared Spectroscopy (FTIR) and Scanning Electron Microscope (SEM); ecological performance tests were carried out to assess the water retention capacity of the bionic soil. The environmental function test includes the detection of the ability of bionic soil to adsorb and solidify heavy metals in the gangue, as shown in Figure 3.

The grouting slurry formula in this study is water: modified bentonite: fly ash: desulphurization gypsum=3:1.3:0.4:0.3. The prepared bionic soil material has good water permeability and air permeability, and its structure is similar to that of rock, which can be closely combined with the surface soil to achieve seamlessness, thus forming a geomorphological feature close to the natural one

Journal of Physics: Conference Series **2941** (2025) 012076 doi:10.1088/1742-6596/2941/1/012076

and effectively taking into account the movement of water and salts, as shown in Figure 4.

Figure 3. Sample preparations.

Figure 4. No side limit resistance strength tester.

4.2. Research on Heavy Metal Accumulation and Precipitation in the Gangue

We collect gangue samples from different locations, dry, crush and sieve the samples to obtain uniform analysis samples; use atomic absorption spectroscopy (AAS) or X-ray fluorescence spectroscopy (XRF) and other methods to determine the content of heavy metal elements (such as lead, cadmium, chromium, arsenic, etc.) in the gangue. The successive extraction method was used to determine the accumulation forms of heavy metal elements in the gangue, including exchangeable state, carbonate-bound state, iron and manganese oxide-bound state, organic-bound state and residue state. [13] The contents of heavy metal elements in different forms were analyzed to assess their environmental risks and bioavailability, as shown in Table 1.

According to the content of precipitated heavy metal elements in the leachate of coal gangue under simulated gangue-soil-groundwater conditions, analyze the influencing factors and migration law of heavy metal elements in coal gangue, and study the characteristics of the precipitation-migration law of heavy metal elements in coal gangue under gangue-soil-groundwater conditions [14,15].

Table 1. Coal gangue heavy metal related indicators and test methods.

Object of characterization	Characterisation reference standard	Remarks
Heavy metal content and accumulation form	Refer to GB/T 16658-2007 "Determination of chromium, cadmium and lead in coal" and GB/T 16659-2008 "Determination of mercury in coal".	Design of the dissolution experiment of each heavy metal element in coal gangue

Dissolution law of heavy metal elements	According to HJ/T 299-2007 "Solid waste leaching toxicity method Sulfuric acid nitric acid method" HJ/T 300-2007 "Solid waste leaching toxicity leaching method Acetic acid buffer solution method" HJ 557-2010 "Solid waste leaching toxicity leaching method Horizontal oscillation method" GB/T 34230-2017 "Coal and gangue leaching experimental method" GB 18599- 2020 "General industrial solid waste leaching experiment" GB 18599- 2020 "General industrial solid waste leaching experiment" GB T 16659-2008 "Coal and gangue leaching experiment" GB T 16659-2008 "Determination method of mercury in coal". 2020 "General industrial solid waste storage and landfill pollution control standards".	Design of gangue leaching experiments of various heavy metal elements to identify the solid waste attribution of industrial solid waste
Migration law of heavy metal elements	GB 5086.1-1997 Solid Waste Leaching Toxicity Leaching Method Flip Method HJ 557-2010 Solid Waste Leaching Toxicity Leaching Method Horizontal Oscillation Method	Design of leaching experiments of each heavy metal element of gangue

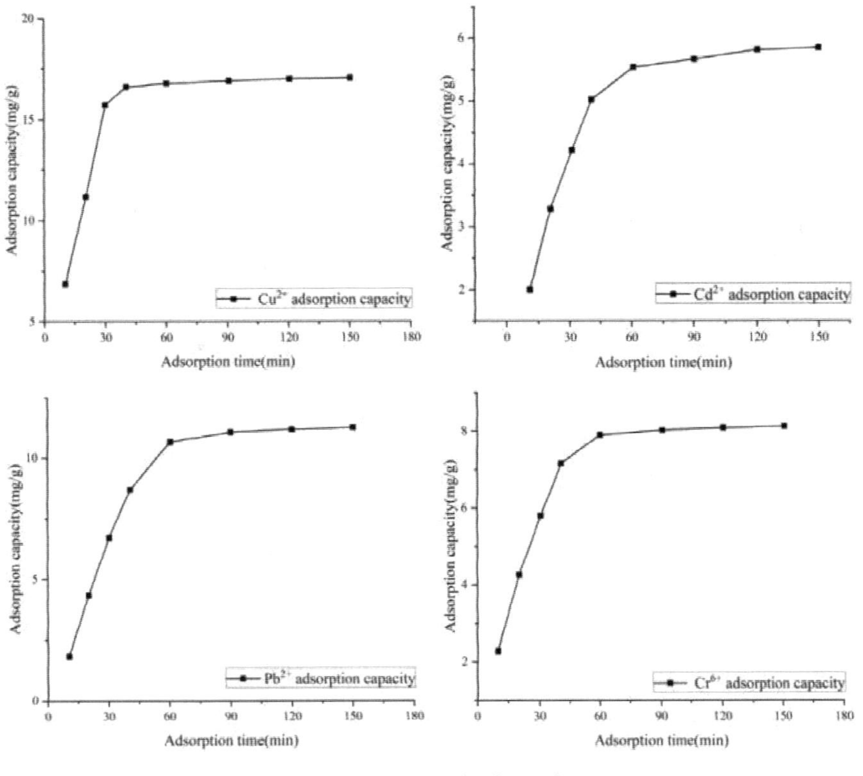

Figure 5. Heavy metal adsorption test.

Characterization of the IBA sand fraction samples by XRD revealed that quartz, calcite, ettringite and goethite were the main crystalline phases, with quartz dominating. SEM-EDX analysis showed that the grains consisted of a variety of fragments with different compositions and structures. These fragments have different shapes and textures, some with vesicles some without, some with sharp edges, some rounded, and some with cracks. Trace metal particles range in size from 1 to 10 μm, as shown in Figure 5. Solid phases with higher melting points (called refractory phases) contain debris,

Journal of Physics: Conference Series **2941** (2025) 012076 doi:10.1088/1742-6596/2941/1/012076

industrial glass, or metal and can be distinguished by the absence of vesicles and sharp edges. Metals may be part of these fragments and may be present as metals, oxides, sulfides, carbonates, chlorides, or silicates. Depending on the melting point, the metal may also be melted. The melted particles should show vesicles and blunt edges. In addition, multiple bonding between different metals has been found. The chemical composition of the IBA samples is also very complex. As can be seen from the XRD spectra of the raw IBA, the main mineral phase is not metallic and about 50 wt% of the mineral phase is dominated by SiO2. In addition, there are significant amounts of calcium, aluminum and iron. In terms of trace elements, six metals are present at levels above 500 ppm, namely barium, chromium, copper, lead, strontium and zinc. Interestingly, the levels of copper and zinc are very close to each other at 2720 ppm and 2811 ppm respectively. As for the rare earth elements (REE), although they are detected, their levels seem to be very low. Lanthanum, cerium and neodymium appear to be the rare earth elements with the highest concentrations in the analyzed samples, as shown in Figure 6.

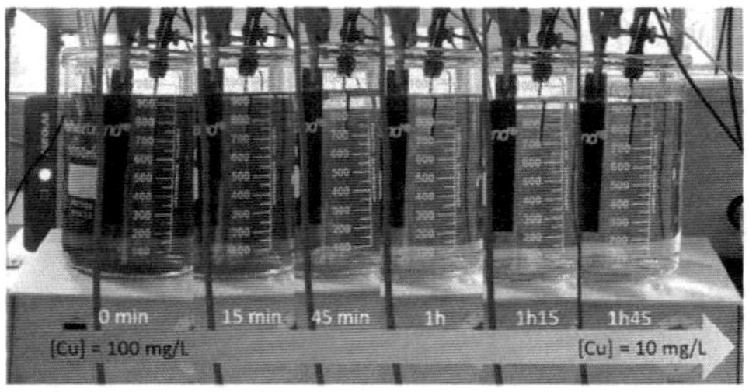

Figure 6. Electrodeposition experiments of heavy metals.

4.3. Research on the Selection of Landfill Auxiliary Materials

Land remediation applicable raw materials screening. We analyze the basic structural properties and physical and chemical properties of the raw materials, plant landscape plants on the surface mulch after solidification of the cementitious materials, carry out reclamation of the coal mining subsidence area, prepare topsoil with the main raw materials and auxiliary raw materials to be selected in the project, and assess the nutrient characteristics and fertilizer quality of the raw materials, including pH, organic matter content, all-nitrogen content, all-phosphorus content, all-potassium content, hydrolysis of nitrogen content, effective phosphorus content, and quick-acting potassium content.

Comprehensive performance comparison of materials prepared based on different raw materials. According to the type and composition of raw materials in different preparation schemes, based on the price and transport cost of raw materials, as well as the composition and cost of the preparation process, evaluate the economy of the preparation of filling cementitious materials, and combined with the slurry's technical indicators and environmental effects, screen the evaluation parameters applicable to the preparation and application of filling cementitious materials, and establish a multilevel mathematical evaluation system, quantitative evaluation of different slurry's usability, and the preferred filling cementitious materials preparation. The raw material composition and preparation scheme of filling cementitious materials are selected to provide a theoretical basis for the process design and on-site construction of harmless treatment of gangue and land remediation of coal mining subsidence area in the mining area, as shown in Table 2.

Table 2. Performance indexes and characterization methods of filling materials.

Characterisation objects	Characterization method
Stability	Evaluate the stability of the slurry by the water secretion rate and the degree of stratification.
Rheology	Determine the yield stress and viscosity coefficient of the slurry by mortar rheometer.
Filling behaviour	Slurry filling test box, slurry filling gangue void experiment, to examine the

gangue particle size distribution, slurry concentration, pumping pressure, and other parameters on the filling degree of completion, combined with the slurry in the void solidification time and degree of shrinkage.

Table 3. Comprehensive performance indexes and evaluation methods of filling materials.

Evaluation objects	Evaluation method
Economy	Depending on the type and composition of the raw material, based on the price and transport costs of the raw material, as well as the composition and costs of the preparation process
Evaluation parameters	Screening of evaluation parameters suitable for the preparation and application of filling gels, taking into account the technical specifications of the slurry and the environmental effects.
Availability	Optimization of the composition of the raw materials and the preparation of filling gels by establishing a multi-level mathematical evaluation system and quantitatively evaluating the availability of different slurries.

Preparation and performance regulation of slurry filling materials. Completion of the preparation and performance control of slurry filling materials, mainly including six aspects. First, to complete the preparation of slurry filler materials and the composition of structural properties characterization: combined with the screening of materials to be used, the use of physical mixing methods for the preparation of fillable slurry materials [16]. Ensure that all the additive materials are from a wide range of sources, cost-controllable, good economy, and in line with the actual engineering application [13]. Secondly, complete the study of the additive program of the slurry material: the screening of additives should follow the principle of non-toxic, non-polluting and green environmental protection. Third, complete the characterization of the filling performance of the slurry material: the filling performance of the slurry material should be combined with the evaluation of the comprehensive experimental effect, as shown in Table 3.

Fourth, the completion of the slurry material curing/stabilization of gangue heavy metal performance characterization: slurry material curing gangue heavy metal sequestration effect needs to be evaluated by experimental characterization results of its performance. Fifth, the completion of the characterization of the flame retardant properties of the slurry material: the flame retardant properties of the slurry material need to be evaluated by experimental characterization results [17]. Sixth, complete the slurry material multi-performance synergistic regulation mechanism: comprehensive slurry materials applied to different grades of gangue filling and grouting effect evaluation, combined with its curing/stabilization of gangue heavy metal performance and flame retardant properties, taking into account the cost and economic benefits of the application of the material, put forward the slurry material grouting backfill and reclamation of the multi-performance synergistic regulation mechanism, clear the mechanism of the application of the material and the application of the long-term stability, as shown in Figure 7 and Figure 8.

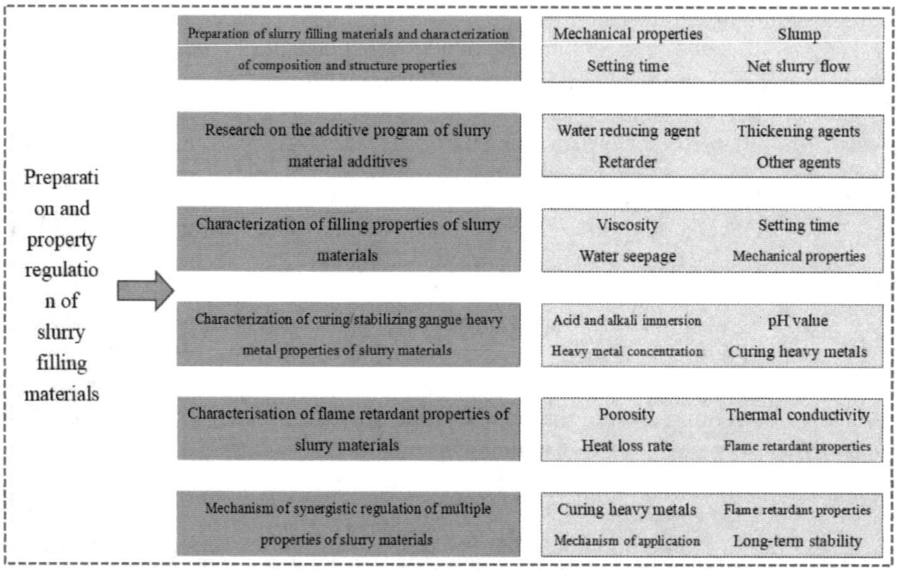

Figure 7. Schematic diagram of the preparation of the slurry filling material and the multi-performance.

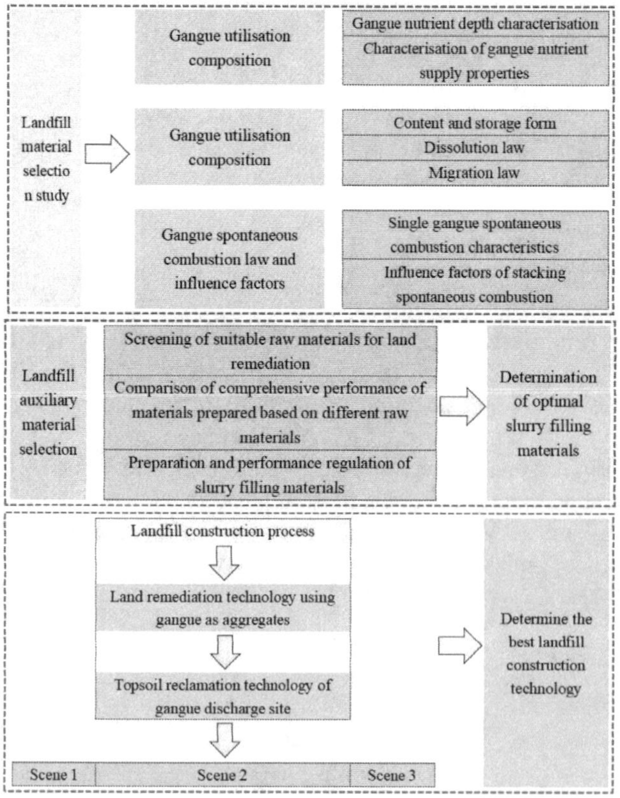

Figure 8. A schematic diagram of the research contents on the landfill disposal mode of the gangue disposal site.

4.4. Gangue Landfill Construction Technology Programme

Land remediation technology with gangue as the main aggregate. We improve the land remediation technology with gangue as the main aggregate [14], combined with the gangue stockpile situation and slurry properties to optimize the construction process optimization. Combined with the laboratory to determine the slurry material by a reasonable filling and grouting process applied to the actual engineering scene, the grouting process has a layered filling and grouting and mixing and filling two kinds of technology.

The working equipment is determined by utilizing the different properties of the raw materials and additives. After confirming the feasibility of the process under laboratory test conditions, the process will be extended to on-site engineering and construction, as shown in Figure 9.

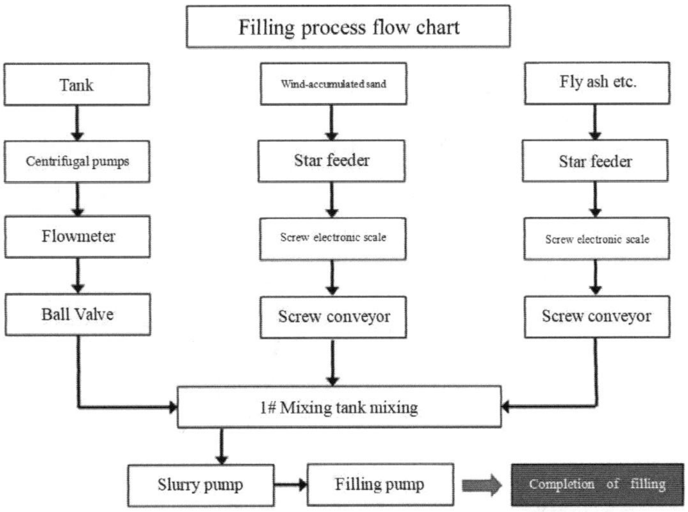

Figure 9. Filling and grouting process system diagram.

Application program of land remediation technology. We take the traditional landfill process as a control, combined with laboratory simulation filling data, and compare the harmless disposal effect of gangue under 3 different scenarios of construction conditions on-site. Compare the two construction perspectives of the seepage control process and slurry filling process, and argue and screen out the best process by setting up 6 monitoring wells and long-cycle monitoring of surface soil and groundwater, and integrating the materials, construction process, economic benefits and monitoring data under the application scenarios [15].

(1) Application of gangue ecological backfilling scenario based on bottom impermeability

Combined with the comparison of the traditional gangue field laying process, in the gangue backfill laying based on bentonite and solid waste improvement of the bottom impermeable layer, concerning the "coal gangue landfill field technical regulations" DB 14 /T 1114-2015 and other land reclamation standards for the gangue backfill in layers, through the reservation of two monitoring wells for shallow groundwater and topsoil overburden physicochemical properties of the monitoring for the evaluation of the effectiveness of the impermeable layer of the application of this scenario [18]. The effect of the impermeable layer is evaluated under the application of this scenario, as shown in Figure 10.

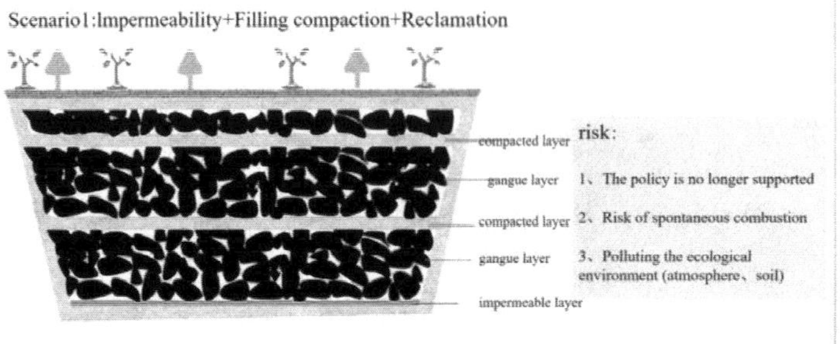

Figure 10. Schematic diagram of improved imvious layer laying process of traditional gangue yard landfill.

(2) Based on the gangue harmless disposal technology after backfilling reclamation program

Combined with the comparison of the traditional gangue field laying process, do not lay the bottom anti-seepage layer. Directly adopt the gangue grouting process for gangue backfill. By reserving 2 monitoring wells for shallow groundwater and topsoil physicochemical property monitoring, the evaluation of the harmless disposal effect of coal gangue under the application of this scenario is carried out, as shown in Figure 11.

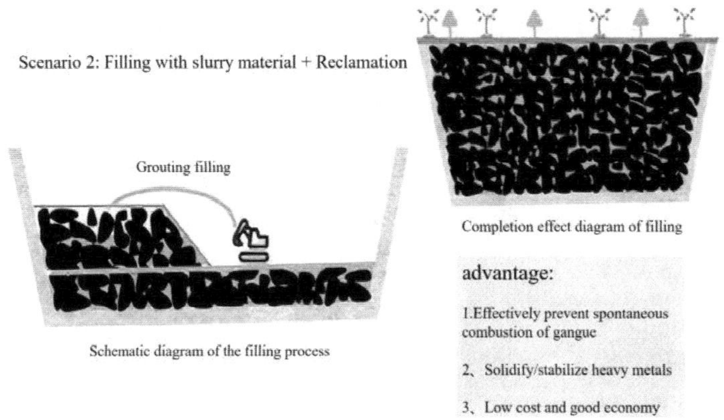

Figure 11. Schematic diagram of slurry material filling and reclamation process.

(3) Based on the anti-seepage + harmless disposal backfill reclamation program

On-site laying of improved impermeable layer, at the same time on the gangue grouting process for gangue backfill, through the buried monitoring wells to observe based on the bottom layer of impermeable and gangue grouting filling combined process effect evaluation, judge the harmless disposal of coal gangue on the environment. By reserving two monitoring wells for monitoring the physical and chemical properties of shallow groundwater and topsoil overburden, the evaluation of the effect of harmless disposal of coal gangue under the application of this scenario is carried out, as shown in Figure 12.

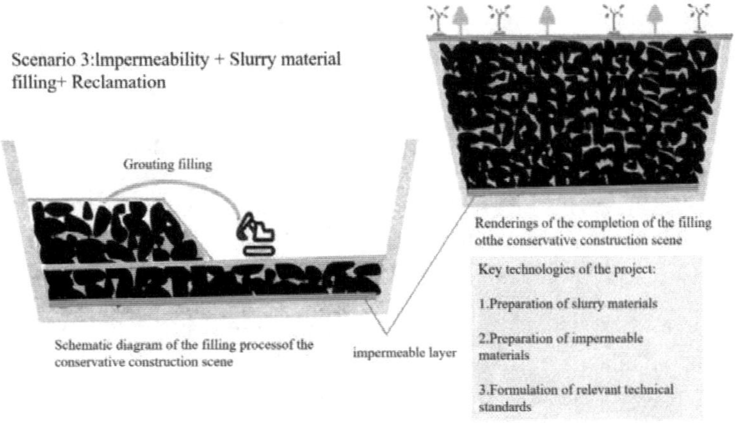

Figure 12. Schematic diagram of anti-seepage + slurry material filling and reclamation process.

5. Economic and Social Benefits

5.1. Economic Benefits

This project is aimed at promoting the construction of a new gangue discharge site in the Shuilianta mining area, taking the research and development of "low-cost, harmless, resourceful" coal gangue treatment technology as the core, and at the same time constructing demonstration projects to provide theoretical and technical support for the construction of new gangue discharge site in Shendong mining area. The project explores the synergistic use of technology and engineering practice suitable for coal gangue resourcefulness, combines the characteristics of gangue and the ecological environment in Shendong Mining Area, carries out ecological engineering, builds an integrated engineering demonstration area with regional characteristics, and provides technological and theoretical guidance for harmless treatment of gangue. Preliminary estimation is that if a gangue field with a capacity of 13.5 million tonnes is constructed by ground reclamation and field building, the treatment cost is about 20 yuan /tonne. In this project, based on the analysis of the physical and chemical properties of coal-based solid waste and soil properties, it is used as the main raw material for the preparation of slurry filler materials and applied to coal gangue reclamation, with low cost of raw materials and simple preparation process. Based on this technology, try to build a slurry material remediation gangue mountain program, it is expected that the cost can be reduced to about 18 yuan/ton. If calculated according to the annual processing capacity of 3 million tonnes, it can create an economic value of about 6 million yuan per year.

5.2. Social Benefits

China's western region has many coal-producing areas, and the coal gangue produced by them cannot be fully realized harmless treatment and resource utilization due to the limitation of disposal technology, economic cost and transportation radius. A large amount of coal gangue piled up in the gangue field, resulting in surface water pollution, land occupation spontaneous combustion and other environmental problems. With the strengthening of national environmental protection policies, the construction of permanent coal gangue dumps has been explicitly prohibited, and the effective treatment of coal gangue has become a key task in the ecological construction of mining areas. Given the current situation of massive accumulation of coal gangue and insufficient in-situ dissipation technology in the western coal-producing areas, this project will study the slurry filling treatment of coal gangue and land reclamation technology system, aiming at realizing the harmless, resourceful and economic treatment of coal gangue, and promoting the green and safe application of the whole life cycle of coal, resource recovery and recycling of the regional resources. The project will provide a technology demonstration platform for regional coal resources and ecological environmental protection, coordination and green development, demonstrate corporate social responsibility, enhance the corporate social image, and provide reference and guidance for the treatment of coal gangue in western mining areas.

6. Conclusion

With the new "Solid Waste Law" on solid waste resource treatment put forward higher requirements, the direct landfill of coal gangue without harmless treatment will no longer be approved. The traditional gangue filling and mining technology has a high cost and low utilization rate, which makes it difficult to realize the comprehensive resource recovery of coal gangue. Therefore, exploring low-cost, harmless and resourceful coal gangue treatment technology, applying it to reclaim coal mining subsidence areas, and formulating technical standards have become an urgent and important task in the utilization of mineral resources and ecological protection in Shendong mining area. In summary, the main conclusions of this paper are as follows:

(1) The gangue's usable components and heavy metal dissolution law are analyzed.

(2) Based on the actual situation of the company, combined with advanced landfill seepage control technology at home and abroad, designed a suitable grouting material bionic soil for Shendong Shuilianta Mine; put forward the idea of the grouting process which is mainly based on layer-by-layer and sub-layer filling and grouting.

(3) The research and development of a new type of landfill disposal technology for gangue fields has been completed, giving the selection of landfill materials and landfill construction techniques based on the harmlessness of gangue. Typical gangue as the main research object, analysis of the use of wind-deposited sand, cemented loess, fly ash and desulfurisation gypsum and other substances used in the preparation of filling slurry materials, based on the slurry material performance regulation and mechanism of research and analysis to determine the filler gangue weight metal curing effect and spontaneous combustion prevention and control of the effect of the addition of different admixtures to determine the effect of filling gangue weight metal curing and spontaneous combustion prevention and control evaluation.

(4) We determine the synergistic control mechanism of slurry materials, build a demonstration project of harmless disposal of coal gangue filling and grouting, and give the design of a harmless disposal project of coal gangue in the Shuilianta coal mine.

(5) Through the research, the scientific research results of gangue disposal are transformed into practical engineering technology, the best construction technology is explored according to the on-site engineering test, and the optimal technical and economic decisions are formulated.

Acknowledgments

This work was financially supported by the National Energy Group 2030 Major Pilot Project "Research and Engineering Demonstration on Harmless Coal Gangue Land Remediation and Reclamation Technology" (E210100485) fund.

7. References

[1] Wang Xiaojun. Research on coal gangue disposal and its application in China's mining areas [J]. Mining technology, 2024,24(03):308-312.

[2] WANG Gang, CHEN Xuechang, CHENG Weimina, et al. Multi-scale characterization of coal pore and fractures and its influence on permeability—taking 14 large coal bases in China as examples[J/OL]. Journal of Chongqing University: 1-17 [2024-03-12].

[3] Tian Li, Yu Xiaomeng, Qin Jin. Research Progress in Utilization of Coal Gangue Resources [J]. Journal of Hebei University of Environmental Engineering,2020,30(5):31-36.

[4] CHANG Jiwen, DU Genjie, DU Jianlei, et al. Current Situation of the Comprehensive Utilization of Coal Gangue in China and the Related Problems and Recommendations [J]. China Environmental Protection Industry, 2022(8):13-17.

[5] GUO Yanxia, ZHANG Yuanyuan, CHENG Fangqin. Industrial development and prospect about

comprehensive utilization of coal gangue[J]. CIESC Journal, 2014,65(07): 2443-2453.
[6] WANG Guoqiang, ZHAO Huahong, WU Daoxiang, et al. Gangue sanitary landfill and recovering ecological environment in Huaibei and Huainan mining area[J]. Journal of China Coal Society,2001(04):428-431.
[7] Liu Feng. Design and application of layered overburden filling process for coal gangue in Shangyuquan Mine[D]. China University of Mining and Technology,2022.
[8] SUN Youbin, WANG Zhenhua. Practice of Comprehensive Treatment of Gangue Mountain in Lvlin Chemical Energy Coal Preparation Plant[J].Energy and Energy Conservation, 2021, No.189(06):79-80.
[9] TANG Zhitao,XIA Yunkai. Analysis of dry separation effect on water-washed reject in Coal Preparation Plants in Shendong Mining Area[J]. Coal Processing & Comprehensive Utilization, 2023, No.282(01):41-44+49.
[10] WANG Ping, FU Zhanyong, WANG Fei, et al. Research progress of the application of coal-based solid waste for soil remediation[J/OL].Clean Coal Technology:1-16.
[11] PAN Shuo, HU Yuqing, WANG Haiyan, et al. Research on comprehensive utilization technology and standard system of coal gangue[J/OL]. China Mining Magazine:1-11.
[12] JIANG Chenyu, DU Qianhui, ZHANG Junhui, et al. Research on Pollution Release and Solidification Treatment of Coal Gangue Subgrade Fill Material[J/OL].Journal of Changsha University of Science & Technology(Natural Science),1-11.
[13] Hao Dengfeng, Liu Weijuan, Xia Dashuai, et al. Study on pressure test of the pipelines filled with gangue slurry into collapse zone[J]. China Energy and Environmental Protection, 2023,45(12):296-302.
[14] Wei Yuedong. Study on Dynamic Reclamation Scheme of Gangue Hill in Production Mine[J] Metal Mine, 2019(10): 140-146.
[15] Shen Qi. Analyze the Influence of Coal Gangue Landfill on Environment and the Countermeasures[J]. Shanxi Chemical Industry, 2024,44(03):247-248+2
[16] Ziejewska C, Bak A, Hodor K H M. Eco-Friendly Coal Gangue and/or Metakaolin-Based Lightweight Geopolymer with the Addition of Waste Glass[J]. Materials, 2023, 16(17). DOI:10.3390/ma16176054.
[17] Sitarz M, Figiela B, Ach M, et al. Mechanical Response of Geopolymer Foams to Heating—Managing Coal Gangue in Fire-Resistant Materials Technology[J]. Energies, 2022, 15. DOI:10.3390/en15093363.
[18] Ashfaq M, Moghal A A B, Basha B M. Reliability-Based Design Optimization of Chemically Stabilized Coal Gangue[J]. Journal of Testing and Evaluation: A Multidisciplinary Forum for Applied Sciences and Engineering, 2022. DOI:10.1520/JTE20210176.

Study on the Treatment Efficiency of Micro-Polluted River Water by Using Fluid Mechanics and Wave Propagation Characteristics

Ying Liu[1,*], Siyang Wang[1], Shuo Xu[2], Lizhu Chen[2], Xuejiao Li[2], Hongxu Liu[2]

[1] Changchun University of Architecture and Civil Engineering, Changchun, Jilin, China.
[2] Jilin Province Hongxing Construction Group Co., LTD, Changchun, Jilin, China.
Email: *YingLiuccjzxy@126.com;

Abstract. This study designed an innovative algorithm based on fluid mechanics and combined it with model simulation to systematically analyze the treatment effect of micro-polluted river water. By constructing a mathematical model of water flow motion, key fluid mechanics characteristics such as energy attenuation and wave velocity change during wave propagation are accurately captured. The removal efficiency of micro-pollutants under different fluid conditions is quantitatively analyzed, further revealing the role of wave propagation characteristics in promoting treatment efficiency. During the simulation process, the system was operated under a variety of boundary conditions and parameter settings to obtain the treatment effect under different working conditions. The results show that the algorithm effectively improves the removal rate of harmful substances in micro-polluted river water under the condition of fine control of fluid mechanics parameters, especially under high-frequency wave propagation conditions, the treatment efficiency is significantly improved, and the removal rate can reach more than 85%. This paper also provides accurate data analysis, including pressure changes, velocity distribution, and energy conversion efficiency during wave propagation, which verifies the reliability of the algorithm and the practical application potential of the model.
Keywords. Micro-polluted river water treatment; fluid mechanics; wave propagation characteristics; efficiency analysis.

1. Introduction

In recent years, with the acceleration of urbanization and industrialization, the problem of micro-polluted river water has become increasingly serious. This type of water usually contains low concentrations of organic pollutants, nutrients, microorganisms and other pollutants. Traditional water treatment methods have limited effects on them and is difficult to achieve economical and effective purification [1]. Therefore, seeking new treatment technologies to improve the purification efficiency of micro-polluted river water has become a research hotspot. Water treatment methods based on fluid mechanics and wave propagation characteristics show good potential. Reference [2] proposed using wave characteristics to enhance the mass transfer process of pollutants in water bodies, effectively accelerating the degradation rate of pollutants, thereby improving the treatment effect to a certain extent. This study introduced a dynamic model of fluid waves to achieve the formation of tiny turbulent structures in the water body, increase the contact frequency between water flow and pollutants, and optimize the reaction efficiency. However, the application effect of this method in a low flow rate environment is still not ideal, and it is impossible to achieve effective removal of micro-pollutants in static water bodies. Reference [3,4] explored the energy transfer mechanism in water treatment and controlled the fluid movement in the water body by optimizing the wave propagation rate. The research results show that controlling the wave velocity can significantly affect the diffusion path and rate of pollutants in water, reduce energy consumption in the water purification process, and thus improve treatment efficiency. Although this method has advantages in terms of energy consumption, it has high requirements for wave speed control and there is uncertainty in the effect under different pollutant types and concentration changes. In addition, based on the turbulence model in fluid mechanics, reference [5] proposed a water treatment method that uses turbulence to enhance the dispersion and diffusion of pollutants. This method introduces periodic oscillation waves into the water body to produce irregular turbulence effects on the microscopic scale of pollutants, thereby accelerating the diffusion and decomposition of pollutants. However, experimental data show that when the pollutant concentration is low, the improvement of treatment efficiency by this method is relatively limited, and it is limited by specific hydraulic conditions and lacks practicality. Reference [5,6] proposed a multi-stage wave propagation system based on vibration waves to simulate the effects of different frequencies and amplitudes on pollutant removal. The results show that by adjusting the vibration frequency and transmission path of the wave, the pollutant removal effect can be enhanced to a certain extent. However, this study mainly focuses on experimental conditions and lacks effective theoretical model support, so it cannot be promoted in a large range of water treatments [6].

This study first constructs a fluid mechanics model to analyze the energy distribution, wave speed changes, and turbulence effects on pollutant diffusion and decomposition during wave propagation. Secondly, a set of efficient wave propagation algorithms is designed to achieve precise control of parameters such as fluid velocity, wave velocity, and wave amplitude to optimize the treatment effect [7]. Finally, through simulation experiments, the reliability and applicability of the model are verified under different fluid conditions and pollutant concentrations, and the feasibility of this method in practical applications is further explored [8].

2. Analysis of Fluid Mechanics Characteristics of Micro-Polluted River Water Treatment

2.1. *Overview of Basic Theories of Fluid Mechanics*

The application of fluid mechanics theory in micro-polluted water treatment covers turbulence, boundary layer effects, and mixing mechanisms. Turbulent characteristics usually refer to the complex and turbulent motion of fluids at high Reynolds numbers, which is a very common flow form in water treatment [7]. The vortex structure in turbulence can significantly enhance the mixing and mass transfer capacity of the fluid, allowing pollutants to contact the treatment medium more effectively, thereby improving the removal efficiency. In the treatment system, turbulence is often controlled and utilized by designing specific flow rates and flow directions [9]. The mixing mechanism refers to the material exchange between different concentration layers formed during the flow of the fluid. This mechanism plays a key role in the oxidation and degradation reactions in the treatment of micro-polluted water.

2.2. *Fluid Mechanics Behavior in Biological Contact Oxidation Ponds*

In biological contact oxidation ponds, the flow state of the fluid is affected by many factors, including pollutant concentration, temperature, and the setting of fluid mechanics parameters. The biological contact oxidation pond decomposes pollutants into lower concentrations of organic matter or even inorganic matter through the interaction of aeration and microbial membranes. Therefore, the analysis of fluid behavior in oxidation ponds is the basis for improving pollutant removal efficiency [10].

2.2.1. Pollution Concentration and Flow State. The change in pollutant concentration will significantly affect the flow state of the fluid in the oxidation pond. Under high concentrations of pollutants, the viscosity of the fluid increases, which easily leads to poor flow. This phenomenon is particularly obvious in some high-pollution load environments [11]. Therefore, when designing the fluid, it is necessary to adjust the dynamic characteristics of the fluid in combination with the pollution concentration to ensure the uniform distribution of the fluid throughout the pond, thereby improving the contact efficiency between the pollutants and the biofilm.

2.2.2. Effect of Temperature on Fluid Mechanics. Temperature also has a significant effect on fluid behavior, especially in low-temperature environments, the viscosity of the fluid increases, the fluidity decreases, and the transmission efficiency of pollutants is affected. This is particularly prominent in water treatment in cold seasons. Low temperature leads to a decrease in the activity of microorganisms, thus affecting the degradation rate of pollutants. In addition, under low-temperature conditions, the fluidity and turbulence characteristics of the water body will also decrease, affecting the overall mass transfer efficiency of the oxidation pond. Some studies have shown that by properly adjusting the aeration volume or increasing the fluid disturbance in the water body, the negative impact of low temperature on fluid flow can be alleviated to a certain extent.

3. Application of Wave Propagation Characteristics in Micro-Polluted River Water Treatment

3.1. Introduction to Wave Propagation Theory

Wave propagation refers to the propagation form of vibration or disturbance in the medium, including sound waves, shock waves, etc. The propagation of waves in the fluid can effectively stimulate the disturbance and mass transfer effect of the fluid, which has a wide range of application potential in the treatment of micro-polluted river water. The rational use of different types of wave propagation characteristics is of great significance for the efficient removal of pollutants. Assuming the frequency of wave propagation is f, the wave speed is c, and the wavelength is λ, its basic propagation relationship can be expressed as:

$$f = \frac{c}{\lambda} \tag{1}$$

By regulating the frequency and speed of waves, the propagation form of waves can be changed, thereby affecting the diffusion behavior of pollutants in water bodies. The propagation of waves not only causes the periodic movement of fluid particles but also changes the distribution state of pollutants at a microscopic scale, thereby providing technical support for improving the efficiency of micro-polluted river water treatment.

3.2. Effect of Wave Propagation on Pollutant Mass Transfer

Wave propagation has a significant driving effect on pollutant mass transfer in water bodies. Fluctuations can promote the migration of pollutants from the fluid phase to the solid phase medium and accelerate the reaction process. Studies have found that under different water quality and temperature conditions, the mass transfer efficiency of wave propagation on pollutants is different. Generally speaking, higher wave frequencies and amplitudes can enhance the diffusion of pollutants, but too high frequencies may cause eddies or backflows in the fluid, affecting the overall mass transfer effect. Assuming that the pollutant concentration in the water body is C and its diffusion coefficient under the influence of fluctuations is D, the enhancement effect of wave propagation on pollutant mass transfer can be expressed by the following equation:

$$\frac{\partial C}{\partial t} = D\nabla^2 C + S_w \tag{2}$$

S_w represents the influence of wave force on concentration gradient. The S_w term introduced by wave propagation can be understood as the contribution of wave force to the diffusion process of pollutants. When the temperature rises, the mass transfer effect of wave propagation is enhanced, which can be described by the relationship equation between wave frequency f and temperature T:

$$f(T) = f_0 \cdot e^{\alpha T} \tag{3}$$

Where f_0 is the reference frequency and α is the temperature influence factor. This relationship shows the regulating effect of temperature on wave frequency, thereby affecting the mass transfer capacity of waves at different temperatures.

3.3. Wave Propagation Optimization Method to Improve Treatment Efficiency

3.3.1. Optimization of Wave Propagation Frequency. Wave frequency is an important factor in controlling the diffusion of pollutants. Assuming the target removal rate is R, its dependence on wave frequency can be expressed by the following formula:

$$R = R_0 + \beta f^{\gamma} \tag{4}$$

Where R_0 is the baseline removal rate, β and γ are the frequency influence coefficient and gain coefficient respectively. This model shows that increasing the wave frequency can improve the removal rate, but too high a frequency may have a counter-effect. Therefore, it is crucial to determine the appropriate frequency range through experiments.

3.3.2. Optimization of Wave Propagation Amplitude. The effect of wave amplitude on mass transfer rate can be expressed by amplitude A. Assume that the relationship between pollutant removal efficiency E and wave amplitude is:

$$E = E_0 + \delta A^2 \tag{5}$$

Where E_0 is the base efficiency and δ is the amplitude influencing factor. The increase in amplitude can accelerate the contact between pollutants and the reaction medium and improve the removal efficiency, but too high an amplitude may induce eddy currents and affect stability.

3.3.3. Shear Force Regulation Based on Wave Propagation Characteristics. The shear force generated during wave propagation can effectively improve the mass transfer effect of pollutants. Assume that the relationship between shear force τ and wave frequency and amplitude is:

$$\tau = \eta f A \tag{6}$$

Where η is the viscosity coefficient of the fluid. Appropriate regulation of shear force not only helps disperse pollutants but also prevents excessive accumulation of biofilm, thereby maintaining efficient operation of the oxidation pond.

3.3.4. Energy Equation of Wave Propagation Force. In the process of wave propagation, energy conversion efficiency is an important optimization factor. Assume that the energy conversion efficiency η_E of wave propagation varies with wave frequency and amplitude as follows:

$$\eta_E = \frac{\kappa f A}{1+\sigma f^2} \tag{7}$$

Where κ and σ are adjustment parameters. Under appropriate wave frequency and amplitude conditions, the highest energy conversion efficiency can be achieved, thereby reducing energy waste and improving treatment effects.

3.3.5. Wave Equation for Pollutant Diffusion Rate. The effect of wave propagation on the pollutant diffusion rate v can be expressed by the following wave equation:

$$v = v_0(1 + \theta \sin(\omega t)) \tag{8}$$

Where v_0 is the basic diffusion rate, θ is the wave amplitude influencing factor, and ω is the angular frequency of the wave. This equation shows that by controlling the angular frequency and amplitude of the wave, the diffusion rate of pollutants can be periodically regulated.

Journal of Physics: Conference Series **2941** (2025) 012077 doi:10.1088/1742-6596/2941/1/012077

3.3.6 Pollutant Concentration Decay Model. Assuming that the change of pollutant concentration with time t follows the exponential decay law and is affected by wave propagation, the model of its concentration $C(t)$ can be expressed as:

$$C(t) = C_0 e^{-\lambda t} + \xi A f \tag{9}$$

Where C_0 is the initial concentration, λ is the attenuation coefficient, and ξ is the wave control factor. This model shows that wave propagation can accelerate the attenuation of concentration, thereby improving treatment efficiency.

4. Application Examples of Biological Contact Oxidation Ponds in Micro-Polluted Water Treatment

Experimental Design and Data Analysis: Firstly, we control the flow rate and wave frequency of the fluid in the biological contact oxidation pond, and generate different turbulence and wave propagation conditions by adjusting the aeration and circulation flow. Secondly, set different pollutant concentrations and temperatures to evaluate the effects of fluid mechanics and wave propagation characteristics on pollutant removal efficiency. Finally, collect water samples to detect indicators such as residual pollutant concentration in water, energy consumption, and system stability to evaluate the performance of different algorithms. Table 1 shows the comparison of different algorithms in COD removal efficiency. Table 2 shows the comparison of different algorithms in energy consumption.

Table 1. Comparison of different algorithms in COD removal efficiency.

	Removal rate (%)			
Parameter settings	Traditional method	Optimization algorithm A	Optimization algorithm B	The algorithm in this paper
Low concentration, low temperature	65	70	75	85
Medium concentration, medium temperature	60	68	72	82
High concentration, high temperature	55	65	70	80

Table 2. Comparison of energy consumption of different algorithms (unit: kWh/m³).

Parameter settings	Traditional methods	Optimization Algorithm A	Optimization algorithm B	This article's algorithm
Low concentration, low temperature	1.2	1	0.9	0.7
Medium concentration, medium temperature	1.5	1.3	1.1	0.8
High concentration, high temperature	1.8	1.5	1.3	1

The proposed algorithm is significantly better than traditional methods and other optimization algorithms in terms of removal efficiency and energy consumption control under low temperature and low concentration conditions. Especially under high concentration and high-temperature conditions, the removal efficiency of the proposed algorithm is about 10% higher than that of optimization algorithm B, and the energy consumption is reduced by 33% compared with optimization algorithm A.

Under low-concentration and low-temperature conditions, the removal rate of the proposed algorithm is significantly better than other algorithms, especially in the later stage of the experiment, with higher stability and better adaptability (Figure 1).

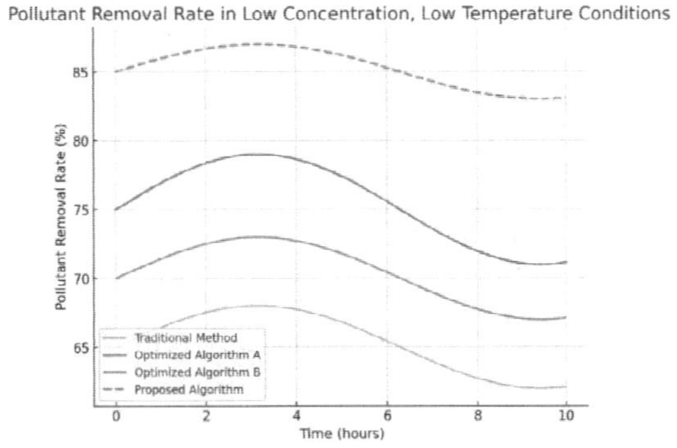

Figure 1. Pollutant removal rate simulation curves of different algorithms under low concentration and low-temperature conditions.

Under medium concentration and medium temperature conditions, the curve of the algorithm in this paper is always higher than the removal rate of other algorithms, especially in the initial stage, the optimization of wave propagation characteristics effectively enhances the pollutant mass transfer effect (Figure 2).

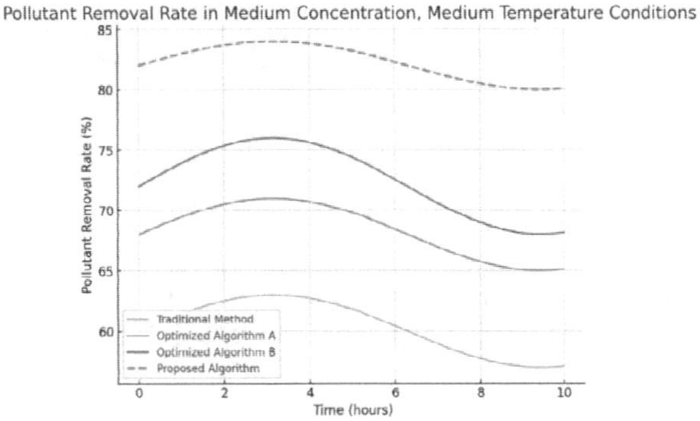

Figure 2. Pollutant removal rate simulation curves of different algorithms under medium concentration and medium temperature conditions.

Under high temperature and high concentration conditions, the proposed algorithm significantly reduces energy consumption. This result shows that the proposed algorithm performs well in the frequency and flow rate control of wave propagation, which not only ensures efficient removal of pollutants but also effectively controls energy consumption (Figure 3).

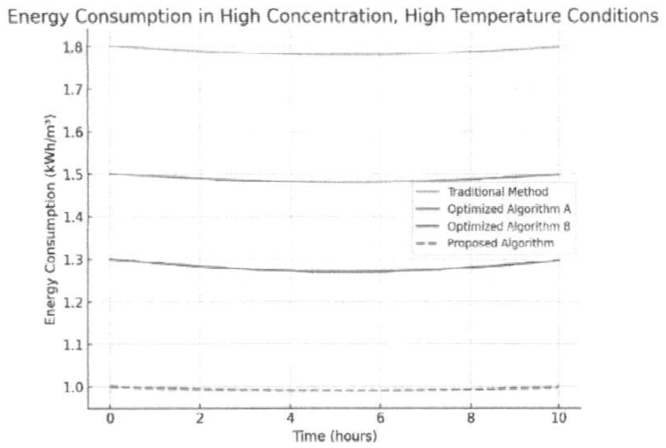

Figure 3. Energy consumption simulation curves of different algorithms under high concentration and

high-temperature conditions.

The proposed algorithm shows superior removal efficiency, low energy consumption, and good system stability under different pollutant concentrations and temperature conditions. Under low-temperature conditions, the proposed algorithm still maintains a high removal rate and system stability, further verifying the applicability of the coordinated optimization of fluid mechanics and wave propagation characteristics in the treatment of micro-polluted water. The proposed algorithm based on the optimization of fluid mechanics and wave propagation characteristics is significantly superior to traditional methods and other optimization algorithms in terms of removal efficiency, energy consumption control, and system stability.

5. Conclusion

This paper conducts an in-depth study on the fluid mechanics and wave propagation characteristics in the treatment of micro-polluted river water, aiming to improve the treatment efficiency. By constructing a fluid mechanics model and designing a corresponding algorithm, this paper effectively analyzes the influence of wave propagation characteristics on the removal effect of pollutants. The results show that the role of wave propagation characteristics in the treatment of micro-polluted river water is significant, especially under the conditions of controlling flow rate, pressure change, and fluctuation frequency, the treatment efficiency has been significantly improved. The simulation results verify the reliability of the proposed algorithm and model. In the process of wave propagation, by reasonably controlling the wave velocity and the transfer of fluid energy, the dispersion and degradation of pollutants can be accelerated, and more efficient treatment of micro-polluted water can be achieved. For example, under high-frequency wave propagation conditions, the pollutant removal rate reaches more than 85%, which is much higher than the treatment effect under traditional static conditions. In addition, the energy attenuation characteristics during wave propagation and the proper control of fluid pressure can help improve the overall treatment efficiency and reduce energy consumption. This method demonstrates the application potential of fluid mechanics parameter optimization in water treatment system design and provides a feasible technical path for the future treatment of micro-polluted water bodies.

Acknowledgments
This work was supported by the funds:

(1) 2024 Department of Education of Jilin Province Science and Technology Research Project Outstanding Youth Project "Research on the Efficiency of Biological Contact Oxidation of Modified Combined Fillers in Removing Nitrogen and Phosphorus in Micro-polluted River Water in Northern China" (JJKH20241677KJ).

(2) 2024 Jilin Province Higher Education Scientific Research Project "Research on 5E+ Hybrid Teaching Model under the Background of Digital Transformation" (JGJX24D0777).

(3) 2024 Jilin Province Higher Education Scientific Research Project "Research on Comprehensive Design Experimental Projects Relying on Professional Discipline Competitions" (JGJX24B49).

(4) 2024 Department of Education of Jilin Province Science and Technology Research Project General Project "Research on Sludge Biochar Catalyzed Persulfate in-situ Reduction Technology of Sludge Cell Breakage" (JJKH20241674KJ).

(5) 2024 Department of Education of Jilin Province Science and Technology Research Project Outstanding Youth Project "Research on the Efficiency of NZVI/BC Activated PMS in Treatment of Tail Water from Urban Sewage Plants" (JJKH20241679KJ).

(6) 2024 Jilin Province Higher Education Research Project "Exploration of Research Paths for Practical Teaching and Innovation and Entrepreneurship Education Reform in Private Colleges and Universities" (JGJX24D0766).

6. References

[1] Zhao Dandan, Liu Jiafu, Liu Jiping, & Du Huishi. A review of the main studies on permafrost degradation in the cold regions of Northeast China. Science Technology and Engineering. 22(30) (2022) 13151-13161.

[2] Huang Dong, Zheng Yanan, Xing Zhilin, Yang Tianyu, Huang Bingbing, & Wu Heng. Research progress and prospects of enhanced domestic sewage treatment by adding solid carbon sources. Journal of Chongqing University of Technology (Natural Science). 38(6) (2024) 298-304.

[3] Liu Jian, Guan Yurong, Wang Huihao, Shao Zhenbao, & Xu Xingqian. Current status and prospects of research on chemical mechanical effects of water and soil. Science Technology and Engineering. 23(10) (2023) 4019-4032.

[4] Liu Yang, Zhao Lixin, Zhang Shuang, Liu Lin, & Xu Baorui. Research and Application of Filtering-swirl Coupling Technology in Heterogeneous Separation. Journal of Mechanical Engineering.58(4) (2022)120-154.

[5] Zou Wenqi, Wang Jiamin, Wang Jianfeng, Zhang Yuxiang, Wei Xiaoxiao, Tang Qingfeng, & Gao Xia. Environmental impact assessment and key characterization techniques of polymer materials throughout their life cycle. Polymer Bulletin. 37(8) (2023) 1011-1023.

[6] Liu Yang, Zhao Lixin, Zhang Shuang, Liu Lin, & Xu Baorui. Research and Application of Filtering-swirl Coupling Technology in Heterogeneous Separation. Journal of Mechanical Engineering.58(4) (2023) 120-154.

[7] Du Shengnan, Xu Jianan, Xiong Fei, Cai Xuyi, Ge Pengbiao, Zhang Liuliu, & Li Juanying. Performance evaluation of different immobilized microbial particles in degrading pollutants in aquaculture tailwater. Journal of Shanghai Ocean University. 33(3) (2024) 682-691.

[8] Wang Xinjie, Wang Yining, Zhao Jian, Liu Shengdong, Xia Xinghui, & Li Yang. Research progress on the impact of river water and sediment movement on the migration of microplastics. Chinese Environmental Science. 42(2) (2022) 863-877.

[9] Paiva A R, Veras T B, Barcellos R L, et al. Evidence of protective effects on aquifer recharge from polluted tropical rivers: An analysis of hyporheic meiofauna and sediments[J]. River Research and Applications, 2022, 38(2):345-357.

[10] Sun X L, Phuc H D, Okamoto R, et al. A 30-year follow-up study in a former cadmium-polluted area of Japan: the relationship between cadmium exposure and β2-microglobulin in the urine of Japanese people[J]. Environmental Science and Pollution Research, 2022, V11(2): 1-7.

[11] Noszczynska M, Pacwa-Plociniczak M, Bondarczuk K Z. The microbial removal of bisphenols in aquatic microcosms and associated alteration in bacterial community[J]. Environmental Science and Pollution Research, 2023, 30(36): 85292-85304.

Study on Eutrophication, Phosphorus Pollution and Morphological Analysis of Separated Lakes

Xiaozhen Liu*, Chen Yang, Luqiang Zhou, Zhiyu Ding, Dengbiao Jiang, Jia Fei

[1] Key Laboratory of Environment and Resource Utilization of Poyang Lake, Ministry of Education, Nanchang University, Nanchang 330031, China.
[2] School of Resources & Environment, Nanchang University, Nanchang 330031, China.
Email: *liuxiaozhen@ncu.edu.cn

Abstract. After urban lakes are divided into "ponds", they are more affected by human beings and more prone to eutrophication. In this paper, five "ponds" in a university park in Nanchang were collected as research objects, and samples of lake water quality and sediment, as well as aquatic plants and animals on ecological floating beds were collected. Relevant indicators such as ammonia nitrogen, TP and sediment phosphorus were analyzed in the water, and the relationship between phosphorus form analysis and water eutrophication was discussed. Results: On the surface, severe eutrophication occurred in the lakes, and the water quality was Class V, which was related to external input and internal pollution of sediments. The main influencing factors of environmental pollution involved TP, ammonia nitrogen, etc. The TP content of one reservoir was seriously polluted, while the other three reservoirs were moderately polluted. The higher the pollution level of organophosphorus in sediments, the higher the content of OP in sediments. Fe/Al-P accounted for the majority of inorganic phosphorus in the four reservoirs. Fe/Al-P is active phosphorus and the main cause of eutrophication in ponds. Suggestions: Improve the drainage and diversion of lake water, clean up underwater silt, and cultivate submerged plants.
Keywords. Urban lake; eutrophication; sediment; phosphorus; morphological analysis.

1. Introduction

As an important part of the ecosystem, the functions of urban lakes cover the roles of water catchment and flood storage, climate regulation, eco-tourism, water supply, and citizens' recreation. The water quality of lakes is closely related to incoming water, such as surface runoff inflow, sewage discharge from neighbouring residents and industrial production wastewater discharge [1,2]. As the type and quantity of pollutant discharges increase significantly, eutrophication of lake water quality seriously affects the safety of water bodies in the region [3]. Nanchang old city such as Donghu District, Qingshan Lake District has a lot of natural lakes in history, due to the process of urbanisation since the nineties of last century, the original water system of city lakes gradually has been divided, and even filled. The lake in the old campus of a university in Nanchang, with a total water area of 12,200 m^2, has also been divided into five 'ponds'; at present, the water source of the 'ponds' mainly comes from natural precipitation, rainy-season surface runoff, and domestic sewage discharge, which makes it a very good site for studying water environmental pollution (figure 1).

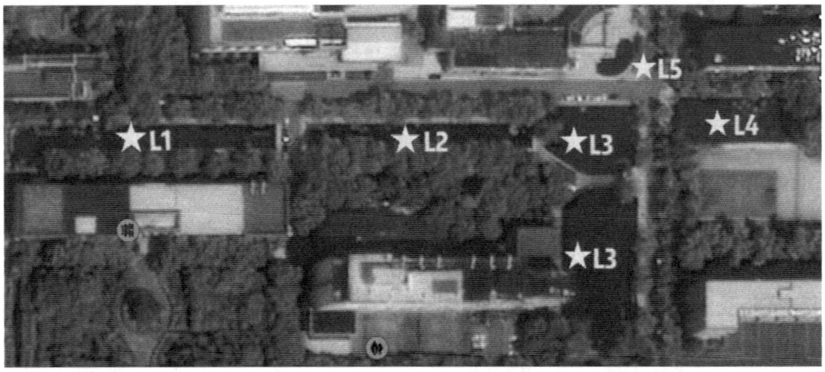

Figure 1. Lake water system map of the old campus of a university in Nanchang.

Compared with lakes, 'water ponds' belonging to small micro-water bodies have a small volume and small environmental capacity, and are susceptible to the influence of people's daily life, and the dilution and degradation capacity of nutrient inputs from exogenous sources is weaker [4], and the nitrogen and phosphorus nutrient salts imported for a long period of time are gradually deposited at the bottom of lake, and phosphorus content in the water body and sediment is affected by water body eutrophication in different degrees [5]. The dual effects of exogenous input and endogenous release make the water ponds highly susceptible to eutrophication and outbreaks of algal blooms.

The division 'ponds' of of university lake was designed and implemented in the past with ecological floating beds, aeration and other measures, but onsite investigation found that the eutrophication degree of water body is still very high, especially in the summer the eutrophication phenomenon is more prominent. The cause of eutrophication and various forms of phosphorus content in the sediment has not been clarified. In this paper, water quality indicators and sediments of the five 'ponds' of university campus, based on onsite research combined with experimental research and analysis, has been performed to explore the endogenous release of phosphorus in the aquatic environment, phosphorus in various forms and eutrophication of ponds, with a view to provide scientific basis for the eutrophication management of small and micro water bodies.

2. Materials and Methods

2.1. Study Site

On-site investigation of the old campus found that 'ponds' eutrophication degree is high, the previous implementation of ecological floating bed wetland is not effective. One of the 'ponds' is almost a black stinking water with a lot of large fish died. The entire lake is divided into five main 'ponds', the surrounding buildings have different functions.The most serious eutrophication pond is next to the student dormitory, the other two more serious eutrophication ponds are around the food laboratory research building, the main lake area was connected by a bridge around the classrooms. Ecological floating beds occupy approximately 6 per cent of the total water body area, and the aeration facilities are not action.

2.2. Field Sampling and Pre-Processing

On 16 Sept. 2023, water samples were taken in each of the five ponds (L1, L2, L3, L4, L5) of the old campus lakes in accordance with the Technical Guidelines for Environmental Impact Assessment of Water and Wastewater Monitoring and Analysis Methods for the Surface Water Environment (HJ2.3-2018), and the water sampling location should avoid inlets or outlets of the water pipelines. A portable sediment sampler [6] was also used to collect bottom sediment at multiple points in four ponds (L1, L2, L3, L4). The collected samples were brought back in clean polythene sealed bags. The water samples were measured in time; the sediment was dried in an oven, tried through a 100-mesh sieve and then labeled in plastic bags for measurement.Five representative plant samples, plantain, wild tumbleweed, great wolf loosestrife, amphibious polygonum and bluebells were collected from the ecological floating beds and brought back to the laboratory in clean polythene ziplock bags and stored at 4°C. Three common animal samples, small fish, river shrimp and snail,were collected. The plant and animal samples were dried, ground and sieved and then labelled in plastic bags for testing. On 17 Nov. 2023, water samples were collected from the five ponds for seasonal comparative analysis.

2.3. Sample Analysis Methods

2.3.1. Experimental Instruments. SHA-C water bath constant temperature oscillator, PU-1900 ultraviolet spectrophotometer, eppendorf5430 centrifuge, etc.

2.3.2. Water Quality Determination. Ammonium molybdate spectrophotometry was used to analyse the phosphorus content in the water samples, and alkaline potassium persulfate digestion UV spectrophotometry (National Standard HJ636-2012) was used to determine the total nitrogen content in the water samples. Nano reagent spectrophotometry (National Standard HJ535-2009) was used to determine the content of ammonia nitrogen in water. Determination of chemical oxygen demand in water by dichromate method (Environmental Standard 828-2017).

2.3.3. Sediment Determination. The determination of phosphorus morphology in sediments was based on the SMT method for phosphorus morphology in freshwater sediments developed by the European Committee for Standard Tests [7], and combined with the sediment phosphorus graded extraction method [8], a seven-step extraction method was comprehensively chosen to morphologically analyse the phosphorus in the sediments. In the experimental method, the inorganic phosphorus was classified into easily exchangeable phosphorus Ex-P, iron and aluminum-bound phosphorus Fe/Al-P, closed storage phosphorus Oc-P, authigenic calcium phosphorus ACa-P, detrital phosphorus De-P and organic phosphorus OP [3]. The phosphorus content was determined by molybdenum-antimony antimony spectrophotometry after the phosphorus was extracted. The method was improved according to this experiment: 1) Iron- and aluminum-bound phosphorus are active phosphorus, and the combined extraction of iron-bound phosphorus and aluminum-bound phosphorus will be analyzed; 2) In order to avoid the influence of different extracts on the experimental results, the samples after centrifugation were filtered with 0.45 μm filter membrane and then measured; 3) The residue was washed and centrifuged with 1 mol/L NaCl solution at the end of each step of the extraction and next step of the extraction was carried out. After each extraction step, the residue was washed with 1 mol/L NaCl solution and centrifuged before next extraction step.

2.3.4. Determination of Plant and Animal Samples. The ammonium molybdate spectrophotometric method (National Standard 11893-89) was used to analyse the total phosphorus content in plant and animal samples.

3. Results and Analyses

3.1. Analysis of Sediment Phosphorus Morphological Content

Organic phosphorus and inorganic phosphorus in the sediments of the four sampling sites accounted for about half of the total; Ex-P and Fe/Al-P are reactive phosphorus, which can well reflect the potential release capacity of sediment phosphorus content [9]. The composition of inorganic phosphorus was relatively large in Fe/Al-P, 24.12%, 27.18%, 37.64%, and 29.29%, respectively; Ex-P accounted for a relatively small proportion, 0.84%, 1.36%, 0.51%, and 0.51%, respectively (table 1, table 2, and figure 2), which indicates that the contribution of Ex-P to the eutrophication situation of the ponds was low; Fe/Al-P was the main cause of eutrophication, Fe/Al-P is reactive phosphorus, which is also released into the water body due to the reaction of iron and aluminum oxides [3], leading to the increase of phosphorus content in the water body and eutrophication of the water body. The content of Oc-P was in the middle of the range, which was 11.39%, 10.70%, 10.51%, and 6.88%, respectively, and Oc-P is the confined phosphorus, which is the phosphorus that is externally wrapped with a layer of Fe_3O_4 film of phosphorus salts [10], it is difficult to release under normal conditions, and has a small impact on the phosphorus content in the water body; ACa-P and De-P content accounted for an even smaller proportion, ACa-P content range of 4.66-5.74%, De-P content range of 1.69-2.80%, both of which have a high stability in the sediment, usually have a small impact on the phosphorus content in the water body and contribute less to the eutrophication of the water body [3].

Table 1. Phosphorus content in sediment with different forms of phosphorus.

Phosphorus form(mg/kg)	L1	L2	L3	L4
Ex-P	9.302	13.140	3.721	4.767
Fe/Al-P	268.721	261.791	273.256	272.674
Oc-P	126.840	103.050	76.300	64.030
ACa-P	63.953	70.930	57.209	43.372
De-P	31.226	26.650	17.974	15.698
IP	493.411	473.643	424.419	398.450
OP	620.543	489.535	301.550	532.558
TP	1113.953	963.178	725.969	931.008

Table 2. Percentage of phosphorus content in different forms in sediment.

Phosphorus form (%)	L1	L2	L3	L4
Ex-P	0.84	1.36	0.51	0.51
Fe/Al-P	24.12	27.18	37.64	29.29
Oc-P	11.39	10.70	10.51	6.88
ACa-P	5.74	7.36	7.88	4.66
De-P	2.80	2.77	2.48	1.69
IP	44.29	49.18	58.46	42.80
OP	55.71	50.82	41.54	57.20

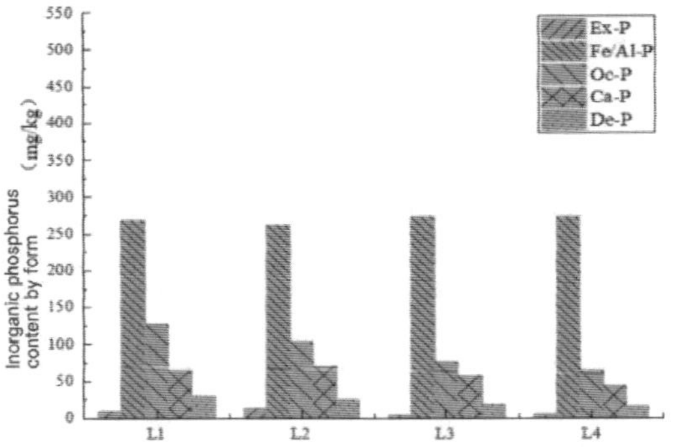

Figure 2. The content of various forms of inorganic phosphorus in sediments

The total phosphorus content in the sediments at the sampling sites ranged from 725.969 to 1113.953 mg/kg, with a mean value of 933.527 mg/kg. According to the existing literature [11,12], it is considered to be uncontaminated when w(TP) ≤ 500 mg/kg in the sediments, and moderately contaminated when w(TP) ≤ 500 mg/kg ≤ w(TP) ≤ 1000mg/kg is moderately polluted, and w(TP) ≥ 1000mg/kg is severely polluted. Among the four sampling points in this experiment, the sediment TP content of L1 was so high (1113.953mg/kg), which was seriously polluted, and the sediment TP content of the remaining three points were among 500mg/kg ~1000mg/kg, which were moderately polluted. The content of organic phosphorus (OP) in the sediments accounted for a relatively large proportion, 55.71%, 50.82%, 41.54% and 57.20%, respectively (figure 3), and the higher the degree of contamination, the higher the content of OP in the sediments of water bodies [13]. The four ponds sampled were all ornamental ponds in the university campus, and the sources of pollution were mainly from domestic sewage discharges from dormitories and laboratory wastewater discharges from food majors, and the water bodies were contaminated by severe exogenous inputs leading to an increase in the content of organic phosphorus (OP).

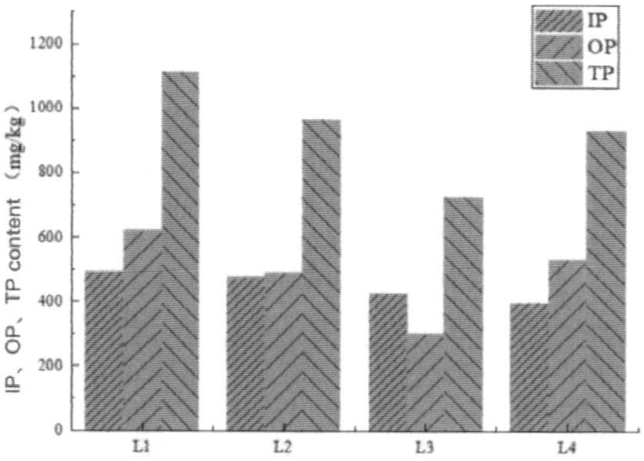

Figure 3. The content of inorganic phosphorus, organic phosphorus, and total phosphorus in the sediments

In aquatic environment, sediments are important reservoirs of phosphorus, which can accept phosphorus from the water column and release phosphorus into the overlying water for biological use under certain conditions [14]. The proportion of reactive phosphorus in total sediment phosphorus is relatively small in lakes with better water quality, while the corresponding increase in reactive phosphorus content in the sediments of heavily polluted urban lakes indicates that eutrophication of the water body is closely related to the content of various forms of phosphorus in the sediments [8]. At the same time, the poor circulation characteristics also led to the gradual deposition of nutrient salts at the bottom inside the reservoir, releasing a variety of nutrients over time.

3.2. Analysis of Water Quality Indicators

The ammonia nitrogen, total phosphorus, total nitrogen and COD_{Mn} data of the water samples collected from each point on 16 September were measured as follows (table 3):

Table 3. Water quality test results of reservoir in September (mg/kg).

sampling location	NH_3-N	TP	TN	COD_{Mn}
L1	0.581	0.162	2.555	10.6
L2	2.358	0.181	5.468	12.4
L3	2.714	0.185	5.451	13.3
L4	2.354	0.189	5.549	15.8
L5	3.019	0.211	5.848	13.7

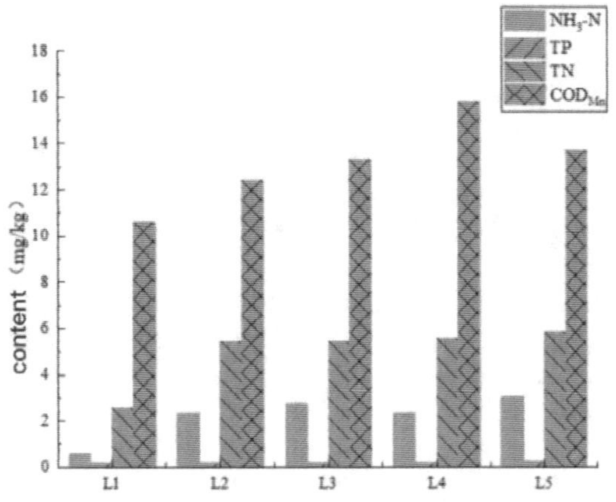

Figure 4. Contents of water quality indicators in ponds.

As can be seen from figure 4, in addition to L1's NH_3-N content in the II - III water standards, the remaining several ponds of NH_3-N content are more than V water standards, and NH_3-N content are accounted for about 50% of the content of TN; several ponds of TP content are more than IV water standards, L5 more than the V water; L1's TN content is more than the V water, L2, L3, L4, L5's TN content is far more than the V water standards; L1, L2, L3, L4, L5 COD_{Mn} content are more than Class V water standards. Compared to urban lakes, as a small micro-water body 'pond' own volume is small, poor connectivity with the outside water body [15], the dilution of nutrient input from external sources is weak, can not well buffer, dilution and sedimentation of input nitrogen and phosphorus nutrients, nutrient salts within the water body is very easy to accumulate, the causes of eutrophication of the water body are diverse [5]. Thus, effective management of eutrophication in these ponds is urgent.

In mid-November 2023, it rained continuously, and after a sun day, on 17 November, the second water quality sampling was carried out, and the measured data were as follows (table 4):

Table 4. Content of water quality indicators in ponds(mg/kg).

sampling location	NH_3-N	TP	TN	COD_{Mn}
L1	0.925	0.015	3.421	9.7
L2	2.575	0.554	5.635	14.5
L3	2.353	0.192	5.471	13.6
L4	2.519	0.305	5.546	15.2
L5	0.347	0.083	1.919	10.2

Compared with the first data, L1 showed an increase in NH_3-N concentration by 59.21% to 0.925, a decrease in TP concentration by 90.74% to 0.015, an increase in TN concentration by 33.90% to 3.421, and an increase in COD_{Mn} concentration by 8.49% to 9.7; the significant decrease in TP concentration could be attributed to the fact that rainfall water was fed into the pond in the form of surface runoff, which was exchanged with phosphorus.L5 s NH_3-N decreased by 88.51% to 0.347, TP concentration decreased by 60.66%, TN concentration decreased by 67.19%, and COD_{Mn} concentration decreased by 25.55% to 10.2; L2's TP increased by 206.08% to 0.554 mg/L, which may be related to the discharge of wastewater from the food laboratory test; L4's TP increased by 62.23% to 0.305, which may be related to dormitory domestic wastewater discharge; the rest other data had little change. In conclusion, the complex changes in water quality parameters were related to exogenous water inflow and endogenous drivers: (1) laboratory wastewater discharge, surface runoff during rainy season bringing plant residues or humus in the neighbourhood; and (2) rainwater harvesting hydrodynamics disturbing floating sediments, and enhanced exchange between the water column and sediments.

3.3. Analysis of Plant Samples

The phosphorus content in stems and leaves of several plant samples sampled in this study was lower than the mean value of phosphorus content of wetland plants in China [16] (leaves: 1.85 mg/g, stems: 1.71 mg/g), and plantain had the lowest phosphorus content, with phosphorus content of 0.312 mg/g in the leaves and 0.408 mg/g in the stems, while great wolf loosestrife had the highest phosphorus content, with phosphorus content of 0.838mg/g in leaves and 0.962mg/g in stems.The study suggests that the phosphorus content of the plants in the ecological floating bed artificial wetland is lower than that of the plants originating from the terrestrial wetland, which provides a basis for the subsequent management, and the proposed cultivation of submerged plants is conducive to the uptake of phosphorus and the reduction of phosphorus content in the water.

3.4. Analysis of Animal Samples

As can be seen from the test data (figure. 5), aquatic animals are rich in phosphorus, but the species varies greatly, with shrimp meat the highest 4607.752mg/kg, followed by small fish, which may be related to the food chain and biological habits. The symbiotic conditions of aquatic plants and aquatic animals can effectively improve the transparency of the water, and play a role in the removal or purification of the components of N, P, SS, and organic matter [17] (table 5).

Table 5. Phosphorus content in plant samples.

samples	amphibious polygonum	canna	windmill grass	bidens frondosa	bluet
stem	0.736	0.408	0.751	0.962	0.684
leaf	0.655	0.312	0.663	0.838	0.703

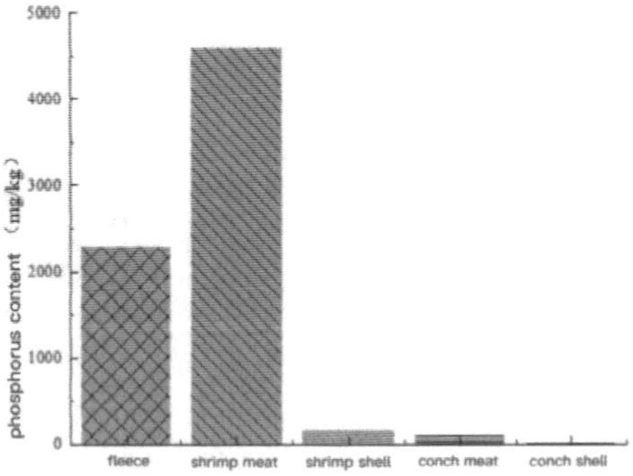

Figure 5. Phosphorus content in animal samples.

4. Conclusion

The serious eutrophication of the lakes in the old campus is related to exogenous input (inadequate domestic sewage network and laboratory food wastewater discharge) and endogenous pollution (pollution of sediments); the water body is of class V water quality, and the main environmental pollution influencing factors involve TP and ammonia nitrogen. Among phosphorus pollution, the sediment TP content of pond L1 reached 1113.953 mg/kg, which was severely polluted. The percentage of organic phosphorus (OP) content in the sediments was also large. And the higher the degree of pollution, the higher the OP content in the sediments in the water bodies. Fe/Al-P accounted for the majority of inorganic phosphorus in the composition of inorganic phosphorus; Ex-P accounted for a very small portion of the total; Fe/Al-P was the main cause of eutrophication of the water pond.

The old campus lake silt has not been cleaned up for more than 30 years, the silt stock is about 12,000 cubic metres, it is recommended to clean up the underwater silt. At the same time, large areas of water cultivation of submerged plants, appropriate fish, shrimp, snails and other aquatic animals, is conducive to the absorption of phosphorus and suspended particles in the sediment, increase the transparency of the water body, reduce the content of phosphorus in the water.

Improve the construction of domestic sewage and rainwater pipeline network around the lakes of the old campus, and do a good job of intercepting and diverting sewage; organic wastewater from laboratories and campus sewage are included in the urban domestic sewage pipeline network.

The old campus has a high degree of greening, plant residues or humus will flow into the lake with the surface runoff in the rainy season. The old campus can be designed to sink two-stage grid sedimentation tank, rainwater through the sedimentation tank and then flow into the lake, which can significantly reduce the pollution of particulate organic matter.

5. References

[1] Huisman J, Codd G A and Paerl H W 2018 Cyanobacterial blooms *Nature Reviews Microbiology*. **16(8)** 471-83.

[2] Ismail W R, Omar M A, Rahaman Z A A 2023 Hydrology, water quality and trophic state of Pergau Reservoir, Kelantan, Malaysia *Water Science*. **37(1)** 131-50.

[3] Zhu G W， Gao G and Qin B Q 2003 *Advances in Water Science*. **14(06)** 714-9.

[4] Birch S and MeCaskie 1999 *Hydrobiologia*. 395:365-378.

[5] Yao J P, Wang G Q and Xue B L 2019 Assessment of lake eutrophication using a novel multidimensional similarity cloud model *Journal of Environmental Management*. **248** 109259.

[6] Liu X Z， Jin J and Yang X Y 2007 Studies on the distribution of perpendicularity pollution of Cu element of sediments in Wucheng section of Ganjiang River *Journal of Nanchang Hangkong University*. **21(3)** 80-84.

[7] Ruban V, Lopez-Sanchez J F and Pardo P 1999 Selection and evaluation of sequential extraction procedures for the determination of phosphorus forms in lake sediment *Journal of Environmental Monitoring*. **1(1)** 51-6.

[8] Zhu G W, Qin B Q and Gao G 2004 Fractionation of phosphorus in sediments and its relation with soluble phosphorus contents in shallow lakes located in the middle and lower reaches of Changjiang River, China *Acta Scientiae Circumstantiae*. **(03)** 381-8.

[9] Jin X C 2008 The key scientific problems in lake eutrophication studies. *Acta Scientiae Circumstantiae*. **(01)** 21-23.

[10] Zhang Y, Yan W M, Yang Y Q 2017 An analysis of phosphorus speciation in sediments of typical small and medium sized cultivated lakes *China Rural Water and Hydropower* **(12)** 99-103.

[11] Li R, Wang S R and Xiao S B 2015 Sediments phosphorus forms and loading in the lakes of the mid-lower reaches of the Yangtze River and Yunnan Plateau, China *China Environmental Science*. **35(06)** 1831-9.

[12] Zhong J C, Liu G F and Fan C X 2009 Environmental effect of sediment dreging in lake(I): the role of sediment dredging in reducing internal phosphorous release *Journal of Lake Sciences*. **21(01)** 84-93.

[13] Liao J Y, Wang S R and Yang S W 2010 Characteristics of organic phosphorus in different forms of lake sediments from east plain region *Research of Environmental Sciences*. **23(09)** 1142-50.

[14] Wu X D, Chang F M and Wu Bin, Sun H J and Zhong W J 2023 Forms of sedimentary phosphorus in the South Yellow Sea and the implication to regional eutrophication trend *Marine Geology & Quaternary Geology*. **43(05)** 106-18.

[15] Yang N， Wang C Y and Xu Y Y 2021 Research status and development trend of the treatment technology of black odor small and micro-water body *Industrial Water Treatment*. **41(05)** 15-21.

[16] Hu W F， Zhang W L and Zhang L H 2014 Stoichiometric characteristics of nitrogen and phosphorus in major wetland vegetation of China *Chinese Journal of Plant Ecology*. **38(10)** 1041-52.

[17] Lv Z J, Inanori R H and Inamori Y H 2009 Analysis of purification function with and without macro animal under temperature changing condition in submerged aquatic plant system *Engineering Journal of Wuhan University.* **42(6)** 723-725.

Optimization of ICP-MS Method for Determination of Praseodymium (Pr) in Vegetables

Guangjun Huang[1] and Xiuhua Zhang[1, 2*]

[1] College of Food and Quality Engineering, Nanning University, Nanning 530200, Guangxi, China.
[2] Guangxi Minsheng Zhongjialian Testing Co., Ltd, Nanning 530009, Guangxi, China.
Email: *763034352@qq.com, 372024483@qq.com

Abstract. Taking praseodymium (Pr) in vegetables as the research object, the microwave digestion/graphite digestion pretreatment methods of Pr in vegetables were compared. Combined with inductively coupled plasma mass spectrometry (ICP-MS), the influence of microwave digestion and graphite digestion sampling amount, digestion time, digestion temperature, carrier gas flow rate of ICP-MS inductively coupled plasma mass spectrometer, torch position and other factors on the determination results was investigated. The optimum reaction conditions were determined by orthogonal experiment as follows: sample volume 0.5 g, digestion temperature 120°C, microwave digestion time 30 min, carrier gas flow 1.001 L/min, torch tube position -1.840 mm. The reliability of the optimized method was verified. The detection limit changed from 0.09177 ug/kg to 0.06043 ug/kg, the linear regression coefficient increased from 0.09990 to 0.999972, and the recovery rate increased from 95.7% to 109.6% to 96.0% to 106.9%.
Keywords. ICP-MS; Vegetables; Pr; Method optimization.

1. Introduction

Low doses of rare earth elements can promote the growth of biological organisms. In particular, praseodymium (Pr) and erbium (Er) in rare earth elements have a significant impact on biological growth, which can increase the yield of various crops, improve varieties, and achieve obvious economic benefits. The widespread use of rare earth elements as micro-fertilizers and feed additives has led to increasingly serious pollution of rare earth elements in the environment, which is likely to cause enrichment of the food chain and further enter the human body. In recent years, the detection of rare earth elements has also received increasing attention [1-3].

For the detection of rare earth elements in food, such as "GB/T 5009.94-2012 National food safety standard-Determination of rare earth elements in plant foods" [4], microwave digestion was used for pretreatment. However, it is difficult to digest under microwave digestion conditions or the reproducibility and stability of the data still have some room for optimization. The content of rare earth elements in vegetables is generally low. The key factors for the determination of rare earth elements in samples are the selection of sample pretreatment methods and the determination of analytical conditions [5].

Inductively coupled plasma mass spectrometry has the characteristics of simple spectrum, less interference, high sensitivity, low detection limit, wide linear range, simple operation, and easy simultaneous analysis of multiple elements. It has become the fastest growing inorganic trace element in recent years [6,7].

2. Materials and Methods

2.1. Materials and Reagents

Nitric acid (Merck, Germany), hydrogen peroxide (Guanghua, Guangdong), rare earth element standard solution (1000 ug/kg, O2SI, USA), the above reagents are AR.

2.2. Instruments and EQUIPMENT

Icecap Inductively Coupled Plasma Mass Spectrometer (Thermo); mars6 microwave digestion instrument (CEM); bHW-09C graphite electrothermal digestion instrument (Beijing Botong Company); mL204 electronic balance (sense 0.1 mg and 1 mg) (MettlerToledo).

2.3. Test Method

2.3.1. Microwave Digestion Method for Sample Pretreatment. The samples were weighed and placed in a sealed polytetrafluoroethylene digestion tank, and 5 ml HNO_3 was added to the microwave digester for digestion. The digestion conditions were programmed at 120°C, 140°C and 180 °C, the heating time was controlled at 5min, and the constant temperature time was controlled at 5min. After the digestion was completed, the inner wall was washed with HNO3, and the digestion solution was transferred together with the washing solution to a 50 ml volumetric flask for constant volume backup. At the same time as reagent blank.

2.3.2. Graphite Digestion Method for Sample Pretreatment. A sample was taken and placed in a sealed Teflon digestion vessel. 5 ml of HNO_3 was added and the digestion was carried out in a microwave digestion apparatus. The digestion conditions were 120°C, 140°C and 180°C, the heating time was controlled at 5min, and the constant temperature time was controlled at 5min. After the digestion is complete, the inner wall is washed by HNO_3, and the digestion solution together with the washing solution is transferred to a 50ml volumetric bottle for constant volume reserve. At the same time as reagent blank.

2.3.3. ICP-MS Determination. Set the measurement conditions: Setting of measurement conditions: The sample tube was placed in the tuning liquid, and the internal standard tube was placed in pure water when the temperature in the atomization room drops to 2°C. The sampling depth, sampling cone position, carrier gas flow rate, charging flow rate, emission power were set to ensure that no collision occurs when the mass number 7 count rate > 2000,89 count rate > 8000,205 count rate > 6000, oxide value < 156/140, double charge value < 1.0% (70/140). The lens voltage and impact gas flow can be adjusted to ensure that the count rate for mass number 56 was < 3000, and the count rate for 89 was > 2000.

After the rare earth element internal standard indium (115In) were selected, the concentration parameter of the standard curve was input, the sample tube was inserted into pure water, and the internal standard tube was inserted into the end of the internal standard liquid, and the sample liquid was determined after the mass number was stable. The content of Pr element in samples was calculated by standard curve method.

3. Analysis of Factors Affecting Pr Content

3.1. Effect of Dissolution Method on Pr Content Determination

For the same sample, two different dissolution methods, microwave dissolution and graphite dissolution, were used under the same conditions of sample weight, dissolution time, dissolution temperature, carrier gas flow rate, and torch position, to investigate the influence of dissolution method on the determination of Pr element in vegetables by ICP-MS.

The standard detection limit (ug/kg) of Pr element in the national standard GB 5009.94-2012 "Food

Safety National Standard for the Determination of Rare Earth Elements in Plant Foods" is Pr≤0.2, and the quantitative limit (ug/kg) is Pr≤0.7. The recovery rate was compared between the two dissolution methods by paired sample T-test, as shown in table 1.

Table 1. Recovery of Pr content by different digestion methods.

Digestion method	Recovery rate (%)	Mean	Standard deviation	Mean standard deviation	T-value	P-value
Microwave digestion	103.1, 98.6, 105.5, 100.7, 104.4, 104.0	102.722.58		1.05	-0.38	0.719
Graphite digestion	95.3, 101.8, 107.0, 103.4, 106.8, 105.9	103.374.45		1.82		

The feasibility of the dissolution method was examined by the paired sample T-test using the sample recovery rate, which generally ranges from 90% to 110%. The recovery rate range of microwave dissolution was 98.6% to 104.4%, and the recovery rate range of graphite dissolution was 95.3% to 107.0%. The P value is 0.719, which is greater than 0.05, indicating that there is no significant difference between the microwave dissolution method and the graphite dissolution method. Microwave dissolution has the advantages of fast heating, high temperature rise, strong dissolution ability, low blank position, and effective avoidance of volatile loss and sample contamination, as well as shortening the dissolution time. Therefore, the microwave dissolution method was chosen for sample pretreatment in subsequent factor investigation.

3.2. Effect of Sample Weight on the Determination of Pr Content

Using microwave digestion with consistent digestion time, digestion temperature, and other factors, the influence of sample weight on the determination of Pr content was investigated.

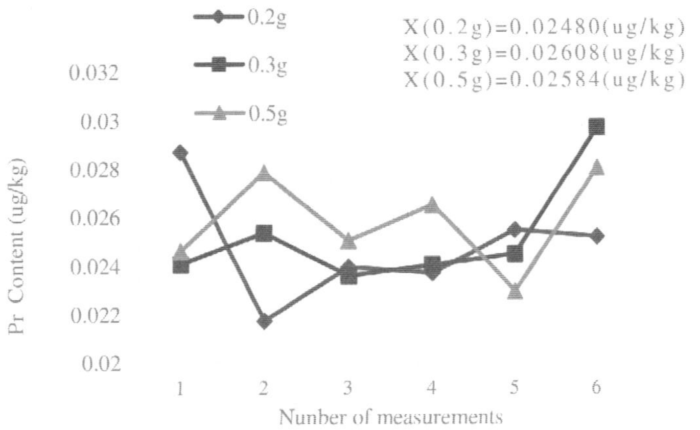

Figure 1. Measurement of Pr content in different weights. Number of measurements

Table 2. Different content of Pr to the sample weight analysis of test results.

	Degree of freedom	Adj SS	Adj MS	F-value	P-value
Sample weight	2	0.000003	0.000001	0.16	0.854
Error	6	0.000052	0.000009		
Total	8	0.000054			

As shown in figure 1, when the sample weight ranges from 0.2 to 0.5 g, the corresponding determined Pr content ranges from 0.02480 to 0.02608 ug/kg. A P-value greater than 0.05 indicates that the influencing factor has no significant effect on the determination results, whereas a P-value less than 0.05 suggests a significant impact. With a P-value of 0.854 in table 2, it is initially judged that different sample weights have no significant effect on the Pr content. However, considering that generally, a larger sample weight results in a smaller absolute error, a sample weight of 0.3 g is determined for use.

3.3. *Impact of Digestion Temperature on the Determination of Pr Content*

Using microwave digestion, with consistent factors such as sample size and digestion method, the influence of digestion temperature on the determination of Pr content was investigated.

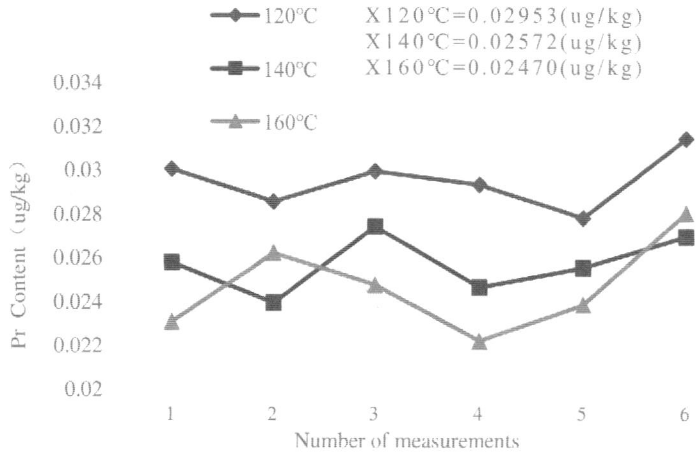

Figure 2. Determination of Pr content at different digestion temperatures.

Table 3. Analysis of Pr content at different digestion temperatures.

	Degree of freedom	Adj SS	Adj MS	F-value	P-value
Dissolution temperature	2	0.000039	0.000019	9.41	0.014
Error	6	0.000012	0.000002		
Error	8	0.000051			

As shown in figure 2, the digestion temperature ranged from 120 to 160°C, corresponding to a Pr content determination range of 0.02470 to 0.02953 ug/kg. In table 3, the P-value of 0.014 is less than 0.05, indicating that, for the same sample under consistent conditions of sample weight, digestion time, and digestion method, the digestion temperature has a significant impact on the determination of Pr content.

3.4. *Impact of Digestion Time on the Determination of Pr Content*

Using microwave digestion, with consistent factors such as sample weight, digestion method, and digestion temperature, the influence of digestion time on the determination of Pr content was investigated.

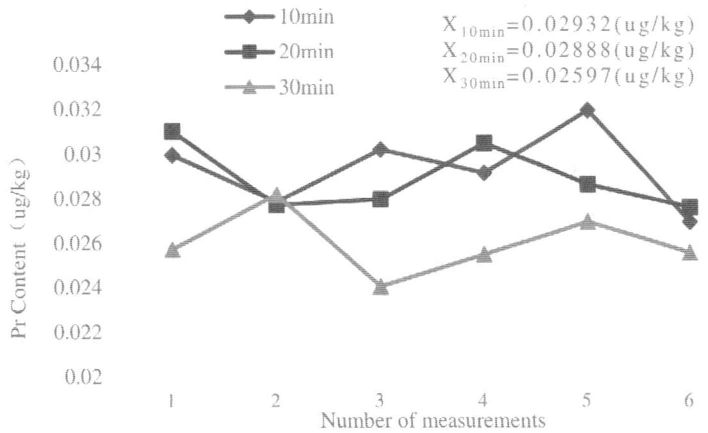

Figure 3. Determination of Pr content at different digestion times.

Table 4. Analysis of Pr content at different digestion time.

	Degree of freedom	Adj SS	Adj MS	F-value	P-value
Dissolution temperature	2	0.000020	0.000010	3.19	0.114
Error	6	0.000019	0.000003		
Total	8	0.000039			

As shown in figure 3, the digestion temperature range is 10 to 30 minutes, corresponding to a measured range of Pr content from 0.02597 to 0.02932 ug/kg. The P-value in table 4 is 0.114, which is greater than 0.05. Therefore, it is concluded that for the same tested sample, under consistent conditions of sample weight, digestion temperature, and digestion method, the digestion time has no significant effect on the determination of Pr content.

3.5. Impact of ICP-MS Carrier Gas Flow Rate on the Determination of Pr Content

The sample carrier gas, also known as nebulizing gas, creates a negative pressure through the nebulizer nozzle, causing the sample solution to form an aerosol under this negative pressure. The aerosol enters the high-temperature plasma along with the carrier gas, and the volume of carrier gas significantly affects the signal intensity. The impact of the carrier gas flow rate on signal intensity and signal-to-noise ratio was investigated, with the results shown in figure 4.

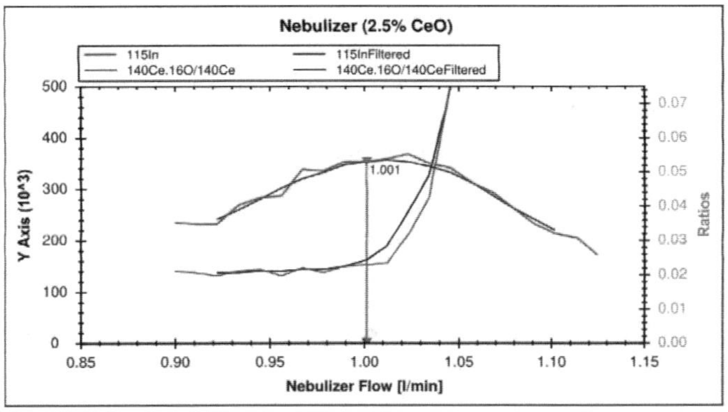

Figure 4. Effect of gas flow rate on signal strength and signal-to-noise ratio.

As illustrated in figure 4, within a certain range, the degree of sample ionization increases with the carrier gas flow rate, leading to a noticeable enhancement in signal intensity. The peak is reached at a carrier gas flow rate of 1.001 L/min. When the flow rate exceeds 1.001 L/min, the signal intensity decreases significantly due to the instant drop in plasma temperature caused by excessive carrier gas flow, which limits the residence time of the sample aerosol in the plasma, resulting in reduced ionization efficiency and compromised signal intensity. Therefore, it is concluded that the optimal condition for the determination of Pr content is achieved at a carrier gas flow rate of 1.001 L/min.

3.6. Impact of ICP-MS Torch Position on Pr Content Measurement

During ICP-MS measurement, adjusting and optimizing the torch position (X-axis, Y-axis, Z-axis) can optimize the ionization efficiency of target ions, reduce the formation of interfering ions, and enhance the sensitivity of measurement results. Additionally, changes in the torch position also influence the measurement results of isotope ratios to a certain extent.

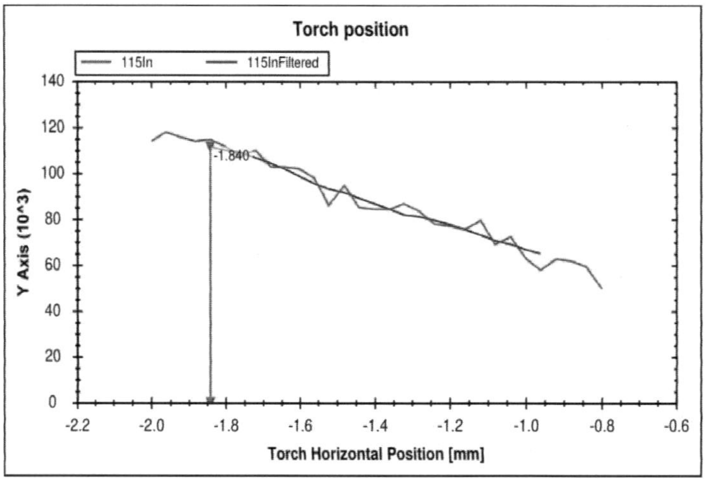

Figure 5. Effect of tube position on signal strength and signal-to-noise ratio.

The impact of torch position along different axes on the measurement signal intensity is illustrated in figure 5. As shown in the figure, when the sampling depth is relatively small (<-1.840 mm), the ion signal intensity decreases significantly. Conversely, when the sampling depth is relatively large (>-1.840 mm), the signal intensity exhibits an unstable trend. The optimal torch position is determined to be -1.840 mm.

3.7. Orthogonal Experiment

Based on the investigation of individual factors such as sample weight, digestion time, and digestion temperature, this experiment employs a three-factor, three-level orthogonal table to seek the optimal experimental conditions. The results are presented in table 5.

Table 5. Orthogonal test table.

Number	Sample weight (g)	Digestion temperature (°C)	Digestion time (min)	Content (ug/kg)
1	0.2	120	10	0.02480
2	0.2	140	20	0.02230
3	0.2	160	30	0.02596
4	0.3	120	20	0.03014
5	0.3	140	30	0.02695

Number	Sample weight (g)	Digestion temperature (°C)	Digestion time (min)	Content (ug/kg)
6	0.3	160	10	0.02558
7	0.5	120	30	0.03123
8	0.5	140	10	0.02741
9	0.5	160	20	0.02588
K1	0.07306	0.08617	0.07779	
K2	0.08267	0.07666	0.07832	
K3	0.08452	0.07742	0.08414	
k1	0.02435	0.02872	0.02593	
k2	0.02756	0.02555	0.02611	
k3	0.02817	0.02581	0.02805	
R	0.00382	0.0317	0.00212	

The magnitude of the range R reflects the primary and secondary order of influence of factors on the test indicators. According to table 5, the primary and secondary order of influence of each factor on Pr content is sample weight, digestion temperature, and digestion time. The optimal conditions are a sample weight of 0.5 g, a reaction temperature of 120°C, and a microwave digestion reaction time of 30 minutes. The comparison before and after method optimization is conducted under these optimal conditions.

4. Reliability Verification After Method Optimization

The purpose of reliability verification of analytical methods is to determine whether the internal quality of samples is effectively controlled and whether the methods are scientific and reasonable. This experiment verifies whether the reliability analysis methods meet the objectives and requirements of the Pr content testing project. Reliability verification involves analyzing precision, limit of detection (LOD), recovery rate, linear range, etc., one by one.

4.1. Precision

Six different standard curve concentrations were selected, and for each concentration, the sample was injected three times according to the established method to record the average response value and calculate the relative standard deviation (RSD).

Table 6. Precision of before and after optimization.

Number		Pr					
1	Concentration before optimization	0.943	1.531	5.099	10.111	20.409	50.332
2	Relative (ug/L) Standard deviation before optimization (RSD%)	8.1	5.7	2.2	1.2	4.3	1.9
3	Concentration after optimization	0.968	1.337	5.160	10.054	20.182	49.799
4	Relative (ug/L) Standard Deviation after optimization (RSD%)	5.6	7.1	5.8	1.3	1.2	1.9

As shown in table 6, Before optimization, the RSD ranged from 1.2% to 8.1%, while after optimization, it ranged from 1.2% to 5.6%. The range significantly decreased after optimization, with relatively stable accuracy. The test results are accurate and reliable, demonstrating good consistency.

4.2. Limit of Detection (LOD)

A standard blank sample was injected 11 times consecutively, and the standard deviation of the 10 samples was calculated. The corresponding concentration (reaction value obtained from the standard curve) was taken as the LOD of the method. The same sample was weighed, digested, and diluted to a fixed volume of 50 ml. The LOD of the method is shown in table 7.

Table 7. Analysis of detection limit before and after optimization.

Number	Pr	Optimized blank CPS value	Optimized blank CPS value	Standard deviation of the blank value before optimization(SD)(ug/kg)	Method detection limit before optimization	Standard deviation of the blank value after optimization (SD)	Method detection limit after optimization (ug/kg)
1	brank1	0.010822801	0.006517992				0.06043
2	brank1	0.010289494	0.005506005				
3	brank1	0.011113812	0.003008826				
4	brank1	0.010435184	0.004145448				
5	brank1	0.011608313	0.003206142				
6	brank1	0.009133486	0.001892199				
7	brank1	0.010982204	0.001600163				
8	brank1	0.008214055	0.002124986	0.03	0.09177	0.02	
9	brank1	0.005647683	0.003129009				
10	brank1	0.005340628	0.002368101				
11	brank1	0.009133486	0.001892199				
12	brank1	0.010982204	0.001600163				
13	brank1	0.008214055	0.002124986				
14	brank1	0.005647683	0.003129009				
15	brank1	0.005340628	0.002368101				

The standard detection limit (ug/kg) for the element Pr is Pr≤0.2, and the quantitation limit (ug/kg) is Pr≤0.7. After optimization, the standard deviation of parallel blank values is relatively small, with a method detection limit of 0.06043 ug/kg < 0.9177 ug/kg. The optimized detection limit is lower than the standard limit, indicating to a certain extent that the optimized measurement method has higher accuracy. The low value of the detection limit broadens the application scope of the method.

4.3. Recovery Rate

A standard solution of rare earth elements with a Pr content close to that of the sample was added to the sample, which was then transferred to a digestion vessel for microwave digestion and subsequently transferred to a 50 ml volumetric flask for volume adjustment and storage. The signal intensity was measured using inductively coupled ICP-MS, and the Pr element content was determined using the standard curve method. The spiked recovery rates are shown in table 8.

Table 8. Analysis of recovery rate before and after optimization.

Sample spike concentrati-on ug/kg	Sample weight before optimiza-tio (g)	Sample weight after optimizati-on (g)	Average before optimizati-on (n=7) ug/kg	Average after optimiza-tio (n=7) ug/kg	Recovery rate before optimizatio-n %	Recovery rate after optimizatio-n %	Standard deviation before optimizati-on RSD %	Standard deviation after optimizati-on RSD %
0.75	0.2134	0.1978	16.31	15.17	108.2	100.6	3.5	4.6
0.75	0.2033	0.2105	14.44	15.12	95.7	100.3	2.8	5.2
0.75	0.1985	0.2002	16.50	15.99	109.5	106.1	2.2	3.1
0.75	0.2006	0.1896	16.09	14.71	106.7	97.6	3.5	2.2
0.75	0.1977	0.2115	15.98	14.48	106.0	96.0	1.6	4.9
0.75	0.2012	0.2056	16.52	16.11	109.6	106.9	1.6	4.1

0.75	0.1996	0.2117	14.49	15.76	96.1	104.5	2.1	2.6

As shown in table 8, before optimization, the recovery rate of the praseodymium element in vegetables was approximately 95.7% to 109.6%. After optimization of the pretreatment process (including digestion), the recovery rate of the praseodymium element in vegetables was around 96.0% to 106.9%. Compared to the pre-optimization rates, the differences were relatively small, indicating a more stable performance. This demonstrates that the optimized measurement method has good accuracy and can be used in practical determination processes.

4.4. Linear Range

A series of 50 ml standard solutions of rare earth elements with concentrations of 1 ppb, 2 ppb, 5 ppb, 10 ppb, 20 ppb, and 50 ppb were prepared. A standard curve was plotted with concentration as the abscissa and counts per second (CPS) as the ordinate.

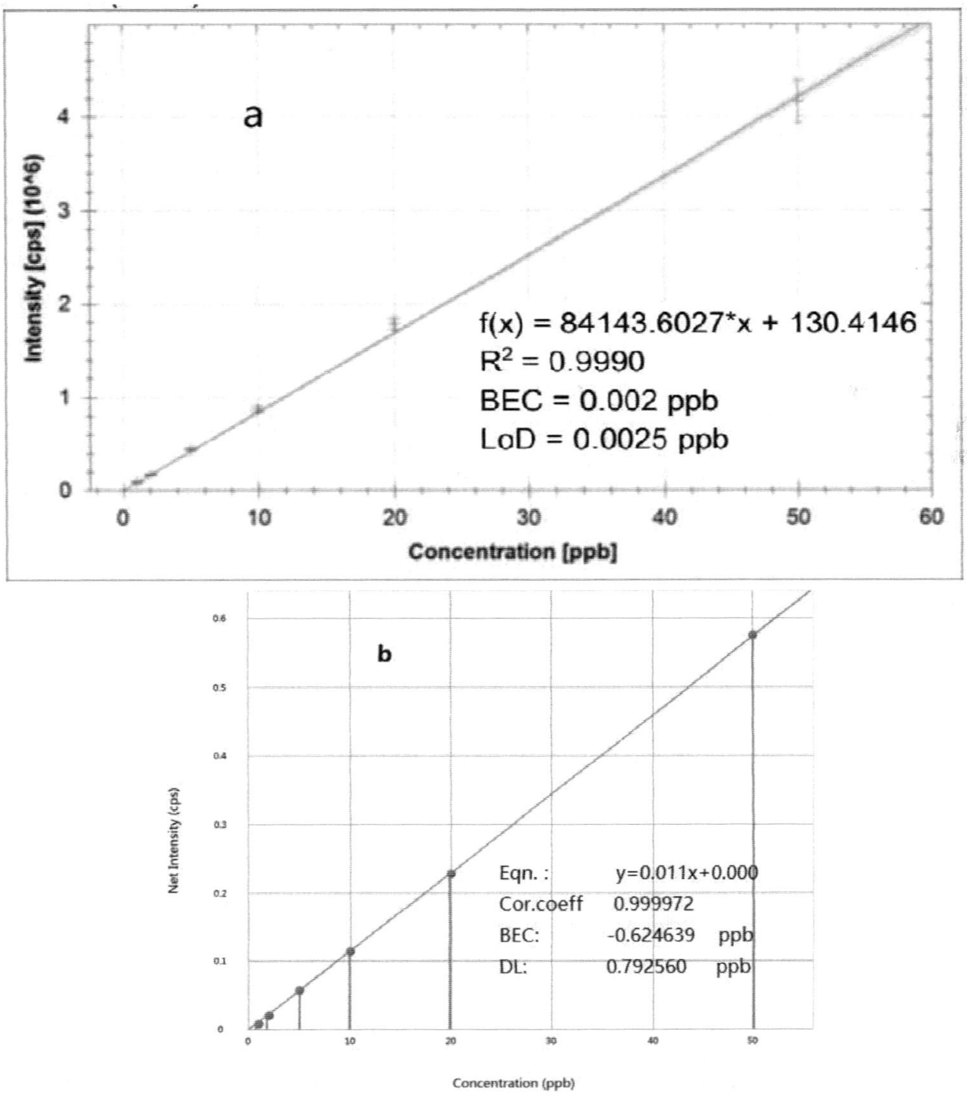

Figure 6. Comparison of standard curves before and after optimization (a before optimization and b after optimization).

Table 9. Determination of Pr element concentration after optimization.

Element	Standard Curve1 (ug/L)	Standard Curve2 (ug/L)	Standard Curve3 (ug/L)	Standard Curve4 (ug/L)	Standard Curve5 (ug/L)	Standard Curve6 (ug/L)
Pr	0.704	1.784	4.990	9.900	19.866	50.089

As shown in figure 6 and table 9, Before optimization, the standard curve equation was y = 84143.6027x + 130.4146, with a linear correlation coefficient R^2 of 0.9990. After optimization, the equation became y = 0.011x + 0.000, with a linear correlation coefficient R^2 of 0.999972. The better the fitting regression effect, the more stable the linear range of the Pr element within its detection limit and quantification limit.

5. Conclusion

By investigating factors such as the sampling amount for microwave digestion, digestion temperature, digestion time, carrier gas flow rate, and torch position, the optimal pretreatment conditions were obtained. These conditions included a sample weight of 0.5 g, a digestion temperature of 120°C, a microwave digestion time of 30 minutes, a sample carrier gas flow rate of 1.001 L/min, and a torch position of -1.840 mm. These conditions are suitable for the determination of praseodymium in vegetables using ICP-MS. The sample weight was 0.5 g, and the microwave digestion system consisted of 5.0 mL of nitric acid, which was then cooled, volume-adjusted, and stored. After optimizing the method, the ICP-MS detection limit for rare earth elements in vegetables subjected to microwave digestion was 0.06043 ug/kg, with a recovery rate of approximately 96.0%to 106.9%, and a correlation coefficient reaching 0.999972. The results indicate that the optimized method is characterized by simple pretreatment, high sensitivity, and low detection limits, making it capable of accurately monitoring trace amounts of Pr in vegetables.

Acknowledgements

This work was supported by Improvement of basic scientific research ability of young and middle-aged teachers in colleges and universities in Guangxi in 2022(2022KY1788); 2021 Professor's Cultivation Projection of Nanning University (2021JSGC05); Nanning University 2022 Practical Teaching Demonstration Course Project (2022XJYYS15): "Food Comprehensive Inspection Course Design".

6. References

[1] Li Z 2018 The influence of rare earth elements Pr, Nd, and Er on the submerged fermentation of russula polysaccharides *Food Science* (12) 312-316.

[2] Chu Z 2019 Study on the Content of Rare Earth Elements in Grain, Vegetables, and Fruits from a Light Rare Earth Mining Area in Shandong Shandong University.

[3] Gong X, Xu X, Wang H, et al. 2018 Analysis of nutritional components and rare earth elements in wild lycium barbarum leaf tea from rare earth regions *Modern Preventive Medicine* (09) 1031-1032.

[4] GB 5009.94-2012, National Food Safety Standard - Determination of Rare Earth Elements in Plant-based Foods.

[5] Xu J, Xiong X, et al. 2020 Determination of 16 rare earth elements in rice by inductively coupled plasma mass spectrometry *Journal of Food Safety and Quality* **11** (20) 7293-7297.

[6] Chen H, Liang P, et al. 2021 Application of inductively coupled plasma atomic emission spectrometry/mass spectrometry in the analysis of trace elements and species in traditional Chinese medicines *Spectroscopy and Spectral Analysis* (06) 1019-1024.

[7] Li Y, Tian H, Yu H, et al. 2022 Determination of total arsenic and inorganic arsenic in health food ingredients from different regions *Science and Technology of Food Industry* **43** (12) 10-17.

Application Research of Green Chelating Agents in the Remediation of Heavy-Metal-Contaminated River Sediments

Hongyun Tian[1], Pan Luo[1], Jing Shuai[1], Linna Xie[2], Tingting Xi[2] and Yongbing Huang[2,*]

[1] China First Metallurgical Group Co.,Ltd., Wuhan 430081, Hubei, China.
[2] Wuhan University of Technology, Wuhan 430000, Hubei, China.
Email: 296106509@qq.com; 297152748@qq.com; j.shuai@foxmail.com; linnaxie@whut.edu.cn; 329834@whut.edu.cn; *hyb732019@163.com

Abstract. This study aimed to utilize chelating agents for the remediation of heavy metal (HM)-contaminated river sediments. The removal efficiency and mechanisms of the green chelating agents glutamic acid (GLDA) and methyl glycine diacetic acid (MGDA) were analyzed under various conditions. The research findings indicate that the removal efficiency for Cd is GLDA> EDTA=MGDA, the removal efficiency for Cu is MGDA=GLDA> EDTA, and for Ni, the removal efficiency is MGDA> GLDA> EDTA. The green chelating agents GLDA and MGDA can remove HMs from river sediments efficiently without harming the environment, serving as alternatives to EDTA. Chelating agents achieve this effect by altering the speciation of HMs in river sediments, with system pH being the primary influencing factor. The supernatant treated with Na_2S contained HMs at levels compliant with national discharge standards, making it suitable for direct discharge.
Keywords. Chelating agents; environmental sustainability; heavy metals; leachate treatment; riverbed sediment remediation.

1. Introduction

In recent years, the sediment in China's major rivers, lakes, and other water bodies has exhibited varying degrees of HM pollution, posing ecological and environmental contamination risks [1]. HM contamination in riverbed sediments mainly originates from the discharge of urban domestic sewage and industrial wastewater, such as that from garment factories, printing facilities, leather shoe manufacturing plants, and furniture factories. The cumulative effect of substantial industrial wastewater discharge leads to the enrichment of HMs in natural water bodies and riverbed sediments, resulting in pollution. The risk assessment of HM pollution in riverbed sediments involves various HMs, including Hg, Cd, Cr, Pb, Mn, Cu, Zn, Ni, Co, and As, among others [2]. Certain heavy metals such as Cd, Cu, and Ni are particularly harmful and toxic to both human health and ecological environments [3]. HMs cannot be degraded in living organisms or the environment. When the sediment environment undergoes changes, heavy metals in the sediment may undergo redissolution and redistribution, leading to their rerelease into the water, their absorption by aquatic flora and fauna, and their subsequent entry into the food chain and accumulation in the human body. A substantial level of HMs can lead to poisoning in humans [4]. Therefore, it is imperative to take effective and environmentally friendly measures to remove HMs from sediment.

The study area is surrounded by a large accumulation of construction and domestic waste. Natural precipitation leads to the leaching of metals such as Cd, Cu, and Ni from solid waste into water bodies, where they are absorbed and amassed by riverbed sediments, causing pollution.

Currently, practical engineering techniques for controlling sediment pollution in domestic and international rivers mainly fall into three categories: physical, chemical, and biological methods. One effective application of the chemical method is the use of chelating agents to remediate HM pollution in riverbed sediments. Among these, using chelating agents for washing is considered a cost-effective, fast, and efficient remediation technique for soil contaminated with heavy metals, and it is regarded as a forward-looking remediation technology. Chelating agents serve to enclose metal ions within their molecular structure, creating stable complexes (figure 1). This process converts heavy metals from their mineral state to their aqueous form, thereby allowing them to be removed. Chelating agents display differing chelation affinities for various metal ions based on factors such as time, concentration, pH, and other environmental conditions [5].

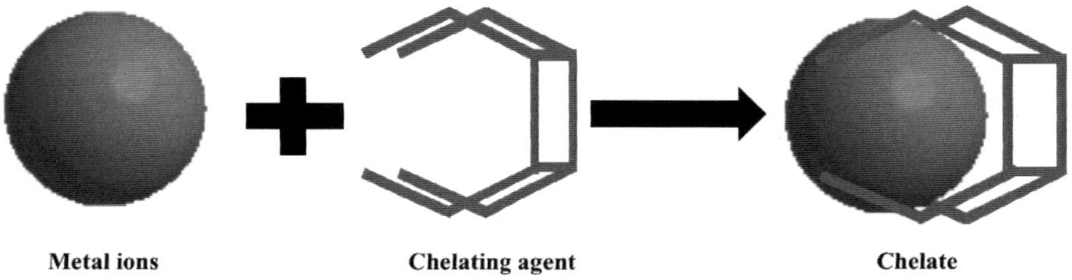

Figure 1. Simulation of chelation mechanisms.

The most well-known chelating ligand is EDTA, a man-made ligand widely used in various applications including water softening, textile manufacturing, cosmetic formulations, food processing, polymer production, metal plating, gas sweetening, pulp and paper manufacturing, and soil and water decontamination processes [6,7]. However, its drawbacks include its limited biodegradability, its persistence in the environment, and its ability to mobilize toxic metals from soil and sediments and stimulate algal growth in surface waters by enhancing the availability of nutrients [6], thus contributing to secondary pollution [8].

In order to improve environmental sustainability, it is meaningful to use biodegradable chelating agents to replace EDTA [9,10]. GLDA is a chelating agent developed by researchers in the late 20th century. Its production primarily utilizes a biogenic carbon source, ensuring high efficiency and energy savings during the manufacturing process. GLDA is also biodegradable. Research has shown that when GLDA is used, over 60 % degrades within 28 days [11]. This results in a significantly smaller ecological footprint for GLDA compared to other conventional chelating agents. The smaller the ecological footprint, the less impact it has on the environment [12]. MGDA is a new type of amino carboxylate chelating agent introduced by BASF [13]. Some studies have shown [14] that MGDA has good biodegradability and lower cost. Table 1 indicates that both GLDA and MGDA possess environmentally friendly characteristics and are safe for humans and the environment. Taken together, GLDA and MGDA do not cause secondary pollution, so they are classified as green chelating agents, laying the foundation for environmentally friendly pollution control. Figure 2 is a schematic illustration of the molecular structures of EDTA, GLDA, and MGDA. The molecular formula of EDTA contains four carboxyl groups, GLDA's formula contains four carboxyl groups, and MGDA's formula contains three carboxyl groups.

Table 1. Comparison of biodegradable properties of GLDA and MGDA with other chelating agents or additives.

Chelating agents	Powerful chelating agent	Biodegradable	Security	Environmentally friendly
EDTA	√	×	√	×
NTA	√	√	×	×
Phosphates	√	Inorganicsalts	×**	×***
Organophosphates	×	×	×	×
GLDA	√	√	√	√
MGDA	√	√	√	√

Note: ** Eutrophication; *** Large ecological footprint.

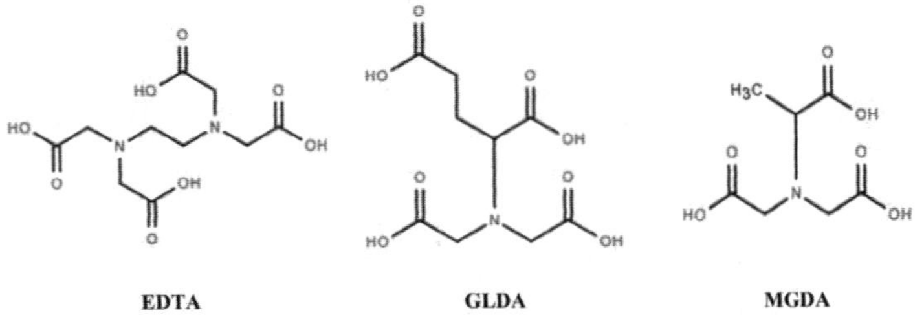

Figure 2. Structural formulae of three chelators.

Recently, there has been significant interest in the use of chelating agents for removing heavy metals from soil. This is because they not only possess high biodegradability, cost-effectiveness, and low toxicity but also contribute to increased organic carbon and nutrient content in the soil after incorporation. However, no prior studies have reported their application in the removal of heavy metals from riverbed sediments. Given the significant differences in the chemical composition and distribution of HM species between soil and sludge, this disparity may lead to variations in the efficiency and mechanism of HM leaching. Therefore, this study provides a theoretical foundation for MGDA and GLDA as environmentally friendly chelating agents for the removal of heavy metals from riverbed sediments.

In this study, the removal efficiency and impact of GLDA and MGDA on the heavy metals Cd, Cu, and Ni in riverbed sediments were investigated with the following objectives: (1) To determine the most suitable removal conditions, including chelating agent pH, concentration ratio, washing duration, and number of cycle rinses. (2) To analyze changes in the speciation of heavy metals using the BCR sequential extraction method and study the mechanism of chelating agents in the removal of the heavy metals Cd, Cu, and Ni. (3) To assess the potential of the green chelating agents GLDA and MGDA as alternatives to EDTA. With further research and exploration to optimize efficiency and performance, biodegradable chelating agents hold significant potential for major applications in the field of environmental remediation.

2. Materials and Methods

2.1. Experimental Materials

Instruments: THZ-C constant-temperature oscillator, ZEEnit700P atomic absorption spectrometer, TG16-II high-speed centrifuge, RK/XZN-100 vibrating milling machine, etc.

Reagents: ethylenediaminetetraacetic acid (EDTA) (Aladdin Biochemical), N, N-bis (carboxymethyl) glutamic acid (GLDA) (AkzoNobel), methyl glycine diacetic acid (MGDA) (TCI), nitric acid (AR), hydrochloric acid (AR), sulfuric acid (AR), hydrofluoric acid (AR), sodium hydroxide, cadmium standard solution, copper standard solution, nickel standard solution, etc.

Riverbed Sediment Samples: The required riverbed sediment samples used in the experiment were collected from a riverbed in Wuhan, Hubei Province, China.

2.2. Experimental Method

(1) Processing the riverbed sediment samples

Analysis of the physicochemical properties of the riverbed sediment samples involved dehydration, air-drying, and removal of non-degradable materials such as plant roots and stones, followed by grinding through a 100-mesh sieve and sealed preservation. Prepared Cd $(NO_3)_2$ aqueous solution, Ni $(NO_3)_2$ aqueous solution, and Cu $(NO_3)_2$ aqueous solution were added to the treated riverbed sediment samples. The solid content in the aqueous solution consists of 270 mg of $Cd(NO_3)_2$, 2530 mg of $Cu(NO_3)_2$, and 3500 mg of $Ni(NO_3)_2$. Every 3 days, an undetermined amount of deionized water was added to ensure microbial activity in the sediment. The samples were equilibrated for two weeks and blank controls were simultaneously set up.

(2) Detection methods for HMs in riverbed sediments

An amount of 0.5 g of the treated riverbed sediment sample was weighed out and then 6 mL of HNO_3, 3 mL of HCl, and 2 mL of HNO_3-$HClO_4$-HF were added in a fume hood. The digested sample was analyzed using atomic absorption spectroscopy to determine the total concentrations of Cd, Cu, and Ni in the riverbed sediment sample. The riverbed sediment samples were digested using a microwave oven, and the content of HMs was measured using atomic absorption spectroscopy. The test solution included three parallel samples and one blank sample. The experimental results were obtained by averaging the values from the three parallel samples.

(3) Morphologic analysis of HMs in riverbed sediments

The BCR (Bureau Communautaire de Référence) sequential extraction method was employed to fractionate HMs into five forms: water-soluble, acid-soluble, reducible, oxidizable, and residual. This was carried out to assess the changes in HM speciation in the riverbed sediment samples.

(4) Experimental design for factors influencing the removal efficiency of HMs in riverbed sediments.

(a) Influence of leaching time on the removal efficiency of HMs in riverbed sediments: Several 0.5g riverbed sediment samples were taken and placed in 50 mL centrifuge tubes. Without adjusting the system pH, chelating agents were added in a molar ratio of 2:1 with respect to HMs Cd, Cu, and Ni. The centrifuge tubes were placed in a thermostatic shaker and the shaking time gradients were set at 1 h, 2 h, 4 h, 8 h, 16 h, 24 h, and 48 h. The shaking conditions were set at 160 rpm/min. After completion of the shaking process, the samples were centrifuged for 20 minutes (5500 rpm). The supernatant was collected to measure the content of the HMs Cd, Cu, and Ni, and the removal efficiencies were calculated accordingly.

(b) Influence of pH on the removal efficiency of HMs in riverbed sediments: Several 0.5g riverbed sediment samples were taken and placed in 50mL centrifuge tubes. A chelating agent was added in a molar ratio of 2:1 with respect to HMs Cd, Cu, and Ni. The leaching time was adjusted to its optimal values. The pH gradient of the system was adjusted to 1, 3, 5, 7, 9, 11, and 13 using HCl and NaOH. Constant temperature was maintained while shaking, the samples were centrifuged, and the supernatant was collected to measure the content of Cd, Cu, and Ni. Their respective removal efficiencies were then calculated.

(c) Influence of chelating agent concentration on the removal efficiency of HMs in riverbed sediments: Several 0.5g riverbed sediment samples were taken and placed in 50 mL centrifuge tubes. The system pH and leaching time were adjusted to their optimal values. Chelating agents were added with HMs in molar ratios of 1:1, 2:1, 3:1, 4:1, 5:1, 6:1, 7:1, and 8:1. Constant temperature was maintained while shaking, the samples were centrifuged, and the supernatant was collected to measure the content of Cd, Cu, and Ni. Their respective removal efficiencies were then calculated.

(d) Influence of leaching frequency on the removal efficiency of HMs in riverbed sediments: Several 0.5g riverbed sediment samples were taken and placed in 50 mL centrifuge tubes. The system pH, chelating agent concentration, and leaching time were adjusted to their optimal values. Three consecutive leaching frequencies were performed and the removal efficiency for each frequency was calculated. The experimental procedure is as follows (figure 3).

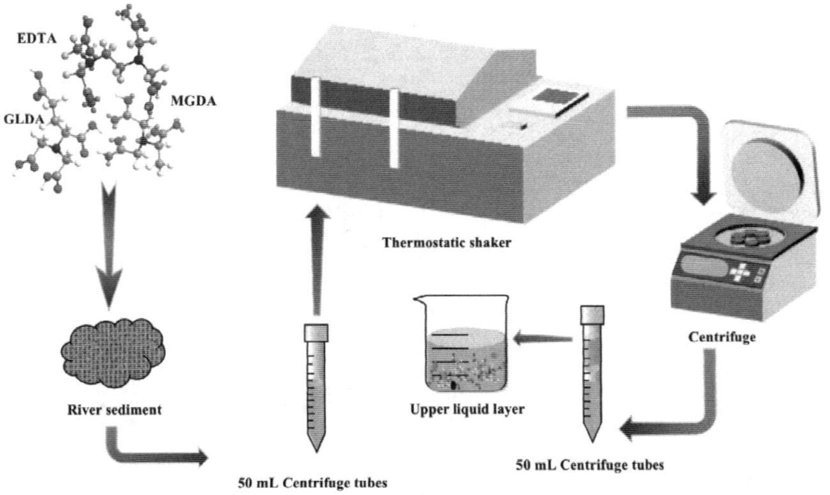

Figure 3. Experimental procedure.

(5) Methods for the treatment of leachate

An amount of 25 ml of the waste leachate obtained after chelating agent leaching was collected. Two parallel sets were prepared for each type of solution. An atomic absorption spectrometer was used to measure the content of Cd, Cu, and Ni in the leachate before treatment. Another two sets of solutions were then taken, with two parallel samples in each set, and these were placed in 25 ml centrifuge tubes. The leachate pH was adjusted to 11 using a NaOH solution. An amount of 1g of solid Na_2S was added to each centrifuge tube and the tubes were gently shaken until the solid Na_2S was completely dissolved [15]. The centrifuge tubes were placed in a thermostatic shaker and the shaking time was set to 24 hours at 160 rpm/min. After completion, the precipitate in the centrifuge tubes was filtered using filter paper, and the filtrate was collected. An atomic absorption spectrometer was used to measure the content of Cd, Cu, and Ni in the leachate after treatment.

(6) Analysis

The removal percentage (RE %) of heavy metal was calculated using the following equation:

$$RE\% = \frac{C_1V_1}{C_2V_2} \times 100\%$$

where C_0 (mg/L) and V_0 (mL) represent the concentration of heavy metals and the volume of the original soil digestion solution, and C_1 (mg/L) and V_1 (mL) represent the concentration of heavy metals and the volume of the extraction solution, respectively.

The experimental data were analyzed using Excel 2019. The figures were drawn with Origin2018.

3. Results and Discussion

3.1. Results of the Physicochemical Properties Analysis of the Original Sludge

Before conducting leaching with biodegradable chelating agents, the basic physicochemical properties of the riverbed sediment samples and the morphologic HMs present in them were determined. The test results are presented in table 2 and figure 4. The results indicate that the predominant morphologies of HMs in the riverbed sediment are reduced and oxidized states, followed by residual and weak acid extraction states.

Table 2. Basic physicochemical properties of the original river sediments.

Types of sludge	pH	Water content	Organic matter content (g/ kg)	Cd (mg/kg)	Cu (mg/kg)	Ni (mg/kg)
Riverbed sediment	6.2	99.2%	126	145	1367	1890

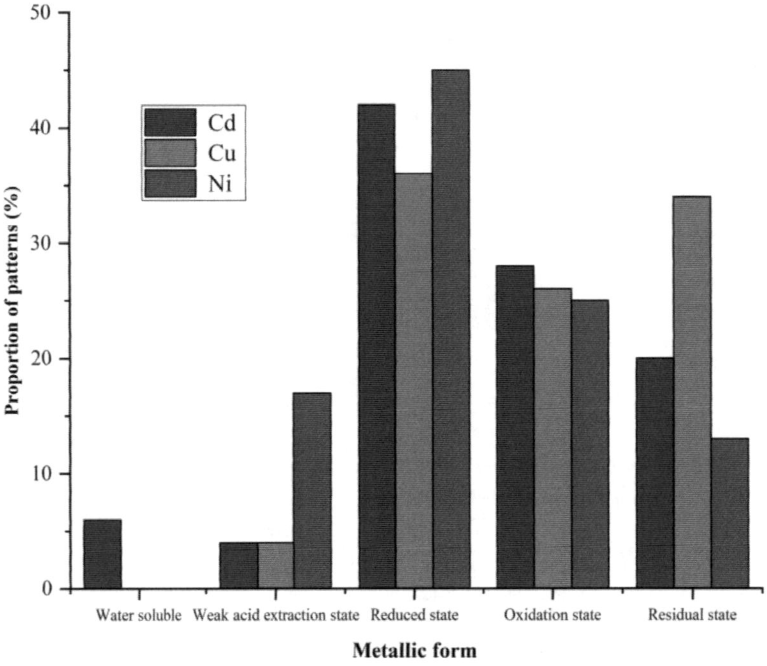

Figure 4. Morphological distribution of HMs in the original river sediments.

3.2. Influence of Leaching Time

Figure 5 shows the leaching kinetics of Cd, Cu, and Ni from the riverbed sediment. The results show that during the leaching process from 0 to 8 hours, GLDA and MGDA exhibited the fastest growth in removal efficiency for HMs. Between 8 and 16 hours, the removal efficiency of GLDA and MGDA stabilized, while EDTA continued to exhibit an increasing removal rate for Cd and Cu. Beyond 16 hours, although some of the heavy metals show fluctuating trends, they all remain within equilibrium values. Based on the economic feasibility principle, the optimal reaction times for EDTA, GLDA, and MGDA were determined to be 16 hours, 8 hours, and 8 hours, respectively.

A further increase in the reaction time had little effect on the overall removal. This might be due to the equilibrium between the water-soluble and insoluble forms of HMs being reached. Experimental research shows that a slower removal time is not the main factor affecting the removal efficiency of heavy metals. The different removal efficiencies of the three chelating agents for different HMs may be related to the stability constants of the metals. For the same type of complex with the same number of coordinating ligands, a higher stability constant indicates a more stable complex [16].

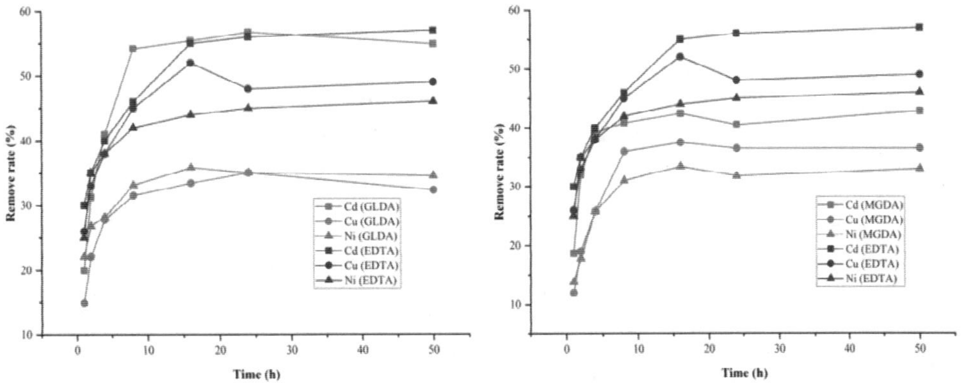

Figure 5. The influence of leaching time on the removal efficiency of HMs.

3.3. Influence of pH

The removal of Cd, Cu, and Ni using EDTA, GLDA, and MGDA was studied at different pH values and the results are presented in figure 6. The extract efficiency decreased with the increase in the pH value of the chelators, which is similar to other scholars' results [16,17]. At pH < 3, EDTA exhibits higher removal rates for Cd, Cu, and Ni compared to GLDA and MGDA. In the pH range of 3-7, all three chelating agents have approximately equal removal rates for Ni. However, in terms of removal rates for Cd, GLDA outperforms EDTA. MGDA's removal efficiency for Cd, Cu, and Ni declines at a slower rate. At pH values exceeding 7, the removal rates for Cd, Cu, and Ni using all three chelating agents gradually decrease, with an overall average removal rate falling below 40%. Under acidic conditions, chelating agents showed the highest removal efficiency for HMs. When the system pH < 3, a significant amount of acid is consumed. Therefore, the optimal pH of 3 is chosen. It can be inferred that under the optimal leaching time and pH conditions, the removal rates for Cd are GLDA > EDTA > MGDA, the removal rates for Cu are EDTA > MGDA > GLDA, and the removal rates for Ni are MGDA > EDTA > GLDA.

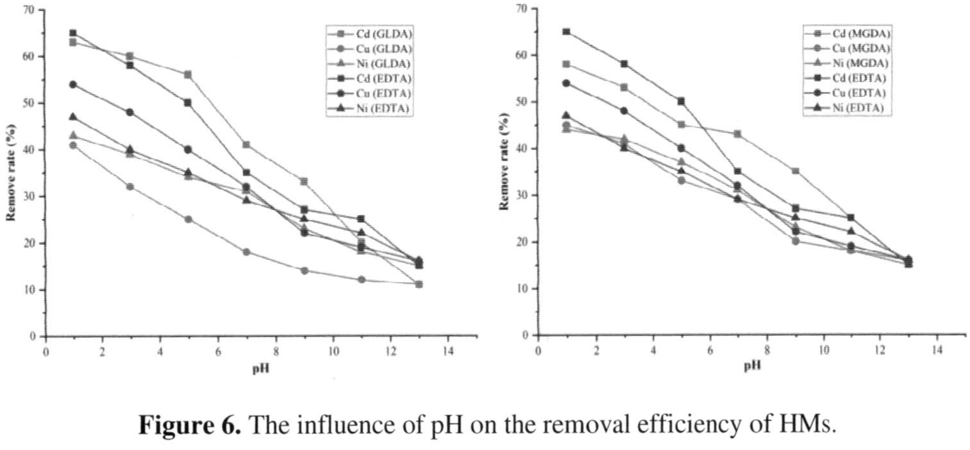

Figure 6. The influence of pH on the removal efficiency of HMs.

The higher removal efficiency of chelating agents under acidic conditions may be due to the acidic environment. Under acidic conditions, as the pH increases, the number of protons in the solution decreases, leading to a weakening of the acidic effect and a subsequent decrease in the removal efficiency. Alternatively, under alkaline conditions, HM ions tend to form precipitates, which can affect the removal efficiency [18]. Based on the results, it can be concluded that the system pH is one of the significant factors affecting the removal of HMs.

3.4. Influence of Chelating Agent Concentration

The data in figure 7 indicate that the removal efficiency of different chelating agents for HMs varies with changes in concentration, and their effects on different types of HMs also differ. The removal efficiency of HMs is positively correlated with chelating agent concentration within a certain range, reaching a plateau and stabilizing after reaching a certain concentration. The removal efficiency for Cd, Cu, and Ni was EDTA>GLDA. When the molar ratio of MGDA to HMs is 7: 1, it has a better removal efficiency for Cu and Ni than EDTA. Taking into consideration the above factors, the optimal concentrations selected are mol EDTA: mol HMs =4:1; mol GLDA: mol HMs =4: 1; and mol MGDA: mol HMs =6: 1.

The chelating agent concentration reaching a plateau and stabilizing after a certain value may be attributed to the dose of chelant being sufficient to meet the soil requirements. Alternatively, it could be due to the reaction of other metal ions and non-metal ions present in the riverbed sediment with the chelating agent, resulting in the consumption of some of the chelating agents.

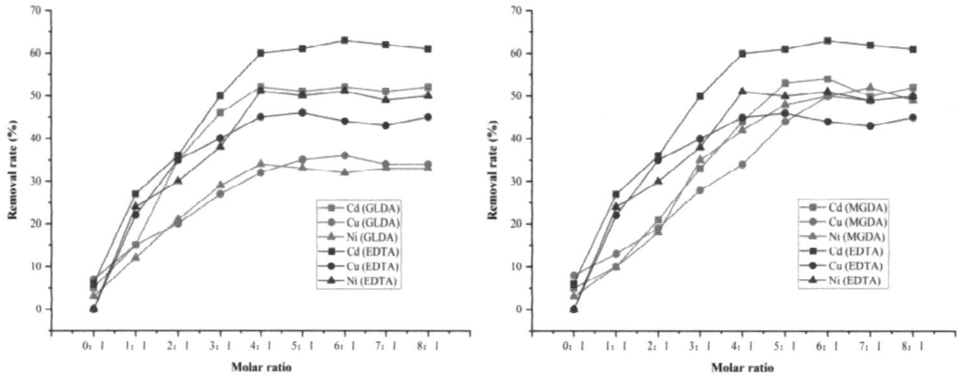

Figure 7. The influence of chelating agent concentration on the removal efficiency of HMs.

3.5. Influence of Leaching Frequency

The riverbed sediment was subjected to multiple consecutive leaching frequencies using three different chelating agents, and the leaching results are shown in figure 8. Single leaching showed the highest HM removal rate. After duplicate leaching, the HM removal efficiency of the chelating agents significantly increased. However, after triplicate leaching, the effect became less pronounced. Under the optimal leaching time, pH, concentration, and triplicate leaching conditions, the removal efficiencies of EDTA for Cd, Cu, and Ni are 78%, 63%, and 80%, respectively; the removal efficiencies of GLDA for Cd, Cu, and Ni are 73%, 60%, and 78%, respectively; and the removal efficiencies of MGDA for Cd, Cu, and Ni are 72%, 60%, and 81%, respectively. Ni exhibits relatively high sensitivity to changes in the leaching frequency. Considering that EDTA does not degrade easily and following the principle of environmental sustainability, duplicate leaching is chosen as the optimal leaching frequency for EDTA, while triplicate leaching is the optimal leaching frequency for GLDA and MGDA.

Some studies have shown that multiple washing may be effective since it can remove the easier (e.g., exchangeable) and more difficult phases (e.g., carbonate and reducible). Alternatively, it could be due to the cumulative effect of leaching frequency, which allows for further extraction of HMs trapped in micropores, resulting in a more thorough reaction.

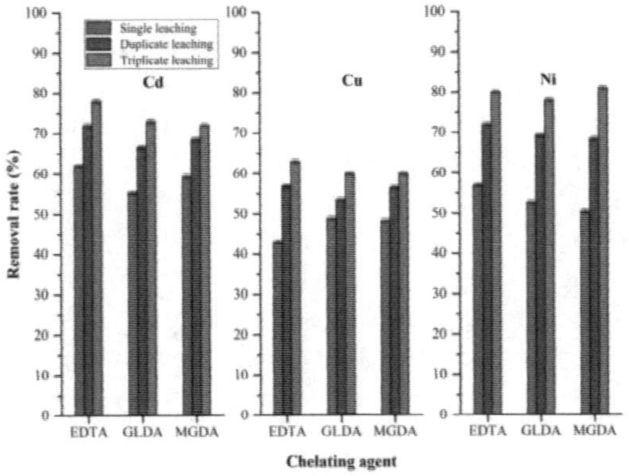

Figure 8. The influence of leaching frequency on the removal efficiency of HMs.

3.6. Analysis of the Morphologic Changes in HMs before and after Leaching

According to the BCR sequential extraction results shown in figure 9, it can be observed that the chelating agents EDTA, GLDA, and MGDA rinse the heavy metals Cd, Cu, and Ni, causing changes in their speciation. After washing, the proportions of water-soluble, weak-acid-extractable, oxidizable, and reducible metal ions all decrease, with the residual fraction becoming the predominant form. EDTA exhibits higher chelating capability compared to GLDA and MGDA, with Cd showing higher levels of residual oxidation and reduction states in comparison to Cu and Ni. This indicates that chelating agents have an impact on the speciation of HMs in riverbed sediment, leading to the release of metals existing in weak-acid-extractable, oxidizable, and reducible states, transforming them into more stable residual forms and thereby achieving removal.

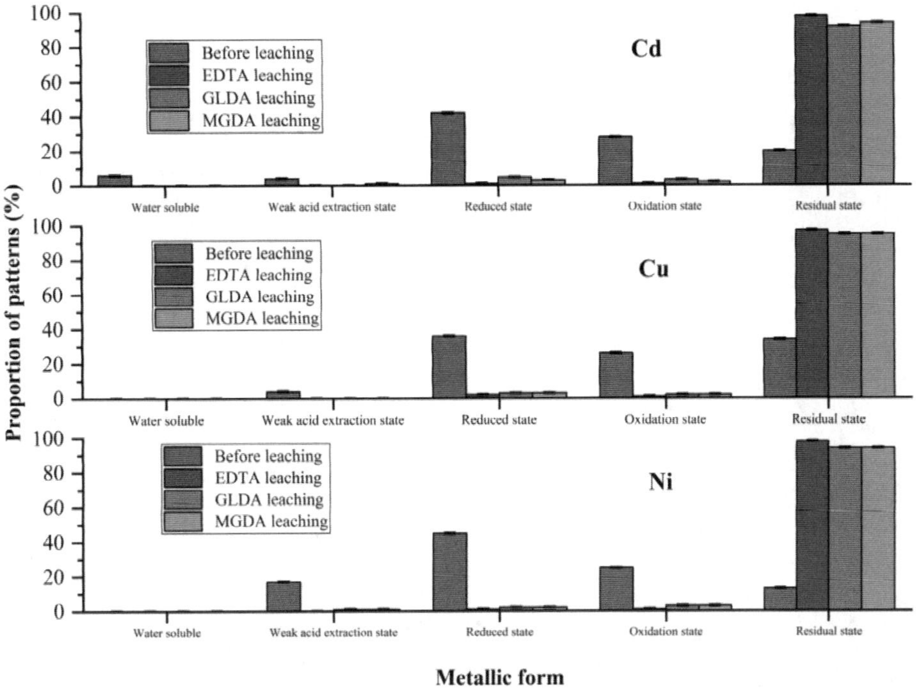

Figure 9. Variation in heavy metal distribution in sludge before and after the chelating agent treatments.

3.7. Treatment of Leachate

Research has shown that adding Na_2S to the leaching solution can result in the formation of more stable sulfide precipitates with heavy metals, thus converting the water-soluble heavy metal ions into precipitates that can be subsequently recovered through filtration. Considering the non-biodegradable nature of the traditional chelating agent EDTA, the experiments primarily focused on leachates obtained after washing with GLDA and MGDA, while leachates produced using EDTA were entrusted to a specialized waste recovery agency for disposal. After the addition of Na_2S, filtration, and calculation, the removal rates of Cd, Cu, and Ni in the leachate are 97.8%, 96.5%, and 98.6%, respectively. After precipitation treatment, the concentrations of Cd, Cu, and Ni in the leachate are as follows: 0.01 mg/L, 0.08 mg/L, and 0.12 mg/L. These meet the safety discharge standards in China ('Comprehensive Discharge Standard for Wastewater—GB8978-1996') and can be directly discharged. Table 3 shows the comparison between the detection values of HMs in leachate and Chinese standard.

Table 3. The detected values of HMs in the leachate and the Chinese standard.

HMs	Detection values	Standard
Cd	0.01mg/L	0.1mg/L
Cu	0.08mg/L	5.0mg/L
Ni	0.12mg/L	1.0mg/L

4. Conclusion

The biodegradable chelating agents studied here have shown promising effects in the remediation of heavy-metal-contaminated riverbed sediment, and under specific conditions they can serve as alternatives to traditional chelating agents, enhancing the sustainability of environmental remediation processes. The optimal room-temperature usage conditions for the three chelating agents are provided in table 4; under these conditions, the removal efficiency for Cd is higher for GLDA (73%) > EDTA (72%) =MGDA (72%), the removal efficiency for Cu is higher for MGDA (60%) =GLDA (60%) > EDTA (58%), and the removal efficiency for Ni is higher for MGDA (81%) > GLDA (78%) > EDTA (72%).

Table 4. The optimal usage conditions for the three chelating agents.

Chelating agent	Time	pH	Concentration	Frequency
EDTA	16 h	3	4:1	2
GLDA	8 h	3	4:1	3
MGDA	8 h	3	6:1	3

5. References

[1] Zhang J, Guo X Y, Zeng Y and Deng J C 2019 Spatial Distribution and Pollution Assessment of Heavy Metals in River Sediments from Lake Taihu Basin *Environmental Science and Pollution Research* **40(5)** 2202-2210.

[2] Chen M, et al. 2015 Research progress of risk assessment of heavy metals pollution in water body sediments *Ecology and Environmental Sciences* p. 1069-1074.

[3] Yu B, et al. 2000 The removal of heavy metal from aqueous solutions by sawdust adsorption — removal of copper *Journal of Hazardous Materials* **80(1)** p. 33-42.

[4] Nagajyoti P C, Lee K D and Sreekanth T V M 2010 Heavy metals, occurrence and toxicity for plants: a review *Environmental Chemistry Letters* **8(3)** p. 199-216.

[5] Wang X, et al. 2015 Heavy metal chemical extraction from industrial and municipal mixed sludge by ultrasound-assisted citric acid *Journal of Industrial and Engineering Chemistry* **27** p. 368-372.

[6] Chen Y, et al. 2022 A critical review on EDTA washing in soil remediation for potentially toxic

elements (PTEs) pollutants. *Reviews in Environmental Science and Bio/Technology* **21(2)** p. 399-423.
[7] George T and Brady M F 2020 Ethylenediaminetetraacetic acid (EDTA) *Europe PMC.*
[8] Beiyuan J, et al. 2018 Combined application of EDDS and EDTA for removal of potentially toxic elements under multiple soil washing schemes *Chemosphere* **205** p. 178-187.
[9] Raynie D E 2020 Innocuous and less hazardous reagents *The Royal Society of Chemistry* **4** p. 92-113.
[10] Pinto I S S, Neto I F F and Soares H M V M 2014 Biodegradable chelating agents for industrial, domestic, and agricultural applications—a review *Environmental Science and Pollution Research* **21(20)** p. 11893-11906.
[11] Kołodyńska D 2013 Application of a new generation of complexing agents in removal of heavy metal ions from different wastes *Environmental Science and Pollution Research* **20(9)** p. 5939-5949.
[12] Wu Q, et al. 2015 Effective removal of heavy metals from industrial sludge with the aid of a biodegradable chelating ligand GLDA *Journal of Hazardous Materials* **283** p. 748-754.
[13] Almubarak T, Ng J H and Nasr-El-Din H 2017 A review of the corrosivity and degradability of aminopolycarboxylic acids *Offshore Technology Conference.*
[14] Pandey S K and Bhattacharya T 2018 Effect of two biodegradable chelates on metals uptake, translocation and biochemical changes of Lantana Camara growing in fly ash amended soil *International Journal of Phytoremediation* **20(3)** p. 214-224.
[15] Zheng X, et al. 2020 Selective removal of heavy metals from simulated wastewater from leaching soil and recovery of eluent *Journal of Environmental Science* **40(3)** p. 995-1003.
[16] Suanon F, et al. 2016 Heavy metal removal from sludge with organic chelators: Comparative study of N, N-bis(carboxymethyl) glutamic acid and citric acid *Journal of Environmental Management* **166** p. 341-347.
[17] Dikinya O and Areola O 2010 Comparative analysis of heavy metal concentration in secondary treated wastewater irrigated soils cultivated by different crops *International Journal of Environmental Science & Technology* **7(2)** p. 337-346.
[18] Wu Q, et al. 2015 Removal of heavy metal species from industrial sludge with the aid of biodegradable iminodisuccinic acid as the chelating ligand *Environmental Science and Pollution Research* **22(2)** 1144-1150.

Investigation on the Application of Carbon Dioxide Power Generation Cycles in Solar Energy Heating Utilization

Lei Zhang[1,*], Yang Jiao[2], Jian Wang[1], Lei Wang[1], Zheng Liu[1], Yuduo Lu[1] and Lei Dong[3]

[1] Datang International Power Generation Co., LTD. Beijing Gaojing Thermal Power Branch., Beijing, 1000043, Beijing, China
[2] China Datang Technology Innovation Co., Ltd. Xiongan,Hebei,china
[3] China Datang Group Science and Technology Research Institute Co., Ltd. , Beijing, 1000043, Beijing, China
Email: *Zhlei1996@126.com; jiaoyang@china-cdt.com: 1485939238@qq.com; 22409585@qq.com; liuzheng70051l@sina.com; lyd006966@163.com; donglei@cdt-kxjs.com

Abstract. In the domain of solar thermal energy utilization, the power cycles that utilize carbon dioxide as the working fluid predominantly encompass the transcritical Rankine cycle and the supercritical Brayton cycle. This study employs MATLAB programming to compute and examine the thermal efficiencies of these two cycles across a spectrum of solar collector temperatures ranging from 200 to 1000°C and carbon dioxide working fluid pressures from 10 to 40 MPa. At elevated temperatures, the thermal efficiencies of both cycles augment with the escalation in working fluid pressure; however, at reduced temperatures, the alterations in thermal efficiencies of the two cycles are converse. Furthermore, beyond a specific temperature, the thermal efficiency of the supercritical Brayton cycle consistently surpasses that of the transcritical Rankine cycle. By means of fitting, the relational expressions for the thermal efficiencies of the two cycles being equivalent were ascertained. In light of practical circumstances, it is proposed that the supercritical Brayton cycle should be adopted when the solar collector temperature exceeds 350°C, whereas the transcritical Rankine cycle is more appropriate below 350°C.
Keywords. Carbon dioxide; transcritical Rankine cycle; supercritical Brayton cycle; cycle thermal efficiency; solar thermal utilization.

1. Introduction

Solar energy, driven by concerns over resources and the environment, has garnered considerable attention in both application and research [1,2]. Within the realm of solar thermal power generation, the exploration and utilization of carbon dioxide as a working fluid in power cycles are primarily centered on the transcritical Rankine cycle and the supercritical Brayton cycle.

Shu and colleagues [3] introduced an enhanced transcritical Rankine cycle utilizing carbon dioxide, which integrates a preheater and a regenerator. This advanced system demonstrates superior thermal efficiency when contrasted with the traditional cycle, rendering it especially apt for solar thermal applications. Li and associates [4] delved into the power characteristics of the transcritical carbon dioxide Rankine cycle and refined the design of the gas heater's inner-threaded tube. Jahar Sarkar [5] posits that the supercritical CO_2 Rankine cycle holds distinct advantages in terms of compactness and environmental friendliness when juxtaposed with both steam Rankine and organic Rankine cycles.

The supercritical carbon dioxide Brayton cycle has gained extensive application in the domain of solar thermal energy utilization [6,7]. Fahad A [8] undertook a comparative analysis of the thermodynamic performances of five distinct types of supercritical carbon dioxide cycles: the fundamental Brayton cycle, the recuperated Brayton cycle, the repressurized Brayton cycle, the precompressed Brayton cycle, and the two-stage expansion Brayton cycle. This investigation revealed that the repressurized Brayton cycle achieves the highest thermal efficiency within the context of solar thermal applications. Rajinesh Singh [9] performed a power analysis of the Brayton cycle, utilizing solar energy as the primary heat source. Pardeep Garg [10] explored concentrated solar power generation employing the supercritical carbon dioxide Brayton cycle, discovering that carbon dioxide sustains high efficiency even at temperatures below 800K. Various scholars have also engaged in theoretical and system simulations of the supercritical carbon dioxide Brayton cycle, examining fluctuations in its performance and their implications [11-15].

Given the diverse array of collector types and the substantial temperature fluctuations inherent in solar thermal energy applications, it is imperative to delineate the applicable ranges for the transcritical Rankine and supercritical Brayton cycles. Wang [16] carried out a comparative exergy analysis of the transcritical Brayton cycle and the performance of the recompression supercritical CO_2 cycle within the context of the organic Rankine cycle, yet failed to specify a definitive range. There exists a paucity of scholarly work conducting comparative analyses of these two cycles.

Hence, this study utilizes MATLAB programming, incorporating the REFPROP function, and applies the first and second laws of thermodynamics to assess the thermal efficiencies of the transcritical Rankine and supercritical Brayton cycles across varying heating temperatures and cycle pressures. The objective is to furnish a theoretical framework for power generation design in the realm of solar thermal energy utilization.

2. The operational procedures for the two CO2 power cycles are outlined below

Considering the myriad configurations of both cycles, this research concentrates on contrasting the basic theoretical cycles. This implies that neither of the two CO_2 power cycles incorporates additional processes such as reheat and regeneration.

Figure 1 illustrates the T-s diagram of the transcritical Rankine cycle. Within the solar collector, CO_2 is elevated to point 1 while in a supercritical state. Subsequently, it enters the expander, undergoing expansion to reach point 2, during which shaft work is produced. Following the extraction of work, the working fluid is cooled within the condenser, reaching point 3'. The CO_2, now in a subcritical state after cooling, is then compressed back to a supercritical state at point 4'. It re-enters the solar collector, thereby completing one cycle.

Figure 2 depicts the T-s diagram of the supercritical Brayton cycle. Analogous to the transcritical Rankine cycle, supercritical CO_2 is heated within the solar collector to reach point 1. It then undergoes expansion to point 2, where it executes external work. Contrary to the Rankine cycle, in the cooler, the exhaust gas, following work extraction, is cooled to a saturated vapor state, not a saturated liquid state. This vapor is subsequently compressed to point 4 and recycled back into the solar collector for further heating.

As illustrated in figures 1 and 2, the operational sequences of both cycles involve four key stages: heating, expansion, cooling, and compression. The key differentiation between these cycles is observed in the state of the working fluid after the cooling process. In the Brayton cycle, the CO_2 working fluid consistently remains in a gaseous phase, primarily existing in a supercritical state. In contrast, the Rankine cycle, due to its inclusion of both subcritical and supercritical states, is classified as a transcritical cycle.

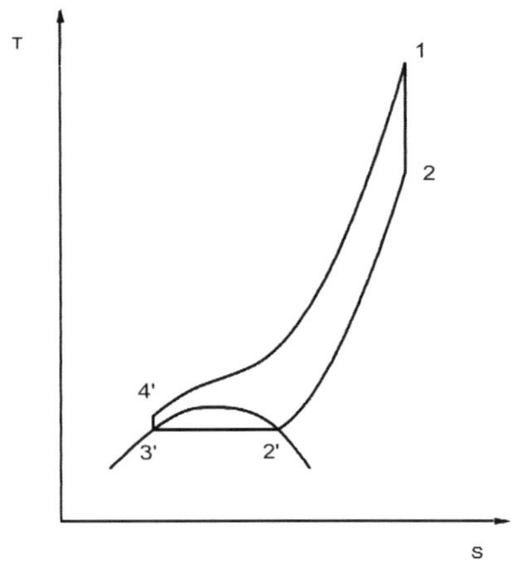

Figure 1. Theoretical T-s Diagram for Transcritical Rankine cycle.

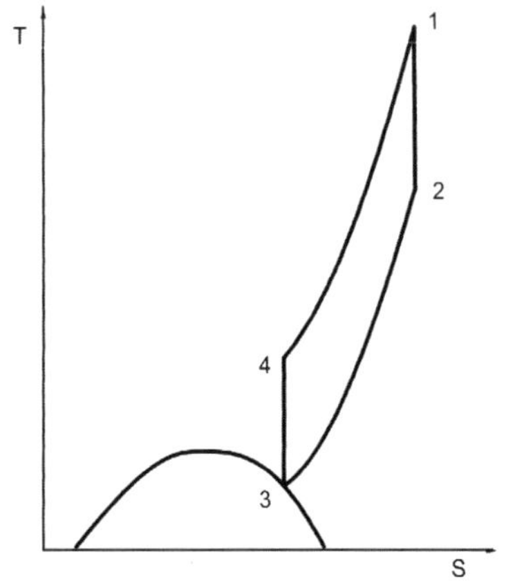

Figure 2. Theoretical T-s Diagram for Brayton cycles.

3. Calculation Model and Initial Conditions

The electrical power output, $P_{e,r}$ and $P_{e,b}$, for the transcritical Rankine cycle and the supercritical Brayton cycle can be expressed as follows:

$$P_{e,(r,b)} = m_{(r,b)} \times (h_1 - h_2) \times \eta_e \tag{1}$$

The compression power consumed during the pressurization process for both cycles, $P_{c,r}$ for the transcritical Rankine cycle and $P_{c,b}$ for the supercritical Brayton cycle, can be expressed respectively as:

$$P_{c,r} = \frac{m_r \times (h_4' - h_3')}{\eta_c} \tag{2}$$

$$P_{c,b} = \frac{m_b \times (h_4 - h_3)}{\eta_c} \tag{3}$$

The heat absorbed by the two cycles in the solar collector, Q_r for the transcritical Rankine cycle and Q_b for the supercritical Brayton cycle, can be expressed respectively as:

$$Q_r = m_r \times (h_1 - h_4) \tag{4}$$

$$Q_b = m_b \times (h_1 - h_4') \tag{5}$$

Then, the thermal efficiencies η_r for the transcritical Rankine cycle and η_b for the supercritical Brayton cycle can be expressed respectively as:

$$\eta_{(r,b)} = \frac{P_{e,(r,b)} - P_{c,(r,b)}}{Q_{(r,b)}} \times 100\% \tag{6}$$

In the comparative calculation, it is presumed that both cycles draw 1000 kW of heat from the solar collector. Considering actual requirements and the prevailing research landscape, the calculation range for the solar collector temperature is established between 200 and 1000°C, with a heating temperature increment of 10°C. The cooling water temperature is set at 15°C [7], with a heat exchange temperature differential of 5°C. The adiabatic efficiencies for both the compressor and expander are assumed to be 0.9 [7]. The pressure range for the working fluid CO_2 in both cycles is set between 10 and 40 MPa.

Journal of Physics: Conference Series **2941** (2025) 012081 doi:10.1088/1742-6596/2941/1/012081

4. Result and Analysis

4.1. Transcritical Rankine Cycle

Figure 3 depicts the thermal efficiency of the transcritical Rankine cycle. As evident from the figure, when the heating temperature remains constant, the thermal efficiency progressively rises with the escalation in working fluid pressure. However, the rate at which the thermal efficiency increases tends to diminish as the working fluid pressure continues to rise.

Under a constant working fluid pressure, two distinct scenarios emerge: At lower working fluid pressures, as the heating temperature escalates, the thermal efficiency initially climbs but subsequently declines, signifying the existence of a maximum value for the cycle; At higher working fluid pressures, as the heating temperature rises, the thermal efficiency continually increases.

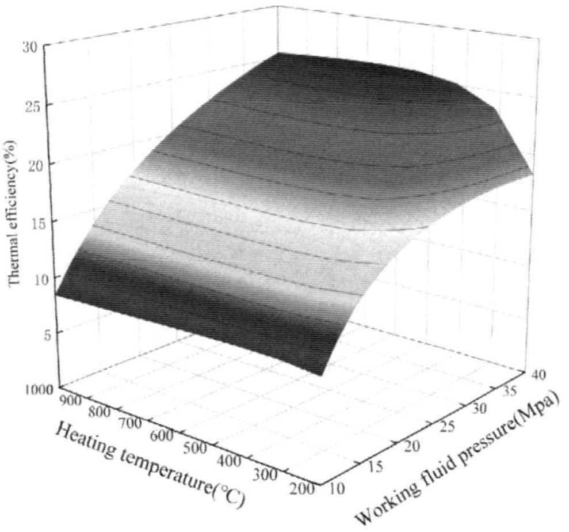

Figure 3. Thermal efficiency for transcritical Rankine cycles.

4.2. Supercritical Brayton Cycle

Figure 4 illustrates the thermal efficiency of the supercritical Brayton cycle. When the heating temperature is held constant, the thermal efficiency tends to increase in conjunction with the escalation in working fluid pressure. However, this trend does not persist at lower heating temperatures.

Under a constant working fluid pressure, two scenarios are observed: At lower working fluid pressures, as the heating temperature increases, the thermal efficiency first increases and then decreases, indicating the presence of a maximum efficiency point for the cycle; At higher working fluid pressures, the thermal efficiency continually rises in tandem with the heating temperature.

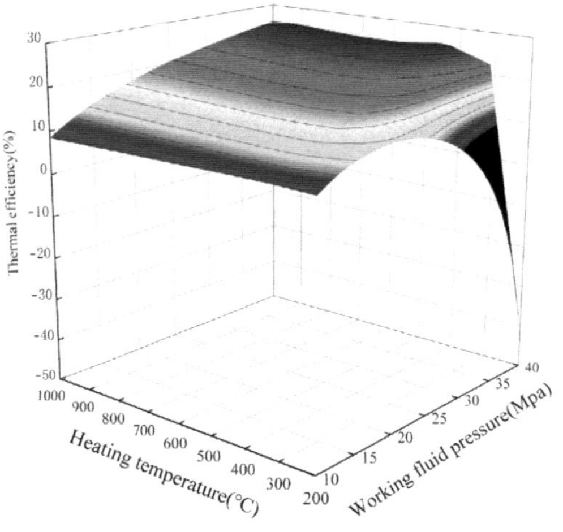

Figure 4. Thermal efficiency for Brayton cycles.

When the heat source temperature is low, a continuous increase in the working fluid pressure may lead to a gradual rise in thermal efficiency, followed by a steep decline into negative values. This signifies that the cycle no longer produces external work but instead demands energy input. This scenario occurs because the efficiencies of the compressor and expander are less than 1. At low temperatures and high pressures, the losses in the compressor and expander are pronounced, and the work generated by the expander is inadequate to offset the energy consumed by the compressor.

4.3. The Difference in Thermal Efficiency between Two Cycles

$\Delta\eta_t$ represents the discrepancy in thermal efficiency between the transcritical Rankine cycle and the supercritical Brayton cycle, defined as follows::

$$\Delta\eta_t = \eta_r - \eta_b \tag{7}$$

As depicted in figure 5, at lower heating temperatures and higher pressures, the thermal efficiency of the transcritical Rankine cycle surpasses that of the supercritical Brayton cycle. When comparing the thermal efficiencies at low temperatures in figures 3 and 4, it is evident that the thermal efficiency of the transcritical Rankine cycle transitions more smoothly, whereas the supercritical Brayton cycle exhibits more pronounced fluctuations. This discrepancy arises from the fact that at lower temperatures, the power consumption of the compressor exerts a more considerable influence on the supercritical Brayton cycle. In such conditions, opting for the transcritical Rankine cycle is more beneficial.

Journal of Physics: Conference Series **2941** (2025) 012081 doi:10.1088/1742-6596/2941/1/012081

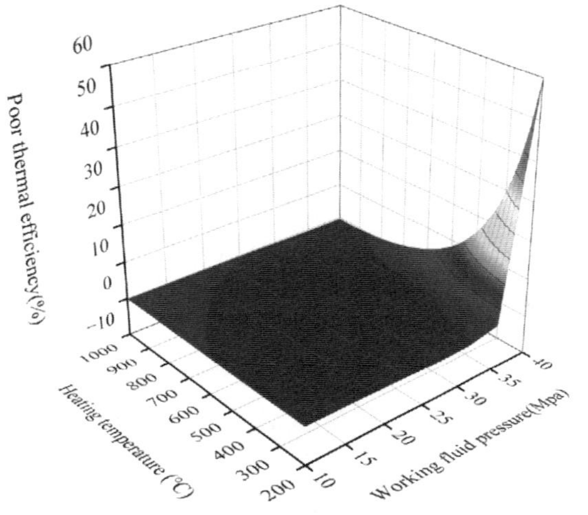

Figure 5. The difference in thermal efficiency between the transcritical Rankine cycles and Brayton Cycles.

Figures 6(a) and 6(b) depict the trends in the discrepancy of thermal efficiencies between the two cycles with varying pressure, respectively, at solar collector temperatures spanning from 200°C to 500°C and from 600°C to 900°C. For easier observation, the curve for T=200°C is presented with a maximum pressure of 25 MPa. As observable from the figures, with the increase in pressure, the difference in thermal efficiencies initially diminishes and then amplifies, signifying the existence of a minimum value.

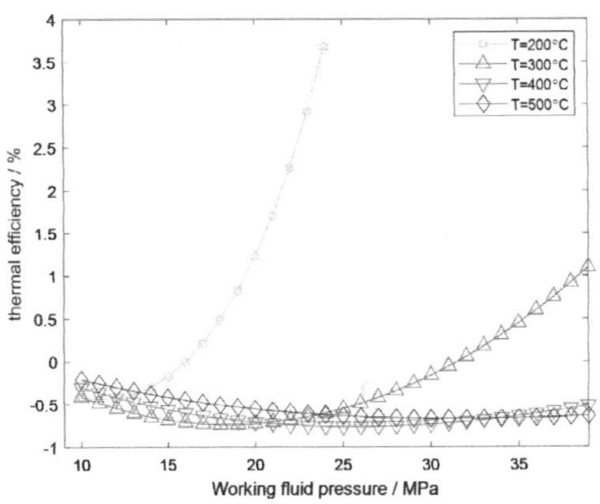

(a) Solar energy collection temperature ranges from 200°C to 500°C.

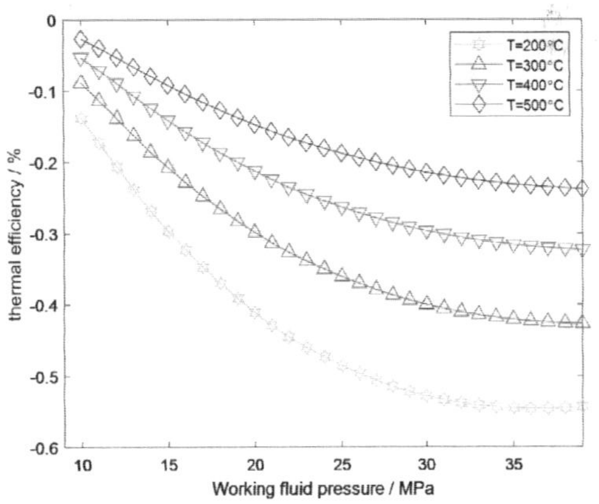

(b) Solar energy collection temperature ranges from 600°C to 900°C.

Figure 6. Differences between the thermal efficiencies of the transcritical Rankine cycles and Brayton Cycles.

Within the computational scope of this paper, the parameters at which the minimum difference in thermal efficiencies between the two cycles is observed are presented in figure 7. In the high-temperature range, an escalation in heating pressure results in an augmentation of the absolute value of the difference in net cycle efficiencies, emphasizing a more significant divergence between the two cycles. However, the minimum difference in net cycle efficiencies emerges within the mid-temperature range.

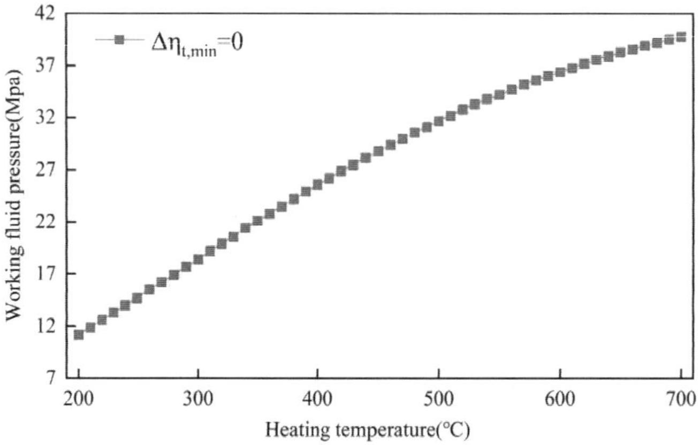

Figure 7. The minimum thermal efficiency for the two cycles.

Within the mid-temperature range, the trend in the net efficiency difference between the cycles undergoes a transformation: as the heating pressure intensifies with a constant heating temperature, the net efficiency difference initially attains a minimum value before progressively increasing, potentially even approaching zero and becoming positive (that is, the net efficiency of the transcritical Rankine cycle surpasses that of the Brayton cycle). When the heating pressure is held constant, an increase in heating temperature results in an initial decrease followed by an increase in the net efficiency difference; the higher the heating pressure chosen, the greater the heating temperature at which the minimum value appears.

The presence of a minimum difference signifies the point at which the Brayton cycle holds the greatest advantage over the transcritical Rankine cycle. As depicted in figure 6, despite the existence of a minimum value for the thermal efficiency difference between the two cycles within the computational range, when the solar collector temperature exceeds 400°C, the thermal efficiency of the transcritical Rankine cycle consistently remains lower than that of the supercritical Brayton cycle, irrespective of variations in the pressure of the CO_2 working fluid.

When the difference in cycle thermal efficiency is zero, employing either of the two cycles will yield identical results, as depicted in the calculations presented in figure 8. If the operating parameters are located above and to the left of the $\Delta\eta_t$=0 curve, it is appropriate to opt for the transcritical Rankine cycle. Conversely, if they fall below this curve, it is recommended to select the supercritical Brayton cycle.

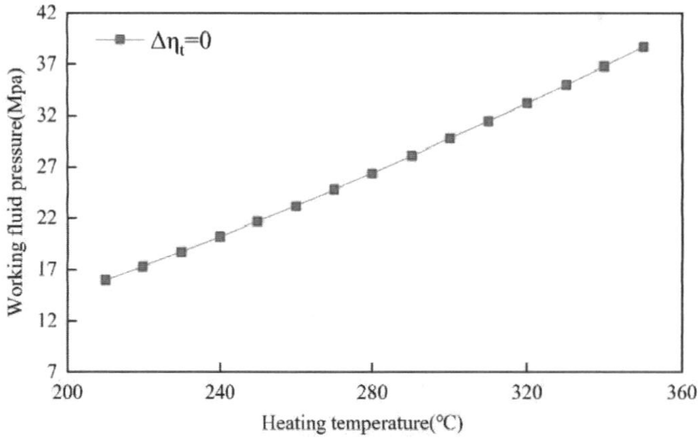

Figure 8. Parameters for no difference in thermal efficiency between the two cycles.

Through fitting, the operating parameters for the $\Delta\eta t=0$ curve can be determined as follows:

$$P = 240.11 \times T^2 + 32628 \times T \tag{8}$$

From equation (8), when the solar collector temperature is 350°C, the corresponding pressure of the CO_2 working fluid approaches 40 MPa, a level that is seldom attained in practical power cycles.

Consequently, as a carbon dioxide power cycle, when the temperature of the solar collector exceeds 350°C, it is advantageous to employ the supercritical Brayton cycle; whereas within the range below 350°C, the transcritical Rankine cycle is more appropriate.

5. Conclusion

In this study, a comparative examination of the thermal efficiencies of the transcritical Rankine cycle and the supercritical Brayton cycle, both utilizing carbon dioxide as the working fluid, was undertaken across a spectrum of solar collector temperatures from 200 to 1000°C and carbon dioxide working fluid pressures from 10 to 40 MPa. The following conclusions were derived:

The thermal efficiency trends of the transcritical Rankine cycle and the supercritical Brayton cycle with carbon dioxide exhibit similarity in the high-temperature and high-pressure regions. However, in the low-temperature and low-pressure domains, the efficiency trends of the two cycles are converse.

The discrepancy in thermal efficiencies between the two cycles is not linear, and at temperatures exceeding 400°C, irrespective of variations in the initial parameters of the working fluid, the efficiency difference consistently remains negative. This indicates that at these temperatures, the thermal efficiency of the supercritical Brayton cycle is invariably superior to that of the transcritical Rankine cycle.

This study has formulated an equation that correlates the parameters at which the efficiencies of the two cycles are equivalent. In light of practical considerations, for solar collectors like troughs and tower collectors, which typically operate at temperatures above 350°C, the supercritical Brayton cycle is advisable. Conversely, for small-scale, low-temperature solar thermal utilization systems, such as solar flat plate collectors, the transcritical Rankine cycle is more appropriate.

6. References

[1] Yang, Y P, Wu, J J, and Hou H J. 2016 Analysis of Thermo-Structural Coupling Transient Characteristics of Trough Solar Collector Tubes. *Acta Energiae Solaris Sinica* **37(12)** 3132-3136.

[2] Ju X, Xu C, Hu Y, et al. 2017 A review on the development of photovoltaic/concentrated solar power (PV-CSP) hybrid systems *Solar Energy Materials & Solar Cells* **161(3)**:305-327.

[3] Shu G, Shi L, Tian H, et al. 2016 An improved CO_2 -based transcritical Rankine cycle (CTRC) used for engine waste heat recovery *Applied Energy* **176(16)** 171-182.

[4] Li Z H, Tang G L, Wu Y X, et al. 2016 Improved gas heaters for supercritical CO_2 Rankine cycles: Considerations on forced and mixed convection heat transfer enhancement *Applied Energy*, **178(18)** 126-141.

[5] Sarkar J 2015 Review and future trends of supercritical CO_2, Rankine cycle for low-grade heat conversion[J]. *Renewable & Sustainable Energy Reviews* **48(8)** 434-451.

[6] Coco-Enríquez L, Muñoz-Antón J, Martínez-Val J M 2015 Integration between direct steam generation in linear solar collectors and supercritical carbon dioxide Brayton power cycles[J]. *International Journal of Hydrogen Energy* **40(44)**:15284-15300.

[7] Garg P, Kumar P, Srinivasan K 2015 A trade-off between maxima in efficiency and specific work output of super- and trans-critical CO_2 Brayton cycles *Journal of Supercritical Fluids* **98(3)** 119-126.

[8] Al-Sulaiman F A, Atif M 2015 Performance comparison of different supercritical carbon dioxide Brayton cycles integrated with a solar power tower *Energy* **82(6)** 61-71.

[9] Singh R, Miller S A, Rowlands A S, et al. 2013 Dynamic characteristics of a direct-heated supercritical carbon-dioxide Brayton cycle in a solar thermal power plant *Energy*, **50(1)** 194-

Journal of Physics: Conference Series **2941** (2025) 012081 doi:10.1088/1742-6596/2941/1/012081

204.
[10] Garg P, Kumar P, Srinivasan K 2013 Supercritical carbon dioxide Brayton cycle for concentrated solar power *Journal of Supercritical Fluids*, **76(4)**:54-60.
[11] Chen Y N, Zhang Y F, Liu, W J, et al. 2017 Simulation Study of Supercritical Carbon Dioxide Fired Power Generation System Thermal Power Generation **46(2)** 22-27.
[12] Iverson B D, Conboy T M, Pasch J J, et al. 2013 Supercritical CO_2 Brayton cycles for solar-thermal energy *Applied Energy* **111(4)** 957-970.
[13] Manente G, Lazzaretto A. 2014 Innovative biomass to power conversion systems based on cascaded supercritical CO_2 Brayton cycles *Biomass & Bioenergy* **69(10)** 155-168.
[14] Moullec Y L. 2013 Conceptual study of a high efficiency coal-fired power plant with CO_2 capture using a supercritical CO_2 Brayton cycle *Energy* **49(1)** 32-46.
[15] Hu L, Chen D, Huang Y, et al. 2015 Investigation on the performance of the supercritical Brayton cycle with CO_2-based binary mixture as working fluid for an energy transportation system of a nuclear reactor *Energy*, **89(9)** 874-886.
[16] Li M, Wang J, Li S, et al. 2014 Thermo-economic analysis and comparison of a CO_2 transcritical power cycle and an organic Rankine cycle *Geothermics* **50(2)** 101-111.

AUTHOR INDEX